THE THEORY OF
ELASTIC WAVES AND WAVEGUIDES

NORTH-HOLLAND SERIES IN
APPLIED MATHEMATICS AND MECHANICS

EDITORS:

H. A. LAUWERIER

*Institute of Applied Mathematics
University of Amsterdam*

W. T. KOITER

*Laboratory of Applied Mechanics
Technical University, Delft*

VOLUME 22

**NORTH-HOLLAND PUBLISHING COMPANY
AMSTERDAM · NEW YORK · OXFORD**

THE THEORY OF
ELASTIC WAVES AND WAVEGUIDES

by

JULIUS MIKLOWITZ

Division of Engineering and Applied Science
California Institute of Technology
Pasadena, California

1978

NORTH-HOLLAND PUBLISHING COMPANY
AMSTERDAM · NEW YORK · OXFORD

© North-Holland Publishing Company 1978

All rights reserved. No part of this publication may be reproduced, stored in a retrieval system, or transmitted, in any form or by any means, electronic, mechanical, photocopying, recording or otherwise, without the prior permission of the copyright owner.

North-Holland ISBN: 0 7204 0551 3

Published by:
NORTH-HOLLAND PUBLISHING COMPANY
AMSTERDAM · NEW YORK · OXFORD

Sole distributors for the U.S.A. and Canada:
Elsevier North-Holland, Inc.
52 Vanderbilt Avenue
New York, NY 10017

Library of Congress Cataloging in Publication Data

Miklowitz, Julius, 1919–
 The theory of elastic waves and waveguides.

 (North–Holland series in applied mathematics and mechanics)
 Includes bibliographical references.
 1. Elastic waves. 2. Boundary value problems.
I. Title.
QA935.M626   531'.33   76-54637
ISBN 0 7204 0551 3

PRINTED IN HUNGARY

To Gloria, Paul and David

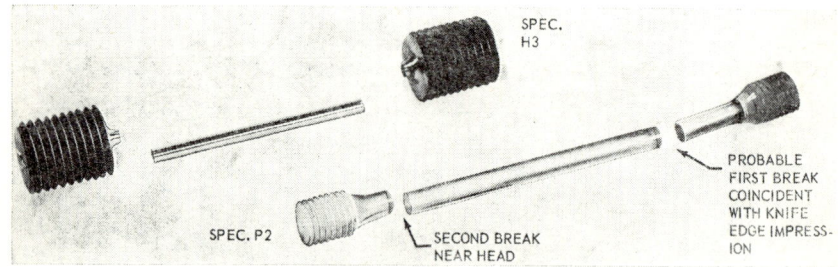

Second fracture due to unloading waves from the first.

PREFACE

The primary objective of this book is to give the reader a basic understanding of waves and their propagation in a linear elastic continuum. The studies presented here, of elastodynamic theory and its application to fundamental boundary value problems, should prepare the reader to tackle many physical problems of modern general interest in engineering and geophysics, and particular interest in mechanics and seismology.

The book focuses on transient wave propagation reflecting the strong interest in this topic exhibited in the literature and my research interest for the past twenty years. Chapters 5–8, and part of 2 are exclusively on transient waves, bringing to the reader a detailed physical and mathematical exposition of the fundamental boundary value problems in the subject. The approach is through the governing partial differential equations with integral transforms, integral equations and analytic function theory and applications being the tools. Transient waves in the infinite and semi-infinite medium and waveguides (rods, plates, etc.) are covered, as well as pulse diffraction problems. Chapters 3 and 4 with their extensive discussions of time harmonic waves in a half space, two half spaces in welded contact and waveguides are of interest *per se*. They are also important as necessary background for the later chapters.

The book will also serve as a reference source for workers in the subject since many important works are involved in the presentation. Many others are cited, but I make no claim to an extensive literature search since time precluded that. In this connection my survey covers the literature through 1964 (see reference [4.4] at the end of Chapter 4).

I found my way into this subject long ago and quite accidentally. In experiments with plexiglas tension specimens, preliminary ones in an investigation of dynamic stress-strain properties, a few of the specimens in these static tests broke suddenly and in a brittle manner in two places. The *frontispiece* p. VI) depicts this phenomenon (also for high-speed tool steel). Simple wave analysis showed the second fracture was created through a series of reflections of the unloading wave from the ends of the remaining elastic cantilever, the source being the first fracture [details in my paper, Journal of Applied Mechanics, *20* (1953) 122–130]. Needless to say this interesting phenomenon dramatizes in a simple way the severe damage that can be created by more complicated unloading (and loading) elastic waves, for example in earthquakes.

The book had its beginnings in a first year graduate course on elastic waves I initiated at the California Institute of Technology in the late fifties. In its present form the course (a full academic year, three lectures a week) draws on a good share of the material presented in this book. Prerequisites for the course have been introductory courses in the theory of elasticity and complex variables. Chapter 1 helps in this since it presents a brief introductory treatment of elasticity. Further, later material involving integral transforms and analytic function theory and applications is quite self-contained.

A one semester or a two quarter course on elastic wave propagation can be based on chapters 2 to 5 with selected material from the beginning of Chapters 6 and 7. Each Chapter has exercises, some problems and proofs primarily designed to involve the reader in the text material.

I would like to thank my colleagues Professors Thomas K. Caughey, James K. Knowles, Eli Sternberg and Theodore T. Y. Wu for reading certain of the chapters and making helpful suggestions. Similar acknowledgment is extended to Professor W. Koiter and my former graduate students Dr. David C. Gakenheimer and Professor Richard A. Scott. It will also become apparent to the reader that my graduate students have made a substantial input to the book for which I am grateful. Last but not least I should like to thank Mrs. Carol Timkovich and Mrs. Joan Sarkissian for their excellent and patient typing of the manuscript, and Cecilia S. J. Lin for her outstanding art work appearing in the major share of the figures herein.

Lastly, let me say I have found working in elastic wave propagation more than exciting. I hope that my book conveys this to the reader and, in particular, leads other young people into the subject with the same fascination that I found in it.

Pasadena, California
J. MIKLOWITZ

TABLE OF CONTENTS

Preface	**VII**
Introduction	**1**
Purpose of the book	1
The early history of the subject	2
Fundamental representations of elastic waves	2
Half space	4
Two half spaces in contact	6
Waveguides	6
Impact	9
Wave diffraction	10
Modern work and reading	10
Contents of present book	12
Other books on elastic waves and related subjects	16
References	17
1. Introduction to linear elastodynamics	**19**
1.1. Introduction; Description of deformation and motion	19
1.2. Tensor notation	20
1.3. Analysis of stress	21
1.3.1. Body and surface forces	21
1.3.2. Components of stress	22
1.3.3. Transformation of the components of stress	24
1.3.4. The stress quadric and the principal stresses	26
1.3.5. Maximum shear stress	28
1.4. Analysis of strain	30
1.4.1. Introduction	30
1.4.2. Finite deformation	31
1.4.3. Infinitesimal strain	32
1.4.4. Relative displacements in neighborhood of point P	33
1.4.4.1. The nature of the rotations	34
1.4.4.2. The nature of the strains	35
1.4.5. Transformation of components of strain	36
1.4.6. The strain quadric and the principal strains	37
1.4.7. The dilatation	38

1.5. Stress-strain relations	38
1.5.1. The generalized Hooke's law	38
1.5.2. Hooke's law for the homogeneous, isotropic, elastic medium	39
1.5.3. Elastic constants for the isotropic case	41
1.6. Dynamic equilibrium; stress equations of motion	42
1.7. Displacement equations of motion	45
1.8. The fundamental boundary-initial value problems of elastodynamics	46
1.9. The superposition principle	47
1.10. The principle of conservation of energy	48
1.11. Uniqueness of solution	50
1.12. Further contributions on the uniqueness of solutions	51
1.13. The Graffi elastodynamic reciprocal theorem	52
1.14. Exercises	55
References	56

2. The fundamental waves of elastodynamics and their representations   57

2.1. Introduction	57
2.2. Fundamental body waves and governing wave equations	57
2.2.1. Equivoluminal and dilatational displacement waves	57
2.2.2. Dilatation and rotation waves	58
2.2.3. The Lamé solution of the displacement equations of motion	58
2.2.4. Helmholtz resolution of a vector	60
2.2.5. Completeness of the Lamé solution	61
2.2.6. Gauge conditions	62
2.3. Types of body waves and governing equations	63
2.3.1. One-dimensional plane waves; D'Alembert's solution	63
2.3.2. Three-dimensional plane waves	65
2.3.3. Spherically symmetric waves	67
2.3.4. Axially symmetric and nonaxially symmetric waves	68
2.4. Body wave generation of waves peculiar to boundaries	68
2.5. Time harmonic body waves	69
2.5.1. Helmholtz equation	69
2.5.2. Time harmonic axially symmetric waves	70
2.6. Propagation of surfaces of discontinuity	71
2.6.1. Kinematical conditions	71
2.6.2. Dynamical conditions	73
2.6.3. Wavefront velocities	74
2.6.4. Wavefronts, characteristics and rays	75
2.6.5. Further comments	78
2.7. Wave motion due to body forces	78
2.7.1. Introduction	78
2.7.2. Basic singular solutions; retarded potentials	80
2.7.3. Elastodynamic solution	84
2.7.4. Two-dimensional radiation problems	85
2.7.5. Basic singular solutions of elastodynamics	86
2.7.5.1. Point load problem	86
2.7.5.2. Center of compression	91
2.7.5.3. Center of rotation	92
2.7.5.4. Point load time harmonic source	93

2.7.5.5. Retarded potentials corresponding to a time harmonic source; steady-state solution	94
2.8. Solution of boundary-initial value problems; integral representations	94
2.9. Initial value or Cauchy problems	96
2.9.1 One-dimensional initial value problem and related characteristics	96
2.9.1.1. Initial displacement condition	98
2.9.1.2. Initial velocity condition	99
2.9.1.3. Discontinuities in initial values	101
2.9.2. The three- and two-dimensional initial value problems	103
2.9.3. The method of characteristics in one-dimensional problems	107
2.9.3.1. First hyperbolic theorem; initial value problem and numerical integration method	110
2.9.3.2. Second hyperbolic theorem; characteristic initial value problem and numerical integration method	112
2.9.3.3. Boundary-initial value problems and numerical integration method	112
2.10. Exercises	116
References	118

3. Reflection and refraction of time harmonic waves at an interface   119

3.1. Reflection of P and SV waves of plane strain from the boundary of an elastic half space	119
3.1.1. Mixed conditions on boundary	123
3.1.1.1. Lubricated-rigid boundary	123
3.1.1.2. In-plane constrained boundary	125
3.1.2. Free boundary	125
3.1.2.1. Reflection of a P wave	125
3.1.2.2. Reflection of an SV wave	131
3.1.3. Elastically restrained boundary	133
3.1.4. Free boundary special cases	135
3.1.4.1. Normal incidence	135
3.1.4.2. Grazing incidence	136
3.1.4.3. SV wave incident at $\beta = \pi/4$	138
3.1.4.4. Total mode conversion	139
3.1.4.5. Reflection of SV waves at critical angles of incidence; total reflection	139
3.1.4.6. Reflection of wave pairs	145
3.1.4.7. Rayleigh surface waves	146
3.1.5. Rigid boundary	152
3.2. Reflection of SH waves from the boundary of an elastic half space	152
3.2.1. Free boundary	155
3.2.2. Rigid boundary	156
3.3. Reflection and refraction of P and SV waves at an interface	156
3.3.1. The solid-solid interface; general solution	156
3.3.2. Solid-solid interface; special cases	162
3.3.2.1. Normal incidence	162
3.3.2.2. Grazing incidence	163
3.3.2.3. Reflection at critical angles of incidence; total reflection	163
3.3.2.4. Stoneley interface waves	165

3.3.3. The fluid-solid interface	168
3.4. Reflection and refraction of *SH* waves at an interface	169
3.4.1. General solution	169
3.4.2. Special cases	173
3.5. Exercises	174
References	177

4. Time harmonic waves in elastic waveguides — 178

4.1. Waves in an infinite plate in plane strain	178
4.1.1. Mixed conditions on plate faces	181
4.1.1.1. Modes of propagation, frequency, and phase velocity spectra	182
4.1.1.2. Wave groups and stationary phase	186
4.1.1.3. Spectral analysis of wave train (or steady) propagation	188
4.1.1.4. Relation between group velocity and the velocity of energy transmission	192
4.1.2. Elastically restrained plate faces	194
4.1.3. Traction free plate faces; Rayleigh–Lamb frequency equation	196
4.1.3.1. General character of Rayleigh–Lamb frequency spectrum; corresponding modes and waves	197
4.1.3.2. Further on real- and imaginary-wave number segments and corresponding modes and waves	201
4.1.3.3. Further on complex-wave number segments and corresponding modes and waves	207
4.2. *SH* waves in an infinite plate	209
4.3. Love waves	211
4.4. Waves in an infinite elastic rod of circular cross section	214
4.4.1. Axially symmetric torsional waves	217
4.4.2. Axially symmetric compressional waves	220
4.4.3. Nonaxially symmetric or flexural waves	223
4.5. Waves in circular cylindrical shells and layered media; literature	226
4.6. Exercises	226
References	229

5. Integral transforms, related asymptotics, and introductory applications — 231

5.1. Introduction	231
5.2. Fourier integral theorem	231
5.3. Laplace transform	232
5.4. Further properties of the Laplace transform and its inverse	237
5.4.1. Uniqueness of $\bar{f}(p)$	237
5.4.2. Linearity of integral operator	237
5.4.3. Transforms of derivatives of $f(t)$	237
5.4.4. Convolution of two functions	238
5.4.5. Inverse of $\bar{f}(p)/p$	239
5.4.6. Inverse of $\exp(-\alpha p)\bar{f}(p)$	239
5.4.7. Laplace transforms of some functions $f(t)$	240

5.4.7.1. Transform of t^{k-1}, $k>0$	240
5.4.7.2. Transform of Dirac delta function $\delta(t)$	240
5.4.7.3. Laplace-transform tables	241
5.5. Bilateral Laplace transform	241
5.6. Exponential Fourier transforms	242
5.6.1. Exponential Fourier transform with real argument	242
5.6.2. Exponential Fourier transform with complex argument	245
5.7. Fourier sine and cosine transforms	246
5.8. Hankel transforms and properties	248
5.9. Asymptotic expansions	250
5.9.1. The nature of asymptotic expansions	251
5.9.2. Poincaré's definition of an asymptotic expansion	252
5.9.3. Some properties of asymptotic expansions	253
5.10. Asymptotic expansions of integrals	255
5.10.1. Types of integrals; critical points	256
5.10.2. Asymptotic expansion of Laplace integrals	256
5.10.2.1. Integration by parts	257
5.10.2.2. Watson's lemma	257
5.10.2.3. Asymptotic expansion of $\bar{f}(p)$; short time (wavefront) approximation	257
5.10.2.4. Long time approximation; solution of the static problem	258
5.10.3. Asymptotic expansion of Fourier integrals	261
5.10.4. Laplace's method	262
5.10.5. Method of steepest descents	264
5.10.6. Method of stationary phase	271
5.11. Cavity source problems	277
5.11.1. Spherical cavity subjected to sudden uniform pressure	277
5.11.2. Circular cylindrical cavity subjected to sudden uniform pressure	282
5.11.3. Short time (wavefront) and long time approximations in the cylindrical cavity problem	289
5.12. Exercises	291
References	296

6. Transient waves in an elastic half space — 298

6.1. Introduction	298
6.2. Plane strain problems	298
6.2.1. Lamb's problem for the surface normal line load source	299
6.2.1.1. Exact inversion by Cagniard-de Hoop method	302
6.2.1.2. Evaluation of exact solution for the surface response	314
6.2.1.3. Evaluation of exact solution for the response at the plane of symmetry	318
6.2.1.4. Wavefront approximations in the Cagniard-de Hoop method	319
6.2.1.5. Wavefront approximations by the method of steepest descents	323
6.2.2. Lamb's problem for the buried line dilatational source	329
6.3. Axially symmetric problems	332

6.3.1. The Lamb problems for the surface and buried vertical
　　　　　　point load sources   323
　　　　　　6.3.1.1. Inversion for surface displacements in problem
　　　　　　　　　　of surface normal point load source; numerical
　　　　　　　　　　evaluation   336
　　　　　　6.3.1.2. Numerical evaluation of the interior solution,
　　　　　　　　　　wavefront behaviors, for surface normal
　　　　　　　　　　point load source   340
　　　　　　6.3.1.3. Inversion for surface displacements in problem
　　　　　　　　　　of buried vertical point load source; numerical
　　　　　　　　　　evaluation and wavefront behaviors   344
　　6.4. A nonaxisymmetric problem; the suddenly applied normal point
　　　　load that travels on the surface   347
　　　　6.4.1. Statement of the problem and formal Laplace transformed
　　　　　　　solution   347
　　　　6.4.2. Preliminaries in the exact inversion by Cagniard-de Hoop
　　　　　　　method   351
　　　　6.4.3. Exact inversion for the interior and supersonic load
　　　　　　　motion; dilatational wave   352
　　　　6.4.4. Wavefront approximations   361
　　6.5. Exercises   362
　　References   365

7. Transient waves in elastic waveguides   367

　　7.1. Approximate theories and one dimensional problems   367
　　　　7.1.1. Approximate theories for axially symmetric compressional
　　　　　　　waves in a rod   367
　　　　　　7.1.1.1. Hamilton's principle   369
　　　　　　7.1.1.2. Love-Rayleigh rod   371
　　　　7.1.2. Approximate theories for flexural (nonaxially symmetric)
　　　　　　　waves in a rod or beam   374
　　　　　　7.1.2.1. Bernoulli–Euler or elementary bending theory   374
　　　　　　7.1.2.2. Timoshenko bending theory   377
　　　　7.1.3. Boundary-initial value problems based on approximate
　　　　　　　theories   382
　　　　　　7.1.3.1. Longitudinal impact problem based on
　　　　　　　　　　Love–Rayleigh rod theory; exact solution   382
　　　　　　7.1.3.2. Long time-far field approximation in the
　　　　　　　　　　longitudinal impact problem   387
　　　　　　7.1.3.3. Problems based on the Bernoulli–Euler and
　　　　　　　　　　Timoshenko bending theories; exact solutions   394
　　　　　　7.1.3.4. Long-time response for sudden shear load
　　　　　　　　　　on an infinite Timoshenko beam   404
　　7.2. Problems for the infinite plate in plane strain   409
　　　　7.2.1. Excitation of plate by two symmetric normal line loads;
　　　　　　　formal solution, numerical evaluation and approximations   409
　　　　　　7.2.1.1. Inversion of spatial transform first   411
　　　　　　7.2.1.2. Inversion of time transform first   425

7.2.2. Excitation of plate by two antisymmetric normal line loads and by a single normal line load	429
7.3. Edge load problems for the semi-infinite waveguide	430
7.3.1. Introduction	430
7.3.2. Plate in plane strain with mixed edge conditions; formal solutions	432
7.3.2.1. Longitudinal impact problem	432
7.3.2.2. Mixed pressure shock problem	434
7.3.2.3. Mixed edge conditions; problems with antisymmetric excitation	435
7.3.2.4. Formal solutions for other waveguides	435
7.3.3. Plate in plane strain with mixed edge conditions; approximate solutions	436
7.3.3.1. Long-time and/or far-field approximations	436
7.3.3.2. Short-time–near-field, wavefront approximations	438
7.3.4. Plate in plane strain with nonmixed edge conditions; formal long-time solutions and their inversions	444
7.3.4.1. Inversion integral forms	444
7.3.4.2. Boundedness condition; integral equations for the edge unknowns	446
7.3.4.3. Problem A: Nonmixed pressure shock; formal long-time solution	449
7.3.4.4. Problem A: Nonmixed pressure shock; long-time solution	455
7.3.4.5. Problem B: Nonmixed line load; formal long-time solution	457
7.3.4.6. Problem B: Nonmixed line load; long-time solution	463
7.3.4.7. Problems involving nonmixed edge displacements; comments	465
7.4. Axially symmetric problems for the infinite plate	466
7.4.1. Excitation by two symmetric normal point loads	466
7.4.2. Mixed edge condition problem; sudden normal displacement on circular cavity wall	471
7.4.3. Excitation by a time-dependent thermal field; numerical evaluation of modal solution	474
7.5. Exercises	475
References	483

8. Pulse scattering by half-plane, cylindrical and spherical obstacles — 485

8.1. Introduction; Wave features in a scattering problem	485
8.2. Plane-elastic pulse diffraction by half-plane obstacles	487
8.2.1. Introduction	487
8.2.2. Diffraction of a plane horizontally polarized shear pulse by a traction free half plane	488
8.2.2.1. Statement of the problem	488
8.2.2.2. Formal solution of the problem	489
8.2.2.3. Inversion of the formal solution and discussion of its nature	492

- 8.2.3. Diffraction of a plane horizontally polarized shear pulse by a rigid half plane — 496
- 8.2.4. Diffraction of a plane dilatational pulse by a traction free half plane — 496
  - 8.2.4.1. Statement of the problem — 496
  - 8.2.4.2. Formal solution of the problem — 498
  - 8.2.4.3. Factorization of the function $F(\zeta)$ — 502
  - 8.2.4.4. Inversion of formal solution and discussion of its nature — 504
- 8.3. Elastic pulse scattering by cylindrical and spherical obstacles — 517
  - 8.3.1. Introduction — 517
  - 8.3.2. Scattering of an elastic pulse by a circular cylindrical cavity — 518
    - 8.3.2.1. Line load source; general features of the wave system — 518
    - 8.3.2.2. Friedlander's representation of solution — 520
    - 8.3.2.3. Normal line load source on cavity wall; formal solution — 521
    - 8.3.2.4. Normal line load source on cavity wall; exact inversion — 524
    - 8.3.2.5. Normal line load source on cavity wall; Rayleigh waves and the long time-far field solution — 529
    - 8.3.2.6. Diffraction of plane compressional pulse by cavity; formal solution — 536
    - 8.3.2.7. Diffraction of plane compressional pulse by cavity; exact inversion — 539
    - 8.3.2.8. Diffraction of plane compressional pulse by cavity; numerical evaluation of solution — 542
    - 8.3.2.9. Diffraction of plane compression pulse by cavity; Fourier series solution, comparison of the two methods and results — 545
    - 8.3.2.10. Diffraction of plane compressional pulse by cavity; approximations and comments — 550
  - 8.3.3. Scattering of a plane compressional pulse by a circular cylindrical elastic inclusion; a brief discussion — 556
    - 8.3.3.1. The problem — 556
    - 8.3.3.2. The literature; methods and results — 556
  - 8.3.4. Diffraction of an elastic pulse by a spherical cavity; a brief discussion — 560
    - 8.3.4.1. The literature — 560
    - 8.3.4.2. Method of solution; results — 560
- 8.4. Exercises — 568
- References — 571

Supplementary reading — 573

- On text material — 573
- On additional effects — 576
- References — 578

Author Index — 581

Subject Index — 586

INTRODUCTION

Purpose of the book

This book is intended to give the reader a basic understanding of waves and their propagation in a linear elastic continuum. Elastodynamic theory, and its application to fundamental problems, are developed here. They underlie the approaches to many physical problems of modern general interest in engineering and geophysics, and particular interest in mechanics and seismology. The challenge in most of these problems stems from the complicated wave reflection, refraction and diffraction processes that occur at a boundary or interface in the continuum. This complexity evidences itself in the partial mode conversion of an elastic wave upon reflection from a traction-free or rigid boundary which converts, for example, compression into compression and shear. When there is a neighboring parallel boundary (forming then a waveguide), the so-created waves undergo multiple reflections between the two boundaries. This leads to dispersion, a further complicating geometric effect, which is characterized by the presence of a characteristic length (like the thickness of a plate). In the case of time-harmonic waves, dispersion leads to a frequency or phase velocity dependence on wavelength, and is responsible for the change in shape of a pulse as it travels along a waveguide. As the title of this book indicates, a healthy share of the material presented will focus on waveguide problems and hence a detailed study of elastic wave dispersion.

It will become apparent in studying the various topics here that obtaining solutions to elastodynamic problems depends strongly on having the appropriate mathematical techniques. It follows that in addition to the analysis of these solutions, the mathematical techniques per se form a natural part of our studies. In particular an understanding of these techniques, and in turn creating still others, lays the ground work for furthering our knowledge in the present subject.

The early history of the subject

The study of elastic wave propagation had its origin in the age-old search for an explanation of the nature of light. In the first half of the nineteenth century light was thought to be the propagation of a disturbance in an elastic aether. As pointed out in Love's interesting historical introduction of the theory of elasticity [1.2,p.7][1], the researches of Fresnel (1816) and Thomas Young (1817) showed that two beams of light, polarized in planes perpendicular to one another, do not interfere with each other. Fresnel concluded that this could be explained only by transverse waves, i.e., waves having displacement in direction normal to the direction of propagation. Fresnel's conclusion gave the study of elasticity a powerful push, in particular attracting the great mathematicians Cauchy and Poisson to the subject.

Fundamental representations of elastic waves

By late in the year 1822 Cauchy (cf. [1.2,p.8]) had discovered most of the elements of the classical theory of elasticity, including the stress and displacement equations of motion[2]. In 1828 Poisson presented his important first *mémoire* [2.1] (published in 1829) on numerous applications of the general theory to special problems. An addition to this *mémoire* [2.2] disclosed that Poisson was the first to recognize that an elastic disturbance was in general composed of both types of fundamental displacement waves, the dilatational (longitudinal) and equivoluminal (transverse) waves. His work showed that every sufficiently regular solution of the displacement equation of motion can be represented by the sum of two component displacements, the first being the gradient of a scalar potential function and the second representing a solenoidal field, where the potential function and

---

[1] Use will be made of bracketed numbers to identify references throughout the book. The references will be found only at the end of the chapter in which they occur first. An exception is this Introduction which also draws on many references appearing in later chapters, e.g., [1.2], the second reference of Chapter 1.

[2] According to Love [1.2, p. 6] Navier (1821) was the first to derive the general equations of equilibrium and vibration of elastic solids.

solenoidal displacement satisfy wave equations having the dilatational and equivoluminal wave speeds, respectively.

Poisson's general solution does not involve the vector potential appropriate to the solenoidal displacement component. Such a solution, i.e., one using both a scalar and vector potential, was apparently first given by Lamé in 1852 [2.5]. Thus through the efforts of Poisson and Lamé it was shown that the general elastodynamic displacement field is represented as the sum of the gradient of a scalar potential and the curl of a vector potential, each satisfying a wave equation. Since its inception this representation of the displacement field has been the core of most advances made through the solution of boundary value problems in linear elastodynamics, the obvious appeal being the wealth of knowledge that exists concerning solutions of the wave equation. The question of completeness of Lamé's solution was raised by Clebsch (1863)[I.1], but his proof was inconclusive. A rigorous completeness proof was given in 1892 by Somigliana [I.2] and subsequently by Tedone (1897) [I.3] and Duhem (1898) [I.4]. In 1885 Neumann [1.7] gave the proof of the uniqueness for the solutions of the three fundamental boundary-initial value problems for the finite elastic medium.

Important early investigations on the propagation of elastic waves were those contributed by Poisson (1831), Ostrogradsky (1831) and Stokes (1849) on the isotropic infinite medium [1.2,p.18]. Poisson and Ostrogradsky solved the initial value problem by synthesis of simple harmonic solutions obtaining the displacement at any point and at any time in terms of the initial distribution of displacement and velocity. Stokes pointed out that Poisson's resulting two waves were waves of the dilatation and rotation. Cauchy (1830) and Green (1839) investigated the propagation of a plane wave through a crystalline medium, obtaining equations for the velocity of propagation in terms of the direction of the normal to the wavefront [1.2, pp. 18, 299]. In general the wave surface (a surface bounding the disturbed portion of the medium) was shown to have three sheets corresponding to the three values of the wave velocity. In the case of isotropy two of the sheets are coincident, and all of the sheets are concentric spheres. The coincident ones correspond to transverse plane waves (in modern nomenclature SV and SH, vertically and horizontally polarized shear waves, respectively), and the third the dilatational wave (the P wave). Exploiting the strain-energy function, Green also showed that for a particular form of this function the wave surface is made up of a sphere representing the dilatational wave and two sheets corresponding to equivoluminal waves. Christoffel (1877) [1.2, pp. 18, 295-299] discussed the propagation of a surface of discontinuity through an elastic medium. He showed the surface moved normally to itself with a velocity that is determined, at any point, by the direction of the normal to the surface, *à la* same law that holds for plane waves propagated in that direction.

Investigation of elastic wave motion due to body forces was first carried out by Stokes (1849) [2.15], and later by Love (1903) [2.16]. On the basis of wave equations on the dilatation and rotation, and Poisson's integral formula (for the solution of the three dimensional initial value problem of the scalar potential), Stokes was the first to derive the basic singular solution for the displacements generated by a suddenly applied concentrated load at a point of the unbounded elastic medium. Love made an independent exhaustive study, solving the point load problem with the aid of retarded potentials. He showed that Poisson's integral formula yields correct results for a quantity only when it is continuous at its wavefront, hence invalidating Stokes' results for the dilation and rotation with (admissible) singular wavefronts. Love confirmed Stokes' solution, gave corrected expressions for the dilatation and rotation when they are singular, and added considerably to the interpretation of the solution. Love's work contained still another important part. This was his extension of Kirchhoff's well-known integral representation (1882) [cf. 2.18] for the potential governed by the inhomogeneous wave equation to one for the displacement in elastodynamics. In recent years this representation has found particular usefulness in wave diffraction problems.

Half space

In 1887 Rayleigh [3.8] made the very important finding of his now well-known surface wave. This wave is generated by a pair of plane harmonic waves, dilatational and equivoluminal (P and SV), in grazing incidence at the surface of an elastic half space. The resultant wave is not plane since it decays exponentially into the interior of the half space. It travels parallel to the surface with a wave speed that is slightly less than that of the equivoluminal body (interior of medium) wave. Rayleigh's wave is a core disturbance in elastodynamic problems involving a traction free surface.

Lamb (1904) [6.1] was the first to study the propagation of a pulse in an elastic half space. The paper was a major advance, one of prime importance in seismology. In it Lamb treated four basic problems, the surface normal line and point load sources, and the buried line and point sources of dilatation. He derived his solutions through Fourier synthesis of the steady propagation solutions. For the surface source problems Lamb evaluated the surface displacements (horizontal and vertical) which showed that the response was composed of a front running dilatational wave, followed by the equivoluminal and Rayleigh surface waves. Lamb also brought forth the important fact that in the far field (from the source) the largest disturbance was the Rayleigh surface wave. He noted the nondispersive nature of this Rayleigh wave, and in the case of the point-load excitation that it decayed as the inverse of the square root of the radial coordinate, a property typical of two-dimensional wave propagation. Other later studies of note on Lamb's problem were those of Nakano (1925, 1930) [I.5, I.6] and Lapwood (1949) [6.8] who investigated Lamb's formal solutions involving internal sources as integrals. Nakano showed the Rayleigh wave does not appear at places near the source. Lapwood treated the step input case. Russian work on Lamb's problem began in the early 1930's. The notable works by Sobolev (1932, 1933), Smirnov and Sobolev (1932), Nariskina (1934) and Schermann (1946) focused on the response for the half space interior. Smirnov and Sobolev gave a fundamentally new method for attacking the problem and other elastodynamic problems (not involving a characteristic length) based on similarity solutions. In 1949 Petrashen generalized this new method, employing Fourier integrals and contour integration, which enabled him to separate the Rayleigh wave from the terms in the solution representing the dilatational and equivoluminal waves[3]. In 1916 Lamb [I.8] extended his work to the cases of impulsive line and point sources traveling in a fixed direction with constant velocity.

The ingenious technique of Cagniard for solving transient wave problems of the half space, and two half spaces in contact, came along in 1939 [3.19]. The technique uses the Laplace transform on time, with spatial variables as parameters. The Laplace transformed solution to a problem is then an integral (e. g., Fourier) containing these parameters and the Laplace transform parameter, which through certain integrand transformations results in the Laplace integral operator (Carson's integral equation). This is then solved for the inverse Laplace transform (the solution) by inspection. As we shall

---

[3] Further detail on these Russian works may be found in Goodier [I.7] and Ewing et al [3.11].

see in this book, Cagniard's method is basic to much of the modern work in transient elastodynamic problems.

Two half spaces in contact

The reflection and refraction of plane harmonic waves from a planar interface between two-welded half spaces was first studied by Knott in 1899 [1.9]. Walker (1919) [1.10] treated the reflection of such waves from a free planar boundary of a half space, i.e., the special case of Knott's problem when one of the half spaces is a vacuum. Jeffreys (1926) [1.11], [1.12], Muskat and Meres (1940) [3.15] and Gutenberg (1944) [3.4] elaborated on Knott's work, the latter two works evaluating numerically energy ratios of reflected and refracted seismic waves for a variety of half space combinations (e.g., fluid-solid). In 1911 Love [3.17] in the course of an investigation on the effect of a surface layer on the propagation of Rayleigh waves found another wave of the same type. For short wavelengths compared to the thickness of the layer Love showed a modified Rayleigh wave existed with velocity dependent on the properties of both media. Later Stoneley (1924) [3.18] showed that this generalized Rayleigh wave had a motion which was greatest near the interface. It is now referred to as the Stoneley wave. Cagniard's exhaustive study of the problem of impulsive radiation from a point source in a space composed of the two-welded elastic solid half spaces was first published in 1939 [3.19]. It was a work of major importance in seismology.

Waveguides

The study of elastic waveguides had its beginnings in the subject of vibrations of elastic solid bodies with the simplest one-dimensional approximate theories being developed first. Euler (1744) and Daniel Bernoulli (1751) derived the governing partial differential equation for the flexural (lateral) vibration of bars (or rods) by variation of a strain-energy function, and then determined the normal modes and the frequency equation for all types of end conditions (combinations of free, clamped and simply supported ends) [1.2, p. 4]. Navier (1824) derived the basic approximate equation for extensional (longitudinal) vibrations [1.2, p. 25]. Chladni (1802) investigated these modes of vibration experimentally, as well as those of extensional (longitudinal) and torsional vibrations. Earlier Chladni (1787) published his experimental results on nodal figures of vibrating plates which were a challenge to theoreticians of that era. In 1821 Germain published the partial differential equation for the flexural vibrations in plates [1.2, p. 5].

In his 1829 *mémoire* Poisson [2.1] showed that the theory of vibrations of thin rods was covered by the exact equations of motion of linear elasticity. Poisson assumed the rod was a circular cylinder of small cross section, and expanded all quantities in powers of the radial coordinate in the section. When terms above the fourth power of the radius were neglected, the exact equations yielded the approximate ones for flexural vibrations, which were identical with those of Euler mentioned earlier. Similarly, the equation for extensional (longitudinal) vibrations was that found earlier by Navier. The analogous equation for torsional vibrations was obtained first by Poisson in the work being discussed.

The exact theory work of Pochhammer (1876) [4.8] for the general vibrations of an infinitely long circular cylinder with a traction free lateral surface was a major advance, one that underlies much of the exact and approximate theory modern research on steady and transient wave propagation in the elastic rod. Using separation of variables r, θ, z and t, the radial and circumferential sectional coordinates, axial coordinate and time, respectively, Pochhammer solved the exact displacement equations of motion. His displacements were represented by an infinite harmonic (in z and t) wave train with amplitude being a product of sinusoidal (in θ) and Bessel (in r) functions. Both the wave train and its amplitude were parametrically dependent on the frequency and wave number. Making use of the conditions of a traction free cylindrical surface, Pochhammer obtained the frequency equations (frequency as a function of wave number) for extensional, torsional and flexural wave trains. (Superposition of two trains of waves traveling in opposite directions along the cylinder gives the steady free vibrations of the infinite cylinder).

Pochhammer also derived the first and second approximations to the lowest branch of the frequency equation for extensional waves, and the first approximation to the lowest branch of the frequency equation for flexural waves. Through the years these have been a guide in the construction and use of approximate wave theories for the rod. Analogous work for the infinite plate in plane strain with traction free faces was carried out by Rayleigh [I.13] and Lamb [I.14] in 1889 for time-harmonic straight crested waves. Since their writing, the Pochhammer and Rayleigh–Lamb frequency equations with their infinite number of branches (roots) have been the subject of many studies, an almost complete understanding of these spectra being achieved only recently.

Lamb (1917) [1.15] was first to analyze the lowest symmetric and antisymmetric modes of propagation in the plate pointing out that for high frequency and short waves they become Rayleigh surface waves. To improve the elementary theory for extensional vibrations (and waves) in a rod, Rayleigh (1894) [7.1] obtained a correction to the frequency equation based on consideration of the radial inertia of the rod element. Love (1927) [1.2, p. 428] derived the equation of motion and end (boundary) conditions for this theory. The importance of the theory stems from the fact that its dispersion relation models exactly the Pochhammer second approximation to the lowest branch of the frequency equation for extensional waves mentioned earlier. Similary, Rayleigh in 1894 [7.1, §§ 161, 162] corrected the Bernoulli–Euler flexural wave theory for the thin rod (the dispersion relation of which models Pochhammer's first approximation to the lowest branch of the frequency equation for flexural waves) by considering the rotatory inertia of the rod element. Subsequently, Timoshenko (1921, 1922) [7.3] showed it was equally important to take account of shear deformation of the element. His approximate theory, accounting for both effects, has played the greater role in modern work on flexural waves in a rod.

Another major advance in waveguides, of prime importance to seismology, was Love's finding in 1911 [3.17, pp. 160–165] of his now well known Love waves. As pointed out in Ewing et al. [3.11, §§ 4.5] the first long-period seismographs, which measured horizontal motion only, exhibited large transverse components in the main disturbance of an earthquake. Love's work showed the waves involved were SH waves confined to a superficial layer of an elastic half space.

Pulse propagation in an elastic waveguide involving dispersion had its beginnings in wave group analysis. In his interesting early monograph on the propagation of disturbances in dispersive media, Havelock [4.1] points out that Hamilton as early as 1839, in his work on the theory of light, investigated the velocity of propagation of a finite train of waves in a dispersive medium. However, the work in the form of short abstracts was overlooked until the early 1900's. Russell (1844) seems to have been the first to observe the wave group phenomenon noting that in water, individual waves moved more quickly than the group as a whole. Stokes (1876) is usually credited with setting down the first analytical expression for group velocity, and Rayleigh with subsequent development. Kelvin's group method of approximating integral representations of dispersive waves came along in (1887) in his work on water waves. This was an important advance, which is now known as the method of stationary phase. Later, Lamb (1900, 1912) presented enlightening graphical methods in the wave group concept. We leave to Havelock's monograph discussions of the contributions of Reynolds (1877), Gibbs (1886), Havelock (1908, 1910), Green (1909), Sommerfeld (1912, 1914), Brillouin (1914) and others to the theory and applications of wave group analysis to a variety of physical fields including elasticity. One will find the references on all of the foregoing contributions to wave group analysis in Havelock's book [4.1].

Impact

As Love [1.2, pp. 25-26] points out many early studies were concerned with the phenomena of impact when two bodies collide, interest initially being in the collision of two rods as a system involving longitudinal waves. It was studied first by Poisson (1833), and later by Saint-Venant (1867). The results of these investigations did not agree satisfactorily with experiment, hence it appeared the impact phenomena could not be described by longitudinal wave theory. In 1882 Voigt suggested that the impacting rods should be thought of as separated by a transition layer with the geometric shape of the interface having an influence on the impact process. His correction led to a little better agreement with experiment. Hertz (1882) was more successful in his treatment of two bodies pressed together. He assumed that the strain produced in each body was a local statical effect, produced gradually and subsiding gradually, and found the duration of impact and the size and shape of the parts that come into contact. They compared favorably with experiment. Later, Sears (1908, 1912) [1.2, p. 440] conducted an extensive investigation of the problem with experiments on longitudinal impact of metal rods with rounded ends, and proposed a theory assuming the ends of the rods come into contact according to Hertz's theory, whereas away from the ends the earlier longitudinal wave theory of Saint-Venant applies. Sears' theory was confirmed by experiment. Further experiments on impact were carried out by Hopkinson (1905) [1.2, p. 117].

Other related problems were treated by longitudinal wave theory. The longitudinal impact of a large body upon one end of a rod was treated by Sébert and Hugoniot (1882), Boussinesq (1883) and Saint-Venant (1883)

[1.2, p. 26]. In 1930 Donnell [I.16] extended the solution to the case of a conical rod. Saint-Venant also solved several other problems by vibration theory that involved a body striking the rod transversely. As Love points out [1.2, p. 26] the problem of a transverse load traveling along a string (modeling a train crossing a bridge) was first treated by Willis (1849) who wrote the differential equation for the problem neglecting the inertia of the wire. Stokes (1849) solved the equation. The importance of the inertia was brought out later, Phillips (1855) and Saint-Venant (1883) writing more complete solutions. Further contributions to the impact problem are discussed in the survey by Goldsmith [I. 17].

Wave diffraction

The study of the diffraction of an elastic wave from an obstacle, like the beginning general studies of elastic wave propagation, had its origin in the elastic solid (aether) theory of light. Famous early works on the diffraction of light waves were the paper by Stokes (1849) mentioned earlier, which treated the diffraction of light by an aperture in a screen, and a series of papers by Rayleigh beginning in 1871 on the diffraction of light by small particles. These works and the further progress in studies on elastic wave diffraction are discussed in an interesting history in the book by Pao and Mow [8.11] on the topic.

Modern work and reading

Interesting in the history of the contributions to this subject is the relatively fallow period lying between the work of the classical elasticians in the nineteenth century and early twentieth century, and modern work which has been expanding at an increasing rate since World War II days. Aside from the fact that the subject offers intrigue and challenge there are several practical reasons for the modern expansion. One of the strongest, at least from the mechanics and engineering point of view, has been the continually growing need for information on the performance of structures subjected to high rates of loading. In geophysics the expanding research activity in elastic waves has also had strong underlying practical reasons such as the need for more accurate information on earthquake phenomena and improved prospecting techniques. Further, seismologists have been concerned with the nuclear detection problem. Developments in the related fields of acoustics and electromagnetic waves, and in applied mathematics in general, have also influenced the interest and progress made in the study of elastic waves. Last but not least, the electronic computer has been of considerable influence. As in many other fields, it has given numerical information for otherwise intractable problems.

The vastness of the modern contributions to the subject of waves in a linear, homogeneous, isotropic elastic medium precludes presenting an abstracted report on them here. The bulk of these studies were, and continues to be, concerned with problems involving boundaries and dispersion. The survey by Miklowitz [4.4] gives a fairly complete coverage of the pertinent literature to 1965. It shows that extended information exists now on transient wave propagation in the elastic half space. Through integral transforms, the Cagniard–deHoop inversion technique, similarity solutions, related asymptotics and other analytical and experimental methods, solutions for most cases of Lamb's problem (surface and buried sources of most types including traveling loads) have been derived and evaluated. Advances were also made on the problem of a buried spherical cavity source in the half space. Concern over underground protective construction created new interest in wave diffraction and scattering by an obstacle in the half space, hence in the related infinite medium problems involving cylindrical and spherical cavities. Some gains were also made in our understanding of elastic wave propagation in a wedge.

Concerning waveguides, extended information exists now on the frequency equations governing extensional, flexural and torsional waves in the infinite elastic rod, plate and cylindrical shell. Recent efforts have established the character of the higher real branches of the frequency spectrum (real frequency vs. real wave number), and the existence and character of the imaginary and complex branches of the spectrum (real frequency vs. imaginary and complex numbers) for these waveguides. This information, basic to transient excitation, and multi-integral transform and other methods, have produced integral solutions for various edge excited semi-infinite waveguides based on both the exact and approximate theories. Evaluation of the solution through asymptotics and numerical integration have produced a significant amount of information on the response of these waveguides.

Important advances have been made in the theory and solution of problems on the diffraction of a plane pulse by a semi-infinite plane boundary (a slit or rigid barrier) in the infinite elastic solid. In two-dimensional diffraction and related crack problems, methods (1) using similarity solutions with and without solutions of integral equations of the Wiener–Hopf-type, and (2) involving the Laplace transform, Wiener–Hopf-type integral equations and the Cagniard–deHoop inversion technique, have been productive. Related interesting advances were made in the study of finite plane cracks and obstacles using other integral equation techniques. In addition, three-dimensional integral representations for the displacement and acceleration vectors have been used in particular problems. Further detail is left to [4.4] which also contains references to earlier surveys.

Contents of present book

Chapter 1 of the present book is entitled *Introduction to linear elastodynamics*. Since the theory of elastic waves and waveguides is based on the classical theory of elasticity, the chapter sets down from the latter the definition of basic quantities, governing equations, the fundamental problems and the uniqueness of their solutions. The treatment presents what is need from the classical theory to attain the objectives of our subject.

Chapter 2, entitled *The fundamental waves of elastodynamics and their representations*, is concerned with certain basics of integrating the elastodynamic displacement equations of motion, and analyzing the general nature of wave solutions so found. The chapter begins with a treatment of body waves, i.e., interior medium waves, dilatational and equivoluminal, and the Lamé solution of the displacement equations of motion which is comprised of both waves. Types of these waves, plane, cylindrical etc., their symmetries and time nature are then discussed. A treatment of propagation of surfaces of discontinuity, and related wavefronts, characteristics and rays follows. An important class of problems, those due to body force disturbances (interior disturbances due to an external source, e.g., gravity), are then discussed at length, followed by a treatment of the one-, three- and two-dimensional initial value problems. The chapter concludes with a study of the method of characteristics for one-dimensional initial value and boundary-initial value problems.

Chapter 3, is entitled *Reflection and refraction of time harmonic waves at an interface*. It presents an extensive study of wave reflection, refraction and generation at a single planar interface for mostly plane waves, harmonic (sinusoidal) in time and (two-dimensional) space. It will be seen that these relatively simple waves (P, SV and SH) bring out basic information on reflection and refraction phenomena, and new waves peculiar to the interface, that are general properties of the more complicated time-dependent waves studied later in the book. The chapter begins with consideration of wave reflection from the boundary of an elastic half space, i.e., the interface is one between two half spaces, one elastic and the other a vacuum. The more general case of two half spaces in welded contact is treated later in the chapter. In both of these problems, all special cases are treated in detail, e.g., normal and grazing incidence, reflection and refraction at critical angles, total reflection, reflection and refraction of wave pairs and their generation of Rayleigh surface and Stoneley interface waves.

Chapter 4, entitled *Time harmonic waves in elastic waveguides*, is a natural extension of Chapter 3. Here we introduce a second planar boundary parallel to the surface of an elastic half space creating an infinite plate or layer. Now the P, SV and SH waves, studied in Chapter 3, reflect from boundary to neighboring boundary, generally (in the P and SV wave cases) undergoing mode conversion at each reflection, and progressing along the length of the plate. The neighboring parallel boundaries are in effect guiding the waves along the plate. This example of a waveguide, as well as others, e.g., rod, cylindrical shell and layered elastic solid, have the common feature of two or more parallel boundaries which introduce one or more characteristic lengths into a problem. These characteristic lengths lead to wave dispersion which is characterized by a dependence of frequency on wavelength. We study in detail the plate in plane strain and the rod, drawing on the modern works of Mindlin, Onoe and coworkers and noting the other waveguides can be treated similarly.

Chapter 5, entitled *Integral transforms, related asymptotics and introductory applications*, paves the way for solving the fundamental time-dependent boundary value problems of the subject carried out in Chapters 6, 7 and 8. As the title indicates we set down here the basics of integral transforms and related asymptotics and the beginnings of their applications in our subject. Starting with the Fourier integral theorem, the theory and properties of the (one-sided) Laplace transform, the bilateral (or two-sided) Laplace transform, the exponential Fourier transforms (of real and complex argument), the Fourier sine and cosine transforms and the Hankel transforms are developed. Then, after a brief introduction to asymptotic expansions and their properties, we discuss asymptotic expansions of integrals and, in particular, the Laplace and Fourier integrals of prime interest to our subject. In this, detailed discussions are given of Laplace's method, the method of steepest descents, the method of stationary phase and the asymptotics of the Laplace transform, which form the tools for long and short time (after the wavefront arrival) approximations in wave problems. Lastly, the problems of spherical and cylindrical cavity sources in the infinite solid are treated. Contour integrations produce the exact solutions and asymptotics the short- and long-time approximations.

Chapter 6, entitled *Transient waves in an elastic half space* treats the basic boundary value problems for the half space through modern integral transform methods. The plane-strain (Lamb's) problems for the surface normal line load source and buried line dilatational source are treated by the Cagniard–deHoop method. In the first of these problems, wavefront approximations are worked out by two methods, a special method used with the Cagniard–deHoop technique and the method of steepest descents. For Lamb's axially symmetric problems for the surface and buried vertical point loads we follow Pekeris' work and method, related to Cagniard's method, but independently developed. Lastly, the chapter presents part of a complete exact solution and its derivation for the problem of the suddenly applied normal point load that travels on the surface of the half space. The solution by Gakenheimer and Miklowitz represents the first application of the Cagniard–deHoop method to a nonaxisymmetric problem.

Chapter 7, entitled *Transient waves in elastic waveguides*, a natural extension of Chapter 4 with its addition of the time variable, enables us to discuss more physically realistic waveguide problems. In effect, here we learn the techniques for integrating over the frequency spectra set down in Chapter 4. The chapter begins with a discussion of approximate theories and one-dimensional problems. Derivation of the classical approximate theories for extensional and flexural waves in a thin rod (Love–Rayleigh, Bernoulli–Euler and Timoshenko theories) is carried out by Hamilton's principle, and boundary-initial value problems based on these theories are solved exactly and approximately using the Laplace transform, contour integration and asymptotics. Problems for the infinite plate in plane strain follow, being solved by a technique given by Lloyd and Miklowitz involving a double integral transform (Laplace on time, exponential Fourier on propagation coordinate). Contour integrations in the planes of the two transform parameters (related to frequency and wave number) lead to a direct correspondence between the component parts of the frequency spectrum and the individual integrals comprising a transient wave solution. This permits a study to be made of component waves in the solution through numerical evaluation of the related integrals. Such evaluations, and related approximations obtained with the aid of the method of stationary phase, are discussed.

Edge load problems for the semi-infinite waveguide form the next topic of Chapter 7. Such problems with their basic corner difficulties have been solved only recently by Skalak (1957), Folk et al. (1958) and DeVault and Curtis (1962) for the rod with mixed edge conditions (a mixture of stress and displacement components specified), and by Miklowitz (1969) and Sinclair and Miklowitz (1975) for the plate in plane strain with nonmixed edge conditions (stress or displacement components specified). The former class of problems are separable. They are solved directly through a double integral transform technique. The latter class of problems are nonseparable. They require in addition to a double transform, integral equations on the edge unknowns, and a boundedness condition on the solution to solve the integral equations. Both techniques and their applications to these problems are examined in detail with long time-far field and short time-near field approximations being deduced. A discussion of related work for other waveguide problems, including axially symmetric ones, is also presented in the chapter.

Chapter 8, entitled *Pulse scattering by half-plane, cylindrical and spherical obstacles*, treats elastodynamic scattering problems that are related to the classical ones in optics, acoustics and electromagnetic waves. The methods presented, however, are modern ones involving integral transforms, integral quations and certain other associated techniques. The first half of the chapter is devoted to cases of diffraction of a plane-elastic pulse by a half-plane scatterer. They are mixed boundary value problems. Specifically, we reat the diffraction of a plane SH-pulse (a Sommerfeld-type problem) by a traction-free half plane, followed by the analogous, but more complicated, case involving the P-pulse. The Laplace transform of the solution in each case is obtained as the solution of a set of dual integral equations, reduced to algebraic ones by a Wiener–Hopf technique inherent in a procedure presented by Clemmow (1951) in related electromagnetic wave problems. Inversion is accomplished by the Cagniard–deHoop method, essentially following earlier work by deHoop (1958). Finally the wave systems in each are discussed.

The second part of the chapter is concerned with diffraction of an elastic pulse from circular cylindrical and spherical scatterers. Such problems have only recently been attacked, and the treatment here reflects this in the methods and results presented. The cylindrical cavity scatterer is treated first by an integral transform method given by Miklowitz (1963, 1966) and Peck and Miklowitz (1969) incorporating Friedlander's (1954) representation of the solution (used in related acoustics problems). Friedlander's representation of the solution is a series having terms of wave form corresponding to propagation in the circumferential direction (about the cavity). This representation is very accurate for early times (not necessarily the earliest) at a station. Two problems are treated, the suddenly applied, normal line load, and the plane wave impingement, on the cavity wall. For long time in the far field (in θ) it is shown that Rayleigh waves are predominant in both problems. For the second problem an exact inversion for the shorter times and the near field, in the form of integrals over modes of propagation, show that the lowest modes (those with smallest imaginary wave numbers) for dilatation, equivoluminal and Rayleigh waves predominate. The technique and results of related earlier work by Baron, Matthews and Parnes (1961, 1962), using a Fourier series technique for the second problem, are also discussed and compared with the foregoing method and results of Miklowitz and Peck. Further related works on approximations are reviewed for the rigid and elastic cylindrical scatterers. Finally a brief review of work is presented on wave scattering by the circular cylindrical elastic inclusion with resultant wave focusing, and wave diffraction from a spherical cavity.

Lastly a section on *Supplementary Reading* is presented. Its purpose is to guide the reader to other important works that are (1) natural extensions of the text material, and (2) on topics dealing with additional effects in the linear elastic medium not treated in the book, because of limitations on time and space, e.g., waves in anisotropic media.

Other books on elastic waves and related subjects

Some brief remarks are in order on some other books in this subject as well as those on related subjects. The book by Kolsky [3.1] published in 1952 contains an introduction to elastic waves. The book by Ewing et al. [3.11], published in 1956 and oriented toward seismology, is a comprehensive treatment of elastic waves with material and extensive references on most topics. Of note also in seismology are the books by Bullen [I.18] published in 1954, Cagniard, first published in French in 1939, and translated and revised by Dix and Flinn in 1962 [3.19], and Brekhovskik [1.19] the translated Russian counterpart of [3.11] published in 1960. More recent comprehensive treatments of elastic waves are the books by Achenbach [2.14] published in 1973, Eringen and Suhubi [1.20] and Graff [1.21] published in 1975.

The following books on special topics in elastic waves are of note: Redwood [3.3] on waveguides, Viktorov [3.12] on Rayleigh and Lamb (plate) waves, Pao and Mow [8.11] on wave diffraction and Auld [1.22] on the theory of waves in a piezoelectric-elastic solid and its application to problems in scattering, waveguides and resonators.

As for related topics the following books will be of interest: On acoustics, the classic treatise by Lord Rayleigh [7.1, both volumes], Friedlander's book [2.10] dealing with modern mathematical techniques for solving problems involving sound pulse reflection and diffraction, and the book by Morse and Ingard [1.23] on theoretical acoustics. On optics, we reference, of course, Sommerfeld [3.7], for waves in water, Stoker [1.24], and for electromagnetic waves, Jones [2.18]. On methods in wave propagation, for dispersive waves we have already referenced Havelock's monograph [4.1]. The recent book of Brillouin [4.2] on this topic is also of note. Finally, we reference two books important to mathematical methods in basic wave phenomena. The first is the Courant–Hilbert volume [2.20] on partial differential equations (by Courant), in particular its chapters on hyperbolic equations in two or more independent variables, and its discussions of the theory of characteristics and rays that are fundamental to wave propagation phenomena. The second is the recent comprehensive treatment by Whitham [1.25] of the theory of linear and nonlinear waves, with applications drawn from acoustics, optics, water waves and gas dynamics.

References

[I.1.] A. Clebsch, *Journal für Reine und Angewandte Mathematik* **61** (1863), 195.
[I.2.] C. Somigliana, *Atti Reale Accad. Linc. Roma,* Ser 5, **1** (1892), 111.
[I.3.] O. Tedone, *Mem. Reale Accad, Scienze Torino,* Ser 2, **47** (1897), 181.
[I.4.] P. Duhem, *Mém. Soc. Sci. Bordeaux,* Ser. V, **3** (1898), 316.
[I.5.] H. Nakano, *Japan Journal of Astronomy and Geophysics* **2** (1925), 233–326.
[I.6.] H. Nakano, *Geophysics Magazine (Tokyo)* **2** (1930), 189–348.
[I.7.] J. N. Goodier, *The Mathematical Theory of Elasticity, Surveys in Applied Mathematics I,* John Wiley and Sons, Inc., New York (1958), 1–47.
[I.8.] H. Lamb, *Philosophical Magazine* [6] **33** (1916), 386–399, 539–548.
[I.9.] C. G. Knott, *Philosophical Magazine* [5] **48** (1899), 64–97.

[I.10.] G. W. Walker, *Philosophical Transactions of the Royal Society (London) A* **218** (1919), 373–393.
[I.11.] H. Jeffreys, *Monthly Notices of the Royal Astronomical Society: Geophysics Supplement* **1** (1926), 321–334.
[I.12.] H. Jeffreys, *Proceedings of the Cambridge Philosophical Society* **22** (1926), 472–481.
[I.13.] Lord Rayleigh, *Proceedings of the London Mathematical Society* **20** (1888–1889), 225–234.
[I.14.] H. Lamb, *Proceedings of the London Mathematical Society* **21** (1889–1890), 85.
[I.15.] H. Lamb, *Proceedings of the Royal Society of London A* **93** (1917), 114–128.
[I.16.] L. H. Donell, *Transactions of the American Society of Mechanical Engineers* **52** (1930), 153–167.
[I.17.] W. Goldsmith, *Impact, The Collision of Solids.* In: Applied Mechanics Surveys, eds. H. Abramson, H. Liebowitz, J. M. Crowley and S. Juhasz, Spartan Books, Washington D. C. (1966), 785–802.
[I.18.] K. E. Bullen, *An Introduction to the Theory of Seismology,* 2nd Edition. Cambridge University Press (1953).
[I.19.] L. M. Brekhovskikh, *Waves in Layered Media, Applied Mathematics and Mechanics* **6**. Academic Press, New York (1960).
[I.20.] A. C. Eringen and E. S. Suhubi, *Elastodynamics Volume 2: Linear Theory.* Academic Press, New York (1975).
[I.21.] K. F. Graff, *Wave Motion in Elastic Solids.* Ohio State University Press, Columbus, Ohio (1975).
[I.22.] B. A. Auld, *Acoustic Fields and Waves in Solids, Volumes 1, 2.* John Wiley and Sons, New York (1973).
[I.23.] P. M. Morse and K. U. Ingard, *Theoretical Acoustics.* McGraw–Hill Book Company, New York (1968).
[I.24.] J. J. Stoker, *Water Waves.* Interscience Publishers, Inc., New York (1957).
[I.25.] G. B. Whitham, *Linear and Nonlinear Waves.* John Wiley and Sons, New York (1974).

CHAPTER 1

INTRODUCTION TO LINEAR ELASTODYNAMICS

1.1. Introduction; Description of deformation and motion

The theory of elastic waves and waveguides is based on the classical theory of elasticity. This chapter therefore sets down from the latter the definition of basic quantities, governing equations, and problems and the uniqueness of their solutions. The treatment presents just the material that is needed to attain the objectives of our subject. Recommended references for the reader are the books by Sokolnikoff [1.1, Chapters 1, 2 and 3], Love [1.2, Chapters I, II, III and VII], and Nowacki [1.3, Chapter 1]. The historical introduction to the theory of elasticity in Love's book is very interesting. In reading it one is impressed by the array of great mathematicians of the 17th, 18th and 19th centuries, who found intrigue in the subject. In the introductory chapter we mentioned those early contributions pertaining to the history of elastic wave propagation. In this chapter we will mention some of the other early contributions to classical elasticity theory. In general however, we will depend on Love's historical introduction to earmark most of them.

There are two basic ways of describing *deformation and motion* in continuum mechanics. They are known as the Lagrangian, or material, and Eulerian, or spatial, descriptions. The *Lagrangian description uses the coordinates of a material point or particle, in its undeformed position, and time, as independent variables.* Assuming, for example, rectangular *Cartesian coordinates* x_1, x_2, x_3, denoting them collectively as x_i ($i = 1, 2, 3$), or the *position vector* x, *with axes fixed in space*, a later *deformed position of the particle* would be given by $x'_i = x'_i(x, t)$. In the *Eulerian description the coordinates of the particle in the deformed position, and time, are taken as independent variables*, i.e., $x_i = x_i(x', t)$, *the undeformed position of the particle is a function of its deformed position and time.* As Sokolnikoff [1.1, § 11], for example, shows, the two descriptions coalesce when the deformation is infinitesimal. Therefore, *in this book*, since we will be dealing exclusively with the linear elastic medium which, as we shall see later, is restricted to infinitesimal strains, the natural and simpler *Lagrangian description will be used*. It follows that the analysis of the stress (§ 1.3) can be done on the basis of the undeformed medium described by the coordinates x_i. This then gives consistency of the coordinates in the material that follows; the analysis of strain (§ 1.4), stress-strain relations (§ 1.5), dynamic equilibrium (§ 1.6), and the related later work in this chapter.

1.2. Tensor notation

Tensor notation permits a compact expression to be written for the equations of mathematical physics that also indicates the form natural laws should take. In particular, Cartesian tensors have been of value in writing the theory of mechanics of a continuous medium, and we will work with these here. A vector F in our Cartesian coordinates x_i (a quantity with magnitude and direction which adds according to the parallelogram law) has the components F_i which are orthogonal projections of F on the coordinate axes. We shall refer to F as the *vector F_i*. Similarly we may have a set of nine quantities such as a_{ij} ($i, j = 1, 2, 3$). Use will be made of the *summation convention* which states that a repeated subscript implies summation over all values the subscript can take, e.g., $a_{ij}x_j = a_{i1}x_1 + a_{i2}x_2 + a_{i3}x_3$. Note that $a_{ij}x_j = a_{ik}x_k$, where i plays the role of a free subscript, and j and k the role of dummy subscripts. Use will be made of the *Kronecker delta* defined as

$$\delta_{ij} = \begin{cases} 1, & \text{for } i=j, \\ 0, & \text{for } i \neq j. \end{cases}$$

Permutation symbols are defined as follows:

$$c_{ijk} = \begin{cases} 1, & \text{if } i, j, k \text{ are an even permutation of } 1, 2, 3, \\ -1, & \text{if } i, j, k \text{ are an odd permutation of } 1, 2, 3, \\ 0, & \text{otherwise.} \end{cases}$$

A permutation is just an interchange of two subscripts, e.g., two permutations of 231 produce 123. If X and Y are two vectors, their *scalar product* is given by $X \cdot Y = X_1 Y_1 + X_2 Y_2 + X_3 Y_3 = X_i Y_i$, and their *vector product* by $(X \times Y)_i = c_{ijk} X_j Y_k$.

1.3. Analysis of stress

1.3.1. Body and surface forces

The forces acting on a body are divided into two groups; body and surface forces. Consider the arbitrary closed region of volume V in a body occupying the region R in the space x_i shown in fig. 1.1. $P(x)$ is a point

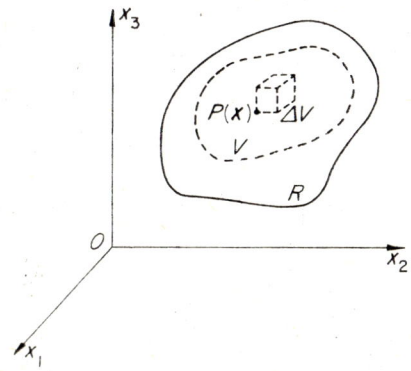

Fig. 1.1. Body R, enclosed volume V and points P in the space x_i.

in V, and ΔV is the element of volume at P. The total body force acting on ΔV, which is created by a source external to the body (like gravity), is taken equal to $\tilde{X}_i$. We define the *body force per unit volume at P* as

$$X_i = \lim_{\Delta V \to 0} \frac{\tilde{X}_i}{\Delta V},$$

assuming this limit to exist and to be independent of the choice of ΔV. Then

$$B_i = \int_V X_i \, dV \tag{1.1}$$

is the total body force acting on V. We note also that the total moment about the origin O, of the body force acting on V is

$$M_i = \int_V c_{ijk} x_j X_k \, dV. \tag{1.2}$$

Surface forces are the forces acting across any surface in the body including its boundary. These forces are due to reactions between adjacent particles. As shown in fig. 1.2 we let S be a surface in V, with P a general point on S, and ΔS an element of S containing P. We call one side of S $(+)$, and the other $(-)$, and consider the surface forces exerted across ΔS by particles on (+) side on particles on (−) side. These forces are equivalent to a single force $\tilde{T}_i$ at P, and a couple $\tilde{G}_i$. We define the *force per unit area* exerted across S at P by particles on the (+) side on particles on (−) side as

$$T_i = \lim_{\Delta S \to 0} \frac{\tilde{T}_i}{\Delta S}.$$

T_i is called the *stress vector*. We note that it depends on the position and orientation of the plane on which it acts, i.e., $T_i = T_i(x, l)$ where l is the unit

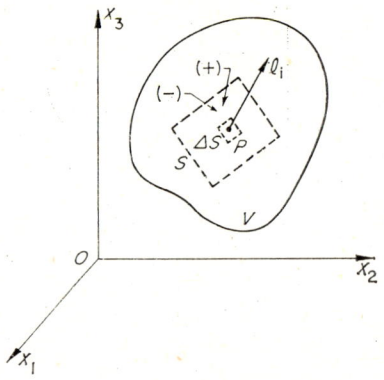

Fig. 1.2. Surface S with element ΔS at P.

normal vector to S at P in the (+) direction, which can also be represented by l_i, the *direction cosines* in the normal direction. It follows that the total surface force acting over the surface S is given by

$$S_i = \int_S T_i dS. \tag{1.3}$$

We further assume that

$$\lim_{\Delta S \to 0} \frac{\tilde{G}_i}{\Delta S} = 0.$$

1.3.2. Components of stress

Let us draw three planes through the general point $P(x)$ parallel to the coordinate axes, as shown in fig. 1.3. We assume the positive side of plane $x_1 = $ constant is the side from which x_1 increases. The stress vector at P for the plane $x_1 = $ constant is σ_{1j}. Similarly we have the stress vector at P for the plane $x_2 = $ constant as σ_{2j}, and for $x_3 = $ constant, σ_{3j}. Hence at P we have nine *components of stress* σ_{ij}; σ_{11}, σ_{22}, σ_{33} are the *normal stresses*, and others like σ_{12}, σ_{23}, etc., the *shear stresses*. The normal stress σ_{11}, for example, is positive for tension, and negative for compression in the x_1 direction.

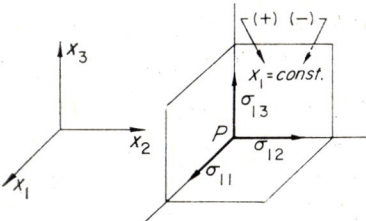

Fig. 1.3. Stress components of the stress vector σ_{1j} at P for the plane $x_1 =$ constant.

We can determine T_i in terms of σ_{ij} and l_i. Consider an element of area at P. l_i are the direction cosines of the positive normal to this element. T_i, σ_{ij}, and X_i are, at P, respectively, the stress vector for this element, the stress components, and the body force per unit volume. We draw a tetrahedron as shown in fig. 1.4 with a corner at P, three faces perpendicular to the

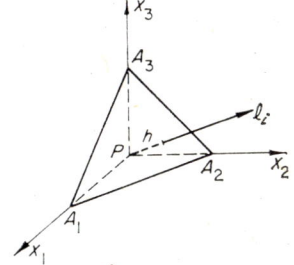

Fig. 1.4. Tetrahedron with corner at P.

cordinate axes, and the fourth face $A_1 A_2 A_3$ perpendicular to l_i, a small stance h from P. Now the average stress vectors over the faces $A_1 A_2 A_3$, $A_2 A_3$, $PA_3 A_1$, and $PA_1 A_2$, are respectively, $T_i + \varepsilon_i$, $\sigma_{1i} + \varepsilon_{1i}$, $\sigma_{2i} + \varepsilon_{2i}$, and $\sigma_{3i} + \varepsilon_{3i}$, and the average body force on the tetrahedron is $X_i + \Delta_i$, where

$$\lim_{h \to 0} (\varepsilon_i, \varepsilon_{ji}, \Delta_i) = 0.$$

Taking the area of face $A_1 A_2 A_3 = S$, and that of $PA_2 A_3$, $PA_3 A_1$, and $PA_1 A_2$, respectively as S_1, S_2, and S_3, we can write $S_i = Sl_i$. Using this relation, the total surface force on the tetrahedron can be written as $S[T_i + \varepsilon_i - l_j(\sigma_{ji} + \varepsilon_{ji})]$. The total body force on the tetrahedron is $V(X_i + \Delta_i)$, where $V = Sh/3$ is its volume. Equilibrium therefore requires that

$$S[T_i + \varepsilon_i - l_j(\sigma_{ji} + \varepsilon_{ji})] + (Sh/3)(X_i + \Delta_i) = 0.$$

Letting $h \to 0$ this reduces to

$$T_i = \sigma_{ji} l_j. \tag{1.4}$$

Later we will prove that σ_{ij} is a tensor of second order, and that it is symmetric, i.e., $\sigma_{ij} = \sigma_{ji}$, which we now assume.

The normal stress σ_n at P is the component of T_i in the direction of l_i. That is, $\sigma_n = T_i l_i$, which using (1.4), with $\sigma_{ji} = \sigma_{ij}$, gives

$$\sigma_n = \sigma_{ij} l_i l_j. \tag{1.5}$$

1.3.3. Transformation of the components of stress

To calculate the components of stress referred to a new set of rectangular coordinates, we consider the two sets x_i and x'_i having a common origin, as shown in fig. 1.5. We denote the direction cosines a_{ij} of the angles between

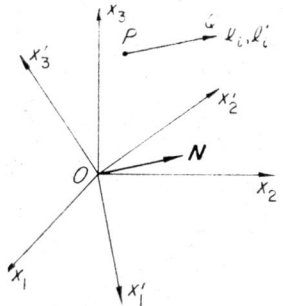

Fig. 1.5. General unit vector in the two spaces x_i, x'_i.

these two sets of axes as follows:

	x_1	x_2	x_3
x'_1	a_{11}	a_{12}	a_{13}
x'_2	a_{21}	a_{22}	a_{23}
x'_3	a_{31}	a_{32}	a_{33}

It follows we have the transformation

$$x'_i = a_{ij} x_j, \tag{1.6}$$

where a_{ij} satisfies the condition

$$a_{ij} a_{ik} = \delta_{jk}, \tag{1.7}$$

representing the orthogonality of the x_i set of axes. We obtain the inverse transformation for x_i by multiplying (1.6) by a_{ik}, and using (1.7), with the result

$$x_i = a_{ji}x'_j, \qquad (1.8)$$

where similarly here a_{ji} statisfies the condition $a_{ji}a_{ki}=\delta_{ik}$, representing the orthogonality of the x'_i set of axes.

Again taking P as the general point in body R, we let PQ be the general unit vector whose components are l_i and l'_i in the respective coordinate systems x_i and x'_i, as shown in fig. 1.5. ON there is the unit vector parallel to PQ. The coordinates of N relative to x_i and x'_i are l_i and l'_i, respectively. It follows from (1.6,) (1.8) that

$$l'_i = a_{ij}l_j, \qquad l_i = a_{ji}l'_j, \qquad (1.9)$$

which state the *laws of transformation of a vector*. That is, a quantity that transforms according to the laws (1.9) is called a first order tensor (a vector in the context of tensor analysis).

Now let σ_{ij} and σ'_{ij} be the stress components at P relative to the unprimed and primed system respectively. According to (1.5) the normal stress at P, corresponding to the direction PQ, is

$$\sigma_n = \sigma_{ij}l_il_j = \sigma'_{ij}l'_il'_j. \qquad (1.10)$$

Substituting for l_i and l_j from the second of (1.9), (1.10) gives

$$\sigma'_{ij}l'_il'_j = \sigma_{ij}a_{mi}l'_m a_{nj}l'_n,$$

which by interchanging m and i, and n and j on the right can be written as

$$(\sigma'_{ij} - a_{im}a_{jn}\sigma_{mn})l'_il'_j = 0. \qquad (1.11)$$

Since (1.11) holds for arbitrary l'_i, and the quantity in parenthesis is independent of l'_i, this quantity vanishes and we have

$$\sigma'_{ij} = a_{im}a_{jn}\sigma_{mn},$$

which, according to (1.6), can also be written as

$$\sigma'_{ij} = x'_{i,m}x'_{j,n}\sigma_{mn}, \qquad (1.12)$$

where we have used the tensor notation $x'_{i,m}$ for $\partial x'_i/\partial x_m$. Eq. (1.12) states the *law of transformation of a second order tensor*. That is, a quantity that transforms according to the law (1.12) is called a second order tensor. Hence our stress components σ_{ij} are components of a second order tensor.

1.3.4. The stress quadric and the principal stresses

We now introduce the coordinates ξ_i with origin at P, and axes parallel to those of x_i, as shown in fig. 1.6. σ_{ij} are the stress components at P. The

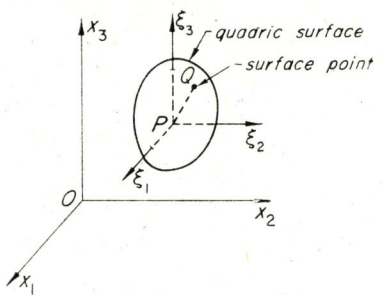

Fig. 1.6. Cauchy's stress quadric.

second degree (in ξ_i) equation

$$\sigma_{ij}\xi_i\xi_j = \pm k^2, \tag{1.13}$$

where k is a constant, is a quadric surface. In particular it is the so-called *stress quadric of Cauchy*. The center of the quadric is at P, and Q is a general point on the quadric with coordinates ξ_i. The coordinates of Q satisfy (1.13). The quadric represents a geometric definition of stress at the point P, through the following two properties:

(1) As Q moves over the quadric, the normal stress corresponding to the direction PQ varies inversely as r^2, where r is the length of PQ depicted in fig. 1.7. This is proved by first noting the direction cosines of PQ are $l_i = \xi_i/r$. Substituting these in (1.5), and using (1.13) gives $\sigma_n = \pm k^2/r^2$.

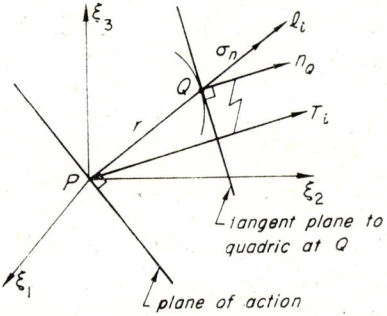

Fig. 1.7. Relation between stress vector at P and normal to stress quadric at Q.

(2) The stress vector at P corresponding to the direction PQ is parallel to the normal to the stress quadric at Q. This is proved by first noting the stress vector at P, corresponding to the direction PQ, is $T_i = \sigma_{ij}\xi_j/r$. The direction components of the normal to the quadric at Q, n_Q shown in fig. 1.7, are $\partial F/\partial \xi_i$, where $F = \sigma_{kl}\xi_k\xi_l$. It follows that $\partial F/\partial \xi_i = \sigma_{il}\xi_l + \sigma_{ki}\xi_k$. Changing l to j, in the first term, and k to j in the second term, and recalling that $\sigma_{ij} = \sigma_{ji}$, this equation reduces to $\partial F/\partial \xi_i = 2\sigma_{ij}\xi_j = 2rT_i$. Hence, the normal n_Q and T_i are parallel.

Every quadric surface has three principal axes. They intersect at its center, are mutually perpendicular, and also pierce the surface of the quadric orthogonally. The directions of the principal axes of the stress quadric are called the *principal directions of stress*. The planes containing these directions in pairs are called the *principal planes of stress*.

Consider again the direction l_i at P. We have the following properties:

(1) The normal stresses at P, corresponding to the principal directions of stress at P, are extrema. Proof follows from property (1) of the quadric.

(2) The stress vectors at P, corresponding to the principal directions of stress at P, lie along the respective principal directions. Proof follows from property (2) of the quadric.

The normal stresses at P, corresponding to the principal directions of stress at P, are called the *principal stresses*, denoted by σ_i. If we choose the coordinate axes x_i parallel to the principal directions of stress at P, then at P, σ_{ij} is as follows:

$$\sigma_{11} = \sigma_1, \quad \sigma_{12} = 0, \quad \sigma_{13} = 0,$$
$$\sigma_{21} = 0, \quad \sigma_{22} = \sigma_2, \quad \sigma_{23} = 0,$$
$$\sigma_{31} = 0, \quad \sigma_{32} = 0, \quad \sigma_{33} = \sigma_3.$$

This discloses the important fact that the principal planes of stress are free of shear stress.

To determine the principal stresses we suppose l_i denotes a principal direction. Then, if we let σ be the corresponding principal stress, we have $T_i = \sigma l_i$, and from (1.4), with $\sigma_{ji} = \sigma_{ij}$, it follows that

$$(\sigma_{ij} - \sigma \delta_{ij})l_j = 0. \tag{1.14}$$

This set of three homogeneous equations for the unknown directions l_j, has a nonvanishing solution if, and only if, the determinant of the coefficients is zero, i.e.,

$$|\sigma_{ij}-\sigma\delta_{ij}|=0. \tag{1.15}$$

Expanding (1.15), one finds

$$|\sigma_{ij}-\sigma\delta_{ij}|=-\sigma^3+I_1\sigma^2-I_2\sigma+I_3=0,$$

where

$$I_1 = \sigma_1+\sigma_2+\sigma_3,$$
$$I_2 = \sigma_1\sigma_2+\sigma_2\sigma_3+\sigma_3\sigma_1,$$
$$I_3 = \sigma_1\sigma_2\sigma_3.$$

This cubic has the three real roots σ_1, σ_2, and σ_3, the principal stresses. Since the principal stresses characterize the state of stress at a point, they must be independent of the choice of coordinate system. It follows that I_1, I_2 and I_3 must be invariant with respect to an orthogonal transformation of coordinates. They are known as the first, second and third *stress invariants*. When the principal axes are not parallel to the coordinate axes, expansion of (1.15) shows

$$I_1 = \sigma_{11}+\sigma_{22}+\sigma_{33},$$
$$I_2 = \sigma_{11}\sigma_{22}+\sigma_{22}\sigma_{33}+\sigma_{33}\sigma_{11}-\sigma_{12}^2-\sigma_{23}^2-\sigma_{31}^2,$$
$$I_3 = \sigma_{11}\sigma_{22}\sigma_{33}+2\sigma_{12}\sigma_{23}\sigma_{31}-\sigma_{11}\sigma_{23}^2-\sigma_{22}\sigma_{31}^2-\sigma_{33}\sigma_{12}^2.$$

To find the direction cosines corresponding to σ_1, we set $\sigma=\sigma_1$ in (1.14), and solve for the l_i, which determine principal directions associated with σ_1. Likewise this procedure is repeated for the other principal directions corresponding to σ_2 and σ_3.

1.3.5. *Maximum shear stress*

As is shown in fig. 1.8, the stress vector T_i at P, for the direction l_i, can

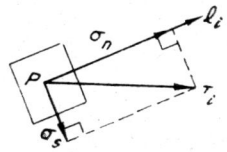

Fig. 1.8. Stress vector T_i and its rectangular components, normal stress σ_n and shear stress σ_s.

be decomposed into the rectangular components σ_r, the normal stress, and σ_s which we define as the *shear stress* in the plane of action at P. We wish to determine the maximum shear stress and its plane of action, which are important quantities in stress analysis. We note first that

$$\sigma_s^2 = T_i T_i - \sigma_n^2.$$

Now choosing the coordinate axes parallel to the principal directions of stress, and using (1.4), (1.5), this relation becomes

$$\sigma_s^2 = \sigma_i^2 l_i^2 - \sigma_i l_i^2 \sigma_j l_j^2, \tag{1.16a}$$

where σ_i are the principal stresses at P. We assume at the outset that $\sigma_1 > \sigma_2 > \sigma_3$. In order to determine the values of l_i that make σ_s a maximum, we can make use of the method of Lagrange multipliers. σ_s will be an extremum when

$$\frac{\partial}{\partial l_k}(\sigma_s^2 + \lambda l_i l_i) = 0, \qquad k = 1, 2, 3, \tag{1.16b}$$

where λ is the Lagrange multiplier. Equations (1.16b), and the relation $l_i l_i = 1$, are four equations for the l_i and λ, corresponding to an extremum σ_s^2. Substituting (1.16a) in (1.16b), and carrying out the differentiation for the three values of k, we find (1.16b) reduces to the three equations

$$l_j [\sigma_j^2 - 2\sigma_i l_i^2 \sigma_j + \lambda] = 0, \qquad j = 1, 2, 3. \tag{1.16c}$$

Not all the l_j's can vanish simultaneously since we have $l_i l_i = 1$. First, the we have solutions of (1.16c) when one of the l_j's does not vanish but the other two do. They are

$$\begin{array}{lll} l_1 = \pm 1, & l_2 = 0, & l_3 = 0; \\ l_1 = 0, & l_2 = \pm 1, & l_3 = 0; \\ l_1 = 0, & l_2 = 0, & l_3 = \pm 1. \end{array} \tag{A}$$

We also have cases where one of the l_j's vanish but the other two do not. Consider the case $l_1 = 0$, $l_2 \neq 0$ and $l_3 \neq 0$. This case reduces to the solution of the second and third equations of (1.16c) ($j = 2, 3$) and the equation $l_2^2 + l_3^2 = 1$. Subtracting the second of these equations from the first, and substituting the third in this, we find $l_2^2 = \frac{1}{2}$, where use was made of $\sigma_2 - \sigma_3 \neq 0$. It follows that $\lambda_3^2 = \frac{1}{2}$. Thus the three solutions of this type are

$$l_1 = 0 \qquad l_2 = \pm \frac{\sqrt{2}}{2}, \qquad l_3 = \pm \frac{\sqrt{2}}{2};$$

$$l_1 = \pm \frac{\sqrt{2}}{2}, \qquad l_2 = 0, \qquad l_3 = \pm \frac{\sqrt{2}}{2}; \qquad \text{(B)}$$

$$l_1 = \pm \frac{\sqrt{2}}{2}, \qquad l_2 = \pm \frac{\sqrt{2}}{2}, \qquad l_3 = 0.$$

For the case when no l_j vanishes we have all three equations in (1.16c). Subtracting these in pairs they reduce to

$$\sigma_1 + \sigma_2 = 2\sigma_i l_i^2,$$

$$\sigma_2 + \sigma_3 = 2\sigma_i l_i^2,$$

$$\sigma_3 + \sigma_1 = 2\sigma_i l_i^2$$

which are consistent only if $\sigma_1 = \sigma_2 = \sigma_3$. But $\sigma_1 > \sigma_2 > \sigma_3$ was imposed earlier so we have no solutions of the present type. Our possible solutions are (A) and (B). (A) gives the principal directions of stress which we have already pointed out correspond to zero shear stress (a minimum shear stress in the present context). The direction corresponding to the maximum shear stress must therefore be contained in the three solutions (B). To find the shear stresses, say τ_i, corresponding to these solutions, we substitute each of them into (1.16a) with the results

$$\tau_1 = \tfrac{1}{2}|\sigma_2 - \sigma_3|, \qquad \tau_2 = \tfrac{1}{2}|\sigma_3 - \sigma_1|, \qquad \tau_3 = \tfrac{1}{2}|\sigma_1 - \sigma_2|.$$

Since $\sigma_1 > \sigma_2 > \sigma_3$, τ_2 *is the maximum shear stress*. It may be seen from the l_i sets (B) that each of the corresponding stresses τ_i act on a surface element that contains one principal axis and bisects the angle between the other two. The maximum shear stress τ_2 therefore acts on the surface elements that contain the σ_2 axis and bisect the angle between the σ_1 and σ_3 axes.

1.4. Analysis of strain

1.4.1. Introduction

A body is strained, or deformed, whenever the distances between its particles are changed. Let us consider an unstrained body occupying our region R, as depicted in fig. 1.9. After straining, the body occupies the

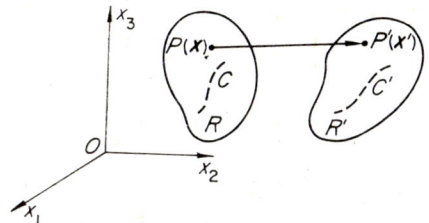

Fig. 1.9. The unstrained and strained bodies.

region R', the general point $P(x)$ having been strained into the positition $P'(x')$, where $x' = (x'_1, x'_2, x'_3)$. There exist relations of the form

$$x'_i = x'_i(x). \tag{1.17}$$

To preserve continuity of the body in deformation, the $x'_i(x)$ must be continuous functions (no dislocations can occur). We assume further that (1.17) has a unique continuous inverse

$$x_i = x_i(x'), \tag{1.18}$$

so that we have a one-to-oneness in the transformation of points in R to those in R'. If in addition to assuming the continuity of the functions $x'_i(x)$ and $x_i(x')$ we further demand that these functions have continuous first derivatives, then their Jacobian exists and is necessarily positive.

1.4.2. Finite deformation

Consider now, in fig. 1.9, the curve C in the region R, which deforms into the curve C' in region R'. With ds the differential arc length along C, and ds' that along C', we have relations

$$(ds)^2 = dx_i dx_i, \qquad (ds')^2 = dx'_i dx'_i. \tag{1.19}$$

From (1.17) we can write

$$dx'_i = x'_{i,j} dx_j. \tag{1.20}$$

It follows from (1.19), (1.20) that

$$(ds')^2 - (ds)^2 = 2\eta_{jk} dx_j dx_k, \tag{1.21}$$

where

$$\eta_{jk} = \tfrac{1}{2}(x'_{i,j} x'_{i,k} - \delta_{jk}),$$

are a set of strain components. We now introduce the *displacement components* $u_i(x)$, defined by

$$u_i(x) = x'_i(x) - x_i. \tag{1.22}$$

Note the displacements, $u_i(x)$ are necessarily continuous, since x'_i and x_i were required to have this property earlier (see the discussion in connection with eqs. (1.17), (1.18)). From (1.22) we have

$$x'_{i,j} = \delta_{ij} + u_{i,j}.$$

It follows from this relation that η_{jk} in (1.21) can be written as

$$\eta_{jk} = \tfrac{1}{2}(u_{j,k} + u_{k,j} + u_{i,j} u_{i,k}). \tag{1.23}$$

which are the finite strain components.

1.4.3. Infinitesimal strain

By requiring the displacement gradients in (1.23) to be small, i.e.,

$$|u_{i,j}| \ll 1, \tag{1.24}$$

(1.23) reduces to

$$\eta_{ij} \simeq \varepsilon_{ij} = \tfrac{1}{2}(u_{i,j} + u_{j,i}), \tag{1.25}$$

he *infinitesimal strain components* of classical elasticity theory. We see herefore that (1.24), through its linearization of η_{ij}, is fundamental to this heory. From (1.21) we see that ε_{jk} governs the change in elemental arc ength during deformation. If $\varepsilon_{jk} = 0$, then $ds' = ds$, which means linear elements do not undergo deformation. In this case the body can still be displaced as a rigid body, i.e., (1.17) may represent a rigid body rotation, or translation, or both, in some cases.

Consider a line element with $ds = dx_1$, and $dx_2 = dx_3 = 0$. In this case, according to (1.21), we have

$$ds' = (1 + 2\varepsilon_{11})^{\frac{1}{2}} dx_1. \tag{1.26}$$

The *extension*, or *elongation per unit length*, of the line element in the x_1 direction is defined by

$$e_1 = \frac{ds' - ds}{ds} = \frac{ds' - dx_1}{dx_1}. \tag{1.27}$$

Substituting (1.26) into (1.27) gives

$$e_1 = (1 + 2\varepsilon_{11})^{\frac{1}{2}} - 1. \tag{1.28}$$

Corresponding definitions exist for e_2 and e_3, the extensions in the x_2 and x_3 directions, respectively. Now since strain ε_{11} is small compared to 1 according to (1.24), a binomial series expansion of the radical in (1.28) shows $e_1 \simeq \varepsilon_{11}$, and similarly we would have $e_2 \simeq \varepsilon_{22}$ and $e_3 \simeq \varepsilon_{33}$.

1.4.4. Relative displacements in neighborhood of point P

As depicted in fig. 1.10 we now introduce local coordinates ξ_i and ξ'_i

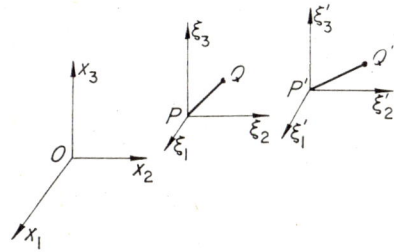

Fig. 1.10. Coordinate systems ξ_i and ξ'_i.

with origins at $P(x)$ and $P'(x')$, respectively, and axes parallel to x_i. We consider further the point $Q(\boldsymbol{\xi})$ near $P(x)$, and the point $Q'(\boldsymbol{\xi}')$ into which Q strains. Now using (1.22) the coordinates of Q' can be written as

$$x'_i + \xi'_i = x_i + \xi_i + u_i(x + \boldsymbol{\xi}). \tag{1.29}$$

Since Q is near P, we can expand the u_i here in a Taylor's series about P with the result

$$\eta_i = \xi'_i - \xi_i = (u_{i,j})_P \xi_j + O(\xi_i \xi_i). \tag{1.30}[1]$$

η_i represents the displacement of point Q relative to P, where the ( )$_P$

---

[1] The order symbols O and o are defined as follows: If as x tends to a limit, $\varphi(x)$ tends to 0 or ∞, and $f(x)/\varphi(x)$ is bounded, we write $f(x) = O[\varphi(x)]$, or that $f(x)$ is the same order of magnitude as $\varphi(x)$. If $f(x)/\varphi(x) \to 0$ as $\varphi(x) \to 0$, then $f(x) = o[\varphi(x)]$.

indicates evaluation at P. Now neglecting the second order term, (1.30) can be written as

$$\eta_i = u_{i,j}\xi_j = \tfrac{1}{2}(u_{i,j}+u_{j,i})\xi_j + \tfrac{1}{2}(u_{i,j}-u_{j,i})\xi_j, \tag{1.31}$$

where we have dropped the subscript P in (1.30), since evaluation of $u_{i,j}$ will always be made there. The first coefficient of ξ_j is recognized as the infinitesimal strains ε_{ij} defined in (1.25). The second coefficient defines the components of the *infinitesimal rotation*

$$\omega_{ij} = \tfrac{1}{2}(u_{i,j}-u_{j,i}), \tag{1.32}$$

which are skew-symmetric since $\omega_{ij} = -\omega_{ji}$. It follows that (1.31) can be written as

$$\eta_i = \varepsilon_{ij}\xi_j + \omega_{ij}\xi_j. \tag{1.33}$$

The elongation per unit length of PQ can be written by making use of (1.30). The elongation is given by

$$e = \frac{r'-r}{r}, \tag{1.34}$$

where r and r' are the lengths of PQ and P'Q', respectively. Now $r'^2 = \xi'_i\xi'_i$, and $r^2 = \xi_i\xi_i$, which using (1.30), lead to

$$r'^2 - r^2 = 2\varepsilon_{ij}\xi_i\xi_j,$$

for infinitesimal strains. It is easy to show from this, and (1.34), that

$$e = \varepsilon_{ij}l_i l_j. \tag{1.35}$$

1.4.4.1. The nature of the rotations. It has already been pointed out that if $\varepsilon_{ij} = 0$, the displacement must be a rigid body one. Under this condition (1.33) reduces to $\eta_i = \omega_{ij}\xi_j$. It is clear, therefore, from (1.33) that η_i has been decomposed into two displacements, one due to pure deformation ($\omega_{ij}=0$) and one to rigid body motion ($\varepsilon_{ij}=0$). From (1.32) we see $\omega_{11}=\omega_{22}=\omega_{33}=0$, and $\omega_{12} = -\omega_{21}$, $\omega_{23} = -\omega_{32}$, and $\omega_{31} = -\omega_{13}$. Redefining the existing ω_{ij} as

$$\omega_1 = -\omega_{23}, \qquad \omega_2 = -\omega_{31}, \qquad \omega_3 = -\omega_{12}, \tag{1.36a}$$

(1.33) can be expanded into

$$\begin{aligned}\eta_1 &= -\omega_3\xi_2 + \omega_2\xi_3,\\ \eta_2 &= -\omega_1\xi_3 + \omega_3\xi_1,\\ \eta_3 &= -\omega_2\xi_1 + \omega_1\xi_2.\end{aligned} \tag{1.36b}$$

Eqs. (1.36b) represent the displacements η_i due to the infinitesimal rotations ω_i about the axes ξ_i. To show this we consider the displacements due to ω_1 about the ξ_1 axis. With reference to fig. 1.11 we see the displacement due

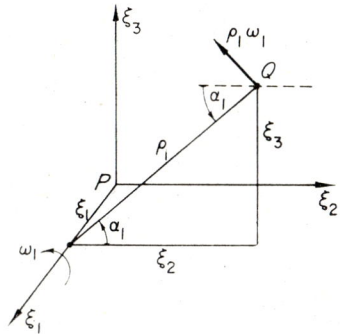

Fig. 1.11. Displacement due to rotation ω_1.

to ω_1 has the magnitude $\varrho_1\omega_1$ and lies in the plane through Q, perpendicular to the ξ_1 axis. The components of this displacement are $(0, -\omega_1\xi_3, \omega_1\xi_2)$, since $\cos \alpha_1 = \xi_2/\varrho_1$ and $\sin \alpha_1 = \xi_3/\varrho_1$. Note that these components yield the ω_1 contributions to the η_i in (1.36b). Similarly the reader can easily show that the components due to ω_2 are $(\omega_2\xi_3, 0, -\omega_2\xi_1)$, and to ω_3 are $(-\omega_3\xi_2, \omega_3\xi_1, 0)$. Adding the three gives the η_i in (1.36b). Finally we should take account of a possible rigid body translation of the point Q, say $\hat{u}_i$. Totally, then, we could have at Q, the displacements $u_i = \hat{u}_i + \eta_i$.

1.4.4.2. *The nature of the strains.* Returning to fig. 1.10, consider the line element PQ in the space ξ_i there, where Q is close to P again. We have already shown, where PQ lies parallel to the coordinate axes, for infinitesimal strains, that $\varepsilon_{11} = e_1$, $\varepsilon_{22} = e_2$, and $\varepsilon_{33} = e_3$, the e_i's being the respective

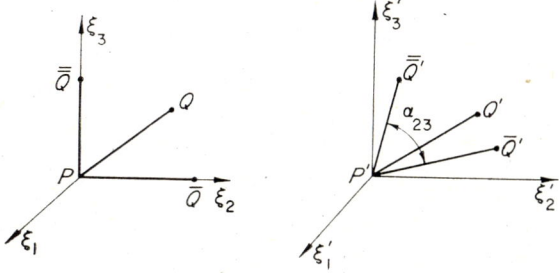

Fig. 1.12. General straining of the line element PQ.

elongations per unit length. Consider now the general case depicted in fig. 1.12, where PQ strains into $P'Q'$, the latter occupying the space ξ'_i. If we take Q to $\bar{Q}$ on the ξ_2 axis, the local coordinates are $(0, \bar{\xi}_2, 0)$. Eqs. (1.30) then show the local coordinates of $\bar{Q}'$ are $[u_{1,2}\bar{\xi}_2, \bar{\xi}_2(1+u_{2,2}), u_{3,2}\bar{\xi}_2]$. It follows that $(P'\bar{Q}')^2 = \bar{\xi}_2^2[1 + \text{higher order terms}]$, according to the condition (1.24), and hence that the direction cosines of $P'\bar{Q}'$ are $(u_{1,2}, 1, u_{3,2})$ to the first order. If we now take Q to $\bar{\bar{Q}}$ on the ξ_3 axis, with local coordinates $(0, 0, \bar{\xi}_3)$, and proceed as in the above case, we find that the direction cosines of $P'\bar{\bar{Q}}'$ are $(u_{1,3}, u_{2,3}, 1)$ to the first order. Now if we let α_{23} be the angle between the lines $P'\bar{Q}'$ and $P'\bar{\bar{Q}}'$, then

$$\cos \alpha_{23} = u_{2,3} + u_{3,2} + [\text{terms of higher order}],$$

or

$$\cos \alpha_{23} \simeq 2\varepsilon_{23}.$$

Now if we let β_{23} represent the decrease in angle between the original elements $P\bar{Q}$ and $P\bar{\bar{Q}}$, that has occurred through deformation, we have $\beta_{23} = (\pi/2) - \alpha_{23}$. It follows that $\sin \beta_{23} = \cos \alpha_{23} = 2\varepsilon_{23}$, and for the small angles β_{23} we are dealing with, $\varepsilon_{23} = (\tfrac{1}{2})\beta_{23}$. It is clear from this result we can represent the deformation in the $\xi_2\xi_3$ plane as shown in fig. 1.13. The original

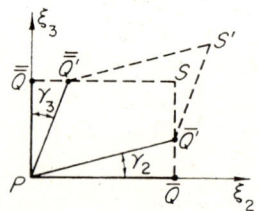

Fig. 1.13. Shear strain.

lines $P\bar{Q}$ and $P\bar{\bar{Q}}$ are angularly displaced by $\gamma_2 = u_{3,2}$ and $\gamma_3 = u_{2,3}$, respectively, and the rectangular element $P\bar{Q}S\bar{\bar{Q}}$ strains into the parallelogram $P\bar{Q}'S'\bar{\bar{Q}}'$. Since we are discussing pure deformation the angles γ_2 and γ_3 are equal. The deformation is caused by a sliding or shearing of elements parallel to the x_1x_2-plane in the x_2 direction, and those parallel to the x_1x_3-plane in the x_3 direction. Hence ε_{23} is called a *shear strain*. It is easy to see ε_{12}, and ε_{31} have the same basic nature.

1.4.5. Transformation of components of strain

Again here we have the line element PQ and the situation in fig. 1.5, with the associated eqs. (1.6)–(1.9). We have here, however, the strain components at P which are ε_{ij} and ε'_{ij}, relative respectively to the unprimed and primed coordinates. Here the elongation per unit length plays the analogous role to σ_n in (1.10), i.e., from (1.35)

$$e = \varepsilon'_{ij} l'_i l'_j = \varepsilon_{ij} l_i l_j, \tag{1.37}$$

and analogously to eqs. (1.10)–(1.12), (1.37) leads to

$$\varepsilon'_{ij} = x'_{i,m} x'_{j,n} \varepsilon_{mn}, \tag{1.38}$$

which shows the strain components ε_{ij} are the components of a second order tensor.

1.4.6. The strain quadric and the principal strains

The analogy with stress à la § 1.3.4 carries over to the strain. We have fig. 1.6 again with ε_{ij} the strains at P. The strain quadric is given by $\varepsilon_{ij}\xi_i\xi_j = \pm k^2$, where the constant k^2 is taken to be infinitesimal. We have two similar properties: (1) the elongation per unit length of PQ varies inversely as r^2, and (2) the displacement of Q relative to P, and due to pure strain only, is parallel to the normal to the strain quadric at Q. Proofs of these properties follow the same lines as the corresponding ones in § 1.3.4.

Associated with the principal axes and planes of the strain quadric we have the so-called *principal directions* and *planes of strain*. It follows that the principal directions of strain are mutually perpendicular and so are the principal planes of strain. Also it follows from property (1) here, that the elongations per unit length of elements, aligned with the principal directions of strain, are extrema. They are the *prinicpal strains* e_1, e_2 and e_3. From property (2) it follows that if PQ is in a principal direction of strain, the displacement of Q relative to P, and due to pure strain only, is in the direction of PQ. Hence *principal directions of strain are orthogonal before and after straining*.

If we choose the coordinates axes x_i parallel to the principal direction of strain at P, then ε_{ij} there is

$$\begin{array}{lll} \varepsilon_{11} = e_1, & \varepsilon_{12} = 0, & \varepsilon_{13} = 0, \\ \varepsilon_{21} = 0, & \varepsilon_{22} = e_2, & \varepsilon_{23} = 0, \\ \varepsilon_{31} = 0, & \varepsilon_{32} = 0, & \varepsilon_{33} = e_3. \end{array}$$

The determination of e_i, and their respective principal directions l_i, follows the same procedure as that used for σ_i and their directions (cf. (1.14), (1.15) and discussion there). Here we treat

$$(\varepsilon_{ij} - e\delta_{ij})l_j = 0,$$
$$|\varepsilon_{ij} - e\delta_{ij}| = 0,$$

and likewise have three strain invariants of the same form as I_1, I_2 and I_3.

1.4.7. The dilatation

Consider the volume element $dV = d\xi_1 d\xi_2 d\xi_3$, a rectangular parallelepiped at P with edges along the coordinate axes ξ_i (see fig. 1.10) which are parallel to the principal directions of strain at P. Suppose the volume element strains into another rectangular parallelepiped with respective faces parallel to those of the original, then the volume of this strained parallelepiped is $dV' = (1+e_1)(1+e_2)(1+e_3)d\xi_1 d\xi_2 d\xi_3$. The *dilatation* is defined as $\Delta = (dV' - dV)/dV$, the *increase in volume per unit volume*. Expanding the right hand side of this relation, and neglecting higher order terms, shows that

$$\Delta = e_1 + e_2 + e_3 = \varepsilon_{11} + \varepsilon_{22} + \varepsilon_{33} = u_{i,i}, \tag{1.39}$$

where we have used the fact that Δ is the first strain invariant.

1.5. Stress-strain relations

1.5.1. The generalized Hooke's law

In 1676, Robert Hooke published his law that in an elastic bar force is proportional to extension. For a one dimensional specimen (a rod) in a tension test this law evidences itself in a stress-strain relation of the form $\sigma = Ee$, a linear relation between stress $\sigma = F/A$ and the elongation per unit length e. F is the applied force, A the original cross sectional area of the specimen. The proportionality constant E is known as *Young's modulus* or the *modulus of elasticity*. The law holds up to a point known as the proportional limit, beyond which there is observable permanent or anelastic deformation.

Previous sections here showed that the stress tensor σ_{ij} determines the state of stress at a point in a continuous medium, and likewise the strain tensor ε_{ij} determines the state of strain. A natural generalization of Hooke's law then is to assume a functional relation between σ_{ij} and ε_{ij} in the medium, i.e., $\sigma_{ij} = T_{ij}(\varepsilon_{11}, \varepsilon_{22}, \cdots)$, where it is assumed that the temperature is fixed, and σ_{ij} is zero when the ε_{ij} is zero, which implies that the unstrained body is unstressed. If the T_{ij} functions are expanded in a power series in ε_{ij}, and linear terms only are retained, we get the *generalized Hooke's law*

$$\sigma_{ij} = c_{ijkl}\varepsilon_{kl}, \qquad (i,j,k,l=1,2,3), \qquad (1.40)$$

which is due to Cauchy. This law is basic to all work in linear elasticity. The c_{ijkl} in (1.40) are the *elastic coefficients*, which in general vary from point to point in the elastic body. If they are independent of the position, the body is said to be *elastically homogeneous*. We assume this is the case in our work here. The c_{ijkl} in (1.40) are 81 in number but, since $\sigma_{ij} = \sigma_{ij}$ and $\varepsilon_{kl} = \varepsilon_{lk}$, $c_{ijkl} = c_{jikl} = c_{ijlk}$, and these coefficients reduce to 36. We further assume the elastic coefficients satisfy the symmetry relation $c_{ijkl} = c_{klij}$ which reduce the c_{ijkl} to 21. Then the function $\hat{U}$ defined by

$$\hat{U} = \tfrac{1}{2}c_{ijkl}\varepsilon_{ij}\varepsilon_{kl}, \qquad (1.41)$$

has the property

$$\frac{\partial \hat{U}}{\partial \varepsilon_{ij}} = \sigma_{ij}. \qquad (1.42)$$

We then say (1.40) becomes the general linear elastic stress-strain relations. $\hat{U}$ in (1.41) is the *strain energy per unit volume*, which is usually referred to as the *strain energy density* or *strain energy density function*. We assume that it is a positive definite function. The 21 independent elastic coefficients c_{ijkl} define the general anisotropic body, that is, a body in which elastic properties at a point are different in different directions. In our work here we assume the elastic body to be *isotropic* at each of its points, i.e., *the elastic properties of the body are identical in all directions at each point*. In this case, as will be shown in the next section, the independent elastic constants reduce to just two.

1.5.2. Hooke's law for the homogeneous, isotropic, elastic medium

We assume isotropy at the general point P in an elastic body. At P there are principal directions of stress and strain. One can show that (1.40), together with the assumption of isotropy, implies that the principal directions of stress and strain coincide at P. Taking this fact for granted one can proceed as follows. Consider the two sets of coordinates x_i and x_i' at P. x_i are a general set, but x_i' are parallel to the principal directions. σ_{ij} and ε_{ij} are the stress and strain components at P relative to the x_i axes, and σ_{ij}' and ε_{ij}'

are those relative to the x_i' axes. It follows that $\sigma_{ij}' = \varepsilon_{ij}' = 0$, for $i \neq j$, at P, and hence from (1.40) that

$$\sigma_{11}' = c_1 \varepsilon_{11}' + c_2 \varepsilon_{22}' + c_3 \varepsilon_{33}'.$$

Since we have isotropy, $c_2 = c_3$, and this relation becomes

$$\sigma_{11}' = \lambda \Delta' + 2\mu \varepsilon_{11}',$$

where Δ' is the dilatation relative to the primed coordinates, $c_2 = \lambda$, and $c_1 - c_2 = 2\mu$. λ and μ (defined in § 1.5.3) are known as the *Lamé constants*. With similar arguments for σ_{22}' and σ_{33}' we have, compactly,

$$\sigma_{ij}' = \lambda \Delta' \delta_{ij} + 2\mu \varepsilon_{ij}'. \tag{1.43}$$

Assume now we have a direction at P defined by l_i and l_i'. Multiplying (1.43) by $l_i' l_j'$ it becomes

$$\sigma_{ij}' l_i' l_j' = \lambda \Delta' (l_i' l_j' \delta_{ij}) + 2\mu \varepsilon_{ij}' l_i' l_j'. \tag{1.44}$$

Now since the normal stress at P, $\sigma_n = \sigma_{ij}' l_i' l_j' = \sigma_{ij} l_i l_j$, and the corresponding elongation per unit length, $e = \varepsilon_{ij}' l_i' l_j' = \varepsilon_{ij} l_i l_j$, and since the dilatation is an invariant, (1.44) can be written as

$$(\sigma_{ij} - \lambda \Delta \delta_{ij} - 2\mu \varepsilon_{ij}) l_i l_j = 0.$$

Now since l_i is an arbitrary direction, and the coefficients here are independent of the l_i, these coefficients must vanish giving the *stress-strain relations*

$$\sigma_{ij} = \lambda \Delta \delta_{ij} + 2\mu \varepsilon_{ij}, \tag{1.45}$$

for the homogeneous, isotropic, linear elastic medium. Since these equations are symmetric in i, j there are six of them in (1.45), which hold at all points of the elastic body. Setting $j = i$ in (1.45), one obtains

$$\sigma_{ii} = I_1 = (3\lambda + 2\mu)\Delta, \tag{1.46}$$

the simple relation between the first invariants of stress and strain. Using this relation in (1.45), the inverse relation for ε_{ij} is found to be

$$\varepsilon_{ij} = -\frac{\lambda \delta_{ij}}{2\mu(3\lambda + 2\mu)} I_1 + \frac{1}{2\mu} \sigma_{ij}. \tag{1.47}$$

Assuming we have finite stresses, in order to have finite strains from (1.47) we must have nonvanishing μ and $3\lambda + 2\mu$.

1.5.3. Elastic constants for the isotropic case

The Lamé constants can be written in terms of the readily measurable constants, the modulus of elasticity E, and Poisson's ratio v. In a case of one dimension (a rod), with a uniform axial stress the only nonvanishing stress component, it is easy to show that

$$E = \frac{\mu(3\lambda + 2\mu)}{\lambda + \mu}, \qquad v = \frac{\lambda}{2(\lambda + \mu)}. \tag{1.48}$$

From these two expressions one finds

$$\mu = \frac{E}{2(1+v)}, \qquad \lambda = \frac{vE}{(1+v)(1-2v)}. \tag{1.49}$$

Both E and v can be obtained from a one-dimensional tension test for a particular material. As we noted in § 1.5.1, E is the slope of the stress-strain curve. *Poisson's ratio* v is defined as the ratio of the *lateral contraction per unit length to the longitudinal extension per unit length*.

If we consider a case of simple shear, where only one component of stress, say σ_{23}, acts on four faces of a unit cube, as shown in fig. 1.14, then (1.47)

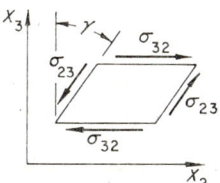

Fig. 1.14. Simple shear.

gives $\varepsilon_{23} = \sigma_{23}/2\mu$. By definition the *shear modulus* (also called the modulus of elasticity in shear or modulus of rigidity) is the ratio of the shear stress to the shear strain (angle γ in fig. 1.14), i.e., $\sigma_{23}/2\varepsilon_{23} = \sigma_{23}/\gamma = \mu$. Since the shear stress and strain would have the same sign, the shear modulus μ is positive.

Now consider an elastic body subjected to a uniform hydrostatic pressure p. In this case it can be shown that everywhere in the body $\sigma_{ij} = -p\delta_{ij}$ holds. We define the *bulk modulus* or modulus of compression $\hat{k}$ as the ratio of the pressure to the decrease in volume per unit volume, i.e., $\hat{k} = p/(-\Delta)$. Using (1.46), (1.49), it is easy to show

$$\hat{k} = \lambda + \frac{2}{3}\mu = \frac{E}{3(1-2v)} = \frac{I_1}{3\Delta}, \tag{1.50}$$

where in the last relation here use has been made of (1.46). Since $\hat{k}$, and therefore $3\lambda+2\mu$, are positive constants for all materials, v must be less than $\tfrac{1}{2}$.

Our above considerations of μ and $3\lambda+2\mu$ show they are restricted by

$$0<\mu<\infty, \qquad 0<3\lambda+2\mu<\infty. \tag{1.51a}$$

Using the first of (1.49), and the second relation in (1.50), it is easily shown that (1.51a) are equivalent to the restrictions

$$E>0, \qquad -1<v<\tfrac{1}{2}. \tag{1.51b}$$

It is useful to have (1.45), (1.47) in terms of the pair (E, v). Using (1.49), we have

$$\sigma_{ij}=\frac{E}{1+v}\left(\frac{v}{1-2v}\Delta\delta_{ij}+\varepsilon_{ij}\right). \tag{1.52}$$

Using the last relation of (1.50), we have from (1.52)

$$\varepsilon_{ij}=\frac{1}{E}[(1+v)\sigma_{ij}-vI_1\delta_{ij}]. \tag{1.53}$$

For the usual cartesian coordinates $(x_1, x_2, x_3)=(x, y, z)$ we can write from (1.53) the well-known strain-stress relations

$$\begin{aligned}
\varepsilon_x &= \frac{1}{E}[\sigma_x-v(\sigma_y+\sigma_z)], \\
\varepsilon_y &= \frac{1}{E}[\sigma_y-v(\sigma_z+\sigma_x)], \\
\varepsilon_z &= \frac{1}{E}[\sigma_z-v(\sigma_x+\sigma_y)], \\
\varepsilon_{xy} &= \frac{1+v}{E}\sigma_{xy}, \quad \varepsilon_{yz}=\frac{1+v}{E}\sigma_{yz}, \quad \varepsilon_{zx}=\frac{1+v}{E}\sigma_{zx}.
\end{aligned} \tag{1.54}$$

1.6. Dynamic equilibrium; stress equations of motion

To derive the equations of dynamic equilibrium we again consider the arbitrary closed region V, with surface S, in the body R in the space x_i (see fig. 1.1), now at a given time t. The corresponding displacements, at time t, of points or particles in R, that in the unstrained state (defined

Ch. 1, § 1.6] DYNAMIC EQUILIBRIUM; STRESS EQUATIONS OF MOTION 43

to be that at uniform temperature and zero displacement) were at x_i, are defined by $u_i(x, t)$. The corresponding so-called *particle velocities* and *accelerations* are du_i/dt and d^2u_i/dt^2, respectively, and since x_i are fixed in space, independent of time, $du_i/dt = \partial u_i/\partial t = \dot{u}_i$, and $d^2u_i/dt^2 = \partial^2 u_i/\partial t^2 = \ddot{u}_i$. It follows that the *inertia force* acting on the volume element ΔV is $-\varrho \Delta V \ddot{u}_i$, where ϱ is the *mass density* of the body, taken here to be a constant. Hence the total inertia force acting on V is given by

$$J_i = -\int_V \varrho \ddot{u}_i dV . \tag{1.55}$$

We can now establish dynamic equilibrium of V in each of the coordinate directions by applying *D'Alembert's principle* to the forces expressed by (1.1), (1.3), and (1.55), i.e., using (1.4) we have

$$\int_S \sigma_{ji} l_j dS + \int_V X_i dV - \int_V \varrho \ddot{u}_i dV = 0 . \tag{1.56}$$

Divergence Theorem. *Let V be a bounded region, the boundary S of which is one or more simple closed surfaces, each consisting of a finite number of smooth surface elements (a smooth surface element is one with a continuously turning normal). The separate surface elements are not tangent to one another, and intersect in at most edges (forming smooth curves) and vertices (formed by the intersection of two or more non-tangent edges). Now let F_j, and the first order partial derivatives of F_j, be continuous in the closed region $V + S$. It can then be proved that*

$$\int_V F_{j,j} dV = \int_S F_j l_j dS . \tag{1.57}$$

This is the divergence theorem. Kellogg [1.4] gives a thorough proof and discussion of the theorem. We note that V can be a multiply connected region. Kellogg points out that the continuity of the first order derivatives of F_j on S can be relaxed since $F_{j,j}$ need only have integrable singularities at the boundary of V (cf. [1.4, § 11, p. 119]). A shorter treatment on the theorem may be found in Sternberg and Smith [1.5]. By applying the Divergence Theorem to the first integral, (1.56) may be written as

$$\int_V (\sigma_{ji,j} + X_i - \varrho \ddot{u}_i) dV = 0 . \tag{1.58}$$

Now this holds for every choice of V. Hence if we assume the integrand here is continuous on R, (1.58) can only vanish if the integrand vanishes at every point in space and time (x, t) in the body R. Therefore

$$\sigma_{ji,j} + X_i = \varrho \ddot{u}_i. \tag{1.59}$$

This is the *stress equation of motion* valid for every point of body R.

We must also consider equilibrium of V for the rotational motion about the origin of coordinates O. This requires the sum of the moments about O due to the surface, body, and inertia forces on V to be zero, hence

$$\int_S c_{ijk} x_j l_m \sigma_{mk} \mathrm{d}S + \int_V c_{ijk} x_j (X_k - \varrho \ddot{u}_k) \mathrm{d}V = 0, \tag{1.60}$$

where we have used (1.2), and c_{ijk}, the permutation symbol defined in § 1.2 By using (1.57) of the divergence theorem again on the first integral here (1.60) becomes

$$\int_V c_{ijk} [\delta_{jm} \sigma_{mk} + x_j (\sigma_{mk,m} + X_k - \varrho \ddot{u}_k)] \mathrm{d}V = 0, \tag{1.61}$$

where δ_{jm} is the Kronecker delta defined in § 1.2. Now (1.59) makes the term in parenthesis in (1.61) zero, and summing over the m's in the remaining term reduces the latter to

$$\int_V c_{ijk} \sigma_{jk} \mathrm{d}V = 0.$$

Again assuming continuity of the integrand here on R, and since V is arbitrary

$$c_{ijk} \sigma_{jk} = 0, \tag{1.62}$$

at every point in space and time (x, t) in R. Expansion of (1.62) readily shows it reduces to $\sigma_{ij} = \sigma_{ji}$, i.e., the stress tensor is symmetric and the nine components at the point P reduce to six, which we assumed earlier. Using this last result in (1.4), (1.59) yields

$$T_i = \sigma_{ij} l_j, \tag{1.63}$$

$$\sigma_{ij,j} + X_i = \varrho \ddot{u}_i, \tag{1.64}$$

both of which have resulted from equilibrium considerations. These relations are attributed to Cauchy who had found them by 1822. In many problems T_i and X_i are known. So the six components of stress must satisfy the three equations (1.63) at the surface of the solid, and the three equations (1.64) in the interior. It is evident that these equations are not sufficient for the determination of the stress state, even if the acceleration terms in (1.64) were known. At this point, therefore, in order to simplify our problem we appeal to the *constitutive law* for the solid being treated, i.e., *the stress-strain relation* (1.45) for the *homogeneous, isotropic, linear elastic medium*.

1.7. Displacement equations of motion

Substituting (1.28) for the strain ε_{ij} in (1.45), and in turn the latter for the stress σ_{ij} in the stress equation of motion (1.64), gives

$$\lambda \Delta_{,j} \delta_{ij} + \mu(u_{i,jj} + u_{j,ij}) + X_i = \varrho \ddot{u}_i,$$

which, after summing over the j's in the first term, interchanging the differentiation order in the third term on the left hand side,[2] and using (1.39), reduces to

$$(\lambda + \mu) u_{j,ji} + \mu u_{i,jj} + X_i = \varrho \ddot{u}_i, \qquad (1.65a)$$

the *displacement equations of motion*, where $u_{i,jj}$ is the *Laplacian of u_i*. Of importance and convenience in problems involving curvilinear coordinates is the vector equivalent of (1.65a), which by inspection of the latter is easily seen to be

$$(\lambda + \mu) \nabla (\nabla \cdot \boldsymbol{u}) + \mu \nabla^2 \boldsymbol{u} + \boldsymbol{X} = \varrho \ddot{\boldsymbol{u}}, \qquad (1.65b)$$

where $\boldsymbol{u}$ is the *displacement vector*, $\boldsymbol{X}$ the *body force* (per unit volume) vector, $\nabla (\ )$ the gradient of $(\ )$, and $\nabla \cdot \boldsymbol{u}$ the *divergence* of $\boldsymbol{u}$, ∇ being the *vector differential operator "del"*. The *Laplacian* of the vector $\boldsymbol{u}$, $\nabla^2 \boldsymbol{u}$ is given by

$$\nabla^2 \boldsymbol{u} = \nabla (\nabla \cdot \boldsymbol{u}) - \nabla \times \nabla \times \boldsymbol{u}, \qquad (1.66a)$$

---

[2] Such an interchange is valid if the two second-order derivatives involved are continuous functions of postion. We assume this here, and in similar situations henceforth.

where $\nabla \times \boldsymbol{u}$ is the *curl* of $\boldsymbol{u}$.[3] Using this relation in (1.65b) gives another useful form of the *vector displacement equation of motion*

$$(\lambda+2\mu)\nabla(\nabla \cdot \boldsymbol{u}) - \mu \nabla \times \nabla \times \boldsymbol{u} + \boldsymbol{X} = \varrho \ddot{\boldsymbol{u}} . \qquad (1.66b)$$

Equations (1.65a), or the vector equivalents (1.65b), (1.66b), are a linear system of hyperbolic partial differential equations for the dependent variables $u_i(x, t)$. These equations together with suitable initial and boundary conditions form the base for writing boundary value problems in this subject. The discussion of the basic problems in dynamic elasticity follows and demonstrates this.

1.8. The fundamental boundary-initial value problems of elastodynamics

There are three fundamental boundary-initial value problems in elastodynamics, which are the analogs of those in elastostatics. In defining them we now assume that the X_i in the displacement equations of motion (1.65) are known in V^4 for all $t \geq t_0$, the *initial time*. In all three problems we wish to determine the displacements $u_i(x, t)$ that satisfy (1.65a) in V, for $t > t_0$, and satisfy the *initial conditions*

$$u_i(x, t) = u_{i0}(x) , \qquad \dot{u}_i(x, t) = v_{i0}(x) , \qquad \text{for } t = t_0 \text{ in } V+S , \qquad (1.67)$$

where u_{i0} and v_{i0} are prescribed functions. The problems differ in the boundary conditions that $u_i(x, t)$ must satisfy:

First problem

$$u_i(x, t) = U_i(x, t) \text{ on } S, \text{ for } t > t_0 , \qquad (1.68)$$

where U_i is a prescribed function.

Second problem

$$T_i(x, t) = F_i(x, t) \text{ on } S, \text{ for } t > t_0 , \qquad (1.69)$$

where F_i is a prescribed function.

---

[3] For further reading on these vector operators, the author suggests the book by Phillips [1.6].
[4] We assume from here on in our work that V represents the body R.

Third problem

$$u_i(x, t) = U_i(x, t) \text{ on } S_1 \brace T_i(x, t) = F_i(x, t) \text{ on } S_2, \quad \text{for } t > t_0, \qquad (1.70)$$

where $S_1 + S_2 = S$, and U_i and F_i are prescribed functions. The third problem is referred to as a *mixed boundary value problem*, having displacements specified on part of the boundary (S_1) and stresses on the rest (S_2). In the second and third problems where T_i is prescribed on S, or part of S, the equilibrium equations (1.63) must be satisfied there. Then through the components of stress on the right hand side of (1.63), and the stress-strain relations (1.45), one can write these boundary conditions in terms of u_i by using (1.25), (1.39).

We will not consider existence of solutions to these basic problems here. However, uniqueness will be set down for them within the scope of the classical theorem due to Neumann. This will be treated shortly with the aid of the material contained in the next two sections.

1.9. The superposition principle

The governing equations (1.64), (1.45), (1.25), (1.39), and (1.65a,b) (1.66b) are all *linear*, i.e., they are all composed of *linear operators L* which satisfy the basic law $L(\alpha u + \beta v) = \alpha L(u) + \beta L(v)$, where u and v are variables and α and β are constants. In this context a linear partial differential equation is defined as $L(u) = F$, where F is given and may be zero. Note that linearity of our equations requires there be no terms involving products of unknown functions or their derivatives. It follows that the *superposition principle* applies to the fundamental problems given by (1.65a) and (1.67)–(1.70). Consider the second problem, for example. We first assume that the elastic body is acted upon by body force X_i' and surface stress T_i'. The corresponding stresses, strains, and displacements are σ_{ij}', ε_{ij}' and u_i', respectively. Now assume that a second set, X_i'' and T_i'', act, with corresponding σ_{ij}'', ε_{ij}'' and u_i''. If both sets $X_i' + X_i''$, and $T_i' + T_i''$ are applied to the body, then it follows from the superposition principle that the corresponding stresses, strains and displacements are $\sigma_{ij}' + \sigma_{ij}''$, $\varepsilon_{ij}' + \varepsilon_{ij}''$ and $u_i' + u_i''$, respectively. The applicability of the principle to the other two problems follows from similar arguments.

1.10. The principle of conservation of energy

Let us now assume that body V is in the unstrained state at $t=t_0$, strains being specified with respect to this state. Essentially following Sokolnikoff's treatment [1.1, § 26], first we define $W(t)$ as the *work done by the external forces* (*body and surface*) acting on V from $t=t_0$ to t. To calculate $W(t)$ we first find the work $\mathrm{d}W$ done in the time interval $(t, t+\mathrm{d}t)$. The general point P, at x_i before deformation, will have at any time t the coordinates $x_i+u_i(x,t)$. In the interval $(t, t+\mathrm{d}t)$ the displacement is therefore $\dot{u}_i \mathrm{d}t$. It follows that the work done by the body forces on the volume element $\mathrm{d}V$ at P in $(t, t+\mathrm{d}t)$ is $X_i \dot{u}_i \mathrm{d}V \mathrm{d}t$. Similarly the surface forces do the work $T_i \dot{u}_i \mathrm{d}S\, \mathrm{d}t = \sigma_{ij} l_j \dot{u}_i \mathrm{d}S\, \mathrm{d}t$, and we can therefore write the expression

$$\frac{\mathrm{d}W(t)}{\mathrm{d}t} = \int_S \sigma_{ij} l_j \dot{u}_i \mathrm{d}S + \int_V X_i \dot{u}_i \mathrm{d}V, \qquad (1.71)$$

the *time rate of change of the work* being done on V, which is sometimes referred to as the *power input*. Using (1.57) of the divergence theorem on the first integral in (1.71), this equation becomes

$$\frac{\mathrm{d}W(t)}{\mathrm{d}t} = \int_V [\sigma_{ij,j} \dot{u}_i + \sigma_{ij} \dot{u}_{i,j} + X_i \dot{u}_i] \mathrm{d}V. \qquad (1.72)$$

Using (1.64) we can write the sum of the first and last terms of the integrand in (1.72) in another important form, i.e.,

$$(\sigma_{ij,j} + X_i) \dot{u}_i = \varrho \ddot{u}_i \dot{u}_i.$$

Using $\sigma_{ji} = \sigma_{ij}$, the second integrand term in (1.72) also has another important form, i.e.,

$$\sigma_{ij} \dot{u}_{i,j} = \tfrac{1}{2} \sigma_{ij} \dot{u}_{i,j} + \tfrac{1}{2} \sigma_{ji} \dot{u}_{j,i} = \sigma_{ij} \dot{\varepsilon}_{ij}.$$

On this basis (1.72) becomes

$$\frac{\mathrm{d}W(t)}{\mathrm{d}t} = \int_V [\varrho \ddot{u}_i \dot{u}_i + \sigma_{ij} \dot{\varepsilon}_{ij}] \mathrm{d}V. \qquad (1.73)$$

Now if we note that the kinetic energy of the body V at time t is given by

$$K(t) = \int_V \tfrac{1}{2} \varrho \dot{u}_i \dot{u}_i \mathrm{d}V = \int_V \widehat{K} \mathrm{d}V, \qquad (1.74)$$

Ch. 1, § 1.10] THE PRINCIPLE OF CONSERVATION OF ENERGY 49

where $\dot{K}$ is the *kinetic energy per unit volume* or *kinetic energy density*, then

$$\frac{dK(t)}{dt} = \frac{d}{dt}\int_V \tfrac{1}{2}\varrho \dot{u}_i \dot{u}_i \, dV = \int_V \varrho \ddot{u}_i \dot{u}_i \, dV. \qquad (1.75)$$

For this differentiation with respect to parameter t to be valid it is sufficient to require that the integrand functions in both integrals here be continuous in x_i and t. The last integral in (1.75) is identical to the first part of the integral in (1.73), hence the latter is equal to dK/dt.

To identify the second part of the integral in (1.73) we note, using the property (1.42) of the strain energy density function $\hat{U}(\varepsilon_{ij})$, that

$$\frac{\partial \hat{U}}{\partial t} = \frac{\partial \hat{U}}{\partial \varepsilon_{ij}} \dot{\varepsilon}_{ij} = \sigma_{ij}\dot{\varepsilon}_{ij}. \qquad (1.76)$$

Hence, from (1.76) the second part of the integral in (1.73)

$$\int_V \sigma_{ij}\dot{\varepsilon}_{ij} \, dV = \int_V \frac{\partial \hat{U}}{\partial t} \, dV = \frac{d}{dt}\int_V \hat{U} \, dV = \frac{dU}{dt}, \qquad (1.77)$$

where U is the *potential energy of deformation* or *strain energy* of the body V at time t. Therefore

$$\frac{dW}{dt} = \frac{dK}{dt} + \frac{dU}{dt}. \qquad (1.78)$$

Since heat energy is neglected here, (1.78) is a form of the *law of conservation of energy*, i.e., the *time rate of change of the work done on V by the body and surface forces is equal to the sum of the time rate of change of the kinetic and potential energies*. Integrating (1.78) between $t = t_0$ and t, noting the unstrained state at $t = t_0$ so that $K = U = 0$ there, we obtain $W(t) = K(t) + U(t)$. Using (1.71), this last equation can be written as

$$W(t) = \int_{t_0}^{t} \left[\int_V X_i \dot{u}_i \, dV + \int_S T_i \dot{u}_i \, dS \right] dt' = K(t) + U(t). \qquad (1.79)$$

Equation (1.79) is known as the *energy identity*.

Using Clapeyron's form for $\hat{U}$

$$\hat{U} = \tfrac{1}{2}\sigma_{ij}\varepsilon_{ij},$$

and (1.45) for our isotropic case, we can show that

$$\hat{U} = \tfrac{1}{2}\lambda\Delta^2 + \mu\varepsilon_{ij}\varepsilon_{ij}, \qquad (1.80)$$

which is obviously a function of ε_{ij} only. Further by expanding this it can be shown to satisfy (1.42). It also can be shown that U in eq. (1.80) is positive definite as was assumed earlier for the general expression (1.41) from which it stems. We make use of this property in the next section.

1.11. Uniqueness of solution

Let us assume that each of the problems stated in § 1.8 has two solutions, given by u_i' and σ_{ij}' and u_i'' and σ_{ij}'', which correspond to identical body forces, and satisfy identical initial and boundary conditions. Since the superposition principle, as discussed in § 1.9, is applicable, it is clear that

$$u_i = u_i' - u_i'', \quad \sigma_{ij} = \sigma_{ij}' - \sigma_{ij}'', \quad X_i = 0, \qquad (1.81)$$

satisfy the governing equations (1.64) or (1.65a) in V, for $t > t_0$. From (1.67) initial conditions on u_i for all three problems are

$$u_i = \dot{u}_i = 0, \qquad (1.82)$$

in $V+S$, at $t = t_0$. From (1.68) and (1.69) boundary conditions on u_i for the first and second problems are respectively

$$u_i = 0,$$
$$T_i = 0, \qquad (1.83)$$

on S, for $t > t_0$. From (1.70) the analogous conditions for the third problem are

$$u_i = 0,$$
$$T_i = 0, \qquad (1.84)$$

respectively on S_1 and S_2, for $t > t_0$.

Returning to (1.79) we see the first integral there is zero since $X_i = 0$. The second integral there is also zero for all three problems because of (1.83) and (1.84). From (1.79) we therefore have

$$K + U = 0. \qquad (1.85)$$

Now both K and U are positive definite, therefore from (1.85), $K = U = 0$, for $t \geq t_0$. This last result means that $\dot{u}_i = 0$, and $\varepsilon_{ij} = 0$, in V for $t \geq t_0$. The first of these results means that $u_i(x, t) = f(x)$ at most, but the second result says that $u_i(x, t)$ can only be a constant, i.e., this solution would be a rigid body displacement. Imposing the first of (1.82), however, shows the constant must be zero and therefore the solution for each of the three problems is $u_i(x, t) = 0$ in V, for $t \geq t_0$. It follows from (1.81) that

$$u'_i = u''_i$$
$$\sigma'_{ij} = \sigma''_{ij}$$

and we therefore have a unique solution for each of the three problems. The proof is due to Neumann [1.7].

1.12. Further contributions on the uniqueness of solutions

As we have seen the uniqueness proof, given in the previous section, was based on a bounded region V. The corresponding uniqueness question for the infinite region arises naturally since problems involving the infinite or semi-infinite medium are fundamental to elastodynamics. As will be shown later, the displacement equations of motion (1.65), (1.66b) are hyperbolic, governing a disturbance with a finite velocity at its front. Hence, in effect, at any particular finite time the disturbed region is always finite, provided the loading is over a finite region, and we might expect that the classical proof could be extended to the infinite domain.[5] Recently Wheeler and Sternberg [1.9] have extended, in this sense, Neumann's classical uniqueness proof. To do this they establish and use a generalized energy identity, i.e., a generalization of (1.79), where t is replaced by $\tau(x)$, effecting not only the upper limits of integrals on the left hand side, but the integrals for $K(t)$ and $U(t)$ on the right hand side [see (1.74) and (1.77)]. For any fixed time, $\tau(x)$ defines an integration over a bounded region. The method in [1.9] was suggested by Zaremba's treatment of related uniqueness questions involving the scalar wave equation [references on Zaremba's and related work are given in [1.9]]. Wheeler [1.10] showed the method of [1.9] could be extended to the uniqueness question for the infinite anisotropic body.

---

[5] Stratton [1.8] presents similar arguments on the analogous electromagnetic wave uniqueness question.

As has been shown in this introduction to classical elastodynamics, continuity is imposed on the displacements and stresses (hence strains). It follows that Neumann's uniqueness proof, which is based on these stringent conditions, is not applicable to problems involving discontinuous loadings. The elastic wave propagation literature, of course, exhibits a great array of problems involving such loadings, and in general the corresponding solutions found beg the uniqueness question. It should be pointed out that recently uniqueness theorems for the second boundary-initial value problem, with certain discontinuous loadings, have been set down. One such theorem is presented in the work by Wheeler and Sternberg [1.9]. They treat cases involving time-dependent concentrated loads at fixed points of application in the interior or on the surface of the elastic solid. Brockway [1.11] treats the cases in which the displacement fields, (1) are continuous, while the stress fields have jump discontinuities, and (2) can have generalized first derivatives, and hence stress fields that have square-integrable singularities. Brockway points out that for (1) the uniqueness can be extended to the first and third boundary-initial value problems as well.

1.13. The Graffi elastodynamic reciprocal theorem

The present theorem sets down a general reciprocal expression relatine the *dynamic equilibrium states* of a homogeneous, isotropic, linear elastic body under different applied loads and initial data. *The theorem,* due to Graffi [1.12], *may be stated as follows*: Let $u_i(x, t)$, $\varepsilon_{ij}(x, t)$, $\sigma_{ij}(x, t)$ and $u'_i(x, t)$, $\varepsilon'_{ij}(x, t)$, $\sigma'_{ij}(x, t)$ *be the displacement, strain and stress corresponding, respectively, to the body and surface forces* $X_i(x, t)$, $T_i(x, t)$ *and* $X'_i(x, t)$, $T'_i(x, t)$, *as well as the initial data* $u_{i0}(x)$, $v_{i0}(x)$ *and* $u'_{i0}(x)$, $v'_{i0}(x)$ *for* $x \in V$. *Then, for* $t > 0$

$$\int_S T_i * u'_i \, dS + \int_V [X * u'_i + \varrho(u_{i0} \dot{u}'_i + v_{i0} u'_i)] \, dV$$

$$= \int_S T'_i * u_i \, dS + \int_V [X' * u_i + \varrho(u'_{i0} \dot{u}_i + v'_{i0} u_i)] \, dV , \quad (1.86)$$

where $T_i * u'_i$ and like quantities are convolutions [cf. (1.92)]. Graffi proved the theorem by drawing on Betti's reciprocal theorem of elastostatics (cf. [1.1 § 109]) and the Laplace transform. Our proof here is essentially that of Graffi's.

Consider the dynamic equilibrium states of the elastic body for the two systems u_i, ε_{ij}, σ_{ij}, X_i, T_i and u'_i, ε'_{ij}, σ'_{ij}, X'_i, T'_i. Then according to (1.63), (1.64) we have

$$T_i(x,t) = \sigma_{ij} l_j(x,t), \quad \sigma_{ij,j}(x,t) + X_i(x,t) = \rho \ddot{u}_i(x,t),$$
$$T'_i(x,t) = \sigma'_{ij} l_j(x,t), \quad \sigma'_{ij,j}(x,t) + X'_i(x,t) = \rho \ddot{u}'_i(x,t). \tag{1.87}$$

The (one-sided) Laplace transform on time t is now applied to all of the foregoing time-dependent quantities.[6] This reduces them to dependence on x and the Laplace transform parameter p, e.g., $T_i(x,t) \to \bar{T}_i(x,p)$ where the bar denotes the Laplace transform of the quantity.

We first calculate the transform of the work that is done by the *unprimed* forces X_i, T_i, on the primed displacement u'_i. It is given by

$$\int_S \bar{T}_i(x,p)\bar{u}'_i(x,p)dS + \int_V (\bar{X}_i - \rho \bar{\ddot{u}}_i)\bar{u}'_i dV$$
$$= \int_S \overline{\sigma_{ij} l_j} \bar{u}'_i dS - \int_V \bar{\sigma}_{ij,j} \bar{u}'_i dV, \tag{1.88}$$

where we have made use of the transforms of (1.87). Note argument (x, p), common to all transformed quantities in (1.88), has been shown in the first integral only for convenience. Now using (1.57) of the divergence theorem, (1.88) becomes

$$\int_S \bar{T}_i \bar{u}'_i dS + \int_V (\bar{X}_i - \rho \bar{\ddot{u}}_i)\bar{u}'_i dV = \int_V (\bar{\sigma}_{ij}\bar{u}'_i)_{,j} dV - \int_V \bar{\sigma}_{ij,j} \bar{u}'_i dV$$
$$= \int_V \bar{\sigma}_{ij} \bar{u}'_{i,j} dV. \tag{1.89}$$

Making use of the transforms of Hooke's law (1.45) and the strain-displacement relations (1.25), (1.39) the integral on the right hand side of (1.89) becomes

$$\int_V (\lambda \bar{\Delta} \bar{\Delta}' + \mu \bar{u}_{i,j} \bar{u}'_{i,j} + \mu \bar{u}_{j,i} \bar{u}'_{i,j}) dV. \tag{1.90}$$

It is obvious that the first two terms of the integrand here are symmetric in

---

[6] The Laplace transform is introduced in Chapter Five for its important later use in solving boundary-initial value problems in our subject. Here we need only some elementary rules of this transform which the reader can find in Chapter Five.

the primed and unprimed variables. Further, interchanging i and j in the last term, it becomes $\mu \bar{u}_{i,j} \bar{u}'_{j,i} = \mu \bar{u}'_{j,i} \bar{u}_{i,j}$. It follows that the integral (1.90) is symmetric in its primed and unprimed variables and, therefore, from (1.89) we have

$$\int_S \bar{T}_i \bar{u}'_i dS + \int_V (\bar{X}_i - \varrho \bar{\bar{u}}_i) \bar{u}'_i dV = \int_S \bar{T}'_i \bar{u}_i dS + \int_V (\bar{X}_i - \varrho \bar{\bar{u}}'_i) \bar{u}_i dV . \quad (1.91)$$

It remains to invert (1.91) to the time domain. Note there that all integrand terms are products of two Laplace transformed quantities. It is well known that the inverse of such a product is given by the *convolution* of the two inverse functions, defined by

$$f(x, t) = f_1 * f_2(x, t) = \int_0^t f_1(x, t-\tau) f_2(x, \tau) d\tau , \quad (1.92)$$

(cf. ch. 5, § 5.4.4). Further, making use of the initial conditions (1.67) for the fundamental boundary-initial value problems and the given initial data of our present theorem, the transforms of the accelerations in (1.91) take the form

$$\bar{\bar{u}}_i(x, p) = p^2 \bar{u}_i(x, p) - p u_{i0}(x) - v_{i0}(x) , \quad (1.93)$$

based on the transform formula for the second time derivative (5.23). Substituting (1.93) into (1.91), and inverting the latter with the aid of (1.92), yields (1.86), which was to be proved. *Note when both the primed and unprimed initial data take on zero values, only the convolution terms survive in* (1.86). Other special properties of Graffi's theorem are discussed in Gurtin [1.13], as well as the literature on the theorem and its use.

In a later paper Graffi [1.14] gave a proof of his theorem through direct use of the properties of the convolution integral. Wheeler and Sternberg [1.9] extended Graffi's theorem to the infinite body. In turn Wheeler [1.10] extended this to the infinite anisotropic body. Of note also is the critique on the use of the reciprocal theorem in seismology by Knopoff and Gangi [1.15], when the medium is inhomogeneous and anisotropic. Lastly, deHoop [1.16] has derived a reciprocity theorem for an inhomogeneous anisotropic, linear viscoelastic medium.

Interesting applications of Graffi's theorem were given by DiMaggio and Bleich [1.17] and Payton [1.18]. In the note by DiMaggio and Bleich interest is in the elastodynamic half space problems involving fixed loads. They show, for example, through the theorem use can be made of existing data for the vertical surface displacement due to a buried step point vertical load (obtained by Pekeris and Lifson, cf. [6.10]) to obtain data for the interior vertical displacement field due to a surface step point vertical load. Payton's interest was in moving point loads. For example, he used the reciprocity theorem, and Pekeris' solution [6.9] for the sudden point normal load on the surface of the half space, to get the solution for the displacement field produced by the moving load.

In the work [1.16] deHoop uses his reciprocity theorem to derive an integral representation of the Kirchhoff type (cf. (2. 117) and § 2. 8) for the displacement vector.

Further examples of reciprocity arise later in this book.

1.14. Exercises

1.1. A semi-infinite elastic rod is struck uniformly on its end ($x=0$) at $t=0$ by a large mass traveling with velocity v_0. Assuming the one-dimensional wave equation

$$\partial^2 u(x, t)/\partial x^2 = \ddot{u}/c^2, \qquad c^2 = E/\varrho$$

governs (an approximation), what are the initial ($t=0$) and boundary ($x=0$) conditions for the problem? Which class of fundamental problems does this one fit into? Show that you can make the boundary and initial conditions for the problem consistent. Give your arguments.

1.2. (a) Derive (1.80) of the text. Show that this form of $\hat{U}(\varepsilon_{ij})$ satisfies (1.42).

(b) Show $\hat{U}$ is a positive definite quadratic form.

1.3. (a) Given the problem depicted in the sketch, a two-dimensional

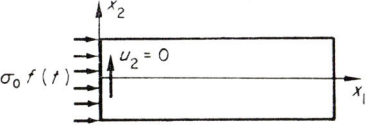

layer suddenly loaded on one edge ($x_1=0$) with the stress $\sigma_{11}(0, x_2, t) = \sigma_0 f(t)$ (σ_0 a constant, f the time input), along with vanishing lateral displacement u_2 there and the rest of the surfaces stress free, prove the problem has a unique solution. Assume we have plane strain such that $u_3 = 0$.

(b) Now consider a similar problem with the edge conditions of part (a) replaced by $\dot{u}_1(0, x_2, t) = v_0 f(t)$, $\sigma_{12} = 0$. Prove this problem has a unique solution. *These problems are known as those with mixed edge conditions* (at $x_1=0$).

References

[1.1.] I. S. Sokolnikoff, *Mathematical Theory of Elasticity*, 2nd Edition. McGraw–Hill Book Company, Inc., New York (1956).
[1.2.] A. E. H. Love, *Mathematical Theory of Elasticity*, 4th Edition. Dover Publications, Inc., New York (1944).
[1.3.] W. Nowacki, *Dynamics of Elastic Systems*. John Wiley and Sons, Inc., New York (1963).
[1.4.] O. D. Kellogg, *Foundations of Potential Theory*. Frederick Unger Publishing Company, New York (1929), Chapter IV.
[1.5.] W. J. Sternberg and T. L. Smith, *The Theory of Potential and* Spherical Harmonics. University of Toronto Press, Toronto, Canada (1946), pp. 48–52.
[1.6.] H. B. Phillips, *Vector Analysis*. John Wiley and Sons, Inc., New York (1933).
[1.7.] F. Neumann, *Vorlesungen über die Theorie der Elastizität der festen Körper und des Lichtäthers*. B. G. Teubner, Leipzig (1885).
[1.8.] J. A. Stratton, *Electromagnetic Theory*, First Edition. McGraw–Hill Book Company, Inc., New York (1941), pp. 487–488.
[1.9.] L. T. Wheeler and E. Sternberg, *Archive for Rational Mechanics and Analysis* **31** (1968), 51–90.
[1.10.] L. T. Wheeler, *Quarterly of Applied Mathematics* **XXVIII** (1970), 91–101.
[1.11.] G. S. Brockway, *Archive for Rational Mechanics and Analysis* **48** (1972), 213–244.
[1.12.] D. Graffi, *Memorie della Accademia delle Scienze, Bologna, Series 10*, **4** (1946, 1947), 103–109.
[1.13.] M. E. Gurtin, *The Linear Theory of Elasticity, Handbook Der Physik VI a/2, Mechanics of Solids*, Eds. S. Flugge, C. Truesdell, Springer–Verlag, Berlin (1972), 1–295.
[1.14.] D. Graffi, *Atti Accademia Scienze, Bologna (11)* **10** (1963), 33–40.
[1.15.] L. Knopoff and A. F. Gangi, *Geophysics* **24** (1959), 681–691.
[1.16.] A. T. deHoop, *Applied Science Research*, **16** (1966), 39–45.
[1.17.] F. L. DiMaggio and H. H. Bleich, *Journal of Applied Mechanics* **26** (1959), 678–679.
[1.18.] R. G. Payton, *Quarterly of Applied Mathematics* **21** (1963–64), 299–313.

CHAPTER 2

THE FUNDAMENTAL WAVES OF ELASTODYNAMICS AND THEIR REPRESENTATIONS

2.1. Introduction

The present chapter is concerned with certain basics of integrating the displacement equations of motion (1.65), or (1.66b), and analyzing the general nature of the wave solutions so found. As we have already pointed out, for finite time, assuming a source over a finite region of body V, such solutions are necessarily restricted to a bounded spatial domain because of the hyperbolicity of the displacement equations of motion. It follows that the disturbance can be thought of as propagating in the unbounded medium. Such a disturbance is necessarily simpler, because it is free of boundary effects such as reflection, refraction, diffraction and dispersion. Hence, the waves comprising this disturbance are referred to as *body waves* to distinguish them from *surface* or *interface waves* generated at, and propagating along a boundary. It is clear, however, since the displacement equations of motion underlie all elastodynamic problems, body waves play a role in all solutions. Physically, this is just saying that these waves will undergo reflection, refraction, and diffraction at a boundary or interface in an elastic medium.

2.2. Fundamental body waves and governing wave equations

2.2.1. Equivoluminal and dilatational displacement waves

Consider the displacement equations of motion (1.65b), setting the body forces $X=0$. Now we set the dilatation $\Delta = \nabla \cdot \boldsymbol{u} = 0$ which gives

$$\nabla^2 \boldsymbol{u} = \ddot{\boldsymbol{u}}/c_s^2, \quad c_s^2 = \mu/\varrho, \qquad (2.1)$$

a wave equation for the displacements u governing *equivoluminal waves*, in which deformation consists of shear and rotation only (dilatation, or volume change, is zero). c_s is the *speed of propagation* of these waves.

Next we set the rotation

$$\omega = \nabla \times u/2 = 0. \tag{2.2}$$

Substituting (2.2) in (1.66a) gives $\nabla^2 u = \nabla(\nabla \cdot u)$, and using this in (1.65b, $X=0$) the latter reduces to

$$\nabla^2 u = \ddot{u}/c_d^2, \quad c_d^2 = (\lambda + 2\mu)/\varrho. \tag{2.3}$$

Equation (2.3) is a wave equation for u governing *dilatational waves*, in which deformation consists of dilatation, or volume change, only (involving shear, but no rotation). c_d is the *speed of propagation* of these waves. Eqs. (2.1), (2.3) disclose the fact that the displacement equations of motion, say (1.65b, $X=0$), govern two fundamentally different types of displacement waves, dilatational and equivoluminal.

2.2.2. Dilatation and rotation waves

It is not difficult to show that the displacement equations of motion also contain wave equations on the dilatation $\Delta = \nabla \cdot u$ and the rotation ω. Taking the divergence of (1.66b, $X=0$), with an appropriate derivative interchange, gives

$$\nabla^2 \Delta = \ddot{\Delta}/c_d^2. \tag{2.4}$$

Similarly, taking the curl of (1.66b, $X=0$) gives, with the aid of the middle factor rule from vector analysis,

$$\nabla^2 \omega = \ddot{\omega}/c_s^2. \tag{2.5}$$

Hence, (2.4) and (2.5) show that the displacement equations of motion govern waves of the dilatation and the rotation.

2.2.3. The Lamé solution of the displacement equations of motion

Poisson, as his classical work in 1829 disclosed, was the first to recognize that a disturbance was, in general, composed of both types of the fundamental displacement waves, the dilatational and equivoluminal waves discussed in § 2.2.1. Poisson's work is contained in his mémoire on the equilibrium and motion of elastic bodies [2.1], and an addition to this mémoire [2.2]. Sternberg [2.3] discusses Poisson's work, and related work of others on the integration of the displacement equations of motion (1.65). A review of some of the discussion in [2.3] will be helpful to our development here. As pointed out there, in [2.1] Poisson gave a class of particular solutions to (1.65b, $X=0$), based on the assumption that the displacement vector u is the gradient of a potential function, i.e., $u = \nabla \varphi$ (hence, these are the solutions (2.3)). In [2.2] Poisson showed that every sufficiently regular solution of (1.65b, $X=0$) can be represented by

$$u(x, t) = u'(x, t) + u''(x, t), \quad u' = \nabla \varphi, \quad \nabla \cdot u'' = 0, \quad (2.6a)$$

where $\varphi(x, t)$, and $u''(x, t)$ satisfy the wave equations

$$\nabla^2 \varphi = \ddot{\varphi}/c_d^2, \quad \nabla^2 u'' = \ddot{u}''/c_s^2. \quad (2.6b)$$

Therefore Poisson showed that the complete solution of (1.65b, $X=0$) can be expressed as a superposition of a dilatational and an equivoluminal motion, associated with the wave speeds c_d and c_s, respectively. Poisson's proof is summarized by Todhunter and Pearson [2.4].

Poisson's general solution (2.6) does not involve the vector potential appropriate to the solenoidal displacement component u''. Such a solution, i.e., one using both a scalar and vector potential, was apparently given first by Lamé [2.5] in 1852. Lamé states that every *displacement vector field* of the form

$$u = \nabla \varphi + \nabla \times \psi, \quad (2.7a)$$

satisfies (1.65b, $X=0$), provided $\varphi(x, t)$ and $\psi(x, t)$ are solutions of

$$\nabla^2 \varphi = \ddot{\varphi}/c_d^2, \quad \nabla^2 \psi = \ddot{\psi}/c_s^2, \quad (2.7b)$$

which is easily verified by substitution. Equations (2.7) show that φ propagates with the speed c_d, and ψ with the speed c_s. When solutions are found to (2.7b) for φ and ψ, u can be obtained from (2.7a), and this is a solution to the displacement equations of motion. The advantage of using (2.7b) is obviously the wealth of knowledge that exists concerning solutions of the wave equation, which carries over to our problems in elastic waves. We should, however, point out in spite of this advantage, sometimes direct use of the displacement equations of motion has advantages. This is usually because one can work with lower order derivatives. We shall see examples of this in our later work here. We note that the dilatation and rotation corresponding to (2.7a) are, respectively,

$$\Delta = \nabla \cdot \boldsymbol{u} = \nabla^2 \varphi, \qquad \boldsymbol{\omega} = \nabla \times \boldsymbol{u}/2 = \nabla \times \nabla \times \boldsymbol{\psi}/2. \qquad (2.8)$$

2.2.4. Helmholtz resolution of a vector

Important to our further development of the theory in this chapter is a theorem due to Helmholtz on the resolution or decomposition of a vector field. Phillips [1.6, § 83] has a discussion of the theorem, and its proof, which we draw on. Consider the vector $\boldsymbol{F}$ which is piecewise continuously differentiable in the finite closed region V. Consider also the vector

$$\boldsymbol{W}(x) = -\frac{1}{4\pi} \int_V \frac{\boldsymbol{F}(x')}{|x-x'|} \, dV_{x'}, \qquad (2.9)$$

where with cartesian coordinates, for example, we have

$$dV_{x'} = dx'_1 dx'_2 dx'_3, \quad \text{and} \quad |x-x'| = [(x_1-x'_1)^2 + (x_2-x'_2)^2 + (x_3-x'_3)^2]^{\frac{1}{2}},$$

which is associated with each point $P(x)$ of space. It can be shown that the so defined $\boldsymbol{W}(x)$ satisfies the vector Poisson's equation

$$\nabla^2 \boldsymbol{W} = \boldsymbol{F}(x), \qquad (2.10a)$$

at interior points where $\boldsymbol{F}(x)$ is continuous, and the vector Laplace's equation

$$\nabla^2 \boldsymbol{W} = 0, \qquad (2.10b)$$

at points exterior to V (cf. [1.6, § 58]). Now expanding the left hand side of (2.10a) by using (1.66a), we have

$$\boldsymbol{F}(x) = \nabla(\nabla \cdot \boldsymbol{W}) - \nabla \times \nabla \times \boldsymbol{W}, \qquad (2.11)$$

and introducing into (2.11) the scalar function $f(x)$, and vector function $G(x)$, defined by

$$f(x) = \nabla \cdot W, \qquad (2.12a)$$

$$G(x) = -\nabla \times W, \qquad (2.12b)$$

(2.11) becomes

$$F(x) = \nabla f + \nabla \times G. \qquad (2.13)$$

The functions f and G are everywhere continuous and differentiable at interior points where F is continuous (cf. [1.6, § § 56–59]). Equation (2.13) expresses Helmholtz's theorem, namely that under the regularity conditions imposed on F it can be resolved into the two vectors ∇f and $\nabla \times G$. Given the vector F, we note (2.9) defines $W(x)$. Then (2.12) can be used to find f and G, hence the resolution of F given by (2.13). We note from (2.12b) that

$$\nabla \cdot G = 0. \qquad (2.14)$$

To extend (2.13) to an infinite domain, one must impose certain regularity conditions on $F(x)$ at infinity.

2.2.5. Completeness of the Lamé solution

The question of completeness of the Lamé solution (2.7) was raised by Clebsch. As Sternberg points out in [2.3], *Clebsch presented the following completeness theorem: Let $u(x, t)$ be a particular solution of (1.65b, $X=0$) in a region of space V for $t_1 < t < t_2$. Then there exists a scalar function $\varphi(x, t)$, and a vector function $\psi(x, t)$, such that $u(x, t)$ is given by (2.7a), with $\nabla \cdot \psi = 0$, and φ and ψ respectively satisfy the first and second of (2.7b).* We should note at this point that (2.7a), together with $\nabla \cdot \psi = 0$, is an example of Helmholtz resolution of a vector field discussed in § 2.2.4.

Clebsch's proof was inconclusive. Later, Somigliana [2.6] proved the theorem rigorously. Still later, Tedone and Duhem gave alternate proofs. Sternberg elucidates Duhem's proof in [2.3]. We follow Somigliana. Making use of the displacement equation of motion (1.66b), for the case of $X=0$, and integrating (1.66b) twice with respect to time, gives

$$u = c_d^2 \nabla \int_0^t \int_0^\tau \nabla \cdot u \, d\eta d\tau - c_s^2 \nabla \times \int_0^t \int_0^\tau \nabla \times u \, d\eta d\tau, \quad (2.15)$$

where we have imposed the quiescent initial conditions $u(x, 0) = 0$, and $\dot{u}(x, 0) = 0$. Now defining φ and ψ as

$$\varphi = c_d^2 \int_0^t \int_0^\tau \nabla \cdot u \, d\eta d\tau,$$

$$\psi = -c_s^2 \int_0^t \int_0^\tau \nabla \times u \, d\eta d\tau, \quad (2.16)$$

2.15) gives (2.7a). To complete the proof therefore, we have only to show hat φ and ψ satisfy (2.7b), since $\nabla \cdot \psi = 0$ is built into the second of (2.16). Differentiating (2.16) twice with respect to time gives

$$\ddot{\varphi} = c_d^2 \nabla \cdot u, \quad \ddot{\psi} = -c_s^2 \nabla \times u. \quad (2.17)$$

Using the first of (2.8) in the first of (2.17), it is clear that φ statisfies the first of (2.7b). Now from the second of (2.8), we find

$$\nabla \times u = \nabla \times \nabla \times \psi = -\nabla^2 \psi + \nabla(\nabla \cdot \psi).$$

But $\nabla \cdot \psi = 0$, hence, from this last expression substituting $-\nabla^2 \psi$ for $\nabla \times u$ in the second of (2.17), shows that ψ satisfies the second of (2.7b). This completes the proof of Clebsch's theorem. It is not difficult to take into account nonvanishing initial conditions and body force in the proof, representing these quantities in (2.16) through Helmholtz resolution of a vector. This is left to the exercises.

2.2.6. Gauge conditions

As has already been noted, u given by (2.7a) satisfies (1.65b, $X=0$), provided (2.7b) hold, even if $\nabla \cdot \psi = 0$ does not hold, i.e., we have Lamé's solution. On the other hand the condition $\nabla \cdot \psi = 0$, an additional constraint, is a natural one. It may be seen that the three components of the displacement in (2.7a) are given in terms of four scalar functions, φ, and the three components of ψ. The most general vector field requires specification of only three scalar functions of position (one for each component of the vector). The condition $\nabla \cdot \psi = 0$ is the means by which the four scalar functions in (2.7a) are reduced to three independent ones.

$\nabla \cdot \psi = 0$ is sometimes referred to as a *gauge condition*, a name stemming from the literature in electromagnetic theory. The invariance of fields such as the displacement u here, being physically measurable quantities, to changes in (the non-unique) potentials representing them, is called *gauge invariance* [cf. Morse and Feshbach, 2.7, pp. 206–207, 210–212]. Consider the given vector potential ψ satisfying $\nabla \cdot \psi = 0$. If we form the new vector potential $\psi' = \psi + \nabla g$, with g a scalar function, it is easy to see that u is invariant to the change in vector potentials, i.e., $\nabla \times \psi' = \nabla \times \psi + \nabla \times \nabla g = \nabla \times \psi$. Now we note, however, that $\nabla \cdot \psi' = \nabla \cdot \psi + \nabla \cdot \nabla g = \nabla^2 g = f$, where f is a nonvanishing scalar function. Hence, assuming ψ satisfies the second of (2.7b), either $\nabla \cdot \psi = 0$, or $\nabla \cdot \psi = f \neq 0$, may be used as the gauge condition. Note that in the latter case also, the four scalar functions in u, given in (2.7a), are reduced to the necessary three independent ones. Applying the divergence operator to the second of (2.7b) we find

$$\left(\nabla^2 - \frac{1}{c_s^2}\frac{\partial^2}{\partial t^2}\right)\nabla \cdot \psi = 0, \tag{2.18}$$

which generates the two gauge conditions discussed above, i.e., either $\nabla \cdot \psi = 0$, the trivial solution of (2.18), or $\nabla \cdot \psi$ is the nonvanishing solution of (2.18). We shall exhibit, in our later work here, examples of ψ that fit into one or the other of these gauge conditions.

2.3. Types of body waves and governing equations

2.3.1. One-dimensional plane waves; D'Alembert's solution

Assume we have but one displacement component $u_1 = u$, a function of $x_1 = x$ and t only, then wave equations, such as (2.1) or (2.3), become

$$\frac{\partial^2 u(x,t)}{\partial x^2} = \ddot{u}(x,t)/c^2 \tag{2.19}$$

where c is the speed of the waves governed by the one-dimensional wave equation (2.19), which would be, in the case of (2.1) and (2.3), respectively c_s and c_d. Now let

$$\eta = x - ct, \qquad \zeta = x + ct,$$

from which

$$u(\eta, \zeta) = u[\eta(x, t), \zeta(x, t)] = u(x, t).$$

It follows, using the chain rule, that

$$\frac{\partial^2 u}{\partial x^2} = \frac{\partial^3 u}{\partial \eta^2} + 2 \frac{\partial^2 u}{\partial \eta \partial \zeta} + \frac{\partial^2 u}{\partial \zeta^2},$$

$$\ddot{u} = c^2 \left(\frac{\partial^2 u}{\partial \eta^2} - 2 \frac{\partial^2 u}{\partial \eta \partial \zeta} + \frac{\partial^2 u}{\partial \zeta^2} \right),$$

and substituting these equations in (2.19), the latter reduces to

$$\frac{\partial^2 u(\eta, \zeta)}{\partial \eta \partial \zeta} = 0. \tag{2.20}$$

Successive partial integrations of (2.20) yield the solution for (2.19)

$$u(\eta, \zeta) = u(x, t) = \int f(\eta) d\eta + G(\zeta),$$

or

$$u(x, t) = F(x - ct) + G(x + ct). \tag{2.21}$$

The solution was given by D'Alembert in 1747. Conditions for its validity are that F and G be twice differentiable (in order that u satisfies (2.19), but otherwise they can be arbitrary. To determine F and G specifically, one must employ the initial conditions of the problem. We will treat such a problem in § 2.9.1.

The general solution (2.21) is composed of the two waves, $F(x-ct)$ and $G(x+ct)$ which are one-dimensional plane waves propagating in the positive and negative x-directions, respectively. F and G are plane waves, since their arguments remain constant during propagation, defining planes that have their normal in the direction of wave propagation, and rendering F and G constant over such planes. We can therefore state the following general definition of a *plane wave: A solution of the wave equation represents a plane wave, if the solution is constant over planes which have their normal in the direction of wave propagation.*

2.3.2. Three-dimensional plane waves

We begin our analysis of body waves, governed by Lamé's solution (2.7), hence (1.65), by treating general plane waves. By virtue of their simplicity, we will find these waves useful in formulating further theory, and in solving problems involving more complicated disturbances. We can construct a plane wave solution of the three-dimensional wave equation, say arbitrary $\alpha(\mathbf{x}, t)$, by drawing on the definition given in the last section. Consider the plane in the space x_i, show in fig. 2.1, defined by $x_j l_j = d$, where

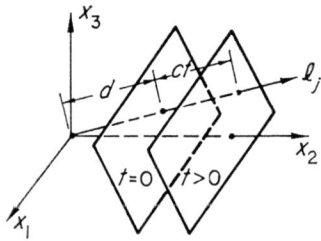

Fig. 2.1. Propagation of a three-dimensional plane wave.

l_j is its outward normal, and d its distance from the origin (along the normal). If we associate the position of this plane with $t=0$, then for $t>0$, assuming the plane moves with speed c in the direction of l_j, it propagates a distance ct. Hence, the equation of the plane in its new position is $x_j l_j = d + ct$, or $x_j l_j - ct = d$. Hence, if we let

$$\alpha(\mathbf{x}, t) = \alpha(x_j l_j - ct), \qquad (2.22)$$

we have a function satisfying the definition of a plane wave, since (2.22) wi. satisfy the wave equation in three space, having propagation speed c The argument of α, $x_j l_j - ct$, gives the relationship between x_j and t, and is generally called the phase of the wave. Hence, in (2.22), α is constant, because it is associated with the planes of constant phase d. Later, we will see that there are more general surfaces of constant phase.

For general plane waves φ and ψ_k, of the form (2.22), it can be shown that they must propagate with speeds c_d and c_s, respectively, in order to satisfy (2.7b). Consider the plane waves

$$\varphi = \varphi(x_j l_j - ct), \qquad (2.23)$$

from which it follows that

$$\nabla^2 \varphi = [\varphi' \cdot (x_j l_j - ct)_{,i}]_{,i}, \qquad (2.24)$$

where the prime indicates differentiation of φ with respect to its argument. Equation (2.24) can be reduced to

$$\nabla^2 \varphi = (\varphi' \delta_{ij} l_j)_{,i} = \varphi'_{,i} l_i = \varphi'' l_i l_i = \varphi''. \tag{2.25}$$

From (2.23) we have $\ddot{\varphi} = c^2 \varphi''$, and substituting this, and (2.25), in the first of (2.7b) gives $\varphi'' = (c/c_d)^2 \varphi''$, which can only be satisfied if $c = c_d$, i.e., *plane waves of φ must propagate with the speed c_d in the space x_i*. A similar procedure shows that *plane waves of ψ_k must propagate with the speed c_s*. Noting that the cartesian form of (2.7a) is

$$u_i = \varphi_{,i} + c_{ijk} \psi_{k,j}, \tag{2.26}$$

it follows from this equation that *plane waves of u_i will have the character of the plane waves of φ and ψ_k*, i.e., it will be composed in general of plane waves that propagate with c_d and c_s.

Through use of (2.26) we can examine the nature of the u_i plane waves. For simplicity, and without the loss of generality, we choose the direction of propagation as that along the positive x_1 axis. Then $l_1 = 1$, $l_2 = l_3 = 0$, and from (2.23) the corresponding plane waves of φ and ψ_k must be

$$\varphi = \varphi(x_1 - c_d t), \qquad \psi_k = \psi_k(x_1 - c_s t), \tag{2.27}$$

since we have just learned that in such waves the speed of propagation must be c_d and c_s, respectively. Consider u_1, which is the displacement component in the propagation direction. From (2.26), (2.27)

$$u_1 = \varphi_{,1} + c_{1jk} \psi_{k,j} = \varphi_{,1} + \psi_{3,2} - \psi_{2,3} = \varphi'(x_1 - c_d t), \tag{2.28}$$

therefore, u_1 is a plane dilatation wave, traveling with speed c_d, and is directed along the propagation direction, or normal to its plane, as indicated by the leading wave in fig. 2.2.

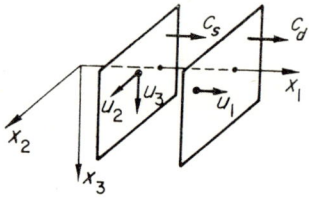

Fig. 2.2. Plane-dilational and-equivoluminal displacement waves.

Similarly,

$$u_2 = \varphi_{,2} + \psi_{1,3} - \psi_{3,1} = -\psi'_3(x_1 - c_s t), \tag{2.29}$$

therefore, u_2 is a plane equivoluminal wave, traveling with speed c_s, and is directed along its plane in the x_2-direction, normal to the propagation direction x_1. Finally

$$u_3 = \varphi_{,3} + \psi_{2,1} - \psi_{1,2} = \psi_2'(x_1 - c_s t), \tag{2.30}$$

is also a plane equivoluminal wave, traveling with speed c_s, but is directed along its plane in the x_3-direction, normal to the propagation direction x_1. Figure 2.2 also shows these slower traveling equivoluminal waves (note $c_d > c_s$). It should be noted that u_1, u_2 and u_3, given in (2.28)–(2.30), form a system of waves that propagate in the x_1-direction, but are independent of each other. It is of interest to point out that if we consider the $x_1 x_2$ plane as horizontally oriented, we can identify the u_1, u_2 and u_3 waves with *the plane harmonic P (compressional), SH (horizontally polarized shear) and SV (vertically polarized shear) waves of seismology*, respectively. We will note, as we progress with the theory, the importance of plane waves, separately, and in combinations, in helping us understand much more involved elastic wave propagation phenomena.

2.3.3. *Spherically symmetric waves*

In the spherical coordinate space (r, θ, η), if wave motion does not vary with the angular coordinates θ and η, we have spherically symmetric waves. Noting that $\partial/\partial \theta = \partial/\partial \eta = 0$, $u_\theta = u_\eta = 0$ here, Lamé's solution reduces to

$$u = \frac{\partial \varphi}{\partial r},$$
$$\nabla^2 \varphi = \frac{1}{r^2} \frac{\partial}{\partial r}\left(r^2 \frac{\partial \varphi}{\partial r}\right) = \ddot{\varphi}/c_d^2, \tag{2.31}$$

where u represents the only nonvanishing displacement component u_r. If φ is set equal to φ'/r in the second of (2.31), the latter becomes

$$\frac{\partial^2 \varphi'}{\partial r^2} = \ddot{\varphi}' c_d^2. \tag{2.32}$$

Noting that $\varphi' = r\varphi$, and that (2.32) has the form of (2.19), it follows from (2.21) that

$$\varphi(r, t) = (1/r)[F(r - c_d t) + G(r + c_d t)]. \tag{2.33}$$

The first term in (2.33) represents a wave traveling outward from $r = 0$, and the second, one traveling inward toward $r=0$. Analogously to the planes of constant phase in plane waves, the arguments of F and G here determine *spherical surfaces of constant phase*. Later in this chapter, and in Chapter 5 we shall treat problems involving spherically symmetric body waves.

2.3.4. Axially symmetric and nonaxially symmetric waves

In the cylindrical coordinate system (r, θ, z), if wave motion does not vary with the angular coordinate θ so that $\partial/\partial\theta = 0$, we have waves that are symmetric with respect to the z-axis. Assuming displacement component u_θ vanishes and if we further impose plane strain, such that $u_z = 0$, $\partial/\partial z = 0$, the Lamé solution (2.7) reduces to

$$u = \frac{\partial \varphi}{\partial r},$$
$$\nabla^2 \varphi = \frac{1}{r}\frac{\partial}{\partial r}\left(r\frac{\partial \varphi}{\partial r}\right) = \ddot{\varphi}/c_d^2, \tag{2.34}$$

where u represents the only nonvanishing displacement component u_r. Solutions φ of (2.34) represent *circular cylindrical surfaces of constant phase*. We will see examples of these surfaces, in § 2.5.2 and later sections of this chapter, and in Chapter 5.

If one assumes now that u_z is nonvanishing, we still have axially symmetric waves, but this case is more complicated, with two displacements involved. We will deal with problems of this type in Chapters 4, 6 and 7.

In the cylindrical coordinate system when variation with respect to θ is involved, one is dealing with nonaxially symmetric waves. We will treat such waves in problems of wave diffraction from cylindrical obstacles in Chapter 8. Another type of nonaxially symmetric wave problem, based on a cartesian coordinate system, is that of a point load traveling on the surface of a half space. This will be treated in Chapter 6. Needless to say, these more complicated problems require use of the full Lamé solution 2.7), and study of more inherently complicated wave surfaces.

2.4. Body wave generation of waves peculiar to boundaries

In later chapters we shall observe the strong role played by plane body waves in bounded media. For example, we shall see in Chapter 3 (and related Chapter 6), on reflection and refraction of time harmonic waves at an

Ch. 2, § 2.5] TIME HARMONIC BODY WAVES 69

interface, that a pair of plane body waves, P and SV, in grazing incidence at a free plane boundary, generates the so-called Rayleigh surface wave. This "new" wave, which travels along the boundary, is no longer plane, its amplitude decaying exponentially into the interior. Further, in Chapter 4 (and related Chapter 7), on time harmonic waves in elastic waveguides, for an infinite plate (or layer) in plane strain, we shall see that interference between plane body waves, after reflection from the plate faces, creates other *symmetric or antisymmetric (with respect to the plate mid-plane) cylindrical wave surfaces of constant phase*. These travel along the length of the plate. Similarly, in a circular cylindrical rod, *axially symmetric and nonaxially symmetric (with respect to the rod axis) surfaces of constant phase* are generated at the lateral surface of the rod, and they propagate along the length of the rod.

2.5. Time harmonic body waves

2.5.1. Helmholtz equation

Consider the three dimensional wave equation

$$\nabla^2 \eta(x, t) = \ddot{\eta}(x, t)/c^2, \qquad (2.35a)$$

where c is the speed of propagation. If we assume that

$$\eta(x, t) = \hat{\eta}(x) \exp(i\omega t), \qquad (2.35b)$$

where $i = \sqrt{-1}$, and ω is the *circular frequency* (which from here on wil be called *frequency*), and substitute (2.35b) into (2.35a), the latter becomes

$$\nabla^2 \hat{\eta}(x) + \varkappa^2 \hat{\eta}(x) = 0, \qquad (2.35c)$$

where $\varkappa = \omega/c$ *is the wave number*, equal to $2\pi/L$, where L is the wave lengt Equation (2.35c) is known as *Helmholtz's equation*. It underlies studies based on time harmonic waves. These waves provide a simple means of bringing out important physical information on wave propagation without the complexities introduced by time dependency. Most of its usefulness stems from wave representations for long time, i.e., where a steady state analysis is, in many instances, sufficient.

2.5.2. Time harmonic axially symmetric waves

As an example of time harmonic body waves we treat those governed by (2.34). Indentifying the second of (2.34) with (2.35a), and using (2.35b), we find the corresponding Helmholtz equation (2.35c) to be the equation governing the ordinary Bessel functions of order zero. It follows that the general solution of this equation is

$$\varphi(r, t) = \hat{\varphi}(r) \exp(i\omega t) = [A(\omega)J_0(\omega r/c_d) + B(\omega)Y_0(\omega r/c_d)] \exp(i\omega t), \quad (2.36a)$$

where J_0, Y_0 are, respectively, *the ordinary Bessel functions of first and second kinds of order zero*. The real part of the right hand side of (2.36a) represents a *standing, or nonpropagating wave*, i.e., the magnitudes at points along the wave change with time, but the wave does not travel. Making use of the relations

$$J_0(x) = \tfrac{1}{2}[H_0^{(1)}(x) + H_0^{(2)}(x)],$$

$$Y_0(x) = \frac{1}{2i}[H_0^{(1)}(x) - H_0^{(2)}(x)],$$

(2.36a) can be written alternately as

$$\varphi(r, t) = [A'(\omega)H_0^{(1)}(\omega r/c_d) + B'(\omega)H_0^{(2)}(\omega r/c_d)] \exp(i\omega t), \quad (2.36b)$$

where $H_0^{(1)}$, $H_0^{(2)}$, are the *Hankel functions of the first and second kinds of order zero*. Approximating (2.36b) for r large, utilizing the leading terms of the asymptotic expansions for the Hankel functions, given in Erdelyi et al. [2.8, pg. 85], it becomes

$$\varphi(r, t) \simeq \left(\frac{2c_d}{\pi\omega r}\right)^{\frac{1}{2}} \left\{ A''(\omega) \exp\left[i\omega\left(t + \frac{r}{c_d}\right)\right] + B''(\omega) \exp\left[i\omega\left(t - \frac{r}{c_d}\right)\right]\right\}. \quad (2.37)$$

The real part of the right hand side of (2.37) exhibits the traveling nature of φ, the first term there being an harmonic wave train propagating inward toward $r=0$, and the second a similar train propagating outward from $r=0$. Assuming A'', B'' real and equal, then (2.37), taken as a whole, still exhibits the *standing wave nature* of φ, since we then have *two wave trains of equal frequency and amplitude, but traveling in opposite directions. The sum of two such waves has spatial nodal points, which characterizes a standing wave.* It should be noted that (2.37) exhibits *circular cylindrical surfaces of constant phase*, which was a property of φ pointed out in § 2.3.4.

2.6. Propagation of surfaces of discontinuity

We have already imposed on the displacements $u_i(x, t)$ the requirement that they be continuous functions of position x at any time t (physically this amounts to a condition that rules out dislocations in the medium). As we shall see no such condition need be imposed on the space and time derivatives of the displacements, i.e., the velocities and strains, and hence the stresses. From the physical point of view, generally one can expect these quantities to be continuous. However, many situations, involving very sudden loadings in the interior or on the surface of an elastic solid, will result in sudden variations of a propagating particle velocity or stress over a very small interval of space and time, and these can be quite closely approximated by finite jumps based on the linear elasticity model. The strong interest in the propagation of surfaces of discontinuities (wavefronts) stems from *approximate representations* that one can obtain for field quantities at, and just behind, the wavefront. In particular, this is important for the more involved geometries, where exact solutions are more difficult to obtain. We shall observe this in our work in the later chapters of this book. On the other hand the topic is basic to wave propagation studies and, as we shall see, one gains greater insight into even the simpler transient problems, through wavefront analysis.

Love [1.2, § § 205-207] sets down the basic kinematical and dynamic conditions that must hold at a propagating surface of discontinuity in an elastic solid. He also shows that for the homogeneous, isotropic solid, a dilatational wavefront propagates with the velocity c_d, and an equivoluminal wavefront with velocity c_s. The theory was discussed recently by H. Keller [2.9] in work on the propagation of stress discontinuities in inhomogeneous, isotropic, elastic media. Further, the subject has a strong relation to wavefront analysis in geometrical optics and acoustics. Modern accounts of this may be found in Friedlander [2.10] and J. Keller [2.11].

2.6.1. Kinematical conditions

Consider a surface of discontinuity S, propagating in the unbounded medium. The situation is shown in fig. 2.3, for a fixed instant of time. It is assumed that S propagates into the undisturbed region (2), leaving a disturbed region (1) behind it. It is also assumed that S moves normal

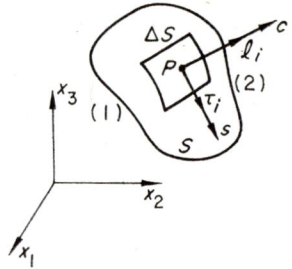

Fig. 2.3. Position of propagating surface of discontinuity at time t.

to itself with velocity c, i.e., each point $P(x)$ of S propagates with velocity c, along the outward (toward (2)) unit normal vector l_i to S at that point. We take s to be any direction in the tangent plane to S at $P(x)$, with τ_i the unit vector along s. It is clear then, that region (1) will have the nonvanishing displacements $(u_i)_1$ and (2) the vanishing displacements $(u_i)_2$. Since we must have continuity of the displacements everywhere in space for all time, it follows that they must vanish on S for all t, i.e.,

$$u_i(x, t) = 0, \qquad \text{for } x \text{ on } S, \qquad \text{and } 0 < t < \infty. \tag{2.38}$$

Referring to fig. 2.3, from (2.38) it follows for any time that

$$\frac{du_i}{ds} = u_{i,j} \tau_j = 0, \tag{2.39}$$

for all directions s satisfying

$$\tau_j l_j = 0. \tag{2.40}$$

Equations (2.39), (2.40) are compatible, since at all points of S,

$$u_{i,j} = \frac{\partial u_i}{\partial l} l_j \tag{2.41}$$

where l is the distance from S measured along the normal l_j. Now (2.38) must hold for an incremental motion of S, corresponding to $x_j' = x_j + c l_j \triangle t$, $t' = t + \triangle t$, i.e., (2.38) must be satisfied to the first order in $\triangle t$, when x_j', t' are substituted for x_j, t. It follows, therefore, that for every point S, we must have

$$\frac{du_i}{dt} = \lim_{\triangle t \to 0} \left[\frac{u_i(x, t') - u_i(x, t)}{\triangle t} \right] + \lim_{\triangle t \to 0} \left[\frac{u_i(x', t) - u_i(x, t)}{\triangle x_j} \right] \cdot \frac{dx_j}{dt} = 0, \tag{2.42}$$

Ch. 2, § 2.6] PROPAGATION OF SURFACES DISCONTINUITY 73

where $\Delta x_j = x'_j - x_j = cl_j \Delta t$. Upon taking the limits in (2.42), it gives

$$\dot{u}_i = -cu_{i,j}l_j, \tag{2.43}$$

which, using (2.41), gives

$$\dot{u}_i = -c\frac{\partial u_i}{\partial l}. \tag{2.44}$$

From (2.44), (2.41) we find

$$\dot{u}_i l_j = -cu_{i,j}. \tag{2.45}$$

The equations (2.43)–(2.45) represent three forms of *the kinematical relations, between finite jumps in the time and space derivatives of* u_i, *that must hold at all points of the propagating surface S*. The derivatives in them must be evaluated at time t from the displacements $(u_i)_1$, i.e., those on that side of S having a disturbance.

As we shall see later in our work, commonly one finds a wavefront moving into a region which has already been disturbed by a faster traveling wave. A very common case is that when both the dilatational and equivoluminal waves are present. For this type of situation we replace $u_i(x, t) = 0$ in (2.38), with

$$[u_i(x, t)] = (u_i)_2 - (u_i)_1 = 0, \tag{2.46}$$

where the *brackets are standard notation for a dynamic jump across a surface of the quantity within*, $(u_i)_2$ being evaluated on the (2) side of S, and $(u_i)_1$ on the (1) side. In (2.46) the jump in u_i is zero since again we must have u_i continuous everywhere in space x_i at time t. It is clear that with $[u_i]$ in (2.46) replacing u_i in (2.38), we would proceed exactly as in derivation of (2.43)–(2.45). Hence, we have, corresponding to (2.46), the jump conditions

$$[\dot{u}_i] = -c[u_{i,j}]l_j,$$
$$[\dot{u}_i] = -c\left[\frac{\partial u_i}{\partial l}\right], \tag{2.47}$$
$$[\dot{u}_i]l_j = -c[u_{i,j}].$$

2.6.2. Dynamical conditions

The dynamical conditions, which hold at the moving surface of discontinuity S, are determined by considering the momentum changes of a thin slice of the medium adjacent to S, and the corresponding impulse-momentum equation. Visualize in fig. 2.3, a small prismatic element of the slice with base ΔS, lateral sides composed of the normals to the edges of ΔS (out of the figure), and an upper end surface, parallel to ΔS a distance $c\Delta t$ away. The element, therefore, lies between the new position of the wave front, and its former position S (in. fig. 2.3), a time Δt earlier. In the time Δt this element passes from state (2) to state (1), that is, from a state of rest and zero strain to one of motion and strain corresponding to the displacements u_i. This change is brought about by the resultant force acting across ΔS, which is given by $-(T_i)_1\Delta S = -(\sigma_{ij})_1 l_j \Delta S$. The corresponding impulse is obtained by multiplying this force by Δt. It follows the *impulse-momentum equation* is

$$\varrho \Delta S c \Delta t (\dot{u}_i)_1 = -(\sigma_{ij})_1 l_j \Delta S \Delta t,$$

which reduces to

$$(\sigma_{ij})_1 l_j = -\varrho c (\dot{u}_i)_1. \tag{2.48}$$

Equations (2.48) hold at all points of S. If initially there is a disturbance in region (2), and jumps in $\dot{u}_i$ exist across S, equation (2.48) would become

$$[\sigma_{ij}] l_j = -\varrho c [\dot{u}_i], \tag{2.49}$$

which follows from arguments like those leading to (2.47).

2.6.3. Wavefront velocities

With the aid of (1.45), and (1.25), (1.39), we can write (2.48) as

$$\varrho c (\dot{u}_i)_1 = -\{\lambda \delta_{ij}(u_{k,k})_1 + \mu[(u_{i,j})_1 + (u_{j,i})_1]\} l_j, \tag{2.50}$$

which, of course, holds at the surface of discontinuity S. Making use of (2.43), (2.45), we can write (2.50) as

$$(\varrho c^2 - \mu)(\dot{u}_i)_1 = -c(\lambda + \mu)(u_{j,j})_1 l_i. \tag{2.51}$$

Now setting the wavefront dilatation $(u_{j,j})_1 = 0$, we find since $(\dot{u}_i)_1$ may be nonvanishing, that (2.51) yields the *wave velocity* $c = c_s = (\mu/\varrho)^{\frac{1}{2}}$ *for this equivoluminal wavefront*. As we anticipated this agrees with our general results for equivoluminal waves in § 2.2.1.

With the purpose of introducing the infinitesimal rotation we rewrite (2.50) as

$$\varrho c (\dot{u}_i)_1 = -\{\lambda \delta_{ij}(u_{k,k})_1 + 2\mu(u_{i,j})_1 - \mu[(u_{i,j})_1 - (u_{j,i})_1]\} l_j. \tag{2.52}$$

The term in brackets is twice the infinitesimal rotation $(\omega_{ij})_1$, according to (1.32), and if we set this rotation equal to zero, and make use of (2.43), (2.45) again, equation (2.52) becomes

$$(\varrho c^2 - 2\mu)(\dot{u}_i)_1 = \lambda(\dot{u}_k)_1 l_k l_i .\tag{2.53}$$

Multiplying (2.53) by l_i, it reduces to

$$[\varrho c^2 - (\lambda + 2\mu)](\dot{u}_i)_1 l_i = 0 ,\tag{2.54}$$

and since $(\dot{u}_i)_1 l_i$ is, arbitrarily, nonvanishing, we find

$$c = c_d = [(\lambda + 2\mu)/\varrho]^{\frac{1}{2}} ,$$

as we might expect for the velocity of *this irrotational or dilatational wavefront* in agreement with our earlier analysis in § 2.2.1. It is clear from the earlier analysis leading to (2.47), (2.49), that the ( )$_1$ terms in (2.50)-(2.54) can be replaced by [ ], i.e., finite jumps (where region (2) has a previous disturbance) also propagate with velocities c_d, c_s.

2.6.4. *Wavefronts, characteristics and rays*

A propagating surface of discontinuity can be represented by the equation

$$F(x, t) = 0 .\tag{2.55}$$

According to the implicit function theorem, one can write (2.55) in the form

$$F(x, t) = \tau(x) - t = 0 ,\tag{2.56a}$$

or

$$t = \tau(x) ,\tag{2.56b}$$

provided $\partial F/\partial t \neq 0$ (cf. Wilson [2.12], for example), which is certainly satisfied for a propagating surface. *Equations (2.56) can be identified with a characteristic in the theory of the wave equation.* Indeed, as Friedlander points out, in his discussion of wavefronts and characteristics [2.10, ch. 2], *a characteristic is a surface in space which propagates normal to itself with its particular characteristic velocity.* It is known that *discontinuities cannot occur except across such surfaces.* Characteristics can also be identified with *the wavefronts of geometrical optics* (or acoustics). Further, (2.56) can be identified with the *surfaces of constant phase* discussed earlier in this chapter, e.g., the three-dimensional plane dilatational waves of § 2.3.2 are identified with the equation $t = (x_j l_j - d)/c_d$, which is a special case of (2.56b).

As we have noted, *a point on the wavefront or characteristic, always moves normal to this surface.* Hence, *the point describes a curve which is orthogonal to the wavefront* throughout the latter's motion. *Such a curve is called a ray.* The equation governing the rays can be obtained by first noting that along such curves we must have

$$\frac{dx_i}{dt} = cl_i = c\frac{\tau(x)_{,i}}{|\nabla \tau(x)|}, \tag{2.57}$$

where $|\nabla \tau(x)|$ is given by

$$|\nabla \tau(x)| = \left\{ \left[\frac{\partial \tau(x)}{\partial x_1}\right]^2 + \left[\frac{\partial \tau(x)}{\partial x_2}\right]^2 + \left[\frac{\partial \tau(x)}{\partial x_3}\right]^2 \right\}^{\frac{1}{2}}.$$

Now since we have (2.56),

$$\frac{dF}{dt} = \frac{\partial \tau(x)}{\partial x_i}\frac{dx_i}{dt} - 1 = 0, \tag{2.58}$$

hence,

$$\tau(x)_{,i} l_i = 1/c. \tag{2.59}$$

Multiplying the second relation in (2.57) by l_i, and using (2.59), we find that

$$|\nabla \tau(x)| = 1/c, \tag{2.60}$$

and using (2.60) in the definition of l_i in (2.57), it follows that

$$l_i = c\tau(x)_{,i}. \tag{2.61}$$

Substitution of (2.61) into the first equation in (2.57) yields

$$\frac{dx_i}{dt} = c^2 \tau(x)_{,i}, \tag{2.62}$$

which gives the rays. In the process we have generated the *eikonal equation* (2.60) (a name stemming from geometrical optics). It is a nonlinear, first order partial differential equation that $\tau(x)$ must satisfy, and is usually stated as $|\nabla \tau(x)|^2 = 1/c^2$. As shown in [2.9], a knowledge of the rays, leads to a solution of the eikonal equation (2.60), and, together with given amplitudes on the initial wavefront, ultimately to the determination of the propagation of discontinuities in particle velocity, strains, and stresses, and the variation of their amplitudes along the rays.

For our homogeneous solid, since c is independent of position and

Ch. 2, § 2.6]   PROPAGATION OF SURFACES OF DISCONTINUITY   77

hence time, the rays are straight lines. This is easily seen by first differentiating (2.62) with respect to time, and rewriting it as

$$\frac{d^2 x_i}{dt^2} = c^2 \frac{\partial}{\partial x_i}\left[\frac{d\tau(x)}{dt}\right].$$

Using (2.56b), this reduces to

$$\frac{dx_i}{dt} = cl_i = \text{constant}.$$

Hence the corresponding family of wavefronts are parallel surfaces whose common normals are the rays. With a given wavefront at some time t', knowledge of the rays permits the construction of the wavefront at some later time t, i.e., one extends each point of the wavefront at t', a distance $c(t-t')$ along the ray (normal) to this wavefront. The process is known as *Huyghen's construction of wavefronts*, stemming from his work in optics in the 1600's. It embodies what is known as *Huyghen's principle*. Essentially the principle states that at time t the wavefront is the envelope of spheres of radius $c(t-t')$, whose centers are points of the wavefront at the previous time t'. For a homogeneous medium this leads to parallel wavefronts.

Friedlander [2.10, ch. 3 § 6] derives, for the present homogeneous case, expressions for the variation of the wavefront magnitude of a field variable along a ray. Assuming an initial wavefront of surface Σ_0, the position of this front, at time t, is given by $v = ct$. In this position the surface is Σ_v, parallel to, and at a normal distance v from, Σ_0. With principal radii of curvatures for Σ_0 and Σ_v, being respectively, R_0, S_0 and $R_v = R_0 + v$, $S_v = S_0 + v$, the variation of the wavefront magnitude for the three-dimensional case is given by

$$M = \left(\frac{d\Sigma_0}{d\Sigma_v}\right)^{\frac{1}{2}} = \left(\frac{R_0 S_0}{R_v S_v}\right)^{\frac{1}{2}}, \tag{2.63}$$

where $d\Sigma_0$, $d\Sigma_v$, are the elements of area on Σ_0, Σ_v, respectively. For the two-dimensional case

$$M = \left(\frac{R_0}{R_v}\right)^{\frac{1}{2}}. \tag{2.64}$$

From (2.63), for spherical waves, we find that M decays as R_v^{-1}, and from (2.64), for cylindrical waves, M decays as $R_v^{-\frac{1}{2}}$. Note since $v = ct$, that for a rge time from (2.63), M decays as t^{-1}, and from (2.64), as $t^{-\frac{1}{2}}$. We will note, in our later work on transient wave propagation, confirmation of these behaviors.

2.6.5. Further comments

As pointed out in the introduction to § 2.6, the strong interest in the propagation of wavefronts stems from approximate representations that can be obtained for field variables, valid at, and just behind, the wavefront. The representation is usually a series in powers of $[t-\tau(x)]$, and one usually needs only one or two terms. Friedlander [2.10, ch. 3 § 9, ch. 4] treats the subject for geometric acoustics, including applications to reflection problems. Karal and J. Keller [2.13] have presented related work for the homogeneous and inhomogeneous elastic medium. Achenbach [2.14, § 4.6] uses the power series method to treat a problem of axially symmetric torsional waves, excited by a suddenly applied shear stress on the wall of a circular cavity in an infinite plate in plane stress.

We will not enter into such wavefront analysis here, since the author prefers to treat the subject using asymptotic expansions of integrals. This will be carried out in the later chapters of this book.

2.7. Wave motion due to body forces

2.7.1. Introduction

An important class of problems in elastodynamics are those involving disturbances on an interior region of an elastic body, where the sources of such disturbances are external to the body. Examples are forces due to the gravitational or a magnetic field. The problems involve the body f orce X, hence the displacement equations of motion (1.65), (1.66b). To study the body waves emanating from these disturbances, one usually considers the initially quiescent unbounded elastic solid, subjected, at time $t=0$, to generally time-dependent sources. On this basis, a point body force, or distributed body forces over a finite region, will generate waves that propagate outward from these sources. Such problems are sometimes called *radiation problems*, but not exclusively, since radiation is also a property of waves generated by surface sources on a body.

Ch. 2, § 2.7] WAVE MOTION DUE TO BODY FORCES 79

Early investigations of wave motion, generated by body forces, were carried out in the classical works on elastic wave propagation by Stokes [2.15] in 1849, and Love [2.16] in 1904. Stokes was the first to derive the basic singular solution for the displacements generated by a suddenly applied concentrated load at a point of the unbounded elastic medium. Stokes derived his solution using wave equations on the dilatation and the rotation, i.e., equations (2.4), (2.5), and Poisson's integral formula (for the solution of initial value problems governed by the three dimensional wave equation; given by (2.146) to be discussed in § 2.9.2) to integrate these equations. In [2.16] Love pointed out that this formula yields correct results for a quantity only when it is continuous at its wave front, hence ruling out its usefulness in representing the dilatation and rotation when these quantities have (admissible) wavefront discontinuities. In order to clear up the point, and its effects on the important results of Stokes, Love made an exhaustive study of the initial value problem for arbitrary initial values, and related wave front discontinuities (we will discuss the initial value problem later). In the course of this study he solved the point problem by a means independent of that used by Stokes, i.e., essentially by using retarded potentials, and equation (2.83) (to be discussed later). Love confirmed Stokes' solution for the point load problem, and added considerably to its interpretation. He also derived corrected expressions for the dilatation and rotation given by Stokes, for the case where the initial velocity was nonvanishing at the boundary of the initially disturbed region of the elastic medium.

In our treatment here, which is based on Lamé's solution (2.7), we will draw mostly on Love's work [2.16], an abstracted version of which appears in his book [1.2, §§ 212–213]. An alternate treatment, also based on Lamé's solution, is given by Achenbach [2.14, §§ 3.6–3.10, 3.12]. Eason et al [2.17] treat the subject with a direct attack on the displacement equations of motion, using a quadruple Fourier transform method. A treatment for the related electromagnetic field may be found in Jones [2.18].

At the outset one expresses X, through Helmholtz resolution, as

$$X = \varrho(c_d^2 \nabla \Phi + c_s^2 \nabla \times \Psi). \tag{2.65}$$

Substitution of (2.7a), and its counterpart (2.65), into the displacement equations of motion, (1.65b) or (1.66b), reduces the latter to the inhomogeneous wave equations

$$\nabla^2 \varphi - \ddot{\varphi}/c_d^2 = -\Phi(x, t), \tag{2.66a}$$

$$\nabla^2 \psi - \ddot{\psi}/c_s^2 = -\Psi(x, t). \tag{2.66b}$$

A general solution of (2.66) is composed of the solution to the homogeneous equations of (2.66), plus particular solutions, representing, through Φ and Ψ, the effects of the given distribution of body forces. The general solution of (2.66) constitutes the solution for u through (2.7a).

2.7.2. Basic singular solutions; retarded potentials

Ultimately we seek a solution in the case where the body forces X are nonvanishing within a finite volume V, and vanish outside of V. The situation is depicted in fig. 2.4, where x locates the general point of of the unbounded

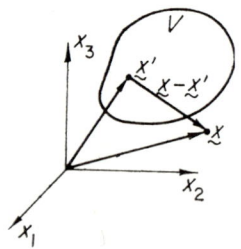

Fig. 2.4. Region V of body forces;

elastic solid, x' the general point within V (at which body forces act), and $|x-x'|$ is the distance between the two points. In the present context let us first solve (2.66a), subject to the initial conditions

$$\varphi(x, 0) = \dot{\varphi}(x, 0) = 0. \tag{2.67}$$

With Φ representing a distribution of sources one can proceed by first getting a solution for a point source, i.e., the *basic singular solution*, and later synthesizing such solutions for each point of the Φ distribution.

To carry out this procedure we will make use of the Dirac delta function, a well-known generalized function. In one dimension the delta function $\delta(x-x')$ has the properties

Ch. 2, § 2.7]   WAVE MOTION DUE TO BODY FORCES   81

$$\delta(x-x')=0 \quad, \quad x \neq x',$$

(2.68)

$$\int_a^b h(x')\delta(x-x')dx' = \begin{cases} h(x), & x \in (a,b)^1, \\ 0, & x \notin (a,b). \end{cases}$$

All the derivatives of the delta function $\delta'(x-x')$, $\delta''(x-x')$, ... $\delta^{(n)}(x-x')$, exist (a prime indicating differentiation with respect to the argument), having similar properties to (2.68), except that the integrals there become

$$\int_a^b h(x')\delta^{(n)}(x-x')dx' = \begin{cases} h^{(n)}(x), & x \in (a,b), \\ 0, & x \notin (a,b). \end{cases}$$

(2.69)

Similarly the delta function in three dimensions $\delta(x-x')$ is defined by

$$\delta(x-x')=0, \quad x \neq x',$$

(2.70)

$$\int_V h(x')\delta(x-x')dV_{x'} = \begin{cases} h(x), & x \in V, \\ 0, & x \notin V, \end{cases}$$

where V is a sufficiently smooth region (cf. Divergence Theorem, Ch. 1 § 1.6, p. 43) and $dV_{x'}$ is an element of volume V in the x' coordinates. Derivatives of the delta function for three dimensions are defined like in (2.69). For cartesian coordinates we have

$$\delta(x-x') = \delta(x_1-x_1')\delta(x_2-x_2')\delta(x_3-x_3'),$$

(2.71a)

$dV_{x'} = dx_1'dx_2'dx_3'$, and

$$|x-x'| = [(x_1-x_1')^2 + (x_2-x_2')^2 + (x_3-x_3')^2]^{\frac{1}{2}}.$$

For spherical coordinates, we have

$$\delta(x-x') = \delta(r-r')\delta(\theta-\theta')\delta(\eta-\eta')/r'^2 \sin \theta,$$

(2.71b)

---

[1] The notation $x \in (a,b)$ means that x is a member of the set (a,b) defined as $a < x < b$. $x \notin (a,b)$ means x is not a member of the set (a,b).

as shown in [2.18, § 1.15]. Interchanging r and r', and then letting $r'=0$, in (2.71b), i.e., the point source is at the origin, it reduces to

$$\delta(x-x') = \delta(r)/4\pi r^2, \tag{2.71c}$$

which can be proved by showing (2.71c) satisfies the appropriate form of (2.70).

We assume, then, a point source, which for convenience may be taken at the origin of coordinates, i.e., $x'=x=0$. Since we have radiation from a point, the disturbance will be a spherically symmetric one, emanating from the source at $r=0$, where $r=(x_1^2+x_2^2+x_3^2)^{\frac{1}{2}}$. As pointed out in [2.14], the basic singular solution $\hat{\varphi}(r,t)$, representing radiation from a point, must satisfy the flux condition

$$\lim_{\varepsilon \to 0} \int_S \frac{\partial \hat{\varphi}}{\partial r} dS = F(t), \tag{2.72}$$

where S is the surface of a sphere of radius $r=\varepsilon$, and $F(t)$ is a function of time that measures the magnitude of the radiation. For $r>0$, $\hat{\varphi}(r,t)$ must satisfy the homogeneous equation of (2.66a), i.e., the second of (2.31), and the homogeneous initial conditions (2.67). Alternately one can find $\hat{\varphi}(r,t)$ by solving the inhomogeneous wave equation (2.66a), for the present point source, namely

$$\frac{1}{r^2}\frac{\partial}{\partial r}\left(r^2 \frac{\partial \hat{\varphi}}{\partial r}\right) - \ddot{\hat{\varphi}}/c_d^2 = F(t)\frac{\delta(r)}{4\pi r^2}, \tag{2.73}$$

where use has been made of (2.71c). For cartesian coordinates (2.73) would become

$$\nabla^2 \hat{\varphi} - \ddot{\hat{\varphi}}/c_d^2 = F(t)\delta(x). \tag{2.74}$$

The solution $\hat{\varphi}(r,t)$ is found by appealing to the first of these formulations, i.e., the solution of the homogeneous equation of (2.73), which is given by (2.33). Since we have a radiation problem (outward traveling energy), the second term there is inadmissible. We therefore choose the first term in the form of

$$\hat{\varphi}(r,t) = -\frac{1}{4\pi r} F(t-r/c_d), \tag{2.75a}$$

to represent the solution, where

$$F(t-r/c_d)=0, \text{ for } t<r/c_d. \tag{2.75b}$$

Equation (2.75b) stems from the hyperbolicity of the homogeneous equation of (2.73). Direct substitution of (2.75) into (2.72) shows that the former satisfies the latter. Hence, the basic singular solution, (2.75), is the solution to the problem. It therefore is the solution of the inhomogeneous equation (2.73) also[2]. The wave nature of (2.75) is clear. It governs the motion of a particle at a station r, showing that this motion begins at the wave arrival time $t=r/c_d$, and has the same duration as that of the source at $r=0$. This source, of course, is due to an external mechanism.

In order to extend our solution to the distribution of sources $-\Phi(x, t)$, i.e., the solution of (2.66a), we proceed as in [2.14, § 3.6.1]. Assume the point source is located at $x=x' \neq 0$. Then the governing equation is

$$\nabla^2 \hat{\varphi} - \ddot{\hat{\varphi}}/c_d^2 = F(t)\delta(x-x'), \tag{2.76}$$

instead of (2.74). It follows from our foregoing discussion the solution of (2.76) is

$$\hat{\varphi}(x, t; x') = -\frac{1}{4\pi|x-x'|} F\left(t - \frac{|x-x'|}{c_d}\right). \tag{2.77}$$

Now we assume $F(t)=\delta(t-t')$, i.e., a delta function applied at t'. From (2.77) the corresponding solution is

$$G(x, t; x', t') = -\frac{1}{4\pi|x-x'|} \delta\left(t-t'-\frac{|x-x'|}{c_d}\right). \tag{2.78}$$

The function G in (2.78) is known as the Green's function[3] for the unbounded

---

[2] This can be proved through an independent solution of (2.73). A neat solution was suggested by my colleague Professor T. K. Caughey. Its derivation has been made into an exercise.

[3] Reading on the theory of Green's functions and their construction, in connection with the scalar wave equation, may be found in Friedlander's book [2.10, ch. 5].

region. It represents the response at the station x at time t due to an impulsive unit point source applied at the point x' at time t'. The response due to the distribution of sources $-\Phi(x, t)$ can be obtained by summing the responses due to each point of the distribution. It follows that $\varphi(x, t)$ is the integral over t' and x' of the product $-\Phi(x', t')G(x, t; x,' t')$

$$\varphi(x, t) = \frac{1}{4\pi} \int_0^t dt' \int_V \frac{\Phi(x', t')}{|x-x'|} \delta(t-t'-|x-x'|/c_d) dV_{x'}. \qquad (2.79)$$

Using (2.68), it is easy to see (2.79) reduces to

$$\varphi(x, t) = \frac{1}{4\pi} \int_{V_d} \frac{\Phi(x', t-|x-x'|/c_d)}{|x-x'|} dV_{x'}, \qquad (2.80)$$

where V_d is defined by the sphere $|x-x'| \leq c_d t$, having its center at x. Equation (2.80) shows that to evaluate the potential φ, at x at time t, due to sources at x', we must sum the values of these sources at the earlier times $t-|x-x'|/c_d$. Therefore the effects of the sources take the finite time $|x-x'|/c_d$ to reach the station x. This nature of (2.80) has led to its name, *retarded potential*. Equation (2.80) is usually written in the briefer form

$$\varphi(x, t) = \frac{1}{4\pi} \int_{V_d} \frac{[\Phi]}{|x-x'|} dV_{x'}, \qquad (2.81)$$

where the brackets indicate the Φ at x' is evaluated at time $t-|x-x'|/c_d$. Equation (2.79) may be verified as the solution of (2.66a) by direct substitution in the latter. Since (2.66b) can be solved in the same way (2.66a) was, we have for its solution

$$\psi(x, t) = \frac{1}{4\pi} \int_{V_s} \frac{[\Psi]}{|x-x'|} dV_{x'}, \qquad (2.82)$$

where the brackets indicate Ψ at x' is evaluated at time $t-|x-x'|/c_s$, and V_s is defined by the sphere $|x-x'| \leq c_s t$, having its center at x.

2.7.3. Elastodynamic solution

It follows from (2.7a), and (2.81), (2.82), that the *elastodynamic solution for the displacements* u, due to general body force X excitation, is

$$u(x,\,t) = \frac{1}{4\pi}\left[\int_{V_d} \nabla\,\frac{[\Phi]}{|x-x'|}\,dV_{x'} + \int_{V_s} \nabla\times\frac{[\Psi]}{|x-x'|}\,dV_{x'}\right], \qquad (2.83)$$

where ∇-operator is with respect to the x_i coordinates.

2.7.4. Two-dimensional radiation problems

In two-dimensional radiation problems we are dealing with line sources. As shown in Achenbach [2.14, § 3.7], one can also solve the inhomogeneous equation for such sources by superposition of impulses. Again one considers first the scalar potential φ. From (2.66a), (2.67) the governing equations for the two dimensional case, are

$$\frac{\partial^2 \varphi}{\partial x_1^2} + \frac{\partial^2 \varphi}{\partial x_2^2} - \ddot{\varphi}/c_d^2 = -\Phi(x_1,\,x_2,\,t)\,, \qquad (2.84)$$

and

$$\varphi(x_1,\,x_2,\,0) = \dot{\varphi}(x_1,\,x_2,\,0) = 0\,. \qquad (2.85)$$

The basic singular solution here is governed by the equation

$$\frac{\partial^2 \hat{\varphi}}{\partial x_1^2} + \frac{\partial^2 \hat{\varphi}}{\partial x_2^2} - \ddot{\hat{\varphi}}/c_d^2 = F(t)\delta(x_1-x_1')\delta(x_2-x_2')\,, \qquad (2.86)$$

and the initial conditions (2.85). The solution of (2.86) can be constructed by linear superposition of the three-dimensional solution (2.77), i.e.,

$$\hat{\varphi}(x,\,t;\,x') = -\frac{1}{4\pi}\int_{-\infty}^{\infty} \frac{F(t-|x-x'|/c_d)}{|x-x'|}\,dx_3'\,, \qquad (2.87)$$

where $|x-x'|$ is defined in (2.71a). Imposing the fact that $F(t-|x-x'|/c_d)=0$, for $t<|x-x'|/c_d$, the limits of integration in (2.87) are actually $x_3'=x_3 \pm(c_d^2 t^2-\varrho^2)^{\frac{1}{2}}$, where $\varrho=[(x_1-x_1')^2+(x_2-x_2')^2]^{\frac{1}{2}}$. It is easy to see that the integrand in (2.87) is symmetric about x_3, hence, the integral there may be written as twice that between the limits $x_3-(c_d^2 t^2-\varrho^2)^{\frac{1}{2}}$ and x_3. Then with the change in variable, $s=t-[\varrho^2+(x_3-x_3')^2]^{\frac{1}{2}}/c_d$, (2.87) reduces to

$$\hat{\varphi}(x_i, t; x_i') = -\frac{c_d}{2\pi} \int_0^{t-\varrho/c_d} \frac{F(s)ds}{[c_d^2(t-s)^2 - \varrho^2]^{\frac{1}{2}}}, \quad (2.88)$$

for $t > \varrho/c_d$. It should be noted that (2.88) is independent of x_3 as one would expect for a two-dimensional problem.

As in the case of the previously discussed three-dimensional problem if $F(t)$ is taken as the delta function $\delta(t-t')$, (2.88) reduces to

$$G(x_i, t; x_i', t') = -\frac{c_d H(t-t'-\varrho/c_d)}{2\pi[c_d^2(t-t')^2 - \varrho^2]^{\frac{1}{2}}}, \quad (2.89)$$

the two-dimensional Green's function, where $H(\tau)$ is the Heaviside step function defined as

$$H(\tau) = 0, \quad \text{for } \tau < 0,$$
$$= 1, \quad \text{for } \tau > 0. \quad (2.90)$$

The function G represents the response along the line x_1, x_2, $-\infty < x_3 < \infty$, due to an impulsive unit line source applied along the line x_1', x_2', $-\infty < x_3' < \infty$. It follows through linear superposition of the solutions (2.89) that the solution to (2.84), (2.85) is

$$\varphi(x_1, x_2, t) = \frac{c_d}{2\pi} \iiint_C \frac{\Phi(x_1', x_2', t')}{[c_d^2(t-t')^2 - \varrho^2]} dx_1' dx_2' dt', \quad (2.91)$$

where C is a cone in the (x_1', x_2', t')-space defined by

$$0 \leq t' \leq t, \quad \varrho^2 \leq c_d^2(t-t')^2. \quad (2.92)$$

it is clear that the corresponding solution for Ψ can be obtained from 2.66b) in the same way.

2.7.5. Basic singular solutions of elastodynamics

2.7.5.1. *Point load problem.* It was already pointed out in § 2.7.1 that the solution for the present problem was given by Stokes [2.15], which later was confirmed by Love [2.16]. Essentially, we follow the latter's treatment here. The problem is depicted in fig. 2.5. A concentrated, time-dependent,

Ch. 2, § 2.7]   WAVE MOTION DUE TO BODY FORCES   87

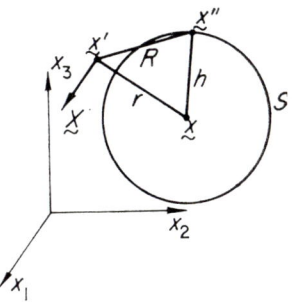

Fig. 2.5. Point load problem.

body force $X(x, t)$ is assumed to act at the point x' of the unbounded medium, in the direction of the x_1 axis. It is therefore given by

$$X(x, t; x') = f(t)\delta(x - x')i_1 ,\qquad(2.93)$$

where $f(t)$ is the time-dependent magnitude of the force, $\delta(x - x')$ is given in (2.71a), x is a field point of a finite closed region V containing x', which can approach x', and i_1 is the cartesian unit vector in the x_1 direction. To calculate the potentials of the disturbance Φ, and Ψ in (2.65), corresponding to (2.93), we draw on the Helmholtz resolution relations (2.9), (2.12). Hence we have

$$W(x, t; x') = -\frac{f(t)i_1}{4\pi} \int_V \frac{\delta(x'' - x')}{|x - x''|} dV_{x''} ,$$

which using (2.70) reduces to

$$W(x, t; x') = -f(t)i_1/4\pi r ,\qquad(2.94)$$

for $x' \in V$, and zero otherwise, where r is $|x - x'|$ given in (2.71a). Applying (2.12) to (2.94) we find

$$\Phi = -\frac{f(t)}{4\pi\varrho c_d^2} \frac{\partial r^{-1}}{\partial x_1} ,\qquad(2.95a)$$

and

$$\Psi = \frac{f(t)}{4\pi\varrho c_s^2}\left(\frac{\partial r^{-1}}{\partial x_3} i_2 - \frac{\partial r^{-1}}{\partial x_2} i_3\right) .\qquad(2.95b)$$

Having (2.95) we can now make use of the retarded potentials (2.81), (2.82) to complete the solution to the problem. Noting that x' is the fixed point of application of the load, (2.81) can be rewritten in the form

$$\varphi(x, t; x') = \frac{1}{4\pi} \int_{V_d} \frac{[\Phi]}{h} dV_{x''}, \qquad (2.96)$$

where V_d is the sphere $h = |x - x''| \leq c_d t$, having its center at x, $[\Phi]$ being the value of Φ at point x'' at the time $t - h/c_d$. From the form of (2.95a), it can be seen, therefore, that $[\Phi]$ in (2.96) must be

$$[\Phi] = -\frac{f(t - h/c_d)}{4\pi \varrho c_d^2} \frac{\partial R^{-1}}{\partial x_1''}, \qquad (2.97)$$

where, as shown in fig. 2.5, R is $|x'' - x'|$. Integration of (2.96) is carried out by noting that the sphere V_d can be partitioned into thin sheets having spherical surfaces with x as center. One such surface is S in fig. 2.5, which has the radius h and surface element dS. Since $f(t - h/c_d)$ is constant over S, it follows that (2.96), incorporating (2.97), can be written as

$$\varphi(x, t; x') = -\frac{1}{16\pi^2 \varrho c_d^2} \int_0^\infty \frac{f(t - h/c_d)}{h} dh \int_S \frac{\partial R^{-1}}{\partial x_1''} dS. \qquad (2.98)$$

The integral over S in (2.98) may be likened to that occurring in potential theory, representing the component of attractive force at x' of a uniform distribution of surface mass density on S. The force vanishes at points internal to S, and at external points it is the same as if the whole mass distribution were concentrated at the center x of S (cf. [1.5], ch. 1, art. 2, and exercise 1, art. 3). It follows that

$$\int_S \frac{\partial R^{-1}}{\partial x_1''} dS = \begin{cases} 0, & \text{for } h > r \\ 4\pi h^2 \partial r^{-1}/\partial x_1, & \text{for } h < r. \end{cases} \qquad (2.99)$$

From the first of (2.99), it can be seen that the upper limit of the integral on h in (2.98) can be replaced by r. Therefore, using the second of (2.99) in (2.98), after the change in variable $h = c_d t'$, the latter becomes

$$\varphi(x, t; x') = -\frac{1}{4\pi \varrho} \frac{\partial r^{-1}}{\partial x_1} \int_0^{r/c_d} t' f(t - t') dt'. \qquad (2.100)$$

A similar procedure, based on (2.82) and (2.95b), gives $\psi_1(x, t; x') = 0$, and

$$\psi_2(x, t; x') = \frac{1}{4\pi\varrho} \frac{\partial r^{-1}}{\partial x_3} \int_0^{r/c_s} t' f(t-t') dt', \qquad (2.101)$$

and

$$\psi_3(x, t; x') = -\frac{1}{4\pi\varrho} \frac{\partial r^{-1}}{\partial x_2} \int_0^{r/c_s} t' f(t-t') dt'. \qquad (2.102)$$

The displacement components corresponding to (2.83), are easily derived from (2.100)–(2.102), with $\psi_1 = 0$. The resulting expression for u_i is

$$u_i^1(x, t; x') = \frac{1}{4\pi\varrho} \left\{ \frac{\partial^2 r^{-1}}{\partial x_i \partial x_1} \int_{r/c_d}^{r/c_s} t' f(t-t') dt' + \frac{\delta_{i1}}{c_s^2 r} f\left(t - \frac{r}{c_s}\right) \right.$$

$$\left. + \frac{1}{r} \frac{\partial r}{\partial x_i} \frac{\partial r}{\partial x_1} \left[\frac{1}{c_d^2} f\left(t - \frac{r}{c_d}\right) - \frac{1}{c_s^2} f\left(t - \frac{r}{c_s}\right) \right] \right\}, \qquad (2.103)$$

where we have used the "super one" to indicate the direction of the load, and where the simplification in the form of u_i^1 stems from noting that $\nabla^2 r^{-1} = 0$. Displacements u_i^2 and u_i^3 corresponding to the body forces in the direction of x_2, x_3, respectively, can be written by considering the symmetries of these displacements with respect to u_i^1. Equation (2.103) leads then, to the compact expression

$$u_i^k(x, t; x') = \frac{1}{4\pi\varrho} \left\{ \frac{\partial^2 r^{-1}}{\partial x_i \partial x_k} \int_{r/c_d}^{r/c_s} t' f(t-t') dt' + \frac{\delta_{ik}}{c_s^2 r} f\left(t - \frac{r}{c_s}\right) \right.$$

$$\left. + \frac{1}{r} \frac{\partial r}{\partial x_i} \frac{\partial r}{\partial x_k} \left[\frac{1}{c_d^2} f\left(t - \frac{r}{c_d}\right) - \frac{1}{c_s^2} f\left(t - \frac{r}{c_s}\right) \right] \right\}, \qquad (2.104)$$

for the three displacement sets ($k = 1, 2, 3$), each of which corresponds, respectively, to the force in the x_k direction. Addition of the three sets will yield the displacements corresponding to a force in an arbitrary direction. A simple change in the variable of integration in (2.104) and carrying out the derivatives there, yields the form

$$u_i^k(x, t; x') = \frac{1}{4\pi\varrho} \left\{ \left[\frac{3(x_i - x_i')(x_k - x_k')}{r^3} - \frac{\delta_{ik}}{r} \right] \int_{1/c_d}^{1/c_s} t' f(t - rt') dt' \right.$$

$$\left. + \frac{\delta_{ik}}{c_s^2 r} f\left(t - \frac{r}{c_s}\right) + \frac{(x_i - x_i')(x_k - x_k')}{r^3} \left[\frac{1}{c_d^2} f\left(t - \frac{r}{c_d}\right) - \frac{1}{c_s^2} f\left(t - \frac{r}{c_s}\right) \right] \right\}.$$

$$(2.105a)$$

Making use of Hooke's law (1.45), together with (1.25) and (1.39), one can calculate the stresses corresponding to (2.105a). Letting $x'=0$, (2.105a) gives $u_i^k(x, t; 0)$, corresponding to the body force applied at the origin of coordinates. $u_i^k(x, t; 0)$, and the corresponding $\sigma_{ij}^k(x, t; 0)$, are given in Wheeler and Sternberg [1.9]. It is clear from (2.105a) that each of the displacement components, due to a suddenly applied point body force in the interior of an elastic solid, is comprised of first a dilatational and a equivoluminal body wave (the non-integral terms in (2.105a)), which propagate with the speeds c_d and c_s, respectively. Secondly, there is, interestingly, a body wave disturbance, the parts of which propagate with speeds c over the continuous range $c_s \leq c \leq c_d$ (the integral term in (2.105a)).

Applying (1.39) for the dilatation to (2.104) or (2.105a), one finds

$$\Delta^k = \frac{1}{4\pi\rho c_d^2} \frac{\partial}{\partial x_k}\left[\frac{1}{r} f\left(t - \frac{r}{c_d}\right)\right], \qquad (2.105b)$$

for the dilatation Δ^k, corresponding to the force in the x_k direction ($k=1$, 2, 3). Similarly, applying the rotations (1.36a), (1.32) to (2.103), one finds

$$\omega_1^1 = 0,$$
$$2\omega_2^1 = \frac{1}{4\pi\rho c_s^2} \frac{\partial}{\partial x_3}\left[\frac{1}{r} f\left(t - \frac{r}{c_s}\right)\right], \qquad (2.105c)$$
$$2\omega_3^1 = -\frac{1}{4\pi\rho c_s^2} \frac{\partial}{\partial x_2}\left[\frac{1}{r} f\left(t - \frac{r}{c_s}\right)\right],$$

where "super one" refers to a load acting along the x_1 direction. When a load is in the x_2 or x_3 direction, calculation of the corresponding rotations is facilitated through use of the general relations

$$\omega_i^k = -\omega_k^i, \quad \omega_k^k = 0, \qquad (2.105\,d)$$

obtained through the symmetries of ω_i^2, ω_i^3 with respect to ω_i^1.

Love investigates (2.105a) further, through (2.103), by considering the special case in which $f(t)$ is a ramp function, i.e.,

$$f(t) = \begin{cases} 0 & \text{for } t < 0, \\ X_1 t / \rho \varepsilon & \text{for } 0 < t < \varepsilon, \\ X_1 / \rho & \text{for } t > \varepsilon. \end{cases}$$

The corresponding solution u_i^1 has four time domains of interest, corresponding to arrivals of the wavefronts, and the ramp rise time ε, i.e.,

$r/c_d < t < (r/c_d) + \varepsilon$, $(r/c_d) + \varepsilon < t < r/c_s$, $r/c_s < t < (r/c_s) + \varepsilon$, and $t > (r/c_s) + \varepsilon$. For the first domain the solution is

$$u_i^1(x, t; x') = \frac{1}{4\pi\varrho} \left\{ \frac{\partial^2 r^{-1}}{\partial x_i \partial x_1} \left[\frac{(t - r/c_d)^2 (t + 2r/c_d)}{6\varepsilon} \right] + \frac{\partial r}{\partial x_i} \frac{\partial r}{\partial x_1} \left(\frac{t - r/c_d}{c_d^2 \varepsilon r} \right) \right\}. \tag{2.106}$$

It can be seen that the displacements are continuous at the wave front $t = r/c_d$, agreeing with our earlier general requirements, i.e., equations (2.38). It is clear, however, that if the input $f(t)$ had been the step function (2.90), the second term in (2.106) would also be a step, hence violating the continuity condition on the displacements. Now taking first derivatives of (2.106) in time and space, the second term there leads to finite jumps in these quantities at the wave front $t = r/c_d$. These, however, are admissible, satisfying the kinematical relations (2.44). Analogously, for the later traveling wavefronts, i.e. arrivals at $t = (r/c_d) + \varepsilon$, $t = r/c_s$, and $t = (r/c_s) + \varepsilon$, the displacements satisfy (2.46), and the kinematical relations (2.47) are satisfied. Therefore, Love's solution satisfies the necessary continuity conditions on the displacements, and kinematical conditions on the velocities and strains, at the wavefronts, provided the input function $f(t)$ is continuous. One should also note from the forms (2.105b), (2.105c) for the dilatation and rotations, that these quantities also exhibit finite jumps at $t = r/c_d$, and $t = r/c_s$, respectively, for the ramp function input.

2.7.5.2. *Center of compression.* Love [2.16, §§ 11–13] shows that other types of singular points can be created by assuming two or more of the point load type act as the loads tend towards coalesence. Consider, for example, a force $f(t)/h$ acting at the origin in the positive direction of the x_1-axis, and an equal and opposite force acting at the point $(h, 0, 0)$. The displacements, due to such a pair of forces, are derived through a limiting procedure, h going to zero with $f(t)$ fixed. To carry this out we first note in (2.95) that the forcing function $f(t)$, for the point load problem, need only be replaced by $f(t)/h$ here, with the limit process for the present two forces being implied. It follows the displacement for the present case is given by

$$\hat{u}_i^1(x, t; \mathbf{0}) = \lim_{h \to 0} \left[\frac{u_i^1(x, t; h, 0, 0) - u_i^1(x, t; \mathbf{0})}{h} \right] = \frac{\partial u_i^1(x, t; \mathbf{0})}{\partial x_1},$$

where $u_i^1(x, t; \mathbf{0})$ is given by (2.105a), with $k = 1$. The singularity creating $\hat{u}_i^1$ is referred to as a *double force without moment*, which in this case is associated with the x_1 axis. It is clear we can similarly derive $\hat{u}_i^2$, $\hat{u}_i^3$, i.e., the displacements due to double forces without moment along the x_2 and x_3 axes. Adding the three we get

$$\hat{u}_i = \frac{\partial u_i^1}{\partial x_1} + \frac{\partial u_i^2}{\partial x_2} + \frac{\partial u_i^3}{\partial x_3}. \tag{2.107}$$

But from (2.105a) we have $u_i^k = u_k^i$, using this in (2.107) gives

$$\hat{u}_i = \frac{\partial u_1^i}{\partial x_1} + \frac{\partial u_2^i}{\partial x_2} + \frac{\partial u_3^i}{\partial x_3} = u_{k,k}^i = \Delta^i, \tag{2.108}$$

in accord with (1.39), i.e., the displacement $\hat{u}_i$ is the dilatation corresponding to a point force acting at the origin in the x_i direction. It follows from (2.105b) that

$$\hat{u}_i(x, t; 0) = \frac{1}{4\pi\varrho c_d^2} \frac{\partial}{\partial x_i}\left[\frac{1}{r}f\left(t - \frac{r}{c_d}\right)\right]. \tag{2.109}$$

The singularity creating $\hat{u}_i$ is referred to as a *center of compression*.

2.7.5.3. Center of rotation. Consider now a force $f(t)/h$ acting at the origin in the positive direction of the x_1-axis, and an equal and opposite force acting at the point $(0, h, 0)$. Using the limit procedure, as in the previous section, the displacements due to this pair of forces are found to be

$$\hat{u}_{i3}^1(x, t; 0) = -\frac{\partial u_i^1(x, t; 0)}{\partial x_2},$$

where $u_i^1(x, t; 0)$ is given by (2.105a), with $k = 1$. The singularity creating $\hat{u}_{i3}^1$ is referred to as a *double force with moment*, the moment being about the x_3-axis, where the subscript 3 has been used to indicate this. The singularity is also related to the directions of the forces, i.e., the x_1-axis in this case. We add to the above pair, a pair composed of a force $-f(t)/h$ acting at the origin in the negative direction of the x_2-axis, and an equal and opposite force acting at $(h, 0, 0)$. The two moments are about the same axis, and are of the same sign, with the directions of the force pairs being at right angles to one another. The limit process then produces the displacements

$$\hat{u}_{i3}(x, t; 0) = \frac{\partial u_i^1}{\partial x_2} - \frac{\partial u_i^2}{\partial x_1}, \tag{2.110}$$

Ch. 2, § 2.7] WAVE MOTION DUE TO BODY FORCES 93

due to these two double forces with equal moments. Using $u_i^k = u_k^i$ in (2.110), it becomes

$$\hat{u}_{i3}(x, t; 0) = (u_{1,2}^i - u_{2,1}^i) = -2\omega_3^i, \qquad (2.111)$$

where we note that these rotations are about the x_3-axis only. The singular point here is referred to as a *center of rotation* about the x_3-axis. Since $\omega_3^3 = 0$, we see that $\hat{u}_{33} = 0$. The remaining two displacement components, $\hat{u}_{13}, \hat{u}_{23}$, are the corresponding rotations in (2.111), evaluated from displacements of the point load solution (2.105a) for, respectively, point loads at the origin in the x_1 and x_2-directions, i.e., $k = 1, 2$ there. The rotation $-2\omega_3^1$, hence $\hat{u}_{13}$, is given in (2.105c). The expression for $-2\omega_3^2$, hence $\hat{u}_{23}$, is given by

$$\hat{u}_{23} = -2\omega_3^2 = -\frac{1}{4\pi\rho c_s^2} \frac{\partial}{\partial x_1} \left[\frac{1}{r} f\left(t - \frac{r}{c_s}\right) \right].$$

Love also discusses, in [2.16, §§ 12,13], generalizations of the basic double forces treated in this and the previous section.

2.7.5.4. Point load time harmonic source It is instructive to examine the nature of the displacement solution (2.105a) for the point load problem, in the special case of a time harmonic input. We assume $f(t) = f_0 \cos \omega t$. Then the integral term in (2.105a) gives

$$\int_{1/c_d}^{1/c_s} t' f(t - rt') \mathrm{d}t' = \frac{f_0}{\omega^2 r^2} \left[\cos \omega \left(t - \frac{r}{c_s}\right) - \cos \omega \left(t - \frac{r}{c_d}\right) \right.$$

$$\left. - \frac{\omega r}{c_s} \sin \omega \left(t - \frac{r}{c_s}\right) + \frac{\omega r}{c_d} \sin \omega \left(t - \frac{r}{c_d}\right) \right]. \qquad (2.112)$$

Hence, we see by (2.112), that the solution (2.105a) is composed of just two simple harmonic wave trains, propagating with the distinct velocities c_d and c_s, respectively. It follows that the present input leads to a degenerate solution, since it does not yield a part composed of disturbances that propagate with all velocities in the continuous range $c_s \leq c \leq c_d$, as we found for the more general input $f(t)$.

2.7.5.5. Retarded potentials corresponding to a time harmonic source; steady-state solution If, at the outset of a body force problem, we had a time harmonic source, such as that in the previous section, it would lead to special forms of the retarded potentials, which evaluated for large time, yield the steady-state solution. Consider (2.66a) again, with the source $-\Phi(x, t) = -g(x)\cos \omega t = -\text{Re}[g(x)\exp(i\omega t)]$, where Re [ ] indicates the real part of [ ]. Substitution of this in (2.80) gives the solution

$$\varphi(x, t) = \frac{1}{4\pi} \text{Re } e^{i\omega t} \int_{V_d} \frac{g(x')}{|x-x'|} e^{-i\omega|x-x'|/c_d} dV_{x'}, \quad (2.113)$$

where V_d is the sphere $|x-x'| \leq c_d t$, with similar integrals for $\Psi(x, t)$. Since the sources are confined to the finite domain V, it follows that for large enough time, V_d may be replaced by V, a region independent of t. Hence, for large time, (2.113) reduces to

$$\varphi(x, t) = \frac{1}{4\pi} \text{Re}\left[e^{i\omega t} \int_V \frac{g(x')}{|x-x'|} e^{-i\omega|x-x'|/c_d} dV_{x'} \right], \quad (2.114)$$

which together with similar integrals for $\Psi(x, t)$, form the steady-state solution for the body force problem. The special case of the point source is treated by Achenbach [2.14, § 3.12.1]. One replaces $g(x')$ with $g(x''; x') = \delta(x''-x')$ (cf. fig. 2.5), and using (2.70), we have from (2.114)

$$\varphi(x, t; x') = \frac{1}{4\pi} \text{Re } e^{i\omega t} \int_V \frac{\delta(x''-x')}{|x-x''|} e^{i\omega|x-x''|/c_d} dV_{x''}$$
$$= \frac{\cos[\omega(t-|x-x'|/c_d)]}{4\pi|x-x'|}. \quad (2.115)$$

This steady-state solution is a wave, harmonic in time and space, propagating outward in $|x-x'|$. The solution applies to the fixed station x, provided $c_d t$ is larger than the radius of the smallest sphere, with center at x, which includes the source point x'.

2.8. Solution of boundary-initial value problems; integral representations

The solution of boundary-initial value problems in elastodynamics has been advanced mainly through direct attacks on the governing system of equations, i.e., the partial differential equations of motion, governing the potentials or displacements, and the boundary and initial conditions. Integral transforms and related analytic function theory have been powerful tools in producing solutions for these problems. This will be evident in the later chapters of this book, where we use such tools to obtain solutions to many of the boundary-initial value problems of our subject. A brief discussion of an alternate attack, however, namely that involving integral representations, is in order. We begin by stating a well-known integral representation, obtained by Kirchhoff in 1882, for the potential $\varphi(x, t)$ governed by the inhomogeneous wave equation

$$\nabla^2 \varphi(x, t) - \ddot{\varphi}(x, t)/c^2 = -g(x, t) \,. \tag{2.116}$$

As the derivation in Stratton [1.8, § 8.1] or Jones [2.18, § 1.17] shows, for a sufficiently smooth domain V, with boundary S, the integral representation for φ is given by

$$\varphi(x, t) = \frac{1}{4\pi} \int_V \frac{[g]}{|x-x'|} \, dV_{x'} + \frac{1}{4\pi} \int_S \left\{ \frac{1}{|x-x'|} \left[\frac{\partial \varphi}{\partial n'} \right] \right.$$
$$\left. -[\varphi] \frac{\partial}{\partial n'} \frac{1}{|x-x'|} + \frac{1}{c|x-x'|} [\dot{\varphi}] \frac{\partial |x-x'|}{\partial n'} \right\} dS_{x'} \,, \tag{2.117}$$

where n' is the outward unit normal to S, and x lies inside V. Further, as in the definition of the retarded potential (2.81), the brackets indicate the quantity within, at x', is evaluated at the retarded time $t-|x-x'|/c$. The volume integral in (2.117), like (2.80), represents the contribution to φ from sources inside V. The surface integral gives the contribution from the sources outside V. If φ and its derivatives are known over S, the field φ is completely determined within V. It is, however, not possible to assign both φ and $\partial \varphi/\partial n'$ arbitrarily on S. For, as Jones [2.18] points out, suppose φ was known on S for all time, then the terms in the surface integral could be evaluated except the one involving $\partial \varphi/\partial n'$, giving an expression for φ. For this expression to reproduce the values of φ on S, then $\partial \varphi/\partial n'$ would have to satisfy an integral equation, which, in general, would not have an arbitrary solution.

In [2.6, §§ 14–17] Love extended Kirchhoff's integral representation to one for the displacement in elastodynamics, cf. [1.2], pg. 303 which also gives a statement of the representation. To do this he made use of the displacement solution (2.105a) for the point body force, and the elastostatic reciprocity theorem due to Betti, treating inertia forces as body forces, cf. § 1.13, (1.91). A similar independent derivation of Love's integral representation was recently given by deHoop [2.19]. Still more recently Wheeler and Sternberg [1.9] provided a more direct and explicit proof of Love's integral representation for the displacement field, through use of Graffi's elastodynamic reciprocity theorem (cf. § 1.13). Further they also derived a corresponding one for the stress field. The integral representation for the displacement field involves the tractions and displacements on the boundary S, and a term which gives the contribution to the displacement from the body forces within V. In [2.19] deHoop showed that this representation has usefulness in elastodynamic diffraction problems. Through it, certain problems of diffraction of a pulse by a half plane are reducible to the solution of certain integral equations, which can be handled by the Wiener-Hopf technique.

Since the displacement integral representation involves both the tractions and the displacements on the surface S of the body, it is not suited to solve the first (surface displacements only specified) or second (surface tractions only specified) fundamental boundary-initial value problem of elastodynamics [cf. ch. 1 § 1.8, eqs. (1.68), (1.69)]. However, in [1.9], Wheeler and Sternberg derived integral representations for solutions of these problems, eliminating from the displacement integral representation, the surface tractions for the first problem, and the surface displacement for the second problem. A detailed treatment of integral representations is given by Achenbach [2.14].

2.9. Initial value or Cauchy problems

Fundamental to any study of wave propagation phenomena are a class of problems known as *initial value, or Cauchy problems*. In them, certain nonvanishing data are specified at the initial time, say $t=0$, usually over a finite region of the infinite medium. The literature on the Cauchy problem is vast. Elaborate treatments are given by Courant and Hilbert [2.20] and Hadamard [2.21]. We begin our treatment with the initial value problem for one spatial dimension.

2.9.1. One-dimensional initial value problem and related characteristics

Here we return to solution (2.21) of the one-dimensional wave equation, and assume it must satisfy the *general initial conditions*

$$u(x, 0) = f(x), \qquad \dot{u}(x, 0) = g(x), \qquad -\infty < x < \infty, \qquad (2.118)$$

We can identify this problem with that of the *transverse displacements u of a stretched elastic string*, under constant tension T. The wave speed in this case is $(T/\varrho)^{\frac{1}{2}}$. Substituting (2.21) into the first of (2.118), we find

$$F(x)+G(x)=f(x). \tag{2.119}$$

Differentiating (2.21) with respect to t, and then substituting this into the second of (2.118), one finds

$$c[G'(x)-F'(x)]=g(x), \tag{2.120}$$

where the primes indicate differentiation with respect to the arguments of these functions. Differentiating (2.119) and adding this to (2.120), we obtain

$$G'(x)=\tfrac{1}{2}f'(x)+\frac{1}{2c}g(x).$$

Integrating this over the x domain $-\infty<x<\infty$, one obtains

$$G(x)=\tfrac{1}{2}f(x)+\frac{1}{2c}\int_{-\infty}^{x}g(x')\mathrm{d}x'+k_1. \tag{2.121}$$

And from (2.119), using (2.121), we find

$$F(x)=\tfrac{1}{2}f(x)-\frac{1}{2c}\int_{-\infty}^{x}g(x')\mathrm{d}x'-k_1. \tag{2.122}$$

Substituting (2.121) and (2.122), into (2.21), the latter becomes

$$u(x,t)=\tfrac{1}{2}[f(x-ct)+f(x+ct)]+\frac{1}{2c}\int_{x-ct}^{x+ct}g(x')\mathrm{d}x', \tag{2.123}$$

the solution to the posed initial value problem. This solution can be verified by direct substitution into (2.19) and (2.118).

The general nature of the solution (2.123) is easily understood if one assumes at the outset, that the initial disturbances $f(x)$ and $g(x)$ are limited to the arbitrary interval $x_1<x<x_2$, being zero outside of this interval. Since the interval is arbitrary, there is no loss of generality in our assumption. It follows, therefore, that

$$x_1 \leq \begin{Bmatrix} x+ct \\ x-ct \end{Bmatrix} \leq x_2,$$

and from these, that

$$t \leq \frac{x_2-x}{c},$$
$$t \leq \frac{x-x_1}{c}. \tag{2.124}$$

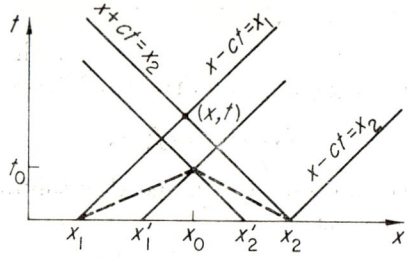

Fig. 2.6. Solution domain for one-dimensional initial value problem.

Equation (2.123) therefore points out that $u(x, t)$ is uniquely determined in the triangular region, given by (2.124), by the initial data $f(x)$ and $g(x)$. Figure 2.6 depicts this region. The lines $x - ct = x_1$ and $x + ct = x_2$ are *characteristics* of the governing one-dimensional wave equation (2.19). They are particular cases of (2.56b) discussed in § 2.6.4. Since $u(x, t)$ in (2.123) depends only on the initial data on the interval $x_1 \leq x \leq x_2$, and these disturbances cannot travel faster than c, we see that, at t_0, a point such as x_0 in fig. 2.6 could not have received a signal from x_1 or x_2, as the dashed lines there have speeds greater than c associated with them. Fig. 2.6 shows the pair (x_0, t_0) has a new inner triangle, governing the disturbances it can receive, defined by $x_1' < x < x_2'$, and the characteristics passing through x_1' and x_2'. Hence we see two families of characteristics are involved in determining $u(x, t)$ in (2.123). This equation, and fig. 2.6, show the contributions to $u(x, t)$, due to the initial displacement $f(x)$, arise from propagations of the atter to (x, t), along characteristics, at speed c. On the other hand we see contributions to $u(x, t)$, due to the initial velocity $g(x)$, arise from propagations of the latter from anywhere on $x_1 \leq x \leq x_2$, i.e., at all speeds from zero to, and including, c. Clearly, along the characteristics, f is constant, and propagates at the maximum speed c in the positive and negative x-directions. A pair of the characteristics pass through each point (x, t). The interval $x_1 < x < x_2$ is called the *domain of dependence* of (x, t). The $t-x$ plane, between the characteristics $x + ct = x_2$ and $x - ct = x_2$, is called the *domain of influence* of the point $(x_2, 0)$, for example. Further insight into the nature of our problem, and solution (2.123), is obtained by treating special cases.

2.9.1.1. Initial displacement condition. Assume that

$$u(x, 0) = f(x), \qquad g(x) = 0.$$

Equation (5.123) therefore reduces to

$$u(x, t) = \tfrac{1}{2}[f(x-ct) + f(x+ct)]. \tag{2.125}$$

Sokolnikoff and Redheffer [2.22, ch. 6 § 3] give an instructive example. Assume $f(x)$ is triangular. Fig. 2.7 depicts the wave propagation that ensues.

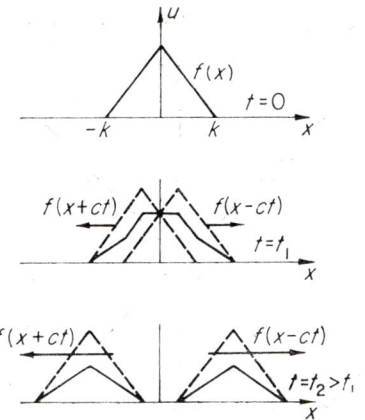

Fig. 2.7. Propagation of displacement generated by an initial disturbance in displacement.

From (2.125), and the figure, we see all parts of these waves move with speed c, hence propagation is along the characteristics in the $t-x$ plane, as we noted earlier. Note also these waves are plane, and are consistent with the discussion given in § 2.3.1.

2.9.1.2. Initial velocity condition. Assume that

$$f(x) = 0, \qquad \dot{u}(x, 0) = g(x).$$

Equation (2.123) therefore reduces to

$$u(x, t) = \frac{1}{2c} \int_{x-ct}^{x+ct} g(x') dx'. \tag{2.126}$$

In this case waves can be of infinite length even though $g(x)$ is of finite extent. We consider another instructive example treated by Sokolnikoff and Redheffer. In the example it is assumed that

$$g(x) = \begin{cases} 1 & \text{for } |x| < k, \\ 0 & \text{for } |x| > k. \end{cases} \tag{2.127}$$

Substitution in (2.126) gives

$$u(x,\ t) = \frac{1}{2c}\int_{x-ct}^{x+ct} dx' \ . \qquad (2.128)$$

Although (2.128) is simple in form, care must be exercised in interpreting the limits of integration for the different domains of x and t. The $t-x$ plane gives the needed guidance. Figure 2.8 shows such a plane for the present

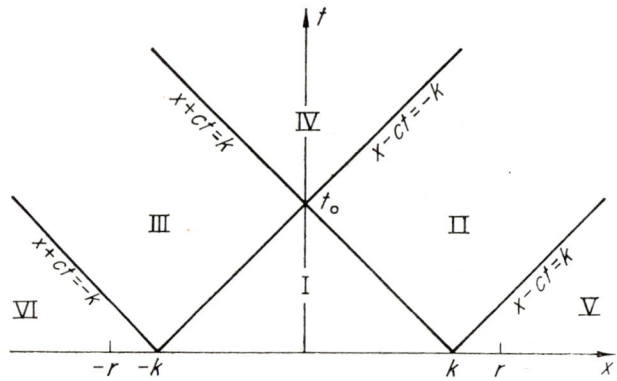

Fig. 2.8. Solution domains in the $t-x$ plane.

problem. For region I we see (2.128) gives

$$u(x,\ t) = \frac{1}{2c}\int_{x-ct}^{x+ct} dx' = \frac{1}{2c}[x+ct-(x-ct)] = t, \qquad (2.129)$$

obviously composed of two waves, propagating with velocity c, that are images of each other in $x = 0$. In (2.129) the characteristics determining the domain of dependence of a point $(x,\ t)$ (within I) must be in accord with those described in fig. 2.6. For example, the vertex $(0,\ t_0)$, has all of $-k \le x \le k$ as its domain of dependence, and $u(0,\ t_0) = t_0 = k/c$, according to (2.129) and fig. 2.8, which is the maximum displacement in this region. For regions II and III, (2.128) gives

$$u(x,\ t) = \frac{1}{2c}\int_{x-ct}^{k} dx' = \frac{1}{2c}[k-(x-ct)], \qquad (2.130)$$

and

$$u(x,\ t) = \frac{1}{2c}\int_{-k}^{x+ct} dx' = \frac{1}{2c}[x+ct+k], \qquad (2.131)$$

respectively. Here again we have waves that move with velocity c. Figure 2.8 shows that in (2.130), $-k \leq x-ct \leq k$, and in (2.131), $-k \leq x+ct \leq k$, so $u(x,t) \leq k/c$. Further (2.130) and (2.131) show u is linear in x and t, in II and III. Points in region IV would have characteristics passing through them which pass through $\pm r$ on the x-axis in fig. 2.8. However, from (2.127), $g(x)=0$ for $|x|>k$, and hence (2.128) again applies, giving for this region

$$u(x,t) = \frac{1}{2c}\int_{-k}^{k} dx' = \frac{k}{c} = u(0, t_0), \qquad (2.132)$$

for all time $t \geq t_0$. Hence we see everywhere in region IV we have the maximum displacement, and this is a static quantity. Clearly, everywhere in regions V and VI, $u(x,t)=0$, since all points (x,t) there have domains of dependence over which $g(x)=0$. It follows that the outer characteristics $x-ct=k$ and $x+ct=-k$, in fig. 2.8, correspond to the positive and negative traveling leading edges of the disturbance $u(x,t)$. Our analysis shows this disturbance would develop and propagate as shown in fig. 2.9. Note that in the last sketch here, t_1 is arbitrarily larger than t_0, (x_1, t_1) being the point of intersection of the line $t=t_1$ (which could be plotted in fig. 2.8) with the characteristic $x-ct=-k$. We note the fronts of the disturbance (leading oblique lines in fig. 2.9) move with speed c. The region behind these fronts corresponds to the final static displacement k/c, governed by (2.132). Note that we would have the same general nature for this region for any function $g(x)$, provided its integral (2.132) does not vanish. This precludes, for example, $g(x)$ being an odd function. For such a function we would have, for $t \geq t_0$, two separated pulses traveling with velocity c along the string, one in the positive, and the other in the negative x-direction, like in the initial displacement problem.

2.9.1.3. Discontinuities in initial values. Suppose the initial values of $f(x)$ in the solution (2.123) for $u(x,t)$ have a discontinuity at $x=x_0$, $x_1 < x_0 < x_2$. Clearly then $u(x,t)$ in (2.123) would be discontinuous along the characteristics $x+ct=x_0$, and $x-ct=x_0$. Now differentiating (2.123) with respect to t, we have

$$\dot{u} = \frac{c}{2}[f'(x+ct)-f'(x-ct)] + \tfrac{1}{2}[g(x+ct)+g(x-ct)], \qquad (2.133)$$

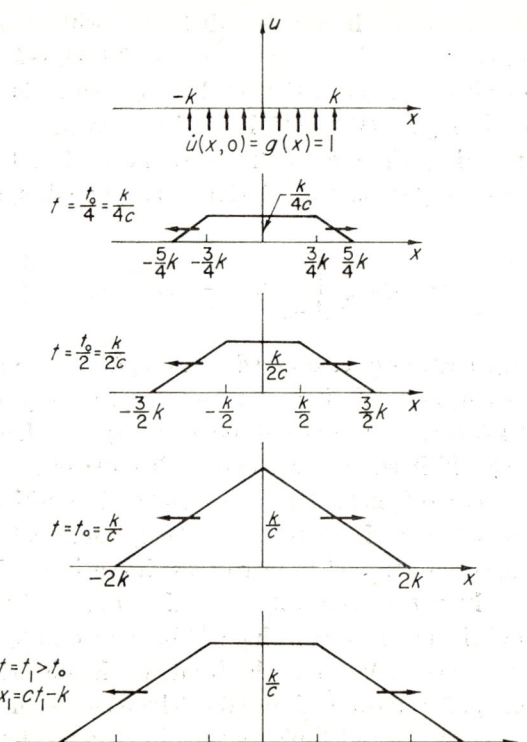

F 2.9. Propagation of displacement generated by an initial disturbance in velocity (after Sokolnikoff and Redheffer).

which, at $t=0$, gives

$$\dot{u}(x, 0) = g(x). \qquad (2.134)$$

Similarly

$$\frac{\partial u}{\partial x} = \tfrac{1}{2}\, f'(x+ct)+f'(x-ct)] + \frac{1}{2c}[g(x+ct)-g(x-ct)], \qquad (2.135)$$

s, at $t=0$,

$$\frac{\partial u(x, 0)}{\partial x} = f'(x), \qquad (2.136)$$

so similarly (2.134) and (2.136) show that a discontinuity at $x=x_0$ in the nitial values of $g(x)$, or $f'(x)$, generates, respectively, a discontinuity in $\dot{u}$ or $\partial u/\partial x$, both of which propagate along the characteristics $x+ct=x_0$, and $x-ct=x_0$.

It is also of interest to consider the case treated by Weinberger [2.23, § 3], where u has a discontinuity in its directional derivative, $\partial u/\partial x + \dot{u}/c$, in the direction of the characteristic $x-ct=$ constant, at some point (x_0, t_0). Forming this directional derivative from (2.133) and (2.135), we have

$$\frac{\partial u}{\partial x} + \frac{\dot{u}}{c} = f'(x+ct) + \frac{1}{c}g(x+ct) \ . \tag{2.137}$$

It follows from (2.137) that $f'(x)+(1/c)g(x)$ must be discontinuous at $x = x_0 + ct_0$. It also follows then from (2.137), that the directional derivative $\partial u/\partial x + \partial u/c\partial t$ has the same discontinuity all along the characteristic $x+ct = x_0 + ct_0$. The situation is analogous for the other characteristic. Hence, a discontinuity in the derivative in the direction of one characteristic propagates along the other of the pair through the discontinuity point (x_0, t_0).

2.9.2. The three- and two-dimensional initial value problems

In three dimensions the initial value problem for the wave equation is stated as follows:

$$\nabla^2 \varphi(\mathbf{x}, t) - \ddot{\varphi}/c^2 = 0 \ , \qquad -\infty < x_i < \infty \ , \qquad t > 0, \tag{2.138}$$

$$\varphi(\mathbf{x}, 0) = f(\mathbf{x}) \ , \qquad \dot{\varphi}(\mathbf{x}, 0) = g(\mathbf{x}) \ , \qquad -\infty < x_i < \infty \ . \tag{2.139}$$

As Love [2.24, § 4] shows, we can derive the solution from Kirchhoff's integral representation (2.117). Assume V is a sphere, having the radius $r = |\mathbf{x} - \mathbf{x}'| = ct$, with $\mathbf{x}$ at the center, and $\mathbf{x}'$ on the boundary S of the sphere. On this basis (2.117) reduces to

$$\varphi(\mathbf{x}, t) = \frac{1}{4\pi} \iint \left\{ \frac{1}{r}\left[\frac{\partial \varphi}{\partial r}\right] + \frac{[\varphi]}{r^2} + \frac{[\dot{\varphi}]}{cr} \right\} dS \ , \tag{2.140}$$

where we note there is no volume integral here since (2.138) is a homogeneous equation, and the integration is over the surface $r=ct$. Now at $r=ct$, using (2.139), we find

$$[\varphi] = \varphi(x', t-r/c) = \varphi(x', 0) = f(x'),$$

$$[\dot{\varphi}] = \dot{\varphi}(x', 0) = g(x'), \qquad (2.141)$$

that is, $[\varphi]$ and $[\dot{\varphi}]$ are the initial values of φ and $\dot{\varphi}$ along the surface $r = ct$. It follows from (2.140), (2.141) that the second and third terms of the former can be written as

$$\bar{f} = \frac{1}{4\pi c^2 t^2} \iint f(x') \, dS,$$

$$t\bar{g} = \frac{1}{4\pi c^2 t} \iint g(x') \, dS, \qquad (2.142)$$

respectively, where the bars indicate these are the average values of f and g over $r = ct$, which is easily seen by inspection of (2.142). The first term in (2.140) can also be expressed in terms of $\bar{f}$ by first noting that $\bar{f}$ is a function of t. Let us call it $\bar{f} = F(t)$, our aim being to calculate its derivative $F'(t)$. We can write

$$F(t') - F(t) = \frac{1}{4\pi} \left[\iint (f)_{r=ct'} \, d\omega - \iint (f)_{r=ct} \, d\omega \right], \qquad (2.143)$$

where $d\omega = dS/c^2 t^2$ is the elemental solid angle subtended at x by element dS of the sphere, and $t' > t$. Further, from the theorem of the mean we have

$$(f)_{r=ct'} - (f)_{r=ct} = c(t'-t) \left(\frac{\partial f}{\partial r} \right)_{r=r'}. \qquad (2.144)$$

where r' is a value of r in the range $ct < r' < ct'$. It follows using (2.143), (2.144) that

$$F'(t) = \lim_{t' \to t} \frac{F(t') - F(t)}{t' - t} = \frac{c}{4\pi} \iint \left(\frac{\partial f}{\partial r} \right)_{r=ct} d\omega = \frac{c}{4\pi c^2 t^2} \iint \left[\frac{\partial \varphi}{\partial r} \right] dS,$$

or

$$F'(t) = \frac{\partial \bar{f}}{\partial t} = \frac{1}{4\pi t} \iint \frac{1}{r} \left[\frac{\partial \varphi}{\partial r} \right] dS. \qquad (2.145)$$

Ch. 2, § 2.9]   INITIAL VALUE OR CAUCHY PROBLEMS   105

It follows from (2.142), (2.145) that (2.140) becomes

$$\varphi(x, t) = \frac{\partial}{\partial t}(t\bar{f}) + t\bar{g},$$

or

$$4\pi c^2 \varphi(x, t) = \frac{\partial}{\partial t}\left[\frac{1}{t}\iint_{r=ct} f(x')dS\right] + \frac{1}{t}\iint_{r=ct} g(x')dS. \quad (2.146)$$

The solution (2.146) is due to Poisson.

In the *two-dimensional initial value problem* we seek the solution of (2.138) again, where now $\nabla^2 \varphi = \varphi_{,ii}$, subject to the initial conditions

$$\varphi(x, 0) = f(x_1, x_2), \qquad \dot{\varphi}(x, 0) = g(x_1, x_2), \qquad -\infty < x_i < \infty.$$

As pointed out in Courant and Hilbert [2.20, ch. 3 § 4] one can obtain the solution φ from that for the three-dimensional problem (2.146). One uses the *method of descent* which in effect eliminates the dependence on x_3 (in the present nomenclature). Consider the second integral in (2.146). We can write it as

$$\varphi_g(x, t) = t\bar{g} = \frac{t}{4\pi c^2 t^2}\iint_{r=ct} g(x_1+\xi_1, x_2+\xi_2, x_3+\xi_3)dS, \quad (2.147)$$

where

$$\xi_1 = x'_1 - x_1 = ct \sin\theta \cos\tau$$

$$\xi_2 = x'_2 - x_2 = ct \sin\theta \sin\tau$$

$$\xi_3 = x'_3 - x_3 = ct \cos\theta$$

where ξ_1, ξ_2, ξ_3 are rectangular coordinates, and r, θ, τ corresponding spherical coordinates, with origin at the sphere center x. Now imposing x_3 independence on (2.147), and noting $dS = c^2 t^2 \sin\theta d\theta d\tau$, we can write (2.147) as

$$\varphi_g(x_1, x_2, t) = \frac{t}{4\pi c^2 t^2}\iint_{r=ct} g(x_1+\xi_1, x_2+\xi_2)c^2 t^2 \sin\theta d\theta d\tau. \quad (2.148)$$

We can write the integral in (2.148) completely in terms of the independent variables ξ_1, ξ_2 by noting $dS = c^2t^2 \sin\theta d\theta d\tau = \sec\theta \cdot d\sigma$, where $d\sigma$ is the projection of the spherical surface element dS on the x_1x_2-plane. Since $\sec\theta = ct/(c^2t^2 - \varrho^2)^{\frac{1}{2}}$, where $\varrho = (\xi_1^2 + \xi_2^2)^{\frac{1}{2}}$ is the distance from the center of $d\sigma$ to the projection of the center of the sphere x on the x_1x_2-plane, it follows (2.148) reduces to

$$\varphi_g(x_1, x_2, t) = \frac{1}{2\pi c} \iint_{\varrho \leq ct} \frac{g(x_1+\xi_1, x_2+\xi_2)}{(c^2t^2-\varrho^2)^{\frac{1}{2}}} d\xi_1 d\xi_2, \qquad (2.149)$$

where we have used the fact that $d\sigma$ can be taken as $d\xi_1 d\xi_2$. Similarly, from the first integral in (2.146), we find

$$\varphi_f(x_1, x_2, t) = \frac{1}{2\pi c} \frac{\partial}{\partial t} \iint_{\varrho \leq ct} \frac{f(x_1+\xi_1, x_2+\xi_2)}{(c^2t^2-\varrho^2)^{\frac{1}{2}}} d\xi_1 d\xi_2. \qquad (2.150)$$

It may be observed the integrations in (2.149), (2.150) are over the circular area $\varrho \leq ct$ in the $\xi_1\xi_2$ plane. It follows the solution for φ is given by

$$\varphi(x_1, x_2, t) = \varphi_f + \varphi_g. \qquad (2.151)$$

For our elastodynamic case φ in (2.146), (2.151) can be identified with the dilational wave potential function in the Lamé solution (2.7), where $c = c_d$. The components of the equivoluminal wave potential function Ψ would also have representations of the form (2.146), (2.151) with $c = c_s$. Hence (2.8) can be written for the displacement solution $u(x, t)$ for the two- and three-dimensional initial value problems.

Comparing (2.146), (2.151), one sees that in the *three-dimensional case* (2.146) the *domain of dependence of a point in space x at time t* consists of just the *surface of a sphere of radius ct around x*, whereas in the *two-dimensional case* (2.151), the *domain of dependence is the entire interior and boundary of the circle of radius ct around x*. It is clear from this comparison, that two-dimensional initial disturbances, like one-dimensional ones (cf. § 2.9.1), travel with maximum speed c, but can also travel at speeds less than c, however, three-dimensional ones always travel at the speed c. It follows that the *domain of dependence* in the *three-dimensional initial value problem obeys Huyghen's principle* (cf. § 2.6.4), but this domain in the corresponding two- or one-dimensional problem does not. This accounts for the fact that sharply defined signals of light or sound, which are essentially three-dimensional in nature, are equally sharp in their transmission, whereas corresponding two-dimensional signals, e.g., a surface disturbance in water, may "ring" indefinitely at a station away from the source. It is known that solutions of wave equations for odd space dimensions of 5, 7, 9,···, also satisfy Huyghen's principle, but the corresponding ones for the even space dimensions do not. In summary, therefore, Huyghen's principle is satisfied by solutions of wave equations for only odd-numbered space dimensions, not including one. Further detail may be found in [2.20, ch. 3 § 6].

2.9.3. The method of characteristics in one-dimensional problems

Useful theorems exist, associated with hyperbolic partial differential equations of second order in two independent variables, and their characteristics, which enable one to integrate the equations by exploiting the coordinates along the characteristics. The associated integration techniques are referred to as the *method of characteristics*. It is applicable to both linear and nonlinear partial differential equations in two independent variables. The literature in fluid and solid mechanics evidences the usefulness of the method of characteristics in wave propagation problems and hence a discussion of it is in order here. The underlying theory is discussed extensively in Courant and Hilbert [2.20, ch. 5, §§ 1, 5, 6] and Garabedian [2.25, ch. 3 § 4, ch. 4 § 4].

We consider the general partial differential equation

$$A \frac{\partial^2 \varphi}{\partial x^2} + 2B \frac{\partial^2 \varphi}{\partial x \partial y} + C \frac{\partial^2 \varphi}{\partial y^2} = F, \qquad (2.152)$$

where A, B, C, and F are functions of x, y, φ, $\partial \varphi / \partial x$, and $\partial \varphi / \partial y$. Equation (2.152) is a quasi-linear equation, being linear in the highest order derivatives of φ. Let

$$r = \frac{\partial^2 \varphi}{\partial x^2}, \qquad s = \frac{\partial^2 \varphi}{\partial x \partial y}, \qquad \frac{\partial^2 \varphi}{\partial y^2} = t. \qquad (2.153)$$

Substituting (2.153) in (2.152), the latter becomes

$$Ar + 2Bs + Ct = F. \qquad (2.154)$$

Now consider the integral surface, or solution, φ of (2.154) that we seek. Suppose φ is given along curve δ_0, the intersection of φ and the coordinate plane $x=0$, as shown in fig. 2.10, together with the tangent planes to φ

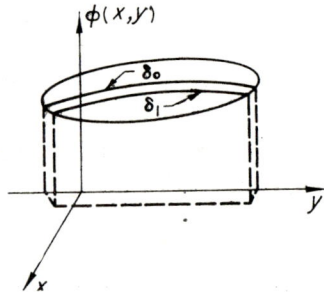

Fig. 2.10. Curve δ_0 of given data and neighboring curve δ_1.

through δ_0. Then we know the normals of these planes and therefore the direction cosines p, q and -1 of these, along δ_0, where

$$p = \frac{\partial \varphi}{\partial x}, \qquad q = \frac{\partial \varphi}{\partial y}.$$

We want to find φ at some neighboring line, say δ_1. It is obtained simply by adding to φ along δ_0,

$$d\varphi = \frac{\partial \varphi}{\partial x} dx + \frac{\partial \varphi}{\partial y} dy = pdx + qdy, \qquad (2.155)$$

which can be computed from the given values of p and q along δ_0. Now we have q along the line δ_1 also, but not p. So to proceed further we need

$$dp = \frac{\partial p}{\partial x} dx + \frac{\partial p}{\partial y} dy = rdx + sdy, \qquad (2.156)$$

which means we need r and s along δ_0. A third equation is given by

$$dq = \frac{\partial q}{\partial x} dx + \frac{\partial q}{\partial y} dy = sdx + tdy, \qquad (2.157)$$

and hence the linear system of equations for r, s, and t along δ_0, is, from (2.154), (2.156) and (2.157),

$$Ar+2Bs+Ct=F,$$
$$dx \cdot r + dy \cdot s = dp,$$ (2.158)
$$dx \cdot s + dy \cdot t = dq,$$

where the dots indicate multiplication. Two possibilities arise for every, point on δ_0. Either the determinant of the system

$$D = \begin{vmatrix} A & 2B & C \\ dx & dy & 0 \\ 0 & dx & dy \end{vmatrix} = A\left(\frac{dy}{dx}\right)^2 - 2B\frac{dy}{dx} + C \neq 0,$$

and r, s, and t are uniquely determined by the given quantities on δ_0, and the partial differential equation (2.154), or

$$D = A\left(\frac{dy}{dx}\right)^2 - 2B\frac{dy}{dx} + C = 0,$$ (2.159)

which has the solutions

$$\frac{dy}{dx} = \frac{B \pm (B^2 - AC)^{\frac{1}{2}}}{A} = \begin{Bmatrix} \alpha \\ \beta \end{Bmatrix},$$ (2.160)

where $\alpha \neq \beta$, α corresponding to the (+), and β to the (−) sign. Equation (2.160) gives, therefore, two solution (directions dy/dx) for proceeding with the calculation of φ on δ_1, as the sketch in fig. 2.11 shows. Equation

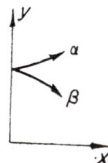

Fig. 2.11. Characteristic directions.

(2.159) is known as the *characteristic condition* for a solution of (2.158) and α and β are the *characteristic directions*. When the characteristic condition (2.159), and hence (2.160), holds, it implies that a linear relation exists between the left hand sides, and therefore between the right hand sides of (2.158), for each of (2.160). The latter are further conditions that must be met if φ, p and q are to be extended to points on δ_1. The later expressions, (2.163) and (2.164), we use, are such conditions.

From (2.160), with $A \neq 0$,

$$B^2 - AC > 0$$

gives two real characteristics. This is the hyperbolic case of (2.152).[4] As we would expect, from our discussion of the initial value problem in § 2.9.1, the one dimensional wave equation (2.19) is an example. We note that in this example A, B and C are constants, hence from (2.160), α and β lead to the paths

$$y = \alpha x + c_1,$$
$$y = \beta x + c_2,$$

which can be identified with the characteristics exhibited in figs. 2.6 and 2.8.

The means by which we can apply this theory of characteristics is based on two fundamental *hyperbolic theorems* associated with (2.152).

2.9.3.1. First hyperbolic theorem; initial value problem and numerical integration method. The *first hyperbolic theorem* can be stated as follows: If φ, and its two first order derivatives, $\partial \varphi / \partial x$ and $\partial \varphi / \partial y$, are given along an open non-characteristic curve, then the solution for φ is uniquely determined in a quadrangle of characteristics about this curve. The theorem governs initial value problems. To apply this theorem we note, using the first of (2.160), that along the α-characteristic, (2.156) and (2.157) become

$$dp = (r + \alpha s) dx,$$
$$dq = (s + \alpha t) dx.$$
(2.161)

From (2.161) we can write

$$dp + \beta dq = [r + (\alpha + \beta)s + \alpha \beta t] dx,$$
(2.162)

but (2.160) shows $\alpha \beta = C/A$, and $\alpha + \beta = 2B/A$, hence, using also the first of (2.158), (2.162) reduces to

$$dp + \beta dq = \left(r + \frac{2B}{A} s + \frac{C}{A} t \right) dx = \frac{F}{A} dx.$$
(2.163)

---

[4] $B^2 - AC = 0$, gives one real characteristic. This is the parabolic case of (2.152), with the one-dimensional heat equation being an example. When $B^2 - AC < 0$, no solutions exist. This is the elliptic case of (2.152), with Laplace's equation, in two dimensions, as an example.

Similarly, along the β-characteristic direction, we have

$$dp + \alpha dq = \frac{F}{A} dx. \tag{2.164}$$

Assume now we have φ, p and q given along an open non-characteristic curve ab, as sketched in fig. 2.12. Points 1 and 2 on ab are necessarily points

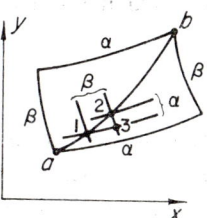

Fig. 2.12. An open non-characteristic curve and intersecting characteristics.

with given information. Point 3 is a point of intersection of α- and β-characteristics. Assuming infinitesimal lengths are straight lines, we can write the difference expressions

$$y_3 - y_1 = \alpha_1(x_3 - x_1),$$

$$y_3 - y_2 = \beta_2(x_3 - x_2),$$

which can be solved for x_3, y_3, since $\alpha_1 \neq \beta_2$. Therefore, from (2.155)

$$\varphi_3 = \varphi_1 + p_1(x_3 - x_1) + q_1(y_3 - y_1). \tag{2.165}$$

From (2.163) we can write

$$p_3 - p_1 + \beta_1(q_3 - q_1) = \frac{F_1}{A_1}(x_3 - x_1), \tag{2.166}$$

and from (2.164)

$$p_3 - p_2 + \alpha_2(q_3 - q_2) = \frac{F_2}{A_2}(x_3 - x_2), \tag{2.167}$$

which can be solved for p_3 and q_3, if $\alpha_2 \neq \beta_1$. This completes the solution for point 3, and we can proceed similarly for other points. We see that the solution must be contained in a quadrangle, whose sides are α and β characteristics that intersect a and b. We also note that if our grid size of characteristics becomes too large we can fill in with more characteristic lines with resulting better accuracy.

Note that the initial value problem we treated in § 2.9.1 is a simple special case of this theorem. That is, u and $\dot{u}$ were given along the non-characteristic curve $t=0$, and hence $\partial u/\partial x$ was too, from u, and the solution was determined in a quadrangle of characteristics about $t=0$ (which, because ≥ 0, was a triangle). We could have employed the present numerical method in this problem, but the analytical method used in § 2.9.1 is preferable in this simple problem.

2.9.3.2. Second hyperbolic theorem; characteristic initial value problem and numerical integration method. The *second hyperbolic theorem* can be tated as follows: *If φ is given along two intersecting characteristic curves, then the solution for φ is uniquely determined in a quadrangle of characteristics within the given two.* The theorem governs the so-called *characteristic initial value problem* (cf. [2.20, ch. 5 §§ 5,6] or [2.25, ch. 4 § 4]). This problem arises in the interaction of waves traveling toward one another. Garabedian [2.25, pp. 516–518 and fig. 62] discusses an instructive example in gas dynamics, namely that of a gas initially at rest in a section of a pipe between a pair of pistons, located say at $x=x_0$ and $x=x_1$. The pistons are suddenly (at $t=0$) pulled away from (or pushed toward) one another. The characteristic initial value problem arises a short time t_0 later, when the two simple waves so generated meet at the center of the pipe section. Derivation of the numerical integration scheme for applying this theorem is left to the exercises. Sokolnikoff and Redheffer [2.22, ch. 6 § 25] give simple uniqueness proofs for φ in the two hyperbolic theorems when the linear one-dimensional wave equation governs.

2.9.3.3. Boundary-initial value problems and numerical integration method.

Consider the problems of one-dimensional wave propagation in a half space $x \geq 0$, where the medium is at rest for $t<0$, excitation being due to a suddenly applied, at $t=0$, uniform stress $\sigma_x = \sigma$, or velocity $v = \dot{u}$, along $x=0$. In boundary value problems it is best to base the method of characteristics on the *first order partial differential equations* that govern, rather than the second order equation (2.152), because related expressions for the *boundary conditions are simpler.* A reduction of the general second order equation (2.152) to the equivalent first order system is discussed in Courant and Hilbert [2.20, pp. 423–424].

We can write (2.152) as

$$G(x, y, \varphi, p, q, r, s, t) = 0 , \tag{2.168a}$$

where

$$p = \partial\varphi/\partial x, \; q = \partial\varphi/\partial y , \; r = \partial^2\varphi/\partial x^2 , \; s = \partial^2\varphi/\partial x \partial y , \; t = \partial^2\varphi/\partial y^2 . \tag{2.168b}$$

(2.168a,b) represent a quasi-linear system of first order partial differential equations for the unknown functions φ, p, q, r, s, t. In the present problems identifying φ with displacement u and y with time t', (2.168a) is the displacement equation of motion

$$\partial^2 u(x, t')/\partial x^2 - \ddot{u}(x, t')/c_d^2 = 0 . \tag{2.169a}$$

Then from the first two equations of (2.168b) we have

$$p = \partial u/\partial x = \sigma/(\lambda + 2\mu) , \qquad q = \dot{u} = v . \tag{2.169b}$$

Now substituting (2.169b) in (2.169a) gives

$$\partial\sigma(x, t')/\partial x = \varrho v(x, t') \tag{2.170a}$$

and substituting (2.169b) in $\dot{p} = s$ [the fourth of (2.168b)], we find

$$\dot{\sigma}(x, t') = (\lambda + 2\mu)\partial v(x, t')/\partial x . \tag{2.170b}$$

(2.170) become the first order system of equations for the present problems, where we can now write just t for t' the time. It may be observed that (2.170a) is the stress equation of motion (1.64), and (2.170b) essentially the stress-strain relation (1.45). Because of the simple nature of this system it could have been written more directly.

The characteristics for the equations (2.170) are obtained from the system of equations (2.170), and (2.156), (2.157), which using the former in the latter reduce the latter to

$$\mathrm{d}x \cdot \partial\sigma/\partial x + (\lambda + 2\mu)\mathrm{d}t \cdot \partial v/\partial x = (\lambda + 2\mu)\mathrm{d}p ,$$

$$\mathrm{d}t \cdot \partial\sigma/\partial x + \varrho\mathrm{d}x \cdot \partial v/\partial x = \varrho\mathrm{d}q . \tag{2.171}$$

The determinant of this system of equations, set equal to zero, gives the characteristic directions

$$\frac{dt}{dx} = \begin{Bmatrix} \alpha \\ \beta \end{Bmatrix} = \pm \frac{1}{c_d}. \tag{2.172}$$

From (2.171), using (2.172), the characteristic equations

$$d\sigma \mp \varrho c_d dv = 0, \tag{2.173}$$

follow, where the upper and lower signs correspond to the α- and β-directions, respectively. Equations (2.173) in finite difference form are equivalent to (2.166), (2.167), since $F=0$ in the present case. Consider then a case where there is a stress input $\sigma_0(t)$ at $x=0$, which rises smoothly from $t=0$. It follows then, with reference to fig. 2.13, that $\sigma=v=0$ all along

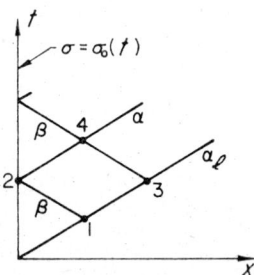

Fig. 2.13. Characteristics for one-dimensional boundary-initial value problem.

the leading characteristic α_l (points 1, 3,...), since σ, v are zero everywhere in the triangular domain $0 \leq t \leq x/c_d$. The stress at the boundary point 2 is given, i.e., $\sigma_2 = \sigma_0(t)$. We calculate v_2 by appealing to the second of (2.173)

$$\sigma_1 - \sigma_2 + \varrho c_d(v_1 - v_2) = 0,$$

from which it follows that $v_2 = -\sigma_0(t)/\varrho c_d$. The variables σ, v at the first interior point 4 can now be calculated by writing (2.173) for the 4-2 and 4-3 increments, along α and β characteristics, respectively, and solving these equations. Evaluation at other points follow similarly.

When the *input function has a jump initially*, from our discussion of the propagation of surfaces of discontinuity in § 2.6 we would expect these jumps to occur across, and propagate along, both the α and β characteristics. Indeed as was proved in § 2.9.1.3, a jump occurring in the direction of one characteristic, say β, propagates along the α characteristic through the discontinuity point. Following Leonard and Budiansky [2.26], we can exploit this fact to calculate the jumps across characteristics and their behavior. Consider the situation in fig. 2.14. We wish to calculate the dis-

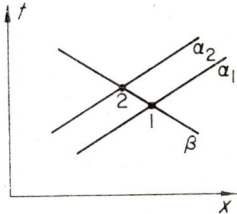

Fig. 2.14. Characteristics involved in calculation of jump in stress.

continuities across α_1, which according to the second of (2.173) is given by

$$\sigma_2 - \sigma_1 + \varrho c_d(v_2 - v_1) = 0,$$

as α_2 approaches α_1, or

$$[\sigma] + \varrho c_d[v] = 0. \qquad (2.174)$$

We note (2.174) agrees with the jump condition (2.49). The variation of the jumps $[\sigma]$, $[v]$ in (2.174), as they propagate along the α characteristics, can be obtained by writing the first of (2.173) for the α_2 and α_1 characteristics, subtracting the latter from the former, and letting α_2 approach α_1. We find

$$d[\sigma] - \varrho c_d d[v] = 0,$$

and using (2.174), this gives

$$[\sigma] = C, \qquad [v] = C/\varrho c_d,$$

where C is a constant, equal to the magnitude of the initial jump in stress. Hence, we see these *jumps remain constant as they propagate in space*. Again our example has been a simple, but instructive, one which would normally be solved in closed form by other means. Chou and Koenig [2.27] have used the above scheme to treat the problem of an axially symmetric cylindrical cavity in an infinite elastic space subjected to uniform internal pressure. We will treat this problem later using the Laplace transform. It should also be pointed out that Ziv [2.28] has given the characteristic equations for both Cartesian and cylindrical coordinates in two-dimensional elastic wave propagation. With the advent of fast computers, it is clear that the method of characteristics may play an increasingly important role in wave propagation work.

2.10. Exercises

2.1. By direct operation on the first of (2.7b) derive (2.4) of the text. Similarly, with the aid of the definition of ω in (2.2), show you can get (2.5) by operating on the second of (2.7b). State your conditions.

2.2. Prove the completeness of the Lamé representation (2.7) for the general case of nonvanishing body force $X(x, t)$ and initial conditions $u(x, 0) = u_0(x)$, $\dot{u}(x, 0) = v_0(x)$. Note X can be written as in (2.65), so that (2.66) replace (2.7c) here, and from (2.7a) we have

$$u_0 = \nabla \varphi_0 + \nabla \times \psi_0, \qquad v_0 = \nabla \dot{\varphi}_0 + \nabla \times \dot{\psi}_0,$$

all based on Helmholz resolution of a vector. The proof follows along the lines of that given in § 2.2.5.

2.3. Consider the propagation of one wavelength of the one-dimensional sinusoidal displacement wave

$$u(x, t) = u_0 \sin \varkappa(x - ct), \qquad \varkappa = 2\pi/L$$

where $\varkappa$, L, c, and u_0 are respectively, the wave number, length, speed and amplitude.

(a) plot the wave as a function of x for the times

$$t = 0, \text{ and } t = t_0 > 0,$$

(b) Prove the wave is plane,
(c) State the period, frequency and phase of the wave.
(d) Give the position of the wavefront for both times,
(e) Rewrite the given expression for u in a form convenient for plotting the wave as a function of t, for a fixed x value. Plot it. Where is the front here?

2.4. Show that the solution of the inhomogenous wave equation (2.73) is given by (2.75) by writing $\delta(r)$ there as $\delta(r-a)$ and exploiting the function

$$\hat{\varphi}^-(r, t) = -\frac{g(t+r/c_d)}{r} + \frac{g(t-r/c_d)}{r}, \quad \text{for} \quad r < a,$$

$$\hat{\varphi}^+(r, t) = \frac{h[t-(r-a)/c_d]}{r}, \quad \text{for} \quad r > a,$$

where $\hat{\varphi}$, $\dot{\hat{\varphi}}$ must be continuous at $r = a$. Ultimately a tends to $0+$. The reader should note further the important condition

$$a^2 \left(\frac{\partial \hat{\varphi}^+}{\partial r} - \frac{\partial \hat{\varphi}^-}{\partial r} \right)_{r=a} = \frac{F(t)}{4\pi}$$

which is obtained by multiplying the governing partial differential equation by r^2 and integrating it over (a^-, a^+), provided it is assumed that $\ddot{\varphi}$ has at most a delta function singularity at $r=a$.

2.5. Derive in detail the steps in § 2.7.4 leading to the solution $\phi(x_1, x_2, t)$ given by (2.91), (2.92) for the two-dimensional radiation problem posed by (2.84), (2.85).

2.6. Derive in detail the steps in § 2.7.5.1 leading to the solution for the displacements $u_i^k(x, t; x')$ given by (2.105a) for the problem of a time dependent point body force acting in the x_k direction ($k=1, 2, 3$).

2.7. In the one-dimensional initial value problem of § 2.9.1.2, involving an initial velocity condition, show that if $g(x)$ in (2.127) is replaced by an odd function, say

$$g(x) = \begin{cases} x, & |x| < k \\ 0, & |x| > k, \end{cases}$$

the resultant disturbance separates, for $t \geq t_0$, into two pulses traveling along the string, one in the positive and one in the negative x-direction. Derive the full solution for the problem and construct a sketch like that of fig. 2.9 for it.

2.8. Determine the characteristics, and derive the solution for the initial value cases

(a) $u=0, \quad \dot{u}=a \sin x,$
(b) $u=a \sin x, \quad \dot{u}=0,$ $\Big\} 0 \leq x \leq \pi.$

Plot the solutions $u(x, t)$ as surfaces.

2.9. Show that a discontinuity in $\partial^2 u/\partial x^2$ is propagated along at least one characteristic through (x_0, t_0) (cf. § 2.9.1.3).

2.10. Prove that when the characteristic condition (2.159), and hence (2.160) holds, it implies that a linear relation exists between the left hand sides, and therefore between the right hand sides of (2.158), for each of (2.160). These right hand side conditions must be met if φ, p and q are to be extended to points on δ_1. (2.163), (2.164) are such conditions.

2.11. Derive the numerical integration method for the second hyperbolic theorem (cf. § 2.9.3.2). As an aid see that for the first hyperbolic theorem in § 2.9.3.1.

References

[2.1.] S. D. Poisson, *Mémoires Académie Science Paris* **8** (1829), 356.
[2.2.] S. D. Poisson, *Mémoires Académie Science Paris* **8** (1829), 623.
[2.3.] E. Sternberg, *Archive Rational Mechanics and Analysis* **6** (1960), 34—50.
[2.4.] I. Todhunter and K. Pearson, *A History of the Theory of Elasticity and of the Strength of Materials, Vol. 1.* Cambridge University Press (1886), 273.
[2.5.] G. Lamé, *Leçons sur la Theorie Mathématique de l'Elasticité des Corps Solides.* Bachelier, Paris (1852).
[2.6.] C. Somigliana, *Atti Real Accad. Linc. Roma,* Ser. 5, **1** (1892), 111.
[2.7.] P. M. Morse and H. Feshbach, *Methods of Theoretical Physics, Parts I and II,* McGraw–Hill Book Company, Inc., New York (1953).
[2.8.] A. Erdelyi, W. Magnus, F. Oberhettinger and F. G. Tricomi, *Higher Transcendental Functions II,* Bateman Manuscript Project. McGraw–Hill Book Company, Inc., New York (1953).
[2.9.] H. Keller, *SIAM Review* **6** (1964), 356–382.
[2.10.] F. G. Friedlander, *Sound Pulses.* Cambridge University Press (1958).
[2.11.] J. Keller, *Journal of Geophysical Research* **68** (1933), 1182–1183.
[2.12.] E. B. Wilson, *Advanced Calculus.* Ginn and Company, New York (1912), 122–124.
[2.13.] F. C. Karal, Jr. and J. B. Keller, *Journal of the Acoustical Society of America* **31** (1959), 694–705.
[2.14.] J. D. Achenbach, *Wave Propagation in Elastic Solids,* North–Holland Publishing Company, Amsterdam (1973).
[2.15.] G. G. Stokes, *Transactions of the Cambridge Philosophical Society* **9** (1849), 1–62.
[2.16.] A. E. H. Love, *Proceedings of the London Mathematical Society, Second Series* **1** (1904), 291–344.
[2.17.] G. Eason, J. Fulton and I. N. Sneddon, *Philosophical Transactions of the Royal Society, London, Series A* **248** (1955–1956), 575–607.
[2.18.] D. S. Jones, *The Theory of Electromagnetism.* The Macmillan Company, New York (1964), 34–45.
[2.19.] A. T. deHoop, *Representation Theorems for the Displacement in an Elastic Solid and Their Application to Elastodynamic Diffraction Theory.* Doctoral Dissertation, Technische Hogeschool, Delft (1958), 14–19.
[2.20.] R. Courant and D. Hilbert, *Methods of Mathematical Physics, V. 2, Partial Differential Equations* by R. Courant. Interscience Publishers, New York (1962), 197–221, 407–766.
[2.21.] J. Hadamard, *Lectures on Cauchy's Problem in Linear Partial Differential Equations.* Dover Publications, New York (1952).
[2.22.] I. S. Sokolnikoff and R. M. Redheffer, *Mathematics of Physics and Modern Engineering.* McGraw–Hill Book Company, Inc., New York (1958).
[2.23.] H. F. Weinberger, *Partial Differential Equations.* Blaisdell Publishing Company, New York (1965).
[2.24.] A. E. H. Love, *Proceedings of the London Mathematical Society, Second Series* **1** (1903), 37–62.
[2.25.] P. R. Garabedian, *Partial Differential Equations.* John Wiley and Sons, Inc., New York (1964).
[2.26.] R. W. Leonard and B. Budiansky, *On Travelling Waves in Beams,* NACA Technical Note 2874 (1953).
[2.27.] P. C. Chou and H. A. Koenig, *Journal of Applied Mechanics* **33** (1966), 159–167.
[2.28.] M. Ziv, *International Journal of Solids and Structures* **5** (1969), 1135–1151.

CHAPTER 3

REFLECTION AND REFRACTION OF TIME HARMONIC WAVES AT AN INTERFACE

At this point in our elastic wave theory it is appropriate to introduce a *single planar interface* in the medium, and study the process of *wave reflection, refraction and generation there*. Attention will focus on mostly plane waves, harmonic (sinusoidal) in time and space. We will find that these relatively simple waves bring out basic information on reflection and refraction phenomena, and new waves peculiar to the interface. Our study begins with the consideration of wave reflection from the boundary of an elastic half space. In this we are assuming the boundary to be an interface between two half spaces, one elastic and the other a vacuum.

3.1. Reflection of *P* and *SV* waves of plane strain from the boundary of an elastic half space

The complexities of general elastic wave reflection from a boundary stem from *mode conversion* there of the body dilatational (*P*) and equivoluminal (*S*) waves, i.e., *P or S into P and S generally, upon reflection, and the generation there of new waves for certain boundaries*. Here we first study this problem, imposing the simplifications of plane strain and plane harmonic waves incident at the boundary. We assume the cartesian space $(x_1, x_2, x_3) = (x, y, z)$, where the corresponding displacement components $(u_1, u_2, u_3) = (u, v, w)$, and as shown in fig. 3.1, a half space defined by $-\infty < x < \infty$, $-\infty < y < \infty$, and $z \geq 0$, with $y > 0$ out of the figure.

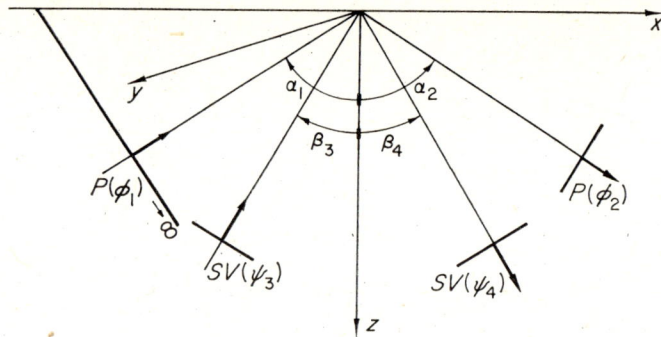

Fig. 3.1. Incident and reflected P and SV waves.

Plane strain imposes $v = \partial/\partial y = 0$, hence $\sigma_{zy} = \sigma_{xy} = 0$[1]. From what we have already learned about the character of general plane displacement waves, it is not surprising that the P and SV waves here are consistent with the plane strain condition of $v = 0$, i.e., the displacements associated with them involve only u and w. Figure 3.2 demonstrates this. Equations (2.26) are

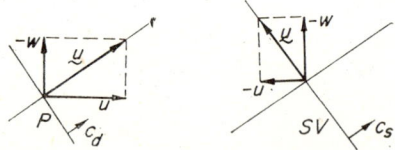

Fig. 3.2. P- and SV-wave displacements.

used to write the displacement-potential relations here. Imposing on these the *plane strain conditions*, that (1) φ and ψ_i can be at most functions of x_1 and x_3, and (2) $u_2 = v$ must vanish, we find it sufficient to set $\psi_1 = \psi_3 = 0$, since these component potentials do not enter into the definition of the

---

[1] A closely related plane problem is that of *plane stress*. A body is in a state of plane stress if there is a family of parallel planes across which no stress exists. In the present coordinate system this is expressed by $\sigma_{2i} = 0$, hence $\sigma_y = \sigma_{zy} = \sigma_{xy} = 0$. It is easy to show that the governing equations of plane stress and plane strain differ only in their constant coefficients. A thin plate under in-plane loads forms a common use of plane stress, e.g., extensional motion of plates (cf. [1.2] p. 497).

nonvanishing displacements $u_1 = u$, and $u_3 = w$. Hence, the displacement-potential relations are

$$u = \frac{\partial \varphi}{\partial x} - \frac{\partial \psi}{\partial z},$$

$$w = \frac{\partial \varphi}{\partial z} + \frac{\partial \psi}{\partial x},$$
(3.1)

with $v = 0$, where we have set $\psi_2 = \psi$.

Shown in fig. 3.1 are the incident P (dilatational) and SV (equivoluminal) waves, φ_1, and ψ_3, respectively, and reflected P and SV waves, φ_2 and ψ_4, respectively, subscripts now on φ, ψ relating only to incidence and reflection. All of these *waves are of infinite beam width* as is indicated in the figure for the incident P wave by the line extending from the boundary to infinity. The *totality of these waves* is given by

$$\varphi = \varphi_1 + \varphi_2 = A_1 \varphi_1^e + A_2 \varphi_2^e,$$

$$\psi = \psi_3 + \psi_4 = A_3 \psi_3^e + A_4 \psi_4^e,$$
(3.2)

where A_1, A_2, A_3, and A_4 are *wave amplitudes*, and

$$\varphi_1^e = \exp\left[i\varkappa_d(l_1 x - n_1 z - c_d t)\right],$$

$$\varphi_2^e = \exp\left[i\varkappa_d(l_2 x + n_2 z - c_d t)\right],$$

$$\psi_3^e = \exp\left[i\varkappa_s(l_3 x - n_3 z - c_s t)\right],$$

$$\psi_4^e = \exp\left[i\varkappa_s(l_4 x + n_4 z - c_s t)\right],$$

where

$$\varkappa_d = \frac{2\pi}{L_d} = \frac{2\pi}{c_d T} = \frac{\omega}{c_d}, \qquad \varkappa_s = \frac{2\pi}{L_s} = \frac{2\pi}{c_s T} = \frac{\omega}{c_s},$$

$\varkappa_d$ and $\varkappa_s$ being *wave numbers*, and L_d and L_s *wavelengths, associated with the P and SV waves*, respectively. T and ω are the *common period and frequency* for all these waves. We note from these last relations that

$$\omega = \varkappa_d c_d = \varkappa_s c_s.$$
(3.3)

The angles in fig. 3.1 are measured from the z axis, α_1 and β_3 being *angles of wave incidence*, and α_2 and β_4 *angles of wave reflection*. Accordingly, the *direction cosines* in (3.2) are $l_1 = \sin \alpha_1$, $n_1 = \cos \alpha_1$, $l_2 = \sin \alpha_2$, $n_2 = \cos \alpha_2$, etc. Note that the direction cosines for the z-direction are negative for incident waves, since along this direction propagation is in the negative z-direction.

It is clear that the plane waves in (3.2) satisfy our differential equations (2.7b). In the most general reflection problem all of these waves will be present. To solve such problems we will, in addition to the displacement-potential relations (3.1), have use for the strain- and stress-potential relations. These can be obtained by first writing the strain-displacement relations for the present case from (1.25) and (1.39)

$$\varepsilon_x = \frac{\partial u}{\partial x}, \qquad \varepsilon_z = \frac{\partial w}{\partial x}, \qquad \varepsilon_y = 0,$$

$$\varepsilon_{zx} = \frac{\partial u}{\partial z} + \frac{\partial w}{\partial x}, \qquad \varepsilon_{xy} = \varepsilon_{zy} = 0, \qquad \Delta = \varepsilon_x + \varepsilon_z, \qquad (3.4)$$

where this is essentially Love's notation [1.2], except that we have dropped the double subscript for the normal strains.[2] The strain-potential relations follow from (3.4), using (3.1),

$$\varepsilon_x = \frac{\partial^2 \varphi}{\partial x^2} - \frac{\partial^2 \psi}{\partial x \partial z}, \qquad \varepsilon_z = \frac{\partial^2 \varphi}{\partial x^2} + \frac{\partial^2 \psi}{\partial z \partial x},$$

$$\varepsilon_{zx} = 2\frac{\partial^2 \varphi}{\partial z \partial x} + \frac{\partial^2 \psi}{\partial x^2} - \frac{\partial^2 \psi}{\partial z^2}, \qquad \Delta = \nabla^2 \varphi. \qquad (3.5)$$

Substitution of (3.5) in (1.45, in Love's notation) gives the stress-potential relations

$$\sigma_x = \frac{\lambda}{c_d^2} \ddot{\varphi} + 2\mu \left(\frac{\partial^2 \varphi}{\partial x^2} - \frac{\partial^2 \psi}{\partial x \partial z} \right),$$

$$\sigma_z = \frac{\lambda}{c_d^2} \ddot{\varphi} + 2\mu \left(\frac{\partial^2 \varphi}{\partial z^2} + \frac{\partial^2 \psi}{\partial x \partial z} \right),$$

$$\sigma_{zx} = \mu \left[\frac{\ddot{\psi}}{c_s^2} + 2 \left(\frac{\partial^2 \varphi}{\partial x \partial z} - \frac{\partial^2 \psi}{\partial z^2} \right) \right], \qquad (3.6)$$

$$\sigma_y = v(\sigma_x + \sigma_z),$$

---

[2] Note $\varepsilon_{zx} = 2\varepsilon_{31} = 2\varepsilon_{13}$, and $\sigma_{zx} = \sigma_{31} = \sigma_{13}$, which from (1.45) gives $\sigma_{zx} = \mu \varepsilon_{zx}$. The normal strains are essentially the same.

Ch. 3, § 3.1]   REFLECTION OF P AND SV WAVES   123

where for convenience we have made use of (2.7b) to express ∇^2 in terms of (¨). We will find these forms particularly useful later also. The totality of the waves in the displacements and stresses may now be obtained by substituting (3.2) in (3.1), and (3.6), respectively, with the results

$$u = i[\varkappa_d(l_1 A_1 \varphi_1^e + l_2 A_2 \varphi_2^e) + \varkappa_s(n_3 A_3 \psi_3^e - n_4 A_4 \psi_4^e)],$$

$$w = i[-\varkappa_d(n_1 A_1 \varphi_1^e - n_2 A_2 \varphi_2^e) + \varkappa_s(l_3 A_3 \psi_3^e + l_4 A_4 \psi_4^e)], \quad (3.7)$$

$$\sigma_z = -\varkappa_d^2[(\lambda + 2\mu n_1^2)A_1 \varphi_1^e + (\lambda + 2\mu n_2^2)A_2 \varphi_2^e] + 2\mu\varkappa_s^2(l_3 n_3 A_3 \psi_3^e - l_4 n_4 A_4 \psi_4^e), \quad (3.8)$$

$$\sigma_{zx} = \mu\{2\varkappa_d^2(l_1 n_1 A_1 \varphi_1^e - l_2 n_2 A_2 \varphi_2^e) - \varkappa_s^2[(l_3^2 - n_3^2)A_3 \psi_3^e + (l_4^2 - n_4^2)A_4 \psi_4^e]\},$$

with similar expressions for σ_x and σ_y. Equations (3.7) and (3.8) are expressions for the displacements, and the two stresses that can occur on the boundary $z=0$. So, depending on the particular boundary conditions defining a problem, two of these equations, hence the four amplitudes A_1, A_2, A_3, and A_4, will be involved at $z=0$. For a particular incident wave, say A_1 or A_3, solutions can then be found for the so-called *reflection coefficients* for the potentials i.e., *ratios of the amplitudes of the reflected waves to the incident wave amplitude*, as a function of incident angle and Poisson's ratio. The cases that follow will demonstrate this.

3.1.1. *Mixed conditions on boundary*[3]

3.1.1.1. *Lubricated-rigid boundary.* In this case

$$w = \sigma_{zx} = 0, \qquad \text{at } z = 0. \quad (3.9)$$

Set $A_3 = 0$, so that we have an incident P wave only. Substituting the second of (3.7) and (3.8) into (3.9), we find

---

[3] Not to be confused with our basic third boundary value problem. The present problems involve one displacement and one stress, both over the entire boundary. Note these types of conditions were involved in exercise 1.3.

$$-\varkappa_d\{n_1 A_1 \exp[i\varkappa_d(l_1 x - c_d t)] - n_2 A_2 \exp[i\varkappa_d(l_2 x - c_d t)]\}$$
$$= -\varkappa_s l_4 A_4 \exp[i\varkappa_s(l_4 x - c_s t)], \qquad (3.10)$$
$$2\varkappa_d^2\{l_1 n_1 A_1 \exp[i\varkappa_d(l_1 x - c_d t)] - l_2 n_2 A_2 \exp[i\varkappa_d(l_2 x - c_d t)]\}$$
$$= \varkappa_s^2(l_4^2 - n_4^2) A_4 \exp[i\varkappa_s(l_4 x - c_s t)].$$

Equations (3.10) will yield a solution for A_2/A_1 and A_4/A_1, independent of x and t, only if we first set

$$\alpha_1 = \alpha_2 = \alpha,$$
$$\varkappa_s(l_4 x - c_s t) = \varkappa_d(l_2 x - c_d t). \qquad (3.11)$$

Noting (3.3), and using the first of (3.11), and $\beta_4 = \beta$, the second of (3.11) reduces to

$$\frac{\sin\alpha}{\sin\beta} = \frac{\varkappa_s}{\varkappa_d} = \frac{c_d}{c_s} = \left[\frac{2(1-v)}{1-2v}\right]^{\frac{1}{2}} = k, \qquad (3.12a)$$

which is similar to *Snell's law of refraction in optics*. Note that (3.11) contains

$$\varkappa = \varkappa_d \sin\alpha = \varkappa_s \sin\beta, \qquad (3.12b)$$

and it follows, since from (3.3) $\omega = \varkappa c$, that

$$c = c_d/\sin\alpha = c_s/\sin\beta. \qquad (3.12c)$$

Equations (3.12b) and (3.12c) show that $\varkappa$ and c, respectively the wave number and wave velocity along the boundary, are the same for both waves. Now using (3.11) and (3.12a) it is easily shown the solution of (3.10) is $A_2 = A_1$, $A_4 = 0$. Similarly with $A_1 = 0$, for the case of an incident *SV* wave, we find the solution is $A_4 = -A_3$, $A_2 = 0$, with $\beta_3 = \beta_4 = \beta$. The two cases are illustrated in fig. 3.3. We see these cases are simple, giving *no mode*

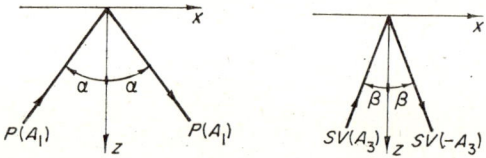

Fig. 3.3. Reflection of *P* and *SV* waves under mixed conditions at boundary.

Ch. 3, § 3.1] REFLECTION OF *P* AND *SV* WAVES 125

conversion, i.e., an incident *P* wave generates just a *P* reflection, and similarly an incident *SV* wave just an *SV* reflection. Note also that α, β are arbitrary and Poisson's ratio is not involved.

3.1.1.2. In-plane constrained boundary. In this case

$$u = \sigma_z = 0, \quad \text{at } z=0. \tag{3.13}$$

This case can be worked out like the previous case in § 3.1.1.1. It gives similar results, i.e., incident *P* wave: $A_3 = A_4 = 0$, $\alpha_1 = \alpha_2 = \alpha$, $A_2 = -A_1$, and incident *SV* wave: $A_1 = A_2 = 0$, $\beta_3 = \beta_4 = \beta$, $A_4 = A_3$.

3.1.2. Free boundary

In this case

$$\sigma_z = \sigma_{zx} = 0, \quad \text{at } z=0. \tag{3.14}$$

3.1.2.1. Reflection of a P wave. Setting $A_3 = 0$, we consider an incident *P* wave. Then substituting (3.8) in (3.14) gives

$$\varkappa_d^2\{(\lambda + 2\mu n_1^2)A_1 \exp[i\varkappa_d(l_1 x - c_d t)] + (\lambda + 2\mu n_2^2)A_2 \exp[i\varkappa_d(l_2 x - c_d t)]\}$$
$$= -2\mu\varkappa_s^2 n_4 l_4 A_4 \exp[i\varkappa_s(l_4 x - c_s t)], \tag{3.15}$$
$$2\varkappa_d^2\{l_1 n_1 A_1 \exp[i\varkappa_d(l_1 x - c_d t)] - l_2 n_2 A_2 \exp[i\varkappa_d(l_2 x - c_d t)]\}$$
$$= \varkappa_s^2(l_4^2 - n_4^2)A_4 \exp[i\varkappa_s(l_4 x - c_s t)].$$

As in our foregoing cases of mixed conditions, we can solve these equations for A_2/A_1 and A_4/A_1, independent of x and t, only through use of (3.11) and (3.12a). Then we find that (3.15) reduce to

$$(\lambda + 2\mu \cos^2 \alpha)(A_1 + A_2) + (\mu k^2 \sin 2\beta)A_4 = 0,$$
$$(\sin 2\alpha)(A_1 - A_2) + (k^2 \cos 2\beta)A_4 = 0. \tag{3.16}$$

Multiplying the first of (3.16) by $\cos 2\beta$, and the second by $\mu \sin 2\beta$, subtracting the latter from the former, and noting that

$$\frac{\lambda}{\mu} + 2\cos^2\alpha = k^2 \cos 2\beta, \tag{3.17}$$

gives

$$\frac{A_2}{A_1} = \frac{\sin 2\alpha \sin 2\beta - k^2 \cos^2 2\beta}{\sin 2\alpha \sin 2\beta + k^2 \cos^2 2\beta}. \tag{3.18}$$

Then from the second of (3.16), using (3.18), we have

$$\frac{A_4}{A_1} = -\frac{2 \sin 2\alpha \cos 2\beta}{\sin 2\alpha \sin 2\beta + k^2 \cos^2 2\beta}. \tag{3.19}$$

Equations (3.18) and (3.19) give the *reflection coefficients for the potentials* for the present case of an incident P wave on the free boundary, which we see are functions of the angle of incidence α and Poisson's ratio ν. No such dependence was exhibited in the cases of mixed boundary conditions treated in § 3.1.1. Note that (3.18) and (3.19) are *independent of wave frequency or length*. It follows that *they hold for general time dependent*

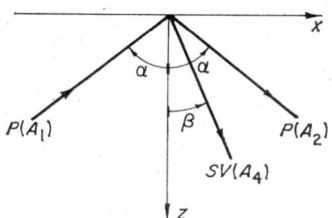

Fig. 3.4. Wave system for *P*-wave incidence.

waves, through a superposition argument. Fig. 3.4 illustrates the wave system we have found.

Inspection of (3.7) and (3.8) shows the reflection coefficients for the displacements and stresses can be obtained from those calculated here for the potentials. From (3.7) we can write for the displacements

$$u = \sum_{j=1}^{4} u_j,$$
$$w = \sum_{j=1}^{4} w_j, \tag{3.20}$$

where

$$u_1 = i\varkappa_d A_1 l_1 \varphi_1^e = U_1 l_1 \varphi_1^e, \qquad u_2 = i\varkappa_d A_2 l_2 \varphi_2^e = U_2 l_2 \varphi_2^e,$$

$$u_3 = i\varkappa_s A_3 n_3 \psi_3^e = U_3 n_3 \psi_3^e, \qquad u_4 = -i\varkappa_s A_4 n_4 \psi_4^e = U_4 n_4 \psi_4^e,$$

and

$$w_1 = -i\varkappa_d A_1 n_1 \varphi_1^e = W_1 n_1 \varphi_1^e, \qquad w_2 = i\varkappa_d A_2 n_2 \varphi_2^e = W_2 n_2 \varphi_2^e,$$

$$w_3 = i\varkappa_s A_3 l_3 \psi_3^e = W_3 l_3 \psi_3^e, \qquad w_4 = i\varkappa_s A_4 l_4 \psi_4^e = W_4 l_4 \psi_4^e.$$

It follows the displacement reflection coefficients can be obtained by forming the appropriate ratios of the above four components of each of the displacements u and w. In the present incident P-wave case, for example, we would have for the *displacement reflection coefficients*

$$\frac{U_2}{U_1} = -\frac{W_2}{W_1} = \frac{A_2}{A_1}, \tag{3.21a}$$

where we have made use of the first of (3.11), and

$$\frac{U_4}{U_1} = \frac{W_4}{W_1} = -k\frac{A_4}{A_1}, \tag{3.21b}$$

where we have used both of (3.11). From (3.8), and similar expressions for σ_x and σ_y, one can, of course, also derive similar expressions for the *stress reflection coefficients*.

Figure 3.5 presents the numerical evaluation of (3.18) and (3.19) for $v = \frac{1}{4}$ ($\lambda = \mu$), and $\frac{1}{3}$.[4] Prominent features of the curves are (1) the maxima of the A_2/A_1 curves, ~ 0.072 at $\sim 69.6°$ for $v = \frac{1}{4}$, and ~ -0.38 at $\sim 65°$ for $v = \frac{1}{3}$, (2) the minima of the A_4/A_1 curves, ~ -0.64 at $\sim 44.9°$ for $v = \frac{1}{4}$, and ~ -0.52 at $\sim 48°$ for $v = \frac{1}{3}$, (3) the fact that at normal incidence the P wave generates no reflected SV wave, and (4) the zeros of A_2/A_1 at $60°$, and $\sim 77.2°$, indicating that *total mode conversion* to an SV wave occurs. Normal incidence, and total mode conversion for the incident P wave, are discussed further in § 3.1.4. Note also the singular point indicated at $\alpha = \pi/2$,

---

[4] The curves for $v = \frac{1}{3}$ were taken from Kolsky [3.1, Fig. 6]. Since Kolsky's curves are for the displacement reflection coefficients, it was necessary to make a simple calculation using (3.21b) for the corresponding A_4/A_1 curve in fig. 3.5.

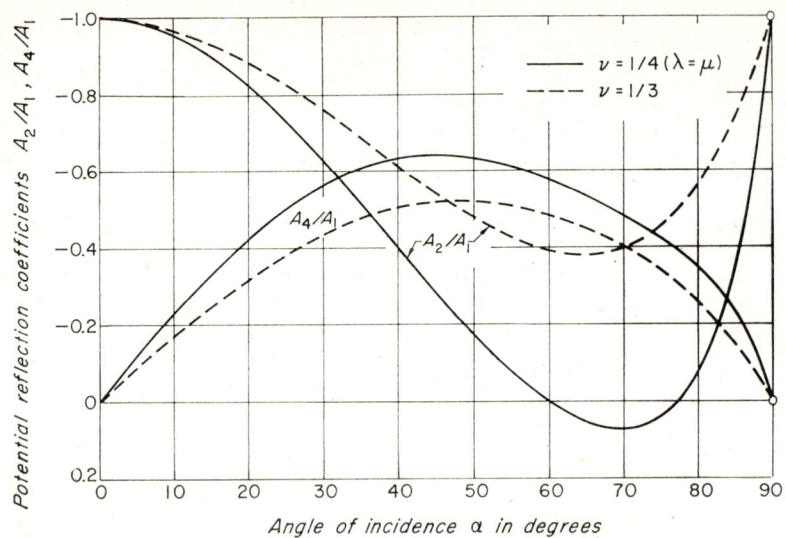

Fig. 3.5. Amplitudes of reflected P and SV waves as a function of angle of incidence for P-wave incidence ($\nu=1/4$, $1/3$) ($\nu=1/3$ after Kolsky).

i.e., grazing incidence of the P wave, which gives $\varphi=\psi=0$, hence zero motion in the present solution. In § 3.1.4.2. it will be shown a nonvanishing solution for this case exists.

Further numerical information on the variation of the P-wave reflection coefficient A_2/A_1 with angle of incidence α, for various Poisson's ratio values, has been obtained by Arenberg [3.2]. Arenberg's results are shown in fig. 3.6.

It is natural to ask *how the incident P-wave energy is partitioned upon reflection between the reflected P and SV waves.* Although these waves are of infinite beam width, as was indicated in fig. 3.1, the relation we seek, involving ratios of energy, e.g., for reflected P wave to incident P wave, is independent of the beam width, except for special cases. Considering, therefore, an incident P wave of finite beam width, which intersects at the free surface an element of unit area, as indicated in fig. 3.7, the incident and reflected P-wave beams will have cross sectional areas of $\cos\alpha$, and the reflected SV-wave beam a cross sectional area of $\cos\beta$. The *total energy density $\hat{E}$ (sum of kinetic and potential energy densities)*, for each of the three waves involved, can be calculated from

$$\hat{E}=\hat{K}+\hat{U}=\tfrac{1}{2}\varrho\dot{u}_i\dot{u}_i+\tfrac{1}{2}\lambda\Delta^2+\mu\varepsilon_{ij}\varepsilon_{ij}, \qquad (3.22)$$

Ch. 3, § 3.1]   REFLECTION OF P AND SV WAVES   129

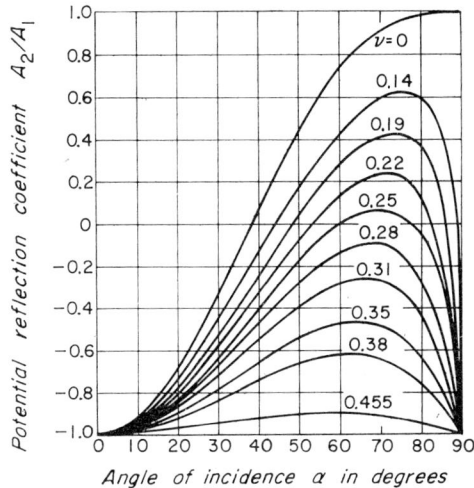

Fig. 3.6. Amplitude of reflected P wave as a function of angle of incidence and Poisson's ratio for P-wave incidence (after Arenberg).

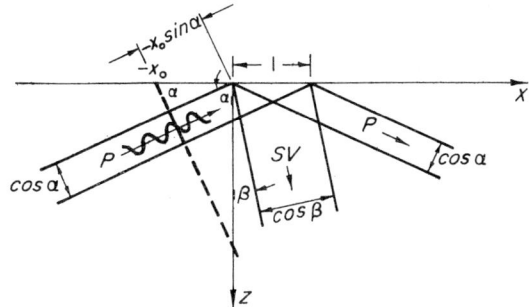

Fig. 3.7. Reflection of P-wave beam.

where we have used (1.74), (1.80). In the present case, for dilatational waves, (3.22) becomes

$$\hat{E}_d = \tfrac{1}{2}\varrho(\dot{w}_d^2 + \dot{u}_d^2) + \tfrac{1}{2}\lambda(\nabla^2\varphi)^2 + \mu\left[\left(\frac{\partial^2\varphi}{\partial x^2}\right)^2 + \left(\frac{\partial^2\varphi}{\partial z^2}\right)^2 + 2\left(\frac{\partial\varphi}{\partial x \partial z}\right)^2\right].$$

To evaluate $\hat{E}_d$ for the incident P wave, the "sub 1" terms in (3.2), (3.7) reduced to real-valued terms for the potential and displacements, are used e.g., $\varphi_1 = A_1 \cos[\varkappa_d(\sin\alpha \cdot x - \cos\alpha \cdot z - c_d t)]$, where A_1 is real. The result is

$$\hat{E}_{d1} = (\lambda + 2\mu)\varkappa_d^4 A_1^2 \cos^2[\varkappa_d(\sin\alpha \cdot x - \cos\alpha \cdot z - c_d t)],$$

half of right hand side coming from $\hat{K}_{d1}$ and half from $\hat{U}_{d1}$, i.e., $\hat{K}_{d1} = \hat{U}_{d1}$. Hence, this expression shows that $\hat{E}_{d1} = 2\hat{K}_{d1}$, and that $\hat{E}_{d1}$ propagates with speed c_d. A similar expression results for $\hat{E}_{d2}$ of the reflected P wave. Related results hold true for equivoluminal wave energy, plane harmonic waves in this case propagating with speed c_s. Generally then for our P and SV waves, we have $\hat{E} = 2\hat{K} = \varrho \dot{u}_i \dot{u}_i$.

The *energy flux* for a particular wave is found by multiplying $\hat{E}$ for the wave by its velocity and beam cross sectional area. Since the energy flux will be *conserved at the free surface*, the energy transmitted along the incident P-wave beam (in. fig. 3.7) must equal the sum of the energies transmitted along the reflected P-and SV-wave beams. It follows we have the *energy balance*

$$\varrho(\dot{u}_1^2 + \dot{w}_1^2)c_d \cos \alpha = \varrho(\dot{u}_2^2 + \dot{w}_2^2)c_d \cos \alpha + \varrho(\dot{u}_4^2 + \dot{w}_4^2)c_s \cos \beta, \quad (3.23)$$

where the "subs 1, 2, and 4", correspond to the so-marked terms in (3.7). Note the energy flux terms here have dimensions of energy per unit time per unit area (of the free surface). Using (3.7) then, (3.23) becomes

$$\varkappa_d^4 c_d^3 A_1^2 \cos \alpha \cos^2 [\varkappa_d (\sin \alpha \cdot x - \cos \alpha \cdot z - c_d t)]$$
$$= \varkappa_d^4 c_d^3 A_2^2 \cos \alpha \cos^2 [\varkappa_d (\sin \alpha \cdot x + \cos \alpha \cdot z - c_d t)]$$
$$+ \varkappa_s^4 c_s^3 A_4^2 \cos \beta \cos^2 [\varkappa_s (\sin \beta \cdot x + \cos \beta \cdot z - c_s t)]. \quad (3.24a)$$

For our infinitely wide beamed waves each of the terms in (3.24a) must be multiplied by infinity, i.e., by ($\overrightarrow{\infty}$). Hence each of these waves carries infinite energy. Since the waves in (3.24a) are time harmonic in the propagation along their beams, for a fixed station (x, z) they vary only with time t within the period $T = 2\pi/\omega$. We can therefore eliminate the remaining time dependence in (3.24a) by using *time averages over the period T* of the $\cos^2[\;]$ waves appearing there. To show this, first we note for any point along the incident P-wave plane (dashed line in fig. 3.7) that $\sin \alpha \cdot x - \cos \alpha \cdot z = -x_0 \sin \alpha$. Hence the time average of the $\cos^2[\;]$ term on the left hand side of (3.24a), over the period T, denoted by $\langle \cos^2[\;] \rangle$, is given by

$$\langle \cos^2 [\varkappa_d (c_d t + x_0 \sin \alpha)] \rangle = \frac{1}{T} \int_t^{t+T} \cos^2 [\varkappa_d (c_d t' + x_0 \sin \alpha)] dt' = \tfrac{1}{2}.$$
$$(3.24b)$$

Since $\varkappa_d(c_d t + x_0 \sin \alpha) = \omega[t - (-x_0 \sin \alpha)/c_d]$, it should be noted that in (3.24b) we are summing over time, the action of a time harmonic wave from its "arrival time" at the fixed station $(-x_0 \sin \alpha)$, which is anywhere along

Ch. 3, § 3.1]   REFLECTION OF P AND SV WAVES   131

the incident P-wave plane (dashed line in fig. 3.7), to a time T later. In effect this is the action of the incident P wave traveling through the station toward the origin, as indicated in fig. 3.7. Similarly the other waves in (3.24a) give the same result as (3.24b). It follows, after normalizing with respect to its left hand side, that (3.24a) becomes

$$1 = \left(\frac{A_2}{A_1}\right)^2 + \frac{\tan \alpha}{\tan \beta}\left(\frac{A_4}{A_1}\right)^2 = \frac{E_2}{E_1} + \frac{E_4}{E_1}. \tag{3.24c}$$

In addition to setting down the partitioning of the incident P-wave energy, as Redwood [3.3, pg. 35] points out, (3.24c) can be used to calculate one of the reflection coefficients from data on the other, e.g., A_4/A_1 from Arenberg's data on A_2/A_1 in fig. 3.6. Gutenberg [3.4] has calculated the values of $(E_2/E_1)^{\frac{1}{2}}$, and $(E_4/E_1)^{\frac{1}{2}}$ as a function of angle of incidence α for several values of Poisson's ratio ν. His curves are reproduced here in fig. 3.8.

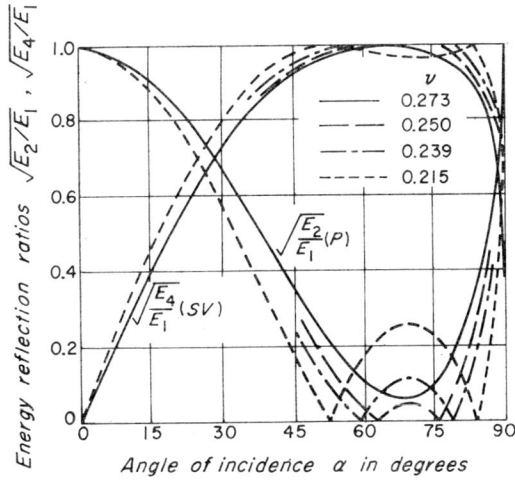

Fig. 3.8. Energy reflection ratios for P-wave incidence (after Gutenberg).

3.1.2.2. Reflection of an SV wave. A similar analysis can be carried out for the incident SV-wave case. The resulting potential reflection coefficients are

$$\begin{aligned}\frac{A_2}{A_3} &= \frac{k^2 \sin 4\beta}{\sin 2\alpha \sin 2\beta + k^2 \cos^2 2\beta}, \\ \frac{A_4}{A_3} &= \frac{\sin 2\alpha \sin 2\beta - k^2 \cos^2 2\beta}{\sin 2\alpha \sin 2\beta + k^2 \cos^2 2\beta},\end{aligned} \tag{3.25}$$

where we note A_4/A_3 is given by the same expression as that for A_2/A_1 in (3.18). Figure 3.9 illustrates the wave system here. It is interesting and

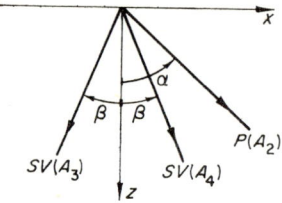

Fig. 3.9. Wave system for SV-wave incidence.

important to note that, in both incident wave cases, we need both types of reflected waves to satisfy the conditions of a stress free surface. This *mode coupling*, between P and SV waves then, is fundamental to this type of boundary surface.

Using (3.20) we can form an analogous set of displacement reflection coefficients for the present case, to those of (3.21) for the incident P wave. We find

$$\frac{U_2}{U_3} = \frac{W_2}{W_3} = \frac{1}{k}\frac{A_2}{A_3},$$
$$\frac{U_4}{U_3} = -\frac{W_4}{W_3} = -\frac{A_4}{A_3}.$$
(3.26)

Again here we could similarly calculate stress reflection coefficients if needed.

Arenberg [3.2] has also presented numerical information on the variation of the SV-wave reflection coefficient A_4/A_3 in (3.26) as a function of the angle of incidence β for various Poisson's ratio values. His results are presented in fig. 3.10. They can be used to calculate U_4/U_3 and W_4/W_3, through the second of (3.26). Prominent features of the curves in fig. 3.10 are (1) the maxima that occur in all, at values of β dependent on ν, (2) that at normal incidence, $\beta=0$, and at $\beta=\pi/4$, the SV wave generates no reflected P wave, in agreement with the first of (3.25), (3) that A_4/A_3 vanishes (hence *total mode conversion* takes place) at two points for each of the curves having the smaller ν values, but not when $\nu=0$, and (4) that there exist *critical incident angles* β, dependent on ν, above which α no longer is a real valued angle according to (3.12a), resulting in *complex reflection coefficients* A_4/A_3 and A_2/A_3. Normal incidence, incidence at $\beta=\pi/4$, total

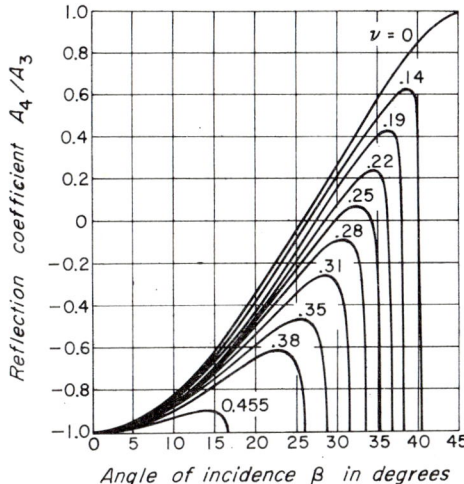

Fig. 3.10. Amplitude of reflected SV-wave as a function of angle of incidence and Poisson's ratio for SV-wave incidence (after Arenberg).

mode conversion, and critical angles of incidence for the SV wave are discussed further in § 3.1.4. Again here we have a singular case for grazing incidence where the motion, according to (3.25), vanishes. The nonvanishing solution for this case is given in § 3.1.4.

The partition of energy in the incident SV wave case can be obtained in the same way as that for the incident P wave just discussed Here we find, for the *energy partition*, the relation

$$1 = \frac{\tan \beta}{\tan \alpha}\left(\frac{A_2}{A_3}\right)^2 + \left(\frac{A_4}{A_3}\right)^2 = \frac{E_2}{E_3} + \frac{E_4}{E_3}. \tag{3.27}$$

Again we point out (3.27) can be used to obtain one reflection coefficient from data on the other, e.g., A_2/A_3 from Arenberg's curves for A_4/A_1 in fig. 3.10. Again here Gutenberg [3.4] has given values for $(E_2/E_3)^{\frac{1}{2}}$ and $(E_4/E_3)^{\frac{1}{2}}$, as a function of angle of incidence β for several values of v. His curves are shown in fig. 3.11.

3.1.3. Elastically restrained boundary. As pointed out by Mindlin [3.5] one can follow the development of coupling between P and SV waves from the *uncoupled*, in the case of mixed boundary conditions, to the *completely coupled* case of the traction free surface, by analyzing the inter-

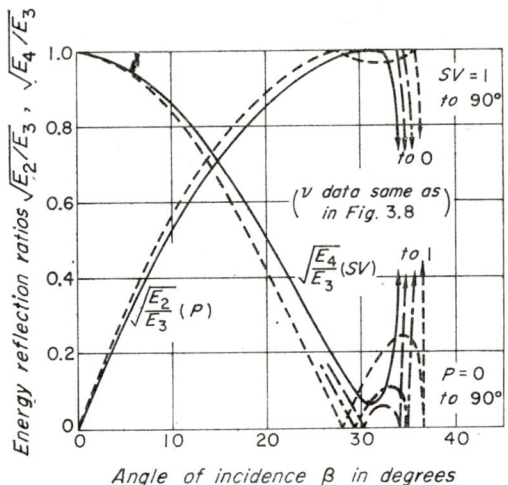

Fig. 3.11. Reflection energy ratios for SV-wave incidence (after Gutenberg).

mediate case of the elastically restrained boundary. The boundary conditions are

$$\sigma_z = -ew, \qquad \sigma_{zx} = 0, \qquad \text{at } z = 0, \qquad (3.28)$$

where e is a parameter (like a spring constant in nature). Parameter e varies from ∞, in which case (3.28) reduce to the mixed conditions ($w = \sigma_{zx} = 0$)

$$\sigma_z = -ew = -\infty \cdot 0 \neq 0, \qquad \sigma_{zx} = 0, \qquad (3.29)[5]$$

to zero, in which case (3.28) reduce to the nonmixed conditions (free surface)

$$\sigma_z = -ew = 0 \cdot w = 0 \quad (w \neq 0), \qquad \sigma_{zx} = 0, \qquad (3.30)$$

For (3.29) the result is $A_4 = -A_3$, $\beta_3 = \beta_4 = \beta$ (see fig. 3.3), and for (3.30) the results are given in equations (3.25) (see fig. 3.9). The reflection coefficients for the general case of $0 < e < \infty$ can be obtained by the process we used in treating the cases in § 3.1.2. The results, given by Mindlin for an incident SV wave, are

---

[5] Note here we have taken w to behave as $c/\infty (c \neq 0)$, and not, say, $c\infty^2$, which would make $\sigma_z = 0$.

Ch. 3, § 3.1] REFLECTION OF P AND SV WAVES

$$\frac{A_2}{A_3} = \frac{1}{1+E_r^2} \cdot \frac{k^2 \sin 4\beta}{\sin 2\alpha \sin 2\beta + k^2 \cos^2 2\beta},$$

$$\frac{A_4}{A_3} = \frac{1}{1+E_r^2} \cdot \frac{\sin 2\alpha \sin 2\beta - k^2 \cos^2 2\beta}{\sin 2\alpha \sin 2\beta + k^2 \cos^2 2\beta} - \frac{E_r^2}{1+E_r^2},$$

(3.31)

where

$$E_r = \frac{ke \cos \alpha}{\mu \varkappa_s (\sin 2\alpha \sin 2\beta + k^2 \cos^2 2\beta)}.$$

Note, through E_r, that (3.31) shows, for a given v, the reflected P-and SV-wave amplitudes are dependent on the incident angle β, as in case of the free boundary, but now also on the stiffness e of the boundary and the wavelength (through $\varkappa_s$) of the incident SV wave. It should be noted in this dependence on e, as it varies from ∞ to 0, the amplitude of the reflected P wave increases from 0, to its value in the first of (3.25), in proportion to $1/(1+E_r^2)$. We will find further use later for the elastically restrained boundary case in our related analysis of elastic waveguides.

3.1.4. Free boundary special cases

As has already been mentioned important special cases arise, in the reflection of P and SV waves from a free boundary, that require further analysis.

3.1.4.1. Normal incidence. Returning to (3.18) and (3.19) for the case of incident P wave, we set $\alpha = 0$, and find $A_2 = -A_1$, $A_4 = 0$, for $A_1 \neq 0$. As shown in fig. 3.12, $P(A_1)$

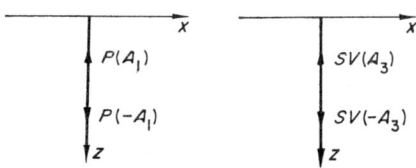

Fig. 3.12. Reflection of P and SV waves at normal incidence.

reflects as the equal in magnitude, but opposite in sign $P(-A_1)$ wave without mode conversion. Likewise for the incident SV wave, (3.25) gives for $\beta = 0$, $A_4 = -A_3$, $A_2 = 0$, for $A_3 \neq 0$, also shown in fig. 3.12.

3.1.4.2. Grazing incidence. Suppose an incident P wave is traveling with its normal parallel to the boundary, i.e., $\alpha = \pi/2$. Equations (3.18), (3.19) then give again

$$A_2 = -A_1, \qquad A_4 = 0, \qquad \text{for } A_1 \neq 0,$$

and using these in (3.2) gives

$$\varphi = \varphi_1 + \varphi_2 = (A_1 - A_1) \exp[i\varkappa_d(x - c_d t)] = 0, \qquad (3.32)$$
$$\psi = \psi_4 = A_4 \psi_4^e = 0,$$

so the wave system vanishes (it is easy to show the displacements and stresses also vanish), and fails to give the result for P-wave grazing incidence. However, Goodier and Bishop [3.6] found the appropriate wave system through a limit procedure. They required A_1 and A_2 to become infinite as α approached $\pi/2$, while the product

$$A_1[(\pi/2) - \alpha] = A_1 \delta \qquad (3.33)$$

remained finite. Now, from (3.12a), we can write $\sin^2[(\pi/2) - \delta] = k^2 \sin^2 \beta$, or

$$\cos^2 \delta = (k^2/2)(1 - \cos 2\beta). \qquad (3.34)$$

Now from (3.34), for small δ, we can write

$$\cos 2\beta = \frac{k^2 - 2}{k^2}\left[1 + \frac{2}{k^2 - 2}\delta^2 + O(\delta^4)\right], \qquad (3.35)$$
$$\sin 2\beta = (1 - \cos^2 2\beta)^{\frac{1}{2}} = \frac{2(k^2-1)^{\frac{1}{2}}}{k^2}\left[1 - \frac{(k^2-2)}{2(k^2-1)}\delta^2 + O(\delta^4)\right].$$

Also

$$\sin 2\alpha = \sin(\pi - 2\delta) = 2\delta + O(\delta^3). \qquad (3.36)$$

Substituting (3.35) and (3.36) in (3.18), (3.19) keeping terms to $O(\delta)$, we find that

$$A_2 \simeq \left[-1 + \frac{8(k^2-1)^{\frac{1}{2}}}{(k^2-2)^2}\delta\right]A_1,$$
$$A_4 \simeq -\frac{4\delta}{k^2 - 2}A_1. \qquad (3.37)$$

Now we also have $\sin \alpha = 1 + O(\delta^2)$, and $\cos \alpha = \delta + O(\delta^3)$, and through (3.12), $\sin \beta = (1/k) + O(\delta^2)$, and $\cos \beta = [(k^2-1)^{\frac{1}{2}}/k] + O(\delta^2)$, so that (3.2), keeping terms to $O(\delta)$, are

$$\varphi = \varphi_1 + \varphi_2 = A_1 \exp\left[i\varkappa_d(x-\delta z - c_d t)\right] + A_2 \exp\left[i\varkappa_d(x+\delta z - c_d t)\right],$$

$$\psi = \psi_4 = A_4 \exp\left\{i\varkappa_s\left[\frac{x}{k} + \frac{(k^2-1)^{\frac{1}{2}}z}{k} - c_s t\right]\right\}.$$
(3.38)

Substituting (3.37) in (3.38), again keeping terms to $O(\delta)$ [from the expansion of $\exp(\pm ik_d\delta z)$] in the latter, and letting $2A_1\delta = A_0$, we find the solution to be

$$\varphi = A_1' \exp\left[i\varkappa_d(x-c_d t)\right] + A_2' i\varkappa_d z \exp\left[i\varkappa_d(x-c_d t)\right],$$

$$\psi = A_4' \exp\left\{i\varkappa_s\left[\frac{x}{k} + \frac{(k^2-1)^{\frac{1}{2}}z}{k} - c_s t\right]\right\},$$
(3.39)

where

$$A_1' = \frac{4(k^2-1)^{\frac{1}{2}}}{(k^2-2)^2} A_0, \qquad A_2' = -A_0, \qquad \text{and} \qquad A_4' = -\frac{2}{k^2-2} A_0,$$

so that the reflection coefficients are

$$\frac{A_2'}{A_1'} = -\frac{(k^2-2)^2}{4(k^2-1)^{\frac{1}{2}}}, \qquad \frac{A_4'}{A_1'} = -\frac{k^2-2}{2(k^2-1)^{\frac{1}{2}}}.$$
(3.40)

The system of waves in (3.39) are, first comprising φ, an incident P wave of the ordinary type, and a reflected P wave, both traveling in the positive x-direction. The amplitude of the latter is, however, z dependent. Hence, this wave is not plane. The reflected SV wave is of the ordinary type. A sketch is given in fig. 3.13.

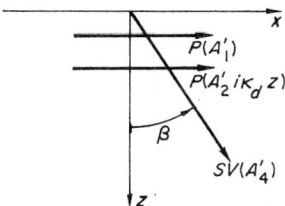

Fig. 3.13. Wave system for grazing incidence of P wave.

At grazing incidence the SV wave generates a similar system of waves. Substituting $\beta = \pi/2$ into (3.25), we find $A_4 = -A_3$, and $A_2 = 0$, provided $A_3 \neq 0$, and hence from (3.2), $\varphi = \psi = 0$, as in the incident P wave case just treated. Goodier and Bishop treated this case too (in a like manner), and found the system of waves

$$\psi = A'_3 \exp\left[i\varkappa_s(x-c_s t)\right] + A'_4 \varkappa_s z \exp\left[i\varkappa_s(x-c_s t)\right],$$
$$\varphi = iA'_2 \exp\left[-(k^2-1)^{\frac{1}{2}}\varkappa_d z\right] \exp\left[i\varkappa_s(x-c_s t)\right],$$
(3.41)

where

$$A'_3 = \frac{4(k^2-1)^{\frac{1}{2}}}{k} A'_0, \qquad A'_4 = -A'_0, \qquad A'_2 = 2A'_0,$$

so that the reflection coefficients are

$$\frac{A'_4}{A'_3} = -\frac{k}{4(k^2-1)^{\frac{1}{2}}}, \qquad \frac{A'_2}{A'_3} = \frac{k}{2(k^2-1)^{\frac{1}{2}}}.$$
(3.42)

The system of waves in (3.41) are, first for ψ, an incident SV wave of the ordinary type, and a reflected SV wave, both traveling in the positive x-direction. The amplitude of the latter is like its analog in the incident P-wave case, i.e., z dependent, and hence the wave is not plane. The reflected P wave also travels along x positive and, note, has the propagation speed c_s. It is a surface wave, its amplitude decaying into the interior of the half space $z>0$, and hence it is not a plane wave either. Fig. 3.14 is a sketch of

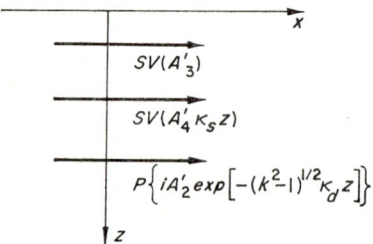

Fig. 3.14. Wave system for grazing incidence of SV wave:

this system. In [3.6], [3.5] some further examples are given of grazing incidence created by reflections and combinations of the wave systems treated here. Grazing incidence plays a basic role in our understanding of wave propagation involving one or more boundaries, and we shall see further evidence of it in our later work.

3.1.4.3. SV wave incident at $\beta = \pi/4$. If we substitute $\beta = \pi/4$ in (3.25), we find $A_2 = 0$, and $A_4 = A_3$, provided $A_3 \neq 0$. Hence, in this case, the SV wave reflects in like sign (or phase), and undergoes no mode conversion. Figure 3.15 demonstrates this.

Ch. 3, § 3.1] REFLECTION OF P AND SV WAVES 139

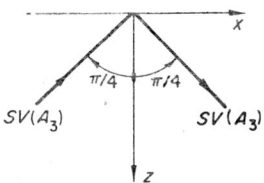

Fig. 3.15. Reflection of SV wave, incident at $\beta = \pi/4$.

3.1.4.4. Total mode conversion. From the solution for the case of P-wave incidence, (3.18), (3.19), we see when

$$\sin 2\alpha \sin 2\beta - k^2 \cos^2 2\beta = 0, \qquad (3.43)$$

that $A_2 = 0$, provided $A_1 \neq 0$. Further, substituting (3.43) in (3.19), we find

$$A_4/A_1 = -\cot 2\beta. \qquad (3.44)$$

For the incident SV-wave case, imposing (3.43) on the second of (3.25) makes $A_4 = 0$, provided $A_3 \neq 0$. Then substituting (3.43) into the first of (3.25), we find

$$A_2/A_3 = \tan 2\beta. \qquad (3.45)$$

We therefore can solve (3.43) and (3.12a) simultaneously to get a range of pairs of α and v, for which an incident P wave converts totally to an SV wave, and a range of pairs of β and v, for which an incident SV wave converts totally to a P wave, the reflection coefficients being given by (3.44) and (3.45), respectively. Figures 3.16 and 3.17 exhibit this. Arenberg [3.2]

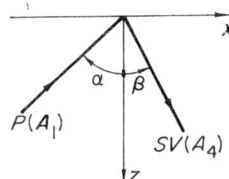 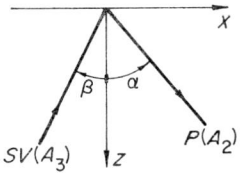

Fig. 3.16. Total mode conversion of a P wave.   Fig. 3.17. Total mode conversion of a SV wave.

has computed the ranges of the pairs α, v and β, v. They are shown in fig. 3.18.

3.1.4.5. Reflection of SV waves at critical angles of incidence; total reflection. Consider (3.12a) again. Since $k > 1$, we have a real value of α as long as

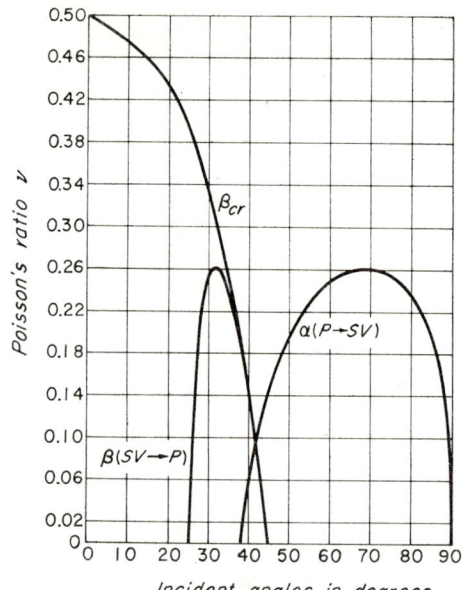

Fig. 3.18. Pairs α, ν and β, ν for total mode conversion, and β_{cr}, ν for total reflection of an incident SV wave (after Arenberg).

$0 \leq \sin \beta \leq 1/k$ or equivalently, $0 \leq \beta \leq \sin^{-1}(1/k) = \sin^{-1}(c_s/c_d)$. Hence there is *no ordinary reflected P wave* in the range

$$\beta_{cr} < \beta < \pi/2, \tag{3.46}$$

where the *critical angle* $\beta_{cr} = \sin^{-1}(1/k)$. For example, with $\nu = 1/4$, $\sin^{-1}(1/\sqrt{3}) < \beta < \pi/2$. *What happens to the motion governed by* (3.2), *when* (3.46) *holds and there is no ordinary reflected P wave?* Assuming for the moment an ordinary reflected P wave, φ in (3.2) can be written as

$$\varphi = \varphi_2 = A_2 \exp\left[i\varkappa(x + \cot \alpha \cdot z - ct)\right]. \tag{3.47}$$

Now from (3.12a)

$$\sin^2 \alpha = k^2 \sin^2 \beta, \tag{3.48}$$

and from this, we find

$$k^2 \cot^2 \alpha = \cot^2 \beta - (k^2 - 1). \tag{3.49}$$

We see from (3.49), when β is in the range defined by

$$0 < \cot^2 \beta < k^2 - 1, \tag{3.50}$$

Ch. 3, § 3.1]   REFLECTION OF P AND SV WAVES   141

which can be shown to be equivalent to (3.46), that $\cot^2\alpha < 0$, showing $\cot\alpha$ is imaginary. From (3.46), and the second of (3.12c), it can be proved that $c_s < c < c_d$ (where c depends on the angle of incidence), hence, within the critical range (3.50) (and the corresponding range for α, through (3.12a)), $\cot\alpha$ and $\cot\beta$ must be taken as

$$\cot\alpha = (\csc^2\alpha - 1)^{\frac{1}{2}} = i[1 - (c/c_d)^2]^{\frac{1}{2}},$$
$$\cot\beta = (\csc^2\beta - 1)^{\frac{1}{2}} = [(c/c_s)^2 - 1]^{\frac{1}{2}}.$$
(3.51)

It follows, using the first of (3.51), that (3.47) becomes

$$\varphi = \varphi_2 = A_2 \exp\left\{-[1-(c/c_d)^2]^{\frac{1}{2}}\varkappa z\right\} \exp\left[i\varkappa(x-ct)\right],$$
(3.52)

which is a surface wave traveling in the positive x-direction, with an amplitude that decays exponentially into the interior $z>0$ (note the latter dictated the positive imaginary form of the first of (3.51)). Note that (3.52) is a case of *grazing reflection*. For the ψ waves we have the ordinary forms given by the second of (3.2), since β is given by (3.46). To calculate the reflection coefficients for the system we first rewrite (3.25) in terms of $\cot\alpha$ and $\cot\beta$, through certain trigonometric identities and (3.48). They are

$$\frac{A_2}{A_3} = \frac{4\cot\beta \cdot [k^2(1+\cot^2\alpha)-2]}{4\cot\alpha\cot\beta + [k^2(1+\cot^2\alpha)-2]^2},$$

$$\frac{A_4}{A_3} = \frac{4\cot\alpha\cot\beta - [k^2(1+\cot^2\alpha)-2]^2}{4\cot\alpha\cot\beta + [k^2(1+\cot^2\alpha)-2]^2},$$
(3.53)

which for the present case, using (3.51), yield

$$A_2/A_3 = \Phi_2 \exp(i\theta_2),$$
$$A_4/A_3 = -\exp(2i\theta_2),$$
(3.54)

where

$$\Phi_2 = \left|\frac{A_2}{A_3}\right| = 4s\left\{\frac{[k^2(1-r^2)-2]^2}{[k^2(1-r^2)-2]^4 + 16r^2s^2}\right\}^{\frac{1}{2}},$$

$$\theta_2 = \tan^{-1}\left\{-\frac{4rs}{[k^2(1-r^2)-2]^2}\right\},$$

and

$$r = [1-(c/c_d)^2]^{\frac{1}{2}}, \qquad s = [(c/c_s)^2-1]^{\frac{1}{2}}.$$

We see from the first of (3.54) that the amplitude A_2 of the reflected P wave is complex. From the second of (3.54) we see the SV wave reflects with equal amplitude but undergoes a change in phase of $\pi + 2\theta_2$. This is a case of "*total reflection*", which has an analog in the reflection and refraction of waves in optics. Figure 3.19 depicts these waves. Arenberg [3.2] gives a curve for the v, β_{cr} pairs, which is shown in our fig. 3.18.

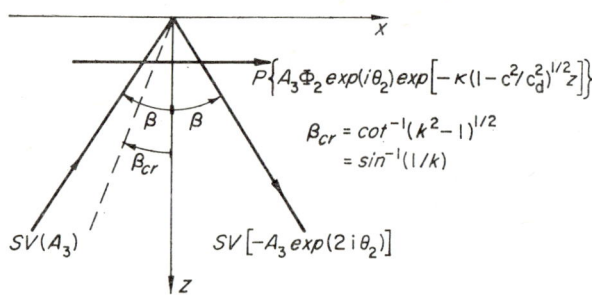

Fig. 3.19. Total reflection of an incident SV wave.

Important in total reflection is an examination of the energy being propagated. In the critical region of β, (3.46), the energy relation (3.27) does not hold, since the latter is based on real angles α. We therefore return to the energy $\hat{E} = \varrho \dot{u}_i \dot{u}_i$, and calculate an energy flux balance similar to (3.27), but now for our special solution (3.54). As we pointed out earlier our waves are of infinite beam width, and it is appropriate here to use this property. On this basis we first write the conservation law for beams of incremental width, i.e.,

$$\varrho[\dot{u}_3^2 + \dot{w}_3^2]c_s \cos\beta \, ds_3 = \varrho[\dot{u}_4^2 + \dot{w}_4^2]c_s \cos\beta \, ds_4 + \varrho[\dot{u}_2^2 + \dot{w}_2^2]c_d \, dz \,, \quad (3.55)$$

where ds_3 and ds_4 are, respectively, the increments of beam widths for the incident and reflected SV waves, and dz that for the reflected P wave in the present solution. As in our development of (3.24) we again use (3.7) and (3.2), but now also (3.52) and (3.54), to expand (3.55) with the result

$$\varrho \varkappa_s^4 c_s^3 A_3^2 \cos^2[\varkappa_s(\sin\beta \cdot x - \cos\beta \cdot z - c_s t)] ds_3$$

$$= \varrho \varkappa_s^4 c_s^3 A_3^2 \cos^2[\varkappa_s(\sin\beta \cdot x + \cos\beta \cdot z - c_s t) + 2\theta_2] ds_4$$

$$+ \varrho \varkappa_d^4 c_d^3 |A_2|^2 \cos^2[\varkappa(x - ct) + \theta_2] \exp\{-2[1 - (c/c_d)^2]^{\frac{1}{2}} \varkappa z\} dz \,.$$

Integrating each of these terms over the beam width (from zero to infinity),

Ch. 3, § 3.1]   REFLECTION OF P AND SV WAVES   143

and again for their time average according to the form (3.24b), yields the energy flux balance

$$\rho\varkappa_s^4 c_s^3 A_3^2(\vec{\infty}) = \rho\varkappa_s^4 c_s^3 A_3^2(\overset{\leftarrow}{\infty}) + \rho\varkappa_d^4 c_d^3 |A_2|^2/2\varkappa[1-(c/c_d)^2]^{\frac{1}{2}}, \quad (3.56)$$

where we recall $(\vec{\infty})$ indicates the limiting beam width. Equation (3.56) points out the important fact that the grazing reflected P wave has finite energy, which propagates parallel to the surface in the direction of x-positive. However, it also shows it can be neglected with respect to the limiting infinite energy in the incident and reflected SV waves, with the results that the incident SV-wave energy all goes into the reflected SV wave. One can interpret this approximately (and more physically) as incident and reflected SV waves of very wide beam widths, and still neglect the P-wave energy. In his treatment of the analogous problem in optics Sommerfeld [3.7] notes similar energy in the grazing reflected wave.

Observing (3.49), we note it is satisfied by

$$\cot\beta = (k^2-1)^{\frac{1}{2}}, \qquad \cot\alpha = 0,$$

corresponding to incidence at β_{cr}. We see from the first of (3.51) the second equation here makes $c = c_d$, and hence r and θ_2 vanish in (3.54). It follows (3.52), with (3.54), yields for this case

$$\varphi = \varphi_2 = \frac{4(k^2-1)^{\frac{1}{2}}}{k^2-2} A_3 \exp\left[i\varkappa_d(x-c_d t)\right],$$

and for the SV-wave amplitude, $A_4 = -A_3$. This wave system is shown in fig. 3.20. For the present case the energy balance (3.56) breaks down, since

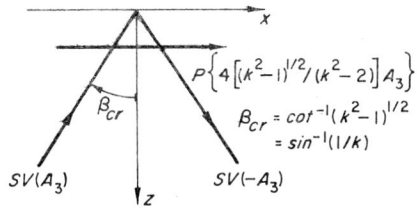

Fig. 3.20. Wave system for SV incidence at β_{cr}.

it is based on $\cot\alpha$ imaginary, and now $\cot\alpha$ is real. It would appear, therefore, that use can be made of (3.27) here. If we rewrite (3.27) in its form before normalization, now taking into account infinitely wide beams, we find

$$\varrho \varkappa_s^4 c_s^3 A_3^2 \cos \beta \cdot (\vec{\infty}) = \varrho \varkappa_s^4 c_s^3 A_4^2 \cos \beta \cdot (\vec{\infty}) + \varrho \varkappa_d^4 c_d^3 A_2^2 \cos \alpha \cdot (\vec{\infty}).$$

Since $\cos \alpha \to 0$ here, the limit of $\cos \alpha \cdot (\vec{\infty})$ by l'Hospital's rule is one. Hence again the P wave has finite energy and we have agreement with (3.56).

It has already been pointed out in our treatment of grazing incidence that, as Goodier and Bishop have shown [3.6], certain of these cases can, through reflection in the z-axis, yield other cases. An important example of this is the solution for P-wave grazing incidence, given here by (3.39) and (3.40), and shown in fig. 3.13. If we reflect the system in fig. 3.13 in the z-axis, and reverse the directions of propagation, we have the system shown in fig. 3.21. Important is the fact that, as the second of (3.39) shows, we have here the case of an incident SV wave at the critical angle β_{cr} [$\beta_{cr} = \cot^{-1}(k^2-1)^{\frac{1}{2}} = \sin^{-1}(1/k)$], for which the system shown in fig. 3.20 was just found. Hence, as Goodier and Bishop noted, and as the comparison

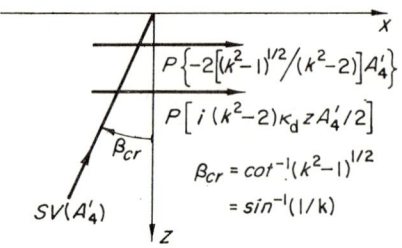

Fig. 3.21. A second wave system for SV incidence at β_{cr}.

of figs. 3.20 and 3.21 shows, there is non-uniqueness of the reflection of an SV wave at this critical angle (since we have not imposed initial conditions or a condition at $z \to \infty$ here, this should not be surprising).

In closing this section it is perhaps worthwhile to point out that, whereas $\beta = \pi/4$ can fall into the critical range for SV-wave incidence, this special angle, treated in § 3.1.4.3 is not critical, since no reflected P wave is involved, i.e., the reflection law (3.12a) is not involved. Substitution of $\beta = \pi/4$ into (3.53) or (3.54) will show that it agrees with the results found in § 3.1.4.3.

3.1.4.6. Reflection of wave pairs. Mindlin [3.5] discusses the general case of reflection of pairs of waves. It is assumed that both the P and SV waves are incident at the half-space boundary. For the free boundary surface, from (3.8), (3.16) and (3.17), we have

$$(A_1+A_2)\cos 2\beta - (A_3-A_4)\sin 2\beta = 0,$$
$$(A_1-A_2)\sin 2\alpha + (A_3+A_4)k^2 \cos 2\beta = 0. \tag{3.57}$$

The solution of (3.57) is, therefore,

$$A_2 = A_1, \qquad A_4 = -A_3,$$
$$A_1/A_3 = \tan 2\beta, \tag{3.58}$$

or

$$A_2 = -A_1, \qquad A_4 = A_3,$$
$$A_1/A_3 = -k^2 \cos 2\beta/\sin 2\alpha. \tag{3.59}$$

Therefore a pair of P and SV waves incident at a free surface, that satisfy $\omega = \varkappa_d c_d = \varkappa_s c_s$, i.e., (3.3), and $\varkappa = \varkappa_d \sin \alpha = \varkappa_s \sin \beta$, i.e., (3.12a), and whose amplitudes satisfy the second of either (3.58) or (3.59), reflect there in their likeness with one of them undergoing a phase change. Figure 3.22 shows

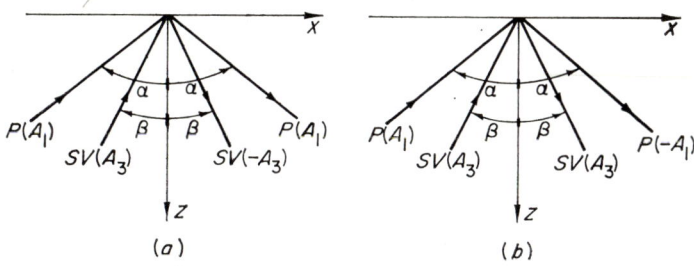

Fig. 3.22. Reflection of wave pairs.

the systems, (a) corresponding to (3.58), and (b) to (3.59). Mindlin points out that at grazing incidence the pairs of waves reduce to combinations of those given by Goodier and Bishop, that we have discussed earlier. He also discusses normal incidence, pointing out that setting $\alpha = 0$, or $\beta = 0$, simple reflections, of the type given in fig. 3.12 here, result. More interesting, however, are the results he obtains for normal incidence, through a limiting process similar to that used by Goodier and Bishop for grazing incidence. Here, however, Mindlin finds waves having linear variation of amplitude, not with depth z from the boundary, but with x along it.

3.1.4.7. Rayleigh surface waves. Rayleigh's discovery in 1887 of his now well-known surface waves demonstrated a fundamental difference between the elastic half space and infinite medium [3.8]. To develop the theory of these waves we first assume that the wave pair φ_1 and ψ_3 of (3.2) are incident at the boundary. These can be written as

$$\varphi = A \exp\left[i\varkappa(x - \cot\alpha \cdot z - ct)\right],$$
$$\psi = B \exp\left[i\varkappa(x - \cot\beta \cdot z - ct)\right], \tag{3.60}$$

where we have let $A_1 = A$, and $A_3 = B$, and have dropped the subscripts on φ and ψ, and where

$$\cot\alpha = \pm\left[(c/c_d)^2 - 1\right]^{\frac{1}{2}}.$$
$$\cot\beta = \pm\left[(c/c_s)^2 - 1\right]^{\frac{1}{2}}. \tag{3.61}$$

For (3.60) to be surface waves, i.e., decaying into the interior $z > 0$, it is necessary to choose the negative imaginary roots from (3.61). That is, we must have $c < c_d$, $c < c_s$, and since $c_d > c_s$, the speed of the Rayleigh surface wave $c = c_R$, must satisfy $c_R < c_s < c_d$. Under these conditions (3.60) become the grazing incidence pair

$$\varphi = A \exp(-rz) \exp\left[i\varkappa(x - c_R t)\right],$$
$$\psi = B \exp(-sz) \exp\left[i\varkappa(x - c_R t)\right], \tag{3.62}$$

where

$$r = \varkappa\left[1 - (c_R/c_d)^2\right]^{\frac{1}{2}}, \qquad s = \varkappa\left[1 - (c_R/c_s)^2\right]^{\frac{1}{2}},$$

are real and positive. Noting that $r = i\varkappa \cot\alpha$, and $s = i\varkappa \cot\beta$, since $c = c_R$, and that we have $\varkappa = \varkappa_d \sin\alpha = \varkappa_s \sin\beta$, we substitute these relations, and (3.62), directly into (3.8) with the results that the stresses corresponding to (3.62) are

$$\sigma_z = \{[(\lambda + 2\mu)r^2 - \lambda\varkappa^2]A \exp(-rz) - i2\mu s\varkappa B \exp(-sz)\} \exp[i\varkappa(x - c_R t)],$$
$$\sigma_{zx} = \mu[-i2r\varkappa A \exp(-rz) - (\varkappa^2 + s^2)B \exp(-sz)] \exp[i\varkappa(x - c_R t)]. \tag{3.63}$$

Now substituting (3.63) into the boundary conditions (3.14), we obtain

$$[(\lambda+2\mu)r^2-\lambda\varkappa^2]A-i2\mu\varkappa sB=0,$$
$$-i2\varkappa rA-(\varkappa^2+s^2)B=0.$$
(3.64)

For a non-trivial solution the determinant of (3.64) must vanish, and this gives

$$(2-k_R^2)^2=4(1-b^2k_R^2)^{\frac{1}{2}}(1-k_R^2)^{\frac{1}{2}},$$ (3.65)

where $k_R^2=c_R^2/c_s^2$, and $b^2=1/k^2$. If we can find a k_R (or c_R) that satisfies this relation, and the surface wave forms of (3.62), this will insure that such waves exist. Squaring (3.65), we find

$$k_R^2[k_R^6-8k_R^4+8k_R^2(3-2b^2)-16(1-b^2)]=0.$$ (3.66)

Now $k_R=0$ means $c_R=0$, for $c_s \neq 0$. Hence, (3.62) becomes time independent. Further from (3.64) we find $A=iB$, since with $c_R=0$, $r=s=\varkappa$. Now from (3.7) we find the corresponding displacements to be

$$u=\varkappa(iA+B)\exp(-\varkappa z)\exp(i\varkappa x),$$
$$w=\varkappa(-A+iB)\exp(-\varkappa z)\exp(i\varkappa x).$$
(3.67)

Clearly then, $A=iB$ makes these displacements vanish, hence they are of no interest. If we set $k_R=1$, i.e., $c_R=c_s$, (3.66) is not satisfied since its left hand side is then equal to 1. Hence, our root, if it exists, must lie in the range

$$0<k_R<1,$$ (3.68)

which corresponds to the restriction $0<c_R<c_s<c_d$, consistent with that imposed earlier. Our problem reduces to finding one real root k_R^2 of the cubic, given in the brackets of (3.66), subject to (3.68), and showing that it satisfies (3.65), whereas the other two roots do not. Hayes and Rivlin [3.9] recently discussed the roots of this cubic equation and their implications. As they point out Somigliana showed there are three real roots k_R^2 for the cubic, provided $v<v_c=0.2637...$, corresponding to $0.3209<b^2<1$. Jeffreys pointed out for this case that only one of these roots satisfied (3.68), the other two roots being greater than one. Now if $v_c \leq v \leq \frac{1}{2}$, corresponding to $0 \leq b^2 \leq 0.3209...$, two of the roots are complex. The one real root here satisfies (3.68). We can state, therefore, that a real and positive k_R exists, satisfying (3.66) and (3.68), for the range $0 \leq v \leq \frac{1}{2}$. Knopoff [3.10] has computed k_R for this range. His results are shown in fig. 3.23.

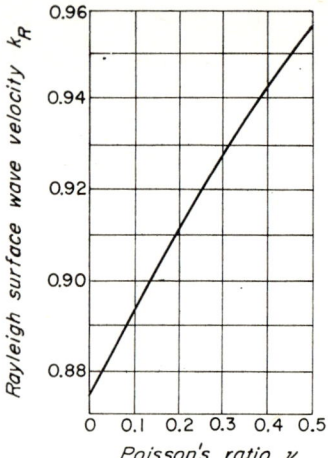

Fig. 3.23. Rayleigh surface wave velocity k_R as a function of Poisson's ratio. (after Knopoff).

On the basis of this information it is not difficult to show that the positive real root k_R of the cubic in (3.66), satisfying (3.68), satisfies (3.65). The other two roots are inadmissible because they violate the real and positive requirement on the radicals that appear in r and s in (3.62), which also appear in (3.65). Consider the range $0 \leq \nu \leq \nu_c$, where the three roots k_{R1}, k_{R2}, and k_{R3}, are such that $0 < k_{R1} < 1$, and $k_{R2} > 1$, $k_{R3} > 1$, e.g., for $\nu = \frac{1}{4}$, $k_{R1} = [2-(2/\sqrt{3})]^{\frac{1}{2}}$, $k_{R2} = [2+(2/\sqrt{3})]^{\frac{1}{2}}$, and $k_{R3} = 2$. In this example direct substitution of k_{R1} into (3.65), will show it satisfies this equation. Since k_{R2} and k_{R3} are greater than one they are inadmissible, i.e., either renders the radical $(1-k_R^2)^{\frac{1}{2}}$ imaginary. For the range $\nu_c \leq \nu \leq \frac{1}{2}$, direct substitution shows the real root satisfying (3.68) satisfies (3.65). The two other roots are complex and hence inadmissible for (3.62). Finally, it is of interest to point out that Achenbach [2.14 § 5.11] has shown, with the aid of the argument principle in complex variables, that (3.65) has only one root, the real k_R satisfying (3.68), in agreement with the discussion here.

Ewing et al. [3.11 § 2-2] give the displacements for the case of $\nu = \frac{1}{4}$ as

$$u = D[\exp(-0.8475\varkappa z) - 0.5773 \exp(-0.3933\varkappa z)] \sin[\varkappa(c_R t - x)], \quad (3.69)$$

$$w = D[-0.8475 \exp(-0.8475\varkappa z) + 1.4678 \exp(-0.3933\varkappa z)] \cos[\varkappa(c_R t - x)],$$

where D is a function of $\varkappa$. They point out that the corresponding particle motion is elliptical retrograde in contrast to the elliptical direct orbit for surface waves on water. On the surface $z=0$ the displacements (3.69) reduce to

$$u = 0.433 D \sin \left[\varkappa(c_R t - x) \right],$$
$$w = 0.62 D \cos \left[\varkappa(c_R t - x) \right], \tag{3.70}$$

so that the ellipse in the u and w is given by

$$\frac{u^2}{(0.433D)^2} + \frac{w^2}{(0.62D)^2} = 1. \tag{3.71}$$

Fig. 3.24 is a sketch of (3.70), or (3.71), for fixed $x_0 = c_R t_0$, showing the

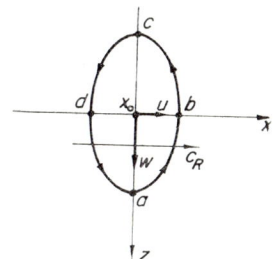

Fig. 3.24. Rayleigh surface wave particle motion.

retrograde nature of the orbit of the surface particle at $(x_0, 0)$. Assuming the motion begins at a, at $t_0 = x_0/c_R$, it progresses counterclockwise through b, c, d and back to a in a complete period. Since (3.70) represent infinite wave trains propagating to the right with speed c_R, all stations x undergo this periodic motion, indefinitely with time. Note that at a and c, $u=0$, with w maxima, and at b and d, $w=0$, with u maxima. From (3.70), and the figure, it can be seen that the maximum amplitude of the vertical displacement is about one and a half times that of the horizontal displacement. In [3.11] it is pointed out that the horizontal displacement vanishes at $z = 0.192 L$, where L is the wavelength $(2\pi/\varkappa)$, changing its sign below this. Viktorov [3.12, pg. 5] presents complete numerical information on the amplitudes of the displacements and stresses, as a function of depth z for $v=0.25$, and 0.34. His curves are shown here in figs. 3.25, 3.26. The displacement curves are plots of the amplitude ratios $\hat{u}/\hat{w}_0$, and $\hat{w}/\hat{w}_0$ versus z/L, where $(\hat{\ })$ indicates the amplitude of the particular displacement, $\hat{w}_0$

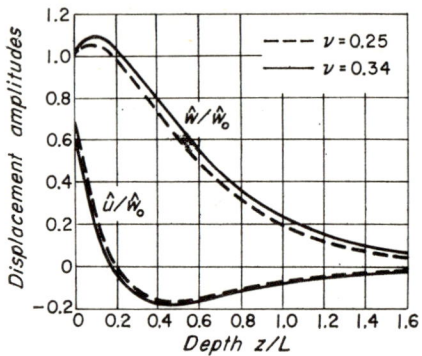

Fig. 3.25. Displacement amplitudes as function of depth (after Viktorov).

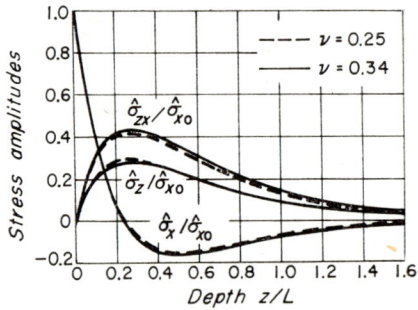

Fig. 3.26. Stress amplitudes as function of depth (after Viktorov).

being the surface vertical displacement. Similarly the stress curves are plots of the amplitude ratios $\hat{\sigma}_x/\hat{\sigma}_{x0}$, $\hat{\sigma}_{zx}/\hat{\sigma}_{x0}$, and $\hat{\sigma}_z/\hat{\sigma}_{x0}$ versus z/L, where $\hat{\sigma}_{x0}$ is $\hat{\sigma}_x$ at the surface. Viktorov's figures show the localization of the motion to a thin layer near the surface of thickness approximately twice the wavelength of the propagating surface waves. It may be noted that the curve $\hat{u}/\hat{w}_0$ in fig. 3.25 exhibits the vanishing of this displacement, and change in sign at $z \simeq 0.2L$, as we mentioned was pointed out in [3.11]. With this change in sign the particle motion would become the reverse of that displayed in fig. 3.24. In [3.11] it is further pointed out the experimental work of Dobrin et al. [3.13] gives good support to the foregoing theory. They determined particle motion with depth below the earth's surface for Rayleigh waves from small explosions. They found the motion is retrograde above 40 feet and direct below. The crossover point was $0.136L$, and the displacements decreased continuously below 40 feet. Interesting is the fact that the experiments were conducted in a region of layered, unconsolidated and semiconsolidated rocks. In spite of the nonhomogeneity the agreement with the theory was good.

Knowles [3.14] studied surface waves of a more general nature than those developed in this section. He considers all possible motions of the half space for which φ and ψ_k have the forms

$$\varphi = \varphi(x_1, x_2) \exp(-ax_3 + i\omega t),$$

$$\psi_k = \psi_k(x_1, x_2) \exp(-bx_3 + i\omega t),$$
(3.72)

where a and b are complex constants with positive real parts, and $\omega = \varkappa c$ is the given frequency of the waves. Assuming that φ and ψ_k satisfy equations (2.7b), and the requirement of vanishing shear stress $\sigma_{\alpha 3} = 0$ ($\alpha = 1, 2$) at $x_3 = 0$, it follows that φ and ψ_3 can be expressed in terms of ψ_1, ψ_2 and c, through the relations

$$\varphi = -\frac{1}{\varkappa}\left(1 - \frac{c^2}{c_d^2}\right)^{-\frac{1}{2}}\left(1 - \frac{c^2}{2c_s^2}\right)\varepsilon_{\alpha\beta}\frac{\partial\psi_\alpha}{\partial x_\beta},$$

$$\psi_3 = \frac{1}{\varkappa}\left(1 - \frac{c^2}{c_s^2}\right)^{\frac{1}{2}}\frac{\partial\psi_\beta}{\partial x_\beta},$$
(3.73)

where

$$a = \varkappa\left(1 - \frac{c^2}{c_d^2}\right)^{\frac{1}{2}}, \qquad b = \varkappa\left(1 - \frac{c^2}{c_s^2}\right)^{\frac{1}{2}},$$
(3.74)

and $\varepsilon_{\alpha\beta}$ is defined by $\varepsilon_{11} = \varepsilon_{12} = 0$, $\varepsilon_{12} = -\varepsilon_{21} = 1$. Further ψ_1 and ψ_2 must satisfy the Helmholtz equation

$$\nabla^2 \psi_\alpha + \varkappa^2 \psi_\alpha = 0,$$

where $\nabla^2 \psi_\alpha = \partial^2/\partial x_1^2 + \partial^2/\partial x_2^2$. The remaining condition of vanishing normal stress $\sigma_{33} = 0$, at $x_3 = 0$, shows that c must satisfy (3.65), i.e., $c = c_R$, and we note (3.68) must be satisfied as discussed earlier. We note that correspondingly, a and b here are respectively r and s in (3.62). Knowles investigates the displacements u_i corresponding to (3.72)–(3.73), and shows that the path of a particle, originally at x_1, x_2, x_3, lies in a plane in the Cartesian space u_i, the orientation of which changes from point to point in the elastic half space. The particle path in this plane is an ellipse. Some special solutions are discussed in [3.14], among which is the case we have treated earlier. It is shown for that case the plane of motion is normal to the boundary of the half space, which agrees with our earlier discussion.

Later in our study of transient excitation of the half space we shall see that the Rayleigh surface wave disturbance predominates at places remote from the source. Indeed, in two dimensional problems, the far-field surface disturbance is found to be the nondecaying, nonchanging, in space and time, Rayleigh wave. This behavior, plus the localization to the surface, has led to strong practical interest in ultrasonic Rayleigh waves. Good examples are their use in the detection of surface and near-surface defects in materials, and in delay lines. They are also used to study properties (mechanical, thermal) of a surface layer of material through measurements of their velocity and attenuation. Viktorov's book [3.12] has its theme in such applications.

3.1.5. Rigid boundary

Another case of interest is the rigid boundary where we have

$$u = w = 0. \tag{3.75}$$

It has a strong analogy to the free boundary and can be worked out similarly. It is left to the exercises.

3.2. Reflection of SH waves from the boundary of an elastic half space

Again assuming the Cartesian half space, and the corresponding displacements set down in § 3.1, the displacements and conditions for the *SH* wave are

$$u_1 = u = 0, \quad u_3 = w = 0,$$
$$u_2 = v = v(x, z, t), \quad \partial/\partial x_2 = \partial/\partial y = 0. \tag{3.76}[6]$$

---

[6] This motion in which v, the displacement normal to the xy plane, is the only nonvanishing one has led to the *alternate name antiplane shear deformation* in contrast to the *in-plane* nature of plane strain deformation. Equations (3.83) show the corresponding antiplane shear strains ε_{xy}, ε_{zy} are the only nonvanishing strains.

Ch. 3, § 3.2] REFLECTION OF *SH* WAVES 153

With arguments similar to those that produced the plane strain displacement-potential relations (3.1), we find using (2.26) and (3.76) that

$$u = u_1 = \frac{\partial \varphi}{\partial x_1} - \frac{\partial \psi_2}{\partial x_3} = 0,$$

$$w = u_3 = \frac{\partial \varphi}{\partial x_3} + \frac{\partial \psi_2}{\partial x_1} = 0, \qquad (3.77)$$

$$v = u_2 = \frac{\partial \psi_1}{\partial x_3} - \frac{\partial \psi_3}{\partial x_1},$$

and that it is sufficient to set $\varphi = \psi_2 = 0$ to satisfy the first two equations here. The two existing potential components ψ_1 and ψ_3, that define v in the third of (3.77), reflect the fact that shearing action on the planes $x_1 =$ constant, and $x_3 =$ constant both contribute to v. These two potentials are related, however, and can be reduced to one. To show this we return to (2.18) and select the gauge condition

$$\frac{\partial \psi_k}{\partial x_k} = \frac{\partial \psi_1}{\partial x_1} + \frac{\partial \psi_2}{\partial x_2} + \frac{\partial \psi_3}{\partial x_3} = 0, \qquad (3.78)$$

that satisfies it. Since $\psi_2 = 0$, (3.78) reduces to

$$\frac{\partial \psi_1}{\partial x_1} = -\frac{\partial \psi_3}{\partial x_3}. \qquad (3.79)$$

Now if we let

$$\psi_1 = \frac{\partial \chi}{\partial x_3}, \qquad \psi_3 = -\frac{\partial \chi}{\partial x_1}, \qquad (3.80)$$

where $\chi(x_1, x_3)$ is still another potential, we see by substituting these into (3.79) the latter is satisfied, i.e., (3.80) establishes the relation between ψ_1 and ψ_3. Now, from (2.7b) and (3.80), we have

$$\begin{pmatrix} \dfrac{\partial}{\partial x_3} \\ \dfrac{\partial}{\partial x_1} \end{pmatrix} (\nabla^2 \chi - \ddot{\chi}/c_s^2) = \qquad (3.81)$$

i.e., both ψ_1 and ψ_3 satisfy (2.7b), provided χ satisfies the same wave equation.

It follows that the totality of the plane harmonic waves in the present case is contained in

$$\chi = \chi_1 + \chi_2 = B_1 \chi_1^e + B_2 \chi_2^e, \tag{3.82}$$

where χ_1 and χ_2 are the incident and reflected waves, respectively, B_1 and B_2 their amplitudes, and

$$\chi_1^e = \exp\left[i\varkappa_s(l_1 x - n_1 z - c_s t)\right],$$

$$\chi_2^e = \exp\left[i\varkappa_s(l_2 x + n_2 z - c_s t)\right],$$

where $l_1 = \sin \gamma_1$, $n_1 = \cos \gamma_1$, etc. The system of waves is shown in fig. 3.27. Now since (3.76) holds here, from (1.25), (1.39) we find

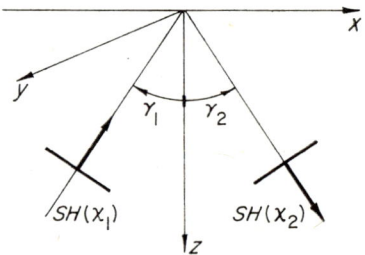

Fig. 3.27. Incident and reflected SH waves.

$$\varepsilon_x = \varepsilon_y = \varepsilon_z = \varepsilon_{zx} = \Delta = 0,$$

$$\varepsilon_{xy} = \frac{\partial v}{\partial x}, \qquad \varepsilon_{zy} = \frac{\partial v}{\partial z}, \tag{3.83}$$

where again we have made use of Love's notation (cf. remarks following (3.4)). The displacement-potential relation for v is obtained by substituting (3.80) into the third of (3.77), with the result

$$v = \frac{\partial^2 \chi}{\partial x_1^2} + \frac{\partial^2 \chi}{\partial x_3^2} = \nabla^2 \chi. \tag{3.84}$$

The strain-potential relations, following from (3.83), using (3.84), are

$$\varepsilon_{xy} = \frac{\partial \nabla^2 \chi}{\partial x}, \qquad \varepsilon_{zy} = \frac{\partial \nabla^2 \chi}{\partial z}. \tag{3.85}$$

Now from (1.45), using (3.85), (3.81) we have the stress-potential relations

Ch. 3, § 3.2]   REFLECTION OF SH WAVES   155

$$\sigma_{xy} = \frac{\mu}{c_s^2} \frac{\partial \ddot{\chi}}{\partial x}, \qquad \sigma_{zy} = \frac{\mu}{c_s^2} \frac{\partial \ddot{\chi}}{\partial z}. \qquad (3.86)$$

It follows using (3.84) (3.81), that the waves in v, corresponding to (3.82), are

$$v = -\varkappa_s^2(B_1\chi_1^e + B_2\chi_2^e), \qquad (3.87)$$

and from (3.86), those in the stresses are

$$\sigma_{xy} = -i\mu\varkappa_s^3(l_1 B_1\chi_1^e + l_2 B_2\chi_2^e),$$
$$\sigma_{zy} = i\mu\varkappa_s^3(n_1 B_1\chi_1^e - n_2 B_2\chi_2^e). \qquad (3.88)$$

We can now consider reflection of these SH waves from the boundary $z=0$.

3.2.1. Free boundary

In this case

$$\sigma_{zy} = 0, \qquad \text{at } z=0. \qquad (3.89)$$

Substitution of the second of (3.88) into (3.89) gives

$$n_1 B_1 \exp\left[i\varkappa_s(l_1 x - c_s t)\right] = n_2 B_2 \exp\left[i\varkappa_s(l_2 x - c_s t)\right],$$

which is satisfied only if $\gamma_1 = \gamma_2 = \gamma$, and $B_2 = B_1$ for any γ, as shown in fig. 3.28. So this is like the mixed boundary cases treated in § 3.1.1, i.e.,

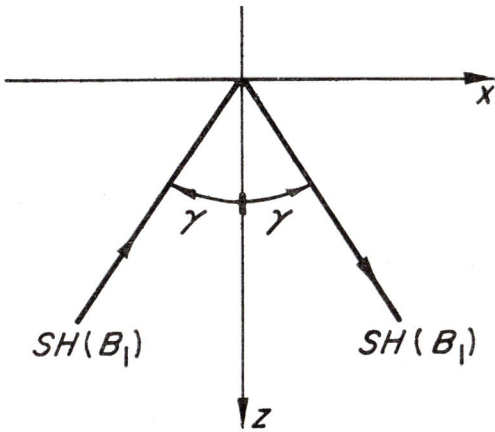

Fig. 3.28. Reflection of an SH wave from a free boundary.

a *P*-wave reflection from a lubricated-rigid boundary, or an *SV* wave from an in-plane constrained boundary. In fact the present case can also be looked at as a mixed conditions case, since $w=0$ is automatically satisfied here. In any event it is interesting that these waves can reflect without generating any mode coupling.

3.2.2. Rigid boundary

In this case

$$v = 0, \qquad \text{at } z = 0. \tag{3.90}$$

Substituting (3.87) into (3.90), we find

$$B_1 \exp\left[i\varkappa_s(l_1 x - c_s t)\right] = -B_2 \exp\left[i\varkappa_s(l_2 x - c_s t)\right],$$

where again this is satisfied only if $\gamma_1 = \gamma_2 = \gamma$, but now with $B_2 = -B_1$ for any γ. Similarly, as in § 3.2.1, this case can be looked at as a mixed conditions case, since $\sigma_z = 0$ is automatically satisfied here. And we find that this case is the analog of the other two reflection cases treated in § 3.1.1, namely, the *SV*-wave reflection from a lubricated-rigid boundary, and the *P*-wave reflection from the in-plane constrained boundary.

It should be pointed out in closing our look at *SH* waves, that we could have done this analysis more simply by drawing directly on v, which is governed by the wave equation (2.1). It is instructive, however, to see these waves in the context of potentials, which has been our theme throughout the work in this chapter. It should be noted that the displacement and stress reflection coefficients are easily calculated from B_2/B_1 with the aid of (3.87), (3.88).

3.3. Reflection and refraction of *P* and *SV* waves at an interface

3.3.1. The solid-solid interface; general solution

Consider the incident wave pair $P(\varphi_1)$, $SV(\psi_3)$, shown in fig. 3.1, but now incident at the interface between two elastic half spaces of different material properties. The reflected waves $P(\varphi_2)$, $SV(\psi_4)$ again occur, and now a new third pair, the refracted $P(\varphi')$, and $SV(\psi')$, traveling in the half space $z<0$ of properties λ', μ', and ϱ'. The situation is depicted in fig. 3.29,

Ch. 3, § 3.3] REFLECTION AND REFRACTION OF P AND SV WAVES   157

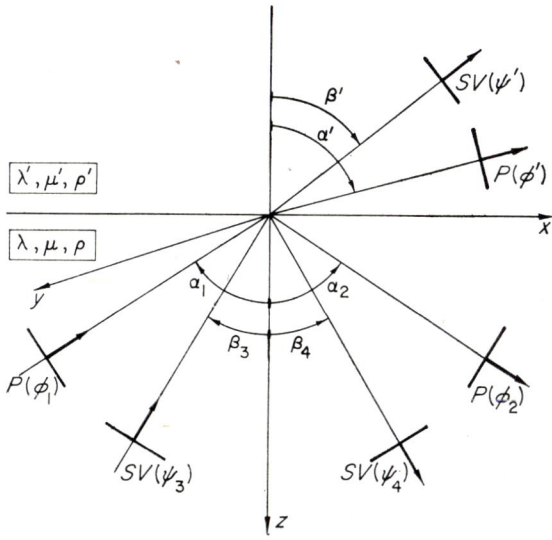

Fig. 3.29. Incident, reflected and refracted P and SV waves.

where coordinates are as they were in fig. 3.1. Equations (3.1)–(3.8) are applicable here for the domain $z>0$, and, in addition, for $z<0$ we have the refracted waves

$$\varphi' = A_{\varphi'} \varphi'^e,$$
$$\psi' = A_{\psi'} \psi'^e, \tag{3.91}$$

where $A_{\varphi'}$ and $A_{\psi'}$ are wave amplitudes, and

$$\varphi'^e = \exp\left[i\varkappa'_d(l_{\alpha'}x - n_{\alpha'}z - c'_d t)\right],$$
$$\psi'^e = \exp\left[i\varkappa'_s(l_{\beta'}x - n_{\beta'}z - c'_s t)\right],$$

where $\varkappa'_d = \omega/c'_d$ and $\varkappa'_s = \omega/c'_s$, hence, $\omega = \varkappa'_d c'_d = \varkappa'_s c'_s$, and α' and β' are the angles of refraction for the φ' and ψ' waves, respectively, measured from the z-axis. It follows the direction cosines in (3.91) are $l_{\alpha'} = \sin \alpha'$, $n_{\alpha'} = \cos \alpha'$, $l_{\beta'} = \sin \beta'$, and $n_{\beta'} = \cos \beta'$.

We assume our half spaces are in welded contact at $z=0$, hence the displacements and stresses must be continuous there, that is,

158 WAVE REFLECTION AND REFRACTION AT INTERFACE [Ch. 3, § 3.3

$$u = u', \qquad w = w',$$
$$\sigma_z = \sigma'_z, \qquad \sigma_{zx} = \sigma'_{zx}. \tag{3.92}$$

Substituting (3.7), (3.8), and the "sub 1 and 3" terms in these equations, written for (3.91), into (3.92), gives

$$\varkappa_d\{l_1 A_1 \exp[i\varkappa_d(l_1 x - c_d t)] + l_2 A_2 \exp[i\varkappa_d(l_2 x - c_d t)]\}$$
$$+ \varkappa_s\{n_3 A_3 \exp[i\varkappa_s(l_3 x - c_s t)] - n_4 A_4 \exp[i\varkappa_s(l_4 x - c_s t)]\}$$
$$= \varkappa'_d l_{\alpha'} A_{\varphi'} \exp[i\varkappa'_d(l_{\alpha'} x - c'_d t)] + \varkappa'_s n_{\beta'} A_{\psi'} \exp[i\varkappa'_s(l_{\beta'} x - c'_s t)],$$

$$-\varkappa_d\{n_1 A_1 \exp[i\varkappa_d(l_1 x - c_d t)] - n_2 A_2 \exp[i\varkappa_d(l_2 x - c_d t)]\}$$
$$+ \varkappa_s\{l_3 A_3 \exp[i\varkappa_s(l_3 x - c_s t)] + l_4 A_4 \exp[i\varkappa_s(l_4 x - c_s t)]\}$$
$$= -\varkappa'_d n_{\alpha'} A_{\varphi'} \exp[i\varkappa'_d(l_{\alpha'} x - c'_d t)] + \varkappa'_s l_{\beta'} A_{\psi'} \exp[i\varkappa'_s(l_{\beta'} x - c'_s t)], \tag{3.93}$$

$$-\varkappa_d^2\{(\lambda + 2\mu n_1^2)A_1 \exp[i\varkappa_d(l_1 x - c_d t)] + (\lambda + 2\mu n_2^2)A_2 \exp[i\varkappa_d(l_2 x - c_d t)]\}$$
$$+ 2\mu\varkappa_s^2\{l_3 n_3 A_3 \exp[i\varkappa_s(l_3 x - c_s t)] - l_4 n_4 A_4 \exp[i\varkappa_s(l_4 x - c_s t)]\}$$
$$= -\varkappa'^2_d(\lambda' + 2\mu' n_{\alpha'}^2)A_{\varphi'} \exp[i\varkappa'_d(l_{\alpha'} x - c'_d t)] + 2\mu'\varkappa'^2_s l_{\beta'} n_{\beta'} A_{\psi'} \exp[i\varkappa'_s(l_{\beta'} x - c'_s t)],$$

$$\mu\{2\varkappa_d^2[l_1 n_1 A_1 \exp[i\varkappa_d(l_1 x - c_d t)] - l_2 n_2 A_2 \exp[i\varkappa_d(l_2 x - c_d t)]]$$
$$- \varkappa_s^2[(l_3^2 - n_3^2)A_3 \exp[i\varkappa_s(l_3 x - c_s t)] + (l_4^2 - n_4^2)A_4 \exp[i\varkappa_s(l_4 x - c_s t)]]\}$$
$$= \mu'\{2\varkappa'^2_d l_{\alpha'} n_{\alpha'} A_{\varphi'} \exp[i\varkappa'_d(l_{\alpha'} x - c'_d t)] - \varkappa'^2_s(l_{\beta'}^2 - n_{\beta'}^2)A_{\psi'} \exp[i\varkappa'_s(l_{\beta'} x - c'_s t)]\}.$$

To obtain solutions of (3.93), for the reflection and transmission (refraction) coefficients, independent of x and t, we must first require that

$$\alpha_1 = \alpha_2 = \alpha, \qquad \beta_1 = \beta_2 = \beta,$$
$$\varkappa = \varkappa_d \sin \alpha = \varkappa_s \sin \beta = \varkappa'_d \sin \alpha' = \varkappa'_s \sin \beta', \tag{3.94a}$$

noting again, that because of a common frequency in all these waves, $\omega = \varkappa_d c_d = \varkappa_s c_s = \varkappa'_d c'_d = \varkappa'_s c'_s = \varkappa c$. The second of (3.94a) contains the refraction alws

Ch. 3, § 3.3] REFLECTION AND REFRACTION OF P AND SV WAVES

$$c = c_d/\sin\alpha = c_s/\sin\beta = c'_d/\sin\alpha' = c'_s/\sin\beta', \tag{3.94b}$$

analogous to (3.12a, c). Under these conditions (3.93) reduce to

$$A_1 + A_2 + \cot\beta \cdot (A_3 - A_4) = A_{\varphi'} + \cot\beta' \cdot A_{\psi'},$$

$$\cot\alpha \cdot (A_1 - A_2) - (A_3 + A_4) = \cot\alpha' \cdot A_{\varphi'} - A_{\psi'}, \tag{3.95}$$

$$\mu[(\cot^2\beta - 1)(A_1 + A_2) - 2\cot\beta \cdot (A_3 - A_4)]$$

$$= \mu'[(\cot^2\beta' - 1)A_{\varphi'} - 2\cot\beta' \cdot A_{\psi'}],$$

$$\mu[2\cot\alpha \cdot (A_1 - A_2) + (\cot^2\beta - 1)(A_3 + A_4)]$$

$$= \mu'[2\cot\alpha' \cdot A_{\varphi'} + (\cot^2\beta' - 1) \cdot A_{\psi'}],$$

where

$$\cot\alpha = [(c/c_d)^2 - 1]^{\frac{1}{2}}, \qquad \cot\beta = [(c/c_s)^2 - 1]^{\frac{1}{2}},$$

$$\cot\alpha' = [(c/c'_d)^2 - 1]^{\frac{1}{2}}, \qquad \cot\beta' = [(c/c'_s)^2 - 1]^{\frac{1}{2}}.$$

By specializing to a single incident wave, of one type or the other, (3.95) can be solved for the four reflection and transmission coefficients. To derive such solutions, first we let

$$A_1 + A_2 = P, \qquad A_3 - A_4 = Q,$$

$$A_1 - A_2 = R, \qquad A_3 + A_4 = S, \tag{3.96}$$

as in the treatment by Ewing et al [3.11 § 3-1]. Now one can solve the first and third of (3.95) for P and Q, and the second and fourth of (3.95) for R and S, obtaining

$$P = p_1 A_{\varphi'} + q_1 A_{\psi'}, \qquad Q = p_2 A_{\varphi'} + q_2 A_{\psi'},$$

$$R = p_3 A_{\varphi'} + q_3 A_{\psi'}, \qquad S = p_4 A_{\varphi'} + q_4 A_{\psi'}, \tag{3.97}$$

where

$$p_1 = e/d, \quad q_1 = f\cot\beta'/d, \quad p_2 = g/d\cot\beta, \quad q_2 = h\cot\beta'/d\cot\beta,$$

$$p_3 = h\cot\alpha'/d\cot\alpha, \quad q_3 = -g/d\cot\alpha, \quad p_4 = -f\cot\alpha'/d, \quad q_4 = p_1,$$

and

$$d = \mu(b+2), \quad e = 2\mu + \mu'b', \quad f = 2(\mu - \mu'), \quad g = \mu b - \mu'b', \quad h = 2\mu' + \mu b,$$

where
$$b = \cot^2 \beta - 1, \qquad b' = \cot^2 \beta' - 1.$$

We assume now just the $P(\varphi_1)$ wave is incident at the interface. Hence setting $A_3 = 0$ in (3.96), and substituting these equations in (3.97), the latter can be written as

$$-\frac{A_2}{A_1} + p_1 \frac{A_{\varphi'}}{A_1} + q_1 \frac{A_{\psi'}}{A_1} = 1,$$

$$\frac{A_4}{A_1} + p_2 \frac{A_{\varphi'}}{A_1} + q_2 \frac{A_{\psi'}}{A_1} = 0,$$

$$\frac{A_2}{A_1} + p_3 \frac{A_{\varphi'}}{A_1} + q_3 \frac{A_{\psi'}}{A_1} = 1,$$

$$-\frac{A_4}{A_1} + p_4 \frac{A_{\varphi'}}{A_1} + q_4 \frac{A_{\psi'}}{A_1} = 0.$$

(3.98)

Equations (3.98) are easily solved if it is first noted that addition eliminates the reflection coefficients A_2/A_1 and A_4/A_1. The results for the potential reflection coefficients are

$$A_2/A_1 = [(p_1 - p_3)(q_2 + q_4) - (p_2 + p_4)(q_1 - q_3)]/D,$$

$$A_4/A_1 = 2(p_4 q_2 - p_2 q_4)/D,$$

(3.99a)

and for the potential transmission coefficients

$$A_{\varphi'}/A_1 = 2(q_2 + q_4)/D,$$

$$A_{\psi'}/A_1 = -2(p_2 + p_4)/D,$$

(3.99b)

where $D = (p_1 + p_3)(q_2 + q_4) - (p_2 + p_4)(q_1 + q_3)$. The wave system we have found for the present case is shown in fig. 3.30.

Ch. 3, § 3.3] REFLECTION AND REFRACTION OF P AND SV WAVES

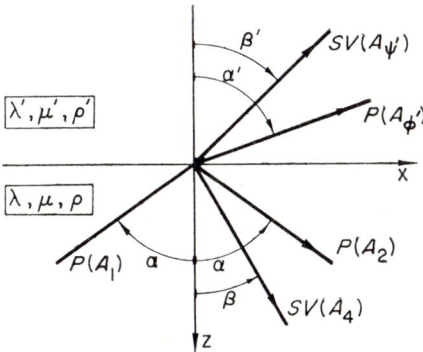

Fig. 3.30. Wave system for P-wave incidence at solid-solid interface.

Assuming now that just the $SV(\psi_3)$ wave is incident at the interface, we set $A_1 = 0$ in (3.96), and substitute these equations in (3.97), which produces a set of equations similar to (3.98). Solution of these equations gives for the potential reflection and transmission coefficients

$$\begin{aligned}
A_2/A_3 &= 2(p_3 q_1 - p_1 q_3)/D \,, \\
A_4/A_3 &= [(p_2 - p_4)(q_1 + q_3) - (p_1 + p_3)(q_2 - q_4)]/D \,, \\
A_{\varphi'}/A_3 &= -2(q_1 + q_3)/D \,, \\
A_{\psi'}/A_3 &= 2(p_1 + p_3)/D \,.
\end{aligned} \quad (3.100)$$

The wave system is the same as that in fig. 3.30, except that the $P(A_1)$ wave, at incident angle α, is replaced by the $SV(A_3)$ wave, at incident angle β. Again using (3.7) and (3.8), and (3.91) here, reflection and transmission coefficients for the displacements and stresses, in terms of those for the potentials in (3.99), (3.100), can be derived [cf. equations (3.20), (3.21)]. Muskat and Meres [3.15] have given extensive tables on the numerical evaluation of the coefficients in (3.99), (3.100), as functions of the incident angles α and β, for several values of the velocity ratio c'_d/c_d and density ratio ϱ'/ϱ, for $\nu = \tfrac{1}{4}$. Expressions here for the partition of energy can be carried out by the same method discussed in § 3.1.2.1. For an incident P wave the partition of energy is governed by

$$1 = \left(\frac{A_2}{A_1}\right)^2 + \frac{\tan \alpha}{\tan \beta}\left(\frac{A_4}{A_1}\right)^2 + \frac{\varrho' \tan \alpha}{\varrho \tan \alpha'}\left(\frac{A_{\varphi'}}{A_1}\right)^2 + \frac{\varrho' \tan \alpha}{\varrho \tan \beta'}\left(\frac{A_{\psi'}}{A_1}\right)^2$$

$$= \frac{E_2}{E_1} + \frac{E_4}{E_1} + \frac{E_{\varphi'}}{E_1} + \frac{E_{\psi'}}{E_1} \,, \quad (3.101a)$$

and for an incident SV wave by

$$1 = \frac{\tan \beta}{\tan \alpha}\left(\frac{A_2}{A_3}\right)^2 + \left(\frac{A_4}{A_3}\right)^2 + \frac{\varrho' \tan \beta}{\varrho \tan \alpha'}\left(\frac{A_{\varphi'}}{A_3}\right)^2 + \frac{\varrho' \tan \beta}{\varrho \tan \beta'}\left(\frac{A_{\psi'}}{A_3}\right)^2$$

$$= \frac{E_2}{E_3} + \frac{E_4}{E_3} + \frac{E_{\varphi'}}{E_3} + \frac{E_{\psi'}}{E_3}. \tag{3.101b}$$

In [3.4] Gutenberg has also given extensive values for the square roots of each of the energy ratios, appearing in (3.101), as a function of incident angle α or β. In each case Gutenberg treats both the incident wave when it occurs in the lower (unprimed material) and upper half space (primed material). Further, each of these are carried out for four different numerical cases of the combination of parameters ϱ/ϱ', c_d/c'_d, c_s/c'_s, v', and v. Gutenberg's curves appear also in [3.11].

3.3.2. Solid-solid interface; special cases

3.3.2.1. Normal incidence. From (3.99) we can derive the coefficients for normal incidence of the P wave. Letting $\alpha \to 0$, it follows from (3.94b) that $c = c_d/\sin \alpha = c_s/\sin \beta = c'_d/\sin \alpha' = c'_s/\sin \beta' \to \infty$. Hence, $\cot \alpha \to c/c_d \to \infty$, $\cot \beta \to c/c_s \to \infty$, $\cot \alpha' \to c/c'_d \to \infty$, and $\cot \beta' \to c/c'_s \to \infty$. It is then easy to show that (3.99) reduce to

$$\frac{A_2}{A_1} = \frac{\varrho' c'_d - \varrho c_d}{\varrho' c'_d + \varrho c_d}, \qquad \frac{A_4}{A_1} = 0,$$

$$\frac{A_{\varphi'}}{A_1} = \frac{2\varrho c'_d}{\varrho' c'_d + \varrho c_d}, \qquad \frac{A_{\psi'}}{A_1} = 0. \tag{3.102}$$

The corresponding nonvanishing displacement coefficients are

$$\frac{W_2}{W_1} = -\frac{A_2}{A_1}, \qquad \frac{W_{\varphi'}}{W_1} = \frac{c_d A_{\varphi'}}{c'_d A_1} = \frac{2\varrho c_d}{\varrho' c'_d + \varrho c_d}, \tag{3.103}$$

the corresponding u components of the displacement being zero in the present case. We see from the first of (3.102), or (3.103), that no wave is reflected ($A_2 = 0$) when $\varrho' c'_d = \varrho c_d$, i.e., the mechanical impedances of the two materials are equal. One would expect this since in effect this condition eliminates the interface. Similarly from (3.100) we can derive the coefficients for normal incidence of an SV wave. Letting $\beta \to 0$ we find

$$\frac{A_4}{A_3} = \frac{\varrho' c'_s - \varrho c_s}{\varrho' c'_s + \varrho c_s}, \qquad \frac{A_2}{A_3} = 0,$$

$$\frac{A_{\psi'}}{A_3} = \frac{2\varrho c'_s}{\varrho' c'_s + \varrho c_s}, \qquad \frac{A_{\varphi'}}{A_3} = 0. \tag{3.104}$$

The corresponding nonvanishing displacement coefficients are

$$\frac{U_4}{U_3} = -\frac{A_4}{A_3}, \qquad \frac{U_{\psi'}}{U_3} = \frac{2\varrho c_s}{\varrho' c_s' + \varrho c_s}, \qquad (3.105)$$

the corresponding w components of the displacement being zero in this case. As might be expected (3.104) and (3.105) are similar in form to those of (3.102) and (3.103).

3.3.2.2. Grazing incidence. To find the coefficients for grazing incidence of a P wave, we let $\alpha \to \pi/2$. We find then, from (3.94b), that $c = c_d = c_s/\sin \beta = c_d'/\sin \alpha' = c_s'/\sin \beta'$. Hence, $\cot \alpha = 0$, $\cot \beta = [(c_d/c_s)^2 - 1]^{\frac{1}{2}}$, $\cot \alpha' = [(c_d/c_d')^2 - 1]^{\frac{1}{2}}$, and $\cot \beta' = [(c_d/c_s')^2 - 1]^{\frac{1}{2}}$. It follows that (3.99) reduce to

$$A_2/A_1 = -1, \qquad A_4 = A_{\varphi'} = A_{\psi'} = 0. \qquad (3.106)$$

Using (3.106) in (3.2) and (3.91), shows the motion vanishes. Equations (3.106) are therefore the analog of (3.32) for the grazing incidence of a P wave on the free surface of an elastic half space.

Similarly for the grazing incidence of an SV wave we let $\beta \to \pi/2$, and find, from (3.94b), that $c = c_s = c_d/\sin \alpha = c_d'/\sin \alpha' = c_s'/\sin \beta'$. Hence, $\cot \beta = 0$, $\cot \alpha = [(c_s/c_d)^2 - 1]^{\frac{1}{2}}$, $\cot \alpha' = [(c_s/c_d')^2 - 1]^{\frac{1}{2}}$, and $\cot \beta' = [(c_s/c_s')^2 - 1]^{\frac{1}{2}}$. It follows that (3.100) reduce to

$$A_4/A_3 = -1, \qquad A_2 = A_{\varphi'} = A_{\psi'} = 0. \qquad (3.107)$$

It follows from (3.106) and (3.107) that one would use the method given in [3.6], presented in § 3.1.4.2, to derive special solutions for P- and SV-wave grazing incidence in the present case of a two-layered solid. In § 3.3.2.4 we shall treat Stoneley interface waves. There we shall see they are composed of two wave pairs, one in grazing incidence, and the other in grazing refraction.

3.3.2.3. Reflection at critical angles of incidence; total-reflection. In § 3.1.4.5 we analyzed the special case of reflection of an SV wave at critical angles of incidence, at the free surface of a half space. As we noted there, when the angles of incidence β were greater than the limiting critical angle β_{cr}, i.e., $\beta_{cr} < \beta < \pi/2$, the reflected P wave traveled along the boundary in grazing reflection, its amplitude decaying exponentially into the interior $z > 0$. Further, we noted that the incident SV wave experienced total reflection, i.e., the reflected SV wave had an amplitude equal to that of the incident wave, undergoing only a phase change. More general phenomena of this type occur at an interface for both P and SV waves.

For the present solid-solid interface we can show that the solution (3.100), for incidence of an SV wave, contains the case of total reflection. Assume our two solids are such that $c_s < c_d < c_s' < c_d'$, an easily realized physical model, in which the primed solid is much stiffer than the unprimed. From (3.94b) we see we have not only the reflection law (3.12a), and its critical range (3.46), but two refraction laws, $\sin \beta' = (c_s'/c_s) \sin \beta$, and $\sin \alpha' = (c_d'/c_s) \sin \beta$, and their respective critical ranges, $\beta_{crs'} < \beta < \pi/2$ and $\beta_{crd'} < \beta < \pi/2$, where $\beta_{crs'} = \sin^{-1}(c_s/c_s')$, and $\beta_{crd'} = \sin^{-1}(c_s/c_d')$. It follows, if we let $c_s < c < c_d < c_s' < c_d'$ (c depending on the angle of incidence), our cotangent functions in (3.95), for the present case, are given by (3.51), and

$$\cot \alpha' = i[1 - (c/c_d')^2]^{\frac{1}{2}},$$
$$\cot \beta' = i[1 - (c/c_s')^2]^{\frac{1}{2}}.$$
(3.108)

Substitution of the first of (3.51) into (3.47), and (3.108) into (3.91), gives for the reflected P wave, (3.52) again (valid for $z > 0$), and for the refracted P and SV waves, respectively,

$$\varphi' = A_{\varphi'} \exp \{[1 - (c/c_d')^2]^{\frac{1}{2}} \varkappa z\} \exp [i\varkappa(x - ct)],$$
$$\psi' = A_{\psi'} \exp \{[1 - (c/c_s')^2]^{\frac{1}{2}} \varkappa z\} \exp [i\varkappa(x - ct)],$$
(3.109)

valid for $z < 0$. For ψ waves we have the ordinary forms in (3.2), since β is real in all three critical ranges involved here. Hence, we see from (3.52) and (3.109), φ, φ', and ψ' decay exponentially away from the interface $z = 0$, φ exhibiting grazing reflection, and φ' and ψ', grazing refraction. One would expect, therefore, total reflection of the incident SV wave. The following calculation of the reflection and transmission (refraction) coefficient proves this. Substituting the cotangent functions (3.51) and (3.108), into (3.100), we find

$$\frac{A_2}{A_3} = \left| \frac{A_2}{A_3} \right| \exp(i\theta_2), \quad \frac{A_4}{A_3} = \exp(2i\theta_2),$$
$$\frac{A_{\varphi'}}{A_3} = \left| \frac{A_{\varphi'}}{A_3} \right| \exp(i\theta_2), \quad \frac{A_{\psi'}}{A_3} = \left| \frac{A_{\psi'}}{A_3} \right| \exp[i(\theta_2 - \pi/2)],$$
(3.110)

where

$$\left| \frac{A_2}{A_3} \right| = \frac{2|\gamma_3 \delta_1 - \gamma_1 \delta_3|}{(E^2 + F^2)^{\frac{1}{2}}}, \quad \left| \frac{A_{\varphi'}}{A_3} \right| = \frac{2|\delta_1 + \delta_3|}{(E^2 + F^2)^{\frac{1}{2}}},$$
$$\left| \frac{A_{\psi'}}{A_3} \right| = \frac{2|\gamma_1 + \gamma_3|}{(E^2 + F^2)^{\frac{1}{2}}}, \quad \theta_2 = \tan^{-1} \frac{E}{F},$$

where
$$E = \delta_4(\gamma_1 + \gamma_3) + \gamma_4(\delta_1 + \delta_3),$$
$$F = \delta_2(\gamma_1 + \gamma_3) - \gamma_2(\delta_1 + \delta_3),$$

where
$$\gamma_1 = \{2\mu - \mu'[2 - (c/c_s')^2]\}/\varrho c^2, \qquad \gamma_2 = \eta/[(c/c_s)^2 - 1]^{\frac{1}{2}},$$
$$\gamma_3 = \tau[1 - (c/c_d')^2]^{\frac{1}{2}}/[1 - (c/c_d)^2]^{\frac{1}{2}}, \qquad \gamma_4 = 2(\mu' - \mu)[1 - (c/c_d')^2]/\varrho c^2,$$
$$\delta_1 = 2(\mu - \mu')[1 - (c/c_s')^2]^{\frac{1}{2}}/\varrho c^2, \qquad \delta_2 = \tau[1 - (c/c_s)^2]^{\frac{1}{2}}/[(c/c_s)^2 - 1]^{\frac{1}{2}},$$
$$\delta_3 = \eta/[1 - (c/c_d)^2]^{\frac{1}{2}}, \qquad \delta_4 = \gamma_1,$$

and where
$$\eta = \{\mu'[2 - (c/c_s')^2] - \mu[2 - (c/c_s)^2]\}/\varrho c^2,$$

and $\tau = \{2\mu' - \mu[2 - (c/c_s)^2]\}/\varrho c^2$. It can be observed, from the A_4/A_3 coefficient in (3.110), that we have total reflection. Further, comparing the reflection coefficients in (3.110) with those in the half space solution (3.54), we see the present case is the direct analog of the half space problem. It is interesting to note the effects of total reflection that are exhibited in Gutenberg's curves, i.e., the absence of real values beyond the critical angle for the square roots of the energy ratios corresponding to the refracted P and SV waves, and the reflected P wave [note this in figs. 3-15c, and 3-16f, g in ref. 3.11].

Energy considerations like those made in § 3.1.4.5 could be made for the present solution. One would expect, from our analysis there, to find the refracted P and SV waves, and the reflected P wave, in the present case, to carry finite energy, and the incident and reflected SV waves infinite energy. It should also be pointed out that reflection at critical angles do not necessarily have to produce total reflection. One can have cases in both P and SV wave incidence where more than one regular reflected or refracted wave occurs (but with complex coefficients involving a phase change). Then the energy divides itself between these waves. A case of this type has been left to the exercises.

3.3.2.4. Stoneley interface waves. Love [3.17], in the course of an investigation on the effect of a surface layer on the propagation of Rayleigh waves, discovered another wave of the same type. For short wavelengths, compared to the layer thickness, Love found a modified Rayleigh wave with velocity dependent on the properties of both media. Its existence required that the shear wave velocities in the two media were nearly equal. Love focused on the disturbance confined to the neighborhood of the surface. Later, Stoneley [3.18] showed that this generalized Rayleigh wave had a motion which was greatest in the neighborhood of the interface, and existed under more relaxed conditions. The wave is now referred to as the *Stoneley wave*. To develop the theory of the Stoneley wave we use basically the same approach as we did for the Rayleigh wave in § 3.1.4.7. Here we again assume the wave pair $P(\varphi_1)$, $SV(\psi_3)$ incident at the interface, but that in addition we have the refracted wave pair $P(\varphi')$, $SV(\psi')$ there [cf. fig. 3.29]. This system of waves can be written as

$$\begin{aligned}\varphi_1 &= A_1 \exp\left[i\varkappa(x - \cot\alpha \cdot z - ct)\right], \\ \psi_3 &= A_3 \exp\left[i\varkappa(x - \cot\beta \cdot z - ct)\right], \\ \varphi' &= A_{\varphi'} \exp\left[i\varkappa(x - \cot\alpha' \cdot z - ct)\right], \\ \psi' &= A_{\psi'} \exp\left[i\varkappa(x - \cot\beta' \cdot z - ct)\right], \end{aligned} \quad (3.111)$$

where the cotangent functions are given in (3.95). To get interface waves, i.e., waves that decay into interiors $z \lessgtr 0$, we must choose these cotangent functions as

$$\begin{aligned} \cot\alpha &= -i[1-(c/c_d)^2]^{\frac{1}{2}}, & \cot\beta &= -i[1-(c/c_s)^2]^{\frac{1}{2}}, \\ \cot\alpha' &= i[1-(c/c_d')^2]^{\frac{1}{2}}, & \cot\beta' &= i[1-(c/c_s')^2]^{\frac{1}{2}}. \end{aligned} \quad (3.112)$$

It follows from (3.112) that we must have $c < c_s$, c_s', c_d, c_d'. Hence the first two waves in (3.111) travel along the interface in grazing incidence, and the second two in grazing refraction, all with a speed c that must be less than all of the body wave speeds in the two media. Since usually $c_s < c_d$, and $c_s' < c_d'$, it follows that c must satisfy

$$c < \bar{c}_s, \quad (3.113)$$

where $\bar{c}_s$ is the smallest of the two equivoluminal body wave speeds c_s and c_s'. Substituting (3.112) in (3.111), the wave system becomes

$$\begin{aligned}\varphi_1 &= A_1 \exp(-rz) \exp\left[i\varkappa(x-ct)\right], \\ \psi_3 &= A_3 \exp(-sz) \exp\left[i\varkappa(x-ct)\right],\end{aligned} \quad (3.114\text{a})$$

in $z \geq 0$, and

$$\begin{aligned}\varphi' &= A_{\varphi'} \exp(r'z) \exp\left[i\varkappa(x-ct)\right], \\ \psi' &= A_{\psi'} \exp(s'z) \exp\left[i\varkappa(x-ct)\right],\end{aligned} \quad (3.114\text{b})$$

in $z \leq 0$, where r and s are defined in (3.62), with c_R replaced by c and

Ch. 3, § 3.3] REFLECTION AND REFRACTION OF P AND SV WAVES 167

$$r' = \varkappa[1-(c/c'_d)^2]^{\frac{1}{2}}, \qquad s' = \varkappa[1-(c/c'_s)^2]^{\frac{1}{2}},$$

all real and positive. Noting again that $\cot \alpha = -ir/\varkappa$, and $\cot \beta = -is/\varkappa$, and in addition that $\cot \alpha' = ir'/\varkappa$, and $\cot \beta' = is'/\varkappa$, we substitute these values for the cotangent functions into (3.95) (our general relations for continuity of the displacements and stresses at the interface) which, when A_2 and A_4 are set equal to zero, reduce to

$$\begin{aligned} \varkappa A_1 - is A_3 - \varkappa A_{\varphi'} - is A_{\psi'} &= 0, \\ r A_1 - i\varkappa A_3 + r' A_{\varphi'} + i\varkappa A_{\psi'} &= 0, \\ \mu(s^2+\varkappa^2)A_1 - i2\mu\varkappa s A_3 - \mu'(s'^2+\varkappa^2)A_{\varphi'} - i2\mu'\varkappa s' A_{\psi'} &= 0, \\ i2\mu\varkappa r A_1 + \mu(s^2+\varkappa^2)A_3 + i2\mu'\varkappa r' A_{\varphi'} - \mu'(s'^2+\varkappa^2)A_{\psi'} &= 0. \end{aligned} \qquad (3.115)$$

The homogeneous equations (3.115) can have a nontrivial solution for the coefficients A_1, A_3, $A_{\varphi'}$, and $A_{\psi'}$, if and only if the determinant of the system vanishes, i.e., if

$$\begin{vmatrix} 1 & -\dfrac{s}{\varkappa} & -1 & -\dfrac{s'}{\varkappa} \\ \dfrac{r}{\varkappa} & -1 & \dfrac{r'}{\varkappa} & 1 \\ \mu\left(2-\dfrac{c^2}{c_s^2}\right) & -2\mu\dfrac{s}{\varkappa} & -\mu'\left(2-\dfrac{c^2}{c'^2_s}\right) & -2\mu'\dfrac{s'}{\varkappa} \\ 2\mu\dfrac{r}{\varkappa} & -\mu\left(2-\dfrac{c^2}{c_s^2}\right) & 2\mu'\dfrac{r'}{\varkappa} & \mu'\left(2-\dfrac{c^2}{c'^2_s}\right) \end{vmatrix} = 0. \qquad (3.116)$$

One solves (3.116) to obtain the speed $c=c_{St}$ of the Stoneley wave. It is not difficult to show that (3.116) is equivalent to Stoneley's result [3.18, Equation (23), pg. 418]. Stoneley also gives an expanded form for (3.116) [3.18, Equation (1), pg. 419]. Much work has gone into investigating the roots of this equation. Cagniard [3.19] has investigated the number and nature of the roots of (3.116) by considering its more general form $D(u)=0$, where u is a complex variable, with the dimensions of *wave slowness* $1/c$. Through a detailed study of $D(u)$, in its mapped plane, and use of the argument principle, Cagniard established there were two zeros of $D(u)$, the complex conjugates $u = \pm i(1/c_{St})$, which correspond to the real roots $c^2 = c_{St}^2$, or $c = c_{St}$, and hence to real Stoneley waves (3.114). These were subject to the condition that the function $p(q)$, where $p=\varrho'/\varrho$ and $q=c_s^2/c'^2_s$, respectively the density and shear wave contrasts, lies in a certain region of the $p-q$

plane. Cagniard computed this region for the case of $\lambda=\mu$, $\lambda'=\mu'$ (or $\nu=\nu'=1/4$), and his results are reproduced here in fig. 3.31. As is exhibited in

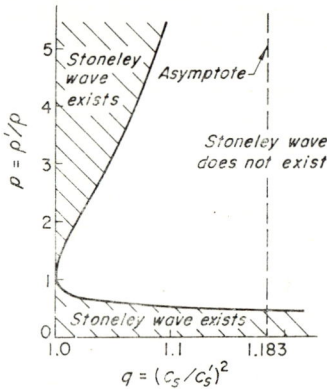

Fig. 3.31. Conditions between the density ratio ϱ'/ϱ and the shear wave ratio $(c_s'/c_s)^2$ for the Stoneley wave to exist ($\lambda=\mu$, $\lambda'=\mu'$, or $\nu=\nu'=1/4$) (after Cagniard).

this figure, and pointed out in [3.19], since the ratio ϱ'/ϱ for all rocks is close to one, the Stoneley wave, in practical cases, can exist only if c_s and c_s' are nearly the same. References to further work on (3.116) may be found Ewing et al [3.11 pp. 112-113]. As pointed out there, Koppe solved (3.116) numerically, and found that $c_{St}>c_R$. On this basis, and using (3.113), the Stoneley wave speed lies in the region $c_R<c_{St}<\bar{c}_s$.

3.3.3. *The fluid-solid interface*

The fluid-solid interface is a special case of the solid-solid interface. It is an important case in seismology, modeling the ocean bottom. We assume the lower medium, $z>0$ in fig. 3.29, is the fluid, and from within it the $P(\varphi_1)$ wave is incident at the interface. Solution of the problem is obtained from (3.99), by letting $\mu \to 0$, and $A_4=0$, since it is assumed that the fluid does not support any shear. From the condition $\mu \to 0$, we have $\varrho c_s^2 \to 0$, which is $c_s \to 0$, $\varrho \neq 0$. It follows for $c>c_s$, that $\cot \beta = [(c/c_s)^2-1]^{\frac{1}{2}} \to c/c_s \to \infty$, and $c_s^2 \cot^2 \beta \to c^2$, but $c_s^2 \cot \beta \to 0$. On this basis (3.99) reduces to

$$A_2/A_1 = \{-\varrho(c^4/c_s'^2) \cot \alpha' + \mu' [(\cot^2 \beta'-1)^2 + 4 \cot \alpha' \cot \beta'] \cot \alpha\}/\hat{D},$$
$$A_{\varphi'}/A_1 = 2\varrho c^2 (\cot^2 \beta'-1) \cot \alpha/\hat{D}, \qquad (3.117)$$
$$A_{\psi'}/A_1 = -4\varrho c^2 \cot \alpha \cot \alpha'/\hat{D},$$

where $\hat{D} = \varrho(c^4/c_s'^2) \cot \alpha' + \mu' [(\cot^2 \beta'-1)^2 + 4 \cot \alpha' \cot \beta'] \cot \alpha$, and the cotangent functions are defined in (3.95). The wave system here is like that in fig. 3.30, except $SV(A_4)$ is no longer present. The previous scheme (3.20), (3.21) can be used to derive the coefficients for the displacements, and stresses, and the energy balance is (3.101a) without the E_4/E_1 and A_4/A_1 term. Ergin [3.16] has computed the square roots of the three energy ratios for the present case, $(E_2/E_1)^{\frac{1}{2}}$, $(E_{\varphi'}/E_1)^{\frac{1}{2}}$, and $(E_{\psi'}/E_1)^{\frac{1}{2}}$, as a function of angle of incidence α. His curves for three different sets of parameters are reproduced here in fig. 3.32. Interesting are the two total reflection points they exhibit ($\alpha_{crd'}$ and $\alpha_{crs'}$ in the nomenclature of § 3.3.2.3 here), corresponding to the "cutting out", first of the refracted P wave, and second the refracted SV wave. In the region $\alpha_{crs'} < \alpha < \pi/2$, $(E_2/E_1)^{\frac{1}{2}} = 1$. Ergin also treats two other cases for the fluid-solid interface, namely, an incident P, and an incident SV wave, in the solid. He gives curves for the square roots of the three energy ratios in these cases also.

As for the special cases here, for normal incidence we have (3.102) and (3.103) again, where now $c_d = (\lambda/\varrho)^{\frac{1}{2}}$ and $c'_d = [(\lambda' + 2\mu')/\varrho']^{\frac{1}{2}}$. Similarly, for graing incidence we again have (3.106) (note in both these cases $A_4 = 0$ stems from the fluid medium). The cases of reflection at critical angles of incidence, and total reflection are left to the exercises. They can be carried out by the techniques described in § 3.3.2.3. In Appendix II of his book [3.19], Cagniard investigated the speed of the Stoneley wave for the fluid-solid interface. Assuming that $z > 0$ is the solid, and $z < 0$ the fluid half space (in fig. 3.29), the corresponding case treated in [3.19] was $c_d^2 = 3c_s^2$, $c_s = 3.84$ km/sec, $c'_d = 1.5$ km/sec, and $\varrho/\varrho' = 2.7$, with the finding that $c_{St} = 1.39$ km/sec, i.e., c_{St} was less than the fluid wave velocity c'_d. It was suggested by Cagniard that this wave be called the *"Scholte wave"*, since he evidently was the first to recognize it.

3.4. Reflection and refraction of SH waves at an interface

3.4.1. General solution

Consider the incident $SH(\chi_1)$ wave shown in fig. 3.27, but now incident at the interface between the two elastic half spaces of different material properties. The reflected $SH(\chi_2)$ wave occurs again, but now in addition we have the new refracted $SH(\chi')$ wave, traveling in the half space $z < 0$ of properties μ' and ϱ'. The situation is shown in fig. 3.33, where the coor-

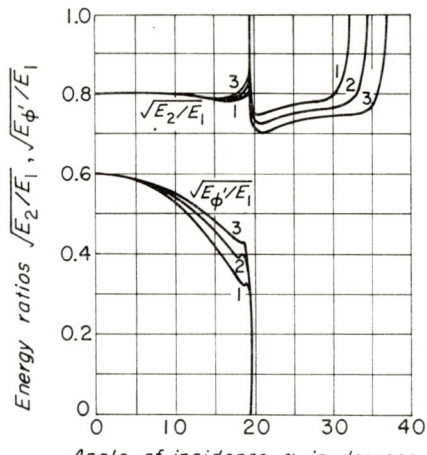

(a) Energy ratios of reflected and refracted P waves

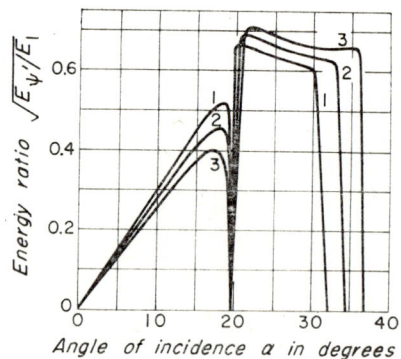

(b) Energy ratio of refracted SV wave

Fig. 3.32. Energy ratios for P-wave incidence in water against a solid (after Ergin)
Data on curves:

Curve	c'_d/c_s	c'_d/c_d	ϱ'/ϱ
1	1.6	3.0	3.0
2	1.7	3.0	3.0
3	1.8	3.0	3.0

dinates are as they were in fig. 3.27. Equation (3.82) is applicable here for $z>0$. In addition, for $z<0$, we have the refracted wave

$$\chi' = B_{\chi'}\chi'^c , \qquad (3.118)$$

Ch. 3, § 3.4] REFLACTION AND REFRACTION OF *SH* WAVES 171

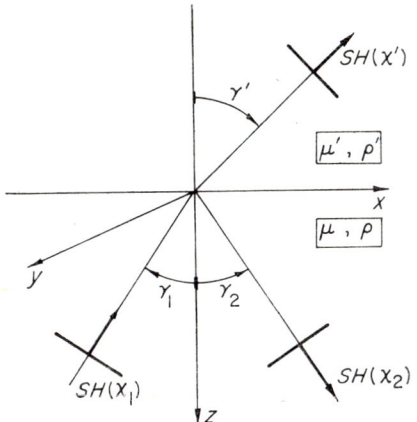

Fig. 3.33. Incident, reflected and refracted *SH* waves.

where $B_{\chi'}$ is the wave amplitude, and

$$\chi' = \exp\left[i\varkappa'_s(l_{\gamma'}x - n_{\gamma'}z - c'_s t)\right],$$

where γ' is the angle of refraction of the χ' wave, measured from the z-axis. It follows the direction cosines in (3.118) are $l_{\gamma'} = \sin \gamma'$ and $n_{\gamma'} = \cos \gamma'$.

Again assuming welded contact of the two half spaces at $z=0$, it follows that the displacement and stress must be continuous there, that is

$$v = v', \qquad \sigma_{zy} = \sigma'_{zy}, \qquad \text{at } z = 0. \tag{3.119}$$

Substituting here for v and σ_{zy} from (3.84), and the second of (3.88), and for v' and σ'_{zy}, from expressions similar to (3.84) and (3.88) based on (3.118), equations (3.119) become

$$\varkappa_s^2 \{B_1 \exp\left[i\varkappa_s(l_1 x - c_s t)\right] + B_2 \exp\left[i\varkappa_s(l_2 x - c_s t)\right]\}$$
$$= \varkappa_s'^2 B_{\chi'} \exp\left[i\varkappa_s'(l_{\gamma'}x - c'_s t)\right], \tag{3.120}$$
$$\mu \varkappa_s^3 \{n_1 B_1 \exp\left[i\varkappa_s(l_1 x - c_s t)\right] - n_2 B_2 \exp\left[i\varkappa_s(l_2 x - c_s t)\right]\}$$
$$= \mu' \varkappa_s'^3 n_{\gamma'} B_{\chi'} \exp\left[i\varkappa_s'(l_{\gamma'}x - c'_s t)\right].$$

As in previous cases, to obtain a solution of (3.120) for B_2/B_1 and $B_{\chi'}/B_1$, independent of x and t, we must first require that

$$\gamma_1 = \gamma_2 = \gamma, \qquad \varkappa = \varkappa_s \sin \gamma = \varkappa'_s \sin \gamma', \tag{3.121}$$

noting again that $\omega = \varkappa_s c_s = \varkappa'_s c'_s = \varkappa c$. The equations for $\varkappa$ in (3.121) contain the refraction laws

$$c = c_s/\sin \gamma = c'_s/\sin \gamma' . \tag{3.122}$$

Under these conditions, (3.120) reduce to

$$B_1 + B_2 - \left(\frac{c_s}{c'_s}\right)^2 B_{\chi'} = 0 ,$$

$$B_1 - B_2 - \frac{\varrho' c_s \cos \gamma'}{\varrho c'_s \cos \gamma} B_{\chi'} = 0 ,$$

which easily yield, for the potential reflection and transmission coefficients,

$$\frac{B_2}{B_1} = \frac{\mu \cos \gamma - \mu'(c_s/c'_s) \cos \gamma'}{\mu \cos \gamma + \mu'(c_s/c'_s) \cos \gamma'} ,$$

$$\frac{B_{\chi'}}{B_1} = \frac{2\varrho\mu' \cos \gamma}{\varrho'[\mu \cos \gamma + \mu'(c_s/c'_s) \cos \gamma']} , \tag{3.123}$$

and displacement coefficients,

$$\frac{V_2}{V_1} = \frac{B_2}{B_1} , \qquad \frac{V_{\chi'}}{V_1} = \left(\frac{c_s}{c'_s}\right)^2 \frac{B_{\chi'}}{B_1} . \tag{3.124}$$

Achenbach [2.14] has computed V_2/V_1 and $V_{\chi'}/V_1$, as a function of angle

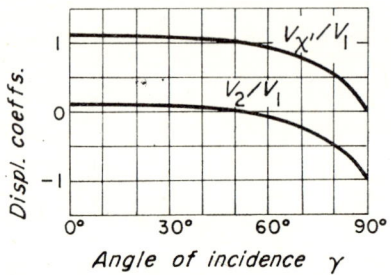

Fig. 3.34. Reflection and refraction coefficients for SH waves as a function of angle of incidence. $\mu'/\mu = 0.64$, $c'_s/c_s = 0.8$. V_2/V_1 reflection coefficient, $V_{\chi'}/V_1$ refraction coefficient (after Achenbach).

of incidence γ. His curves are reproduced in fig. 3.34. The partition of energy here is expressed by

$$1 = \left(\frac{B_2}{B_1}\right)^2 + \frac{\varrho' c_s^2 \tan \gamma}{\varrho c'^2_s \tan \gamma'} \left(\frac{B_{\chi'}}{B_1}\right)^2 .$$

3.4.2. Special cases

The coefficients for normal incidence are obtained by letting $\gamma \to 0$ in (3.122), which shows that γ' also $\to 0$. It follows our general solution (3.123) for the potential reflection and refraction coefficients reduces to

$$\frac{B_2}{B_1} = \frac{\rho c_s - \rho' c'_s}{\rho c_s + \rho' c'_s}, \qquad \frac{B_{\chi'}}{B_1} = \frac{2\rho c'^2_s}{c_s[\rho c_s + \rho' c'_s]}.$$

The corresponding displacement coefficients are

$$\frac{V_2}{V_1} = \frac{B_2}{B_1}, \qquad \frac{V_{\chi'}}{V_1} = \frac{c^2_s}{c'^2_s} \cdot \frac{B_{\chi'}}{B_1} = \frac{2\rho c_s}{\rho c_s + \rho' c'_s}.$$

The coefficients for grazing incidence are obtained by letting $\gamma \to \pi/2$ in (3.123). We find

$$\frac{B_2}{B_1} = -1, \qquad \frac{B_{\chi'}}{B_1} = 0.$$

Such a solution leads to zero motion, as we have noted in previous cases. It follows that one would use the technique in § 3.1.4.2 to derive the appropriate special solution for this case.

Total reflection occurs for the incident SH wave, if γ exceeds its critical angle $\gamma_{cr} = \sin^{-1}(c_s/c'_s)$, i.e., if $\gamma_{cr} < \gamma < \pi/2$. Following the technique given in § 3.3.2.3 for such cases, it is easily shown that in the present case the refracted wave is of the form

$$\chi' = B_{\chi'} \exp\{[1-(c/c'_s)^2]^{\frac{1}{2}}\varkappa z\} \exp[i\varkappa(x-ct)], \qquad (3.125)$$

with $B_{\chi'}$ being complex, and that χ_1 reflects totally, i.e., χ_2 is a regular SH wave except for a phase change. The two coefficients are given by

$$\frac{B_{\chi'}}{B_1} = \left|\frac{B_{\chi'}}{B_1}\right| \exp(i\theta_{\chi'}), \qquad \frac{B_2}{B_1} = \exp(2i\theta_{\chi'}), \qquad (3.126)$$

where

$$\left|\frac{B_{\chi'}}{B_1}\right| = \frac{A}{(B^2+C^2)^{\frac{1}{2}}}, \qquad \theta_{\chi'} = \tan^{-1}\left(-\frac{C}{B}\right),$$

and where

$$A = 2\varrho\mu'[(c/c_s)^2 - 1]^{\frac{1}{2}}/\varrho', \quad B = \mu[(c/c_s)^2 - 1]^{\frac{1}{2}}, \quad C = \mu'[1 - (c/c_s')^2]^{\frac{1}{2}}.$$

One further special case is worth mentioning. From the first of (3.123) we see the incident SH wave is totally transmitted if

$$\mu \cos \gamma - \mu'(c_s/c_s') \cos \gamma' = 0, \quad (3.127)$$

i.e., $B_2 = V_2 = 0$. We substitute from this equation into the second of (3.123) which gives

$$B_{\chi'}/B_1 = (c_s'/c_s)^2. \quad (3.128a)$$

This in turn, when substituted into the second of (3.124), gives

$$V_{\chi'} = V_1. \quad (3.128b)$$

We therefore can solve (3.127) and the second of (3.122), simultaneously, to get a range of pairs γ and the four material constants μ, ϱ, μ', and ϱ', for which an incident SH wave is totally transmitted, the transmission coefficients being given by (3.128). Note the present case is analogous to total mode conversion discussed in § 3.1.4.4.

3.5. Exercises

3.1. Starting with (2.26) derive the displacement-potential relations (3.1) for plane strain.

3.2. In the case of the in-plane constrained boundary for the half space (§ 3.1.1.2), the text shows (3.13) govern. Derive the solutions given there for both the reflection of an incident P and SV wave.

3.3. Determine the stress reflection coefficients in terms of the potential reflection coefficients for the case of an incident P wave at a free boundary treated in § 3.1.2.1 of the text.

3.4. Derive (3.27) of the text, the energy partition relation for the incident SV wave at a free boundary.

3.5. Show if $2A_1\delta = A_0$, (3.37) and (3.38) of the text lead to the system of waves (3.39) and (3.40) for the P-wave grazing incidence problem.

3.6. Derive the solution (3.41) and (3.42) for the system of waves in the SV-wave grazing incidence problem.

3.7. Consider the normal incidence of a wave pair (P and SV), whose amplitudes are given by (3.58), as a limiting case (α and $\beta \to 0$) under the condition that in the limit $A_3 \tan \alpha = A_3'$. Derive the wave system and sketch it.

3.8. The sketch shows a plane harmonic P wave (of plane strain), incident at an angle α on the boundary of an elastic half space, this boundary being

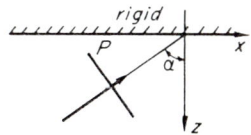

completely constrained (i.e., displacement $u=0$). Derive (1) the expressions for the potential reflection coefficients for the case of

(a) arbitrary incident angle α, and Poisson's ratio for the half space,
(b) normal incidence,
(c) grazing incidence,
(d) total mode conversion,

and (2) the energy partition relation for case (a) above.

3.9. Using the displacement equation of motion (2.1) for equivoluminal waves, show that you can develop the theory for SH waves without using potentials as was done in § 3.2 of the text. Show that your treatment leads to the same solutions as we found in §§ 3.2.1, 3.2.2 for the free and rigid boundaries.

3.10. With reference to fig. 3.29 assume we have P wave incident at the interface. Noting again that $c_s < c_d < c_s' < c_d'$ for our solids, we now let $c_s < c_d < c_s' < c < c_d'$. With the method described in § 3.3.2.3 show for the present case we have critical angles of incidence for the P wave. Then show (3.99) for the present case reduce to

$$\frac{A_2}{A_1} = \left|\frac{A_2}{A_1}\right| e^{i\theta_2}, \qquad \frac{A_4}{A_1} = \left|\frac{A_4}{A_1}\right| e^{i\theta_4},$$

$$\left|\frac{A_{\varphi'}}{A_1}\right| = \left|\frac{A_{\varphi'}}{A_1}\right| e^{i\theta_{\varphi'}}, \qquad \frac{A_{\psi'}}{A_1} = \left|\frac{A_{\psi'}}{A_1}\right| e^{i\theta_{\psi'}}$$

where

$$\left|\frac{A_2}{A_1}\right| = \left(\frac{L^2+M^2}{G^2+H^2}\right)^{\frac{1}{2}}, \qquad \left|\frac{A_4}{A_1}\right| = 2\left(\frac{\gamma_2^2\delta_4^2+\gamma_4^2\delta_2^2}{G^2+H^2}\right)^{\frac{1}{2}},$$

$$\left|\frac{A_{\varphi'}}{A_1}\right| = \frac{2|W|}{(G^2+H^2)^{\frac{1}{2}}}, \qquad \left|\frac{A_{\psi'}}{A_1}\right| = 2\left(\frac{\gamma_2^2+\gamma_4^2}{G^2+H^2}\right)^{\frac{1}{2}}$$

$$\theta_2 = \tan^{-1}\left(-\frac{T}{S}\right), \qquad \theta_4 = \tan^{-1}\left(-\frac{B}{A}\right),$$

$$\theta_{\varphi'} = \tan^{-1}\left(-\frac{H}{G}\right), \qquad \theta_{\psi'} = \tan^{-1}\left(\frac{D}{C}\right)$$

where

$A = \gamma_2\delta_4 G - \gamma_4\delta_2 H$, $\qquad B = \gamma_4\delta_2 G + \gamma_2\delta_4 H$
$C = \gamma_2 G + \gamma_4 H$, $\qquad D = \gamma_4 G - \gamma_2 H$
$S = LG - MH$, $\qquad T = MG + LH$
$L = \gamma_1 W - \gamma_2 X$, $\qquad M = \widehat{\gamma}_3 W + \gamma_4 X$
$G = \gamma_1 W - \gamma_2 V$, $\qquad H = \widehat{\gamma}_3 W - \gamma_4 V$
$V = \delta_1 + \delta_3$, $\qquad W = \delta_2 + \delta_4$, $\qquad X = \delta_1 - \delta_3$

and

$$\gamma_3 = \tau[1-(c/c_d')^2]^{\frac{1}{2}}/[(c/c_d)^2-1]^{\frac{1}{2}}, \qquad \delta_1 = 2(\mu-\mu')[(c/c_s')^2-1]^{\frac{1}{2}}/\varrho c^2$$

$$\delta_2 = \tau[(c/c_s')^2-1]^{\frac{1}{2}}/[(c/c_s)^2-1]^{\frac{1}{2}}, \qquad \delta_3 = -\eta/[(c/c_d)^2-1]^{\frac{1}{2}}$$

with γ_1, γ_2, γ_4, δ_4, η and τ being defined in (3.108) of the text. What waves, if any, decay with distance from the interface, and what form do they take? Do we have total reflection here? How would you expect the energy to partition?

3.11. For the case of the P-wave incidence in water against a solid, discussed in § 3.3.3, we let $c_d < c_s' < c < c_d'$. Discuss the forms of the φ_2, φ' and ψ' waves in this case. Are any of these waves exponentially decaying with distance

z from the interface? Do any of them have complex coefficients? Give your arguments.

3.12. In the case of the P-wave incidence in water against a solid, prove that letting $c_d < c < c'_s < c'_d$ leads to total reflection in the fluid with a phase change for the reflected P wave of $2\theta_2$, where

$$\theta_2 = \cot^{-1}\left\{\frac{\varrho' c_s'^4[(c/c_d)^2 - 1]^{\frac{1}{2}}\{[(2-(c/c'_s)^2]^2 - 4[1-(c/c'_d)^2]^{\frac{1}{2}}[1-(c/c'_s)^2]^{\frac{1}{2}}\}}{\varrho c^4[1-(c/c'_d)^2]^{\frac{1}{2}}}\right\}$$

Discuss the forms φ' and ψ'.

3.13. Derive (3.125) and (3.126), the solution for total reflection of an SH wave at a solid-solid interface.

References

[3.1.] H. Kolsky, *Stress Waves in Solids*. Dover Publications, Inc., New York (1963)
[3.2.] D. L. Arenberg, *Journal of the Acoustical Society of America* **20** (1948), 1–26
[3.3.] M. Redwood. *Mechanical Waveguides*. Pergamon Press, New York (1960).
[3.4.] B. Gutenberg, *Bulletin of the Seismological Society of America* **34** (1944), 85–102.
[3.5.] R. D. Mindlin, Waves and Vibrations in Isotropic Elastic Plates. In: *Structural Mechanics*, J. N. Goodier and N. J. Hoff, eds., Pergamon Press, New York (1960), 199–232.
[3.6.] J. N. Goodier and R. E. D. Bishop, *Journal of Applied Physics* **23** (1952), 124–126.
[3.7.] A. Sommerfeld, *Optics, Lectures on Theoretical Physics IV*. Academic Press, New York (1964), 27–32.
[3.8.] Lord Rayleigh, *Proceedings of the London Mathematical Society* **17** (1887), 4–11.
[3.9.] M. Hayes and R. S. Rivlin, *Zeitschrift für Angewandt Mathematik und Physik* **13** (1962), 80–83.
[3.10.] L. Knopoff, *Bulletin of the Seismological Society of America* **42** (1952), 307–308.
[3.11.] W. M. Ewing, W. S. Jardetzky, and F. Press, *Elastic Waves in Layered Media*. McGraw–Hill Book Company, Inc., New York (1957).
[3.12.] I. A. Viktorov, *Rayleigh and Lamb Waves*. Plenum Press, New York (1967).
[3.13.] M. B. Dobrin, R. F. Simon, and P. L. Lawrence, *Transactions of the American Geophysical Union* **32** (1951), 822–832.
[3.14.] J. K. Knowles, *Journal of Geophysical Research* **71** (1966), 5480–5481.
[3.15.] M. Muskat and M. W. Meres, *Geophysics* **5** (1940), 115–148.
[3.16.] K. Ergin, *Bulletin of the Seismological Society of America* **42** (1952) 349–372.
[3.17.] A. E. H. Love, *Some Problems of Geodynamics*. Cambridge University Press, Cambridge (1911), 165–178.
[3.18.] R. Stoneley, *Proceedings of the Royal Society (London) A* **106** (1924), 416–428.
[3.19.] L. Cagniard, *Reflection and Refraction of Progressive Seismic Waves*, Translated and Revised by E. A. Flinn and C. H. Dix. McGraw–Hill Book Company, Inc., New York (1962), 42–49.

CHAPTER 4

TIME HARMONIC WAVES IN ELASTIC WAVEGUIDES

An infinite elastic plate is in effect a half space that has become bounded in its depth by the introduction of a second parallel boundary. One can easily visualize the *P*, *SV* and *SH* waves, we have just studied, reflecting now from boundary to neighboring boundary, generally, in the *P* and *SV* wave cases, undergoing mode conversion at each reflection, and progressing along the length of the plate. The neighboring parallel boundaries are in effect then guiding the waves along the plate. Rods, cylindrical shells, and a layered elastic solid are other examples of *waveguides*. The common feature is two or more parallel boundaries, which introduces one or more characteristic lengths into the problem, and leads to *wave dispersion*, *characterized in harmonic waves by a dependence of frequency on wave length*. Here we shall study these waves in an infinite elastic plate and rod.

4.1. Waves in an infinite plate in plane strain

Consider the two equal amplitude, and like sign, *P* waves propagating in a plate of thickness 2h, symmetrically with respect to the mid-plane $z=0$, as depicted in fig. 4.1. The waves are both incident and reflected since

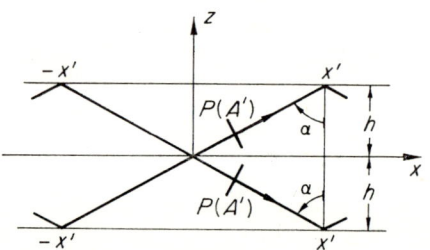

Fig. 4.1. Symmetric *P* waves in an infinite plate.

they propagate from their last reflection points, say $x=-x'$, $z=\pm h$, toward their next points of incidence $x=x'$, $z=\pm h$. The waves therefore account for all the motion within and on the surfaces of the plate over the arbitrary domain $-x' \leq x \leq x'$, hence over the entire plate $-\infty < x < \infty$.

The sum of the P waves is given by

$$\varphi = A' \exp[i\varkappa_d(\sin \alpha \cdot x + \cos \alpha \cdot z - c_d t)]$$
$$+ A' \exp[i\varkappa_d(\sin \alpha \cdot x - \cos \alpha \cdot z - c_d t)], \qquad (4.1)$$

which reduces to

$$\varphi = 2A' \cos(\eta_d z) \exp[i\varkappa(x - ct)], \qquad (4.2)$$

where again here $\varkappa = \varkappa_d \sin \alpha$, $c = c_d/\sin \alpha$, and $\omega = \varkappa c$ are, respectively, the *wave number*, and *wave phase velocity in the propagation direction*, and corresponding *frequency*. η_d is the *wave number in the z, or plate thickness direction*, which can be written in terms of ω, $\varkappa$, and c_d, i.e.,

$$\eta_d = \varkappa_d \cos \alpha = (\varkappa_d^2 - \varkappa^2)^{\frac{1}{2}} = [(\omega/c_d)^2 - \varkappa^2]^{\frac{1}{2}}. \qquad (4.3)$$

It may be noted that φ is symmetric with respect to the plate mid-plane $z=0$, and for a particular $(\omega, \varkappa)$ pair, it can be interpreted, when the phase $\varkappa(x - ct) = $ constant, as a *propagating cylindrical surface of constant phase*.

The antisymmetric counterpart of (4.1) can be created by changing the sign (or phase) of the second wave there. This just changes the signs of the associated displacements as indicated in fig. 4.2. The sum of these waves is

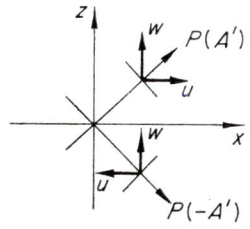

Fig. 4.2. Antisymmetric P waves in an infinite plate.

given by

$$\varphi = 2A'i \sin(\eta_d z) \exp[i\varkappa(x - ct)]. \qquad (4.4)$$

It is clear we can construct the analogous cases for symmetrically and antisymmetrically disposed *SV* waves similarly. It also follows, as Mindlin points out [3.5, p. 208], that *P* and *SV* waves, for both symmetric and antisymmetric cases, can be grouped in pairs as described in ch. 3, § 3.1.4.6. In totality all of these waves are contained in

$$\varphi = f(z) \exp[i\varkappa(x-ct)],$$
$$\psi = g(z) \exp[i\varkappa(x-ct)], \qquad (4.5)$$

where

$$f(z) = A \sin \eta_d z + B \cos \eta_d z,$$
$$g(z) = C \sin \eta_s z + D \cos \eta_s z,$$

where $\varkappa = \varkappa_d \sin \alpha = \varkappa_s \sin \beta$, $c = c_d/\sin \alpha = c_s/\sin \beta$, and $\omega = \varkappa_d c_d = \varkappa_s c_s = \varkappa c$, with η_d being given by (4.3), and η_s by

$$\eta_s = \varkappa_s \cos \beta = (\varkappa_s^2 - \varkappa^2)^{\frac{1}{2}} = [(\omega/c_s)^2 - \varkappa^2]^{\frac{1}{2}}. \qquad (4.6)$$

Note that f and g contain both symmetric and antisymmetric incident waves. We have already pointed out that expressions like (4.1) and (4.2) represent waves that are both incident and reflected. Hence for a symmetric pair, for example, we have, in the nomenclature of (3.2), $2A_2 = 2A_1 = 2A' = B$, and $2A_4 = -2A_3 = C$ as in the first two of (3.58), i.e., the cos $\eta_d z$ term in f, and the sin $\eta_s z$ term in g, represent symmetric dilatational and equivoluminal motion, respectively, which is easily checked by applying (3.1) to (4.5). An antisymmetric pair would therefore be represented by the other two terms in f and g, i.e., $2A_2 = -2A_1 = 2A'i = A$, and $2A_4 = 2A_3 = D$ as in the first two of (3.59). We shall see that these two cases are fundamental to the theory of wave propagation or vibrations in plates, the type of waves involved, symmetric or antisymmetric or both, being dependent on the same features of the loading.

Applying the displacement- and stress-potential relations, (3.1) and (3.6) to (4.5), we find the corresponding nonvanishing displacement and stress waves to be

$$u = [i\varkappa f(z) - g'(z)] \exp[i\varkappa(x-ct)],$$
$$w = [f'(z) + i\varkappa g(z)] \exp[i\varkappa(x-ct)], \qquad (4.7)$$

and

$$\sigma_x = \mu[(2\eta_d^2 - \varkappa_s^2)f(z) - i2\varkappa g'(z)] \exp[i\varkappa(x-ct)],$$

$$\sigma_z = \mu[(\varkappa^2 - \eta_s^2)f(z) + i2\varkappa g'(z)] \exp[i\varkappa(x-ct)],$$

$$\sigma_{zx} = \mu[i2\varkappa f'(z) + (\eta_s^2 - \varkappa^2)g(z)] \exp[i\varkappa(x-ct)], \quad (4.8)$$

$$\sigma_y = v(\sigma_x + \sigma_z),$$

where the prime indicates differentiation with respect to z. We now consider the different types of conditions that can arise on the faces of the plate.

4.1.1. Mixed conditions on plate faces

We have need here only for the case involving lubricated-rigid faces, given by the conditions

$$w = \sigma_{zx} = 0, \qquad \text{at } z = \pm h.$$

Into these equations we substitute the second of (4.7) and the third of (4.8), and in turn $f(z)$ and $g(z)$ from (4.5), which gives the four homogeneous equations, independent of x and t, for the amplitudes A, B, C, and D,

$$\eta_d \cos \eta_d h \cdot A - \eta_d \sin \eta_d h \cdot B + i\varkappa \sin \eta_s h \cdot C + i\varkappa \cos \eta_s h \cdot D = 0,$$

$$\eta_d \cos \eta_d h \cdot A + \eta_d \sin \eta_d h \cdot B - i\varkappa \sin \eta_s h \cdot C + i\varkappa \cos \eta_s h \cdot D = 0,$$

$$2i\varkappa\eta_d \cos \eta_d h \cdot A - 2i\varkappa\eta_d \sin \eta_d h \cdot B + (\eta_s^2 - \varkappa^2) \sin \eta_s h \cdot C$$

$$+ (\eta_3^2 - \varkappa^2) \cos \eta_s h \cdot D = 0, \quad (4.9)$$

$$2i\varkappa\eta_d \cos \eta_d h \cdot A + 2i\varkappa\eta_d \sin \eta_d h \cdot B - (\eta_s^2 - \varkappa^2) \sin \eta_s h \cdot C$$

$$+ (\eta_s^2 - \varkappa^2) \cos \eta_s h \cdot D = 0.$$

The determinant of the system of equations (4.9) is

$$D = 4\varkappa_s^2\eta_d^2 \sin \eta_d h \cos \eta_d h \sin \eta_s h \cos \eta_s h.$$

So we have four non-trivial solutions of (4.9), corresponding to

$$\left\{\begin{matrix} \sin \eta_d h \\ \cos \eta_d h \\ \sin \eta_s h \\ \cos \eta_s h \end{matrix}\right\} = 0, \tag{4.10}$$

where in each we require $\varkappa_s = \omega/c_s \neq 0$ ($\omega=0$ would correspond to zero motion). Consider either of the first two of (4.10). From the first two of (4.9) we have the solutions

$$C = D = 0, \quad \begin{cases} A=0, \quad \sin \eta_d h = 0; \quad \eta_d = m\pi/2h, \quad m=0, 2, 4, \ldots, & (4.11a) \\ B=0, \quad \cos \eta_d h = 0; \quad \eta_d = m\pi/2h, \quad m=1, 3, 5, \ldots. & (4.11b) \end{cases}$$

Analogously from either of the last two of (4.10), and from the last two of (4.9), we have the solutions

$$A = B = 0, \quad \begin{cases} D=0, \quad \sin \eta_s h = 0; \quad \eta_s = n\pi/2h, \quad n=2, 4, 6, \ldots, & (4.12a) \\ C=0, \quad \cos \eta_s h = 0; \quad \eta_s = n\pi/2h, \quad n=1, 3, 5, \ldots. & (4.12b) \end{cases}$$

Equations (4.11) and (4.12) are in agreement with Mindlin's results [3.5].

4.1.1.1. Modes of propagation, frequency, and phase velocity spectra. Note that the solutions (4.11) and (4.12) show that dilatational and equivoluminal motions are uncoupled here. That is, in (4.11) $C=D=0$, means that $g(z)=0$, hence dilatational motion only, and in (4.12) $A=B=0$, means $f(z)=0$, or equivoluminal motion only. We would expect this since these motions were uncoupled in the half space also. The different values of m and n determine *modes of propagation*, with (4.11a) representing symmetric and (4.11b) antisymmetric dilatational, and (4.12a) symmetric and (4.12b) antisymmetric equivoluminal modes. Substitution of (4.11) and (4.12), into (4.5), gives the modal forms for the φ and ψ waves. Likewise (4.7) gives the corresponding displacements. Fig. 4.3 depicts the first few of these modes for φ, ψ, u, and w. With the aid of (4.8) similar plots for the stresses are easily obtained. One can think of a particular mode, with its nodal lines, as being created by the interference of the plane harmonic P or SV waves, as they cross one another in reflecting from the plate faces, in their general propagation along the plate.

Ch. 4, § 4.1]   WAVES IN AN INFINITE PLATE...   183

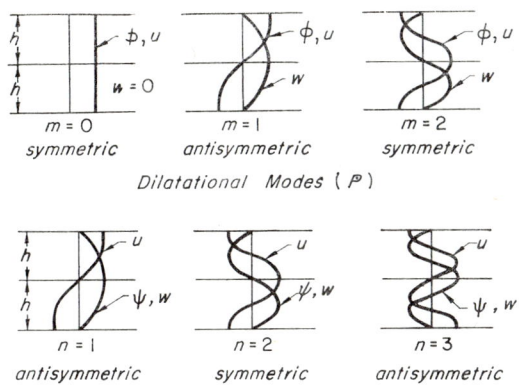

Fig. 4.3. Modes of propagation in an infinite plate with mixed face conditions.

From (4.3) and (4.11), we can write

$$\omega^2/c_d^2 = (m\pi/2h)^2 + \varkappa^2 , \qquad (4.13a)$$

or in dimensionless form as

$$\Omega^2 = \omega^2/\omega_s^2 = k^2(m^2 + \zeta^2) , \qquad (4.13b)$$

where $\zeta = 2h\varkappa/\pi$, and $\omega_s = \pi c_s/2h$. Equations (4.13), commonly called *frequency equations*, disclose the fact that *for each dilatational mode of propagation m there is a continuous spectrum of frequency-wave number pairs.* It is clear from (4.13), and (4.11), there are an *infinite number* of such relations of *branches of this frequency equation.* Note that h is fundamental to the definition of Ω and ζ in (4.13), so that a long or short wave $2\pi/\varkappa$ is only long or short with respect to h, and similarly a low or high frequency ω is only low or high with respect to ω_s, which depends inversely on h. From (4.13a) we can write

$$c^2/c_d^2 = (m\pi/2h\varkappa)^2 + 1 , \qquad (4.14a)$$

or in dimensionless form

$$C^2 = c^2/c_s^2 = k^2[(m/\zeta)^2 + 1] . \qquad (4.14b)$$

(4.14) are the *phase velocity-wave number relations*. These, and (4.13), are sometimes called dispersion relations, the name stemming from the fact that a disturbance governed by such relations changes its shape as it propagates, because of the various phase velocities associated with the different wave-lengths composing the disturbance.

Now, similarly, from (4.6) and (4.12), we find that the frequency and phase velocity spectra for the equivoluminal waves are contained in

$$\Omega^2 = n^2 + \zeta^2, \qquad (4.15a)$$

and

$$C^2 = (n/\zeta)^2 + 1. \qquad (4.15b)$$

If now we require frequency Ω to be real and positive[1] (on physical grounds) we see from both (4.13b) and (4.15a), that ζ can be *real γ or imaginary $i\delta$, but not complex*. We restrict C, in (4.14b) and (4.15b), to real values (again on physical grounds) but here we also impose realness on ζ, because an imaginary ζ corresponds to a standing wave. This is easily seen by examination of the exponential function appearing in (4.5), (4.7), (4.8). With $\zeta = i\delta$ we find

$$\exp\left[i\varkappa(x-ct)\right] = \exp\left[i\zeta(\hat{x}-C\hat{t})\right] = \exp(-\delta\hat{x})\exp(-i\Omega\hat{t}), \qquad (4.15c)$$

where dimensionless $\hat{x} = \pi x/2h$, and $\hat{t} = \pi c_s t/2h$. Eq. (4.15c) means the waves in (4.5), (4.7) and (4.8) are exponentially decaying, in x (or $\hat{x}$), standing waves. In the waveguide literature such waves are referred to as *edge waves*, meaning they decay away from an edge, or a point of load application, say at $x=0$. Edge waves have been observed in both steady and transient wave propagation and vibration experiments. We shall study them further in our work in this chapter on the traction free plate, and also in our later work on transient excitation of waveguides. It may be noted that with ζ real, (4.13b) gives a family of hyperbolas, and (4.15a) a family of equilateral hyperbolas. With ζ imaginary, (4.13b) are ellipses, and (4.15a) circles. Note also that the curves of C are easily computed from the Ω curves or vice

---

[1] It can be negative also since Ω occurs in (4.13b) and (4.15a) as the square. We therefore need only an analysis for Ω positive.

Ch. 4, § 4.1]   WAVES IN AN INFINITE PLATE...   185

versa, since $\Omega = \gamma C$. One can find extensive plots of these spectra for $v = 1/3$ ($k = 2$) in Mindlin's work [3.5, cf. the thin lines in figs. 18, 19]. Figure 4.4

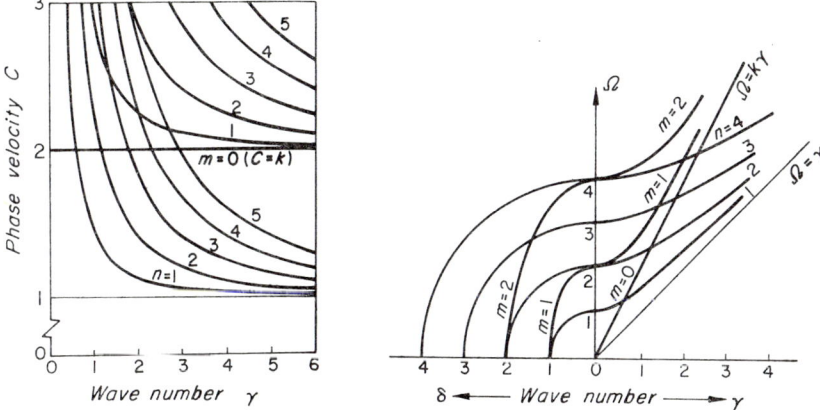

Fig. 4.4. Phase velocity and frequency spectra for the infinite plate with mixed face conditions for $v = 1/3$ (after Mindlin).

shows a few of the lowest branches. We will refer to the real and imaginary wave number parts of a branch of the frequency equation, m or $n = $ constant, as the *real and imaginary wave number segments* of the branch. As may be seen in the frequency spectra plots in fig. 4.4, the branches change their nature at $\zeta = 0$. These points will be called *branch points*. As we shall show in § 4.1.3.1 the slopes $d\Omega/d\gamma$, $d\Omega/d\delta$ there, vanish.

The curves for $m = 0$, i.e., $C = k$, and $\Omega = k\gamma$, in fig. 4.4 represent the *fundamental dilatational mode of the infinite medium*. The *higher m branches* in the figure represent the *anharmonic overtones in an infinite plate, with mixed face conditions*, of this fundamental mode of the infinite medium. The even m represent the symmetric, and odd m the antisymmetric modes. Similarly the curves $C = 1$, and $\Omega = \gamma$, represent the fundamental equivoluminal mode, and n the anharmonic overtones, again with even n symmetric, and odd n antisymmetric. These higher modes are also referred to as thickness modes, i.e. the m modes as thickness-dilatational (or stretch), and n modes as thickness-equivoluminal (or shear) modes. Spectra such as these are invaluable in the information they contain on steady wave propagation. Before analyzing them, however, we introduce the concepts of a wave group and stationary phase, which are most important to a fuller understanding and usefulness of such spectra. Later we shall also see how these concepts can be applied to derive approximate solutions to transient wave problems from these spectra.

4.1.1.2. Wave groups and stationary phase. Havelock, in his interesting monograph on dispersive waves [4.1], points out that Hamilton, as early as 1839, investigated the velocity of propagation of a finite train of waves in a dispersive medium. However, since only short abstracts on this research were published, the work was overlooked until the early 1900's (Havelock reviews Hamilton's main findings in [4.1]). Stokes (1876) is usually credited with setting down the first analytical expression of group velocity, and Rayleigh with subsequent development. Kelvin's group method of approximating integral representations of dispersive waves (1887), now known as the method of stationary phase (which will be discussed in Chapter 5), was a further important advance. We begin our development of the wave group concept with the construction of a simple group, first set down by Stokes. Consider the case of one-dimensional propagation, in which the displacement $u(x, t)$ is the sum of two infinite harmonic wave trains of equal amplitude A, but real wave numbers $\varkappa$ and $\varkappa'$, and phase velocities c and c', differing by only $\Delta\varkappa$ and Δc, respectively. Then,

$$u(x, t) = A \cos [\varkappa(x-ct)] + A \cos \{(\varkappa+\Delta\varkappa)[x-(c+\Delta c)t]\}. \tag{4.16}$$

Equation (4.16) can be expanded into

$$u(x, t) = 2A \cos \left\{ \frac{\varkappa(x-ct)+(\varkappa+\Delta\varkappa)[x-(c+\Delta c)t]}{2} \right\}$$
$$\times \cos \left\{ \frac{\varkappa(x-ct)-(\varkappa+\Delta\varkappa)[x-(c+\Delta c)t]}{2} \right\}. \tag{4.17}$$

Multiplying out the quantities in brackets, taking the incremental quantities to their limit, and discarding higher order terms, (4.17) reduces to

$$u(x, t) \simeq 2A \cos [(\mathrm{d}\varkappa/2)(x-c_g t)] \cos \varkappa(x-ct), \tag{4.18}$$

where

$$c_g = c + \varkappa(\mathrm{d}c/\mathrm{d}\varkappa). \tag{4.19}$$

$u(x, t)$ in (4.18) is a *simple wave group*, and c_g in (4.19) is the *wave group velocity*. It represents at any instant t, a wave train of wave length $2\pi/\varkappa$ whose amplitude varies sinusoidally, slowly with x over a long wavelength of $4\pi/d\varkappa$. Therefore, as this train moves forward, its shape changes. It does, however, have a periodic nature. We see that the amplitude of the wave train moves with c_g, and is in effect an envelope of the train. As pointed out in Brillouin's book [4.2], (4.18) represents a *carrier* with frequency ω and a *modulation* with frequency $c_g d\varkappa/2$. Fig. 4.5 is a sketch of (4.18) for

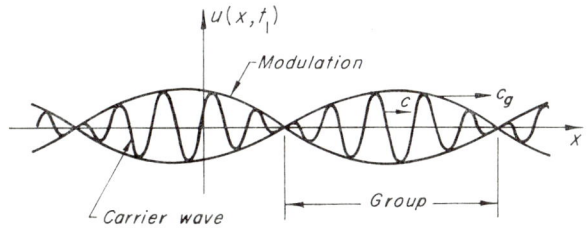

Fig. 4.5. A simple wave group at a fixed time t_1.

some fixed time, say t_1. The group in fig. 4.5 may also be thought of in terms of an observer, traveling on the envelope with velocity c_g, who sees in his own locale approximately simple harmonic motion of fixed wave length and amplitude. When $c_g < c$ the phase waves of the carrier will appear to build up at the back of the group, advance to the front and vanish there, thus as if they were advancing through the group. According to (4.19) we have $dc/d\varkappa < 0$ here. Such a case is referred to as *normal dispersion*. When $c_g > c$, hence $dc/d\varkappa > 0$, we have the converse case, which is referred to as *anomalous disperson*.

We can generalize the analysis to any finite number of component wave trains,

$$u(x, t) = \sum_{n=1}^{N} A_n \cos \varkappa(x - ct) \tag{4.20}$$

provided $\varkappa$ and c vary only slightly over these terms. Now if we require that the phase $\varkappa(x - ct)$ be stationary over the range of $\varkappa$ in these terms, we have

$$d[\varkappa(x - ct)]/d\varkappa = x - [d(\varkappa c)/d\varkappa]t = 0 . \tag{4.21}$$

Now from (4.19) it follows that

$$d(\varkappa c)/d\varkappa = d\omega/d\varkappa = c_g \tag{4.22}$$

and therefore that (4.21) reduces to

$$x - c_g t = 0 . \tag{4.23}$$

Equation (4.23) is called the *condition of stationary phase*. It says that the disturbance at a station x at time t is due to a group of waves, such as those in (4.20), that travels with velocity c_g, i.e., the group is composed of waves having almost equal wavelengths, phase velocities, frequencies, etc., that are in phase at position x at time t. It follows that we can extend these ideas to a much wider range of $\varkappa$, provided we think of the whole as the sum of individual wave groups, each of which is composed of component wave trains that always are confined to a small variation, say, around some mean $\varkappa_0$. We note from (4.22) that c_g is given by $d\omega/d\varkappa$. Therefore, frequency spectra, such as the $\Omega - \gamma$ plots in fig. 4.4, contain all the needed information for a wave group analysis in steady wave propagation. That is, for a cluster or group of waves about $\varkappa_0$, $c(\varkappa_0) = \omega(\varkappa_0)/\varkappa_0$, $c_g(\varkappa_0) = d\omega/d\varkappa_0$, and we have (4.23). Fig. 4.6 depicts this data.

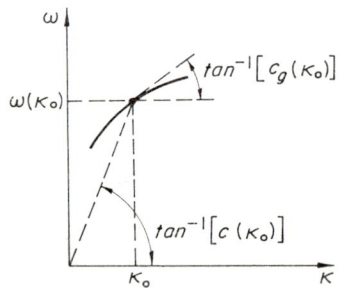

Fig. 4.6. Dispersion data in the $\omega - \varkappa$ plane.

4.1.1.3. Spectral analysis of wave train (or steady) propagation. We return now to (4.13b), (4.14b), and (4.15), and fig. 4.4, to extract from these the nature of the steady wave propagation they govern. First consider $\gamma = 0$ (wavelength infinite). The $\Omega - \gamma$ plots in fig. 4.4, for $m, n \geq 1$, show that here, for every branch, Ω has a minimum. This value is called the *cutoff frequency*, since below it there are no ordinary (Ω-real wave number) propagating waves. From (4.13b) the dilatational cutoff frequencies are given by

$$\Omega = km, \qquad m = \begin{cases} 2, 4, 6 \ldots, \text{ symmetric} \\ 1, 3, 5 \ldots, \text{ antisymmetric} \end{cases}, \qquad (4.24)$$

and from (4.15a), the equivoluminal cutoff frequencies by

$$\Omega = n, \qquad n = \begin{cases} 2, 4, 6 \ldots, \text{ symmetric} \\ 1, 3, 5 \ldots, \text{ antisymmetric} \end{cases}, \qquad (4.25)$$

where we observe the dilatational cutoff frequencies are dependent on Poisson's ratio, but the equivoluminal ones are not. Now from both (4.14b) and (4.15b), and fig. 4.4, we see the phase velocities at cutoff all become infinite. Then since $c = c_d/\sin \alpha$, and $c = c_s/\sin \beta$, it is clear that in cutoff the plane waves, such as those in (4.1), are incident at $\alpha = 0$ and $\beta = 0$, and are independent of x, i.e., these waves propagate normal to the faces, back and forth across the plate, as depicted in fig. 4.7. Hence, the corres-

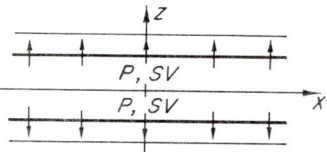

Fig. 4.7. Incident P and SV waves at cutoff in the infinite plate with mixed face conditions

ponding waves in (4.5) are standing waves with respect to x. From (4.13b) with $\zeta = \gamma$, we find, using (4.22),

$$C_g = d\Omega/d\gamma = k^2 \gamma/\Omega .\qquad(4.26)$$

where $C_g = c_g/c_s$, with a similar expression from (4.15a). Hence at cutoff $C_g = d\Omega/d\gamma = 0$. This agrees with the slopes exhibited in fig. 4.4, and the nonpropagating (in x) nature of these waves, both of which were pointed out earlier. Note that since $\Omega = \gamma C$, where $\gamma \to 0$ and $C \to \infty$ at cutoff, the constant values, km and n, given in (4.24) and (4.25), can be interpreted as limits derived from l'Hospital's rule.

Now for $\zeta = \gamma \to \infty$, (4.13b), (4.14b), and (4.15) give, respectively,

$$\Omega = k\gamma, \qquad C = k ,\qquad(4.27a)$$

and

$$\Omega = \gamma , \qquad C = 1 ,\qquad(4.27b)$$

as fig. 4.4 shows. From (4.27a), using (4.26), we find for the dilatational waves

$$C_g = k = C .\qquad(4.28a)$$

Similarly, from (4.27b) for the equivoluminal waves, we find

$$C_g = 1 = C .\qquad(4.28b)$$

This equality to phase velocities is typical of a horizontal tangent in $C - \gamma$ curves, as (4.19) requires. Hence these limiting infinitely high frequency-short waves also have, in addition to their phase velocities, group velocities that are equal to the body wave speeds. Further, since $c = c_d/\sin \alpha$, $c = c_s/\sin \beta$, the values for C in (4.27) show that α and β must tend to $\pi/2$ as a limit, hence the corresponding high frequency-short waves, such as (4.1), have propagation normals that approach the x-axis. The related waves in (4.5) would have $\varkappa = \varkappa_d$ and $c = c_d$ as limits. One can think of the waves, as so short, compared to h, that the latter injects the infinite medium nature into this situation. The lowest branch $m = 0$, corresponding to the dilatational mode of the infinite medium, is an exception, but this mode is non-dispersive, all of its waves satisfying $\Omega = k\gamma$, $C = k$.

Writing the dilatational wavelength in the propagation direction as

$$L = 2\pi/\varkappa = cT = (c_d/\sin \alpha)(2\pi/\omega), \qquad (4.29)$$

with a similar relation for the equivoluminal wavelength, we note for fixed angle of incidence α, as ω increases, L and T decrease. And for a fixed ω, and T, as α increases, L decreases, which is consistent with the limiting values we found, i.e., as $\alpha \to 0$, $L \to \infty$ (cutoff), and as $\alpha \to \pi/2$, $L \to 0$, since $\omega \to \infty$ in the latter. Now note that as m, or n increases, ω increases, and the thickness wavelength $2\pi/\eta_d$, or $2\pi/\eta_s$, decreases. We see, therefore, the wavelengths in the thickness and propagation directions decrease together with increasing ω.

As a means of identifying wave groups on a response record (e.g., displacement as a function of time at a fixed station) Davies [4.3], in his classic work on the elastic rod, employed predominant period-time of occurrence plots. The *predominant period* T_p is defined as the mean period of waves, which are in phase at station x, at time t. This is therefore the stationary phase solution for periods. In dimensionless terms

$$T_p/T_h = 1/\Omega_p, \qquad (4.30)$$

where $\Omega_p = \omega_p/\omega_s$, and $T_h = 4h/c_s$ is the time it takes a wave, traveling with speed c_s, to propagate twice the plate thickness. It is clear that our T and Ω are predominant for each of the groups they define, so we have no further need for the subscript p. Now from the stationary phase condition (4.23) we can write,

$$t/T_x = 1/C_g, \qquad (4.31)$$

where $T_x = x/c_s$ is the time it takes a wave traveling with speed c_s to propagate the distance x. Now (4.30) and (4.31) can be obtained from an $\Omega - \gamma$, or

$C-\gamma$, curve, such as those in fig. 4.4, and if we plot them as ordinate and abscissa, respectively, we have a predominant period-time of occurrence curve. In the present case the simplicity of the dispersion relations makes it easy to sketch such plots. For example, from (4.13b), or (4.26), we have

$$C_{gm} = d\Omega_m/d\gamma = k^2\gamma/\Omega_m, \qquad (4.32)$$

where the m denotes the branch number. Fig. 4.8 shows the $C_g-\gamma$ plots,

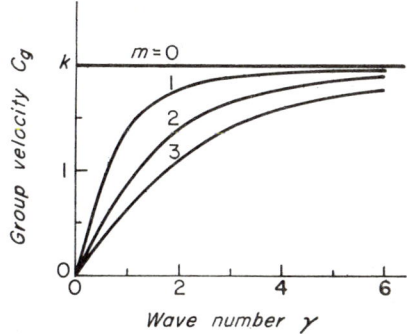

Fig. 4.8. Group velocity spectra for dilatational waves in the infinite plate with mixed face conditions ($v = 1/3$).

obtained from (4.13b), or the plots of this in fig. 4.4, and (4.32). Now using (4.30), (4.31), (4.13b) and (4.32), or for the latter two, the related plots in figs. 4.4, 4.8 ($v = 1/3$), we find the $T/T_h - t/T_x$ curves shown in fig. 4.9.

We note from our dimensionless variables in (4.30), (4.31), and fig. 4.9, that for a certain material, given by E, v and ϱ, multiplying the plate thick-

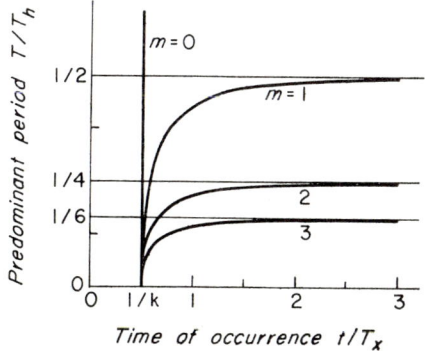

Fig. 4.9. Predominant period-time of occurrence curves for dilatational waves in the infinite plate with mixed face conditions ($v = 1/3$).

ness by a certain factor, multiplies T by the same factor. In turn multiplying the x position by a factor, multiplies t by the same factor.

4.1.1.4. Relation between group velocity and the velocity of energy transmission. Consider the simple group in (4.18) again. Forming $\hat{E}$ from it we find

$$\hat{E} = \varrho \dot{u}^2 = 4\varrho\omega^2 A^2 \cos^2[(d\varkappa/2)(x-c_g t)] \sin^2[\varkappa(x-ct)] + O(d\varkappa).$$

Taking the time average of this expression over several periods T, during which the modulation does not change much, we find

$$\langle \hat{E} \rangle \simeq 2\varrho\omega^2 A^2 \cos^2[(d\varkappa/2)(x-c_g t)],$$

which suggests the time-averaged energy density propagates with the velocity $c_E = c_g = d\omega/d\varkappa$. In Havelock's monograph [4.1] he treats some examples of one-dimensional dispersive waves, drawn from mechanics and electromagnetics. He shows for these examples, and generally for a medium having a strain-energy function, that $c_E = d\omega/d\varkappa = c_g$.

Achenbach [2.14 § 6.4] has shown that the time-averaged energy of SH waves in an infinite plate propagates with $c_E = d\omega/d\varkappa = c_g$. He also gives a proof [2.14 § 6.5] that $c_E = d\omega/d\varkappa = c_g$ holds generally for time harmonic wave propagation in waveguides of constant cross section. It is therefore applicable to the waves of interest in this chapter. We can apply Achenbach's scheme for SH waves to the waves governed by (4.11). It suffices to consider the symmetric dilatational waves governed by (4.11a), the other three cases can be treated similarly. First we calculate $\hat{K}$ and $\hat{U}$, and thus $\hat{E}$ for (4.11a). The results are

$$\hat{K} = \tfrac{1}{2}\varrho\omega^2 B^2 [\varkappa^2 \cos^2 \eta_d z \cos^2 \delta + \eta_d^2 \sin^2 \eta_d z \sin^2 \delta],$$

$$\hat{U} = \tfrac{1}{2}\lambda\varkappa_d^4 B^2 \cos^2 \eta_d z \cos^2 \delta + \mu B^2 [(\varkappa^4 + \eta_d^4) \cos^2 \eta_d z \cos^2 \delta \quad (4.33)$$

$$+ 2\varkappa^2 \eta_d^2 \sin^2 \eta_d z \sin^2 \delta],$$

where $\delta = \varkappa(x-ct)$. Now integrating $\hat{K}$ and $\hat{U}$ in (4.33) over the thickness of the waveguide, and in turn taking their time averages over the period T, we have

$$\langle K \rangle = \frac{1}{T} \int_0^T dt \int_{-h}^{h} [\hat{K} \text{ in } (4.33)] dz = \frac{1}{4} \varrho h \varkappa_d^2 \omega^2 B^2$$

$$\langle U \rangle = \frac{1}{T} \int_0^T dt \int_{-h}^{h} [\hat{U} \text{ in } (4.33)] dz = \frac{1}{4} \varrho h \varkappa_d^2 \omega^2 B^2, \quad (4.34)$$

where the dimensions of $\langle K \rangle$, $\langle U \rangle$ are energy per unit area. From (4.34) we see the total energy per unit area is

$$\langle E \rangle = \langle K \rangle + \langle U \rangle = 2\langle K \rangle. \tag{4.35}$$

Returning to (1.71) we have for the power input in the present case

$$\frac{dW}{dt} = \dot{W} = \int_S \sigma_{ij} l_j \dot{u}_i dS$$

since $X_i = 0$. The time rate of change of work, or power, per unit area is therefore

$$\hat{\dot{W}} = \sigma_{ij} l_j \dot{u}_i = T_i \dot{u}_i. \tag{4.36}$$

The *power* $\hat{\dot{W}}$ represents the *rate at which energy is transmitted per unit time per unit area*; hence it represents the *energy flux across the unit area*. The time average of this energy flux in the present case is therefore $\langle E \rangle c_E$. Hence we have the following relation between $\langle \dot{W} \rangle$ and $\langle E \rangle$

$$\langle \dot{W} \rangle = \langle E \rangle c_E = 2\langle K \rangle c_E, \tag{4.37}$$

where use has also been made of (4.35). The time-averaged power $\langle \dot{W} \rangle$ here is obtained by integrating $\hat{\dot{W}}$ in (4.36) over the thickness of the waveguide, and then taking its time average over the period T, i.e.,

$$\langle \dot{W} \rangle = -\frac{1}{T} \int_0^T dt \int_{-h}^{h} [\sigma_x \dot{u} + \sigma_{xz} \dot{w}] dz. \tag{4.38}$$

The negative sign is to insure a positive valued power, since the products of the stress and velocity are negative. After substituting into (4.38) from (4.7) and (4.8), and carrying out the indicated integrations for our present symmetric dilatational waves, (4.38) reduces to

$$\langle \dot{W} \rangle = \mu h \varkappa^2 \varkappa_s^2 c B^2 / 2, \tag{4.39}$$

which has the dimensions of power per unit length. Substituting the first of (4.34), and (4.39), into (4.37), we find

$$c_E = \langle \dot{W} \rangle / 2\langle K \rangle = c_d^2 / c. \tag{4.40}$$

Now from (4.13a) we find $d\omega/d\varkappa = c_d^2/c$, hence it follows from (4.40) that $c_E = d\omega/d\varkappa = c_g$. As would be expected for the antisymmetric dilatational waves (4.11b), the same result is found.

4.1.2. Elastically restrained plate faces

Following Mindlin [3.5] we now treat the case of a plate with elastically restrained faces. As we shall see this case is important since it discloses how coupling between P and SV waves develops at a boundary as the constraint there is relaxed. As in the analogous half-space case we now have the boundary conditions

$$\sigma_z = \mp ew, \qquad \sigma_{zx} = 0, \qquad \text{at } z = \pm h. \tag{4.41}$$

Substituting f and g of (4.5), with $A = D = 0$, in the second of (4.7), and in the second and third of (4.8), and substituting these expressions in (4.41), we find, for symmetric modes,

$$\mu[B(\varkappa^2 - \eta_s^2)\cos\eta_d h + 2iC\varkappa\eta_s \cos\eta_s h] = e[B\eta_d \sin\eta_d h - iC\varkappa \sin\eta_s h],$$

$$2iB\varkappa\eta_d \sin\eta_d h + C(\varkappa^2 - \eta_s^2)\sin\eta_s h = 0, \tag{4.42}$$

and similarly, but with $B = C = 0$, for antisymmetric modes,

$$\mu[A(\varkappa^2 - \eta_s^2)\sin\eta_d h - 2iD\varkappa\eta_s \sin\eta_s h] = -e[A\eta_d \cos\eta_d h + iD\varkappa \cos\eta_s h],$$

$$2iA\varkappa\eta_d \cos\eta_d h - D(\varkappa^2 - \eta_s^2)\cos\eta_s h = 0. \tag{4.43}$$

Consider first (4.43) for antisymmetric modes. For $e > 0$ (4.43) are satisfied provided

$$\cos\eta_d h = 0, \qquad \cos\eta_s h = 0,$$
$$A/iD = \pm[2\varkappa\eta_s/(\varkappa^2 - \eta_s^2)] = \mp\tan 2\beta. \tag{4.44}$$

If we observe (4.11b) and (4.12b) again, we see (4.44) shows that some roots of (4.43) are identical with some of the roots for the plate with mixed face conditions. Recall that the latter corresponds to $e \to \infty$ (and $w = 0$) in the context of the present case. Since now the $(\omega, \varkappa)$ pairs must be the same for both dilatational and equivoluminal parts, the roots in (4.44) determine $(\omega, \varkappa)$, or $(c, \varkappa)$, pairs that fall on the branches m, n, in fig. 4.4, but at intersections of m odd and n odd only. Now for $e > 0$, $\sin\eta_d h = 0$ and $\sin\eta_s h = 0$ [note (4.11a) and (4.12a)] are not roots of (4.43), since the latter cannot be satisfied by a single ratio of A/iD. Therefore, the branches of (4.43) do not pass through the intersections of the m even and n even branches in fig. 4.4 (note the latter are the branches representing

Ch. 4, § 4.1]   WAVES IN AN INFINITE PLATE...   195

symmetric dilatational and equivoluminal modes for the infinite plate with mixed face conditions).

With $e=0$, however, (4.43) is satisfied by

$$\sin \eta_d h = 0, \qquad \sin \eta_s h = 0,$$

$$\frac{A}{iD} = \mp[(\varkappa^2 - \eta_s^2)/2\varkappa\eta_d] = \pm k^2 \cos 2\beta \csc 2\alpha.$$
(4.45)

Observing (4.11a) and (4.12a) again, we see from (4.45) that it is $e=0$ that makes the branches of (4.43) go through the intersections of the m even and n even branches for the plate with mixed face conditions, in addition to the m odd and n odd found earlier, i.e., (4.44). More generally, for $0 \leq e \leq \infty$, we solve the equations (4.43) for A/iD, where for compatibility of the two equations there, we have

$$\frac{A}{iD} = \frac{2\mu\varkappa\eta_s \sin \eta_s h - e\varkappa \cos \eta_s h}{\mu(\varkappa^2 - \eta_s^2) \sin \eta_d h + e\eta_d \cos \eta_d h} = \frac{(\eta_s^2 - \varkappa^2) \cos \eta_s h}{2\varkappa\eta_d \cos \eta_d h}.$$
(4.46)

The last equality in (4.46) is the frequency equation. Fig. 4.10 here, taken

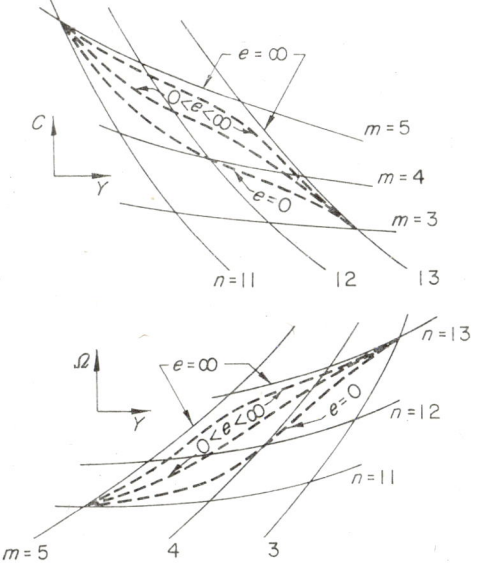

Fig. 4.10. Phase-velocity and frequency spectra of the antisymmetric modes of an infinite plate as influenced by the development of coupling between dilatational and equivoluminal overtones, due to the relaxation of the plate face constraints (after Mindlin).

from Mindlin [3.5], summarizes the influence of the elastic restraint. It shows how a real segment of a branch $\Omega - \gamma$, and $C - \gamma$ branch of (4.46), is affected by the development of coupling between dilatational and equivoluminal overtones, as the boundary restraint is relaxed.

The equations for the case of symmetric modes follow easily from the foregoing analysis because of the closeness in form of (4.42) and (4.43). One can see that simple interchanges of sine and cosine, and odd m, n, with even m, n, in the foregoing analysis, gives the behavior of the branches corresponding to (4.42).

We note from fig. 4.10 that between the intersections of m even with n even, and m odd with n odd, the $C - \gamma$ branch, or $\Omega - \gamma$ real segment of a branch for the infinite plate with mixed face conditions, form upper and lower bounds of both C, or Ω, and γ for the infinite plate with traction free faces. Mindlin points out that this behavior leads to *terracing* of the real segments of the upper branches for the infinite plate with free faces. Figure 4.11, taken from his paper [3.5], demonstrates this for the phase-velocity spectrum.

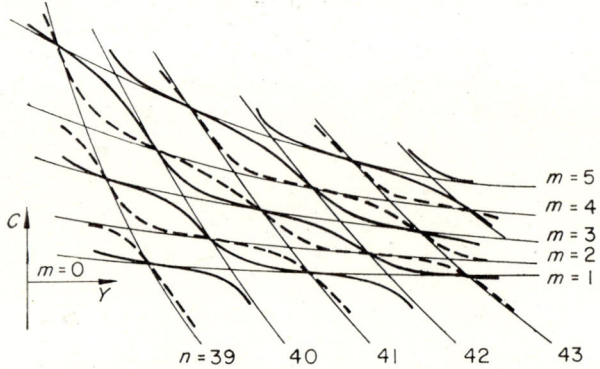

Fig. 4.11. Terrace-like structure of upper portion of phase-velocity spectrum of an infinite plate with traction-free faces (after Mindlin).

4.1.3. *Traction free plate faces; Rayleigh–Lamb frequency equation*

Setting $e = 0$ in (4.42) and (4.43), they give

$$\frac{B}{iC} = \frac{2\varkappa\eta_s \cos \eta_s h}{(\eta_s^2 - \varkappa^2)\cos \eta_d h} = \frac{(\varkappa^2 - \eta_s^2)\sin \eta_s h}{2\varkappa\eta_d \sin \eta_d h}$$

$$\frac{A}{iD} = \frac{2\varkappa\eta_s \sin \eta_s h}{(\varkappa^2 - \eta_s^2)\sin \eta_d h} = \frac{(\eta_s^2 - \varkappa^2)\cos \eta_s h}{2\varkappa\eta_d \cos \eta_d h}.$$

(4.47)

for symmetric and antisymmetric modes, respectively. The equality of the second two terms in each of (4.47) is imposed by the required compatibility of the two equations, in each of (4.42) and (4.43). These equalities may be written in dimensionless form as

$$F(\Omega^2, \zeta^2) = \frac{\tan(\pi\eta'_s/2)}{\tan(\pi\eta'_d/2)} + \left[\frac{4\zeta^2\eta'_s\eta'_d}{(\zeta^2-\eta'^2_s)^2}\right]^{\pm 1} = 0 \qquad (4.48)$$

where $\eta'_s = 2h\eta_s/\pi = (\Omega^2 - \zeta^2)^{\frac{1}{2}}$, and $\eta'_d = 2h\eta_d/\pi = [(\Omega/k)^2 - \zeta^2]^{\frac{1}{2}}$, and where the (+) and (−) are for symmetric, and antisymmetric modes, respectively. Equations (4.48) represent the classical frequency equations, given by Rayleigh and Lamb in 1889, for our present case of time harmonic-straight crested waves in an infinite elastic plate (in plane strain) with free faces. More recent work by Holden, Mindlin, Onoe, and others, has resulted in a comprehensive understanding of the spectra, modes, and waves governed by this equation. The paper by Mindlin [3.5], and a survey by Miklowitz [4.4], have outlines of the contributions made on this.

4.1.3.1. General character of Rayleigh–Lamb frequency spectrum; corresponding modes and waves. Consider the frequency equations (4.48) again. *General roots, or branches,* $\zeta(\Omega)$ of these equations, namely those *composed of segments with real, imaginary, and complex wave numbers,* have been shown to exist through numerical computation and approximations. With the existence of the general root $\zeta(\Omega)$ it is not difficult to show (4.48) has the conjugate root $\bar\zeta(\Omega)$ also, since ζ occurs as ζ^2. For the same reason it follows that $-\zeta(\Omega)$ and $-\bar\zeta(\Omega)$ are also roots, these being reflections of the former two in the origin $\zeta = 0$. So for $\Omega \geq 0$, a complex-wave number segment of a general branch (4.48) has images in both the planes $\gamma = 0$, and $\delta = 0$. Likewise real-, and imaginary-, wave number segments have images in the planes $\gamma = 0$, and $\delta = 0$, respectively. It should also be noted that since Ω occurs as Ω^2 in (4.48), $-\Omega$ corresponds to the reflection of the $\Omega > 0$ spectra, just discussed, in the plane $\Omega = 0$. Fig. 4.12, taken from Mindlin [3.5], depicts these general features of the branches of the Rayleigh–Lamb frequency equation for symmetric modes. The lowest three branches for relatively small $|\zeta|$ are shown. Proceeding from $\Omega = \infty$ to 0, along the third branch, we see it is real (3R) until it reaches a branch point at cutoff ($\zeta = 0$). It then becomes, and remains, imaginary (3I), until a second branch point is reached at the next lowest cutoff frequency. The branch then becomes, and remains, real (3R) until a third branch

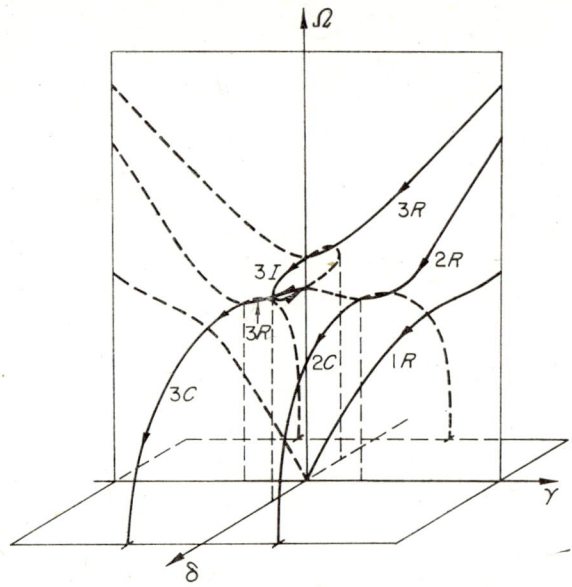

Fig. 4.12. First three branches of Rayleigh–Lamb frequency equation for symmetric modes (solid lines) and images (dashed lines) showing real (R)-, imaginary (I)-, and complex (C)- wave number segments (after Mindlin).

point is reached. This occurs at the minimum in (3R), at a real negative wave number. The branch then proceeds to $\Omega=0$ along a complex-wave number segment (3C). It may be seen, therefore, that the third branch we have been discussing is an analytically continued one, having a one to one correspondence between Ω and ζ, as Ω proceeds from infinity to zero. The production of this *analytically continued* information (in. fig. 4.12) requires a numerical procedure for the segments away from the branch points, and a proper means for determining where the branch points lie, and what changes in the branches occur there. In a report by Mindlin [4.5], on the work of Medick, Onoe, and Mindlin, it is shown that the branch points occur at places along the real and imaginary wave number segments of the branches, where the slopes are zero, i.e., where $\partial\Omega/\partial\gamma=0$ and $\partial\Omega/\partial\delta=0$, respectively. Imaginary and complex segments emanate from these points (which include cutoff points), as fig. 4.12 demonstrates. This is proved in [4.5] (for the real segment case) by first rewriting (4.48) in its real and imaginary parts, and then expanding in powers of δ. Then, retaining first order terms, it follows that

Ch. 4, § 4.1]   WAVES IN AN INFINITE PLATE...   199

$$F(\Omega, \gamma) = 0, \qquad \delta[\partial F(\Omega, \gamma)/\partial \gamma] = 0. \qquad (4.49)$$

Equations (4.49) govern the branches in the $\Omega - \zeta$ space near the real $\Omega - \gamma$ plane. They are satisfied by

$$F(\Omega, \gamma) = 0, \qquad \delta = 0, \qquad (4.50\text{a})$$

and

$$F(\Omega, \gamma) = 0,$$

$$\partial F(\Omega, \gamma)/\partial \gamma = 0, \qquad \delta \neq 0. \qquad (4.50\text{b})$$

Equation (4.50a) represents the real segments, and (4.50b) the complex segments, which emanate from the real segments. Now since along the real segments of the branches

$$dF = \frac{\partial F(\Omega, \gamma)}{\partial \Omega} d\Omega + \frac{\partial F(\Omega, \gamma)}{\partial \gamma} d\gamma = 0,$$

it follows, since $d\Omega/d\gamma = \partial \Omega/\partial \gamma$, that

$$\frac{\partial F(\Omega, \gamma)}{\partial \gamma} + \frac{\partial F(\Omega, \gamma)}{\partial \Omega} \frac{\partial \Omega}{\partial \gamma} = 0. \qquad (4.51)$$

Therefore, if the second of (4.50b) is satisfied, we must have $\partial \Omega/\partial \gamma = 0$, since usually $\partial F(\Omega, \gamma)/\partial \Omega$ does not vanish. This proves that complex segments emanate from the real segments, at points of zero slope of the latter, say at point (Ω^*, γ^*). When the branch points are at cutoff, then $\gamma^* = 0$, and imaginary segments emanate from the real segments, since generally the latter have zero slope at cutoff. There is an exception when two cutoff modes of the same symmetry have the same frequency, i.e., equivoluminal and dilatational modes. Then the slopes of the real segments $\partial \Omega/\partial \gamma$ do not vanish, and no imaginary segment is generated. Further, from (4.50b), (4.51), we must have $\partial F(\Omega, \gamma)/\partial \Omega = 0$. We shall discuss this case further in the next section.

For complex segments emanating from the imaginary segments at points of zero slope in the latter, the proof is similar to the above for real segments. Reference [4.5] points out, however, that whereas a real segment has at most one point of zero slope, certain of the imaginary segments have an infinite number of such points. This can be observed in fig. 4.13 of the next section, taken from reference [4.5].

Concerning the changes in the branches occurring at the branch points, it may be observed in fig. 4.12 that for branches 2 and 3 here (solid lines), the complex and imaginary segments satisfy the condition. Im $\zeta = \delta > 0$. This is a necessary condition if the wave forms $\exp[i\zeta(\hat{x} - C\hat{t})]$ in (4.5) (4.7), and (4.8) [cf. (4.15c)] are to be bounded as $\hat{x} \to \infty$. Such a criterion then could be set down at the outset which would produce the branches 2 and 3 (solid lines) in fig. 4.12 (and all similar higher branches), instead of the dashed line images there (which would represent unbounded waves). It follows that the admissible complex and imaginary segments, which must satisfy Im $\zeta = \delta > 0$, correspond to edge waves, as discussed in § 4.1.1.1 [cf. (4.15c) there] for imaginary segments.

In the foregoing discussion we have been using *"branch points"* to describe a point on a branch at which a change in the nature of the branch occurs, e.g., from a real to an imaginary wave segment. Further, we have referred to a branch (in fig. 4.12) undergoing such changes, and having a one to one correspondence between Ω and ζ, as Ω proceeds from infinity to zero, as an analytically continued one. Later, in our treatment of transient waves in an elastic waveguide, we shall see that these "branch points" in the $\Omega - \zeta$ space correspond truly to branch points in a more general $p - \varkappa$ space where p, the Laplace transform parameter, plays the role of a complex frequency, and $\varkappa$ (or ζ) is again our complex wave number. In that work we will derive solutions through *analytical continuation of branches* $\varkappa_j(p)$ *of the generalized Rayleigh–Lamb frequency equations* $F(p^2, \varkappa^2) = 0$, where, for example, $\varkappa_3(p)$ would correspond to branch 3 in fig. 4.12. Accordingly, we will show (1) that $dp/d\varkappa = 0$ at the branch points, (2) the behavior of the branches near these points is

$$\varkappa_j(p) - \varkappa_j^* \simeq C(p - p^*)^{\frac{1}{2}}, \qquad (4.52)$$

where C is a constant, and (3) we must have Im $\varkappa_j(p) > 0$, to satisfy a boundedness condition on our waves for $x \to \infty$.

Concerning the modes associated with the branches of (4.48), we note with Ω real and positive, and ζ real, imaginary, or complex, that η'_d and η'_s can be as follows:

For ζ *real*:

η'_d, η'_s real, if $\Omega > k\gamma$,

η'_d imaginary, η'_s real, if $\gamma < \Omega < k\gamma$,

η'_d, η'_s imaginary, if $\Omega < \gamma$.

For ζ *imaginary*:

η'_d, η'_s real.

For ζ *complex*:

η'_d, η'_s complex.

Clearly, for these cases, from (4.5), (4.7), and (4.8) we have functions that are trigonometric, hyperbolic, or products of these two, for mode shapes, depending on whether the η's are real, imaginary, or complex, respectively.

As we have seen the analysis leading to (4.5) was based on pairs of *P* and *SV* waves, incident upon and reflecting from the faces of an infinite plate. It follows therefore that the waves, associated with pairs (Ω, ζ) or (C, γ), in (4.48), and fig. 4.13, can be interpreted as being composed of pairs of *P* and *SV* waves incident upon and reflecting between the faces $z = \pm h$ of the free infinite plate.

4.1.3.2. Further on real- and imaginary-wave number segments and corresponding modes and waves. It was pointed out in § 4.1.2, that the *real wave number segments of the frequency spectrum and the phase velocity spectrum of the infinite plate with mixed face conditions, were bounds for the corresponding segments of the Rayleigh–Lamb spectrum* (see fig. 4.10). In Mindlin's report [4.5] use was made of this fact, and the like one for the imaginary wave number segments, to sketch on these bounds the real- and imaginary-wave number segments for a large number of the branches of the Rayleigh–Lamb freguency equations (4.48), for $\nu = 0.31$. They are shown in fig. 4.13, which is a reproduction of fig. 3 in [4.5] (or fig. 19 in [3.5]). In the sketching, aid is obtained from formulas for the coordinates and slopes of the Rayleigh–Lamb segments, at their intersections with the bounds, their ordinates, slopes,

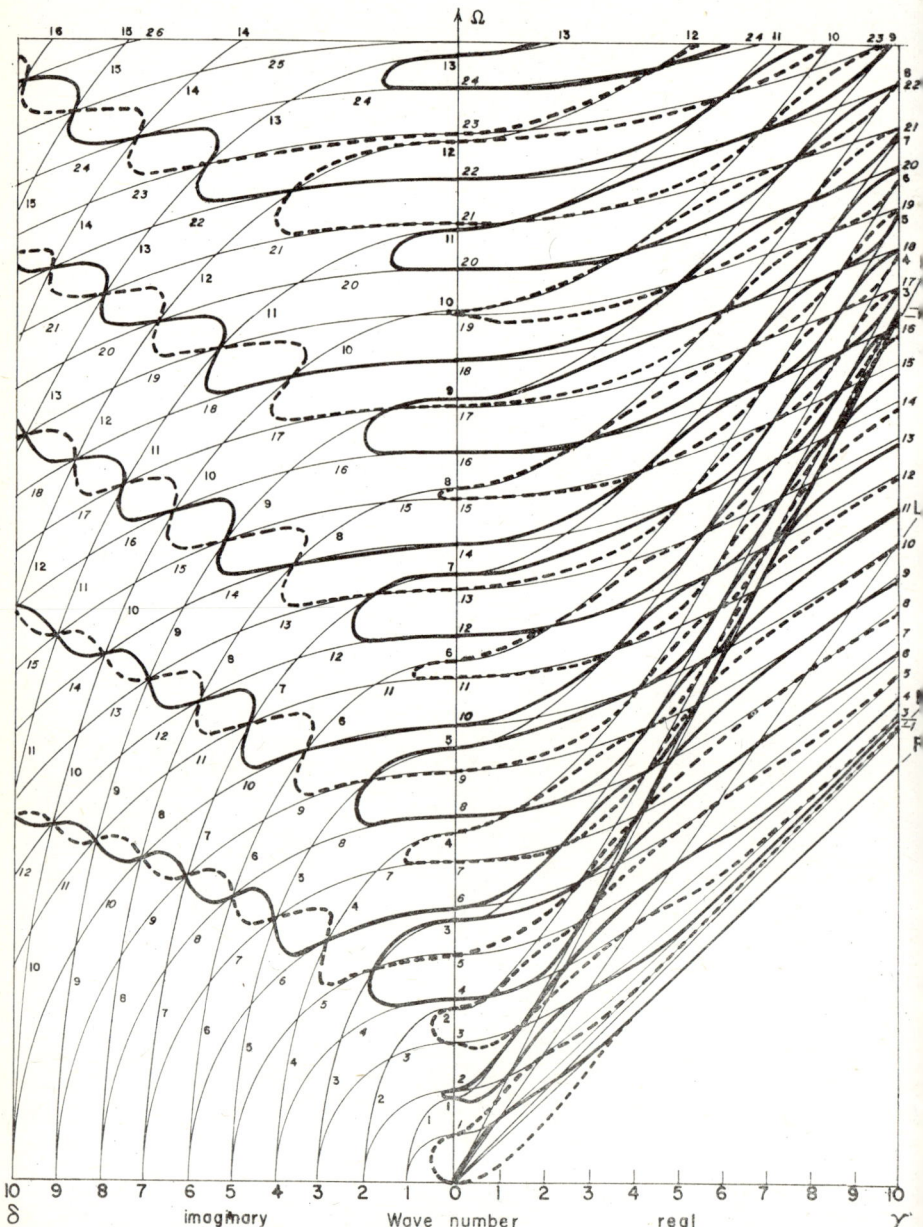

Fig. 4.13. Rayleigh–Lamb frequency spectra for real- and imaginary-wave numbers for $\nu = 0.31$; solid lines for symmetric and dashed for antisymmetric modes (courtesy Mindlin).

and curvatures at cutoff ($\zeta=0$), and their asymptotic behavior at large ζ (γ or δ). These formulas are given in [4.5]. One can construct the corresponding phase velocity spectrum by using the real-wave number segments in fig. 4.13, in conjunction with the relation $C=\Omega/\gamma$. Fig. 18 in [3.5] contains extensive sketches of this spectrum for $v=1/3$.

We have already seen that cutoff is independent of x and associated with normal incidence. The latter, from ch. 3 § 3.1.4.1, governs reflection of P and SV waves with no mode coupling. Hence we would expect the same of (4.48). Imposing $\zeta\to 0$ on (4.48), noting that when $\zeta\to 0$, $\eta'_d=\Omega/k$, and $\eta'_s=\Omega$, where Ω is the non-vanishing cutoff frequency, it reduces to

$$\frac{\tan(\pi\eta'_s/2)}{\tan(\pi\eta'_d/2)}=0,$$

(4.53)

$$\frac{\tan(\pi\eta'_d/2)}{\tan(\pi\eta'_s/2)}=0,$$

for symmetric and antisymmetric modes, respectively. (4.53) can be rewritten as

$$\frac{\sin(\pi\eta'_s/2)\cos(\pi\eta'_d/2)}{\cos(\pi\eta'_s/2)\sin(\pi\eta'_d/2)}=0,$$

(4.54)

$$\frac{\sin(\pi\eta'_d/2)\cos(\pi\eta'_s/2)}{\cos(\pi\eta'_d/2)\sin(\pi\eta'_s/2)}=0,$$

to show that they yield the roots

$$\Omega=k\alpha, \quad \alpha=\begin{cases}1, 3, 5\ldots\text{symmetric modes}\\ 2, 4, 6\ldots\text{antisymmetric modes}\end{cases}, \quad (4.55a)$$

$$\Omega=\beta, \quad \beta=\begin{cases}1, 3, 5\ldots\text{antisymmetric modes}\\ 2, 4, 6\ldots\text{symmetric modes}\end{cases}. \quad (4.55b)$$

Equations (4.55a), (4.55b), represent the frequencies of the dilatational and equivoluminal modes, respectively. They are referred to as *simple thickness-stretch, and -shear modes*, respectively. Note that, as in the case of mixed plate face conditions, the dilatational cutoff frequencies are dependent on Poisson's ratio, but the equivoluminal ones are not. These cutoff frequencies are shown in fig. 4.13 with the numbers to the immediate right of the Ω axis being the β's, and immediate left the α's. Comparing (4.25), (4.55b), and (4.24), (4.55a), as well as figs. 4.4, 4.13, it may be observed that the cutoff frequencies for the two plate cases, mixed and traction-free face conditions, are the same for the equivoluminal, but not the dilatational modes. The cutoff α's and β's are a useful means of ordering the real segments, e.g., those for the first, second, and third symmetric thickness-stretch modes are solid curves in fig. 4.13, marked 1, 3, and 5 to the left of the Ω axis (i.e., $\alpha = 1, 3, 5$). Equations (4.55) point out that the Rayleigh–Lamb equations (4.48) each have an infinite number of branches (or roots), like those shown in fig. 4.12.

In the last section it was pointed out that there are exceptional cases when two real segments of the same symmetry (dilatational and equivoluminal) intersect at cutoff, i.e., they have the same cutoff frequency. In a case of this type the slopes of the real segments $\partial \Omega / \partial \gamma \neq 0$, no imaginary segment is generated, and $\partial F(\Omega/\gamma)/\partial \Omega = 0$. Observing (4.55) we see these cases occur when $k = \beta/\alpha$. An important example is when the first two real segments, for symmetric thickness modes (in fig. 4.13), meet at cutoff, i.e., for $k=2(v=\frac{1}{3})$, and $\beta=2$, $\alpha=1$. As v goes through this value $\frac{1}{3}$, the nature of the spectra changes, and hence $v=\frac{1}{3}$ is a critical value. That is, for $v<\frac{1}{3}$, as fig. 4.13 shows, the first thickness-stretch (dilatational) and first thickness-shear (equivoluminal) real segments (and modes) are separated, with the latter being higher than the former. For $v>\frac{1}{3}$ they interchange their relative positions. Fig. 4.14 demonstrates this. Such sensitivity of the spectra to changes in Poisson's ratio v is discussed in greater detail in Mindlin's report [4.5, § 11].

We should point out here that there are other possible types of motion, associated with normal incidence of wave pairs, with $\zeta = 0$ and Ω's from (4.55). They are of the type we discussed in ch. 3, § 3.1.4.6 and found in exercise 3.8, ch. 3. § 3.5.

In fig. 4.13, of course, real segments with positive values of $\zeta = \gamma$ only are shown. Actually where the slope of a real segment there is negative, this corresponds to a negative value of γ, hence fig. 4.13 shows the reflection

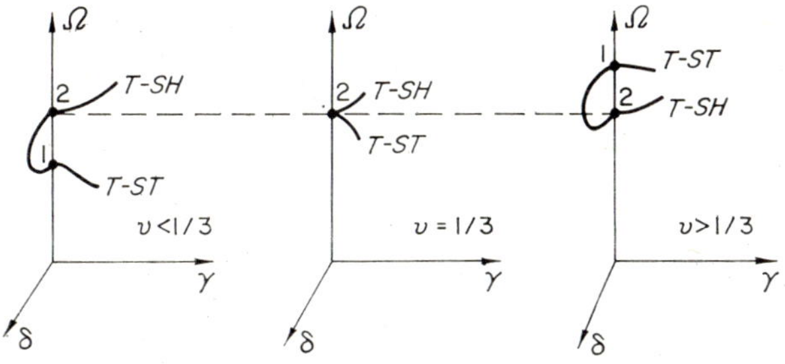

Fig. 4.14. Behavior of the first two real segments for symmetric thickness modes near cutoff and the critical Poisson's ratio value 1/3 (after Mindlin).

in the plane $\gamma=0$ for such segments. These are really the images of the real segments of the branches appearing in fig. 4.12 (dashed lines), e.g., the part of the first real segment for symmetric thickness modes in fig. 4.13, having negative slope (just after cutoff), is really the reflection of $3R$ in $\gamma=0$, the dashed curve in fig. 4.12. This discloses the general feature that curvature for the real segments at cutoff may be positive or negative. For the positive case the phase velocity and group velocity are both positive, but for the negative case the *phase velocity is negative* and the *group velocity positive*, in the vicinity of cutoff. Note that wave groups, corresponding to points at, and near cutoff, and the minimum points on these real segments, will occur at long time, since $d\Omega/d\gamma = C_g \to 0$ at these places.

Mindlin's detailed discussion in [3.5] of the spectra in fig. 4.13, and related modes and waves, is of interest here. As he points out, starting at $\zeta=0$, the ratio of the curvatures of the real or imaginary segment of a branch and bound determines whether the segment starts out above or below the bound. Thereafter, as fig. 4.13 shows, the segment is contained between bounds, crossing them only at successive intersections of bounds m even with n even, and m odd with n odd. As we have noted earlier, at these intersections the corresponding modes satisfy both mixed and traction free conditions on the faces of the infinite plate. The shapes of these modes across the thickness of the plate are characterized by an even number, $2m$ and $2n$, of quarter wavelengths of the dilatational and equivoluminal parts of the displacement, respectively, (cf. fig. 4.3). To identify the wave nature of these modes one compares (4.44) and (4.45), with the last of (3.58) and (3.59), the latter being cases of reflecting pairs of P and SV waves, as we have seen. The antisymmetric modes are predominantly equivoluminal ($|A/iD|<1$) at intersections of m odd with n odd, and predominantly dilatational ($|A/iD|>1$) at intersections of m even with n even. This can be proved by calculating $|A/iD|$ from the first equalities in the second equations of (4.44) and (4.45), using data supplied by the frequency spectra in fig. 4.13, and the first equations in (4.44) and (4.45). The converse holds for the symmetric case, which can be proved in a manner similar to the antisymmetric case. Note that the real and imaginary segments show agreement with this (at the intersections) by their more predominant alignment with the bounds $m=$ constant, when they are predominantly dilatational, and $n=$ constant, when they are predominantly equivoluminal in nature. This is understandable if one recalls that the bounds $m=$ constant, and $n=$ constant, are, respectively, the dilatational and equivoluminal branches defined by (4.11) and (4.12), and shown in fig. 4.4. We note, then that between intersections on a real or imaginary segment, of a certain symmetry, there is a continuous transition from one type of deformation to the other, and from one even number of quarter wavelengths to the next.

Let us consider the nature of the real segments, and modes and waves they govern, further. As γ increases, all these segments, except the lowest symmetric and antisymmetric ones, cross the fundamental one, $\Omega = k\gamma$ ($m=0$), line OD in fig. 4.13 where $c=c_d$. At the intersections, since $c=c_d$, the symmetric and antisymmetric modes correspond to combinations of the waves found by Goodier and Bishop for grazing incidence (cf. ch. 3, § 3.1.4.2), given in [3.5] by equations (41) and (42), and figs. 7 and 8, respectively. The line OD ($m=0$) is a bound for the segments representing symmetric modes, which cross it at its intersections with bounds n even. The slopes of these segments, at these intersections, are all the same, and less than the slope of OD, if $\nu \neq 0$. Hence $c_g < c_d$ at these points, being given by

$$\frac{c_g}{c_d} = \frac{4(k^2-1)+(k^2-2)^2}{4(k^2-1)+k^2(k^2-2)^2}.$$

Note that c_g/c_d is a function of ν only. The line $m=0$ is not a bound for the real segments of antisymmetric modes, and these segments do not cross it at its intersections with bounds n odd. After the first few segments, however, the intersections of bounds, and these segments, with $m=0$ get very close together, and c_g from segment to segment is nearly constant and $<c_d$, for $\nu \neq 0$.

Still further out, all the segments, except that for the lowest antisymmetric mode, cross the line $OL(\Omega=(2)^{\frac{1}{2}}\gamma)$ at its intersesections with bounds $n=$ constant, to which the segments are tangent. The modes are purely equivoluminal at these points, being composed of SV waves having angles of incidence and reflection of $\pi/4$. Hence, they are of the type shown in fig. 3.15 (note $c = c_s/\sin \beta = (2)^{\frac{1}{2}} c_s$ here). These points, $\Omega=(2)^{\frac{1}{2}}n$, $\gamma=n$, correspond to those Lamé found for the modes of vibration of rectangular parallelepipeds. These points, $\Omega=(2)^{\frac{1}{2}}n$, $\gamma=n$, and the points $\Omega=\beta$, $\gamma=0$, corresponding to the thickness shear cutoff modes, are the only points in the frequency spectra for real and imaginary ζ that correspond to purely equivoluminal modes. It follows that they are the only points, for real and imaginary ζ, that are invariant to a change in Poisson's ratio ν. After crossing $\Omega=(2)^{\frac{1}{2}}\gamma$, all the segments, except the lowest ones for symmetric and antisymmetric modes, approach $\Omega=\gamma$ (line OE in fig. 4.13), where the limiting phase velocity $c=c_s$, asymptotically as γ becomes large. The segment for the lowest symmetric mode is the only one that intersects the line OE. At this intersection the mode shape corresponds to a combination of the waves found by Goodier and Bishop for grazing incidence, given in [3.5] by equations (43), and fig. 9. The lowest segments for symmetric and antisymmetric modes approach the line $\Omega=(c/_Rc_s)\gamma$ (line OR in fig. 4.13), where the limiting phase velocity $c=c_R$, asymptotically as γ becomes large. The segment for the symmetric mode approaches OR from above, and that for the antisymmetric mode from below. The velocity ratio c_R/c_s is calculated from (3.65), or (3.66), subject to (3.68). Hence the very highest frequency-very shortest waves, corresponding to the lowest symmetric and antisymmetric modes, are Rayleigh surface waves (cf. ch. 3, § 3.1.4.7).

4.1.3.3. Further on complex-wave number segments and corresponding modes and waves. Fig. 4.12, in its consideration of just the lowest three branches for symmetric modes, exhibits just two complex segments and their images. It has already been pointed out, in § 4.1.3.1, such segments emanate from minima in the real and imaginary segments of the spectra, with the former having at most one such point on each segment, and the latter certain segments with an infinity of such points. These points are evident in fig. 4.13, along with necessary maxima on the imaginary segments. The complex segments emanating from the minima, occurring on the imaginary segments (and to a lesser extent, the real segments) are loops which join the maxima on the imaginary segments in an ordered way to provide a path for analytical continuation of the higher branches of (4.48). The scheme is depicted in fig. 4.15, taken from a paper by Onoe, McNiven, and Mindlin [4.6] on axially symmetric waves in the infinite circular cylindrical rod (to be discussed later in this chapter). Since the spectrum for axially symmetric

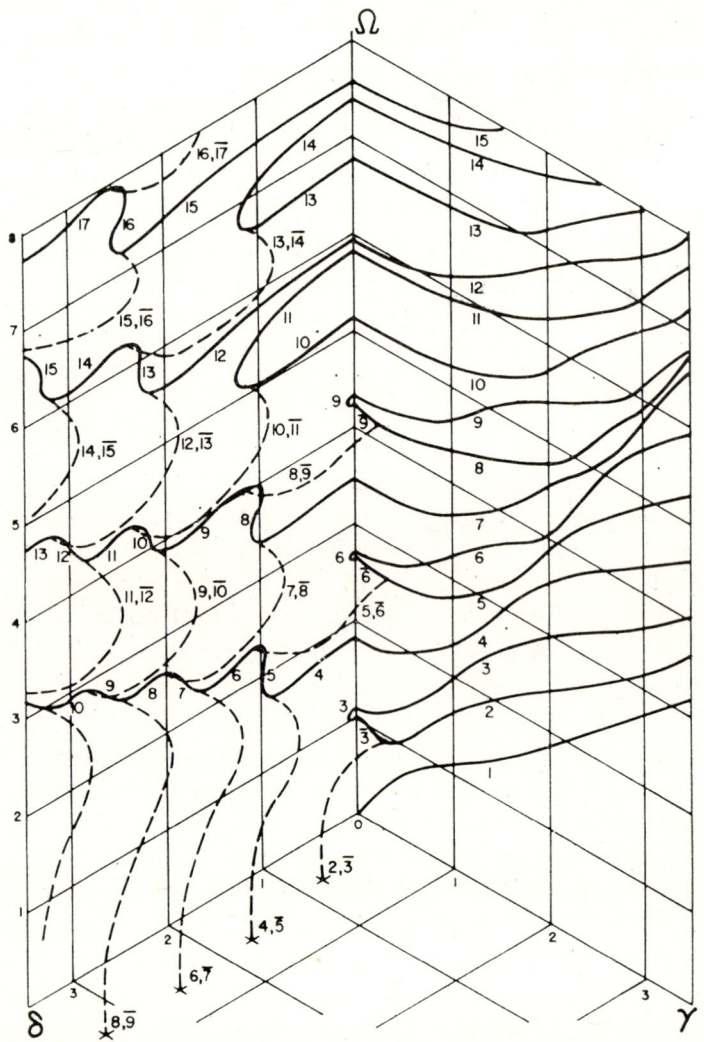

Fig. 4.15. Frequency spectra for axially symmetric waves in an infinite circular cylindrical rod (after Onoe, McNiven, and Mindlin).

waves in the rod is quite similar to that the symmetric waves in a plate we are discussing, fig. 4.15 suffices to bring out the nature of the continuation for the higher branches of the latter. The plot of the spectrum in fig. 4.15 is confined to one octant of $\Omega-\zeta$ space, which suffices because of the symmetries we have already pointed out. Consider the seventh branch, for example, starting with the real segment marked 7 in fig. 4.15. After going through its cutoff point this branch becomes, and remains, imaginary until it reaches the first minimum point. It then continues on a complex loop generated there, marked 7, $\overline{8}$, with $\overline{8}$ indicating this loop is also the image (in the plane, $\gamma=0$) of a similar continuation of the eighth branch. It may then be seen the loop intersects a maximum on the next lowest imaginary segment of like-nature. The branch then proceeds as an imaginary segment, marked 7, until reaching a minimum point, and finally proceeds as a complex segment generated there, marked 6, $\overline{7}$, until it pierces the $\Omega=0$ plane. Note from fig. 4.15 that such piercing points form an infinite set of complex numbers. The analogous sets for the present plate case can be found by approximating $F(\Omega^2, \zeta^2)$ in (4.48), which yields

$$\lim_{\Omega \to 0} \frac{F(\Omega^2, \zeta^2)}{\Omega^2} = \mp \frac{1}{4(1-\nu)\zeta^2 \sinh \pi\zeta} (\sinh \pi\zeta \pm \pi\zeta) , \qquad (4.56a)$$

where the upper signs correspond to symmetric, and lower signs to antisymmetric modes. Hence the corresponding sets are obtained from the roots of the equations

$$\sinh \pi\zeta \pm \pi\zeta = 0 . \qquad (4.56b)$$

It may be noted that these equations are oft occurring in two-dimensional elastostatic problems, which would be expected from our present $\Omega \to 0$ approximation.

4.2. SH waves in an infinite plate

Following the treatment in § 4.1, we consider two equal amplitude, and like sign, *SH* waves propagating in our plate of thickness 2h, symmetrically with respect to the midplane $z=0$, as shown in fig. 4.16. These waves can be written from (3.82), and analogously to the way we constructed (4.5).

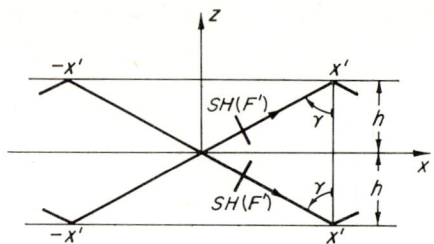

Fig. 4.16. Symmetric *SH* waves in an infinite plate.

We have in totality here, then,

$$\chi = j(z) \exp\left[i\varkappa(x-ct)\right], \qquad (4.57a)$$

where

$$j(z) = E \sin \eta_s z + F \cos \eta_s z .$$

In the nomenclature of (3.82) and fig. 3.28, $2B_2 = 2B_1 = 2F' = F$, and $2B_2 = -2B_1 = 2E' = E$, these being associated with symmetric and antisymmetric waves, respectively, where here $\omega = \varkappa c = \varkappa_s c_s$, $\varkappa = \varkappa_s \sin \gamma$, and $\eta_s = \varkappa_s \cos \gamma = [(\omega/c_s)^2 - \varkappa^2]^{\frac{1}{2}}$. Applying (3.84), (3.81) to (4.57a) gives for the one existing displacement,

$$v = -\varkappa_s^2 j(z) \exp\left[i\varkappa(x-ct)\right], \qquad (4.57b)$$

and for the one existing stress on the plate faces, using the second of (3.86),

$$\sigma_{zy} = -\mu \varkappa_s^2 j'(z) \exp\left[i\varkappa(x-ct)\right] . \qquad (4.58)$$

For the plate with traction free faces, we have from (4.58)

$$j'(\pm h) = 0 ,$$

which is satisfied with

$$E=0, \quad \sin \eta_s h=0, \quad \text{or} \quad \eta_s=n\pi/2h, \qquad n=0, 2, 4,\ldots, \tag{4.59a}$$

or

$$F=0, \quad \cos \eta_s h=0, \quad \text{or} \quad \eta_s=n\pi/2h, \qquad n=1, 3, 5,\ldots, \tag{4.59b}$$

(4.59a), (4.59b) give the symmetric and antisymmetric modes, respectively. Now from η_s we can write

$$\Omega^2=\omega^2/\omega_s^2=n^2+\zeta^2, \tag{4.60}$$

which we recognize is the same as (4.15a), the frequency equation for equivoluminal waves in a plate (in plane strain) with mixed conditions on its faces. It follows that we have the same spectra here as the curves $n=$ constant in fig. 4.4, with, in addition, the important label $n=0$ on the curves $C=1$ and $\Omega=\gamma$ there, as (4.59a) requires. The corresponding lower mode shapes, $n=0, 1, 2,$ and 3, for χ and v would be the same, respectively, as the $m=0$, symmetric dilatational, $n=1, 2,$ and 3, equivoluminal modes for u, in fig. 4.3. Note also that the $C_g-\gamma$ curves here would be the same as in fig. 4.8, with n replacing m, and 1 replacing k there. Similarly the $T/T_h-t/T_x$ curves here would be the same as in fig. 4.9, with again n replacing m, 1 replacing k, and cutoff periods $T/T_h=1/n$ replacing $1/km$ there. The wave features such as normal incidence at cutoff, grazing incidence at the high frequency-short wave limit, etc., that we found for the plate (in plane strain) with mixed face conditions, would carry over here also.

4.3. Love waves

Ewing et al. [3.11, § 4–5] point out that the first long-period *seismographs*, which measured horizontal motion only, exhibited large transverse components in the main disturbance of an earthquake. This early established fact in seismology was explained in 1911 by Love [3.17, pp. 160–165]. It is easily shown that there can be no *SH* surface wave on the free surface of a homogeneous elastic half space. Hence this simple model could not explain the measurements. Love, however, showed the waves involved were *SH* waves, confined to a superficial layer of an elastic half space, the layer having different properties than the rest of the half space. Essentially

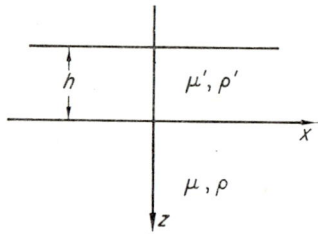

Fig. 4.17. An elastic half space with a superficial layer.

following Love's treatment, we consider first the geometric model shown in fig. 4.17. The origin of our cartesian coordinates (x, y, z) is taken along the interface, with positive y out of the plane of the figure. The constants μ' and ϱ' are, respectively, the shear modulus and material density of the layer. Since we are dealing with SH waves we can draw on relations in §§ 3.2.4.2. It follows only the displacement component along y, $v(x, z, t)$, is nonvanishing. Making use of (4.57b), (4.58), and (4.57a), the displacement v', and stress σ'_{zy}, in the layer can be written as

$$v'(x, z, t) = -\varkappa_s'^2 [E \sin \eta_s' z + F \cos \eta_s' z] \exp [i\varkappa(x-ct)], \quad (4.61)$$

and

$$\sigma'_{zy}(x, z, t) = -\mu' \varkappa_s'^2 \eta_s' [E \cos \eta_s' z - F \sin \eta_s' z] \exp [i\varkappa(x-ct)], \quad (4.62)$$

respectively, where $\eta_s' = (\varkappa_s'^2 - \varkappa^2)^{\frac{1}{2}} = \varkappa[(c/c_s')^2 - 1]^{\frac{1}{2}}$, $\varkappa_s' = \omega/c_s'$, and $c_s' = (\mu'/\varrho')^{\frac{1}{2}}$. In order that the energy be essentially confined to the layer, we write the displacement v and stress σ_{zy} for the half space $z > 0$, as interface waves, decaying exponentially with z into the interior. Hence from (3.87), (3.86), and (3.82), we have for these waves

$$v(x, z, t) = -\varkappa_s^2 B \exp(-i\eta_s z) \exp [i\varkappa(x-ct)] \quad (4.63)$$

$$\sigma_{zy}(x, z, t) = \mu \varkappa_s^2 (i\eta_s) B \exp(-i\eta_s z) \exp [i\varkappa(x-ct)], \quad (4.64)$$

where $\eta_s = -i\varkappa[1-(c/c_s)^2]^{\frac{1}{2}}$. The forms of η_s, η_s' restrict the phase velocity to the range $c_s' < c < c_s$.

On the free surface $z=-h$, the condition $\sigma'_{zy}=0$ applies, and at the interface $z=0$, the continuity conditions $v=v'$, $\sigma_{zy}=\sigma'_{zy}$ must hold. Eqs. (4.61)–(4.64) are substituted into these conditions, producing the homogeneous set of equations for B, E, and F

$$E \cos \eta'_s h + F \sin \eta'_s h = 0,$$

$$\varkappa_s^2 B - \varkappa_s'^2 F = 0,$$

$$\mu \varkappa_s^2 (i\eta_s) B + \mu \varkappa_s'^2 \eta'_s E = 0.$$

Setting the determinant of these equations equal to zero, we find the phase velocity c-wave number $\varkappa$ relation, governing the dispersion of Love waves, to be

$$\tan \eta'_s h = \mu(i\eta_s)/\mu' \eta'_s,$$

or

$$\tan \{[(c/c'_s)^2 - 1]^{\frac{1}{2}} \varkappa h\} = \mu[(1 - (c/c_s)^2]^{\frac{1}{2}}/\mu'[(c/c'_s)^2 - 1]^{\frac{1}{2}}. \quad (4.65)$$

Letting c approach c_s in (4.65), we find that

$$[(c/c'_s)^2 - 1]^{\frac{1}{2}} \varkappa h \to 0, \pi, 2\pi, \ldots, n\pi, \qquad n = 0, 1, 2, \ldots,$$

so we have an infinite number of real roots or branches of (4.65). The lowest branch ($n=0$) therefore has the property of $\varkappa h \to 0$ as $c \to c_s$. We note in this case, since $\omega = \varkappa c \to \varkappa c_s \to 0$, that this limit governs a wave of infinite wavelength and infinitely long period. At the other end of the c range, where $c \to c'_s$, (4.65) yields the roots

$$[(c/c'_s)^2 - 1]^{\frac{1}{2}} \varkappa h \to \pi/2, 3\pi/2, \ldots, (2n+1)\pi/2, \qquad n = 0, 1, 2, \ldots,$$

In this case all of these roots must have the property that $\varkappa h \to \infty$ as $c \to c'_s$. It follows, therefore, for all of the branches, that this limit governs waves that are of infinitely short wavelength and period. The literature on *Love waves*, both in time harmonic and transient wave studies, is vast. Extensive numerical evaluations of (4.65) for different values of the parameters μ/μ',

c_s/c_s' have been carried out. For our purposes the results for the first two branches of the phase- and group-velocity spectra, given by Ewing et al. [3.11, Fig. 4-53], and based partially on data obtained by Stoneley [4.7], will suffice. They are reproduced here in fig. 4.18. Note the agreement in

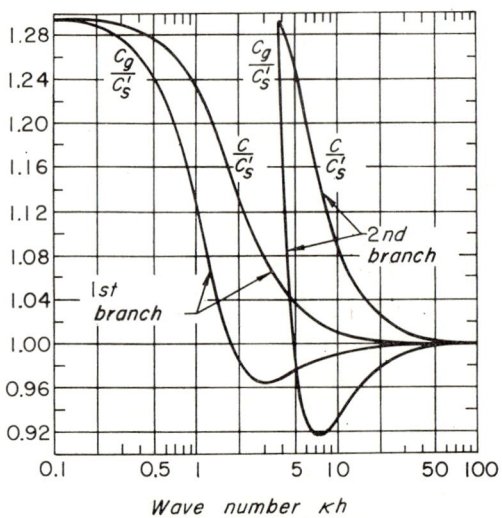

Fig. 4.18. Phase- and group-velocity curves for first- and second-mode Love waves for case $c_s/c_s' = 1.297$ and $\mu/\mu' = 2.159$ (after Ewing, Jardetsky and Press).

the figure with the limits discussed here for $\varkappa \to 0, \infty$. The minimums in the group velocity curves are important too. As we shall see later, they correspond to a higher order stationary phase point and its larger disturbance for long time.

4.4. Waves in an infinite elastic rod of circular cross section

Work on wave propagation in the elastic rod dates back to Pochhammer's classical study published in 1876, predating, therefore, Rayleigh and Lamb's work on the plate in plane strain. On the basis of time harmonic wave trains Pochhammer treated all types, axially symmetric (compressional or longitudinal), non-axially symmetric (flexural or transverse), and axially symmetric torsional waves in the infinite rod of circular cross section and traction free cylindrical surface. The first two types are analogs of the symmetric and antisymmetric waves in the plate in plane strain, respectively. The last type is the analog of *SH* waves in an infinite plate. In Love's treatment [1.2, §§ 199–202] of Pochhammer's work [4.8], one sees that the latter employed the displacement equations of motion for the rod, and through separation, finally governing equations on the dilatation and rotation. He then obtained the frequency equations for the three types of waves, and certain approximations to these equations. A more direct way of getting these results was recently effected by Wong, Miklowitz, and Scott [4.9], as a byproduct of related work on the infinite rod of elliptic cross section. It draws directly on displacement potentials, and the wave equations governing them. We return to Lamé's general solution (2.7) of (1.65b), or (1.66b), with $X=0$. As shown in Morse and Feshbach [2.8, Part II, pp. 1762–1767], the vector wave equation (2.7b) on ψ is satisfied in general cylindrical coordinates by

$$\psi = \nabla \times (\chi z) + \nabla \times \nabla \times (\eta z), \qquad (4.66)$$

where z is the unit vector along the axial cylindrical coordinate z, provided the scalar functions χ, η satisfy the wave equations

$$\left(\nabla^2 - \frac{1}{c_s^2}\frac{\partial^2}{\partial t^2}\right)\binom{\chi}{\eta} = 0. \qquad (4.67)$$

ψ in (4.66) satisfies (2.18) through the gauge condition $\nabla \cdot \psi = 0$. An alternate to (4.66) is given by

$$\psi = \chi z + \nabla \times (\eta z), \qquad (4.68)$$

where again χ, η must satisfy (4.67). ψ in (4.68) satisfies (2.18) through the gauge condition $\nabla \cdot \psi = \partial \chi / \partial z \ne 0$. We will use (4.68) in our development here. To show that (4.67) are necessary conditions for ψ in (4.68) to satisfy (2.7b), we first take the curl of (4.68) with the result

$$\nabla \times \psi = \nabla \times \chi z + \nabla\left(\frac{\partial \eta}{\partial z}\right) - (\ddot{\eta}/c_s^2)z, \qquad (4.69)$$

where we have used the condition (4.67) on η. Now taking the curl of (4.69), we have, using the expansion in (1.66a),

$$\nabla(\nabla\cdot\boldsymbol{\psi})-\nabla^2\boldsymbol{\psi}=\nabla\left(\frac{\partial\chi}{\partial z}\right)-\frac{1}{c_s^2}\frac{\partial^2}{\partial t^2}[\chi z+\nabla\times(\eta z)]. \quad (4.70)$$

where we have satisfied the condition (4.67) on χ. Now since $\nabla\cdot\boldsymbol{\psi}=\partial\chi/\partial z$, substituting this and (4.68) in the right hand side of (4.70), we find it reduces to

$$\nabla^2\boldsymbol{\psi}=\ddot{\boldsymbol{\psi}}/c_s^2, \quad (4.71)$$

which was to be proved. We can now use (4.68) to derive our displacement potential relations.

Fig. 4.19 shows the cylindrical coordinates r, θ, z, and corresponding

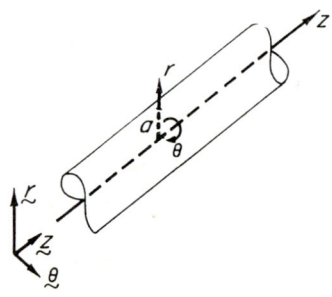

Fig. 4.19. The circular cylindrical rod and its coordinate system.

unit vector triad (r, θ, z), for the infinite rod of circular cross section. The rod is of radius a. Using (2.7a), with (4.68) for $\boldsymbol{\psi}$, and well known expansions for the vector operations in the present coordinates [1.6], we find the general displacement-potential relations for this problem are

$$\begin{aligned} u_r &= \frac{\partial\varphi}{\partial r}+\frac{\partial\chi}{r\partial\theta}+\frac{\partial^2\eta}{\partial z\partial r}, \\ u_\theta &= \frac{\partial\varphi}{r\partial\theta}-\frac{\partial\chi}{\partial r}+\frac{\partial^2\eta}{r\partial z\partial\theta}, \\ u_z &= \frac{\partial\varphi}{\partial z}-\frac{\partial}{r\partial r}\left(r\frac{\partial\eta}{\partial r}\right)-\frac{\partial^2\eta}{r^2\partial\theta^2}, \end{aligned} \quad (4.72)$$

where u_r, u_θ, and u_z are the displacements in the r, θ, and z directions, respectively. The strain-displacement relations are (cf. Love [1.2, p. 56])

$$\varepsilon_r = \frac{\partial u_r}{\partial r}, \qquad \varepsilon_\theta = \frac{1}{r}\left(\frac{\partial u_\theta}{\partial \theta} + u_r\right), \qquad \varepsilon_z = \frac{\partial u_z}{\partial z},$$

$$\varepsilon_{r\theta} = \frac{\partial u_r}{r\partial \theta} + r\frac{\partial}{\partial r}\left(\frac{u_\theta}{r}\right), \qquad \varepsilon_{z\theta} = \frac{\partial u_z}{r\partial \theta} + \frac{\partial u_\theta}{\partial z}, \qquad \varepsilon_{rz} = \frac{\partial u_r}{\partial z} + \frac{\partial u_z}{\partial r}.$$

(4.73)

The stress-strain relations, obtained from Hooke's law, are

$$\sigma_r = \lambda\Delta + 2\mu\varepsilon_r, \qquad \sigma_\theta = \lambda\Delta + 2\mu\varepsilon_\theta, \qquad \sigma_z = \lambda\Delta + 2\mu\varepsilon_z,$$

$$\sigma_{r\theta} = \mu\varepsilon_{r\theta}, \qquad \sigma_{z\theta} = \mu\varepsilon_{z\theta}, \qquad \sigma_{rz} = \mu\varepsilon_{rz}.$$

(4.74)

The strain-potential relations are obtained by substituting (4.72) in (4.73), and clearly from (2.7a) we have

$$\Delta = \nabla^2 \varphi = \frac{\partial^2 \varphi}{\partial r^2} + \frac{1}{r}\frac{\partial \varphi}{\partial r} + \frac{1}{r^2}\frac{\partial^2 \varphi}{\partial \theta^2} + \frac{\partial^2 \varphi}{\partial z^2}.$$

(4.75)

Substitution of the strain-potential relations, and (4.75), in (4.74), gives the stress-potential relations. The stress components σ_r, $\sigma_{r\theta}$, and σ_{rz} must vanish on the cylindrical surface of the elastic rod, $r = a$, for the case of a traction free surface, which we treat in the following sections.

4.4.1. Axially symmetric torsional waves

The simplest case of waves in an elastic rod of circular cross section is that corresponding to axially symmetric torsion. In this case

$$u_r = u_z = 0, \qquad u_\theta = u_\theta(r, z, t), \qquad \partial/\partial\theta = 0.$$

(4.76)

Applying (4.76) to (4.72) shows

$$\varphi = \eta = 0,$$

$$u_\theta = -\frac{\partial \chi}{\partial r}.$$

(4.77)

Applying them again to (4.73) shows the existing strains here are

$$\varepsilon_{r\theta} = r \frac{\partial}{\partial r}\left(\frac{u_\theta}{r}\right), \qquad \varepsilon_{z\theta} = \frac{\partial u_\theta}{\partial z}. \tag{4.78}$$

Substituting the last of (4.77) in (4.78), and then the latter in (4.74), we find

$$\sigma_{r\theta} = -\mu r \frac{\partial}{\partial r}\left(\frac{\partial \chi}{r \partial r}\right), \qquad \sigma_{z\theta} = -\mu \frac{\partial^2 \chi}{\partial r \partial z}, \tag{4.79}$$

as the stress-potential relations. Now as we have just shown, χ must satisfy the wave equation (4.67) in order that (4.68) satisfies the second of (2.7b) (or (4.71)). From the form in (4.75), and the last of (4.76), here the wave equation for χ in (4.67) must take the form

$$\frac{\partial^2 \chi}{\partial r^2} + \frac{1}{r}\frac{\partial \chi}{\partial r} + \frac{\partial^2 \chi}{\partial z^2} = \ddot{\chi}/c_s^2. \tag{4.80}$$

If we assume a solution of the form

$$\chi = Af(r) \exp\left[i\varkappa(z-ct)\right], \tag{4.81}$$

substitution into (4.80) shows $f(r)$ must satisfy the Bessel equation

$$\frac{d^2 f}{dr^2} + \frac{1}{r}\frac{df}{dr} + \eta_s^2 f = 0. \tag{4.82}$$

It follows that

$$\chi = AJ_0(\eta_s r) \exp\left[i\varkappa(z-ct)\right], \tag{4.83}$$

where J_0 is the ordinary Bessel function of the first kind or order zero (cf. ch. 2 § 2.5.1), and we note z is the propagation direction, $\varkappa$ and c again being wave number and phase velocity, respectively, along that direction. η_s is given by (4.6), and again $\omega = \varkappa c$. Note we discarded $Y_o(\eta_s r)$, the ordinary Bessel function of the second kind in our solution (4.83), since we require continuity in the displacement u_θ.

For the case of the rod with a traction free surface we are dealing with, the first of (4.79) must vanish at $r = a$, i.e.,

Ch. 4, § 4.4]   WAVES IN AN INFINITE ELASTIC ROD   219

$$\sigma_{r\theta}\bigg|_{r=a} = -\mu\left[r\frac{\partial}{\partial r}\left(\frac{\partial \chi}{r\partial r}\right)\right]_{r=a} = 0. \tag{4.84}$$

Substituting (4.83) in (4.84), this condition becomes

$$-\mu a\left\{\frac{\mathrm{d}}{\mathrm{d}r}\left[\frac{\mathrm{d}J_0(\eta_s r)}{r\mathrm{d}r}\right]\right\}_{r=a} = 0.$$

This reduces to the roots of

$$\eta_s = 0,$$
$$J_2(\eta_s a) = 0. \tag{4.85}$$

The first of (4.85), from (4.83), corresponds to $f(r) = \text{constant}$, a trivial solution of (4.82) (with $\eta_s = 0$), corresponding to $u_\theta = 0$. Eq. (4.82), with $\eta_s = 0$, also has the singular solution $f(r) = \log r$ corresponding to an infinite displacement, hence inadmissible. Differentiating (4.82), with $\eta_s = 0$, once with respect to r, gives

$$\frac{\mathrm{d}^3 f}{\mathrm{d}r^3} + \frac{1}{r}\frac{\mathrm{d}^2 f}{\mathrm{d}r^2} - \frac{1}{r^2}\frac{\mathrm{d}f}{\mathrm{d}r} = 0, \tag{4.86}$$

and if we can find a solution f to this equation, which reduces the left side of (4.82), with $\eta_s = 0$, to a constant, (4.86) is valid. $f(r) = -r^2/2$ is such a solution, which from (4.83) gives

$$\chi = -\frac{Ar^2}{2}\exp\left[i\varkappa(z-ct)\right].$$

The last of (4.77) shows the corresponding displacement is

$$u_\theta = Ar\exp\left[i\varkappa(z-ct)\right]. \tag{4.87}$$

One should note that this result could have been arrived at more directly by using (2.1). However, it is instructive to see how this special case is handled when potentials are used. Since with $\eta_s = 0$ here, hence $\Omega = \gamma$, (4.87) corresponds to the nondispersive fundamental torsional mode of the infinite medium.

The anharmonic overtones are given by the zeros $\eta_s a = \alpha_n \neq 0$ of the second of (4.85). It follows the frequency equation is

$$\Omega^2 = \alpha_n^2 + \zeta^2, \qquad n = 0, 1, 2, 3\ldots, \qquad \alpha_0 = 0, \tag{4.88}$$

where now $\omega_s = c_s/a$, and $\zeta = a\varkappa$. The α_n values for $n = 1, 2, 3\ldots$ may be found in the Abramowitz and Stegun handbook [4.10], the lowest three being $\alpha_1 = 5.136$, $\alpha_2 = 8.417$, and $\alpha_3 = 11.620$. We note that (4.88) has the same form as (4.60), for SH waves in an infinite plate. In fact, as was pointed out earlier, the present *torsional waves in a rod are the analog of the SH waves in a plate*. One would therefore have curves like those in fig. 4.4, for $n=0$ ($\Omega=\gamma$, $C=1$), $n=1, 2,\ldots$, for $\alpha_0, \alpha_1, \alpha_2\ldots$ here, as well as corresponding ones to those in figs. 4.8, 4.9. Note from (4.88) that the cutoff frequencies are given by $\Omega=\alpha_n$, $n=1, 2, 3\ldots$. The mode shapes for u_θ are the functions of r in (4.87), for $n=0$, i.e., $u_\theta=r$, and in

$$u_\theta = A' J_1(\alpha_n r/a) \exp\left[i\varkappa(z-ct)\right], \tag{4.89}$$

for $n=1, 2, 3,\ldots$, i.e., $u_\theta = J_1(\alpha_n r/a)$, where $A' = -(\alpha_n/a)A$. Eq. (4.89) was obtained from (4.83) using the last of (4.77).

It is of further interest to note that since here

$$\boldsymbol{u} = \nabla \times \chi \boldsymbol{z} = -\frac{\partial \chi}{\partial r}\boldsymbol{\theta} = u_\theta \boldsymbol{\theta}, \tag{4.90a}$$

the analogy with SH waves is contained in

$$\boldsymbol{u} = \nabla \times \psi_1 \boldsymbol{x} = \frac{\partial \psi_1}{\partial z}\boldsymbol{y} = v\boldsymbol{y}. \tag{4.90b}$$

Note that v here is only the first term of the last of (3.77), but this is consistent with ψ_1 being the only nonvanishing component of $\boldsymbol{\psi}$ now. Indeed the frequency equation (4.60) could have been obtained using just ψ_1, and (4.90b). We note that our analogy has the correspondence of r, θ, z and z, $-y$, x, χ and ψ_1, u_θ and $-v$, and hence the surface stresses $\sigma_{r\theta}$ and σ_{zy}. It follows that the wave features found for the SH waves in the plate carry over here.

4.4.2. Axially symmetric compressional waves

In this case we have

$$u_r = u_r(r, z, t), \quad u_z = u_z(r, z, t), \quad u_\theta = 0, \quad \partial/\partial\theta = 0. \tag{4.91}$$

Applying these to (4.72), one finds it is sufficient to let $\chi=0$ to define the displacement-potential relations

$$\begin{aligned} u_r &= \frac{\partial \varphi}{\partial r} + \frac{\partial^2 \eta}{\partial r \partial z}, \\ u_z &= \frac{\partial \varphi}{\partial z} - \frac{\partial}{r\partial r}\left(r \frac{\partial \eta}{\partial r}\right). \end{aligned} \tag{4.92}$$

Ch. 4, § 4.4]   WAVES IN AN INFINITE ELASTIC ROD   221

Applying (4.91) also to (4.73), we find the existing strains to be

$$\varepsilon_r = \frac{\partial u_r}{\partial r}, \qquad \varepsilon_\theta = \frac{u_r}{r}, \qquad \varepsilon_z = \frac{\partial u_z}{\partial z}, \qquad \varepsilon_{rz} = \frac{\partial u_r}{\partial z} + \frac{\partial u_z}{\partial r}. \qquad (4.93)$$

Now substituting (4.92) in (4.93), and then the latter in (4.74), we find the stress-potential relations

$$\sigma_r = \frac{\lambda}{c_d^2} \ddot{\varphi} + 2\mu \left(\frac{\partial^2 \varphi}{\partial r^2} + \frac{\partial^3 \eta}{\partial r^2 \partial z} \right),$$

$$\sigma_{rz} = \mu \left\{ 2 \frac{\partial^2 \varphi}{\partial r \partial z} + \frac{\partial^3 \eta}{\partial r \partial z^2} - \frac{\partial}{\partial r} \left[\frac{\partial}{r \partial r} \left(r \frac{\partial \eta}{\partial r} \right) \right] \right\}, \qquad (4.94)$$

for the stresses that can occur on the boundary $r = a$, where we have used (4.75) (without the term involving θ), and the first of (2.7b). The wave equations governing φ and η are of the form given in (4.80), hence, as we have seen in the case of (4.83), φ and η are of the form

$$\varphi = A J_0(\eta_d r) \exp \left[i\varkappa(z - ct) \right],$$
$$\eta = B J_0(\eta_s r) \exp \left[i\varkappa(z - ct) \right], \qquad (4.95)$$

η_d and η_s being given in (4.3) and (4.6), respectively, and where again $\omega = \varkappa c = \varkappa_s c_s = \varkappa_d c_d$.

The free cylindrical surface of the rod requires that the stresses in (4.94) vanish. Substituting (4.95), in these conditions then, they become

$$\left[-\frac{\lambda \omega^2}{c_d^2} J_0(\eta_d r) + 2\mu \frac{d^2 J_0(\eta_d r)}{dr^2} \right]_{r=a} A + 2\mu i\varkappa \frac{d^2 J_0(\eta_s r)}{dr^2} \bigg|_{r=a} B = 0,$$

$$2i\varkappa \frac{d J_0(\eta_d r)}{dr} \bigg|_{r=a} A - \left\{ \varkappa^2 \frac{d J_0(\eta_s r)}{dr} + \frac{d}{dr} \left[\frac{d}{r dr} \left(r \frac{d J_0(\eta_s r)}{dr} \right) \right] \right\}_{r=a} B = 0. \qquad (4.96)$$

Setting the determinant of the coefficients of A and B in these equations equal to zero, we obtain *Pochhammer's frequency equation*, which can be reduced to the convenient form, given by Onoe et al. in [4.6],

$$(\varkappa^2 - \eta_s^2)^2 \left[\frac{\eta_d a J_0(\eta_d a)}{J_1(\eta_d a)} \right] + 4\varkappa^2 \eta_d^2 \left[\frac{\eta_s a J_0(\eta_s a)}{J_1(\eta_s a)} \right] = 2\eta_d^2 (\eta_s^2 + \varkappa^2). \qquad (4.97)$$

The functions $z J_0(z) / J_1(z)$, basic to (4.97), have been studied at length by Onoe [4.11]. The Rayleigh–Lamb frequency equation (4.48), for straight-crested symmetric waves in an infinite elastic plate, can be written in a form analogous to that in (4.97).

$$(\varkappa^2 - \eta_s^2)^2 \left[\frac{\eta_d h \cos \eta_d h}{\sin \eta_d h} \right] + 4\varkappa^2 \eta_d^2 \left[\frac{\eta_s h \cos \eta_s h}{\sin \eta_s h} \right] = 0. \tag{4.98}$$

The close resemblance between (4.97) and (4.98) dramatically exhibits the fact that the frequency spectra for the infinite rod of cross-sectional radius a, and the infinite plate of half-thickness h, for axially-symmetric and symmetric waves, respectively, are very similar. In particular, when ω and $\varkappa$ are small, (4.97) and (4.98) are practically identical, except for differences in the constants in the power series expansions of the Bessel and trigonometric functions. Holden first pointed out the strong similarity in the rod and plate spectra. Fig. 4.15, which shows the rod spectra for axially symmetric waves, was taken from [4.6], as noted earlier. In it $\Omega = \omega/\omega_s$, $\zeta = \varkappa a/\sigma = \gamma + i\delta$, where $\omega_s = \sigma c_s/a$, and σ is the lowest non-zero root of $J_1(\sigma_n) = 0$, i.e., 3.8317. In retaining our nomenclature ω_s, ζ, γ and δ here, we are replacing $2h/\pi$ appearing in their plate definition by a/σ. Note the similarity between these spectra and those in figs. 4.13, 4.12 for the infinite plate. The large variable asymptotes for the real segments $\Omega = k\gamma$, $\Omega = \gamma$, and $\Omega = (c_R/c_s)\gamma$, are again involved here. The asymptote for the lowest real segment for small Ω and γ, is given by $\Omega = (c_b/c_s)\gamma$ here, whereas for the infinite plate by $\Omega = (c_p/c_s)\gamma$, where $c_b = (E/\varrho)^{\frac{1}{2}}$, and $c_p = [E/\varrho(1-v^2)]^{\frac{1}{2}}$, the so-called *"bar" and "plate" velocities*. We shall see later that these lowest real segments in the neighborhood of the small variable asymptotes form the base for approximate theories of rods and plates. The kind of analysis that we discussed for the spectra and related modes and waves in the infinite plate case, i.e., bounds, etc., is carried out in detail for the rod in [4.6]. An analysis of the thickness modes could be carried out, as we did for the plate, (see fig. 4.3), using the bounds for the present case, and (4.95) and (4.92).

Consider (4.68) again. Expanding, we find

$$\psi = \frac{\partial \eta}{r \partial \theta} r - \frac{\partial \eta}{\partial r} \theta + \chi z. \tag{4.99}$$

Imposing axial symmetry on (4.99), it reduces to

$$\psi = -\frac{\partial \eta}{\partial r} \theta + \chi z. \tag{4.100}$$

If we take account of our analysis in the previous section for the axially symmetric torsion case, observing the absence of χ in (4.92), we can interpret (4.100) as being composed of non-torsional (first term) and torsional (second term) motion. The former, of course, represents the axially symmetric shear and rotation of the present case. It is consistent with the fact that ε_{rz} and ω_θ are the only components of shear strain and rotation that exist. These are measured on the planes $\theta =$ constant to which $\psi = -\dfrac{\partial \eta}{\partial r} \boldsymbol{\theta}$ is perpendicular. In the case of axially symmetric torsion, $\varepsilon_{r\theta}$, $\varepsilon_{z\theta}$, ω_z, and ω_r exist. The strain $\varepsilon_{r\theta}$ and rotation ω_z are measured on the planes $z =$ constant to which $\psi = \chi z$ is perpendicular. The strain $\varepsilon_{z\theta}$ and ω_r are measured on the planes $r =$ constant to which r is perpendicular, however, these quantities are inherent in SH and torsional wave motion, involving the representation of ψ by a single component (cf. ch. 3 § 3.2 and the previous section, last paragraph).

4.4.3. Nonaxially symmetric or flexural waves

This is our most general case. Here we have all the displacements, u_r, u_θ, and u_z, all the strains in (4.73), and stresses in (4.74), and they are all functions of r, θ, z, and t. Eq. (4.72) gives the displacement-potential relations, substitution of these in (4.73), the strain-potential relations, and in turn the latter in (4.74), the stress-potential relations, with the aid of (4.75). Since φ, χ, and η are solutions of wave equations, where now the full Laplacian (4.75) governs, we find, similar to the previous cases treated here, the potentials to be of the form

$$\varphi = A J_n(\eta_d r) \cos n\theta \exp\left[i\varkappa(z - ct)\right],$$
$$\chi = B J_n(\eta_s r) \sin n\theta \exp\left[i\varkappa(z - ct)\right], \qquad (4.101)$$
$$\eta = C J_n(\eta_s r) \cos n\theta \exp\left[i\varkappa(z - ct)\right],$$

where $n = 1, 2, 3 \ldots$, dictates the particular sinusoidal variation in the circumferential direction, and the order of the coupled Bessel function variation in the radial direction, of the potentials. The Y_n's have again been discarded to assure continuity of the displacements at $r = 0$. Eqs. (4.101) really represent an infinite array of possible distributions in the θ direction, because of n. Pochhammer treated the case of $n = 1$, which is the basic flexural case, and of prime interest here. The higher n values, representing higher circumferential modes, find their most prominent use in analysis of wave propagation in cylindrical shells.

A free cylindrical surface now requires that all the stresses there, σ_r, $\sigma_{r\theta}$, and σ_{rz}, vanish. As in the case of (4.96) for compressional waves, this generates the Pochhammer frequency equation for flexural waves. The equation has been a subject of extensive investigation through the years. Notable, very recently, is the work of Pao and Mindlin [4.12], and a second paper by Pao [4.13]. The attack is similar to that applied to the axially symmetric case in [4.6], but more involved. In [4.12] the frequency equation is written in the convenient form

$$J_1(\alpha)J_1^2(\beta)(f_1 O_\beta^2 + f_2 O_\alpha O_\beta + f_3 O_\beta + f_4 O_\alpha + f_5) = 0, \qquad (4.102)$$

where

$$O_x = O_1(x) = xJ_0(x)/J_1(x),$$

is Onoe's function of the first kind of order one [4.11], and

$$f_1 = 2(\beta^2 - \zeta^2)^2, \qquad f_2 = 2\beta^2(5\zeta^2 + \beta^2),$$
$$f_3 = \beta^6 - 10\beta^4 - 2\beta^4\zeta^2 + 2\beta^2\zeta^2 + \beta^2\zeta^4 - 4\zeta^4,$$
$$f_4 = 2\beta^2(2\beta^2\zeta^2 - \beta^2 - 9\zeta^2),$$
$$f_5 = \beta^2(-\beta^4 + 8\beta^2 - 2\beta^2\zeta^2 + 8\zeta^2 - \zeta^4),$$

and $\alpha = \eta_d a$, $\beta = \eta_s a$, $\zeta = \varkappa a = \gamma + i\delta$, and $\omega_s = c_s/a$. Again here we retain our plate nomenclature ω_s, ζ, γ and δ, replacing $2h/\pi$ in them by a. Fig. 4.20 shows the real and imaginary segments for the branches of (4.102) taken from Pao's paper [4.13]. The branches (circled numbers) and two sets of bounds (B_1, B_2) are shown in the figure. Note the close similarity of the lowest two branches, with the corresponding ones in the antisymmetric wave spectra for the infinite plate, shown in fig. 4.13. Note, however, that the imaginary segments show distinct differences with the corresponding ones of the plate spectra at the third branch, and into the higher branches. If (4.101), with $n=1$, are substituted in (4.72), we find the displacements are

$$u_r = U(r) \cos \theta \exp[i\varkappa(z - ct)],$$
$$u_\theta = V(r) \sin \theta \exp[i\varkappa(z - ct)], \qquad (4.103)$$
$$u_z = W(r) \cos \theta \exp[i\varkappa(z - ct)],$$

where

$$U(r) = A \frac{dJ_1(\eta_d r)}{dr} + B \frac{J_1(\eta_s r)}{r} + iC\varkappa \frac{dJ_1(\eta_s r)}{dr},$$
$$V(r) = -\left[A \frac{J_1(\eta_d r)}{r} + B \frac{dJ_1(\eta_s r)}{dr} + iC\varkappa \frac{J_1(\eta_s r)}{r} \right],$$
$$W(r) = i\varkappa A J_1(\eta_d r) + C\eta_s^2 J_1(\eta_s r).$$

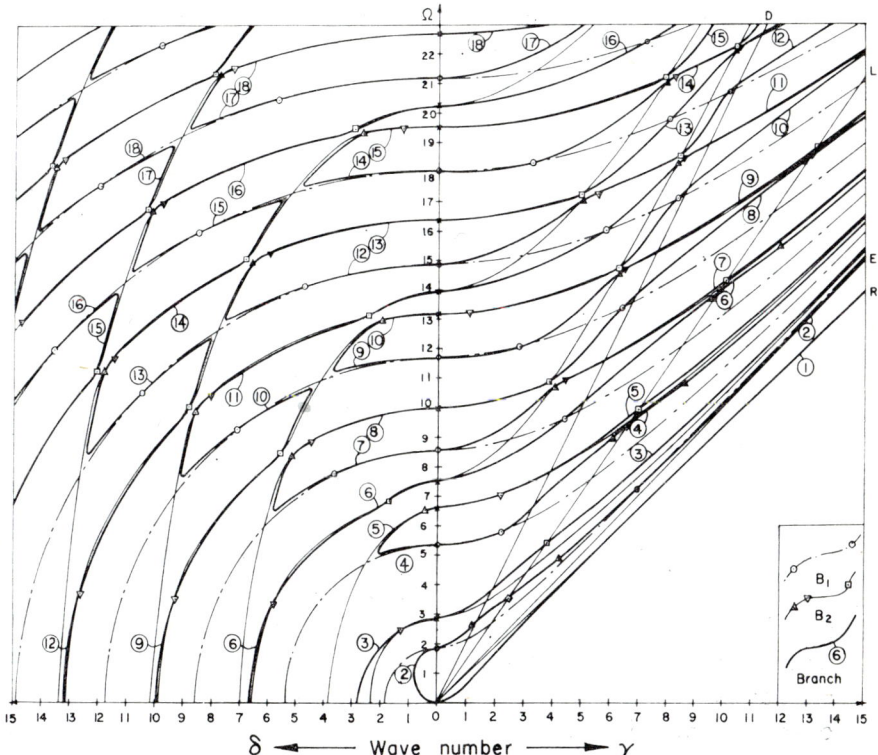

Fig. 4.20. Real and imaginary segments of the Pochhammer frequency spectra for nonaxially symmetric modes ($\nu = 1/3$) (after Pao).

The latter are in agreement with Love's expressions, Eq. (66), pg. 292 in [1.2], if it is noted that Love's B and C correspond to iC and B here, respectively. As Love points out, letting $r=0$ in (4.103), we see $u_r \sin \theta + u_\theta \cos \theta = 0$, hence the motion of points on the axis of the rod takes place in a diametral plane, containing the unstrained position of that axis and the line from which ? is measured. This, coupled with the fact that u_z in (4.103) vanishes at $r=0$, means the motion of points on the axis is necessarily perpendicular to that axis. Hence this motion is transverse, or flexural, in nature.

Finally, observing (4.99) again, we see nonaxially symmetric motion is a very complicated one, with shear and rotation on all coordinate planes, and in addition, the dilatation associated with φ.

4.5. Waves in circular cylindrical shells and layered media; literature

Although the subject media are more complicated examples of waveguides, analysis of them is similar to the preceding work on the plate and rod. A discussion of the literature on the circular cylindrical shell may be found in Miklowitz [4.4], and for layered media in Ewing et al. [3.11] and Brekhovskikh [I. 19].

4.6. Exercises

4.1. Using (4.32) and (4.13b) derive expressions for C_{gm} and $dC_{gm}/d\gamma$. Show that these lead to the set of $C_{gm} - \gamma$ branches shown in fig. 4.8. Again from (I.32), (4.13b), and (4.30), (4.31), derive expressions for $T(t/T_x)/T_h$ and $d(T/T_h)/d(t/T_x)$, and show they lead to the set of T/T_h vs t/T_x branches shown in fig. 4.9. In the latter expressions it is best to treat the $m=0$ branch separately, through the parametric equations (4.30), (4.31).

4.2. Assume in our plane-strain waveguide theory that the infinite plate (of thickness $2h$) is made of an elastic inviscid fluid. What are the displacement-potential relations for this case? State the general form for displacement and stress waves propagating along the plate. Assuming the pressure on the plate faces is zero,
(a) find the corresponding admissible modes of propagation,
(b) sketch the first few modes and discuss their symmetry or antisymmetry w.r.t. the plate mid-plane,
(c) derive the frequency-, phase velocity-, and group velocity-wave number relations, and sketch and discuss the first few branches in each case,
(d) derive the predominant period-time of occurrence relation, and again sketch and discuss the first few branches.

4.3. Consider the infinite beam or rod governed by the Bernoulli-Eule bending theory equation of motion

$$\frac{\partial^4 y(x, t)}{\partial x^4} + \frac{\ddot{y}(x, t)}{d^2} = 0,$$

where y is the transverse deflection of the beam, $d^2 = c_b^2 r_g^2$, where $c_b^2 = E/\varrho$ and r_g is the beam sectional radius of gyration. Assume that a disturbance composed of plane harmonic wave trains of the form $y = A \exp\left[i\varkappa(x - ct)\right]$

is traveling along the beam, say from a source at the origin $x=0$. Derive the $c-\varkappa$, $c_g-\varkappa$ relations. Derive the predominant period-time of occurrence relation $T/T_{r_g}-t/T_x$ for the disturbance and sketch it. With the aid of this relation sketch the disturbance propagating along x-positive, as a function of x for a particular time $t=t_1$, indicating where one would find the longest and shortest wavelengths.

4.4. For the plate with elastically restrained faces, derive the frequency equation for the general case $0 \leq e \leq \infty$ for symmetric waves. Show that this equation, and (4.46), contain the modes (4.11) and (4.12) for the plate with mixed conditions on its faces.

4.5. Note, from (4.51), that along the real segments of the roots or branches of the Rayleigh–Lamb frequency equation (4.48),

$$\frac{d\Omega}{d\gamma} = -\frac{\partial F/\partial \gamma}{\partial F/\partial \Omega}.$$

Using this relation, and suitable approximations from (4.48) for symmetric modes, show that at cutoff $d\Omega/d\gamma=0$, and at and near cutoff, $\gamma(\Omega)$ behaves as

$$\gamma(\Omega) \sim (\Omega-\Omega^*)^{\frac{1}{2}},$$

where Ω^* is the cutoff frequency. Note that this behavior is exhibited in fig. 4.13.

4.6. Assume that the Raleigh–Lamb frequency equations (4.48) have the root $\zeta(\Omega)$. Show then that they also have the roots $\bar{\zeta}(\Omega)$, $-\zeta(\Omega)$ and $-\bar{\zeta}(\Omega)$, and then the roots $\zeta(-\Omega)$, $-\bar{\zeta}(-\Omega)$, $-\zeta(-\Omega)$ and $\bar{\zeta}(-\Omega)$.

4.7. If $m=7$, $n=17$, calculate and show from fig. 4.13 that

$$\left|\frac{A}{iD}\right| < 1,$$

and similarly, if $m=6$, and $n=18$, calculate and show that

$$\left|\frac{A}{iD}\right| > 1,$$

as discussed in § 4.1.3.2. Does it make sense? Explain.

4.8. For a linearly elastic infinite plate (or layer) of thickness h in plane strain, resting on an elastic foundation, the frequency equation is given by

$$(\pi^3 h/8\mu)K_e\eta_d'\Omega^2[\theta_1^2 \tan(\pi\eta_s'/2) - \theta_2 \tan(\pi\eta_d'/2)]$$

$$= \frac{[1+\cos(\pi\eta_d'/2)][1+\cos(\pi\eta_s'/2)]}{\cos(\pi\eta_d'/2)\cos(\pi\eta_s'/2)}[\theta_1^2 \tan(\pi\eta_d'/4) - \theta_2 \tan(\pi\eta_s'/4)]$$

$$\cdot [\theta_1^2 \tan(\pi\eta_s'/4) - \theta_2 \tan(\pi\eta_d'/4)]$$

where η_d', η_s' and Ω are the same as in (4.48),

$$\theta_1 = (\pi^2/4)(2\zeta^2 - \Omega^2), \qquad \theta_2 = -(\pi^4/4)\eta_d'\eta_s'\zeta^2$$

where $\zeta = \gamma + i\delta$ is the complex wave number, and K_e is the foundation spring constant (force per unit area per unit deflection). The equation is due to Das Gupta [4.14] who assumed for boundary conditions that one face of the layer was traction free and the other (the plate-foundation interface) was free of shear stress but subject to normal stress proportional to the displacement of that face.

The most interesting feature of Das Gupta's equation is related to the displacement symmetry it reflects about the base of the foundation. As

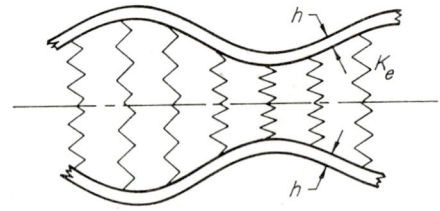

the accompanying sketch shows the problem is equivalent to that of a sandwich of two flat plates separated by an elastic layer.

Show (1) for an infinite spring constant ($K_e \to \infty$), Das Gupta's frequency equation reduces to the Rayleigh–Lamb equation (4.48) for symmetric waves in a plate of $2h$ thickness, and (2) for the other extreme ($K_e = 0$) his equation reduces to that of the frequency equation for a free plate of thickness h. Show also that the cutoff frequencies contained in the given equation are

$$\Omega_n = n = 2, 4, 6 \cdots$$

and the roots of

$$\frac{\pi\Omega_m}{2k} \tan \frac{\pi\Omega_m}{2k} = \frac{K_e h}{k^2\mu}, \qquad m = 0, 1, 2\cdots$$

Prove that the latter cutoff frequencies vary from $\Omega_m = 2km$ to $2k(m+\frac{1}{2})$ as the foundation stiffness K_e increases from zero to infinity. Note when $K_e \to \infty$ then, that the above cutoff frequencies Ω_n and Ω_m agree with the second of (4.55b) and first of (4.55a), respectively, the Rayleigh–Lamb cutoff frequencies for the symmetric modes. The reader will be interested further in the paper by Lloyd and Miklowitz [4.15] which presents more complete frequency spectra for the present problem including complex wave number segments which are generated by the foundation.

4.9. The *SH* waves in the infinite elastic plate have the mode corresponding to $n=0$, whereas in the infinite plate in plane strain no such mode existed. Explain the physics of this.

4.10. Derive (4.90a), (4.90b) and show with some sketches why the discussion about them in the text is true.

4.11. Suppose an infinite elastic plate has conditions of complete restraint at its faces, i.e., displacement $u=0$ at $z=\pm h$. Assuming plane strain, and

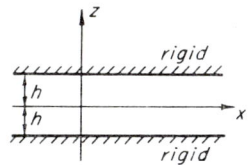

harmonic wave propagation in the x direction, derive the frequency equations for the symmetric and antisymmetric (with respect to $z=0$) modes of propagation. From these equations deduce the cutoff frequencies for both cases, the associated phase velocities and orientation of the plate particle motions.

References

[4.1.] T. H. Havelock, *The Propagation of Disturbances in Dispersive Media*. Cambridge University Press, (1914); reprinted by Stechert–Hafner Service Agency, Inc., New York (1964).

[4.2.] L. Brillouin, *Wave Propagation and Group Velocity*, Academic Press, New York (1960).

[4.3.] R. M. Davies, *Philosophical Transactions of the Royal Society London*, Series A, **240** (1948), 375–457.

[4.4.] J. Miklowitz, *Elastic Wave Propagation*. In: Applied Mechanics Surveys, eds. H. N. Abramson, H. Liebowitz, J. N. Crowley, S. Juhasz. Spartan Books, Washington (1966), 809–839.

[4.5.] R. D. Mindlin, *Mathematical Theory of Vibration of Elastic Plates*. In: Proceedings of Eleventh Annual Symposium on Frequency Control. U. S. Army Signal Corps Engineering Laboratories, Fort Monmouth, New Jersey (1957), 1–40.
[4.6.] M. Onoe, H. D. McNiven and R. D. Mindlin, *Journal of Applied Mechanics* **29** (1962), 729–734.
[4.7.] R. Stoneley, *Bulletin of the Seismological Society of America* **38** (1948), 263–274.
[4.8.] L. Pochhammer, *Journal für Mathematik* **81** (1876), 324–336.
[4.9.] P. K. Wong, J. Miklowitz and R. A. Scott, *Journal of the Acoustical Society of America* **40** (1966), 393–398.
[4.10.] M. Abramowitz and I. A. Stegun, eds. *Handbook of Mathematical Functions*. National Bureau of Standards, Applied Mathematics Series 55 (1964), 409.
[4.11.] M. Onoe, *Formulas and Tables, The Modified Quotients of Cylinder Functions*. Report of the Institute of Industrial Science, University of Tokyo **4** (1955), 216–237.
[4.12.] Y. H. Pao and R. D. Mindlin, *Journal of Applied Mechanics* **27** (1960), 513–520.
[4.13.] Y. H. Pao, *Journal of Applied Mechanics* **29** (1962), 61–64.
[4.14.] S. C. Das Gupta, *Bulletin of the Seismological Society of America* **45** (1955), 115–120.
[4.15.] J. R. Lloyd and J. Miklowitz, *Journal of Applied Mechanics* **29** (1962), 459–464.

CHAPTER 5

INTEGRAL TRANSFORMS, RELATED ASYMPTOTICS AND INTRODUCTORY APPLICATIONS

5.1. Introduction

At this point it is appropriate to begin our study of the basic boundary-initial value problems in elastic wave propagation. As pointed out earlier, integral transform techniques, and related asymptotics, are ideally suited for these problems, and we will set down the fundamentals of these in the first part of this chapter. Later in the chapter we will apply the Laplace transform and related asymptotics to solve the problems of spherical and circular cylindrical cavities in the infinite elastic solid subjected to sudden uniform pressure. Aside from the interest we have in these one-dimensional problems per se, they are important simple vehicles for the development of the integral transform techniques, which will be further developed in the later chapters of this book dealing with geometrically more complicated problems.

5.2. Fourier integral theorem

Exponential integral transforms, such as the Fourier and Laplace, are based on the classical *Fourier integral theorem*. The theorem may be stated as follows: *Let $g(t)$ and its derivative $g'(t)$ be real functions of the real variable t that are piecewise continuous on every finite interval of the t-axis. If $g(t)$ has a finite discontinuity at, say $t=t_d$, then let it be defined there as the mean value of the limits from the right and left*

$$g(t) = \tfrac{1}{2}[g(t_d+) + g(t_d-)]. \qquad (5.1)$$

Further, let $g(t)$ be absolutely integrable over the t-axis, i.e.,

$$\int_{-\infty}^{\infty} |g(t)| \, dt < \infty. \qquad (5.2)$$

Then at every point t the function $g(t)$ is represented by the Fourier integral

$$g(t) = \frac{1}{2\pi} \int_{-\infty}^{\infty} d\varkappa \int_{-\infty}^{\infty} g(\tau) e^{i\varkappa(\tau-t)} d\tau . \qquad (5.3)$$

Churchill [5.1, §§ 50, 51] gives a proof of the theorem. The theorem contains sufficient conditions on $g(t)$ for it to be represented by the Fourier integral (5.3). In particular the absolute integrability conditions (5.2) is severe. Titchmarsh [5.2, § 1.10] discusses the relaxation of this condition. As we progress in our development of integral transforms it will be shown that certain relaxations can be made of the other conditions in the theorem.

5.3. Laplace transform

The *one-sided Laplace transform*, commonly called the *Laplace transform* (as we shall refer to it here), is basic to integral transform methods dealing with problems involving initial conditions. This is because the Laplace transform "asks for" the given initial data. Let us define $g(t)$ as

$$\begin{aligned} g(t) &= e^{-\gamma t} f(t), & t &> 0, \\ &= 0, & t &< 0, \end{aligned} \qquad (5.4)$$

where the latter implies that $f(t)=0$, for $t<0$, and γ is a positive number. Further, we require $f(t)$ to be of exponential order, i.e., $O\left[\exp(\gamma_1 t)\right]$, as $t \to \infty$, where $\gamma_1 < \gamma$. This condition will insure that $g(t)$ in (5.4) satisfies the absolute integrability condition (5.2), for $t \to \infty$, in the Fourier integral theorem of § 5.2. Then provided $f(t)$ satisfies all the other requirements of the theorem so will $g(t)$ in (5.4). It follows that $g(t)$ can be expanded in the form (5.3), and hence substituting (5.4) into (5.3), we have

$$g(t) = e^{-\gamma t} f(t) = \frac{1}{2\pi} \int_{-\infty}^{\infty} e^{-i\varkappa t} d\varkappa \int_{0}^{\infty} f(\tau) e^{-(\gamma - i\varkappa)\tau} d\tau . \qquad (5.5)$$

Letting p be the *complex variable* $=\gamma - i\varkappa$, this reduces to

$$f(t) = \frac{1}{2\pi i} \int_{\gamma-i\infty}^{\gamma+i\infty} \bar{f}(p) e^{pt} dp , \qquad (5.6a)$$

where

$$\bar{f}(p) = \int_{0}^{\infty} f(t) e^{-pt} dt , \qquad (5.6b)$$

where p is the transformation parameter. Equation (5.6b) gives the Laplace transform of $f(t)$, and (5.6a) its inverse $f(t)$. From (5.4) we see that the transform pair (5.6) are defined for $t > 0$. When $f(t)$ has a finite discontinuity it is given by the inversion integral (5.6a) with

$$f(t) = \tfrac{1}{2}[f(t_d+) + f(t_d-)] \tag{5.7}$$

substituted on the left there, which stems from (5.1), (5.4) and (5.5). If $t_d = 0$, taking the second of (5.4) into account, we see that $f(t)$ in (5.7), hence (5.6a), becomes

$$f(t) = \tfrac{1}{2} f(0+). \tag{5.8}$$

Note therefore, that if $f(t)$ is continuous at $t = 0$, $f(0) = 0$. Convergence of the operator integral in (5.6b) stems essentially from the stated exponential order of $f(t)$ at $t \to \infty$. To show this we consider the boundedness of the integral

$$\int_0^\infty |f(t)e^{-pt}| dt = \int_0^\infty e^{-\gamma t} |f(t)| dt, \tag{5.9}$$

where we have used $p = \gamma - i\varkappa$. The integral on the right may be expanded in accord with

$$\int_0^\infty e^{-\gamma t} |f(t)| dt = \int_0^T + \int_T^\infty e^{-\gamma t} |f(t)| dt, \tag{5.10}$$

where the time $T \gg 1$. Since $f(t)$ has at most finite discontinuities, according to our Fourier integral theorem, the first integral on the right hand side of (5.10) converges absolutely. Because of its exponential order at $t \to \infty$, $|f(t)|$ in the second integral satisfies $|f(t)| < K \exp(\gamma_1 t)$, where K is a positive constant. It follows the second integral (5.10)

$$\int_T^\infty e^{-\gamma t} |f(t)| dt < K \int_T^\infty e^{-(\gamma - \gamma_1)t} dt = \frac{K}{\gamma - \gamma_1} \exp[-(\gamma - \gamma_1)T].$$

Since $\gamma = \operatorname{Re} p > \gamma_1$, this last result is a finite positive number, completing the proof that *the operator integral in* (5.6b) *converges absolutely in the half plane* $\operatorname{Re} p > \gamma_1$. Finally we should note that since (5.9) is a statement of the integral in (5.2) for the present case, our proof shows (5.2) has been satisfied.

In applying (5.6) one normally uses the integral operator in (5.6b), and associated operations for derivatives, etc., on the unknown function $f(t)$

to obtain its transform $\bar{f}(p)$. Then provided $\bar{f}(p)$ satisfies certain conditions in the complex p-plane, the inversion integral (5.6a) will yield (and define) $f(t)$, compatible with the conditions already stated for this function. This forms the main use of the pair (5.6) for our purposes, $f(t)$ usually being the unknown dependent variable in a differential equation, and hence the solution we seek. We therefore proceed with setting down the conditions $\bar{f}(p)$ must satisfy.

Following Churchill [5.3, §§ 62, 66] we make use of the Cauchy integral formula writing

$$\bar{f}(p) = \frac{1}{2\pi i} \int_C \frac{\bar{f}(z)}{z-p} \, dz, \qquad (5.11)$$

where p is interior to $C = -L + C_R$, the closed path in the right half z-plane, in its limiting sense $|z| \to \infty$, as depicted in fig. 5.1. To use (5.11) to

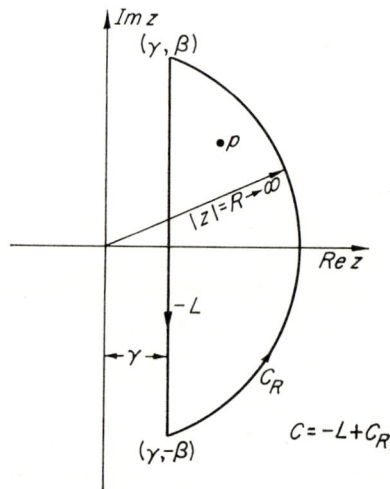

Fig. 5.1. Integration path for the Cauchy integral formula for $\bar{f}(p)$.

calculate $\bar{f}(p)$, $\bar{f}(z)$ must be *analytic* within and on C, i.e.,

$$\bar{f}(z) \text{ analytic in } \operatorname{Re} z \geq \gamma. \qquad (5.12)$$

We add the further requirement that

$$\bar{f}(z) = O(z^{-k}), \qquad |z| \to \infty, \qquad \text{in } \operatorname{Re} z \geq \gamma, \qquad (5.13)$$

where k is a positive number.

Expanding (5.11), we have

$$\bar{f}(p) = -\frac{1}{2\pi i}\left[\lim_{\beta\to\infty}\int_{\gamma-i\beta}^{\gamma+i\beta}\frac{\bar{f}(z)}{z-p}\,dz - \lim_{|z|\to\infty}\int_{C_R}\frac{\bar{f}(z)}{z-p}\,dz\right]. \quad (5.14)$$

Using (5.13), it is easily shown that the second integral vanishes, which stems from

$$\left|\lim_{R\to\infty}\int_{C_R}\right| \leq \lim_{R\to\infty}\frac{\pi}{R^k} = 0.$$

It follows that (5.14) reduces to

$$\bar{f}(p) = \frac{1}{2\pi i}\int_{\gamma-i\infty}^{\gamma+i\infty}\frac{\bar{f}(z)}{p-z}\,dz. \quad (5.15)$$

Now we apply the inversion integral (5.6a) to (5.15) with the result

$$f(t) = \frac{1}{2\pi i}\int_{\gamma-i\infty}^{\gamma+i\infty}\left[\frac{1}{2\pi i}\int_{\gamma-i\infty}^{\gamma+i\infty}\frac{\bar{f}(z)}{p-z}\,dz\right]e^{pt}\,dp. \quad (5.16)$$

Consider the inner integral here. On its path of integration, we have $f(z)$ analytic, according to (5.12), and $\operatorname{Re} p > \operatorname{Re} z = \gamma$. Further, according to (5.13), $|\bar{f}(z)/(p-z)| < M/|z|^{k+1}$ for $|z| \gg 1$, where M is a positive constant, and $k > 0$. It follows that this inner integral converges absolutely and, hence, uniformly with respect to parameter p for $\operatorname{Re} p > \operatorname{Re} z = \gamma$. The latter property means the order of integration in (5.16) can be interchanged so that it can be written as

$$f(t) = \frac{1}{2\pi i}\int_{\gamma-i\infty}^{\gamma+i\infty}\left[\frac{1}{2\pi i}\int_{\gamma-i\infty}^{\gamma+i\infty}\frac{e^{pt}}{p-z}\,dp\right]\bar{f}(z)\,dz. \quad (5.17)$$

The inner integral here is the inverse of the transform $1/(p-z)$ which is $\exp(zt)$. This is easily established by using the operator integral (5.6b), solving this simple integral equation by inspection. By so doing, one finds $\exp(zt)$ under the conditions $t > 0$, and $\operatorname{Re} p > \operatorname{Re} z = \gamma$, consistent with our above conditions. Alternately, one could evaluate the inner integral directly with residue theory. On this basis (5.17) reduces to

$$f(t) = \frac{1}{2\pi i}\int_{\gamma-i\infty}^{\gamma+i\infty}\bar{f}(z)e^{zt}\,dz, \qquad t > 0, \qquad \gamma_1 < \gamma < \infty. \quad (5.18)$$

As shorthand for the path $\gamma - i\infty$ to $\gamma + i\infty$ (the path L in fig. 5.1) we will adopt Br *"the Bromwich contour."*[1] Noting that (5.18) is identical to the inversion integral (5.6a), we have again generated the latter, but importantly, in the process, these *conditions on* $\bar{f}(p)$: In the half plane Re $p \geq \gamma$,

$$\bar{f}(p) \text{ is analytic}, \tag{5.19a}$$

and

$$\bar{f}(p) = O(p^{-k}), \qquad |p| \to \infty, \qquad k > 0. \tag{5.19b}$$

(5.19b) implies that $|\bar{f}(p)| \to 0$, as $|p| \to \infty$, uniformly with respect to arg p for $-\pi/2 < \arg p < \pi/2$. Using this fact and (5.19a) it can be shown that the inversion integral in (5.6a) vanishes for $t < 0$, hence so does $f(t)$, consistent with the second of (5.4). The proof is left to the exercises. This is a very useful property since, as we will see in our problems which involve *hyperbolic equations*, it will enable us to show that the solution is identically zero at a station before the fastest wave arrives there. The pair (5.6) together with the conditions on (1) $f(t)$, given in the Fourier integral theorem, and (2) $\bar{f}(p)$, given by (5.19), form what is known as the *Fourier–Mellon inversion theorem.*

Another important fact emerges from (5.19b), namely that it shows a class of functions, singular at $t = 0$, are admissible for the Laplace transform. These are the functions t^{k-1}, $0 < k < 1$, which have the transforms $\Gamma(k)/p^k$, where Γ is the gamma function, as we shall show in § 5.4.7.1. The functions t^{k-1}, $0 < k < 1$, are absolutely integrable in accord with (5.2), for the present case of $g(t)$ in (5.4). For Re $p > 0$, their transforms $\Gamma(k)/p^k$ satisfy the conditions (5.19). It follows, therefore, that these functions represent a relaxation of the piecewise continuity (a sufficient condition) on $g(t)$ in the Fourier integral theorem of § 5.2, for $t = 0$, $g(t)$ being given by (5.4). It should be pointed out that the functions t^{k-1} are very important to wave propagation studies since, as we shall see, they arise as *wavefront singularities* in hyperbolic problems.

Of further importance to the discussion in this section is a theorem on the convergence of the *operator integral* (5.6b) which is less restrictive than that discussed earlier. It may be stated as follows: If the integral $\bar{f}(p)$ converges for $p = p_0$, then $\bar{f}(p)$ exists, and represents an analytic function

---

[1] The name stems from the fact that the integral in (5.18) was first given in 1916 by the mathematician T. J. I'A. Bromwich and independently by K. W. Wagner.

Ch. 5, § 5.4] FURTHER PROPERTIES OF LAPLACE TRANSFORMS

of p, in the half-plane $\text{Re}\,(p-p_0)>0$. A function $f(t)$ is said to belong to $L(p_0)$ if the integral (5.6b), considered as one over the interval $0 \leq t \leq T$, $T<\infty$, exists for $p=p_0$. It should be noted that the theorem relaxes the absolute integrability requirement on $f(t)$. Widder [5.4, ch. II] gives a proo of this theorem.

5.4. Further properties of the Laplace transform and its inverse

A certain number of further basic properties have to be stated to make our transform method (5.6) workable. Most of these can be derived from (5.6).

5.4.1. Uniqueness of $\bar{f}(p)$

Since $\bar{f}(p)$ is analytic in $\text{Re}\,p \geq \gamma$, it is a unique function of p there. It follows that when $p \to \text{Re}\,p = x$, $\bar{f}(p) \to \bar{f}(x)$ for $x \geq \gamma$, consistent with the fact that the integration rules for the operator integral in (5.6b) are the same for p real or complex.

5.4.2. Linearity of integral operator

By direct substitution of

$$f(t) = Af_1(t) + Bf_2(t), \tag{5.20a}$$

in the operator integral (5.6b) where A and B are constants, it is easil' shown that

$$\bar{f}(p) = A\bar{f}_1(p) + B\bar{f}_2(p), \tag{5.20b}$$

where $\bar{f}_1(p)$ and $\bar{f}_2(p)$ are, respectively, the transforms of $f_1(t)$ and $f_2(t)$

5.4.3. Transforms of derivatives of $f(t)$

Using (5.6b) we can write

$$\bar{f}^{(1)}(p) = \int_0^\infty e^{-pt} f'(t) \, dt, \tag{5.21}$$

where $\bar{f}^{(1)}(p)$ is the Laplace transform of $f'(t)$, the first derivative of $f(t)$. By a parts integration on (5.21) it becomes

$$\bar{f}^{(1)}(p) = \left[e^{-pt}f(t)\right]_0^\infty + p\int_0^\infty e^{-pt}f(t)dt,$$

where if we recall the exponential order of $f(t)$, $|f(t)| < K \exp(\gamma_1 t)$, K is a constant and $\operatorname{Re} p = \gamma > \gamma_1$, this reduces to

$$\bar{f}^{(1)}(p) = p\bar{f}(p) - f(0+), \tag{5.22}$$

where $f(0+)$ is the right hand limit of $f(t)$ as $t \to 0$. We can apply (5.22) to $\bar{f}^{(2)}(p)$ to get

$$\bar{f}^{(2)}(p) = p^2 \bar{f}(p) - p f(0+) - f'(0+), \tag{5.23}$$

the transform of $f''(t)$. This process shows that $f^{(n)}(t)$ has the transform

$$\bar{f}^{(n)}(p) = p^n \bar{f}(p) - \sum_{r=0}^{n-1} p^r f^{(n-r-1)}(0+). \tag{5.24}$$

5.4.4. Convolution of two functions

Let $\bar{f}_1(p)$ and $\bar{f}_2(p)$ be the transforms of $f_1(t)$ and $f_2(t)$, respectively. Then if $\bar{f}(p) = \bar{f}_1(p)\bar{f}_2(p)$ is the transform of $f(t)$,

$$f(t) = \frac{1}{2\pi i}\int_{Br} \bar{f}_1(p)\bar{f}_2(p)e^{pt}dp = \frac{1}{2\pi i}\int_{Br} \bar{f}_1(p)e^{pt}\left[\int_0^\infty e^{-p\tau}f_2(\tau)d\tau\right]dp.$$

The order of integration here can be interchanged because the inner integral converges uniformly in p for $\operatorname{Re} p > \gamma_1$ (proved in § 5.3). We have, therefore,

$$f(t) = \int_0^\infty f_2(\tau)\left[\frac{1}{2\pi i}\int_{Br} e^{p(t-\tau)}\bar{f}_1(p)dp\right]d\tau. \tag{5.25}$$

According to the inversion integral (5.6a), the inner integral here defines $f_1(t-\tau)$ for $t-\tau > 0$, and according to (5.19b), and the discussion following that equation, gives

$$f_1(t-\tau) = 0, \qquad t-\tau < 0.$$

It follows that (5.25) reduces to

$$f(t) = f_2(t) * f_1(t) = \int_0^t f_2(\tau)f_1(t-\tau)d\tau. \tag{5.26a}$$

Ch. 5, § 5.4]   FURTHER PROPERTIES OF LAPLACE TRANSFORMS   239

A simple substitution shows that $f(t)$ may also be written as

$$f(t)=f_1(t)*f_2(t)=\int_0^t f_1(\tau)f_2(t-\tau)d\tau\ . \tag{5.26b}$$

Equations (5.26) *define the convolution of the two functions $f_1(t)$ and $f_2(t)$, which corresponds to the product of their transforms.* They are quite useful in finding the inverse $f(t)$, when the inverses of $\bar{f}_1(p)$ and $\bar{f}_2(p)$, are either known or can be found more easily than $f(t)$ itself. Note that (5.26) show the convolution operation is *commutative*, i.e., $f_1(t)*f_2(t)=f_2(t)*f_1(t)$. The operation is *distributive* w.r.t. addition, and is also *associative*, i.e., $f_1(t)*[f_2(t)+f_3(t)]=f_1(t)*f_2(t)+f_1(t)*f_3(t)$ and $f_1(t)*[f_2(t)*f_3(t)]=[f_1(t)*f_2(t)]*f_3(t)$. Using (5.26b) these are not difficult to prove.

5.4.5. *Inverse of $\bar{f}(p)/p$*

Suppose

$$\bar{F}(p)=\bar{f}(p)/p\ . \tag{5.27}$$

In the nomenclature of the previous section we let $\bar{f}_1(p)=1/p$, and $\bar{f}_2(p)=\bar{f}(p)$. It is easily shown using the operator integral (5.6b), that

$$f_1(t)=H(t)\ , \tag{5.28}$$

where $H(t)$ is the step function defined by (2.90). Since $f_2(t)=f(t)$, substituting this and (5.28) in (2.26a) gives

$$F(t)=\int_0^t f(\tau)d\tau\ . \tag{5.29}$$

Equation (5.29) shows that division of a Laplace transform of a function by p, as in (5.27), corresponds to integration of the function form 0 to t. Equation (5.29) points out further that when $f(0+)=0$, the *inverse operations on $\bar{f}(p)$, multiplication and division by p*, i.e., (5.22) and (5.27), *correspond to inverse operations on $f(t)$, differentiation and integration.*

5.4.6. *Inverse of $\exp(-\alpha p)\bar{f}(p)$*

From (5.6a) we see $F(t)$, the inverse of $e^{-\alpha p}\bar{f}(p)$, $\alpha>0$, is given by

$$F(t)=\frac{1}{2\pi i}\int_{Br}\bar{f}(p)e^{p(t-\alpha)}dp=\begin{cases}0\ , & t<\alpha\ ,\\ f(t-\alpha)\ , & t>\alpha\ .\end{cases} \tag{5.30}$$

We see, therefore, *multiplication of a Laplace transform $\bar{f}(p)$ by $\exp(-\alpha p)$ corresponds to a simple shift of the origin for the function $f(t)$, to $t=\alpha$*.

5.4.7. Laplace transforms of some functions $f(t)$

5.4.7.1. Transform of t^{k-1}, $k>0$. From (5.6b) we have

$$\bar{f}(p) = \int_0^\infty t^{k-1} e^{-pt} dt, \qquad t>0.$$

Substituting $x = pt$ in this integral, one finds

$$\bar{f}(p) = p^{-k} \int_0^\infty e^{-x} x^{k-1} dx = \Gamma(k)/p^k, \tag{5.31}$$

which is an analytic function for Re $p > 0$. We will have particular interest in the function t^{k-1}, $0 < k < 1$, since (as we pointed out in § 5.3) very often it represents the singular time behavior of an elastic disturbance at its front. We note $k=1$ is the *finite jump* $H(t)$ given in (2.90). The range $0 < k < 1$ gives infinite, but *integrable singularities*, the strongest being associated with the lowest admissible k value. The value $k=0$ is not included in our foregoing theory (see 5.19b). It can, however, be included in a special way as is shown in the next section.

5.4.7.2. Transform of Dirac delta function $\delta(t)$. The *Dirac delta function* was defined generally in (2.69). This definition shows $\delta(t)$ is defined by

$$\delta(t) = 0, \qquad t \neq 0,$$
$$\int_{-\infty}^\infty \delta(t) dt = 1. \tag{5.32}$$

According to (5.32), we can consider $\delta(t)$ to be the limit of a rectangular pulse (impulse) $f(t)$, defined by

$$f(t) = \begin{cases} 0, & t<0, \\ \dfrac{1}{h}, & 0<t<h, \\ 0, & t>h, \end{cases} \tag{5.33}$$

as the width h of this pulse shrinks to zero. From (5.6b) the transform of this pulse is given by

$$\bar{f}(p) = \frac{1}{h} \int_0^h e^{-pt} dt = \frac{1}{hp}(1 - e^{-ph}). \qquad (5.34)$$

From (5.34) then, the transform of the delta function is given by

$$\bar{\delta}(p) = \lim_{h \to 0} \bar{f}(p) = 1, \qquad (5.35)$$

which from (5.19b) corresponds to $k=0$. We see therefore that $\delta(t)$ is our strongest singularity, taking the place of $1/t$ which does not have a Laplace transform. Equation (5.33) shows $\delta(t)$ here has an asymmetrical nature about $t=0$, and because of this is sometimes called the asymmetrical delta function. We shall omit the word asymmetrical in our reference to this function. Formally from the inversion integral (5.6a), and (5.35), the inverse $\delta(t)$ is represented by

$$\delta(t) = \frac{1}{2\pi i} \int_{Br} e^{pt} dp. \qquad (5.36)$$

5.4.7.3. Laplace-transform tables. Many other Laplace transform pairs, some of which will be useful in our work can be found in a variety of tables. Those in the book by Churchill [5.3, pp. 458–466] will suffice for our purposes.

5.5. Bilateral Laplace transform

The *two-sided Laplace transform*, commonly referred to as the *bilateral Laplace transform* (as we shall refer to it here), is useful in problems involving functions $f(t)$ defined over all of the real t-axis. It finds use, for example, as we shall see later, in diffraction problems, where there may be need to describe events for times $t<0$. The bilateral Laplace transform may be considered as the sum of two (one-sided) Laplace transforms, i.e.,

$$\bar{f}(p) = \int_{-\infty}^{\infty} e^{-pt} f(t) dt = \int_{-\infty}^{0} e^{-pt} f(t) dt + \int_{0}^{\infty} e^{-pt} f(t) dt,$$

or

$$\bar{f}(p) = \int_{0}^{\infty} e^{pt} f(-t) dt + \int_{0}^{\infty} e^{-pt} f(t) dt. \qquad (5.37)$$

The second integral in (5.37) converges for $\operatorname{Re} p = \gamma > \gamma_1$, say, as we have seen in § 5.3. Now assume the first integral converges for $\operatorname{Re}(-p) = -\gamma > -\gamma_2$ or $\operatorname{Re} p = \gamma < \gamma_2$. Further, if $f(t)$ is such that $\gamma_1 < \gamma_2$, we have the *common strip of convergence in the p-plane*, $\gamma_1 < \operatorname{Re} p = \gamma < \gamma_2$, for the two integrals in (5.37). It follows the transform pair is given by

$$\bar{f}(p) = \int_{-\infty}^{\infty} e^{-pt} f(t) dt, \qquad \gamma_1 < \operatorname{Re} p < \gamma_2,$$

$$f(t) = \frac{1}{2\pi i} \int_{\gamma - i\infty}^{\gamma + i\infty} \bar{f}(p) e^{rt} dp, \qquad \gamma_1 < \gamma < \gamma_2. \tag{5.38}$$

The p domain in the first of (5.38) guarantees the absolute convergence of the operator integral there. The γ domain in the second of (5.38) points out that the Bromwich contour must lie within the strip of convergence for the operator integral. From the manner in which it was derived the bilateral Laplace transform could be expected to have many similarities in its properties to those of the Laplace transform. They can be derived along the lines that we used for the Laplace transform in §§ 5.3, 5.4. Recommended reading on the bilateral Laplace transform, its properties and use is the extensive treatment by van der Pol and Bremmer [5.5].

5.6. Exponential Fourier transforms

5.6.1. *Exponential Fourier transform with real argument*

Equation (5.3) can be written as

$$g(x) = \frac{1}{2\pi} \int_{-\infty}^{\infty} \tilde{g}(\varkappa) e^{-i\varkappa x} d\varkappa, \tag{5.39a}$$

where

$$\tilde{g}(\varkappa) = \int_{-\infty}^{\infty} g(x) e^{i\varkappa x} dx. \tag{5.39b}$$

The function $\tilde{g}(\varkappa)$ in (5.39b) defines the *exponential Fourier transform*, and (5.39a) its *inverse* $g(x)$. Setting $\varkappa = -\varkappa$ in (5.3) one finds the alternate transform pair

$$g(x) = \frac{1}{2\pi} \int_{-\infty}^{\infty} \tilde{g}(\varkappa) e^{i\varkappa x} \mathrm{d}\varkappa \ . \tag{5.40a}$$

and

$$\tilde{g}(\varkappa) = \int_{-\infty}^{\infty} g(x) e^{-i\varkappa x} \mathrm{d}x \ , \tag{5.40b}$$

i.e., the signs in the exponentials in (5.39) may be interchanged with no change in $g(x)$. The pair (5.39) or (5.40) are a convenient pair for reducing a variable, say x, in a problem to a parameter, when the variable is specified over the infinite domain $-\infty < x < \infty$. A half space or an infinite layer or plate are examples. Sufficient conditions on the inverse function $g(x)$ in (5.39a) or (5.40a) have already been stated in the Fourier integral theorem of § 5.2. Concerning the *properties* of the exponential Fourier transform $\tilde{g}(\varkappa)$, we note the following:

i) Since we have absolute integrability of $g(x)$ (5.2), the integral in (5.39b) converges absolutely, and uniformly with respect to $\varkappa$. Hence, $\tilde{g}(\varkappa)$ is a continuous function of $\varkappa$, even if $g(x)$ is discontinuous.

ii) From this absolute integrability of $g(x)$ it also follows, on the basis of the *Riemann–Lebesgue lemma*, that

$$\lim_{\varkappa \to \pm\infty} \tilde{g}(\varkappa) = \lim_{\varkappa \to \pm\infty} \int_{-\infty}^{\infty} g(x) e^{i\varkappa x} \mathrm{d}x = 0 \ . \tag{5.41}$$

A proof of the lemma is given in Weinberger [2.22, p. 316].

iii) In many cases where $g(x)$ is not absolutely integrable, we can derive $\tilde{g}(\varkappa)$ as a *Cauchy principal value* of its defining integral (5.39b), for $\varkappa > 0$, and similarly for $\varkappa < 0$, hence establishing continuity of $\tilde{g}(\varkappa)$ in these domains of $\varkappa$. Generally there is a discontinuity at $\varkappa = 0$. Weinberger [2.23, § 64] gives the proof and some instructive examples.

The exponential Fourier transforms of derivatives of $g(x)$ can be obtained through parts integration of the operator integral (5.39b), and iterations. Suppose $g'(x)$, as well as $g(x)$, satisfies our Fourier integral Theorem in § 5.2. Then the transform of $g'(x)$ is given by

$$\widetilde{g^{(1)}}(\varkappa) = \int_{-\infty}^{\infty} g'(x) e^{i\varkappa x} \mathrm{d}x = [g(x)e^{i\varkappa x}]_{-\infty}^{\infty} - i\varkappa \int_{-\infty}^{\infty} g(x) e^{i\varkappa x} \mathrm{d}x \ . \tag{5.42}$$

Then provided $g(x) \to 0$, as $x \to \pm\infty$, (5.42) reduces to

$$\widetilde{g^{(1)}}(\varkappa) = -i\varkappa \tilde{g}(\varkappa) \ . \tag{5.43}$$

Through iterations of (5.43), we find the transform of the nth order derivative of $g(x)$ is

$$\widetilde{g^{(n)}}(\varkappa)=(-i\varkappa)^n\tilde{g}(\varkappa) .\tag{5.44}$$

Inversion of (5.39a), when the transforms cannot be found in the tables, is accomplished in much the same way as in the Laplace transform, i.e., through methods based on *analytic function theory*. It is important to note that the exponential Fourier transform $\tilde{g}(\varkappa)$ of an absolutely integrable function $g(x)$ need not be absolutely integrable, and because of this the inversion integral (5.39a) is a Cauchy principal value. Weinberger gives a simple example of this, namely, with

$$g(x)=\begin{cases}1, & |x|\leq 1,\\ 0, & |x|>1,\end{cases}$$

which is absolutely integrable, the corresponding transform $\tilde{g}(\varkappa)=2\sin\varkappa/\varkappa$ is not.

To derive the shift rule for the exponential Fourier transform with real argument we follow Weinberger's simple derivation. Consider the function

$$h(x)=g(ax-b) ,$$

where a and b are real constants. If $g(x)$ is absolutely integrable so is $h(x)$, and we have from (5.39b)

$$\tilde{h}(\varkappa)=\int_{-\infty}^{\infty}g(ax-b)e^{i\varkappa x}dx .$$

Letting $ax-b=\xi$, and $a>0$, this integral becomes

$$\tilde{h}(\varkappa)=\frac{1}{a}\int_{-\infty}^{\infty}g(\xi)e^{i\varkappa(\xi+b)/a}d\xi .$$

If $a<0$, a minus sign is introduced through the limits of integration. In either case it follows that

$$\widetilde{g(ax-b)}=\exp(i\varkappa b/a)\tilde{g}(\varkappa/a)/|a| ,\tag{5.45a}$$

which is the *shift rule*. When $a=1$ in (5.45a), we have

$$\widetilde{g(x-b)}=\exp(ib\varkappa)\tilde{g}(\varkappa) ,\tag{5.45b}$$

which says that a translation to the right by b, corresponds to an $\exp(ib\varkappa)$

multiplied exponential Fourier transform. If $b=0$ in (5.45a) we see multiplying x by a constant a corresponds to dividing $\varkappa$ by a and the exponential Fourier transform by $|a|$.

The *convolution* product for the present transform can be obtained much as it was in the Laplace transform [cf. § 5.4.4]. Letting $\tilde{g}_1(\varkappa)$ and $\tilde{g}_2(\varkappa)$ be the transforms of $g_1(x)$ and $g_2(x)$, respectively, and assuming $g_1(x)$, $g_2(x)$ and $g(x)$ are absolutely convergent, then if $\tilde{g}(\varkappa)=\tilde{g}_1(\varkappa)\tilde{g}_2(\varkappa)$ is the transform of $g(x)$ the convolution integral is given by

$$g(x)=\int_{-\infty}^{\infty} g_2(x')g_1(x-x')\mathrm{d}x' . \qquad (5.46)$$

The proof is left to the exercises.

5.6.2. Exponential Fourier transform with complex argument

A transform closely related to the bilateral Laplace transform but with somewhat simpler properties, hence somewhat simpler to use, is the exponential Fourier transform with *complex argument*. The transform pair can be derived from the bilateral Laplace transform pair (5.38) by noting the transformation parameter of the former, say ξ, corresponds to a simple $\pi/2$ rotation of the parameter p of the latter. Hence we write $\xi=ip=\varkappa+i\gamma$ and $p=-i\xi=\gamma-i\varkappa$, and substitute for p in (5.38) with the results

$$\tilde{g}(\xi)=\int_{-\infty}^{\infty} g(x)e^{i\xi x}\mathrm{d}x , \qquad \gamma_1<\mathrm{Im}\ \xi<\gamma_2 ,$$

$$g(x)=\frac{1}{2\pi}\int_{-\infty+i\gamma}^{\infty+i\gamma} \tilde{g}(\xi)e^{-i\xi x}\mathrm{d}\xi , \qquad \gamma_1<\gamma<\gamma_2 . \qquad (5.47)$$

It may be observed that (5.47) have the form of (5.39), and indeed reduce to (5.39) when $\gamma=0$, i.e. when ξ becomes real $\varkappa$. The operational properties, like the transform of a derivative of a function (5.43) hold for the present complex variable ξ, as well as the shift rule (5.45) and the convolution integral (5.46). Proofs are given by Weinberger [2.23, § 73] in which the integrability requirements on the inverse functions and analyticity requirements on the transforms (note the domain $\gamma_1<\mathrm{Im}\ \xi<\gamma_2$) are discussed.

5.7. Fourier sine and cosine transforms

The Fourier sine and cosine transform pairs can be obtained from the exponential Fourier transform pairs as special cases when $g(x)$ *is odd or even*. Equation (5.39b) can be written as

$$\tilde{g}(\varkappa) = \int_{-\infty}^{\infty} g(x)(\cos \varkappa x + i \sin \varkappa x) dx.$$

When $g(x)$ is odd the $\cos \varkappa x$ term in this integral gives a zero contribution. It follows that

$$\tilde{g}(\varkappa) = 2i \int_0^{\infty} g(x) \sin \varkappa x dx = 2i \tilde{g}^s(\varkappa), \qquad (5.48)$$

where $\tilde{g}^s(\varkappa)$ *is the Fourier sine transform*. Its inverse is obtained by first substituting the right hand side of (5.48) into the inversion integral (5.39a) and writing the latter as

$$g(x) = \frac{i}{\pi} \int_{-\infty}^{\infty} \tilde{g}^s(\varkappa)(\cos \varkappa x - i \sin \varkappa x) d\varkappa.$$

By substituting $-\varkappa$ for $\varkappa$ in (5.48) we see $\tilde{g}^s(\varkappa)$ is odd, and therefore the last integral reduces to

$$g(x) = \frac{2}{\pi} \int_0^{\infty} \tilde{g}^s(\varkappa) \sin \varkappa x d\varkappa. \qquad (5.49)$$

Equations (5.48), (5.49) define the *Fourier sine transform pairs*, and we see they are appropriate for use in the half range $0 < x < \infty$.

Similarly when $g(x)$ is even, we find

$$\tilde{g}(\varkappa) = 2 \int_0^{\infty} g(x) \cos \varkappa x dx = 2 \tilde{g}^c(\varkappa), \qquad (5.50)$$

where $\tilde{g}^c(\varkappa)$ *is the Fourier cosine transform*, and

$$g(x) = \frac{2}{\pi} \int_0^{\infty} \tilde{g}^c(\varkappa) \cos \varkappa x d\varkappa \qquad (5.51)$$

for ist inverse. Again the pairs are appropriate for use in $0 < x < \infty$.

Parts integrations again supply the sine and cosine transforms of derivatives. From (5.48) we have

$$\widetilde{g^{(1)s}}(\varkappa) = \int_0^\infty g'(x) \sin \varkappa x \, dx = [g(x) \sin \varkappa x]_0^\infty - \varkappa \int_0^\infty g(x) \cos \varkappa x \, dx \, .$$

Then if $g(x) \to 0$, as $x \to \infty$,

$$\widetilde{g^{(1)s}}(\varkappa) = -\varkappa \widetilde{g}^c(\varkappa) \, . \tag{5.52}$$

And from (5.50)

$$\widetilde{g^{(1)c}}(\varkappa) = \int_0^\infty g'(x) \cos \varkappa x \, dx = [g(x) \cos \varkappa x]_0^\infty + \varkappa \int_0^\infty g(x) \sin \varkappa x \, dx \, , \tag{5.53}$$

and again if $g(x) \to 0$, as $x \to \infty$, (5.53) reduces to

$$\widetilde{g^{(1)c}}(\varkappa) = \varkappa \widetilde{g}^s(\varkappa) - g(0) \, . \tag{5.54}$$

Iteratively, from (5.52), using (5.54), we find

$$\widetilde{g^{(2)s}}(\varkappa) = -\varkappa \widetilde{g^{(1)c}}(\varkappa) = -\varkappa^2 \widetilde{g}^s(\varkappa) + \varkappa g(0) \, , \tag{5.55}$$

and from (5.54), using (5.52)

$$\widetilde{g^{(2)c}}(\varkappa) = \varkappa \widetilde{g^{(1)s}}(\varkappa) - g'(0) = -\varkappa^2 \widetilde{g}^c(\varkappa) - g'(0) \, , \tag{5.56}$$

provided $g'(x) \to 0$, as $x \to \infty$. This process can be continued when transforms of higher order derivatives are needed. It should be pointed out that in these half range transforms and inverses, since we are dealing with special cases of $g(x)$ (i.e., odd and even), and their $\tilde{g}(\varkappa)$ (i.e., $2i\tilde{g}^s(\varkappa)$ and $2\tilde{g}^c(\varkappa)$), the properties we have discussed for $g(x)$, and its higher order derivatives, and their exponential transforms, carry over here. We note, however, continuity of the half range transforms is over the half range $\varkappa \geq 0$ (corresponding to absolute integrability of the inverses). When the inverses do not have absolute integrability, whereas we had a Cauchy principal value defining integral for these in the exponential Fourier transform, here we would have ordinary *improper integrals*, which define half range transforms, continuous over the range $\varkappa > 0$. Similarly, Cauchy principal value defining integrals for the inverses of exponential Fourier transforms correspond to ordinary improper integrals for the half range inverses.

The relations (5.48) and (5.50), defining the relationships between $\tilde{g}(\varkappa)$ and $\tilde{g}^s(\varkappa)$, and $\tilde{g}(\varkappa)$ and $\tilde{g}^c(\varkappa)$ are very important. Tables usually contain an array of the half range pairs. The tables by Oberhettinger [5.6] contain about 1800. Hence, for example, when a problem is solved by the exponential Fourier transform, its inverse can very often be found from that of $\tilde{g}^s(\varkappa)$ or $\tilde{g}^c(\varkappa)$ in the tables. It should also be noted from (5.55) and (5.56), when solving half range problems, that if $g(0)$ is given, say on the boundary ($x=0$), a sine transform is appropriate, whereas if $g'(0)$ is given, the cosine transform fits (i.e., they "ask for" that information). Note also from (5.52) and (5.54) that one differentiation in x corresponds to a transform of the other type. A second differentiation gives homogeneity in each transform as (5.55) and (5.56) show. This mixing, and homogeneity, of transforms is also true for the odd and even higher order derivatives, respectively. Churchill gives expressions for the transforms of the higher order derivatives. He also gives convolution integrals for the sine and cosine transforms [5.3, ch. 12, 13]. Further reading on the Fourier transforms is left to the books by Weinberger [2.23, ch. X], Churchill [5.3, ch. 12, 13] and Sneddon [5.7]. Sneddon treats some interesting applications in his book.

5.8. Hankel transforms and properties

Hankel transforms find their use in problems involving axial symmetry. The suddenly applied point load on the surface of an elastic half space is an important example of such a problem. We shall treat this in Chapter 6. As in the case of the exponential Fourier transform pairs (5.39) we need sufficient conditions on, say $g(r)$, like those in the Fourier integral theorem of § 5.2 to permit expansion in a form analogous to (5.39). With a slight modification these conditions are those given in Watson [5.8, § 14.4] for *Hankel's integral theorem* which may be stated as follows: *Let $g(r)$ be an arbitrary function of real r such that*

$$\int_0^\infty |\sqrt{r}g(r)|dr<\infty, \tag{5.57}$$

and $g(r)$ and $g'(r)$ are piecewise continuous over each of the subintervals of $0\leq r<\infty$. Then, if $\nu\geq-\tfrac{1}{2}$,

$$\int_0^\infty \varkappa\tilde{g}^\nu(\varkappa)J_\nu(\varkappa r)d\varkappa=\tfrac{1}{2}\left[g(r+)+g(r-)\right], \tag{5.58}$$

Ch. 5, § 5.8]   HANKEL TRANSFORMS AND PROPERTIES   249

where

$$\tilde{g}^v(\varkappa) = \int_0^\infty rg(r)J_v(\varkappa r)\mathrm{d}r\,. \tag{5.59}$$

Equation (5.59) is the Hankel transform of order v of g(r), and (5.58) gives its inverse g(r). The piecewise continuity of g(r) and g'(r) is used here (as in the fundamental Fourier integral theorem in § 5.2) instead of the requirement in Watson of limited total fluctuation of g(r) in an interval about r, which is an equivalent statement to bounded variation of g(r). Our conditions certainly fit this. Proof of the Hankel integral theorem is given in Watson [5.8, §§ 14.41–14.5].

As we did for the Fourier transforms similar comments are in order for most of the properties of the Hankel transform. They are quite similar, however, i.e., most of the properties (except for the convolution theorem) stated in §§ 5.6.1, 5.7 having their analogs here. The absolute integrability of $\sqrt{r}g(r)$ required by (5.57), guarantees the operator integral in (5.59) converges absolutely and uniformly, and hence that the Hankel transform $\tilde{g}(\varkappa)$ is a continuous function of $\varkappa$, even if g(r) is discontinuous. Here, we have an analog of the Riemann–Lebesgue lemma which shows

$$\lim_{\varkappa \to \infty} \tilde{g}^v(\varkappa) = 0\,. \tag{5.60}$$

In cases where $\sqrt{r}g(r)$ is not absolutely integrable, here the transforms $\tilde{g}^v(\varkappa)$, continuous over $\varkappa > 0$, may exist through (5.59) as an improper integral. And the inverses given on the left hand side of (5.58) may also exist as improper integrals.

From (5.59) parts integration again gives the transforms of derivatives. For g'(r) we find

$$\tilde{g}'^v(\varkappa) = (v-1)\int_0^\infty g(r)J_v(\varkappa r)\mathrm{d}r - \varkappa\tilde{g}^{v-1}(\varkappa)\,, \tag{5.61}$$

which is based on the *further condition*

$$\lim_{r \to 0, \infty} [rg(r)] = 0\,. \tag{5.62}$$

A simpler expression can be found for (5.61) by using the recursion relation $2vJ_v(x) = x[J_{v-1}(x) + J_{v+1}(x)]$ in the integral there with the result

$$\tilde{g}'^v(\varkappa) = -\frac{\varkappa}{2v}\left[(v+1)\tilde{g}^{v-1}(\varkappa) - (v-1)\tilde{g}^{v+1}(\varkappa)\right]\,, \tag{5.63}$$

which can be used iteratively to get the transforms for higher derivatives. A very useful special case is obtained from (5.63) (or (5.61) for $v=1$),

$$\tilde{g}'^1 = -\varkappa \tilde{g}^0(\varkappa) . \tag{5.64}$$

Of particular value are the transforms of functions containing combinations of derivatives that appear in Bessel type differential equations. Following Sneddon [5.7, § 10] we use parts integration to write

$$\int_0^\infty rg''(r)J_v(\varkappa r)\mathrm{d}r = -\int_0^\infty g'(r)\frac{\mathrm{d}}{\mathrm{d}r}[rJ_v(\varkappa r)]\mathrm{d}r , \tag{5.65}$$

where we have *imposed* $rg'(r) \to 0$ *as* $r \to 0$. Now from (5.65), and a parts integration, we find

$$\int_0^\infty r\left[g''(r) + \frac{1}{r}g'(r)\right]J_v(\varkappa r)\mathrm{d}r = \varkappa \int_0^\infty g(r)\frac{\mathrm{d}}{\mathrm{d}r}[rJ'_v(\varkappa r)]\mathrm{d}r . \tag{5.66}$$

From the Bessel Equation for $J_v(\varkappa r)$.

$$\varkappa \frac{\mathrm{d}}{\mathrm{d}r}[rJ'_v(\varkappa r)] = -r\left(\varkappa^2 - \frac{v^2}{r^2}\right)J_v(\varkappa r) ,$$

and using this in (5.66) it becomes

$$\int_0^\infty r\left[g''(r) + \frac{1}{r}g'(r) - \frac{v^2}{r^2}g(r)\right]J_v(\varkappa r)\mathrm{d}r = -\varkappa^2 \tilde{g}^v(\varkappa) . \tag{5.67}$$

In particular when $v=0$, we have the important special case

$$\int_0^\infty r\left[g''(r) + \frac{1}{r}g'(r)\right]J_0(\varkappa r)\mathrm{d}r = -\varkappa^2 \tilde{g}^0(\varkappa) . \tag{5.68}$$

One can find further reading on the Hankel transform, theory and applications, in Sneddon's book [5.7].

5.9. Asymptotic expansions

When use is made of integral transforms in solving complicated physical problems, one might expect through the inversion integrals and contour integration to generate solutions represented by one or more definite integrals. Evaluation of these integrals for extended exact information for particular parameters (usually space x and time t) in most instances, requires numerical integration. On the other hand sometimes only the value of a definite integral (or integrals) at some limiting value of a parameter is needed, e.g., large x or t. The methods by which a definite integral can be approximated for a limiting value of a parameter are referred to as *asymptotic expansions of integrals*. The literature in elastic waves exhibits very profitable use of these methods in solving problems, and since we will draw on such methods a discussion of them is in order. We begin with a brief treatment of asymptotic expansions and their properties.

5.9.1. The nature of asymptotic expansions

The nature of an asymptotic expansion is usually best brought out by an example. Following Wu [5.9, § 2.1] and McLachlan [5.10, Appendix 3] we consider the integral

$$f(x) = \int_x^\infty \frac{e^{x-t}}{t} \, dt, \tag{5.69}$$

where $x > 0$ and t is real. This integral converges uniformly with respect to x in any closed interval $0 < x \leq x'$. We wish to find its value when x is large. Through repeated integrations by parts we find

$$f(x) = S_n(x) + R_n(x), \tag{5.70a}$$

where

$$S_n(x) = \sum_{k=1}^n (-)^{k-1}(k-1)! x^{-k} = \sum_{k=1}^n u_k, \tag{5.70b}$$

and

$$R_n(x) = (-)^n n! \int_x^\infty \frac{e^{x-t}}{t^{n+1}} \, dt. \tag{5.70c}$$

$S_n(x)$ is the *partial sum*, and $R_n(x)$ the *remainder*, of the expansion (5.70a). The series (5.70) is exact. Its terms and the remainder alternate in sign, so the error committed in representing $f(x)$ by the partial sum $S_n(x)$, that is by neglecting $R_n(x)$, is at most equal to the first neglected term $u_{n+1}(x)$, i.e., from (5.70c) we have

$$|R_n(x)| \leq n! x^{-(n+1)}. \tag{5.71}$$

Consider the ratio of two successive terms in S_n,

$$|u_{n+1}/u_n| = nx^{-1}. \tag{5.72}$$

We see this $\to 0$ as $x \to \infty$, for fixed n, but becomes *unbounded as* $n \to \infty$, *for any x no matter how large*. It follows for a given large x, the moduli of the terms of S_n first decrease for $n \leq x$ to a minimum, and then increase for $n \geq x$. Hence, for a given x, if S_n includes all terms up to but not the smallest, the least possible error is made in representing $f(x)$ by S_n. Taking further terms would increase the error. McLachlan demonstrates this behavoir numerically. Such behavior is typical of an asymptotic series. In the next two sections we discuss the definition and some important properties of an asymptotic expansion of an analytic function, essentially following the Jeffreys and Jeffreys treatment [5.11, ch. 17].

5.9.2. Poincaré's definition of an asymptotic expansion

The usual definition given for an asymptotic expansion is due to Poincaré (1886). It may be stated as follows: If $f(z)$ is an analytic function, and $S(z)$ is the series

$$S(z) = A_0 + A_1/z + A_2/z^2 + \ldots + A_n/z^n + \ldots,$$

having the partial sum

$$S_n(z) = \sum_{k=0}^{n} A_k z^{-k}, \tag{5.73}$$

the series is called an asymptotic expansion of $f(z)$ within a given interval of argument z, if for every n

$$\lim_{|z| \to \infty} z^n [f(z) - S_n(z)] = 0 \tag{5.74}$$

where $f(z) - S_n(z) = R_n(z)$, the remainder after n terms. When this definition has been met we denote the relation between $f(z)$ and its asymptotic expansion by

$$f(z) \sim S(z). \tag{5.75}$$

Equation (5.74) says that $f(z) - S_n(z) = o(z^{-n})$, hence we have

$$f(z) = \sum_{k=0}^{n} A_k z^{-k} + o(z^{-n}), \qquad |z| \to \infty. \tag{5.76}$$

This says the *error* in stopping at n terms is $o(z^{-n})$. Equation (5.75) implies the error is of $O(z^{-n-1})$, a sharper estimate.

An asymptotic expansion can be convergent or divergent. An example of the first is a Laurent's series (power series in $1/z$) which converges in the exterior domain, say $|z| > R$, for it is known that the modulus of the remainder $|R_n|$ after the term z^{-n} is less than $M(|z|/R)^{-n}[(|z|/R) - 1]^{-1}$, for all values of arg z. The error term, therefore, is $O[|z|^{-(n+1)}] = o(|z|^{-n})$. Such a series, however, may need too many terms to be a useful approximation. The alternate would be a divergent series of the type discussed in the previous section, where a few terms does the job. We note that that series is an asymptotic expansion since it meets all the conditions of Poincaré's definition applied to a real function.

5.9.3. Some properties of asymptotic expansions

We discuss here some important properties of asymptotic expansions.

i) *Asymptotic expansions are unique.* To prove this assume we have from (5.74) for the sector $|z| > R$, $\alpha \leq \arg z \leq \beta$,

$$\lim_{|z| \to \infty} z^n [f(z) - S_n(z)] = 0, \qquad (5.77)$$

$$\lim_{|z| \to \infty} z^n [f(z) - T_n(z)] = 0,$$

where

$$T_n = \sum_{k=0}^{n} B_k z^{-k}.$$

It follows from (5.77) that

$$\lim_{|z| \to \infty} z^n \sum_{k=0}^{n} (A_k - B_k) z^{-k} = 0.$$

hence $A_k = B_k$, which was to be proved. As pointed out in [5.11, § 17.024], however, the converse is not true, namely the same expansion in a given region may be an asymptotic expansion of several functions provided their differences $f(z) - g(z)$ satisfy, for every n,

$$\lim_{|z| \to \infty} z^n [f(z) - g(z)] = 0.$$

A case where this happens is when $f(z)-g(z)=e^{-z}$, with arg z in the range $-\pi/4 \leq \arg z \leq \pi/4$.

ii) *Two asymptotic power series valid in the same sector can be multiplied unconditionally to give another asymptotic power series.* Consider the two series

$$f(z) = S_n(z) + o(z^{-n}),$$
$$g(z) = T_n(z) + o(z^{-n}).$$
(5.78)

Now the product

$$S_n(z)T_n(z) = \left(\sum_{k=0}^{n} A_k z^{-k}\right)\left(\sum_{k=0}^{n} B_k z^{-k}\right) = \sum_{k=0}^{n} C_k z^{-k} + o(z^{-n}), \quad (5.79)$$

where $C_k = A_0 B_k + A_1 B_{k-1} + \ldots + A_k B_0$. Hence we have from (5.78), (5.79) for every n

$$f(z)g(z) = S_n(z)T_n(z) + o(z^{-n}) = \sum_{k=0}^{n} C_k z^{-k} + o(z^{-n}), \quad (5.80)$$

which was to be proved. A similar proof can be carried out to establish that *two asymptotic power series can be divided.* That is $S_n(z)/T_n(z)$ yields another asymptotic power series provided $B_0 \neq 0$. It is simple to show *addition of two asymptotic expansions leads to another asymptotic expansion.*

iii) *An asymptotic expansion can be integrated unconditionally.* Again assuming $f(z)$ analytic in the sector $|z| > R$, $\alpha \leq \arg z \leq \beta$, and $|z^n[f(z)-S_n(z)]| < \varepsilon$, we choose a path, from some point z_0 within the sector to ∞, over which arg z is a constant. It follows that

$$\left|\int_{z_0}^{\infty} [f(z) - S_n(z)] dz\right| < \int_{|z_0|}^{\infty} \frac{\varepsilon dr}{r^n} = \frac{\varepsilon}{(n-1)|z_0|^{n-1}},$$

and using this we have

$$\left| |z_0|^{n-1} \left[\int_{z_0}^{\infty} f(z) dz - \int_{z_0}^{\infty} S_n(z) dz\right]\right| < \frac{\varepsilon}{n-1},$$

from which it follows that termwise integration of $S_n(z)$ gives an asymptotic expansion of $\int_{z_0}^{\infty} f(z) dz$.

Next let $|z_0|=|z_1|$ and choose a path from z_0 to z_1 which is a circular arc about $z=0$, hence a path over which modulus of z is a constant. Then since the arc is of length $L<2\pi|z_0|$

$$\left| \int_{z_0}^{z_1} [f(z)-S_n(z)]dz \right| < \frac{\varepsilon L}{|z_0|^n} < \frac{2\pi\varepsilon}{|z_0|^{n-1}}.$$

It follows we have the same result, as in the previous case, where now we have a path of constant modulus z. Since $f(z)$ and $S_n(z)$ have no singularities in the sector, it follows we can termwise integrate $S_n(z)$ over any path in the sector to get the asymptotic expansion of the integral of $f(z)$ over the same path since such a general path can always be represented by the sum of a path over constant arg z and one over constant modulus z, each of which we just proved have such an asymptotic expansion.

iv) Further properties. We note further that asymptotic expansions can also be integrated with respect to a parameter provided that the expansion holds uniformly in the parameter. This is proved in Erdelyi [5.12, § 1.4] for a real parameter. In general it is not permissible to differentiate asymptotic expansions with respect to the variable or parameters. An asymptotic power series can be differentiated according to the following statement: If $f(z)$ in

$$f(z) \sim A_0 + A_1/z + A_2/z^2 + \ldots \text{ to } n \text{ terms, as } z \to \infty,$$

is differentiable, and if $f'(z)$ possesses an asymptotic power series expansion, then

$$f'(z) \sim -A_1/z^2 - 2A_2/z^3 - \ldots \text{to } n-1 \text{ terms, as } z \to \infty.$$

In the case of analytic functions it is not necessary to assume that $f'(z)$ possesses an asymptotic power series expansion. Proofs are given in [5.12, § 1.6]. For further reading on properties of asymptotic expansions reference [5.12, ch. I] is recommended.

5.10. Asymptotic expansions of integrals

There are several methods of obtaining asymptotic expansions of integrals. We shall discuss some of these here, drawing on Erdelyi [5.12] and Wu [5.9].

5.10.1. Types of integrals; critical points

The asymptotic expansion methods are applicable to the integrals of the form

$$\int_a^b f(t)e^{xg(t)}dt, \qquad (5.81)$$

$$\int_a^b f(t)e^{ixh(t)}dt. \qquad (5.82)$$

In (5.81) we have two possibilities, (1) x is a large positive parameter, $g(t)$ is a real function of a real variable t and $f(t)$ may be real or complex, or (2) x is a complex parameter with large modulus, f and g are analytic functions of complex variable t, and the integration is along a path in the complex t-plane. In (5.82) the conditions on x, $h(t)$ and $f(t)$ are the same respectively, as those on x, $g(t)$, and $f(t)$ in (1) above for (5.81). *A point at which the derivative $g'(t)$ or $h'(t)$ vanishes is called a critical point. If there exists no critical point within the limits of integration, then the end points of integration are called critical end points.*

Of particular interest to us will be a special case of (5.81) obtained by setting $g(t) = -t$, and $a = 0$, $b = \infty$, which gives the *Laplace integral*

$$\int_0^\infty f(t)e^{-xt}dt, \qquad (5.83)$$

and a special case of (5.82) obtained by setting $h(t) = t$, with (a, b) mostly a finite real interval, which gives the *Fourier integral*

$$\int_a^b f(t)e^{ixt}dt. \qquad (5.84)$$

It is clear that both (5.83), (5.84) have no critical points other than the end points.

5.10.2. Asymptotic expansion of Laplace integrals

Let us now consider the asymptotic expansion for $x \to \infty$ of the Laplace integral

$$I(x) = \int_0^\infty f(t)e^{-xt}dt. \qquad (5.85)$$

We have already discussed the requirements for convergence of this integral in § 5.3, which we will draw on in our further development here.

5.10.2.1. Integration by parts. As our example in § 5.9.1 demonstrated, repeated integrations by parts is a productive means of obtaining asymptotic expansions of integrals. Under the conditions that $f(t)$ is n times continuously differentiable in $0 \leq t \leq c$ and belongs to $L(x_0)$ for some x_0, then repeated parts integration of (5.85) gives

$$I(x) \sim \sum_{k=1}^{n} f^{(k-1)}(0) x^{-k}, \tag{5.86}$$

uniformly in arg x, *as* $x \to \infty$ *in the sector* S_Δ *given by* $0 < |x| < \infty$ *and* $|\text{arg } x| \leq (\pi/2) - \Delta < \pi/2$. The proof is given in [5.12], [5.9].

5.10.2.2. Watson's lemma. When $f(t)$ has a branch point at $t=0$, then the asymptotic expansion of (5.85) can be obtained by *Watson's lemma*. The lemma is stated as follows: *If* $I(x)$ *(in (5.85)) exists for some* x_0, *and for* $0 < \lambda_1 < \lambda_2 < \ldots$,

$$f(t) \sim \sum_{k=1}^{n} a_k t^{\lambda_k - 1} \quad \text{as} \quad t \to 0, \tag{5.87}$$

then

$$I(x) \sim \sum_{k=1}^{n} a_k \Gamma(\lambda_k) x^{-\alpha_k}, \tag{5.88}$$

uniformly in arg x *as* $x \to \infty$, *in* S_Δ, where S_Δ is defined in § 5.10.2.1., and Γ is the gamma function. The lemma states that the asymptotic expansion (5.88) is obtained by substituting (5.87) into (5.85) and integrating termwise. The proof of the lemma may be found in [5.12], [5.9].

5.10.2.3. Asymptotic expansion of $\bar{f}(p)$; short time (wavefront) approximation.

An important application of Watson's lemma stems from identifying $\bar{f}(p)$ with I(x) in (5.88). It follows that the corresponding terms in (5.87) and (5.88) are pairs, for $|p|$ large and t small. An asymptotic expansion found for an $\bar{f}(p)$ for $|p|$ large, when identified with (5.88) has its inverse in (5.87), i.e., substitution of (5.88) for $\bar{f}(p)$ into (5.6a) will yield (5.87) as a termwise integration. When applied one usually finds that only one or two terms are useful, but in many cases this is all that is needed to disclose the short time, or in our hyperbolic problems, the wavefront character of a solution. We shall apply the technique in the cylindrical cavity source problem later in this chapter, and to other problems in later chapters.

5.10.2.4. Long time approximation; solution of the static problem. The pair (5.87), (5.88) can also be used to derive an important asymptotic expansion for $f(t)$ as an inverse Laplace transform valid for large t. From the expansion we can set down the static solution for problems in a convenient way. Observing the exp (pt) term in the Bromwich integral (5.6a), we note that *for large time t*, on the basis of isolated singularities, *one could expect the singularity of $\bar{f}(p)$, closest to the Br path* (the *extreme* singularity), *to determine $f(t)$ approximately*. We assume that in the neighborhood of this isolated singularity, say a branch point or pole at $p = p_1$, $\bar{f}(p)$ behaves as

$$\bar{f}(p) \simeq (p-p_1)^{-\nu} \sum_{k=0}^{n} a_k (p-p_1)^k, \tag{5.89}$$

where ν is a real number, and where other possible isolated singularities, say at $p = p_i$, $\operatorname{Re} p_i < \operatorname{Re} p_1$, $i = 2, 3, \ldots$, have been neglected. To find $f(t)$ for large time, we appeal to the inversion integral (5.6a) and contour integration, using (5.89). Figure 5.2a shows the completion of Br that we will use,

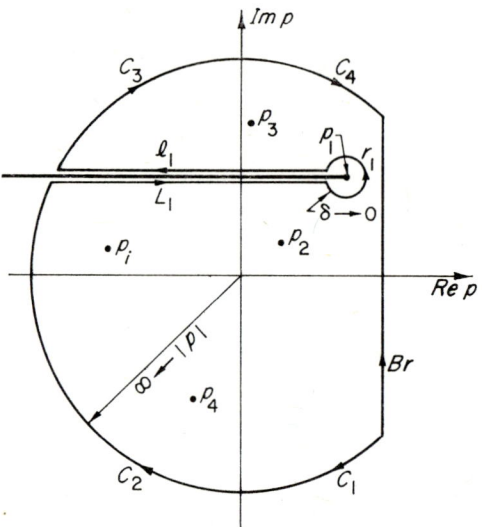

Fig. 5.2a. Contour integration for the derivation of the asymptotic expansion of $f(t)$ for large time.

in its limiting < sense, $|p| \to \infty$ on the paths C_1, C_2, C_3 and C_4, and $\delta \to 0$ on r_1. Assuming a branch point at p_1, with the aid of the Cauchy–Goursat theorem using (5.6a), we can write

$$f(t) = \frac{1}{2\pi i} \int_{Br} \bar{f}(p) e^{pt} dp$$

$$= \frac{1}{2\pi i} \left[\int_{C_1} + \int_{C_2} + \int_{L_1} + \int_{r_1} + \int_{l_1} + \int_{C_3} + \int_{C_4} \right] \bar{f}(p) e^{pt} dp, \quad (5.90)$$

where the chosen single-valued branch of the multivalued function $\bar{f}(p)$, appearing in the Bromwich integral on the left of (5.90) must be analytically continued to and along the paths for the integrals on the right. Since $t > 0$ in (5.90), and $\bar{f}(p)$ here meets the order condition (5.19b), it is not difficult to show, (1) with the aid of *Jordan's inequality*, $2\theta/\pi \leq \sin\theta \leq \theta$, where $\theta = \arg p$, that the sum of the integrals over the paths C_2, C_3 vanishes, and (2) the integrals over the paths C_1, C_4 vanish. The proof is left to the exercises. On this basis (5.90) reduces to

$$f(t) = \frac{1}{2\pi i} \left[\int_{L_1 + l_1} + \int_{r_1} \bar{f}(p) e^{pt} dp \right]. \quad (5.91)$$

To evaluate the integrals over the paths L_1, l_1 along the cut, and r_1, we let

$$p - p_1 = \eta \exp(i\varphi), \quad -\pi < \varphi < \pi, \quad \eta > 0, \quad (5.92)$$

where this complex vector is shown in fig. 5.2b. We note that the definition (5.92) satisfies the requirement that $\operatorname{Re} p > 0$ on the Br path. From it, the single-valued branch of $\bar{f}(p)$ on Br is defined. Continuation of $p - p_1$ in

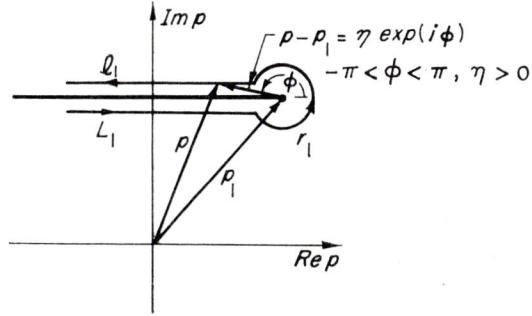

Fig. 5.2b. Definition of complex vector $p - p_1$.

(5.92), and hence $\bar{f}(p)$ in (5.89), to the paths L_1, l_1 and r_1 is simple then, since we have only the one branch point at $p=p_1$. We find on L_1, $p-p_1 = \eta \exp(-i\pi)$, on l_1, $p-p_1 = \eta \exp(i\pi)$, and on r_1, $p-p_1 = \delta \exp(i\varphi)$, $-\pi < \varphi < \pi$. It follows that

$$\int_{L_1+l_1} \bar{f}(p)e^{pt}dp \simeq \frac{e^{p_1 t}}{\pi} \int_0^\infty \sum_{k=0}^n \sin[\pi(v-k)]a_k \eta^{-v+k} e^{-t\eta} d\eta, \quad (5.93)$$

and

$$\lim_{\delta \to 0} \int_{r_1} \bar{f}(p)e^{pt}dp \simeq \frac{e^{p_1 t}}{2\pi} \sum_{k=0}^n a_k \lim_{\delta \to 0} \int_{-\pi}^\pi \delta^{1-v+k} e^{i\varphi(1-v+k)} d\varphi. \quad (5.94)$$

For $v < k+1$, this last integral vanishes. Recalling that t is large in (5.93) and the integrals there are Laplace integrals, we can identify the series in (5.93) with that in (5.87), i.e. η here corresponds to t there, $\sin[\pi(v-k)]a_k$ to a_{k+1}, and $-v+k$ to $\lambda_{k+1}-1$. It follows from (5.88), identifying t here with x there, that for large t

$$f(t) \sim t^{v-1} e^{p_1 t} \sum_{k=0}^n \frac{a_k}{\Gamma(v-k)} t^{-k}.$$

By expanding $\Gamma(v-k)$, we can put this asymptotic expansion in the convenient form

$$f(t) = \frac{t^{v-1} e^{p_1 t}}{v!} \sum_{k=0}^n [v(v-1)(v-2)...(v-k)] a_k t^{-k} + O(e^{p_1 t} t^{v-n-2}). \quad (5.95)$$

The transform pairs (5.89), (5.95) are in agreement with those given by Smith [5.13, § 10.5] who derived them in a somewhat different manner. The pairs work for the extreme singularity when it is a branch point or a pole of $\bar{f}(p)$, since (5.89) represents poles when v is an integer. When v is an integer we violate our condition $v < k+1$ under which (5.94) vanishes. Note, however, the expression just above (5.95) involves $1/\Gamma(z)$, $z=v-k$, which is analytic everywhere in the complex z-plane through the analytic continuation provided by Hankel's contour integral for this function. Hence v can take on integer values. Note, in this case, (5.89) degenerates to $\bar{f}(p) = a_0(p-p_1)^{-v}$, i.e., $k=0$ only. Correspondingly then, (5.90) becomes $f(t) = t^{v-1} \exp(p_1 t)/(v-1)!$, which one can obtain from simple residue theory or the Laplace transform tables. For an extreme singularity that is an essential singularity, one would find the corresponding Laurent's series for $\bar{f}(p)$ and use residue theory to evaluate the asymptotic expansion for $f(t)$.

The solution of the static problem, if it exists, corresponds in our dynamic problems to a limit as $t\to\infty$. From (5.95) we see such a solution would stem from $p_1=0$, $v=1$, which reduces (5.95) to

$$f(t)=a_0, \qquad (5.96a)$$

and therefore (5.89) to

$$\bar{f}(p)=a_0/p. \qquad (5.96b)$$

Equations (5.96) point out, therefore, if the static solution exists, the transform must have a simple pole at the origin, and this must be the extreme singularity. When such a solution exists, then the approximation of $\bar{f}(p)$ for $p\to 0$ will result in (5.96b). We shall apply this technique to find the static solution from the dynamic solution for the cylindrical cavity source problem later in this chapter. We shall also apply it in later chapters.

5.10.3. *Asymptotic expansion of Fourier integrals*

Let us now consider the asymptotic expansion for $x\to\infty$ of the Fourier integral

$$F(x)=\int_a^b f(t)e^{ixt}\,dt. \qquad (5.97)$$

We assume that $f(t)$ is absolutely integrable in (a, b), but it can have branch points in the closed interval (a, b). It follows from the Riemann–Lebesgue lemma that

$$F(x)=\int_a^b f(t)e^{ixt}dt\to 0, \qquad \text{as } x\to\infty, \qquad (5.98)$$

as pointed out in § 5.6, *ii*. For a function $f(t)$ of bounded variation

$$F(x)=\int_a^b f(t)e^{ixt}dt=O(x^{-1}), \qquad \text{as } x\to\infty. \qquad (5.99)$$

The proof may be found in Jeffreys and Jeffreys [5.11, § 14.03].

To obtain more detailed information, the interval (a, b) is divided into sub-intervals so that the branch points, if any, only occur at the end points of each interval. In each of these intervals the asymptotic expansion of

(5.97) is obtained by repeated integrations by parts. The following are useful theorems:

i) If $f(t)$ is n times continuously differentiable for $a \leq t \leq b$, then

$$F(x) = \int_a^b f(t) e^{ixt} dt$$

$$= \sum_{k=1}^n \left(\frac{i}{x}\right)^k \left[e^{iax} f^{(k-1)}(a) - e^{ibx} f^{(k-1)}(b)\right] + o(x^{-n}) \qquad \text{as } x \to \infty .$$

(5.100)

ii) If $f(t)$ has singularities at both end points, and if

$$f(t) = (t-a)^{\lambda-1}(b-t)^{\mu-1} \eta(t), \qquad 0 < \lambda, \mu \leq 1 ,$$

where $\eta(t)$ is n times continuously differentiable for $a \leq t \leq b$, then

$$F(x) = \int_a^b (t-a)^{\lambda-1}(b-t)^{\mu-1} \eta(t) e^{ixt} dt$$

$$= \sum_{k=1}^n [A_k(x) + B_k(x)] + O(x^{-n}) \qquad \text{as } x \to \infty , \qquad (5.101)$$

where

$$A_k(x) = \frac{\Gamma(k+\lambda-1)}{\Gamma(k)} e^{i\pi(k+\lambda-1)/2} x^{-k-\lambda+1} e^{iax} \frac{d^{k-1}}{da^{k-1}} [(b-a)^{\mu-1} \eta(a)] ,$$

$$B_k(x) = \frac{\Gamma(k+\mu-1)}{\Gamma(k)} e^{i\pi(k-\mu-1)/2} x^{-k-\mu+1} e^{ibx} \frac{d^{k-1}}{db^{k-1}} [(b-a)^{\lambda-1} \eta(b)] ,$$

and where $O(x^{-n})$ in (5.101) may be replaced by $o(x^{-n})$ when $\lambda = \mu = 1$. Proof of these theorems may be found in [5.12].

5.10.4. Laplace's method

We now consider integrals having the form (5.81), i.e.,

$$I(x) = \int_a^b f(t) e^{xg(t)} dt . \qquad (5.102)$$

Conditions (1) in § 5.10.1 are assumed to hold, namely that x is a large positive parameter, $g(t)$ is a real function of real t, and $f(t)$ may be real or complex. It is assumed further that $f(t)$ is continuous and $g(t)$ is twice continuously differentiable in the interval (a, b). Now suppose that $g(t)$

has a finite number of maxima in (a, b), with $g(\tau)$ being the largest of these maxima. It follows that the function $\exp[g(t)-g(\tau)]=1$ at $t=\tau$, and that it is less than one elsewhere in (a, b). Hence, for large x, $\exp\{x[g(t)-g(\tau)]\}$ is significant only in a small neighborhood of $t=\tau$, being negligible elsewhere. The behavior of these exponential functions is depicted in fig. 5.3. It follows

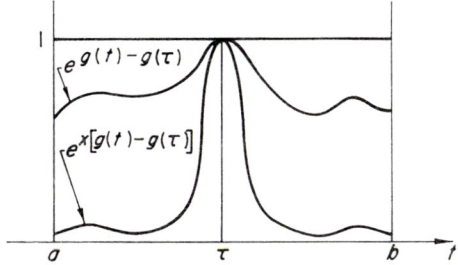

Fig. 5.3. Behavior of exponential functions in Laplace's method for $x \gg 1$.

the major contribution to $I(x)$ for large x comes from the immediate vicinity of $t=\tau$. Since when $g(t)$ has more than one maximum in (a, b) we can always subdivide this interval, we lose no generality in assuming $g(t)$ has only one and this is at $t=\tau$, $a<\tau<b$. Further the case of $\tau=a$ or $\tau=b$ is just a special case in this assumption.

Laplace introduced a new variable, say η defined by

$$\eta = \pm[g(\tau)-g(t)]^{\frac{1}{2}},$$

into (5.102), where the upper sign is for $t>\tau$, and lower for $t<\tau$, with the result that

$$I(x) = e^{xg(\tau)} \int_\alpha^\beta e^{-x\eta^2} f(t) \frac{dt}{d\eta} d\eta, \qquad (5.103)$$

where $\alpha = -[g(\tau)-g(a)]^{\frac{1}{2}} < 0$ and $\beta = [g(\tau)-g(b)]^{\frac{1}{2}} > 0$. It is not difficult to obtain the first term of the asymptotic expansion of the integral in (5.103). With x large, the exponential term there, near $\eta=0$ ($t\to\tau$), dominates the integration so much, that the error made by approximating $f(t)dt/d\eta$ by its value as $\eta\to 0$ is negligible. Therefore the following approximations can be made:

$$f(t) = f(\tau) + f'(\tau)(t-\tau) + \tfrac{1}{2}f''(t-\tau)^2 + O[(t-\tau)^3],$$
$$\eta^2 = g(\tau)-g(t) \simeq -g''(\tau)(t-\tau)^2/2, \qquad (5.104)$$
$$\eta \simeq [-g''(\tau)/2]^{\frac{1}{2}}(t-\tau), \qquad dt/d\eta \simeq [-2/g''(\tau)]^{\frac{1}{2}}.$$

Hence, using (5.104) and a simple change of variable in the resultant integral, (5.103) becomes

$$I(x) \sim f(\tau) e^{xg(\tau)} \left[\frac{-2}{xg''(\tau)} \right]^{\frac{1}{2}} \int_{\alpha\sqrt{x}}^{\beta\sqrt{x}} e^{-\xi^2} d\xi. \qquad (5.105a)$$

Since $x \to \infty$ here, the upper and lower limits $\to \infty$ and $-\infty$, respectively. It follows that

$$I(x) \sim f(\tau) e^{xg(\tau)} \left[\frac{2\pi}{x|g''(\tau)|} \right]^{\frac{1}{2}} \qquad \text{as } x \to \infty. \qquad (5.105b)$$

When $\tau = a$ or b, $I(x)$ is one half the value given in (5.105b).

It may happen that one or more of the leading terms in the Taylor's series for $f(t)$ in (5.104) vanish. In this case one replaces the $f(\tau)$ we used in the above derivation by the first nonvanishing term in the Taylor's series. For example, if $f(\tau) = f'(\tau) = 0$ and $f''(\tau) \neq 0$, we find by a derivation similar to that above, the approximation

$$I(x) \sim \frac{\sqrt{\pi}}{4} f''(\tau) e^{xg(\tau)} \left[\frac{2}{x|g''(\tau)|} \right]^{\frac{3}{2}} \qquad \text{as } x \to \infty. \qquad (5.106)$$

As Wu [5.9, § 2.9] points out one can generalize the procedure leading to (5.105b) to write the expansion corresponding to a general maximum of $g(t)$, which is defined by $g'(\tau) = g''(\tau) = \ldots = g^{(2n-1)}(\tau) = 0$ and $g^{(2n)}(\tau) < 0$, $n = 1, 2, \ldots$, where τ is referred to as the critical point of order $2n-1$. Here we introduce the new variable

$$\eta^{2n} = g(\tau) - g(t) \simeq \frac{1}{(2n)!} g^{(2n)}(\tau)(t-\tau)^{2n}$$

from which it can be shown that

$$I(x) \sim \frac{1}{n} \Gamma\left(\frac{1}{2n}\right) f(\tau) e^{xg(\tau)} \left[\frac{(2n)!}{x|g^{(2n)}(\tau)|} \right]^{\frac{1}{2n}}. \qquad (5.107)$$

Letting $n = 1$ here gives the special case (5.105b).

5.10.5. *Method of steepest descents*

Consider again integrals of the form (5.81), i.e.,

$$I(x) = \int_\alpha^\beta f(t) e^{xg(t)} dt, \qquad (5.108)$$

where conditions (2) in § 5.10.1 are assumed to hold, namely that x is a complex parameter with large modulus, f and g are analytic functions in some domain of the complex variable t-plane, and the path of integration is one that lies in this domain. An asymptotic expansion of (5.108) can be obtained by the well-known *method of steepest descents*, which was originated by Riemann and developed further by Debye. This method has been used extensively in wave propagation problems.

The method depends on some well-known geometrical features of the surfaces representing the real and imaginary parts of an analytic function of a complex variable. Letting

$$t = \xi + i\eta, \qquad xg(t) = u(\xi, \eta) + iv(\xi, \eta),$$

then it may be seen that $|\exp[xg(t)]| = \exp[u(\xi, \eta)]$ will dominate the integrand in (5.108) as $|x| \to \infty$. A plot of this surface over its $\xi\eta$-base plane is called a *relief*. In analytic function theory u, v are known as conjugate harmonic functions. These functions satisfy the Cauchy–Riemann equations

$$\frac{\partial u}{\partial \xi} = \frac{\partial v}{\partial \eta}, \qquad \frac{\partial u}{\partial \eta} = -\frac{\partial v}{\partial \xi}, \qquad (5.109)$$

and Laplace's equation in two dimensions, i.e.,

$$\frac{\partial^2 u}{\partial \xi^2} + \frac{\partial^2 u}{\partial \eta^2} = 0. \qquad (5.110)$$

Using (5.109) it is not difficult to prove that the curves $u = $ const. and $v = $ const. are everywhere orthogonal except where $\partial u/\partial \xi$ and $\partial u/\partial \eta$ both vanish, i.e., where $g'(t) = 0$. The curves $u = $ const. are called *level curves*. Along such curves v, the phase of $\exp(xg)$, changes as rapidly as possible. The curves $v = $ const. are known as *steepest paths*. Along these curves $\exp(xg)$ has a constant phase, but u, and therefore $\exp u$, the modulus of $\exp(xg)$, changes as rapidly as possible. This follows from the orthogonality of the curves of $u = $ const. and $v = $ const., which means maximum $|du/ds|$ occurs when direction s is along a curve of $v = $ const. Since u satisfies Laplace's equation (5.110) it follows that the relief $\exp u$ may have infinities, or *hills* of infinite height, and zeros, or *valley* bottoms. A maximum or minimum point in the relief is not possible. This is easily seen by noting that at such a point, say at $t = \tau$, u would have to satisfy

$$\frac{\partial u}{\partial \xi} = \frac{\partial u}{\partial \eta} = 0, \qquad (5.111)$$

or $g'(\tau) = 0$. The second derivatives corresponding to (5.111) are

$$\frac{\partial^2 u}{\partial \xi^2} < 0, \qquad \frac{\partial^2 u}{\partial \eta^2} < 0,$$

for a maximum, and

$$\frac{\partial^2 u}{\partial \xi^2} > 0, \qquad \frac{\partial^2 u}{\partial \eta^2} > 0,$$

for a minimum. Clearly for either of these cases we cannot satisfy Laplace's equation (5.110). The one other possibility corresponding to (5.111) is the stationary point known as a *saddle point or col*. It has derivatives $\partial^2 u/\partial \xi^2$ and $\partial^2 u/\partial \eta^2$ at the point that are of equal magnitude but of opposite sign. Hence they satisfy (5.110). The saddle points are the zeros of $g'(t)$. If $g'(\tau) = g''(\tau) = \ldots g^{(n)}(\tau) = 0$, and $g^{(n+1)}(\tau) \neq 0$, τ is called a *saddle point of order n*. At a saddle point of order n, $n+1$ level curves intersect at equal angles, each of which is bisected by one steepest path. These $n+1$ intersecting level curves divide the $\xi\eta$-plane into $2(n+1)$ sectors. Half of these sectors are *hills*, where $u > u(\tau)$, and half are *valleys*, where $u < u(\tau)$. The hills and valleys occur alternately about the saddle point τ, being linked by the saddle point. Through every ordinary point t, these possess only one steepest path. In each valley all the steepest paths meet at the bottom, which is a singularity of $xg(t)$ and a zero of exp u. These geometric features are illustrated in fig. 5.4 through the mapping of a simple relief on the $\xi\eta$-plane, namely that for $g(t) = t^2 = \xi^2 - \eta^2 + i2\xi\eta$. The level curves here are $u = x(\xi^2 - \eta^2) = c_1$ and steepest paths $v = 2x\xi\eta = c_2$. Since $g'(t) = 2t = 2(\xi + i\eta)$, we have a saddle point at $t = \tau = 0$, which is at the origin, marked s in fig. 5.4. In the figure the level curves are indicated by dashed lines, and the steepest paths by solid lines with an arrow indicating the direction of steepest descent. It may be noted that these paths intersect at the valley bottoms which are at $\eta \to \pm \infty$. There $u \to -\infty$, hence $xg(t)$ is singular, but exp u vanishes. Since $g''(t) = 2$, the one and only saddle point here is of first order. Note also other features at the saddle point in the figure which agree with our earlier comments, namely (1) there are two level curves ($u = 0$) that intersect at equal angles $\pi/2$, (2) each angle is bisected by one steepest path ($v = 0$), and (3) the two intersecting level curves ($u = 0$) divide the $\xi\eta$-plane into 4 sectors, two hills and two valleys (occurring alternately) that are linked by the saddle point.

The method of steepest descents consists in deforming the path of integration of the integral in (5.108) to make it coincide as far as possible with certain steepest paths. The possibilities are discussed by Erdelyi [5.12, § 2.5]

Ch. 5, § 5.10] ASYMPTOTIC EXPANSIONS OF INTEGRALS 267

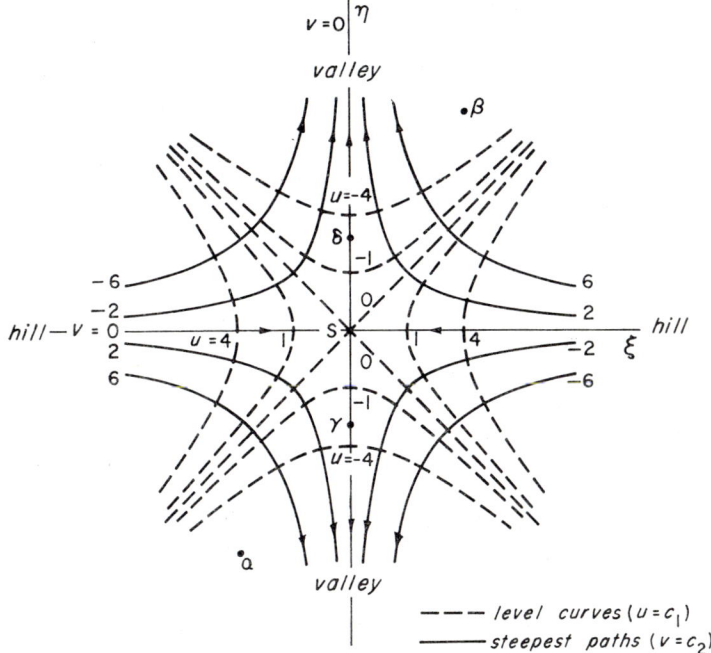

Fig. 5.4. Geometric features of a saddle surface.

and Wu [5.9, § 2.10]. If the limits of integration α and β lie on the same steepest path, then we deform the path of integration to coincide with this steepest path whether it passes through saddle points or not. More often than not α and β will not lie on the same steepest path, then the deformed path may be made up entirely of segments from different steepest paths. Two cases arise, α and β in the same valley, or in different valleys. In the first we can descend from α along a steepest path to a singularity (valley bottom), and then ascend along another steepest path to β. In the second case (see fig. 5.4) as in the first, we descend from α to the singularity, but now climb from this singularity across the saddle point to the valley containing β. Point β is reached through a steepest path to the singularity of that valley from which one can ascend to β along another steepest path. Alternatively parts of the deformed path can be made up of segments that traverse paths of steepest descent, e.g. the path $\alpha\gamma s\delta\beta$, since exp u over $\alpha\gamma$ and $\delta\beta$ will be negligible compared to its values in the neighborhood of s. In a more complicated relief it may be necessary to pass through several saddle points before β can be reached.

In any case since $u = \text{Re}[xg(t)]$ is monotonic along a steepest path, except at saddle points, Laplace's method (§ 5.10.4) can be used to evaluate the integral (5.108) asymptotically. The main contribution to $I(x)$ comes from a short interval (or intervals) of the deformed path, which often is part of the steepest path through the saddle point where u has its largest values. It follows in applying Laplace's method we can expand f and g into Taylor's series for this short interval. We see our evaluation of $I(x)$ is simplified therefore, by keeping large values of u in as short an interval of the deformed path as possible.

We develop the asymptotic expansion of the integral $I(x)$ in (5.108) for the case where x is real, positive and large, and α and β lie in different valleys across a first order saddle point τ. Near $t=\tau$, we have

$$g(t) = g(\tau) + \tfrac{1}{2} g''(\tau)(t-\tau)^2 + \ldots \tag{5.112}$$

We take the direction of the deformed path such that along the path near τ $g''(\tau)(t-\tau)^2$ is real and negative, hence from (5.112)

$$g(t) - g(\tau) = \tfrac{1}{2} g''(\tau)(t-\tau)^2 = -\varrho^2/2, \tag{5.113}$$

where ϱ is real, and our chosen short path is one of steepest descent [since $v(t) = v(\tau) = 0$ in (5.113)]. For t on the path, we write

$$t - \tau = r \exp(i\sigma), \tag{5.114}$$

with r real and small, and σ the angle this vector makes with the real t-axis. Substituting (5.114) into (5.113) gives

$$\varrho^2 = -g''(\tau) r^2 \exp(2i\sigma), \tag{5.115}$$

and it follows from this equation that $g''(\tau) \exp(2i\sigma)$ must be real and negative, i.e.,

$$g''(\tau) = |g''(\tau)| \exp[i(-2\sigma \pm \pi)] \tag{5.116}$$

Substitution of (5.116) into (5.115) gives

$$\varrho = \pm r |g''(\tau)|^{\frac{1}{2}}. \tag{5.117a}$$

Now from (5.113), using (5.114), (5.116) and (5.117a) we also have

$$\frac{dt}{d\varrho} = \pm e^{i\sigma}|g''(\tau)|^{-\frac{1}{2}}. \tag{5.117b}$$

Since $g''(\tau)$ will be known, (5.116) determines the two possible values of σ. They differ by π. As pointed out by Jeffreys and Jeffreys [5.11, § 17.04] in their treatment of the method of steepest descents, to select the correct value of σ one must inspect the behavior of u, v in order to determine the sense in which the path goes through the saddle point. If we select the value of σ that makes r positive at points on the path after passing through the saddle point τ, we have to take the positive sign in (5.117a), and hence (5.117b), as ϱ goes from $-\infty$ to $+\infty$ on the path. On this basis then, the deformed path integral for $I(x)$ can be approximated by

$$I(x) \sim e^{xg(\tau)} \int_{-\varepsilon}^{\varepsilon} e^{-x\varrho^2/2} f(t) \frac{dt}{d\varrho} d\varrho = f(\tau)e^{xg(\tau)+i\sigma} \left(\frac{2}{x|g''(\tau)|}\right)^{\frac{1}{2}} \int_{-\varepsilon\sqrt{x/2}}^{\varepsilon\sqrt{x/2}} e^{-w^2} dw,$$

which using (5.105) gives

$$I(x) \sim f(\tau)e^{xg(\tau)+i\sigma} \left[\frac{2\pi}{x|g''(\tau)|}\right]^{\frac{1}{2}}, \qquad \text{as } x \to \infty. \tag{5.118}$$

It is instructive to return to our example $g(t) = t^2$ again. Here $g'' = 2$, and therefore from (5.116) we find $\sigma = \pm \pi/2$. Now here we have from (5.114) $t = r \exp i\sigma$. We recall that the upper half of the imaginary axis in fig. 5.4 is a path of steepest descent through the saddle point s at $t = \tau = 0$. On this path $t = i\eta$, hence we get agreement between these two expressions for t by setting $\sigma = \pi/2$ and $r = \eta > 0$. It follows this choice of σ is consistent with the positive sign in (5.117a), (5.117b) which led to (5.118), making ϱ an increasing positive number as we proceed along the steepest path toward larger positive $r = \eta$. This example then is consistent with our earlier discussion of how σ is chosen. Our three references here [5.9, 5.11, 5.12] treat examples on the application of the method of steepest descents. We shall use it in a problem involving elastic wave propagation in Chapter 6.

In the treatment above on the method of steepest descents we have assumed that f and g in the integral (5.108) were analytic in some domain of the complex t-plane, the path of integration being in this domain. If, however, $f(t)$ has an isolated singularity in the domain, such that it lies within the closed path of integration consisting of the original path in

(5.108), and the deformed path, then account must be taken of the contribution of the singularity. Assuming the singularity is at $t=t_1$ and sufficiently removed from the saddle point $t=\tau$, the integration about it will contain at most a factor $\exp[u(t_1)]$, which will make it negligible with respect to that from the saddle point $\exp[u(\tau)]$. Hence our treatment leading to (5.118) remains valid, with one or more singularities like this.

When t_1 is close to τ assessment of the contribution is much harder. Van der Waerden [5.14] gives a valuable technique for treating this case. In fact his technique is an *alternate* one for evaluating (5.108) that *does not exploit paths of steepest descent* with the obvious advantage of not having to find them. Van der Waerden applys his technique to (5.108) in the form

$$I(x)=\int_C f(t)e^{xg(t)}dt,$$

where x is real, negative and large, and where at both ends of the contour C, Re $g(t)\to\infty$ (neither is a necessary condition). The new variable of integration g is introducted giving

$$I(x)=\int_C f\frac{dt}{dg}e^{xg}dg=\int_C P(g)e^{xg}dg.$$

Now (1) the contour C in the g-plane must be drawn, which is easy since g is known on C, (2) the branch points and poles of $P(g)$ must be determined, and (3) the integrand is expanded in a power series in the neighborhood of every branch point and pole which is followed by termwise integration. Two kinds of branch points may occur. The first are at points where $dg/dt=0$. These correspond to the saddle points in the classical method of steepest descents which we have just discussed. The others stem from the function $f(g)$, which may also have poles. Only the branch points and poles lying on or to the right of the contour C have to be considered, "to the right of" meaning where Re g has larger values than its values on C. The branch points or poles which lie more to the right than others may be neglected giving the technique further simplicity. Van der Waerden treats the case of a pole near a saddle point. For further detail on his method, and this case, the reader is referred to the quoted paper [5.14]. In the problem in chapter 6 mentioned earlier we shall also show some techniques applicable to cases in which a saddle point lies on a branch cut.

Returning to the classical method of steepest descents we first remark

Ch. 5, § 5.10]   ASYMPTOTIC EXPANSIONS OF INTEGRALS   271

that further reading on the case where a pole is near a saddle point may be found in the report by Cerrillo [5.15]. It is of further interest here to point out the different features of the method of steepest descents when it is applied to waveguide problems. In such problems one finds that $xg(t)$ in (5.108) has a phase form like $i[\varkappa_j(\omega)x - \omega t]$, where ω is the variable of integration, x and t are the space and time variables, and $\varkappa_j(\omega)$ is a particular branch of the frequency equation. It follows that the saddle points are defined by $d\varkappa_j(\omega)/d\omega = t/x$. Hence, for fixed x, they are a function of time t, i.e. their position, say ω_{js}, changes with time. This mobility leads to an approximation of the type (5.118) for the first order saddle point, when two saddle points are sufficiently separated, and to one representing a second order saddle point when two saddle points are close together or collide. Cerrillo [5.15] discusses the latter case which is sometimes called the *extended saddle point method*. Applications to elastic waveguide problems may be found in the works of Folk, et al. [5.16a, b] and Jones and Ellis [5.17a, b]. Brillouin's treatment of a problem on the propagation of light in dispersive media [4.2, ch. III] is also instructive. We shall see in the next section, that the method of stationary phase is a simpler way to deal with waveguide problems.

5.10.6. Method of stationary phase

The *method of stationary phase* is a well-known means for deriving *asymptotic expansions of integrals of the type* (*5.82*). Such integrals commonly arise in solutions of *waveguide problems*, and as we shall see in Chapter 7, the method of stationary phase plays an important role in solving such problems. We have already introduced the underlying concepts of the method in ch. 4, § 4.1.1.2, namely that predominant disturbances at a station x at time t, in a *dispersive medium*, can be interpreted as being due to *wave groups*. The latter are composed of harmonic waves that are in phase at station x at time t, all other harmonic waves cancelling. As we pointed out in ch. 4, § 4.1.1.2 Kelvin's stationary phase method, which we are about to discuss, is based on these concepts. Consider the integral

$$I(x) = \int_a^b f(t) e^{ixh(t)} dt, \tag{5.119}$$

where x is a large, positive variable, $h(t)$ is a real function of the real variable t, and $f(t)$ can be real or complex. If $h(t)$ is continuously differentiable with nonvanishing $h'(t)$ in a $\leq t \leq b$, then (5.119) can be reduced to a Fourier integral by the transformation $h(t) = \eta$. This is easily seen since then (5.119) becomes

$$I(x) = \int_{h(a)}^{h(b)} \frac{f[k(\eta)]}{h'[k(\eta)]} e^{ix\eta} d\eta,$$

where $k(\eta) = t$. Now since f/h' is of bounded variation in a $\leq t \leq b$, according to the Riemann–Lebesgue lemma, this integral is of $O(x^{-1})$ for x large (cf. § 5.10.3). However, when $h'(t)$ vanishes at some point $t = \tau$ in (a, b), then as we shall show, the contribution to $I(x)$ from the neighborhood of the critical point τ is more important than the contributions of the end points. *The point τ where $h'(\tau) = 0$ is called a point of stationary phase.* Now assume τ is the only stationary point in (a, b), and assume $h''(\tau) > 0$. Then the function $h(t) - h(\tau)$, shown in fig. 5.5, vanishes at $t = \tau$ and is positive

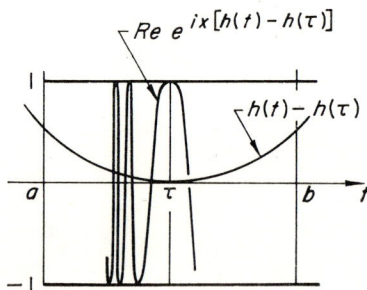

Fig. 5.5. Variation of functions in stationary phase method.

everywhere else. For x large the function $\exp\{ix[h(t) - h(\tau)]\}$, also shown in fig. 5.5, oscillates very rapidly near the critical point τ. It follows the main contribution to $I(x)$ comes from the neighborhood of $t = \tau$.

To derive an asymptotic expansion for $I(x)$ we further assume that $f(t)$ is continuous and $h(t)$ is twice continuously differentiable. Then introducing into (5.119) the new variable of integration w, defined by

$$w^2 = h(t) - h(\tau), \qquad (5.120)$$

we have

$$I(x) = e^{ixh(\tau)} \int_{-\alpha}^{\beta} e^{ixw^2} f(t) \frac{dt}{dw} dw, \qquad (5.121)$$

where $\alpha=[h(a)-h(\tau)]^{\frac{1}{2}}$ and $\beta=[h(b)-h(\tau)]^{\frac{1}{2}}$. Now in the neighborhood of τ, we see from (5.120), with the aid of a Taylor's series for $h(t)$, that

$$w^2 \simeq \tfrac{1}{2}h''(\tau)(t-\tau)^2, \tag{5.122}$$

where we have used $h'(\tau)=0$. It follows that

$$w \simeq [h(t)-h(\tau)]^{\frac{1}{2}} = [h''(\tau)/2]^{\frac{1}{2}}(t-\tau),$$

and hence $dt/dw \simeq [2/h''(\tau)]^{\frac{1}{2}}$, and we can write $f(t) \simeq f(\tau)$. On this basis (5.121) can be written as

$$I(x) = \left[\frac{2}{h''(\tau)}\right]^{\frac{1}{2}} f(\tau) e^{ixh(\tau)} \int_{-\alpha}^{\beta} e^{ixw^2} dw$$

$$-e^{ixh(\tau)} \int_{-\alpha}^{\beta} e^{ixw^2} \left\{ \left[\frac{2}{h''(\tau)}\right]^{\frac{1}{2}} f(\tau) - f(t) \frac{dt}{dw} \right\} dw. \tag{5.123}$$

The first integral in (5.123) can be written as

$$\int_{-\alpha}^{\beta} e^{ixw^2} dw = \int_{-\infty}^{\infty} - \int_{-\infty}^{-\alpha} - \int_{\beta}^{\infty} e^{ixw^2} dw. \tag{5.124}$$

The first integral here gives the well-known result $[\pi/x]^{\frac{1}{2}} \exp(i\pi/4)$, and since $w=0$ (the only critical point here) is outside the range of integration in the second two integrals, they are Fourier integrals and are of $O(x^{-1})$ as $x \to \infty$. Further, the quantity in $\{\ \}$ in the second integral in (5.123) is of bounded variation and hence this Fourier integral is also of $O(x^{-1})$ for $x \to \infty$. Hence (5.123) becomes

$$I(x) = \left[\frac{2\pi}{xh''(\tau)}\right]^{\frac{1}{2}} f(\tau) e^{i[xh(\tau)+(\pi/4)]} + (Ox^{-1}) \qquad as\ x \to \infty. \tag{5.125}$$

Clearly we might have a maximum in w^2 at the point of stationary phase. We would treat this case in an analogous manner noting now that since $h''(\tau)<0$, w^2 must be replaced by $-w^2$, or now

$$w^2 = h(\tau)-h(t) \simeq \tfrac{1}{2}|h''(\tau)|(t-\tau)^2.$$

It is left to the reader to derive the result for this case analogous to (5.125). We have for either case, the asymptotic expansion for x large

$$I(x) = \left[\frac{2\pi}{x|h''(\tau)|}\right]^{\frac{1}{2}} f(\tau) e^{i[xh(\tau)+(\pi/4)\mathrm{sgn}h''(\tau)]} + O(x^{-1}), \tag{5.126}$$

where *sgn* stands for the sign of the quantity following it. If the stationary point occurs at an end point then we see from (5.124) we would have half the value on the right of (5.126). It should also be pointed out that one must obtain a relation like (5.126) for each and every point of stationary phase that occurs. Indeed in our waveguide problems we will encounter integrals where a *continuous distribution of points of stationary phase* occur over the entire range of integration, these stemming from a real segment of a branch of a frequency equation and the continuous array of wave groups they represent. We shall treat an example of this later on in this section.

Here, as in the case of saddle points, we can have stationary phase points of higher order. A points τ at which $h'(\tau) = h''(\tau) = \ldots = h^{(n)}(\tau) = 0$ with $h^{(n+1)}(\tau) \neq 0$ is called a stationary point of order n, $n = 1, 2, 3 \ldots$ As Wu [5.9, § 2.11] points out, for a point of second order ($n=2$) we let

$$w^3 = h(t) - h(\tau) \simeq \tfrac{1}{6} h'''(\tau)(t-\tau)^3 .$$

Then (5.119) becomes

$$I(x) = e^{ixh(\tau)} \int_{-\alpha}^{\beta} e^{ixw^3} f(t) \frac{dt}{dw} \, dw , \qquad (5.127)$$

where $\alpha = [h(a) - h(\tau)]^{\frac{1}{3}}$ and $\beta = [h(b) - h(\tau)]^{\frac{1}{3}}$. Following similar argumens, to those used in deriving (5.126), noting here we have $dt/dw \simeq [6/h'''(\tau)]^{\frac{1}{3}}$, and that we must retain two terms of the Taylor's series for $f(t)$, i.e., $f(t) \simeq f(\tau) + f'(\tau)(t-\tau)$, we find

$$I(x) = \left[\frac{6}{xh'''(\tau)} \right]^{\frac{1}{3}} f(\tau) e^{ixh(\tau)} \int_{-\infty}^{\infty} e^{ixw^3} dw + O(x^{-\frac{2}{3}}) ,$$

which reduces to

$$I(x) = \frac{\Gamma(1/3)}{\sqrt{3}} \left[\frac{6}{xh'''(\tau)} \right]^{\frac{1}{3}} f(\tau) e^{ixh(\tau)} + O(x^{-\frac{2}{3}}) . \qquad (5.128)$$

It is clear that the integral (5.119) is a special case of that in (5.108), where in the former the path of integration is a portion of the real axis of t, and this path is also a level line, i.e., $u = \text{const.} = 0$. It would follow then that the method of steepest descents can be applied to the integral in (5.119). To show this we first note that a saddle point has to satisfy (5.111). Along the real axis since $u = 0$ so is $\partial u/\partial \xi = 0$, and the first of (5.111) is satisfied. Now

Ch. 5, § 5.10]   ASYMPTOTIC EXPANSIONS OF INTEGRALS   275

from the second of the Cauchy–Riemann conditions (5.109), the second of (5.111) is satisfied provided $\partial v/\partial \xi = 0$. Thus in (5.119), where $v(t) = h(t)$, we have saddle points where $v'(\tau) = h'(\tau) = 0$. To apply the method of steepest descents to (5.119) we take t complex and require that

$$ih''(\tau)(t-\tau)^2 = h''(\tau)r^2 \exp\left[2i\left(\sigma + \frac{\pi}{4}\right)\right] = -\varrho^2 \tag{5.129}$$

according to (5.113)–(5.115), i.e., that $ih''(\tau)(t-\tau)^2$ is negative real. Therefore, for $h''(\tau) > 0$, σ must be $\pi/4$, and when $h''(\tau) < 0$, σ must be $-\pi/4$. Substituting these values and $g(\tau) = ih(\tau)$ into (5.118) gives

$$I(x) \sim \left[\frac{2\pi}{x|h''(\tau)|}\right]^{\frac{1}{2}} f(\tau) e^{i[xh(\tau) + (\pi/4)\mathrm{sgn}h''(\tau)]},$$

which agrees with the stationary phase result (5.126).

Kelvin applied his stationary phase method to a one dimensional problem in water waves involving dispersion. His application is instructive here since in particular it brings out the nature of the approximation when it is based on an infinite spectrum of wave groups. We treat an analogous problem to that of Kelvin, an infinite elastic string on an elastic foundation along the x-axis, subjected initially to a spatially limited initial disturbance in the transverse displacement y. The initial conditions for the problem are

$$y(x, 0) = y_0(x), \qquad \dot{y}(x, 0) = 0.$$

Using (5.3) we can express $y_0(x)$ as the Fourier integral

$$y_0(x) = \frac{1}{\pi} \int_0^\infty d\varkappa \int_{-\infty}^\infty y_0(\xi) \cos \varkappa(x-\xi) d\xi.$$

Assuming $y_0(x)$ is an even function, this integral becomes

$$y_0(x) = \frac{2}{\pi} \int_0^\infty U(\varkappa) \cos \varkappa x \, d\varkappa,$$

where $U(\varkappa)$ is the Fourier cosine transform of $y_0(x)$ (we will not exploit this property here, however). After a time $t > 0$ wave trains move out from the initial disturbance a distance ct, hence for any pair (x, t) $y(x, t)$ is given by

$$y(x, t) = \frac{2}{\pi} \int_0^\infty U(\varkappa) \cos \varkappa x \cos \varkappa ct \, d\varkappa$$

$$= \frac{1}{\pi} \int_0^\infty U(\varkappa) \{\cos [\varkappa(x-ct)] + \cos [\varkappa(x+ct)]\} d\varkappa . \tag{5.130}$$

We may observe that our solution (5.130) is composed of two integrals the first involving wave trains that propagate along x positive, and the other equal trains that propagate along x negative. Assuming that $y_0(x)$ is the symmetric delta function, (5.130) reduces to

$$y(x, t) = \frac{1}{2\pi} \int_0^\infty \{\cos [\varkappa(x-ct)] + \cos [\varkappa(x+ct)]\} d\varkappa , \tag{5.131}$$

valid for $x > 0$. This problem may be interpreted as having an infinite number of harmonic wave trains of equal amplitude, in phase at the origin $x = 0$, at time $t = 0$, and cancelling elsewhere. We want to know $y(x, t)$ for any pair (x, t), $x > 0$, $t > 0$. Consider again the more general solution (5.130). The phase velocity $c(\varkappa)$ [or frequency $\omega(\varkappa)$] in our problem is a continuous function over the infinite domain of wave number $\varkappa$, i.e., $0 \leq \varkappa < \infty$, c and ω being real and positive. We can evaluate (5.130) for large time by the method of stationary phase. Consider the integral in (5.130) involving $\cos[\varkappa(x-ct)]$. It is a generalization of the series (4.20), hence the condition of stationary phase (4.23) would be expected to hold. We can identify this integral with (5.119), writing the latter as

$$y(x, t) = \frac{1}{\pi} \operatorname{Re} \int_0^\infty U(\varkappa) e^{ith(\varkappa)} d\varkappa , \tag{5.132}$$

where we have taken $a = 0$, $b = \infty$, $x = t$, $t = \varkappa$, $f(t) = U(\varkappa)$, and $h(t) = h(\varkappa)$, where

$$h(\varkappa) = \varkappa[(x/t) - c(\varkappa)], \qquad x/t \text{ fixed} . \tag{5.133}$$

A stationary point $\varkappa = s$ is given by $h'(s) = 0$, which from (5.133) yields the stationary phase condition

$$h'(s) = (x/t) - c_g(s) = 0 , \tag{5.134}$$

in agreement with (4.23). Equation (5.134) discloses a constraint on the fixity of x/t, namely that it must always be equal to a group velocity c_g at some s value, but it is clear since $c_g(\varkappa)$ is a continuous function [which follows from sufficiently smooth $c(\varkappa)$ or $\omega(\varkappa)$] we will have a continuous array of x/t values, hence a continuous array of wave groups making up our disturbance. It follows our *long time disturbance*, corresponding to the $\cos[\varkappa(x-ct)]$ integral in (5.130), is, from (5.126)

$$y(x, t; s) = \sqrt{2}[\pi t |c_g'(s)|]^{-\frac{1}{2}} U(s) \cos\left\{s[x-c(s)t] + \frac{\pi}{4} \mathrm{sgn}[-c_g'(s)]\right\} + O(t^{-1}),$$

(5.135)

where we have used $h''(s) = -c_g'(s)$ obtained from (5.134). Note that we have groups corresponding to s values in the range $0 \leq \varkappa = s < \infty$. Most convenient use of (5.135) can be made by noting that, for a fixed x station, time t can be represented by the group velocity $c_g(s)$, hence the response y at a station can be calculated by letting s go through monotonic ranges of the $c_g(s)$ relation. Reference [7.22] of Chapter 7 contains an example.

The $\cos[\varkappa(x+ct)]$ integral in (5.130), involving negative traveling harmonic waves in the domain $x>0$, does not have stationary points that satisfy (5.134), since c_g cannot be negative. It follows this integral can be reduced to a Fourier integral and hence contributes a term of $O(t^{-1})$ [cf. (5.119) and following discussion], which is like those already represented by the order condition in (5.135). Hence (5.135) is the long time solution of our problem. We should point out that (5.135) is in agreement with Kelvin's result which may be found in Havelock's monograph [4.1, eq. (46), pg. 35]. Havelock treats several other examples in his Chapter IV.

In closing it should be pointed out that when a second order stationary point occurs, since then $h'(s) = h''(s) = 0$ and $h'''(s) \neq 0$, we see from (5.134) that $h''(s) = -c_g'(s) = 0$. This says we have such a point at a maximum or minimum of the $c_g(\varkappa)$ curve, and the corresponding contribution would be given by (5.128). We note this $O(t^{-\frac{1}{3}})$ term would dominate any contributions from first order points, since they are of $O(t^{-\frac{1}{2}})$ for long time.

5.11. Cavity source problems

5.11.1. Spherical cavity subjected to sudden uniform pressure

We treat the problem of the spherical cavity of radius $r=a$ in the infinite elastic solid, the wall of which is subjected to sudden uniform pressure, say σ_0, an inherently negative magnitude constant. The uniformity of the pressure reduces the problem to one of spherical symmetry with $u_r = u$ therefore, being the only nonvanishing displacement. Making use of (2.31), the problem can be stated as follows:

$$\frac{\partial^2 \varphi(r,t)}{\partial r^2} + \frac{2}{r}\frac{\partial \varphi}{\partial r} = \ddot{\varphi}/c_d^2, \qquad r > a, \qquad t > 0, \qquad (5.136\text{a})$$

$$\varphi(r, 0) = \dot{\varphi}(r, 0) = 0, \qquad r \geq a, \qquad (5.136\text{b})$$

$$\lim_{r \to \infty} \varphi(r, t) = 0, \qquad t > 0, \qquad (5.136\text{c})^2$$

$$\sigma_r(a, t) = \sigma_0 H(t). \qquad (5.136\text{d})$$

Equations (5.136) state, respectively, the governing equation, initial conditions, the radiation condition and the boundary condition at $r = a$. Applying the Laplace transform integral (5.6b) to (5.136a), using (5.136b) and (5.23), we find

$$\frac{d^2 \bar{\varphi}(r, p)}{dr^2} + \frac{2}{r}\frac{d\bar{\varphi}}{dr} = k_d^2 \bar{\varphi}, \qquad (5.137)$$

where $k_d = p/c_d$, and where in transforming spatial derivatives of φ we have used the interchangeability differentiation and integration in (5.6b), e.g.,

$$\frac{d\bar{\varphi}(r,t)}{dr} = \int_0^\infty \frac{\partial \bar{\varphi}(r,t)}{\partial r} e^{-pt} dt = \frac{d}{dr} \int_0^\infty \varphi(r,t) e^{-pt} dt = \frac{d\bar{\varphi}(r,p)}{dr}. \qquad (5.138)$$

When (1) the integrands of the two integrals in (5.138) are piecewise continuous with respect to r and t, and (2) the first integral is uniformly convergent with respect to r, (5.138) is valid (a proof is given in Churchill [5.3, § 15]).

We assume a solution of (5.137) of the form

$$\bar{\varphi}(r, p) = A(p) \exp[-n(p)r]/r. \qquad (5.139)$$

---

[2] A *radiation condition* like this is not actually required for a hyperbolic problem, such as the present one, since it is inherent in the solution of the governing differential equation. However, such a statement at the outset of a wave propagation problem is a useful reminder in the development of the solution.

Substitution of (5.139) into the differential equation (5.137), yields

$$n^2 = k_d^2; \qquad n = \pm k_d, \tag{5.140}$$

respectively the *characteristic equation* of (5.137), and its roots. Now transforming (5.136b) we have

$$\overline{\lim_{r \to \infty} \varphi(r, t)} = \int_0^\infty \lim_{r \to \infty} \varphi(r, t) e^{-pt} dt = \int_0^\infty 0 \cdot e^{-pt} dt = 0, \tag{5.141}$$

and therefore from (5.141), under the same requirements as discussed for (5.138),

$$\overline{\lim_{r \to \infty} \varphi(r, t)} = \lim_{r \to \infty} \int_0^\infty \varphi(r, t) e^{-pt} dt = \lim_{r \to \infty} \bar{\varphi}(r, p) = 0. \tag{5.142}$$

For $Re\, n(p) > 0$, $\varphi(r, p)$ in (5.139) will satisfy the transformed radiation condition (5.142). Anticipating the use of the inversion integral (5.6a) to get $\varphi(r, t)$, we see $n = k_d$, for $Re\, p > 0$, is the root of (5.140) we need, since this will satisfy $Re\, n(p) > 0$. Therefore (5.139) becomes

$$\bar{\varphi}(r, p) = A(p) \exp(-k_d r)/r. \tag{5.143}$$

The first of (2.31)

$$u = \frac{\partial \varphi}{\partial r}, \tag{5.144}$$

is the displacement-potential relation in the present problem. Love [1.2, pg. 56] gives the strain-displacement relations

$$\varepsilon_r = \frac{\partial u}{\partial r}, \qquad \varepsilon_\theta = \varepsilon_\eta = \frac{u}{r}, \tag{5.145}$$

where r, θ, and η are the polar coordinates. It follows from Hooke's law, (5.145) and (5.144), that the existing stress-potential relations are

$$\sigma_r = \lambda \nabla^2 \varphi + 2\mu \frac{\partial^2 \varphi}{\partial r^2},$$
$$\sigma_\theta = \sigma_\eta = \lambda \nabla^2 \varphi + \frac{2\mu}{r} \frac{\partial \varphi}{\partial r}, \tag{5.146}$$

where $\nabla^2 \varphi$ is the left hand side of (5.136a). By substituting (5.143) into the transform of the first of (5.146), and then substituting this in the transform of (5.136d), we find $A(p)$. Then (5.143) gives the transformed solution for φ as

$$\bar{\varphi}(r,p) = \frac{ac_s^2\sigma_0 e^{-k_d(r-a)}}{\mu r p[p^2+bp+c]} = \bar{\varphi}^* e^{-k_d(r-a)}, \tag{5.147}$$

where $b=4c_s^2/ac_d$, and $c=4c_s^2/a^2$. The transformed solution $\bar{\varphi}$ in (5.147) has the typical form for a hyperbolic problem, where wave arrival times, related to a separated out shift operator, $exp\left[-k_d(r-a)\right]$ here, are involved. This term is associated, therefore, with e^{pt} in the inversion integral (5.6a), and hence $\bar{\varphi}^*$ must satisfy the conditions (5.19), in order to be an admissible transform for this inversion integral.

Now $\bar{\varphi}^*$ has simple poles at $p=0$, p_1, and p_2, where

$$p_{1,2} = \alpha[-\beta \pm i(1-\beta^2)^{\frac{1}{2}}], \qquad \alpha = 2c_s/a, \qquad \beta = k^{-1}.$$

Therefore a Br path anywhere in the half plane $Re p > 0$ will satisfy (5.19a). Also

$$\bar{\varphi}^* = O(p^{-3}), \qquad |p| \to \infty, \tag{5.148}$$

and therefore (5.19b) is satisfied. It follows that (5.6a) gives us the formal solution to our problem

$$\varphi(r,t) = \begin{cases} 0, & \tau = t - \dfrac{r-a}{c_d} < 0, \\ \dfrac{c_s^2\sigma_0}{\mu}\left(\dfrac{a}{r}\right) \cdot \dfrac{1}{2\pi i} \displaystyle\int_{Br} \dfrac{e^{p\tau}}{p[p^2+bp+c]} \, dp, & \tau > 0, \end{cases} \tag{5.149}$$

the first of (5.149) stemming from the remarks following (5.19b), cf. exercise 5.1. The time $\tau = 0$ defines the arrival time of the disturbance $\varphi(r, t)$ at the station $(r-a)$. To get a real form for $\varphi(r, t)$ for $\tau > 0$ we must invert the Br integral. Here this integral contains a simple transform and the inverse can be found in the tables. However, the procedure we have set down in arriving at this formal solution (5.149) are the steps one would also consider in other more complicated problems, where usually the inverse cannot be found in the tables. Then the inversion integral serves to state the formal solution, and its inversion can usually be carried out with the aid of methods based on analytic function theory. We shall set down these methods as the need for them arises.

Using the tables, e.g., in [5.3] then, the solution $\varphi(r, t)$ for $\tau>0$ in (5.149) can be written as

$$\varphi(r, t) = \frac{a^2\sigma_0}{4\mu} \cdot \frac{a}{r}\left[1 - \frac{e^{-\beta\tau}}{(1-\beta^2)^{\frac{1}{2}}}\left\{\beta \sin\left[\alpha(1-\beta^2)^{\frac{1}{2}}\tau\right]\right.\right.$$
$$\left.\left.+ (1-\beta^2)^{\frac{1}{2}} \cos\left[\alpha(1-\beta^2)^{\frac{1}{2}}\tau\right]\right\}\right]. \quad (5.150)$$

The solutions for u, and the stresses, can be obtained by substituting the first of (5.149), and (5.150) in (5.144), (5.146). Verification that the first of (5.149), and (5.150), comprise the solution of the stated boundary value problem (5.136), can be easily shown by direct substitution of the former in the latter. In particular, since the solution is bounded in r and t, the initial conditions (5.136b), and the radiation condition (5.136c), are easily shown to be satisfied because of the nature of the first of (5.419), i.e., the identically zero response in $t<(r-a)/c_d$, or $(r-a)>c_d t$. Note also this character of the solution makes it easy to show that all the higher order space and time derivatives of φ, i.e., strains, velocities and stresses, also satisfy the radiation condition (5.136c). This is typical of our hyperbolic problems. As one would expect, (5.150) yields the static solution when $\tau \to \infty$, i.e.,

$$\varphi(r) = \sigma_0 a^3/4\mu r. \quad (5.151)$$

The corresponding displacement and stresses may be obtained by substituting (5.151) in (5.144) and (5.146). Note they decay spatially as r^{-2} and r^{-3}, respectively. Analogously, by imposing $\tau \to 0+$ on (5.150), and the derived solution for u, and the stresses, one finds that at, and just back of, the front of the disturbance,

$$\varphi(r, t) \simeq \frac{c_s^2\sigma_0}{2\mu}\left(\frac{a}{r}\right)\tau^2, \qquad u(r, t) \simeq -\frac{c_s\sigma_0}{k\mu}\left(\frac{a}{r}\right)\tau,$$
$$\sigma_r(r, t) \simeq \sigma_0\left(\frac{a}{r}\right)H(\tau), \qquad \sigma_\theta = \sigma_\varphi = \frac{\lambda}{\lambda+2\mu}\sigma_r(r, t). \quad (5.152)$$

Note that these quantities decay spatially as r^{-1} in agreement with the spatial decay of spherical wavefronts set down in § 2.6.4.

Selberg [5.18] has presented numerical evaluations for the time behavior of the stresses. His results, which are reproduced here in fig. 5.6, are for Poisson's ratio $v = 1/4$ except for one curve [cf. (a) in fig. 5.6, $r/a \to \infty$,

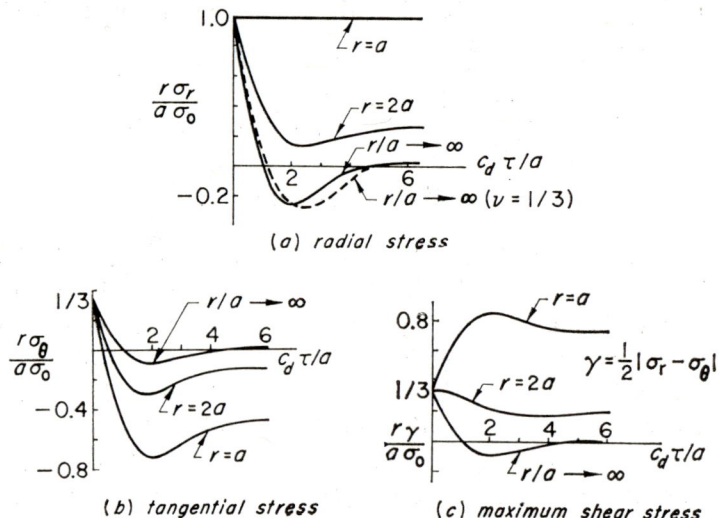

Fig. 5.6. Response of stresses in spherical cavity source problem ($v = 1\ 4$) (after Selberg).

$v = 1/3$]. It may be noted that the response curves, when compared to the static stresses at a station, exhibit a dynamic increase in the tensile tangential stress and a corresponding decrease in the compressive radial stress. In the survey by Miklowitz [4.4] a general discussion is given of work done on the spherical cavity problem, including cases involving spherically non-symmetric sources.

5.11.2. *Circular cylindrical cavity subjected to sudden uniform pressure*

Here we treat a problem closely related to that in the previous section, i.e., that of the circular cylindrical cavity of radius $r = a$ in the infinite elastic solid, the wall of which is again subjected to sudden uniform pressure σ_0. In spite of the problem's close analogy with that of the spherical cavity we shall see that the solution and its derivation are not as simple. Hence, in addition to yielding the wave information in this problem, the derivation of the solution will have instructional value to us. The present problem is one of plane strain in cylindrical coordinates (r, θ, z), hence the requirements that the displacement u_z, and variations in the axial direction of the cavity, must vanish. Since here we have axial symmetry, because of the loading, it follows that (2.34) can be used to write the boundary-initial value problem as

$$\frac{\partial^2 \varphi(r,t)}{\partial r^2} + \frac{1}{r}\frac{\partial \varphi}{\partial r} = \ddot{\varphi}/c_d^2, \qquad r>a, \qquad t>0, \qquad (5.153\text{a})$$

$$\varphi(r,0) = \dot{\varphi}(r,0) = 0, \qquad r \geq a, \qquad (5.153\text{b})$$

$$\lim_{r \to \infty} \varphi(r,t) = 0, \qquad t>0, \qquad (5.153\text{c})$$

$$\sigma_r(a,t) = \sigma_0 H(t). \qquad (5.153\text{d})$$

The transformed equation is

$$\frac{d^2 \bar{\varphi}(r,p)}{dr^2} + \frac{1}{r}\frac{d\bar{\varphi}}{dr} - k_d^2 \bar{\varphi} = 0, \qquad (5.154)$$

the general solution of this being

$$\bar{\varphi}(r,p) = A(p)I_0(k_d r) + B(p)K_0(k_d r), \qquad (5.155)$$

where I_0 and K_0 are, respectively, the modified Bessel functions of the first and second kinds of order zero. The asymptotic expansions for $I_\nu(z)$ and $K_\nu(z)$ are given in Erdelyi, et al. [2.8, formulas (5), (7), pg. 86]. From these we find the behavior for large r for I_0 and K_0 to be

$$I_0(k_d r) \simeq (2\pi k_d r)^{-\frac{1}{2}} \exp(k_d r),$$
$$K_0(k_d r) \simeq (\pi/2k_d r)^{\frac{1}{2}} \exp(-k_d r), \qquad (5.156)$$

certainly valid for $-(\pi/2) < \arg p < \pi/2$, i.e., on a Br path in $\operatorname{Re} p > 0$. Therefore, in order to satisfy the radiation condition (5.153c), i.e., the transform of (5.153c) for $\operatorname{Re} p > 0$, we must set $A(p) = 0$, and (5.155) reduces to

$$\bar{\varphi}(r,p) = B(p)K_0(k_d r), \qquad (5.157)$$

which now, according to the second of (5.156), satisfies the transformed radiation condition.

The displacement-potential relation for the one displacement here is given by the first of (2.34),

$$u = u_r = \frac{\partial \varphi}{\partial r}. \tag{5.158}$$

The strains are [cf. (4.73)]

$$\varepsilon_r = \frac{\partial u}{\partial r}, \qquad \varepsilon_\theta = \frac{u}{r}, \qquad \varepsilon_z = 0, \tag{5.159}$$

and the stress-potential relations

$$\sigma_r = \lambda \nabla^2 \varphi + 2\mu \frac{\partial^2 \varphi}{\partial r^2},$$

$$\sigma_\theta = \lambda \nabla^2 \varphi + \frac{2\mu}{r} \frac{\partial \varphi}{\partial r}, \tag{5.160}$$

$$\sigma_z = \nu(\sigma_r + \sigma_\theta).$$

The boundary condition (5.153d) defines $B(p)$, and using (5.158) we have for the transformed displacement,

$$\bar{u}(r, p) = -\frac{\sigma_0 K_1(k_d r)}{(\lambda + 2\mu) p F(p)}, \tag{5.161}$$

where

$$F(p) = (2/k^2 a) K_1(k_d a) + k_d K_0(k_d a).$$

The singularities in $\bar{u}(r, p)$ are (1) a simple pole at $p = 0$, (2) a logarithmic branch point there too, the common one for the integer order K_ν functions (hence for $\bar{u}$), and (3) possible poles due to the possible zeros of $F(p)$ in the denominator of (5.161). We argue that the latter poles cannot be in the right half p-plane since this would make the solution unbounded in time, and we know a static solution exists for the problem, and must be contained in (5.161). Therefore these poles if they exist, can lie only in the left hand p-plane $\text{Re}\,p < 0$. It follows our Br path can be anywhere in $\text{Re}\,p > 0$. We can now examine our order condition on $\bar{u}(r, p)$ by again using the leading terms of the K_ν functions in (5.161), which all behave like the second of (5.156). For large $|p|$ we find

$$\bar{u}(r, p) \simeq -\frac{c_d \sigma_0}{\lambda + 2\mu} \sqrt{\frac{a}{r}} \frac{e^{-k_d(r-a)}}{p^2} = \bar{u}^*(r, p) e^{-k_d(r-a)}, \tag{5.162}$$

hence $\bar{u}^*$, of $O(p^{-2})$, satisfies our order condition. It follows the formal solution for the present problem is contained in

$$u(r,t) = \begin{cases} 0 & , \quad t < \dfrac{r-a}{c_d}, \\ -\dfrac{\sigma_0}{\lambda+2\mu} \cdot \dfrac{1}{2\pi i} \displaystyle\int_{Br} \dfrac{K_1(k_d r)}{pF(p)} e^{pt} dp & , \quad t > \dfrac{r-a}{c_d}. \end{cases} \quad (5.163)$$

The inverse of the second of (5.163) cannot be found in the tables, so we will therefore use contour integration, and the Cauchy–Goursat theorem to invert it. In a paper dealing with a problem, strongly related to the present one, Miklowitz [5.19] made use of the contour shown in fig. 5.7,

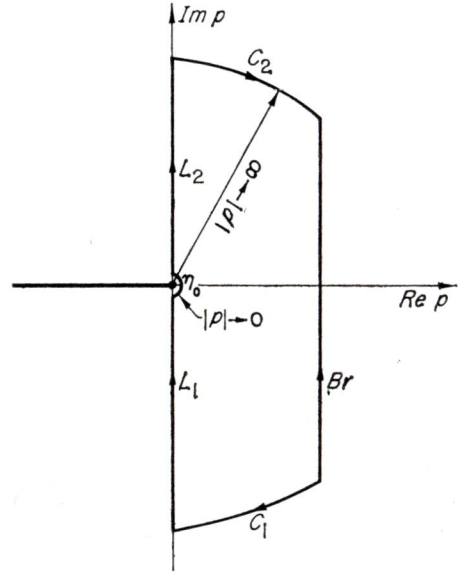

Fig. 5.7. Contour integration in the p-plane for cylindrical cavity source problem.

since it avoids the necessity of seeking the left–half plane (Re $p<0$) zéros of $F(p)$. We make use of the same contour here. In Selberg's treatment [5.18] of the present problem, he inverted the Br integral by first proving that $F(p)$ had just one root lying in the upper half p-plane, and that was in the second quadrant. He then located the pole numerically, and completed a form of Br that enclosed it, in order to derive his solution.

To carry out the contour integration we must first set down a particular branch of the *multivalued* function $K_1(k_d r)/pF(p)$ on Br, that agrees with Re $p>0$ there, i.e., we let

$$p = \eta e^{i\alpha}, \quad -\pi/2 < \alpha < \pi/2. \quad (5.164)$$

The branch cut is taken, as shown in fig. 5.7, along the negative real axis so that the *region of analyticity of the branch*, consistent with (5.164) is defined by

$$-\pi < \alpha < \pi, \qquad \eta > 0. \qquad (5.165)$$

Note also that this branch definition is consistent with that of the asymptotic expansions like the second of (5.156), used to derive (5.157) and (5.162). Then, with reference to fig. 5.7, we would complete Br to the left so that

$$\int_{Br} = \int_{C_1} + \int_{L_1} + \int_{\eta_0} + \int_{L_2} + \int_{C_2} \qquad (5.166)$$

where all integrals involved are the limiting ones indicated in the figure. Clearly, according to the *Cauchy–Goursat theorem*, the integrands on the right must be *analytical continuations* of the branch we have selected on Br [based on (5.164), (5.165)]. We have already pointed out that $\bar{u}^*(r, p)$ satisfies the general order condition (5.19b), hence the contributions from the paths C_1 and C_2 are zero (cf. exercise 5.5). Now on L_1, $p = \eta e^{-i\pi/2} = -i\eta$, and on L_2, $p = \eta e^{i\pi/2} = i\eta$, and since $K_\nu(\bar{z}) = \overline{K_\nu(z)}$ for real ν, where the bar indicates the conjugate of the quantity, i.e., $K_\nu(z)$ satisfies the *principle of reflection*, the contributions of the L_1 and L_2 integrals in (5.166) to $u(r, t,)$ for $t > (r-a)/c_d$, may be combined into

$$u(r, t)|_{L_1+L_2} = -\frac{\sigma_0}{\pi(\lambda+2\mu)} \int_0^\infty \operatorname{Im}\left[\frac{K_1(i\eta r/c_d)e^{it\eta}}{(2/k^2 a)K_1(i\eta a/c_d) + (i\eta/c_d)K_0(i\eta a/c_d)}\right]\frac{d\eta}{\eta}. \qquad (5.167)$$

Now on η_0, $p = \eta e^{i\alpha}$, $-\pi/2 \leq \alpha \leq \pi/2$, and since for small z,

$$K_0(z) \simeq -\ln z, \qquad K_1(z) \simeq 1/z,$$

the contribution of the η_0 integral in (5.166) becomes

$$u(r, t)|_{\eta_0} = -\frac{c_d \sigma_0}{2\pi(\lambda+2\mu)r} \lim_{\eta \to 0} \int_{-\pi/2}^{\pi/2} \frac{e^{-i\alpha} d\alpha}{\eta[2c_d e^{-i\alpha}/k^2 a^2 \eta - (\eta e^{i\alpha}/c_d)\ln(\eta e^{i\alpha}a/c_d)]}, \qquad (5.168)$$

provided r remains finite (which is insured since the second of (5.163) is the solution behind the wave front, i.e., $(r-a) < c_d t$). Now if the limit of the integrand here, as $\eta \to 0$, exists, we are justified in taking the limit inside the integral, and therefore (5.168) reduces to

$$u(r, t)|_{\eta 0} = -\sigma_0 a^2 / 4\mu r . \tag{5.169}$$

Our solution for $t > (r-a)/c_d$ is therefore the sum of the right hand sides of (5.167), (5.169). With the aid of the relation

$$K_\nu(z) = \frac{i\pi}{2} e^{i\nu\pi/2} H_\nu^{(1)}(iz) , \tag{5.170}$$

where $H_\nu^{(1)}$ is the Hankel function of the first kind of order ν, the integrand in (5.167) can be reduced to one containing only the real functions, J_0, J_1, Y_0, Y_1, sine, and cosine. Numerical integration then affords the evaluation with no particular difficulties [5.19]. For η large the integrand in (5.167) behaves as $1/\eta^2$, and for η small like $\sin \eta t/\eta$, hence the integral converges for $r \geq a$, and $t > (r-a)/c_d$. Fig. 5.8 shows the nature of the stress waves governed by the present solution. These results were reproduced from Selberg's paper [5.18]. Selberg took the delay time variable in these plots

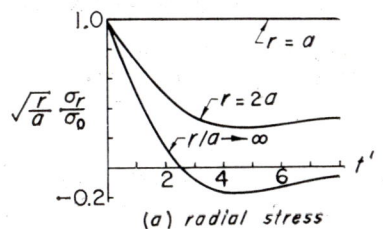

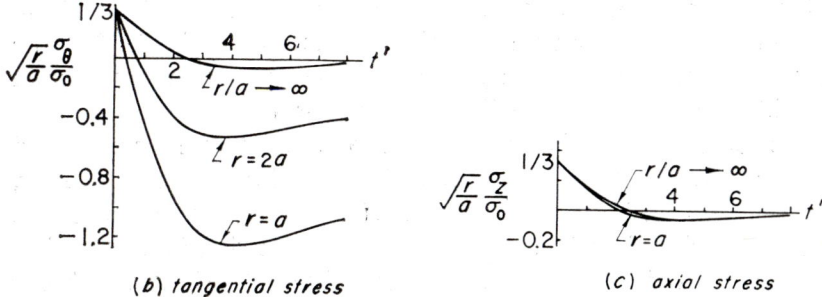

Fig. 5.8. Response of stresses in the cylindrical cavity source problem ($t' = t - (r-a)/a$ based on a unit of time $t = a/c_d$, $\nu = 1/4$) (after Selberg).

$t' = t - (r-a)/a$, based on letting the unit of time t be a/c_d. Poisson's ratio ν was taken as $\tfrac{1}{4}$ ($\lambda = \mu$). The similarity in shapes of the curves for the radial and tangential stresses with the corresponding ones for the spherical cavity problem (cf. fig. 5.6) should be noted. It should also be noted, however, that the latter curves decay faster with r.

Verification of the present solution can also be carried out directly as in the case of the spherical cavity problem. We see from (5.158), that the solution for $\varphi(r, t)$ is obtained by a simple integration of (5.163) over r, setting the constant of integration (a function of p) equal to zero for agreement with (5.157). The solution for $\varphi(r, t)$ for $t < (r-a)/c_d$, is therefore zero, and for $t > (r-a)/c_d$, is the sum of simple integrations of (5.167) and (5.169) over r. To show this solution satisfies (5.153a), direct substitution would be employed. The zero solution for $t < (r-a)/c_d$, easily shows the initial conditions (5.153b), and infinity condition (5.153c), are satisfied. Forming σ_r from the solution for $u(r, t)$ we find

$$\sigma_r(r, t) = \begin{cases} 0, & t < \dfrac{r-a}{c_d}, \\ \dfrac{\sigma_0}{\pi} \int_0^\infty \mathrm{Im}\left[\left\{\dfrac{(2/k^2 r)K_1(i\eta r/c_d) + (i\eta/c_d)K_0(i\eta r/c_d)}{(2/k^2 a)K_1(i\eta a/c_d) + (i\eta/c_d)K_0(i\eta a/c_d)}\right\} e^{it\eta}\right] \dfrac{d\eta}{\eta} \\ \quad + \dfrac{\sigma_0}{2}\left(\dfrac{a^2}{r^2}\right), & t > \dfrac{r-a}{c_d}. \end{cases} \quad (5.171)$$

Hence, at $r = a$, (5.171) reduces to $\sigma_r(a, t) = \sigma_0 H(t)$, which verifies (5.153d). It is important to note that the solution for u is bounded in r and t, but that φ corresponding to (5.169) grows as $\ln r$. We shall prove shortly this term corresponds to one half of the long time (back of wavefront) or static solution, the other half coming from the integral in (5.167). Note that such a term would be expected since it represents the well-known logarithmic potential for two-dimensional problems.

The solution we obtained here is of a form that does not readily lend itself to short time (wavefront) or long time analysis. As we argued earlier, for such cases good use can be made of the asymptotics of the Laplace transform pairs, given in §§ 5.10.2.3, 5.10.2.4, to derive, in a convenient manner, short time (wavefront) and long time approximations directly from the inversion integral(5.6a). In the next section we will apply these techniques to obtain such approximations for the present problem.

5.11.3. Short time (wavefront) and long time approximations in the cylindrical cavity problem

Using the tools set down in §§ 5.10.2.3, 5.10.2.4 we can extract the wave front approximations and static solution from (5.163). Approximating $\bar{u}(r,p)$ in (5.161) for large $|p|$, we find (5.162), of course. To identify $\bar{u}^*$ with the leading term in (5.88) we must have $\lambda_1 = 2$, and

$$a_1 = -\frac{c_d \sigma_0}{\lambda + 2\mu} \sqrt{\frac{a}{r}}.$$

It follows from (5.87) the leading term of our wave front expansion is

$$u(r,t) \simeq -\frac{c_d \sigma_0}{\lambda + 2\mu} \sqrt{\frac{a}{r}} \left(t - \frac{r-a}{c_d} \right). \tag{5.172}$$

The corresponding σ_r and σ_θ are given by

$$\sigma_r = \sigma_0 \sqrt{\frac{a}{r}} H\left(t - \frac{r-a}{c_d} \right),$$
$$\sigma_\theta = \frac{\lambda}{\lambda + 2\mu} \sigma_r. \tag{5.173}$$

Similarly, if we approximate $\bar{u}(r,p)$ in (5.161), for small $|p|$, we find

$$\bar{u}(r,p) \simeq -\frac{a\sigma_0}{2\mu} \left(\frac{a}{r} \right) \frac{1}{p} \tag{5.174}$$

for the leading term. We identify this with (5.96b), where a_0 is everything on the right of (5.174) except $\frac{1}{p}$. It follows from (5.96a) the static solution here is

$$u(r) = -\sigma_0 a^2 / 2\mu r. \tag{5.175}$$

Note that (5.169) is half of (5.175), the other half being wrapped up in (5.167) as was pointed out earlier. Since (5.167) is a Fourier integral one can use the asymptotics of § 5.10.3 to prove this. It is left to the exercises. For the stresses corresponding to (5.175) we find

$$\sigma_r = -\sigma_\theta = \sigma_0 a^2 / r^2, \qquad \sigma_z = 0. \tag{5.176}$$

The results (5.176) are in agreement with Lamé's classical solution for the static problem. From (5.172) and (5.173), we see the displacement and stresses at their wavefronts decay as $r^{-\frac{1}{2}}$ in agreement with the spatial decay we noted for cylindrical wavefronts in § 2.6.4. Further, for long time these quantities decay spatially as r^{-1} and r^{-2}, respectively. Comparing these short time and long time results with those of the spherical cavity case, (5.152) and (5.151), it is clear each of the latter represents more severe spatial decay. In addition to their usefulness in describing physical features of a problem, approximations like those found in this section are an independent check on the numerical work done in evaluating the complete solution.

Mention should be made of further available results on the cavity source problems. Selberg [5.18] gives further numerical results for the cylindrical cavity problem for the exponential decay input $[\sigma_r/\sigma_0]_{r=a} = e^{-kt}$, plotting the radial stress at $r/a \to \infty$ and tangential stress at $r=a$, for a range of k values. He also gives a set of curves for radial, tangential and maximum shear stress, for $r=a$, $2a$ and $r/a \to \infty$, for $k=0.25$. In [5.19] Miklowitz presented numerical results in the cylindrical cavity problem for the tangential stress response to a ramp function input in radial stress, at $r=a$, at a sequence of r stations. This case was studied further by the method of characteristics by Chou and Koenig [2.27] as well as other cases of both the spherical and cylindrical pressurized cavities. Further work in these problems, which quite early was of strong interest in the study of seismic sources, is discussed in the survey by Miklowitz [4.4].

Finally mention should be made of related work carried out by Goodier and Jahsman [5.20], who treated the problems of suddenly applied uniform shear stress and rotary velocity on the cylindrical cavity wall in an infinite plate in plane stress. The problems are the mathematical analog of the compressional case we have discussed. Goodier and Jahsman used an inversion technique set down earlier by Kromm for the pressurized cylindrical cavity problem we have treated. The method draws on the Laplace transform but with a rather special method not using the inversion integral. Instead, integral equations for the unknowns, the shear stress and rotary velocity, are written using the convolution theorem which involved the inverse of $e^p K_2(pr)$ where p, r were dimensionless. This inverse was obtained by inspection from Laplace transform integral for $e^p K_2(pr)$, which was written with the aid of the Schläfli integral representation for $K_2(pr)$. The integral equations were solved numerically. The reader will be interested in the numerical results given in [5.20] for the shear stress and rotary velocity waves in two problems.

5.12. Exercises

5.1. Noting the conditions (5.19) on $\bar{f}(p)$, and the remarks following these equations in the text, show that

$$f(t) = \frac{1}{2\pi i} \int_{Br} \bar{f}(p) e^{pt} dp = 0, \qquad \text{for } t < 0.$$

Hint: You can appeal to contour integration in the right half p–plane and Jordan's lemma.

5.2. (a) Find the exponential Fourier transform (of real argument) $\tilde{g}(\varkappa)$ of the function $\tilde{g}(x) = \exp(-|x|)$.
(b) Invert your result in (a), using the inversion integral and contour integration, which will give a check on the pairs $g(x)$, $\tilde{g}(\varkappa)$.

5.3. With k a positive constant,
(a) find the Fourier sine and cosine transforms $\tilde{g}^s(\varkappa)$, $\tilde{g}^c(\varkappa)$ of $g(x) = \exp(-kx)$,
(b) show that the exponential Fourier transform $\tilde{g}(\varkappa)$ of $g(x) = \exp(-k|x|)$ is twice the Fourier cosine transform $\tilde{g}^c(\varkappa)$ of $g(x) = \exp(-kx)$ and
(c) show that the Fourier cosine transform $\tilde{g}^c(\varkappa)$ of $g(x) = k/(x^2 + k^2)$ is $\pi e^{-k\varkappa}/2$.

5.4. Prove that the convolution integral for the exponential Fourier transform is (5.46) of the text.

5.5. Show that the integrals over the paths C_2, C_3 and those over the paths C_1, C_4, in (5.90) of the text give zero contribution to the inversion there.

5.6. Exercise 4.3 dealt with the predominant period–time of occurrence relation for flexural waves in a beam governed by the Bernoulli–Euler theory. Assume now we have an initial value problem for this theory, in which the initial values are $y(x, 0) = y_0(x) = \delta_s(x)$, $\dot{y}(x, 0) = 0$ where δ_s is the symmetric delta function. [cf. (6.5) ch. 6]. Show that the exact solution for this problem, given by the first of (5.130), yields

$$y(x, t) = (\tfrac{1}{2})(\pi dt)^{-\frac{1}{2}} \cos\left[(x^2/4dt) - (\pi/4)\right] + O(1/t), \qquad t \gg 1,$$

by the method of stationary phase. Show this result agrees with the results of exercise 4.3. To obtain this result you will need parts of the analysis in exercise 4.3. Now using the first of (5.130) again, show that it integrates to the same result just found for $y(x, t)$ without the order term. This shows that the method of stationary phase gives the exact result in the present problem.

5.7. For the problem of the spherical cavity in the infinite elastic solid, prove through *residue theory*, that the second of (5.149) yields (5.150).

5.8. With the aid of the Laplace transform on time t, solve the boundary-initial value problem of the semi-infinite stretched string stated as follows:

$$\partial^2 y(x, t)/\partial x^2 = \ddot{y}/c^2, \qquad x > 0, \qquad t > 0,$$

$$y(x, 0) = \dot{y}(x, 0) = 0, \qquad x \geq 0,$$

$$y(0, t) = f(t), \qquad t \geq 0,$$

$$\lim_{x \to \infty} y(x, t) = 0, \qquad t > 0.$$

Here y is the transverse (small) displacement of the string which initially lies along the x-axis ($x \geq 0$), and speed $c = (T/\varrho)^{\frac{1}{2}}$, where T is the assumed constant tension force in the string ϱ being its material density (mass per unit length). The displacement end ($x = 0$) condition is that the string moves verticaly according to $f(t)$. There is no motion at x infinite.

5.9. With the aid of the Laplace transform on time t, solve the semi-infinite elastic rod boundary-initial value problem stated as follows:

$$\partial^2 u(x, t)/\partial x^2 = \ddot{u}/c_b^2, \qquad x > 0, \qquad t > 0,$$

$$u(x, 0) = \dot{u}(x, 0) = 0, \qquad x \geq 0,$$

$$\lim_{x \to \infty} u(x, t) = 0, \qquad t > 0,$$

$$\partial u(0, t)/\partial x = -\sigma_0 H(t)/E.$$

The theory here is a good approximation for a long thin rod, with u being the axial displacement. The function $H(t)$ is the Heaviside step function, σ_0 the magnitude of the end stress input, and c_b the bar velocity $\sqrt{E/\varrho}$, E being Young's modulus and ϱ the material density. With the aid of the shift operator in § 5.4.6 show also that your result for the semi-infinite rod problem can be extended to the solution of the corresponding problem of the cantilever rod of length l, given by

$$u(x, t) = -\frac{c\sigma_0}{E} \sum_{n=0}^{\infty} (-)^n \left\{ \left[t - \frac{(2n+2)l - x}{c} \right] - \left[t - \frac{2nl + x}{c} \right] \right\}.$$

Note in the cantilever case the fixed end condition $u(l, t) = 0$ replaces the condition at $x \to \infty$ for the semi-infinite rod. Verify your solution for both problems, and for the cantilever case sketch the displacement response $u(l/2, t)$ and $u(0, t)$. Note the latter response can be written as $u(0, t) = (c\sigma_0/E)H\left(\dfrac{2l}{c}, t\right)$, H being the periodic triangular wave function of period $4l/c$. Prove from your work here the transform of this function is $\bar{H}(a, p) = (1/p^2) \tanh(ap/2)$. Churchill [5.3, § 7] discusses other similar functions.

5.10. With the aid of the Laplace transform on time t, consider the boundary-initial value problem for the cantilever rod in torsion stated as follows:

$$\partial^2 \theta(x, t)/\partial x^2 = \ddot{\theta}/c_s^2, \qquad 0 < x < l, \qquad t > 0,$$

$$\theta(x, 0) = x\theta_0/l, \qquad \dot{\theta}(x, 0) = 0, \qquad 0 \leq x \leq l,$$

$$\theta(0, t) = 0, \qquad \partial\theta(l, t)/\partial x = 0, \qquad t > 0.$$

Here θ is the angle of twist. The torsional moment on a rod section is given by $M_t = \mu I \partial\theta/\partial x$, where μ is the Lamé constant, and I the polar moment of interia of the rod section. The theory here corresponds to the *nondispersive torsional mode* (4.87) discussed in § 4.4.1. Note in the problem statement that initially the rod is twisted such that the displacement θ is a linear function of x, the end $x = l$ having the maximum $\theta = \theta_0$. The boundary conditions show that immediately after $t = 0$, $\theta = 0$ at the fixed end $x = 0$ and $\partial\theta/\partial x = 0$ at the now free end $x = l$ (no moment hence no stress). The problem is therefore one of end moment release, hence unloading waves propagating from this source. Show that the angular response at the end $x = l$ is given by

$$\theta(l, t) = \theta_0 \left[1 - \frac{c}{l} H\left(\frac{2l}{c}, t\right) \right],$$

where H is the triangular wave function discussed in exercise 5.9. Sketch the response.

5.11. With the aid of the proper transform on x show that the initial value problem for the transverse displacements $y(x, t)$ in a string stretched along the entire x-axis, given by

$$\partial^2 y(x, t)/\partial x^2 = \ddot{y}/c^2, \qquad -\infty < x < \infty, \qquad t > 0,$$

$$y(x, 0) = F(x), \qquad \dot{y}(x, 0) = 0, \qquad -\infty < x < \infty$$

$$\lim_{|x| \to \infty} [y(x, t), \partial y/\partial x, \dot{y}] = 0, \qquad t \geq 0,$$

with c defined in exercise 5.8, has the solution

$$y(x, t) = (\tfrac{1}{2})[F(x - ct) + F(x + ct)]$$

in agreement with D'Alembert's solution (2.21), (2.125).

5.12. Starting with the formal solution in (5.163) derive in detail the inversion in the text for the cylindrical cavity problem for $t > (r-a)/c_d$, i.e., $u(r, t)$, the sum of the right hand sides of (5.167) and (5.169).

5.13. Using the asymptotics of the Fourier integral in § 5.10.3 of the text prove that the integral (5.167) contributes one half of the static solution.
Hint: You can write the integral in (5.167) in the form $\int_0^\varepsilon + \int_\varepsilon^\infty$, $0 < \varepsilon \ll 1$, to take care of the integrand singularity at $\eta = 0$.

5.14. A thin infinite elastic plate (in plane stress) containing a circular cavity of radius $r = a$ is suddenly excited by a step uniform (in r) velocity

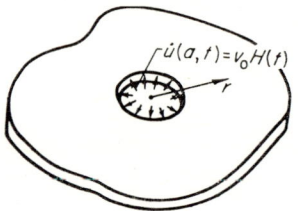

on the cavity wall, as the sketch depicts. The governing equation of motion for the displacement $u(r, t)$ is

$$\frac{\partial^2 u(r, t)}{\partial r^2} + \frac{1}{r} \frac{\partial u(r, t)}{\partial r} - \frac{u(r, t)}{r^2} = \frac{\ddot{u}(r, t)}{c_p^2}$$

where $c_p = [E/(1-v^2)\varrho]^{\frac{1}{2}}$ is the so called "*plate*" wave speed.

(a) Formulate the rest of the boundary value problem, and using a Laplace transform on time t, derive the formal solution for $\dot{u}(r, t)$ i.e., the *Br* inversion integral.

(b) Show that $\bar{\dot{u}}(r, p)$ satisfies the two basic conditions required of a Laplace transform.

(c) Noting that a station here is given by $r-a$, find the wavefront approximation (short time) for the solution, and find the long time (static) solution.

(d) Show the initial conditions, the boundary condition $(r=a)$ and the condition at $r=\infty$ are satisfied by your formal solution in (a), and that the approximations in (c) satisfy the boundary condition.

5.15. Apply the asymptotics of the Laplace transform and its inverse to the *Br* integral in (5.149) to derive (5.151) and (5.152) for the spherical cavity problem. Note the first of (5.149) defines the solution ahead of the wavefront.

5.16. Consider the initial value problem for the axially symmetric transverse displacement $w(r, t)$ of a stretched infinite membrane given by

$$\nabla^2 w(r, t) = \partial^2 w/\partial r^2 + (1/r)\partial w/\partial r = \ddot{w}/c^2, \qquad 0 < r < \infty, \qquad t > 0,$$

$$w(r, 0) = ah(r^2 + a^2)^{-\frac{1}{2}}, \qquad \dot{w}(r, 0) = 0, \qquad 0 \leq r < \infty,$$

$$\lim_{r \to 0, \infty} \{r[w(r, t), \partial w/\partial r]\} = 0, \qquad t > 0,$$

where the wave speed $c = (T/\varrho)^{\frac{1}{2}}$, T representing the constant tension in the membrane (force/unit length) and ϱ the membrane density (mass/unit area). a and h are constants. With the aid of the Hankel transform show the solution is

$$w(r, t) = \frac{h}{\sqrt{2}} \left\{ \frac{1}{\left[\left(1 + \frac{r^2 - c^2 t^2}{a^2}\right)^2 + 4\left(\frac{ct}{a}\right)^2\right]^{\frac{1}{2}}} + \frac{1 + \frac{r^2 - c^2 t^2}{a^2}}{\left(1 + \frac{r^2 - c^2 t^2}{a^2}\right)^2 + 4\left(\frac{ct}{a}\right)^2} \right\}.$$

The derivation is simplified by making use of the integral

$$\int_0^\infty e^{-a\varkappa} J_0(\varkappa r) d\varkappa = (r^2 + a^2)^{-\frac{1}{2}}.$$

5.17. A large mass traveling at velocity v_0 impacts, at time $t=0$, the end $x=0$ of a semi-infinite viscoelastic rod. The rod element is modeled approximately as that of a *Maxwell material*, i.e., a spring and dashpot in series. The boundary-initial value problem may be stated as

$$\partial^2 \sigma(x, t)/\partial x^2 - \ddot{\sigma}/c^2 - (\varrho/\varphi)\dot{\sigma} = 0, \qquad 0 < x < \infty, \qquad t > 0,$$

$$\sigma(x, 0) = \dot{\sigma}(x, 0) = 0, \qquad 0 \le x < \infty,$$

$$\dot{u}(0, t) = v_0 H(t),$$

$$\lim_{x \to \infty} [u(x, t), \sigma(x, t)] = 0, \qquad t > 0,$$

where σ, u are the normal stress and displacement in the axial direction, $c = (E/\varrho)^{\frac{1}{2}}$ and φ is a coefficient of viscosity. Show that the solution of this problem is

$$\sigma(x, t) = \begin{cases} 0, & t < x/c, \\ -v_0 \varrho c e^{-(a/2)t} I_0\{(a/2)[t^2 - (x/c)^2]\}, & t > x/c, \end{cases}$$

where $a = E/\varphi$ and I_0 is the modified Bessel function of the first kind of zeroth order. Note you can assume that the stress equation of motion $\partial \sigma/\partial x = \varrho \ddot{u}$ holds at $x = 0$, and that $u, \dot{u}$ are zero initially. Plot the wavefront behavior for the stations x_1, x_2 and x_3 where $x_3 > x_2 > x_1$. Discuss the long-time behavior of the solution for fixed x. Discuss also the influence of a on the solution.

References

[5.1.] R. V. Churchill, *Fourier Series and Boundary Value Problems*, 2nd Edition. McGraw-Hill Book Company, Inc., New York (1963).

[5.2.] E. C. Titchmarsh, *Introduction to the Theory of Fourier Integrals*, 2nd Edition. Oxford University Press (1948).

[5.3.] R. V. Churchill, *Operational Mathematics*, 3rd Edition. McGraw-Hill Book Company, Inc., New York (1972).

[5.4 [D. V. Widder, *The Laplace Transform*. Princeton University Press (1941).

[5.5.] B. van der Pol and H. Bremmer, *Operational Calculus*, 2nd Edition. Cambridge University Press (1959).

[5.6.] F. Oberhettinger, *Tabellen zur Fourier Transformation*. Springer-Verlag, Berlin (1957).

[5.7.] I. N. Sneddon, *Fourier Transforms*. McGraw-Hill Book Company, Inc., New York (1951).

REFERENCES

[5.8.] G. N. Watson, *Theory of Bessel Functions*, 2nd Edition. Cambridge University Press (1958).

[5.9.] T. Y. T. Wu, *Course Notes on Hydrodynamics of Free Surface Flows*, California Institute of Technology, Pasadena, California (1958).

[5.10.] N. W. McLachlan, *Complex Variable Theory and Transform Calculus*, 2nd Edition. Cambridge University Press (1953).

[5.11.] H. Jeffreys and B. S. Jeffreys, *Methods of Mathematical Physics*, 3rd Edition. Cambridge University Press (1956).

[5.12.] A. Erdelyi, *Asymptotic Expansions*. Dover Publications, Inc., New York (1956).

[5.13.] M. G. Smith, *Laplace Transform Theory*. D. Van Nostrand Company Ltd., London (1966).

[5.14.] B. L. van der Waerden, *Applied Science Research* **B2** (1951–52) 33–45.

[5.15.] M. V. Cerrillo. *An Elementary Introduction to the Theory of the Saddlepoint Method of Integration*. Technical Report 55: 2a, Research Laboratory of Electronics, Massachusetts Institute of Technology, Cambridge, Massachusetts (1950).

[5.16a.] R. Folk, *Time Dependent Boundary Value Problems in Elasticity*, Ph. D. Thesis, Lehigh University, Bethlehem, Pennsylvania (1958).

b. R. Folk, G. Fox, C. A. Shook and C. W. Curtis, *Journal of the Acoustical Society of America* **30** (1958), 552–558.

[5.17a.] O. E. Jones, *Theoretical and Experimental Studies on the Propagation of Longitudinal Elastic Strain Pulses in Wide Rectangular Bars*, Ph. D. Thesis, California Institute of Technology, Pasadena, California (1961).

b. O. E. Jones and A. T. Ellis, *Journal of Applied Mechanics* **30** (1963), 51–60.

[5.18.] H. L. Selberg, *Arkiv för Fysik* **5** (1952), 97–108.

[5.19.] J. Miklowitz, *Journal of Applied Mechanics* **27** (1960), 165–171.

[5.20.] J. N. Goodier and W. E. Jahsman, *Journal of Applied Mechanics* **23** (1956), 284–286.

CHAPTER 6

TRANSIENT WAVES IN AN ELASTIC HALF SPACE

6.1. Introduction

Lamb [6.1] was the first to study the propagation of a pulse in an elastic half space. In his classical work, presented in 1904, Lamb treated four basic problems, the surface normal line and point load sources, and the buried line and point sources of dilatation. He derived his solutions for these problems through Fourier synthesis of the steady propagation solutions. For the surface source problems Lamb evaluated the surface displacements (horizontal and vertical), bringing forth the very important fact that the *largest disturbance in the far field* (from the source) was the *Rayleigh surface wave*. He noted the nondispersive nature of this disturbance, and in the case of the point load excitation, that it decayed ar $r^{-\frac{1}{2}}$, a property typical of two-dimensional propagation we have noted earlier. Ewing et al. [3.11, §§ 2-3 to 2-6] present the essential features of Lamb's methods and solutions. They also discuss the contributions of others to the general problem of the propagation of a disturbance in a half space [3.11, § 2.7]. More recent contributions on the problem (through 1964), which through the years has taken on the name Lamb's problem, have been discussed by Miklowitz [4.4]. Modern *integral transform methods* afford a much more direct and compact way for deriving solutions to the problem. This will be evident in our discussion in the sections that follow.

6.2. Plane strain problems

The present problems are two dimensional, and as such, will lend themselves to treatment by a *double integral transform*. One of these transforms is the Laplace (cf. §§ 5.3, 5.4), introduced to reduce the time variable, as in the one-dimensional problems of Chapter 5. In selecting the second transform, for a spatial variable, the geometry of the medium, and again what information the transform "asks for" at boundaries determines the appropriate choice. The Fourier exponential transform with real argument (cf. § 5.6.1) is suited to the present, and other related two-dimensional problems of the elastic half space.

6.2.1. Lamb's problem for the surface normal line load source

Figure 6.1 depicts the problem. A uniform (with respect to y) normal line load, $-\sigma_0 F(t)$, is suddenly applied to the surface of an elastic half space, $z=0$, at time $t=0$. The function $F(t)$ represents the time behavior

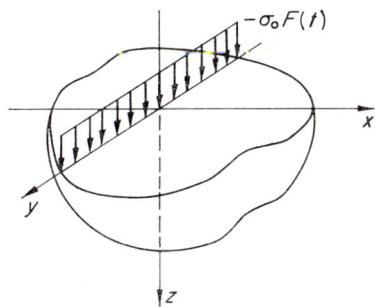

Fig 6.1. Problem of normal line load on surface of half space.

of the loading, and σ_0 is a positive magnitude constant of dimensions force per unit length. We wish to derive and analyze the wave system in the half space (or plane), $-\infty < x < \infty$, $z \geq 0$, that emanates, and propagates, from this source for $t > 0$. The uniformity of the loading in y makes this a plane strain problem ($v=0$, $\partial/\partial y = 0$), and therefore using (2.7b), the boundary-initial value problem may be stated as

$$\left.\begin{array}{l}\nabla^2 \varphi(x, z, t) = \ddot{\varphi}/c_d^2 \\ \nabla^2 \psi(x, z, t) = \ddot{\psi}/c_s^2\end{array}\right\} \text{in } -\infty < x < \infty, \qquad z > 0, \qquad \text{for } t > 0, \qquad (6.1)$$

$$\varphi(x, z, 0) = \dot{\varphi} = \psi = \dot{\psi} = 0, \qquad \text{in } -\infty < x < \infty, \qquad z \geq 0, \qquad (6.2)$$

$$\left.\begin{array}{l}\sigma_z(x, 0, t) = -\sigma_0 \delta_s(x) F(t) \\ \sigma_{zx}(x, 0, t) = 0\end{array}\right\} \text{for } -\infty < x < \infty, \qquad t > 0, \qquad (6.3)$$

$$\lim_{r \to \infty} \left[\varphi(x, z, t), \psi, \partial \varphi/\partial x, \text{etc.}\right] = 0, \qquad \text{for } t > 0, \qquad (6.4)$$

where $r=(x^2+z^2)^{\frac{1}{2}}$, and $\delta_s(x)$ is the *symmetrical delta function*,[1] defined by

$$\delta_s(x) = \begin{cases} 0, & x \leq -\varepsilon, \\ \dfrac{1}{2\varepsilon}, & -\varepsilon < x < \varepsilon, \\ 0, & x \geq \varepsilon, \end{cases} \tag{6.5a}$$

where $\varepsilon \to 0$, and

$$\int_{-\infty}^{\infty} \delta_s(x)\,dx = 1. \tag{6.5b}$$

Applying the Laplace transform to (6.1), using (6.2), and then to this, the Fourier exponential transform (real argument $\varkappa$), using the Laplace transforms of (6.4), reduce (6.1) to

$$\frac{d^2 \bar{\tilde{\varphi}}(\varkappa, z, p)}{dz^2} - (k_d^2 + \varkappa^2)\bar{\tilde{\varphi}} = 0,$$

$$\frac{d^2 \bar{\tilde{\psi}}(\varkappa, z, p)}{dz^2} - (k_s^2 + \varkappa^2)\bar{\tilde{\psi}} = 0, \tag{6.6}$$

where $k_d = p/c_d$, and $k_s = p/c_s$. Solutions of (6.6) are of the form

$$\bar{\tilde{\varphi}} = A(\varkappa, p) \exp(-\eta_d z) + C(\varkappa, p) \exp(\eta_d z),$$

$$\bar{\tilde{\psi}} = B(\varkappa, p) \exp(-\eta_s z) + D(\varkappa, p) \exp(\eta_s z), \tag{6.7}$$

where $\eta_d = (k_d^2 + \varkappa^2)^{\frac{1}{2}}$, and $\eta_s = (k_s^2 + \varkappa^2)^{\frac{1}{2}}$. Now (6.7) must satisfy the double transforms of (6.4). Hence, since $\varkappa$ is real, if we take p real and positive,[2] we can select the *real positive branches* of η_d and η_s, and then require C and D

---

[1] The asymmetrical delta function was defined in § 5.4.7.2, which as we pointed out there will be referred to as the delta function. Equation (2.68) governs both types.

[2] It suffices to consider such values of p since they lead to a unique inverse transform $f(t)$ according to *Lerch's theorem*. That is, if $\bar{f}(p) = \int_0^{\infty} f(t)e^{-pt}dt$, $p \geq p_0$, is satisfied by a function $f(t)$ of bounded variation, this function is unique. The proof is given in Widder [5.4, § 6].

to vanish, reducing (6.7) to

$$\tilde{\varphi}(\varkappa, z, p) = A(\varkappa, p) \exp(-\eta_d z),$$
$$\tilde{\psi}(\varkappa, z, p) = B(\varkappa, p) \exp(-\eta_s z).$$
(6.8)

These satisfy the double transforms of (6.4).

The displacement- and stress-potential relations here are given by (3.1) and (3.6). The double transforms of σ_z and σ_{zx} in (3.6) are

$$\tilde{\sigma}_z(\varkappa, z, p) = \lambda k_d^2 \tilde{\varphi}(\varkappa, z, p) + 2\mu \left(\frac{d^2 \tilde{\varphi}}{dz^2} - i\varkappa \frac{d\tilde{\psi}}{dz} \right),$$
$$\tilde{\sigma}_{zx}(\varkappa, z, p) = \mu \left[k_s^2 \tilde{\psi}(\varkappa, z, p) - 2 \left(i\varkappa \frac{d\tilde{\varphi}}{dz} + \frac{d^2 \tilde{\psi}}{dz^2} \right) \right].$$
(6.9)

The substitution of (6.8) in (6.9), and then substituting this in the double transforms of (6.3), gives two algebraic equations for $A(\varkappa, p)$ and $B(\varkappa, p)$, i.e.,

$$(\lambda k_d^2 + 2\mu \eta_d^2)A + i2\mu\varkappa\eta_s B = -\sigma_0 \bar{F}(p),$$
$$i2\varkappa\eta_d A + (k_s^2 - 2\eta_s^2)B = 0,$$
(6.10)

since $\tilde{\delta}_s(\varkappa) = 1$. Solution of (6.10) shows

$$A(\varkappa, p) = M(k_s^2 + 2\varkappa^2),$$
$$B(\varkappa, p) = M(i2\varkappa\eta_d),$$
(6.11)

where $M = -\sigma_0 \bar{F}(p)/\mu R$, in which

$$R(\varkappa, p) = (k_s^2 + 2\varkappa^2)^2 - 4\varkappa^2 \eta_d \eta_s$$

is Rayleigh's function (cf. (3.65)). Substitution of (6.11) into (6.8) completes the definition of the transformed solutions for the potentials. Applying the double transforms of (3.1) to these solutions we find the transformed solutions for the displacements

$$\tilde{u}(\varkappa, z, p) = -i\varkappa A \exp(-\eta_d z) + \eta_s B \exp(-\eta_s z),$$
$$\tilde{w}(\varkappa, z, p) = -\eta_d A \exp(-\eta_d z) - i\varkappa B \exp(-\eta_s z).$$
(6.12)

Under the conditions stated for η_d, and η_s, after (6.7), we see $\tilde{u}$ and $\tilde{w}$, as Fourier exponential transforms with p and z fixed, are continuous functions of $\varkappa$ in $-\infty<\varkappa<\infty$, provided $R(\varkappa, p)$ in (6.11) does not have zeros in this domain. It can be proved that $R(\varkappa, p)$ does have a pair of simple zeros, corresponding to the Rayleigh surface wave as discussed in § 3.1.4.7. They satisfy (3.65) or (3.66) subject to (3.68), and are off the real axis lying along the imaginary axis in the complex $\varkappa$-plane. Now since $z \geq 0$, it is not difficult to show that

$$\lim_{\varkappa \to \pm \infty} \begin{bmatrix} \tilde{u} \\ \tilde{w} \end{bmatrix} = 0,$$

consistent with (5.41). The proof is left to the exercises. These properties make $\tilde{u}$, and $\tilde{w}$, well-behaved exponential Fourier transforms, i.e., in the sense of i) and ii) in § 5.6.1. Therefore, we use (5.39a) to set down the formal inverse exponential Fourier transforms corresponding to (6.12),

$$\begin{bmatrix} \bar{u}(x, z, p) \\ \bar{w}(x, z, p) \end{bmatrix} = \frac{1}{2\pi} \int_{-\infty}^{\infty} \begin{bmatrix} \tilde{u}(\varkappa, z, p) \\ \tilde{w}(\varkappa, z, p) \end{bmatrix} \exp(-i\varkappa x) \mathrm{d}\varkappa . \qquad (6.13)$$

6.2.1.1. Exact inversion by Cagniard–deHoop method. The inversion of the Laplace transforms $\bar{u}$ and $\bar{w}$ in (6.13) can be carried out by an ingenious method first set down by Cagniard[3] [3.19]. We will use a modified form of it due to deHoop [6.2] which is simpler. Essentially the method employs transformations and contour integrations to warp the integrals in (6.13) into the Laplace integral operators (5.6b), and then the so-produced integral equations for u and w are solved by inspection. Since p is real, and positive, we begin by introducing the real variable $\zeta = \varkappa/k_d$ into the integrals in (6.13), that is, with

$$\varkappa = k_d \zeta, \qquad \mathrm{d}\varkappa = k_d \mathrm{d}\zeta,$$

$$\eta_d = k_d(\zeta^2 + 1)^{\frac{1}{2}}, \qquad \eta_s = k_d(\zeta^2 + k^2)^{\frac{1}{2}}, \qquad (6.14)$$

---

[3] Note use of the Laplace inversion integral (5.6a) is not very suitable to problems such as the present, because one would have a double inversion integral involving branch points of η_d and η_s in the p-plane.

these integrals become

$$\bar{u}(x,z,p) = \frac{\sigma_0 \bar{F}(p)}{2\pi\mu} \int_{-\infty}^{\infty} \left[f_d(\zeta)e^{-pg_d(\zeta)} + f_s(\zeta)e^{-pg_s(\zeta)} \right] d\zeta ,$$

$$\bar{w}(x,z,p) = \frac{\sigma_0 \bar{F}(p)}{2\pi\mu} \int_{-\infty}^{\infty} \left[h_d(\zeta)e^{-pg_d(\zeta)} + h_s(\zeta)e^{-pg_s(\zeta)} \right] d\zeta ,$$

(6.15)

where

$$g_d(\zeta) = (1/c_d)[\eta'_d(\zeta)z + i\zeta x] , \qquad g_s(\zeta) = (1/c_d)[\eta'_s(\zeta)z + i\zeta x] ,$$

$$f_d(\zeta) = i\zeta(k^2 + 2\zeta^2)/R(\zeta) , \qquad f_s(\zeta) = -i2\zeta[\eta'_d(\zeta)\eta'_s(\zeta)]/R(\zeta) ,$$

$$h_d(\zeta) = \eta'_d(\zeta)(k^2 + 2\zeta^2)/R(\zeta) , \qquad h_s(\zeta) = -2\zeta^2 \eta'_d(\zeta)/R(\zeta) ,$$

and

$$R(\zeta) = (k^2 + 2\zeta^2)^2 - 4\zeta^2 \eta'_d(\zeta)\eta'_s(\zeta) ,$$

where the "sub d and s" terms in the integrals of (6.15) are associated with the dilatational and equivoluminal parts of the displacements, respectively, and $\eta'_d(\zeta) = \eta_d(\zeta)/k_d = (\zeta^2 + 1)^{\frac{1}{2}}$ and $\eta'_s(\zeta) = \eta_s(\zeta)/k_d = (\zeta^2 + k^2)^{\frac{1}{2}}$.

The *basic idea* in the Cagniard–deHoop inversion of integrals like those in (6.15) is to let $g(\zeta) = t$, which is real and positive. This deforms the path of integration off the real axis of ζ. Clearly then, consideration must be given the character of f, g, and h in (6.15) in the ζ-plane. The functions f and h have branch points at $\zeta = \pm i$, $\pm ik$, and simple poles at $\zeta = \pm ik/k_R = \pm \zeta_R$ (the zeros of $R(\zeta)$ for the Rayleigh surface wave[4]). The function g has branch points at $\zeta = \pm i$ and $\zeta = \pm ik$. The plane can be cut as shown in fig. 6.2, and the simple multivalued functions $\eta'_d(\zeta)$, $\eta'_s(\zeta)$ can be continued analytically off the real axis where they are real and positive (as we fixed them earlier, i.e., η_d, η_s). If we let

---

[4] As pointed out earlier these correspond to the pair satisfying (3.65) or (3.66), subject to (3.68), e.g., $\pm k_{R1}$ for $\nu = \frac{1}{4}$ given in § 3.1.4.7. It is left to the exercises to show this for the definitions given $\eta'_d(\zeta)$ and $\eta'_s(\zeta)$ for the cut ζ-plane in fig. 6.2.

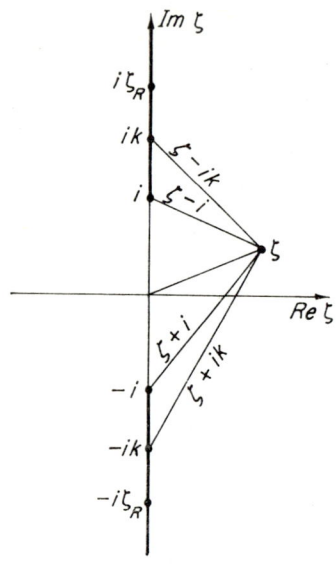

Fig. 6.2. The defining cut ζ-plane for analytic functions f, g and h.

$$\zeta - i = \varrho_1 \exp(i\varphi_1), \qquad -3\pi/2 < \varphi_1 < \pi/2, \qquad \varrho_1 > 0,$$

$$\zeta + i = \varrho_2 \exp(i\varphi_2), \qquad -\pi/2 < \varphi_2 < 3/2\pi, \qquad \varrho_2 > 0,$$

it is easy to show that in the cut lower half plane $-\pi/2 < \arg \eta'_d(\zeta) < \pi/2$. A similar analysis shows the same is true for $\eta'_s(\zeta)$. It follows that

$$\mathrm{Re}\begin{bmatrix} \eta'_d(\zeta) \\ \eta'_s(\zeta) \end{bmatrix} \geqq 0,$$

everywhere in the cut lower half plane, and they $\to \infty$ as $|\zeta| \to \infty$. Further, in the lower half plane, $\mathrm{Im}\,\zeta < 0$. On the basis of this analysis, if we were to integrate the integrals in (6.15) over the completed path shown in fig. 6.3,

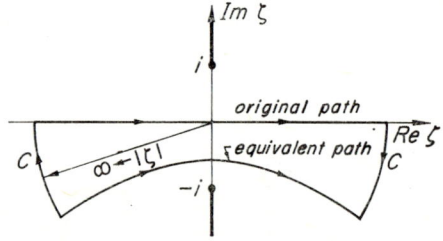

Fig. 6.3. Equivalent path of integration in Cagniard–de Hoop method.

imposing $z \geq 0$, and $x > 0$, an application of Jordan's lemma to the integrals on the large circular paths C shows these integrals vanish. Hence, the integrals in (6.15) (along the original path, the real axis) are equivalent to those on any path in the lower half plane, from infinity in the third quadrant to infinity in the fourth quadrant, passing between $\zeta = 0$ and $\zeta = -i$, as shown in fig. 6.3.

Let us now impose $g_d(\zeta) = t$. As we shall see this leads to a lower half plane contour of the type of the *equivalent path* in fig. 6.3. From (6.15) we have

$$t = (1/c_d)[(\zeta^2 + 1)^{\frac{1}{2}} z + i\zeta x] \,. \tag{6.16}$$

Solving (6.16) for ζ we have

$$\zeta = \frac{-ixc_d t \pm z[c_d^2 t^2 - (x^2 + z^2)]^{\frac{1}{2}}}{x^2 + z^2} \,, \tag{6.17}$$

and introducing polar coordinates $r = (x^2 + z^2)^{\frac{1}{2}}$, and $\theta = \tan^{-1}(x/z)$, where $0 \leq r < \infty$, $-\pi/2 \leq \theta \leq \pi/2$, θ being measured from the z-axis, the new path of integration (6.17) becomes

$$\zeta = \pm[(c_d t/r)^2 - 1]^{\frac{1}{2}} \cos \theta - i(c_d t/r) \sin \theta, \quad r/c_d \leq t < \infty, \tag{6.18}$$

with $[(c_d t/r)^2 - 1]^{\frac{1}{2}} \geq 0$, and $0 \leq \theta \leq \pi/2$, which is a branch of the hyperbola

$$(\operatorname{Im}\zeta/\sin \theta)^2 - (\operatorname{Re}\zeta/\cos \theta)^2 = 1 \,, \tag{6.19}$$

lying in the lower half ζ-plane, as required by our foregoing analysis, with asymptotes $\operatorname{Im}\zeta/\operatorname{Re}\zeta = \pm \tan \theta$. Note that limiting θ to the right quarter plane is not really a restriction in the physical problem because of the symmetry with respect to the z-axis. Now the earliest time in (6.18) is $t = r/c_d$, which gives $\zeta = -i \sin \theta$, the *vertex* of the branch. As t grows, the real parts ($\pm$), and the negative imaginary part grow, until at $t \to \infty$, $\operatorname{Im}\zeta/\operatorname{Re}\zeta \to \pm \tan \theta$, i.e., (6.18) tends to its asymptotes. Figure 6.4 provides a sketch of this *Cagniard path* which is associated with dilatational waves. We see if $0 \leq \sin \theta < 1$, or $0 \leq \theta < \pi/2$, is imposed, the Cagniard path does not intersect the branch cut shown in fig. 6.4. Now with ζ in the fourth quadrant, $-\bar{\zeta}$ (the bar indicating complex conjugate) is in the third quadrant of the ζ-plane, hence integrating the g_d integrals, in (6.15), over the path (6.18), $C_3 + C_4$ in fig. 6.4, we find

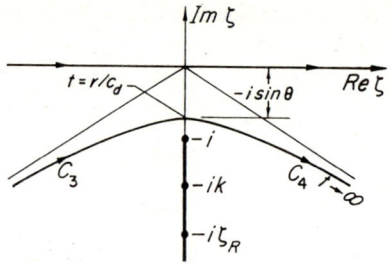

Fig. 6.4. The Cagniard path corresponding to the dilatational waves.

$$\int_{C_3}\begin{bmatrix}f_d(\zeta)\\h_d(\zeta)\end{bmatrix}\exp\left[-pg_d(\zeta)\right]d\zeta+\int_{C_4}\begin{bmatrix}f_d(\zeta)\\h_d(\zeta)\end{bmatrix}\exp\left[-pg_d(\zeta)\right]d\zeta=$$

$$\int_{-C_3}\begin{bmatrix}f_d(-\bar\zeta)\\h_d(-\bar\zeta)\end{bmatrix}\exp\left[-pg_d(-\bar\zeta)\right]d\bar\zeta+\int_{C_4}\begin{bmatrix}f_d(\zeta)\\h_d(\zeta)\end{bmatrix}\exp\left[-pg_d(\zeta)\right]d\zeta. \quad (6.20)$$

Now we note that

$$g_d(-\bar\zeta)=(1/c_d)[\overline{\eta'_d(\zeta)}z-i\bar\zeta x]=\overline{g_d(\zeta)},$$

$$f_d(-\bar\zeta)=-i\bar\zeta(k^2+2\bar\zeta^2)/R(-\bar\zeta)=\overline{f_d(\zeta)}, \quad (6.21)^5$$

$$h_d(-\bar\zeta)=\overline{\eta'_d(\zeta)}(k^2+2\bar\zeta^2)/R(-\bar\zeta)=\overline{h_d(\zeta)},$$

where $R(-\bar\zeta)=(k^2+2\bar\zeta^2)^2-4\bar\zeta^2\overline{\eta'_d(\zeta)\eta'_s(\zeta)}$, i.e., the functions g_d, f_d, and h_d in the third quadrant are conjugates of those in the fourth quadrant, and using these relations, the integrals on the right of (6.20) combine to give the dilatational contributions to the Laplace transformed displacements as

$$\begin{Bmatrix}\bar u_d(x,z,p)\\ \bar w_d(x,z,p)\end{Bmatrix}=\frac{\sigma_0\bar F(p)}{\pi\mu}\int_{C_4}Re\left\{\begin{bmatrix}f_d(\zeta)\\h_d(\zeta)\end{bmatrix}\exp\left[-pg_d(\zeta)\right]d\zeta\right\}. \quad (6.22)$$

Now using (6.18) for ζ along C_4, i.e.,

---

[5] Note that the first of (6.21) involves the real functions $g_d(\zeta)$ and $g_d(-\bar\zeta)$, i.e., real functions of the complex variables ζ and $-\bar\zeta$, both equal to t real and positive.

$$\zeta = \omega_d(t) = [(c_d t/r)^2 - 1]^{\frac{1}{2}} \cos\theta - i(c_d t/r) \sin\theta, \qquad r/c_d \leq t < \infty, \qquad (6.23)$$

and introducing this, and $g_d(\zeta) = t$, in (6.22) it becomes

$$\begin{Bmatrix} \bar{u}_d(x,z,p) \\ \bar{w}_d(x,z,p) \end{Bmatrix} = \frac{\sigma_0 \bar{F}(p)}{\pi\mu} \int_{r/c_d}^{\infty} \mathrm{Re}\left\{ \begin{bmatrix} f_d(\omega_d) \\ h_d(\omega_d) \end{bmatrix} \frac{d\omega_d(t)}{dt} \right\} \exp(-pt) dt. \quad (6.24)$$

The process is basically the same for the g_s integrals in (6.31). We have

$$t = g_s(\zeta) = (1/c_d)[(\zeta^2 + k^2)^{\frac{1}{2}} z + i\zeta x], \qquad (6.25)$$

and therefore find the new path of integration to be

$$\zeta = \pm [(c_s t/r)^2 - 1]^{\frac{1}{2}} k \cos\theta - i(c_d t/r) \sin\theta, \qquad r/c_s \leq t < \infty, \quad (6.26)$$

with $[(c_s t/r)^2 - 1]^{\frac{1}{2}} \geq 0$, $0 \leq \theta \leq \pi/2$. Equation (6.26) is again a branch of a hyperbola, similar to (6.19), lying in the lower half ζ-plane, with asymptotes Im $\zeta / \mathrm{Re}\, \zeta = \pm \tan\theta$. The initial time is now $t = r/c_s$, which gives $\zeta = -ik \sin\theta$ as the vertex of the branch. Hence, provided $0 \leq k \sin\theta < 1$ or $0 \leq \theta < \sin^{-1}(1/k) = \beta_{cr}$, (6.26) does not intersect the cut, and the situation is just the same as in the g_d integrals, i.e., fig. 6.4 applies again, with the simple changes just mentioned. It follows that our fourth quadrant path is given by

$$\zeta = \omega_s(t) = [(c_s t/r)^2 - 1]^{\frac{1}{2}} k \cos\theta - i(c_d t/r) \sin\theta, \qquad r/c_s \leq t < \infty, \quad (6.27)$$

and that the equivoluminal contributions to the Laplace transformed displacements, for this case ($0 \leq k \sin\theta < 1$), are contained in

$$\begin{Bmatrix} \bar{u}_s(x,z,p) \\ \bar{w}_s(x,z,p) \end{Bmatrix} = \frac{\sigma_0 \bar{F}(p)}{\pi\mu} \int_{r/c_s}^{\infty} \mathrm{Re}\left\{ \begin{bmatrix} f_s(\omega_s) \\ h_s(\omega_s) \end{bmatrix} \frac{d\omega_s(t)}{dt} \right\} \exp(-pt) dt.$$
$$(6.28)$$

When $1 < k \sin\theta < k$, or $\sin^{-1}(1/k) \leq \theta < \pi/2$, the Cagniard contour (6.27) will intersect the cut, so this contour must be *supplemented* by a path, say H, that approaches but circumvents this cut and joins the Cagniard contour C_3, C_4, as shown in fig. 6.5.

The lineal paths of H, along the cut, meet the contour C_3, C_4 at $\zeta = -ik \sin\theta$, corresponding to the time $t = r/c_s$, as fig. 6.5 shows. Now along

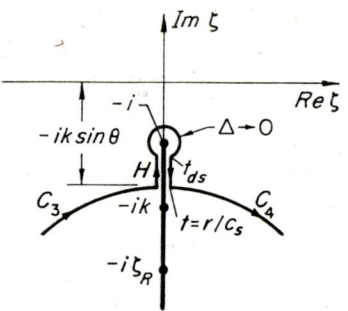

Fig. 6.5. Supplemental path H.

these paths, ζ becomes negative imaginary in the limit (as the cut is approached). Further, $|\zeta|$ decays along these paths from $\zeta = -ik \sin \theta$ toward the ends at the branch point $\zeta = -i$. It follows from (6.27) that these paths are represented by

$$\zeta = \omega_{ds}(t) = i\{[1-(c_s t/r)^2]^{\frac{1}{2}} k \cos \theta - (c_d t/r)\sin \theta\}, \quad (6.29)$$

showing that $t \leq r/c_s$. The minimum time t_{ds} is reached at the branch point $\zeta = -i$, hence substituting the latter in (6.25) we find,

$$t_{ds} = (1/c_d)[(k^2-1)^{\frac{1}{2}} z + x] = (r/c_d)[\sin \theta + (k^2-1)^{\frac{1}{2}} \cos \theta]. \quad (6.30)$$

Letting $\zeta = -i + \Delta \exp(i\delta)$, on the circular path of H, we can assess its possible contribution in the present case. From (6.15) we have for these g_s integrals on this path,

$$\lim_{\Delta \to 0} i\Delta \int_{-\pi/2}^{3\pi/2} \begin{bmatrix} f_s[-i+\Delta \exp(i\delta)] \\ h_s[-i+\Delta \exp(i\delta)] \end{bmatrix} \exp[-pg_s(-i+\Delta e^{i\delta}) + i\delta] d\delta. \quad (6.31)$$

Now as $\Delta \to 0$,

$$g_s(-i+\Delta e^{i\delta}) = (1/c_d)[(k^2-1)^{\frac{1}{2}} z + x] = t_{ds},$$

$$f_s(-i+\Delta e^{i\delta}) \simeq -2(-2i\Delta e^{i\delta})^{\frac{1}{2}}(k^2-1)^{\frac{1}{2}}/(k^2-2)^2,$$

$$h_s(-i+\Delta e^{i\delta}) \simeq 2(-2i\Delta e^{i\delta})^{\frac{1}{2}}/(k^2-2)^2,$$

and therefore the integrals (6.31) vanish like $\Delta^{\frac{3}{2}}$. It follows, again taking into account the conjugate properties of the functions f_s, g_s, and h_s on the lineal paths of H, that this contour yields the further equivoluminal contributions to the Laplace transformed displacements,

$$\begin{Bmatrix} \bar{u}_{ds}(x,z,p) \\ \bar{w}_{ds}(x,z,p) \end{Bmatrix} = \frac{\sigma_0 \bar{F}(p)}{\pi\mu} \int_{t_{ds}}^{r/c_s} \mathrm{Re}\left\{ \begin{bmatrix} f_s(\omega_{ds}) \\ h_s(\omega_{ds}) \end{bmatrix} \frac{d\omega_{ds}(t)}{dt} \right\} e^{-pt} dt, \quad (6.32)$$

where ω_{ds} and t_{ds} are given in (6.29) and (6.30), respectively.

To evaluate the integrals in (6.24), (6.28), and (6.32) we must write the integrands there completely in terms of t. This can be done by making use of the fact that our Cagniard integration paths are all in the fourth quadrant of the ζ-plane where $\mathrm{Re}\,\eta'_{d,s} \geqq 0$ and $\mathrm{Im}\,\eta'_{d,s} \leqq 0$. Therefore from (6.16) and (6.23)

$$\eta'_d(\omega_d) = (\omega_d^2 + 1)^{\frac{1}{2}} = (1/z)(c_d t - i\omega_d x) = (c_d t/r)\cos\theta - i[(c_d t/r)^2 - 1]^{\frac{1}{2}}\sin\theta. \quad (6.33a)$$

Also, we have

$$\eta'_s(\omega_d) = (\omega_d^2 + k^2)^{\frac{1}{2}}. \quad (6.33b)$$

Similarly, from (6.25) and (6.27) we have

$$\eta'_s(\omega_s) = (c_d t/r)\cos\theta - ik[(c_s t/r)^2 - 1]^{\frac{1}{2}}\sin\theta. \quad (6.34a)$$

Also

$$\eta'_d(\omega_s) = (\omega_s^2 + 1)^{\frac{1}{2}}. \quad (6.34b)$$

And from (6.25) and (6.29),

$$\eta'_s(\omega_{ds}) = (c_d t/r)\cos\theta + k[1 - (c_s t/r)^2]^{\frac{1}{2}}\sin\theta. \quad (6.35a)$$

Also

$$\eta'_d(\omega_{ds}) = (\omega_{ds}^2 + 1)^{\frac{1}{2}}. \quad (6.35b)$$

The terms $R(\omega_d)$, $R(\omega_s)$ and $R(\omega_{ds})$, are written from $R(\zeta)$ in (6.15) with the aid of (6.23), (6.27), (6.29), (6.33), (6.34) and (6.35).

Because of the circular cylindrical nature of the dilatational and equivoluminal wave fronts in the present problem, it is natural to seek information on the corresponding cylindrical displacement components u_r and u_θ, since they are directed normal and tangential, respectively, to these fronts. The Laplace transforms for u_r and u_θ are easily written from (6.24), (6.28) and (6.32), with the aid of the expressions

$$u_r = u \sin\theta + w \cos\theta,$$

$$u_\theta = u \cos\theta - w \sin\theta,$$
(6.36)

and (6.15), (6.23), (6.27), (6.29), (6.33a) and (6.35a). The resulting expressions are

$$\bar{u}_r(r, \theta, p) = \bar{u}_{rd} + \bar{u}_{rs} + \bar{u}_{rds}$$

$$= \frac{\sigma_0 \bar{F}(p)}{\pi\mu} \left\{ \int_{r/c_d}^\infty \frac{c_d t}{r} \operatorname{Re}\left[\frac{(k^2 + 2\omega_d^2)}{R(\omega_d)} \cdot \frac{d\omega_d}{dt} \right] e^{-pt} dt \right.$$

$$+ \int_{r/c_s}^\infty k[(c_s t/r)^2 - 1]^{\frac{1}{2}} \operatorname{Re}\left[\frac{-2\omega_s \eta_d'(\omega_s)}{R(\omega_s)} \cdot \frac{d\omega_s}{dt} \right] e^{-pt} dt \quad (6.37)$$

$$+ \int_{t_{d_c}}^{r/c_s} k[1 - (c_s t/r)^2]^{\frac{1}{2}} \operatorname{Re}\left[\frac{-i2\omega_{ds}\eta_d'(\omega_{ds})}{R(\omega_{ds})} \cdot \frac{d\omega_{ds}}{dt} \right] e^{-pt} dt \right\},$$

$$\bar{u}_\theta(r, \theta, p) = \bar{u}_{\theta d} + \bar{u}_{\theta s} + \bar{u}_{\theta ds}$$

$$= \frac{\sigma_0 \bar{F}(p)}{\pi\mu} \left\{ \int_{r/c_d}^\infty [(c_d t/r)^2 - 1]^{\frac{1}{2}} \operatorname{Re}\left[\frac{i(k^2 + 2\omega_d^2)}{R(\omega_d)} \cdot \frac{d\omega_d}{dt} \right] e^{-pt} dt \right.$$

$$+ \int_{r/c_s}^\infty \frac{c_d t}{r} \operatorname{Re}\left[\frac{-i2\omega_s \eta_d'(\omega_s)}{R(\omega_s)} \cdot \frac{d\omega_s}{dt} \right] e^{-pt} dt \quad (6.38)$$

$$+ \int_{t_{d_c}}^{r/c_s} \frac{c_d t}{r} \operatorname{Re}\left[\frac{-i2\omega_{ds}\eta_d'(\omega_{ds})}{R(\omega_{ds})} \cdot \frac{d\omega_{ds}}{dt} \right] e^{-pt} dt \right\}.$$

According to our analysis, (6.37) and (6.38) are valid in the region $0 < r < \infty$, $0 \leq \theta < \pi/2$, the third integral occurring only for $\sin^{-1}(1/k) \leq \theta < \pi/2$. This then is the *transformed solution for the "interior" of the half space*.

For the solution representing the surface response we return to (6.15), and now using (6.36), with $\theta = \pi/2$, we find

$$\bar{u}_r(r, \pi/2, p) = \bar{u}(r, \pi/2, p)$$

$$= \frac{\sigma_0 \bar{F}(p)}{2\pi\mu} \int_{-\infty}^\infty \frac{i\zeta[(k^2 + 2\zeta^2) - 2\eta_d'(\zeta)\eta_s'(\zeta)]}{R(\zeta)} e^{-ipr\zeta/c_d} d\zeta, \quad (6.39a)$$

$$\bar{u}_\theta(r, \pi/2, p) = -\bar{w}(r, \pi/2, p)$$

$$= -\frac{\sigma_0 \bar{F}(p) k^2}{2\pi\mu} \int_{-\infty}^\infty \frac{\eta_d'(\zeta)}{R(\zeta)} e^{-ipr\zeta/c_d} d\zeta. \quad (6.39b)$$

Ch. 6, § 6.2] PLANE STRAIN PROBLEMS 311

Now substituting $\theta = \pi/2$ into (6.23), (6.27) and (6.29) we have $\zeta = -ic_d t/r$, with $r/c_d \leq t < \infty$, $r/c_s \leq t < \infty$ and $t_{ds} \leq t \leq r/c_s$, respectively. Clearly then, this case has the Cagniard contours for the dilatational and equivoluminal wave integrals wrapped around the branch cut shown in fig. 6.4. In the case of the dilatational wave integrals the circular paths at $|\zeta| \to \infty$ are quarter circles, and C_3 goes up to $\zeta = -i$, just to the left of the cut, and C_4 returns just to the right of the cut. It is similar for the equivoluminal wave integrals, taken with the "sub ds" integrals. We see therefore the solution for u_r and u_θ on the surface may be written by first calculating the contributions for the Rayleigh simple pole at $\zeta = -i\zeta_R$ for these displacements, and adding these to the integrals in (6.37) and (6.38), the first two in each case now being

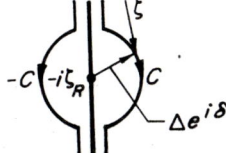

Fig. 6.6. Integration around the Rayleigh pole.

Cauchy principal values. The principal values arise because the Rayleigh singularity occurs along $\zeta = \omega_d = \omega_s = -ic_d t/r$ when $t = r/c_R$, hence occurs along the real time paths of integration.

It is simple to calculate the contributions of the Rayleigh pole. Let $\zeta = -i\zeta_R + \Delta e^{i\delta}$, as shown in fig. 6.6. Again here the conjugate properties of our functions apply. It follows that

$$\bar{u}_{rR}(r, \pi/2, p) = \frac{\sigma_0 \bar{F}(p)}{\pi\mu} \lim_{\Delta \to 0} \int_C \text{Re}\left\{\begin{bmatrix} \text{integrand} \\ \text{in (6.39a)} \end{bmatrix} d\zeta\right\}, \quad (6.40)$$

which, noting that near $\zeta = -i\zeta_R$, $R(\zeta) \simeq \Delta e^{i\delta} R'(-i\zeta_R)$, where $R'(-i\zeta_R) = [dR\, d\zeta]_{\zeta = -i\zeta_R}$, reduces to

$$\bar{u}_{rR}(r, \pi/2, p) = -\frac{k^2(k^2 - 2\zeta_R^2)\sigma_0}{2\mu\zeta_R[R'(-i\zeta_R)/i]} \bar{F}(p) e^{-pr/c_R}, \quad (6.41)$$

where

$$R'(-i\zeta_R)/i = -2[8(k^2-1)\zeta_R^6 - 4k^6\zeta_R^2 + k^8]/\zeta_R(k^2 - 2\zeta_R^2)^2,$$

a real quantity. Noting that the integrand numerator in (6.39a), as $\zeta \to -i\zeta_R$, is real, and the corresponding $\eta'_d(\zeta)$ in (6.39b) is imaginary, we conclude that the contribution from the Rayleigh pole in this case is zero, i.e.,

$$[\bar{u}_\theta(r, \pi/2, p)]_{R.P.} = 0. \tag{6.42}$$

Our *transformed solution for the surface of the half space*, $\theta = \pi/2$, *is therefore*

$$\bar{u}_r(r, \pi/2, p) = [\text{R.H.S. (6.41)}] + \frac{\sigma_0 \bar{F}(p)}{\pi\mu}\left[\sum \begin{array}{c}\text{integrals}\\ \text{in (6.37)}\end{array}\right], \tag{6.43}$$

$$\bar{u}_\theta(r, \pi/2, p) = \frac{\sigma_0 \bar{F}(p)}{\pi\mu}\left[\sum \begin{array}{c}\text{integrals}\\ \text{in (6.38)}\end{array}\right], \tag{6.44}$$

where the first two integrals in each case are Cauchy principal values.

The transformed "interior" solution (6.37) and (6.38), and "surface" solution (6.43) and (6.44) are integral equations for the unknown inverses $u_r(r, \theta, t)$ and $u_\theta(r, \theta, t)$. As deHoop points out these *integral equations* have to be satisfied for all real values of p greater than some fixed positive number. Then the solutions, or inverses, of these equations are unique, according to Lerch's theorem. Recall this was built into $\bar{u}_r$ and $\bar{u}_\theta$ by requiring p to be real and positive, permitting the selection of the real and positive branches of η_d and η_s in (6.8), as pointed out in the discussion after (6.7). Now setting $F(t) = \delta(t)$, so that $\bar{F}(p) = 1$, inspection of (6.37), (6.38) shows the corresponding inverses u_r and u_θ of the interior solution are

$$u_r^\delta(r, \theta, t) = \frac{\sigma_0}{\pi\mu}\left\{\frac{c_d t}{r}\,\text{Re}\left[\frac{(k^2 + 2\omega_d^2)}{R(\omega_d)} \cdot \frac{d\omega_d}{dt}\right] H(t - r/c_d)\right.$$

$$+ k[(c_s t/r)^2 - 1]^{\frac{1}{2}}\,\text{Re}\left[\frac{-2\omega_s \eta'_d(\omega_s)}{R(\omega_s)} \cdot \frac{d\omega_s}{dt}\right] H(t - r/c_s)$$

$$\left. + k[1 - (c_s t/r)^2]^{\frac{1}{2}}\,\text{Re}\left[\frac{-2i\omega_{ds}\eta'_d(\omega_{ds})}{R(\omega_{ds})} \cdot \frac{d\omega_{ds}}{dt}\right][H(t - t_{ds}) - H(t - r/c_s)]\right\}, \tag{6.45}$$

$$u_\theta^\delta(r, \theta, t) = \frac{\sigma_0}{\pi\mu}\left\{[(c_d t/r)^2 - 1]^{\frac{1}{2}}\,\text{Re}\left[\frac{i(k^2 + 2\omega_d^2)}{R(\omega_d)} \cdot \frac{d\omega_d}{dt}\right] H(t - r/c_d)\right.$$

$$+ \frac{c_d t}{r}\,\text{Re}\left[\frac{-2i\omega_s \eta'_d(\omega_s)}{R(\omega_s)} \cdot \frac{d\omega_s}{dt}\right] H(t - r/c_s)$$

$$\left. + \frac{c_d t}{r}\,\text{Re}\left[\frac{-2i\omega_{ds}\eta'_d(\omega_{ds})}{R(\omega_{ds})} \cdot \frac{d\omega_{ds}}{dt}\right][H(t - t_{ds}) - H(t - r/c_s)]\right\}, \tag{6.46}$$

Ch. 6, § 6.2]   PLANE STRAIN PROBLEMS   313

where the super δ on the u's means response to the delta function, and the H functions are steps, defined in (2.90). Equations (6.45) and (6.46) are valid in $0<r<\infty$, $0\leq\theta<\pi/2$, except that the third terms in each occur only in the region $\beta_{cr}\leq\theta<\pi/2$.

At the surface $\theta=\pi/2$, the inverses, from (6.43) and (6.44), are

$$u_r^\delta(r,\pi/2,t)=\begin{bmatrix}\text{R.H.S. of (6.45)}\\ \text{evaluated at }\theta=\pi/2\end{bmatrix}-\frac{k^2(k^2-2\zeta_R^2)\sigma_0}{2\mu\zeta_R[R'(-i\zeta_R)/i]}\delta(t-r/c_R),$$
(6.47)

$$u_\theta^\delta(r,\pi/2,t)=\begin{bmatrix}\text{R.H.S. of (6.46)}\\ \text{evaluated at }\theta=\pi/2\end{bmatrix},$$
(6.48)

valid for $0<r<\infty$.

The solution contained in (6.45) through (6.48) can be extended to the more general time input $F(t)$, originally in the problem, with the aid of the convolution theorem given in (5.26). That is,

$$\begin{bmatrix}u_r^F(r,\theta,t)\\ u_\theta^F(r,\theta,t)\end{bmatrix}=\int_0^t F(t-\tau)\begin{bmatrix}u_r^\delta(r,\theta,\tau)\\ u_\theta^\delta(r,\theta,\tau)\end{bmatrix}d\tau,$$
(6.49)

where u_r^δ and u_θ^δ are given by (6.45) through (6.48).

The solution (6.45)–(6.48) for Lamb's problem of the surface normal line load source of delta function time behavior shows that such a source generates four different types of waves that establish the character of the half-space motion. One of these is a circular cylindrical *dilatational wave*, the front of which radiates out from the source ($r=0$) with speed c_d. The second is a circular cylindrical *equivoluminal wave*, the front of which also radiates out from the source, but with speed c_s. The third is an *equivoluminal wave, commonly referred to as a head or von Schmidt wave*. It is generated by grazing incidence of the circular cylindrical dilatational wave at the surface $z=0$. This plane fronted head wave occurs only in the region $\beta_{cr}\leq\theta\leq\pi/2$ traveling with speed c_s along the normal to the front, which makes the angle β_{cr} with the z-axis. The occurrence of the head wave in this region introduces equivoluminal deformation in front of the cylindrical equivoluminal wave, in effect making the latter two-sided at $r=c_s t$. Lastly there is a *singular Rayleigh surface wave* disturbance which travels out from the source at speed c_R. This disturbance also occurs in the interior where it is non-singular. Generally it is a two-sided disturbance, and hence not a wave in the ordinary sence since it lacks a front. Figure 6.7, exploiting the symmetry of the problem with respect to the z-axis, contains a sketch of these waves, for fixed arbitrary time t, showing the parts they play in the solution (6.45)–(6.48).

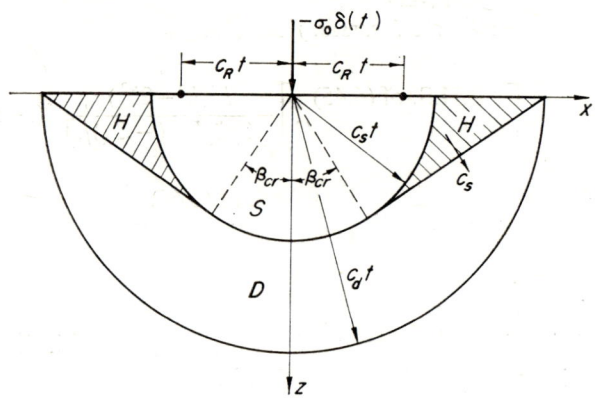

Fig. 6.7. The nature of disturbed region of half space (for fixed time) for surface normal line load source.

The circular cylindrical dilatational wave covers the region D, defined by $0 < r \leq c_d t$. Similarly the circular cylindrical equivoluminal wave covers region S, defined by $0 < r \leq c_s t$. The head waves are confined to the cross hatched regions H, defined on the right by $c_s t \leq r \leq c_d t/(\sin\theta + (k^2-1)^{\frac{1}{2}}\cos\theta)$, and $\beta_{cr} \leq \theta \leq \pi/2$, with a symmetrical definition on the left. Since the lower limit of $\theta = \beta_{cr}$, the situation at the surface point $z=0$, $r=c_d t$, is the transient wave equivalent of the time harmonic P-wave *grazing incidence* solution, given by (3.39) and (3.40), and shown in fig. 3.13. Note, since $z=0$, the reflected P wave there, being linearly dependent on z, vanishes at $z=0$, and here too we have no such wave. The progress of the two-sided Rayleigh wave disturbance is shown on the surface.

6.2.1.2. Evaluation of exact solution for the surface response. It is of interest to evaluate the surface solution (6.47) and (6.48). Since $\theta = \pi/2$, from (6.23), (6.27), (6.29) and (6.30) we have

$$\begin{Bmatrix} \omega_d \\ \omega_s \\ \omega_{ds} \end{Bmatrix} = -\frac{ic_d t}{r}, \quad \begin{cases} r/c_d \leq t < \infty, \\ r/c_s \leq t < \infty, \\ t_{ds} \leq t \leq r/c_s, \end{cases} \quad (6.50)$$

Ch. 6, § 6.2]   PLANE STRAIN PROBLEMS   315

where $t_{ds} = r/c_d$. It follows from (6.50), and (6.33), (6.34) and (6.35), taking into account the fact that Re $\eta'_{d,s} \geq 0$, Im $\eta'_{d,s} \leq 0$, as discussed there, that

$$\eta'_d(\omega_d) = -i[(c_d t/r^2) - 1]^{\frac{1}{2}}, \qquad r/c_d \leq t < \infty,$$

$$\eta'_s(\omega_d) = \begin{cases} k[1 - (c_s t/r)^2]^{\frac{1}{2}}, & r/c_d \leq t < r/c_s, \\ -ik[(c_s t/r)^2 - 1]^{\frac{1}{2}}, & r/c_s \leq t < \infty, \end{cases}$$

$$\left.\begin{array}{l} \eta'_d(\omega_s) = -i[(c_d t/r)^2 - 1]^{\frac{1}{2}} \\ \eta'_s(\omega_s) = -ik[(c_s t/r)^2 - 1]^{\frac{1}{2}} \end{array}\right\}, \qquad r/c_s \leq t < \infty, \qquad (6.51)$$

$$\left.\begin{array}{l} \eta'_d(\omega_{ds}) = -i[(c_d t/r)^2 - 1]^{\frac{1}{2}} \\ \eta'_s(\omega_{ds}) = k[1 - (c_s t/r)]^{\frac{1}{2}} \end{array}\right\}, \qquad r/c_d \leq t \leq r/c_s.$$

The quantities $R(\omega_d)$, $R(\omega_s)$, and $R(\omega_{ds})$ are easily written from (6.50) and (6.51). Now for $r/c_d \leq t < r/c_s$, only the first and third terms of (6.45), (6.46), contribute to the surface solution given in (6.47), (6.48). For the later region $t > r/c_s$ the third terms in (6.45), (6.46) vanish, but now the second terms, as well as the first, must be taken into account. Under these conditions, substitution of (6.50), (6.51) into (6.47), (6.48) reduce the latter to the dimensionless forms

$$\frac{\pi\mu r}{c_s \sigma_0} u_r^\delta(r, \pi/2, t) = \frac{\pi\mu x}{c_s \sigma_0} u^\delta(x, 0, \tau)$$

$$= 2k^3 \left\{ \frac{\tau STU}{R_1} [H(\tau - 1) - H(\tau - k)] - \frac{\pi(2 - k_R^2)^3}{8G} \delta(\tau - \zeta_R) \right\}, \qquad (6.52)$$

$$\frac{\pi\mu r}{c_s \sigma_0} u_\theta^\delta(r, \pi/2, t) = -\frac{\pi\mu x}{c_s \sigma_0} w^\delta(x, 0, \tau)$$

$$= k^3 \left\{ \frac{S^2 T}{R_1} [H(\tau - 1) - H(\tau - k)] + \frac{T}{R_2} H(\tau - k) \right\}, \qquad (6.53)$$

where

$$R_1 = S^4 + 16\tau^4 T^2 U^2, \qquad R_2 = S^2 - 4\tau^2 TV,$$

$$S = k^2 - 2\tau^2, \qquad T = (\tau^2 - 1)^{\frac{1}{2}},$$

$$U = (k^2 - \tau^2)^{\frac{1}{2}}, \qquad V = (\tau^2 - k^2)^{\frac{1}{2}},$$

$$\tau = c_d t/x, \qquad G = 8(k^2 - 1) - 4k^2 k_R^4 + k^2 k_R^6.$$

The surface solution (6.52), (6.53) is in agreement with deHoop's solution for the problem. The left hand sides become dimensionless if σ_0 for the present case ($\delta(t)$ input) is taken as impulse per unit length. Forrestal et al. [6.3] have presented numerical evaluations of this solution for three values of Poisson's ratio; $v=0$, 0.25 and 0.40. Their results are reproduced here in fig. 6.8. It is obvious the solution is dominated by the *Rayleigh wave*

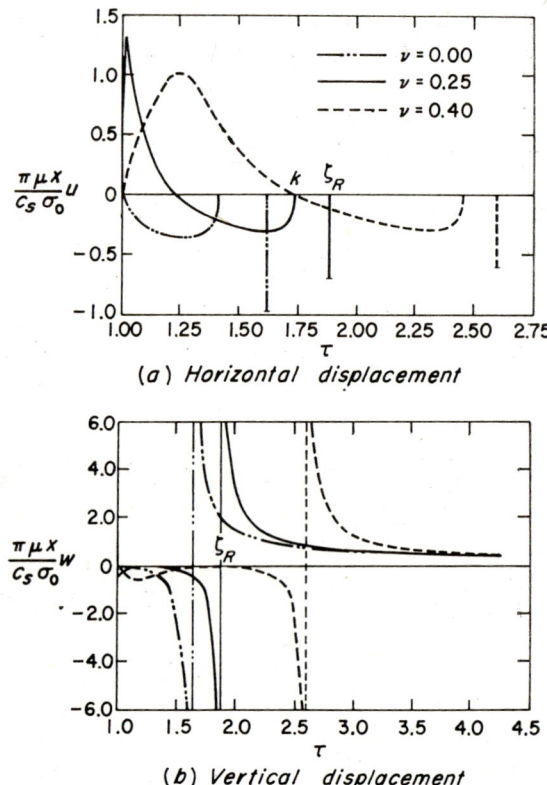

Fig. 6.8. Surface displacements as function of time in Lamb's problem for surface normal line load source (after Forrestal, Fugelso, Neidhardt and Felder).

singularities, which arrive at $\tau = \zeta_R$ (indicated in fig. 6.8 for $v=0.25$ curves). In the case of $u_r^\delta = u^\delta$ this term is explicit in the solution, and we can see it is *nondecaying in space (x) and time, and does not change its shape as it propagates*. The delta function singularities in the figure are represented by vertical straight lines, their lengths representing the "strengths" or coefficients of the delta functions in (6.52).[6]

The corresponding singularity of $u_\theta^\delta = -w^\delta$ is contained in the second term of (6.53) due to the vanishing of R_2 at $t = x/c_R (\tau = \zeta_R)$. The singularity can be determined by expanding R_2 in a Taylor series about $t = r/c_R$. We find $R_2(t) \simeq (t - r/c_R) R_2'(r/c_R)$ which leads to

$$\lim_{t \to (r/c_R)\pm} \left[\frac{\pi \mu r}{c_d \sigma_0} u_\theta^\delta(r, \pi/2, t) \right] = -\frac{k_R^2(1 - b^2 k_R^2)}{Q(k_R)(\tau - \zeta_R)}$$

$$= \frac{\pi \mu r}{c_d \sigma_0} u_{\theta R}^\delta(r, \pi/2, t), \qquad \text{for } \frac{r}{c_R} - \varepsilon < t < \frac{r}{c_R} + \varepsilon, \qquad (6.54)$$

where $\varepsilon \ll r/c_R$ and

$$Q(k_R) = 2\{8[2 - (1 + b^2)k_R^2] + 4(k_R^2 - 2)^3 + (k_R^2 - 2)^4\} / (k_R^2 - 2)^2.$$

We note also that $(\pi \mu r / c_d \sigma_0) u_{\theta R}^\delta(r, \pi/2, t) = -(\pi \mu x / c_d \sigma_0) w_R^\delta(x, 0, t)$. Equation (6.54) shows the Rayleigh surface wave disturbance for $u_{\theta R}^\delta = -w_R^\delta$ behaves as $-1/(t - r/c_R)$, as against the delta function behavior of $u_{rR}^\delta = u_R^\delta$. Note that it, like u_{rR}^δ is nondecaying in space and time and has a nonchanging shape (once it is free of the other disturbances as in the far field). Both are *two-sided phenomena*, as mentioned earlier, $u_{\theta R}^\delta$ being odd, and u_{rR}^δ even, through the arrival time $t = r/c_R$. It should be pointed out that the nondecaying nature of the Rayleigh disturbance here is attributable to the two-dimensional (spatial) nature of this problem, rather than the singular nature of the disturbance, i.e., nonsingular Rayleigh disturbances, resulting from smoother inputs $F(t)$, are also nondecaying and nonchanging in shape in the present problem. This nondecaying nature determines a further important property of the *Rayleigh disturbance*, namely that it *predominates at surface positions remote from the source (far field)* in this basic two-dimensional half space problem. Note, for instance, in (6.52), (6.53), (and fig. 6.8) that the other contributions to the displacements decay like $1/x$ in

---

[6] In [6.3] the Rayleigh wave contribution to (6.52) was calculated differently from the way it was done here and in deHoop's work [6.2], and the resulting coefficients are algebraically different. However, independent numerical calculations, made here for these coefficients in (6.52), agree with those of [6.3] shown in fig. 6.8 The rest of the solution in [6.3] agrees with (6.52), (6.53).

the far field for large time. Lamb was the first to set down these and other important features of the Rayleigh disturbance [6.1]. Using (6.49) one can show that for the step input, $F(t) = H(t)$, $u_{rR}^H(r, \pi/2, t)$ behaves as $H(t - r/c_R)$, which one-sided can be considered odd through $t = r/c_R$, and $u_{\theta R}^H(r, \pi/2, t)$ as $\log |t - r/c_R|$, even through $t = r/c_R$. This is left to the exercises.

6.2.1.3. Evaluation of exact solution for the response at the plane of symmetry. It is also of interest to evaluate the response on the plane of symmetry $\theta = 0$. From (6.36) we have

$$u_r = w, \qquad u_\theta = u. \tag{6.55}$$

Also from (6.23), (6.27)

$$\omega_d(\tau) = (\tau^2 - 1)^{\frac{1}{2}}, \qquad 1 \leq \tau < \infty,$$
$$\omega_s(\tau) = (\tau^2 - k^2)^{\frac{1}{2}}, \qquad k \leq \tau < \infty, \tag{6.56}$$

where now $\tau = c_d t/r = c_d t/z$. Using in addition, (6.33), (6.34) and $R(\zeta)$ (6.15), in the interior solution (6.45), (6.46), we find that

$$\frac{\pi\mu r}{c_s \sigma_0} u_r^\delta(r, 0, t) = \frac{\pi\mu z}{c_s \sigma_0} w^\delta(0, z, \tau) = k \left\{ \tau^2 \left[\frac{k^2 + 2\omega_d^2}{\omega_d R(\omega_d)} \right] H(\tau - 1) \right.$$
$$\left. - 2\tau \left[\frac{\omega_s \eta_d'(\omega_s)}{R(\omega_s)} \right] H(\tau - k) \right\}, \tag{6.57}$$

$$u_\theta^\delta(r, 0, t) = u^\delta(0, z, \tau) = 0.$$

This solution agrees with deHoop's [6.2], and with that in Forrestal et al. [6.3], except the latter seem to have lost the factor k in the second term of (6.57) (cf. their equation [18]). It is simple to show, directly from (6.57), that in the neighborhood of $\tau = 1$,

$$(\pi\mu z/c_s \sigma_0) w_d^\delta(0, z, 1+) \simeq k^{-1}[2(\tau - 1)]^{-\frac{1}{2}}, \tag{6.58a}$$

and that the behavior of the equivoluminal part of the displacement, near $\tau = k+$, is

$$(\pi\mu z/c_s \sigma_0) w_s^\delta(0, z, k+) \simeq -2k^{-\frac{3}{2}}[2(\tau - k)]^{\frac{1}{2}}. \tag{6.58b}$$

Numerical evaluation of (6.57) was carried out in [6.3] for $v=1/4$, which is reproduced here in fig. 6.9.[7]

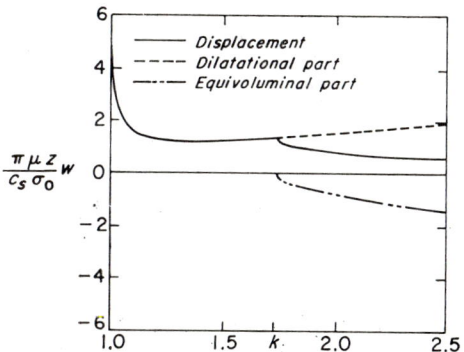

Fig. 6.9. Displacement reponse at plane of symmetry $x=0$ in Lamb's problem for for the surface normal line load source (after Forrestal, Fugelso, Neidhardt and Felder).

6.2.1.4. Wavefront approximations in the Cagniard–deHoop method. The wavefront approximations (6.58) made in the previous section were obtained directly from the exact solution (6.57). This of course can always be done if one has the exact solution to a problem. To carry out a general analysis of wavefronts in the Cagniard—deHoop method without getting the exact solution first, we proceed as in Rosenfeld and Miklowitz [6.4]. Consider $\bar{u}_d(x, z, p)$ in (6.24). Since $\zeta(t)=\omega_d(t)$ [given in (6.23)], $\omega'_d(t) = 1/g'_d(\zeta)$, and (6.24) yields

$$u_d^\delta(x, z, t) = (\sigma_0/\pi\mu)\text{Re}\{f_d[\zeta(t)]/g'_d[\zeta(t)]\}H(t-r/c_d). \qquad (6.59)$$

To approximate u_d^δ near its front we note that there $t=g_d(\zeta)=r/c_d+\varepsilon$, where ε satisfies $0 \leq \varepsilon \ll 1$. The point ζ, corresponding to the minimum time $t=g_d(\zeta)=r/c_d$, is given by (6.23), i.e. $\zeta=-i\zeta_0=-i\sin\theta$. It is easily shown that $g'_d(-i\zeta_0)=0$, and hence $\zeta=-i\zeta_0$ is a saddle point of the relief of $\exp[-pg_d(\zeta)]$ in (6.22), where now p is large. However, we will not, at

---

[7] The factor k error mentioned above is probably typographical, since a check on this numerical evaluation, using the corresponding wavefront expansion (6.58b), shows good agreement.

this point, exploit the method of steepest descents to obtain our approximation of u_d^δ in (6.59).[8] Instead we expand $g_d(\zeta)$ at $\zeta = -i\zeta_0$, along the C_4 path [cf. (6.22)], with the result

$$t = g_d(\zeta) = g_d(-i\zeta_0) + [g_d''(-i\zeta_0)/2](\zeta + i\zeta_0)^2 + \ldots \qquad (6.60)$$
$$\simeq r/c_d + [g_d''(-i\zeta_0)/2](\zeta + i\zeta_0)^2 \ .$$

From (6.60)

$$\zeta + i\zeta_0 \simeq [2(t - r/c_d)/g_d''(-i\zeta_0)]^{\frac{1}{2}} , \qquad (6.61\text{a})$$

and

$$g_d'(\zeta) \simeq g_d''(-i\zeta_0)(\zeta + i\zeta_0) , \qquad (6.61\text{b})$$

from which we find

$$g_d'(\zeta) \simeq [2g_d''(-i\zeta_0)(t - r/c_d)]^{\frac{1}{2}} . \qquad (6.62)$$

It follows from (6.62) and setting $f_d(\zeta) \simeq f_d(-i\zeta_0)$, that (6.59) is approximated by

$$(\pi\mu r/c_s\sigma_0)u_d^\delta(x, z, \tau) \simeq kf_d(-i\zeta_0) \cos\theta/[2(\tau-1)]^{\frac{1}{2}} , \qquad (6.63)$$

where $\tau = c_d t/r$, and

$$f_d(-i\zeta_0) = (k^2 - 2\sin^2\theta) \sin\theta/R(-i\zeta_0) ,$$
$$R(-i\zeta_0) = (k^2 - 2\sin^2\theta)^2 + 4(k^2 - \sin^2\theta)^{\frac{1}{2}}\sin^2\theta \cos\theta ,$$

valid everywhere in the interior $0 < r < \infty$, $0 \leq \theta < \pi/2$, for $r/c_d \leq t < (r/c_d) + \varepsilon$. It may be seen that the approximation breaks down at the surface $\theta = \pi/2$, which we have ruled out. The analogous approximation for w_d^δ in (6.24) is written similarly. It is given by

$$\frac{\pi\mu r}{c_s\sigma_0} w_d^\delta(x, z, \tau) \simeq kh_d(-i\zeta_0) \cos\theta/[2(\tau-1)]^{\frac{1}{2}} , \qquad (6.64)$$

---

[8] The method of steepest descents will be used in the next section, as an alternate means of obtaining wavefront approximations, independent of the Cagniard-deHoop method.

where
$$h_d(-i\zeta_0) = (k^2 - 2\sin^2\theta)\cos\theta/R(-i\zeta_0).$$

The approximations (6.63), (6.64) reduce to the second of (6.57) and (6.58a) when θ is set equal to zero. Using (6.63), (6.64) and (6.36) it is simple to write the corresponding approximations for u_{rd}^δ, $u_{\theta d}^\delta$. *We shall refer to all these wavefronts as regular, the implication being that they are cylindrical wavefronts radiated from the source, with no disturbance* (in this case, dilatational) *occurring before them at a station.* They are associated with Cagniard paths that do not intersect the branch cut.

A similar process can be carried out for the regular equivoluminal wavefronts u_s^δ, w_s^δ. They are the fronts restricted to the θ domain $0 \leq \theta < \beta_{cr}$, which can be obtained from (6.28), (6.27). Their wavefronts are singular behaving as $(\tau-k)^{-\frac{1}{2}}$, except at $\theta = 0$ as would be expected from (6.58b). The derivations are left to the exercises.

Analogously the wavefront approximation for the *head wave* (a reflected wave) is derived from (6.32). Since $\zeta(t) = \omega_{ds}(t)$ (given in (6.29)), $\omega'_{ds}(t) = 1/g'_s(\zeta)$, and (6.32) yields

$$u_{ds}^\delta(x, z, t) = (\sigma_0/\pi\mu)\text{Re}\{f_s[\zeta(t)]/g'_s[\zeta(t)]\}[H(t-t_{ds}) - H(t-r/c_s)], \quad (6.65)$$

where t_{ds} is given by (6.30). We approximate u_{ds}^δ for the time domain $0 \leq t - t_{ds} \leq \varepsilon \ll 1$. The point ζ, corresponding to the minimum time t_{ds}, is $\zeta = -i$, since substitution of $\zeta = -i$ into (6.25) was how the definition (6.30) of t_{ds} was obtained. Hence, we expand $g_s(\zeta)$ at $\zeta = -i$, along the straight line portion of path H to the right of and parallel to the imaginary axis in the fourth quadrant of the ζ-plane, cf. fig. 6.5. We have

$$t = g_s(\zeta) = g_s(-i) + g'_s(-i)(\zeta+i) + \ldots$$
$$\simeq t_{ds} + g'_s(-i)(\zeta+i). \quad (6.66)$$

From (6.66)
$$g'_s(-i) = (t - t_{ds})/(\zeta + i),$$
where
$$g'_s(-i) = i(r/c_d)\{\sin\theta - [\cos\theta/(k^2-1)^{\frac{1}{2}}]\}.$$

Critical to our derivation is the term $(\zeta^2+1)^{\frac{1}{2}}$ which appears in $f_s(\zeta)$ given in (6.15). We note from our earlier analysis of this term (cf. fig. 6.2, and corresponding analysis given after (6.15)) that, on our fourth quadrant path of H near i,

$$(\zeta^2+1)^{\frac{1}{2}} \simeq -i[2(t-t_{ds})/|g'_s(-i)|]^{\frac{1}{2}}, \qquad (6.68)$$

where we have used (6.66). Using (6.68), and the fact that $g'_s(\zeta) \simeq g'_s(-i)$ which is obtained from (6.66), we have

$$\frac{f_s(\zeta)}{g'_s(\zeta)} \simeq \frac{f_s(-i)}{g'_s(-i)} = \frac{2(k^2-1)^{\frac{1}{2}}[2(t-t_{ds})]^{\frac{1}{2}}}{(k^2-2)^2|g'_s(-i)|^{\frac{3}{2}}}.$$

It follows our head wavefront approximation is

$$\frac{\pi\mu r}{c_s\sigma_0} u^\delta_{ds}(x,z,\tau) \simeq \frac{k(k^2-1)^{\frac{1}{2}}}{(k^2-2)^2}\left[\frac{2}{c_d|g'_s(-i)|/r}\right]^{\frac{3}{2}}(\tau-k_{ds})^{\frac{1}{2}}, \qquad (6.69)$$

where $\tau = c_d t/r$, and $k_{ds} = c_d t_{ds}/r$. It may be observed that in the region close to the ray $\theta = \beta_{cr}$, where the head wavefront becomes tangent to the equivoluminal wavefront, the magnitudes of (6.69) become very large since $g'_s(-i)$ approaches zero as $\theta \to \sin^{-1}(1/k)$ (note the corresponding large numbers in fig. 7 of [6.4]. It can be shown through an approximation to the head wave integrals of the exact solution (6.45), (6.46), in which we take $k - k_{ds} = \varepsilon$, hence $\tau = k_{ds} + \varepsilon/2 = k - \varepsilon/2$ (so that we always evaluate at a time half way between the two-sided equivoluminal and head wavefronts in this critical region of θ), that

$$u^\delta_{\theta ds} \sim (k^2\sin^2\theta - 1)^{\frac{1}{2}}/[2k(\tau - k_{ds})]^{\frac{1}{2}}, \qquad u^\delta_{rds} \sim (k^2\sin^2\theta - 1)^{\frac{1}{2}} H(\tau - k_{ds}).$$

So we see from the first of these results that we trade the spatial singularity of $g'_s(-i)$ for one in time, but the latter is admissible since it can be integrated out through a convolution with a slower rising input function (as we remarked in our discussion of the Rayleigh surface wave earlier). We therefore conclude that (6.69) breaks down in this critical region of θ.

We go on to the wavefront approximations for the two-sided equivoluminal waves. In this case, since we are again dealing with the domain $\beta_{cr} < \theta < \pi/2$, the Cagniard path intersects the branch cut, so we have the situation depicted in fig. 6.5. The two-sided equivoluminal wavefront approximation is obtained by expanding about the (saddle) point $\zeta = -i\zeta_0 = -ik\sin\theta$, where $k\sin\theta > 1$. We approximate (6.28), (6.27) for $r/c_s \leq t \leq r/c_s + \varepsilon$, and (6.32), (6.29) for $r/c_s - \varepsilon \leq t \leq r/c_s$. For the first case

Ch. 6, § 6.2]   PLANE STRAIN PROBLEMS   323

$$t = g_s(\zeta) \simeq r/c_s + [g_s''(-i\zeta_0)/2](\zeta + i\zeta_0)^2 ,\qquad (6.70)$$

or

$$\zeta + i\zeta_0 = [2(t - r/c_s)/g_s''(-i\zeta_0)]^{\frac{1}{2}} .\qquad (6.71)$$

From (6.70), (6.71) we have

$$g_s'(\zeta) \simeq g_s''(-i\zeta_0)(\zeta + i\zeta_0) = [2g_s''(-i\zeta_0)(t - r/c_s)]^{\frac{1}{2}} .\qquad (6.72)$$

For the second case

$$\zeta + i\zeta_0 \simeq i[2(r/c_s - t)/g_s''(-i\zeta_0)]^{\frac{1}{2}}$$

and therefore

$$g_s'(\zeta) \simeq i[2g_s''(-i\zeta_0)(r/c_s - t)]^{\frac{1}{2}} .\qquad (6.73)$$

The two-sided wavefront approximation for u_s^δ is therefore given by

$$\frac{\pi\mu r}{c_s\sigma_0} u_s^\delta(x, z, \tau) \simeq \sqrt{k} \cos\theta \begin{cases} \mathrm{Re}[kf_s(-i\zeta_0)][2(\tau - k)]^{-\frac{1}{2}} , & \tau > k , \\ -\mathrm{Re}[ikf_s(-i\zeta_0)][2(k - \tau)]^{-\frac{1}{2}} , & \tau < k , \end{cases}$$

$$(6.74)$$

where

$$kf_s(-i\zeta_0) = \frac{2i \sin\theta \cos\theta (k^2 \sin^2\theta - 1)^{\frac{1}{2}}}{k(1 - 2\sin^2\theta)^2 - 4i\sin^2\theta \cos\theta (k^2\sin^2\theta - 1)^{\frac{1}{2}}} ,$$

and where we have used $g_s''(-i\zeta_0) = rc_s/c_d^2 \cos^2\theta$. We note as θ tends to the critical point $\sin^{-1}(1/k)$, the approximation tends to zero.

6.2.1.5. Wavefront approximations by the method of steepest descents. The foregoing wavefront approximations were based on the Cagniard–de Hoop method of solution in the present problem. It is instructive to apply the method of steepest descents (§ 5.10.5) in the problem, particularly since it does not depend on the Cagniard–deHoop method. Work of this nature was carried out by Knopoff and Gilbert [6.5] for Lamb's problem of the normal point load on the surface (we discuss this problem later), and later by Rosenfeld [6.6] in the present problem. As our point of departure we return to the Laplace transformed solutions for the displacements u, w in (6.15) (which differs only trivially from (6.13)). Consider the dilatational part of $\bar{u}$ given by

$$\bar{u}_d(x, z, p) = \frac{\sigma_0 \bar{F}(p)}{2\pi\mu} \int_{-\infty}^{\infty} f_d(\zeta) e^{-pg_d(\zeta)} \mathrm{d}\zeta ,\qquad (6.75)$$

for p large, since this should yield short time (following the wavefront) behavior. Recalling our discussion in § 5.10.5 here we would write

$$\exp[-pg_d(\zeta)] = \exp[u(\xi,\eta)+iv(\xi,\eta)],$$

where $\zeta = \xi + i\eta$, $\exp u$ is the relief, v is the phase of $\exp[-pg_d(\zeta)]$, $u = $ constant are level curves, and $v = $ constant are steepest paths. A saddle point is given by the root of

$$g'_d(\zeta) = (1/c_d)[z\zeta/\eta'_d(\zeta) + ix]. \qquad (6.76)$$

It is $\zeta = -i\zeta_0$, where $\zeta_0 = \sin\theta$. This makes sense since we have already seen that $\zeta = -i\zeta_0$ corresponds to the minimum time $t = r/c_d$ for dilatational waves in the present problem (cf. fig. 6.4). When it is possible, recall we choose a path of steepest descent through a saddle point. This will work only if the saddle point is not on a branch cut which fig. 6.4 shows is the case here. To find the steepest path we examine the function $g_d(\zeta)$. It may be written as

$$g_d(\eta) = (1/c_d)[(\xi^2 - \eta^2 + 1 + 2i\xi\eta)^{\frac{1}{2}}z + i(\xi + i\eta)x]. \qquad (6.77)$$

Along the imaginary axis, $\xi = 0$, and (6.77) reduces to

$$g_d(\eta) = (1/c_d)[(1-\eta^2)z - \eta x]$$

which is real for $-1 \leq \eta \leq 1$, i.e., $-p\,\text{Im}\,g_d(\zeta) = v = 0$. It follows this portion of the imaginary axis is a path of steepest descent, one, however, on which $g_d(\eta)$ decays symmetrically away from the saddle point leading to increasing values of the relief $\exp[-pg_d(\eta)]$. Hence this path climbs the hills.[9] To find the mate of this ascending path, *we ask is there another path that satisfies $g_d(\zeta) = q$ real (therefore $v = 0$ again) going through the saddle point from valley to valley?* The answer of course is yes since if $q = t$ this path is is the Cagniard path. Note, on it, $\exp[-pg_d(\zeta)]$ is a maximum at the saddle point and vanishes as $g_d(\zeta) = q \to \infty$ (the valley bottoms). We note this path of steepest descent crosses the imaginary axis normal to it at $\eta = -\sin\theta$.

---

[9] This is easily proved by substituting $\eta = -\sin\theta \pm \varepsilon$ in $g_d(\eta)$.

To determine the direction along this path we have from (5.114)

$$\zeta + i\zeta_0 = \varrho \exp(i\sigma), \tag{6.78}$$

where ϱ is real and small and σ is the angle this vector makes with the ξ-axis. According to our discussion after (5.117b), we set $\sigma=0$ in (6.78), since our steepest path at $\zeta = -i\zeta_0$ is parallel to the real ζ-axis, and we wish to have ϱ positive after passing through the saddle point from left to right.

Now $g_d(-i\zeta_0) = r/c_d$ which is a minimum since r/c_d is the minimum time. This is consistent with the fact that $\exp[-pg_d(\zeta)]$ in (6.75) must be a maximum at the saddle point $\zeta = -i\zeta_0$, p being real and positive. With $g_d(-i\zeta_0)$ being a minimum, however, the derivations in Laplace's method in § 5.10.4, based on $g(t)$ there being a maximum, must be carried out for the case of a minimum. The difference is trivial, however, since one has only to replace Laplace's variable $\eta = \pm[g(\tau) - g(t)]^{\frac{1}{2}}$ (just before (5.103)) with $\eta = \pm[g(t) - g(\tau)]^{\frac{1}{2}}$ in the derivation and proceed as before. One finds the resultant approximation (5.105b) is replaced by

$$I(x) \sim f(\tau) e^{-xg(\tau)} [2\pi/xg''(\tau)]^{\frac{1}{2}} \qquad \text{as } x \to \infty, \tag{6.79}$$

since $g''(\tau) > 0$ now. The derivation in the method of steepest descents approximation, which draws on (6.79), changes accordingly. The minus signs on the right of (5.113), (5.115) are dropped, and accordingly $g''(\tau)\exp(2i\sigma)$ there must now be real and positive. It follows that (5.118) must be replaced by

$$I(x) \sim f(\tau) e^{-xg(\tau)+i\sigma} [2\pi/xg''(\tau)]^{\frac{1}{2}} \qquad \text{as } x \to \infty. \tag{6.80}$$

Hence, from (6.80) we have

$$\bar{u}_d(x, z, p) \sim \frac{\sigma_0 \bar{F}(p)}{2\pi\mu} f_d(-i\zeta_0) e^{-pg_d(-i\zeta_0)} \left[\frac{2\pi}{pg_d''(-i\zeta_0)}\right]^{\frac{1}{2}} \qquad \text{as } p \to \infty. \tag{6.81}$$

where

$$f_d(-i\zeta_0) = \frac{(k^2 - 2\sin^2\theta)\sin\theta}{(k^2 - 2\sin^2\theta)^2 + 4(k^2 - \sin^2\theta)^{\frac{1}{2}}\sin^2\theta\cos\theta},$$

$$g_d(-i\zeta_0) = r/c_d, \qquad g_d''(-i\zeta_0) = r/c_d \cos^2\theta.$$

It follows that

$$\bar{u}_d^\delta(x, z, p) \sim \frac{\sigma_0}{2\pi\mu} f_d(-i\zeta_0) e^{-pr/c_d} \left[\frac{2\pi c_d \cos^2\theta}{rp} \right]^{\frac{1}{2}}. \tag{6.82}$$

Now identifying (6.82) with the leading term of the *Watson's lemma expansion* (5.88), we find its Laplace inverse transform (the leading term of (5.87)) to be

$$(\pi\mu r/c_s\sigma_0) u_d^\delta(x, z, \tau) \sim k \cos\theta f_d(-i\zeta_0)[2(\tau-1)]^{-\frac{1}{2}} \tag{6.83}$$

in agreement with (6.63). At the surface ($\theta = \pi/2$) (6.83) fails since it gives zero there. Again here one could also reproduce the corresponding approximation (6.64) for $w_d^\delta(x, z, \tau)$.

A similar treatment can be carried out for the wavefront approximations to the regular equivoluminal waves u_s^δ, w_s^δ for the region $0 \leq \theta < \sin^{-1}(1/k)$ since again here the pertinent path of steepest descent does not intersect the branch cut. We have already mentioned in the last section that these wavefronts behave as the regular dilatational ones, i.e., $\sim (\tau-k)^{-\frac{1}{2}}$, except at $\theta = 0$ where $u_s^\delta = 0$, and $w_s^\delta \sim (\tau-k)^{\frac{1}{2}}$ as (6.58b) shows. Derivation of the regular wavefronts for u_s^δ, w_s^δ are left to the exercises. At $\theta = 0$ ($x = 0$) for w_s^δ, (5.106) is applicable, giving a result in agreement with (6.58b).

For the interior region $\beta_{cr} < \theta < \pi/\hat{\varphi}$ the saddle point $\zeta = -i\zeta_0 = -ik \sin\theta$, corresponding to $g_s(\zeta)$ in (6.15), lies on the branch cut as we have seen in the previous section, and the method of steepest descents must be modified as shown by Knopoff and Gilbert [6.5]. The situation in fig. 6.5 is again pertinent. *For the head wave* we expand $g_s(\zeta)$ at $\zeta = -i$ along the straight line portion of path H to the right of the imaginary axis. Hence, we have from (6.66)

$$g_s(\zeta) \simeq t_{ds} - i\varrho g_s'(-i), \tag{6.84a}$$

where we have used $\zeta + i = -i\varrho$, and

$$\eta_d'(\zeta) \simeq -i(2\varrho)^{\frac{1}{2}}. \tag{6.84b}$$

The time t_{ds} is given by (6.30). Using (6.84), the head wavefront approximation for u is obtained from the equivoluminal part of $\bar{u}$ in (6.15)

$$\bar{u}_s(x, z, p) = \frac{\sigma_0 \bar{F}(p)}{2\pi\mu} \int_{-\infty}^{\infty} f_s(\zeta) e^{-pg_s(\zeta)} d\zeta. \tag{6.85}$$

We find

Ch. 6, § 6.2] PLANE STRAIN PROBLEMS 327

$$\bar{u}_{ds}(x, z, p) = -\frac{\sqrt{2}\sigma_0 \bar{F}(p)}{\pi\mu} \left[\frac{f_s(\zeta)}{\eta'_s(\zeta)}\bigg|_{\zeta=-i}\right] e^{-pt_ds} \int_0^\varepsilon \sqrt{\varrho}\, e^{-p|g'_s(-i)|\varrho}\, d\varrho,$$

where ε is a small positive number, $g'_s(-i)$ is given in (6.67), and

$$\frac{f_s(\zeta)}{\eta'_d(\zeta)}\bigg|_{\zeta=-i} = \frac{-2(k^2-1)^{\frac{1}{2}}}{(k^2-2)^2}.$$

For p very large it follows the main contribution to the integral in (6.86) comes from the vicinity of $\varrho=0$, and hence the integration there may be extended to ∞. Then the simple transformation $p|g'_s(-i)|\varrho=s$ leads to a well-known integral, hence again to the resultant approximation (6.69) for our delta function input.

For the two-sided equivoluminal wave, we again deal with (6.85), expanding now, however, about the saddle point $\zeta=-i\zeta_0=-ik\sin\theta$ on the branch cut. Hence, for $t>r/c_s$, we have from (6.70)

$$g_s(\zeta) \simeq r/c_s + [g''_s(-i\zeta_0)/2](\zeta+i\zeta_0)^2, \qquad (6.87)$$

and we let $f_s(\zeta)\simeq f_s(-i\zeta_0)$. We recognize these approximations as those in the classical steepest descents approximation for the case when the saddle point does not lie on a cut. We discussed this generally in § 5.10.5, and further with regard to the integrals of interest in this section, as well as an application to $\bar{u}_d(x, z, p)$. The deformed path of integration for (6.85) is taken as that from the saddle point $-i\zeta_0$ out a short distance along the path of steepest descent (the Cagniard path) in the fourth quadrant (cf. fig. 6.5). The integral has the form of those in $I(x)$ preceding (5.118), except for changes introduced leading to $I(x)$ in (6.80). Also because of the conjugation involved here the integration is taken only over the interval $(0, \varepsilon)$. It follows (6.80) applies, and we get the form of (6.81) for $\bar{u}_s(x, z, p)$

$$\bar{u}_s(x, z, p) = \frac{\sigma_0 \bar{F}(p)}{2\pi\mu} \text{Re}[f_s(-i\zeta_0)] e^{-pr/c_s} \left[\frac{2\pi}{pg''_s(-i\zeta_0)}\right]^{\frac{1}{2}} \quad \text{as } p\to\infty, \qquad (6.88)$$

where $f_s(-i\zeta_0)$ is given in (6.74), and $g''_s(-i\zeta_0)=rc_s/c_d^2\cos^2\theta$. Therefore, the inverse of (6.88) for the delta function input duplicates the first of (6.74). For $t<r/c_s$, we follow Rosenfeld's treatment [6.6]. Expanding along the lineal part of the path H to the right of the imaginary axis where $\zeta+i\zeta_0=i\varrho$, we find from (6.87) that

$$g_s(\zeta) \simeq r/c_s - \varrho^2 [g_s''(-i\zeta_0)/2]. \tag{6.89}$$

Using (6.89), the approximation to (6.85) becomes

$$\bar{u}_s(x, z, p) \simeq \frac{\sigma_0 \bar{F}(p)}{2\pi\mu} \text{Re}[-if_s(-i\zeta_0)e]^{-pr/c_s} \int_0^\infty e^{pg_s''(-i\zeta_0)/2]\varrho^2} d\varrho, \tag{6.90}$$

where we must now *consider p a negative number* to obtain convergence of the integral in (6.90). The use of negative p will be justified only by observing the results we obtain later on this basis agree with those obtained in the last section by the Cagniard–deHoop method. Under this condition extending the integration in (6.90) from ε to ∞ was permissible, because the major contribution to the integral comes from the vicinity of $\varrho = 0$. Setting $\bar{F}(p) = 1$ for our delta function input, and carrying out the simple integration in (6.90) we find

$$\bar{u}_s^\delta(x, z, p) \simeq \frac{\sigma_0}{2\pi\mu} \text{Re}[-if_s(-i\zeta_0)]e^{-pr/c_s} \left[\frac{2\pi}{(-p)g_s''(-i\zeta_0)}\right]^{\frac{1}{2}}. \tag{6.91}$$

We can identify $e^{-pr/c_s}/(-p)^{\frac{1}{2}}$ in (6.91) with the aid of the Laplace transform

$$\frac{e^{-pr/c_s}}{(-p)^{\frac{1}{2}}} = \int_0^\infty \frac{e^{-(-p)(t-r/c_s)}}{(\pi t)^{\frac{1}{2}}} dt. \tag{6.92}$$

Using the transformation $-(t - r/c_s) = t'$ in this integral, it can be shown that (6.92) becomes

$$\frac{e^{-pr/c_s}}{(-p)^{\frac{1}{2}}} = \frac{1}{\sqrt{\pi}} \int_{-\infty}^{r/c_s} \frac{e^{-pt'}}{(r/c_s - t')^{\frac{1}{2}}} dt'.$$

For $r/c_s - \varepsilon < t \leq r/c_s$ the inverse of $e^{-prc_s}/(-p)^{\frac{1}{2}}$ is therefore $1/[\pi(r/c_s - t)]^{\frac{1}{2}}$. Hence from (6.91) our wavefront disturbance derived here for $t < r/c_s$ duplicates the second of (6.74). In their work Knopoff and Gilbert [6.5] made use of the *Tauberian theorem* associated with the short time asymptotics of the Laplace transform. In the present approximation they also exploited p negative developing a Tauberian theorem to handle it. Further reading on Tauberian theorem smay be found in Widder [5.4, ch. V].

6.2.2. Lamb's problem for the buried line dilatational source

Garvin [6.7] recently derived and evaluated a closed form solution for the surface displacements in the subject problem using Cagniard's method of inversion with certain modifications. With § 6.2.1 to draw on only the *essential differences* in the present problem, and Garvin's treatment, need be brought out here. The problem is depicted in fig. 6.10. The half-space

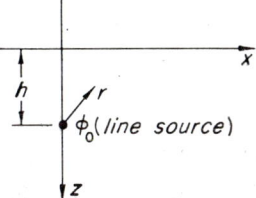

Fig. 6.10. Problem of buried line dilatational source in half space.

experiences a sudden line dilatational source φ_0, on the z-axis a distance h from the origin. We note the governing equations (6.1)–(6.4), with the right hand side of the first of (6.3) now zero, and where now $r = [x^2 + (h-z)^2]^{\frac{1}{2}}$, apply again here. The solution corresponding to the axially symmetric source φ_0 can be represented by its solution from the related infinite medium problem. The transform of this solution is given by (5.157). Using this, (6.8) and the form of (6.13), we find the Laplace transformed potentials in the present problem are

$$\bar{\varphi}(x, z, p) = \bar{\varphi}_0(r, p) + \frac{1}{2\pi} \int_{-\infty}^{\infty} A(\varkappa, p) e^{-\eta_d z - i\varkappa x} d\varkappa , \qquad (6.93)$$

$$\bar{\psi}(x, z, p) = \frac{1}{2\pi} \int_{-\infty}^{\infty} B(\varkappa, p) e^{-\eta_s z - i\varkappa x} d\varkappa ,$$

where

$$\bar{\varphi}_0(r, p) = \bar{F}(p) K_0(k_d r) ,$$

where $\bar{F}(p)$ is the transform of $F(t)$, the time input of the source. We now need an integral expression for $\bar{\varphi}_0(r, p)$ to match the forms of the integrals in (6.93). Garvin gives such an expression developed by a means due basically to Lapwood [6.8]. Essentially Lapwood's derivation begins with a well-known *integral representation* for the K_0 function

$$K_0(k_d r) = \int_0^\infty e^{-k_d [x^2 + (h-z)^2]^{\frac{1}{2}} \cosh w} dw , \qquad r > 0 . \qquad (6.94)$$

But a linear form in x and z is needed to match (6.93) rather than the quadratic one appearing here. At $x=0$ (6.72) becomes

$$K_0[k_d(h-z)] = \int_0^\infty e^{-kd(h-z)\cosh w} \, dw, \quad z<h. \tag{6.95}$$

Now let $k_d \sinh w = \varkappa$, and $k_d \cosh w = \eta_d$ (note these stem from $\eta_d^2 = k_d^2 + \varkappa^2$), and (6.95) becomes

$$K_0[k_d(h-z)] = \int_0^\infty \frac{e^{-(h-z)\eta_d}}{\eta_d} \, d\varkappa, \quad z<h. \tag{6.96}$$

An even trigonometric function of x is now introduced to have a representation of $\bar{\varphi}_0(r, p)$ off the z-axis, which is a solution of the Laplace transformed wave equation on $\bar{\varphi}_0$, and reduces to (6.96) when $x=0$. Such a solution is

$$K_0(k_d r) = \int_0^\infty \frac{e^{-(h-z)\eta_d}}{\eta_d} \cos \varkappa x \, d\varkappa, \quad z<h, \tag{6.97}$$

which must be equivalent to $K_0(k_d r)$ in (6.94). It is easy to show that (6.97) may be written as

$$K_0(k_d r) = \frac{1}{2} \int_{-\infty}^\infty \frac{e^{-(h-z)\eta_d - i\varkappa x}}{\eta_d} \, d\varkappa, \quad z<h. \tag{6.98}$$

Substitution of (6.98) into the first of (6.93) gives the latter the appropriate structure. The right hand sides of (6.93) can then be substituted into the corresponding Laplace transformed stress-potential relations, i.e., the exponential Fourier transform inverses of (6.9). Then in turn we set the resulting integrands of these equal to zero, for $z=0$, to satisfy the free boundary conditions, which determines the unknowns $A(\varkappa, p)$ and $B(\varkappa, p)$ and an explicit term for the source disturbance. The resulting transformed surface displacements given by Garvin are

$$\bar{u}(x, 0, p) = -\frac{4p^2 \bar{F}(p)}{c_s^2} \operatorname{Im} \int_0^\infty \frac{\varkappa \eta_s e^{-h\eta_d - i\varkappa x}}{R(\varkappa, p)} \, d\varkappa,$$

$$\bar{w}(x, 0, p) = 2\bar{w}_0 + \frac{4\bar{F}(p)}{c_s^2} \operatorname{Re} \int_0^\infty \frac{\varkappa^2[p^2 + 2c_s^2\varkappa^2 - 2c_s^2\eta_s\eta_d] e^{-h\eta_d - i\varkappa x}}{R(\varkappa, p)} \, d\varkappa, \tag{6.99}$$

where $\bar{w}_0$ is the transform of the vertical displacement that would result at $(x, 0)$ in an infinite medium. Garvin sets $F(t) = (a/2)H(t)$, where a is a constant of dimensions length squared (area), giving a corresponding radial displacement representative of an earthquake source, a sudden jump at r/c_d followed by gradual incomplete recovery; cf. his fig. 4. The transformed source term is then

$$\bar{\varphi}_0(r,p) = \frac{a}{2p} K_0(k_d r) = \frac{a}{2p} \int_0^\infty e^{-k_d r \cosh w} dw,$$

which, through the transformation $t = (r/c_d) \cosh w$, gives

$$p\bar{\varphi}_0(r,p) = \frac{a}{2} \int_{r/c_d}^\infty \frac{e^{-pt} dt}{(t^2 - r^2/c_d^2)^{\frac{1}{2}}}.$$

Inversion of this by inspection leads to the solution

$$\varphi_0(r,t) = 0, \quad \text{for} \quad t < r/c_d,$$

$$= (a/2) \cosh^{-1}(c_d t/r), \quad \text{for} \quad t > r/c_d,$$

which yields for the radial displacement

$$u_{r0}(r,t) = \frac{\partial \varphi_0}{\partial r} = 0, \quad \text{for } t < r/c_d,$$

$$= -\frac{at}{2r(t^2 - r^2/c_d^2)^{\frac{1}{2}}}, \quad \text{for } t > r/c_d \quad (6.100)$$

with the constant a inherently negative. The inverse of $\bar{w}_0$ in (6.99) is the z-component of u_{r0} in (6.100) evaluated at $z = 0$. Garvin inverts the integrals contributing to $\bar{u}$ and $\bar{w}$ in (6.99) by Cagniard's method much as it was applied in the previous section.

Of particular interest in Garvin's work is the numerical evaluation and analysis of his solution. In particular, note the response plots for $u(x, 0, t)$ and $w(x, 0, t)$, given in his fig. 5, which are reproduced here in fig. 6.11. They show the birth and development of the Rayleigh wave disturbance as the station x/h grows. Note once this Rayleigh pulse takes its shape ($x/h > 20$), it is nondecaying (in x and t), and does not change its shape. Recall this was also the case in the surface source two-dimensional problem discussed in § 6.2.1. The apparent change in shape in fig. 6.11 for the more remote stations is due to a contraction in the real time scale as x/h increases, i.e., $\tau \simeq c_d t/h(x/h)$. Note that w and u are even and odd, respectively, through $t = x/c_R$, and that each is a *smooth pulse*.

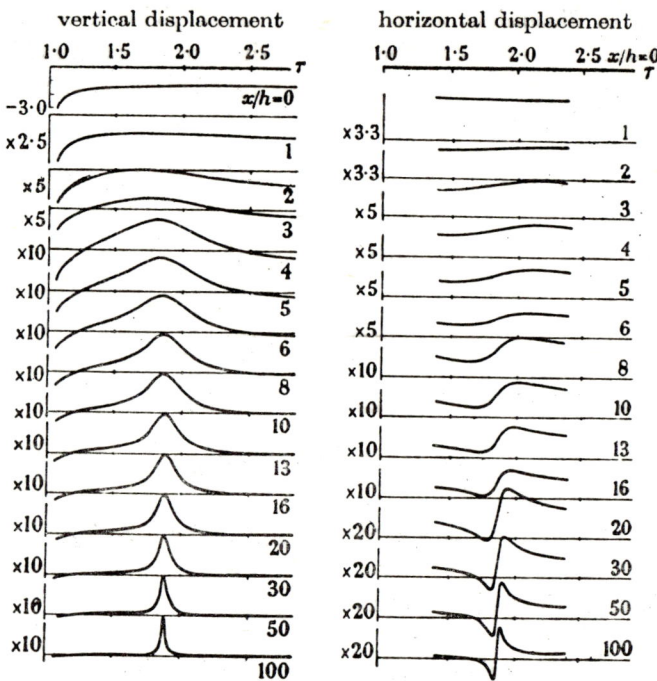

Fig. 6.11. Response of vertical and horizontal surface displacements at various stations x/h for Lamb's problem of buried line dilatational source ($\tau = c_d t/(x^2+h^2)^{\frac{1}{2}}$, $\nu = 1/4$) (after Garvin).

Garvin gives the formal transformed solutions for the displacements $\bar{u}(x, z, p)$, $\bar{w}(x, z, p)$ [cf. his equations (15), (16), together with (20), (21)] which are the analogs of (6.15) of the surface source problem of § 6.2.1. If one so desired they could be exploited further, in the manner we have set down in § 6.2.1, to get the interior solution and corresponding approximations.

6.3. Axially symmetric problems

Lamb's basic problems for the surface and buried point load sources fall into the present category. Modern treatments of these problems along the lines of the techniques prescribed in the foregoing § 6.2 were given by Pekeris [6.9] and Pekeris and Lifson [6.10]. We will obtain his formal solutions somewhat differently by drawing on the *Hankel transforms* in § 5.8 which are natural for these problems, and in line with the general theme here.

6.3.1. The Lamb problems for the surface and buried vertical point load sources

Following Pekeris [6.9] we can write the formal solution for the buried vertical point load source, and obtain from it, as a limit, the special case for the surface normal point load source. The problem is depicted in fig. 6.12.

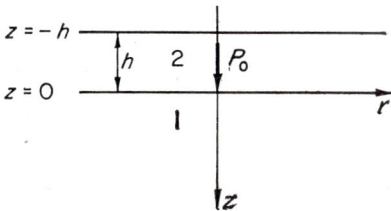

Fig. 6.12. Problem of buried vertical point load source in half space.

The source is a point compressional load P_0 (a negative constant), acting in the z-direction at the origin of the cylindrical coordinate system (r, z), a depth h below the surface. The space variation of the source is specified by the condition that, at the level of the source $(z=0)$, the surface integral of the applied stress $\sigma_{z1} - \sigma_{z2}$ must be equal to P_0, i.e.,

$$2\pi \int_0^\infty (\sigma_{z1} - \sigma_{z2}) r \, dr = P_0 . \qquad (6.101)$$

We let

$$\sigma_{z1} - \sigma_{z2} = \frac{P_0 \delta_s(r)}{\pi r} \qquad (6.102)$$

represent the point source,[10] where $\delta_s(r)$ is the symmetric delta function given in (6.5), but defined here for the domain of r, and hence giving the value $1/2$ on the right of (6.5b) instead of 1. Thus (6.102) is easily shown to

---

[10] Pekeris chose another representation.

satisfy (6.101). The rest of the statement of the boundary value problem is straightforward. We have

$$\left.\begin{array}{l}\nabla^2\varphi(r, z, t) = \ddot{\varphi}/c_d^2 \\ \nabla^2\eta(r, z, t) = \ddot{\eta}/c_s^2\end{array}\right\} \text{in } 0 \leq r < \infty, \qquad -h < z < \infty, \quad \text{for } t > 0, \quad (6.103)$$

$$\varphi(r, z, 0) = \dot{\varphi} = \eta = \dot{\eta} = 0, \qquad 0 \leq r < \infty, \quad -h \leq z < \infty, \quad (6.104)$$

$$\sigma_{z2}(r, -h, t) = \sigma_{zr2}(r, -h, t) = 0, \qquad 0 \leq r < \infty, \quad t > 0, \quad (6.105)$$

$$\left.\begin{array}{l}u_{r1}(r, 0, t) = u_{r2}(r, 0, t), \\ u_{z1}(r, 0, t) = u_{z2}(r, 0, t), \\ \sigma_{zr1}(r, 0, t) = \sigma_{zr2}(r, 0, t), \\ \sigma_{z1}(r, 0, t) - \sigma_{z2}(r, 0, t) = 0, \quad t < 0, \\ = \dfrac{P_0 \delta_s(r) F(t)}{\pi r}, \quad \text{for } t > 0,\end{array}\right\} \text{for } t > 0, \qquad 0 \leq r < \infty, \quad (6.106)$$

$$\lim_{r \to 0, \infty} \{r[\varphi(r, z, t), \eta, \partial\varphi/\partial r, \text{etc.}]\} = 0, \quad \text{for } -h \leq z < \infty, \; t > 0, \quad (6.107)$$

$$\lim_{z \to \infty} \{\varphi(r, z, t), \eta, \partial\varphi/\partial r, \text{etc.}\} = 0, \quad \text{for } 0 \leq r < \infty, \; t > 0, \quad (6.108)$$

where ∇^2 is given by (4.75) without the θ term, and $F(t)$ is the time input function. Note that the first three of (6.106) are required continuity conditions across the interface $z = 0$, as is the last of (6.106) away from $r = 0$ (note $u_\theta = 0$ in this problem).

We now apply the Laplace transform on t to (6.103), using (6.104). To these, on the basis of the Laplace transforms of (6.107), the order zero Hankel transform from (5.59) is applied along with (5.68), which reduces (6.103) to the ordinary differential equations

$$\begin{aligned}\frac{d^2 \bar{\varphi}^{\sim 0}(\varkappa, z, p)}{dz^2} - \eta_d^2 \bar{\varphi}^{\sim 0} &= 0, \\ \frac{d^2 \bar{\eta}^{\sim 0}(\varkappa, z, p)}{dz^2} - \eta_s^2 \bar{\eta}^{\sim 0} &= 0.\end{aligned} \qquad (6.109)$$

The transformed solutions of (6.109) are

$$\begin{aligned}\bar{\varphi}_1^{\sim 0}(\varkappa, z, p) &= A(\varkappa, p) \exp(-\eta_d z), \\ \bar{\eta}_1^{\sim 0}(\varkappa, z, p) &= B(\varkappa, p) \exp(-\eta_s z),\end{aligned} \qquad (6.110)$$

for region 1 in fig. 6.12, where η_d and η_s are again selected as those branches that are real and positive, for p real and positive, to satisfy the conditions on $\bar{\varphi}^{\sim 0}$, and $\bar{\eta}^{\sim 0}$ at large z, given by the double transforms of (6.108). The region 2 transformed solutions are

$$\bar{\varphi}^{\sim 0}_2 = C(\varkappa, p) \exp(-\eta_d z) + D(\varkappa, p) \exp(\eta_d z),$$
$$\bar{\eta}^{\sim 0}_2 = E(\varkappa, p) \exp(-\eta_s z) + F(\varkappa, p) \exp(\eta_s z), \qquad (6.111)$$

since here we have a finite domain in z. The displacement-potential relations are given by (4.92). From the form of these, and from (6.110), (6.111), (5.64) and (5.68), we see the appropriate transformed displacement-potential relations are

$$\bar{u}^{\sim 1}_r = -\varkappa \left(\bar{\varphi}^{\sim 0} + \frac{d\bar{\eta}^{\sim 0}}{dz} \right),$$
$$\bar{u}^{\sim 0}_z = \frac{d\bar{\varphi}^{\sim 0}}{dz} + \varkappa^2 \bar{\eta}^{\sim 0}. \qquad (6.112)$$

From (4.92), (4.93) the stress

$$\sigma_z = \frac{\lambda}{c_d^2} \frac{\partial^2 \varphi}{\partial t^2} + 2\mu \left\{ \frac{\partial^2 \varphi}{\partial z^2} - \frac{\partial}{\partial z} \left[\frac{\partial}{r \partial r} \left(r \frac{\partial \eta}{\partial r} \right) \right] \right\}. \qquad (6.113)$$

The appropriate transforms of this, and σ_{zr} in (4.94) are found to be

$$\bar{\sigma}^{\sim 0}_z = \lambda k_d^2 \bar{\varphi}^{\sim 0} + 2\mu \left(\frac{d^2 \bar{\varphi}^{\sim 0}}{dz^2} + \varkappa^2 \frac{d\bar{\eta}^{\sim 0}}{dz} \right),$$
$$\bar{\sigma}^{\sim 1}_{zr} = -\mu\varkappa \left(2 \frac{d\bar{\varphi}^{\sim 0}}{dz} + \frac{d^2 \bar{\eta}^{\sim 0}}{dz^2} + \varkappa^2 \bar{\eta}^{\sim 0} \right), \qquad (6.114)$$

with the aid of (5.64) and (5.68). Substitution of (6.110) and (6.111), into (6.112) and (6.114), forms the general transformed displacement and stress solutions for each of the regions 1 and 2. Substitution of the latter into the double transforms of the boundary conditions (6.105) and (6.106) then yields six equations for the six unknowns A, B, C, D, E, and F. Pekeris gives their values in his equations (16)–(19) based on $\lambda = \mu$ ($\nu = 1/4$). Using our Hankel inversion integral (5.58) the corresponding Laplace transformed solutions for the surface displacements are given by

$$\bar{u}_z^H(r, -h, p) = \frac{P_0}{2\pi\mu p} \int_0^\infty \varkappa J_0(\varkappa r)[-(2\varkappa^2 + k_s^2)e^{-\eta_d h} + 2\varkappa^2 e^{-\eta_s h}] \frac{\eta_d}{R(\varkappa, p)} d\varkappa ,$$

(6.115)[11]

$$\bar{u}_r^H(r, -h, p) = -\frac{P_0}{2\pi\mu p} \int_0^\infty \varkappa^2 J_1(\varkappa r)[-2\eta_d\eta_s e^{-\eta_d h} + (2\varkappa^2 + k_s^2)e^{-\eta_s h}] \frac{d\varkappa}{R(\varkappa, p)} ,$$

or $F(t) = H(t)$, where $R(\varkappa, p)$ is given in (6.11).

6.3.1.1. Inversion for surface displacements in problem of surface normal point load source; numerical evaluation. For $h \to 0$ (6.115) gives the transformed solution for the surface displacements for the case of the surface normal point load source. Pekeris inverts these by a *technique close in nature to Cagniard's*. It is of interest to bring out the basic features of his technique and solution, and the numerical evaluation of the latter. Consider the transformed vertical surface displacelment which from the first of (6.115) is

$$\bar{u}_z^H(r, 0, p) = -\frac{P_0 k_s}{2\pi\mu c_s} \int_0^\infty \frac{\varkappa \eta_d J_0(\varkappa r)}{R(\varkappa, p)} d\varkappa .$$

(6.116)[12]

Following Pekeris we introduce $\varkappa = k_s \zeta$ (same type transformation as in (6.14)) into (6.116) and it becomes

$$\bar{u}_z^H(r, 0, p) = -\frac{P_0}{2\pi\mu c_s} N(k_s r) ,$$

(6.117)

where

$$N(k_s r) = \int_0^\infty J_0(k_s r \zeta) \zeta m(\zeta) d\zeta ,$$

$$m(\zeta) = \frac{\alpha}{R(\zeta)} ; \qquad R(\zeta) = (2\zeta^2 + 1)^2 - 4\zeta^2 \alpha\beta ,$$

$$\alpha = \left(\zeta^2 + \frac{1}{3}\right)^{\frac{1}{2}} , \qquad \beta = (\zeta^2 + 1)^{\frac{1}{2}} .$$

---

[11] (6.115) are equations (20) and (21) from Pekeris, with our nomenclature. Note his work employs the *p*-multiplied Laplace transform.

[12] This is (22) in Pekeris' work, except for sign. Apparently his sign there is wrong, since (6.116) was derived from the first of (6.115), which does agree with (20) in the quoted work.

To get $N(k_s r)$ into the form of the Laplace integral operator, we first note that the singularities of its integrand are branch points at $\zeta = \pm i\sqrt{3}$, $\pm i$, and simple poles at $\zeta = \pm i\gamma$, where $\gamma = (3+\sqrt{3})^{\frac{1}{2}}/2$ (the latter correspond to the admissible zeros of the Rayleigh function $R(\zeta)$). The complex $\zeta = \xi + i\delta$ plane is cut as shown in fig. 6.13, to make $m(\zeta)$ analytic. The values given

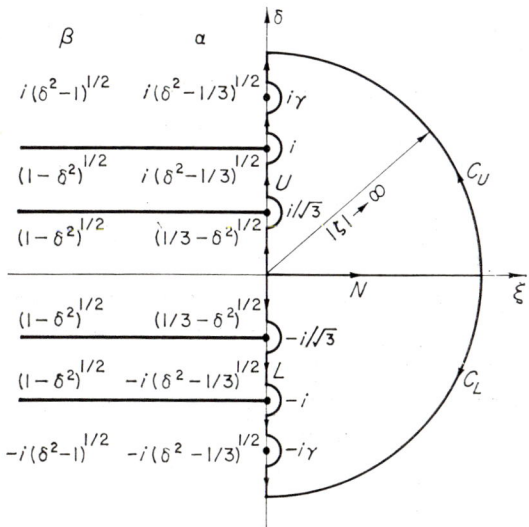

Fig. 6.13. Contour integrations in the ζ-plane for Lamb's problem of surface normal point load source (after Pekeris).

for α and β in the figure are those they have on the imaginary axis to the right of them which are determined by the branch cuts. Verification is left to the exercises.

Now

$$J_0(k_s r \zeta) = \tfrac{1}{2}\left[H_0^{(1)}(k_s r \zeta) + H_0^{(2)}(k_s r \zeta)\right], \tag{6.118}$$

where $H_0^{(1)}$ and $H_0^{(2)}$ are, respectively, the Hankel functions of first and second kinds of order zero. Substitution of (6.118) in $N(k_s r)$ makes the latter the sum of two integrals each of which is over the path N (real $\zeta = \xi \geq 0$) in fig. 6.13. A contour integration for the first integral over the first quadrant path $N + C_U - U$ readily shows this integral is equal to that over the path U, i.e., using the relation $H_0^{(1)}(is) = -(2i/\pi)K_0(s)$,

$$\tfrac{1}{2}\int_0^\infty H_0^{(1)}(k_s r \zeta)\zeta m(\zeta)\,d\zeta = \frac{i}{\pi}\int_0^\infty K_0(k_s r \delta)\delta mi(\delta)\,d\delta, \tag{6.119}$$

since Jordan's lemma shows the integral along C_U contributes nothing.[13] Similarly a fourth quadrant integration over the path $N+C_L-L$ for the second integral shows it is equal to that over the path L, i.e., using $H_0^{(2)}(-is) = (2i/\pi)K_0(s)$,

$$\tfrac{1}{2}\int_0^\infty H_0^{(2)}(k_s r\zeta)\zeta m(\zeta)\,d\zeta = -\frac{i}{\pi}\int_0^\infty K_0(k_s r\delta)\delta m(-i\delta)\,d\delta. \tag{6.120}$$

Since $m(i\delta)$ and $m(-i\delta)$ are conjugate functions, the sum of the integrals on the right hand sides of (6.119) and (6.120) gives

$$N(k_s r) = -\frac{2}{\pi}\,\mathrm{Im}\int_0^\infty K_0(k_s r\delta)\delta m(i\delta)\,d\delta. \tag{6.121}$$

We have already pointed out that $R'(i\gamma)$ is imaginary [cf. (6.41)]. From fig. 6.13 we see, near $\delta=\gamma$, α is also imaginary, and since $K_0(k_s r\delta)$ is real, it is easily seen that the integral in (6.121), taken over the small circle about $\delta=\gamma$, vanishes. Thus the Rayleigh pole here, as in the analogous case of the vertical displacement $u_{\theta R}$ of the line load problem, also gives no contribution [cf. (6.42)]. Further since $m(i\delta)$ is real in $0 \leq \delta < 1/\sqrt{3}$, (cf. fig. 6.13), it follows (6.121) reduces to

$$N(k_s r) = -\frac{2}{\pi}\,\mathrm{Im}\left[\mathrm{P.\,V.}\int_{1/\sqrt{3}}^\infty K_0(k_s r\delta)\delta m(i\delta)\,d\delta\right], \tag{6.122}$$

where we note that the principal value must be taken since R vanishes at $\delta=\gamma$ along the path of integration.

Now noting that p occurs only in $K_0(k_s r\delta)$, its inverse can be obtained nike that for $p\bar{\varphi}_0$ in the derivation leading to (6.100). It follows that the version of (6.117) yields

$$u_z^H(r, 0, t) = \frac{P_0}{\pi^2 \mu r}\,\mathrm{Im}\left[\mathrm{P.\,V.}\int_{1/\sqrt{3}}^\tau \frac{\delta m(i\delta)\,d\delta}{(\tau^2-\delta^2)^{\frac{1}{2}}}\right] \tag{6.123}$$

---

[13] The leading terms of the asymptotic expansions for the H_0 functions for large argument are needed in the proof (cf. Erdelyi et al. [2.8, (1), (2) pg. 85] for them).

where $\tau = c_s t/r$. As δ varies from $1/\sqrt{3}$ to τ, $m(i\delta)$ in (6.123) is evaluated from the values of α and β in fig. 6.13. Pekeris uses partial fraction decomposition to get an algebraic form for (6.123).

A similar analysis gives u_r^H for the surface, and again it is found, as in the line load case, the Rayleigh pole does contribute. Here, however, u_r^H is in terms of elliptic integrals. Pekeris' numerical results for the surface displacements are reproduced here in fig. 6.14.

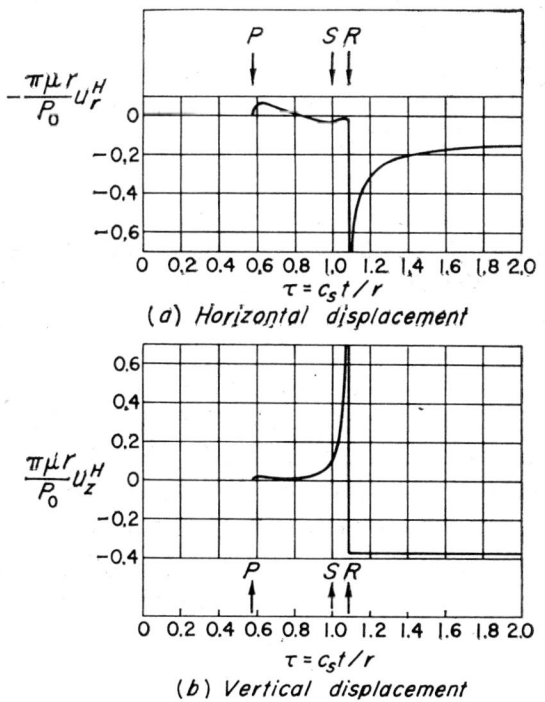

Fig. 6.14. Surface displacements as a function of time in Lamb's problem for surface normal point load source (P, S, R mark arrival times of dilatational, equivoluminal and Rayleigh waves) (after Pekeris).

If one compares the results in fig. 6.14 with those of fig. 6.8 for the line load case, many similar features are seen in spite of the less severe time input in the present case (i.e., $H(t)$ instead of $\delta(t)$). This suggests the point load input leads to stronger singularities at the wave fronts than the line load. Indeed Pekeris finds a $1/(t-r/c_R)^{\frac{1}{2}}$ and a $1/(r/c_R-t)^{\frac{1}{2}}$ behavior in the surface displacements u_r^H, u_z^H, respectively, at the Rayleigh wave arrival time ($\tau=\gamma$) (cf. his eqs. (47) and (66)). The corresponding (step time input) singular terms in the line load case are $\ln|t-r/c_R|$ and $H(t-r/c_R)$. This $1/(t-r/c_R)^{\frac{1}{2}}$ *strengthening of the singularity, from line to point load inputs, is also true of the response at the other wave fronts in the interior.* We shall discuss this further in the next section. It is important to note that here the Rayleigh disturbance is one-sided (recall the two-sided Rayleigh waves in line load case). But note u_{zR}^H becomes singular ahead of r/c_R, i.e., as $t \to (r/c_R)-$. Note also that the Rayleigh disturbance here depends on the spatial variable r, that is, near the arrival time $\tau=\gamma$, both displacements behave as $1/\sqrt{r}$, as is shown by equations (47) and (66) in Pekeris' paper. Note that this is typical for two-dimensional (cylindrical) wavefront propagation as we noted in § 2.6.4.

6.3.1.2. Numerical evaluation of the interior solution, wavefront behaviors, for surface normal point load source. In recent work on the problem of the traveling normal point load on the surface of an elastic solid (to be discussed in § 6.4), Gakenheimer and Miklowitz [6.11] derived the exact solution for the present problem as a special case in which the velocity of the load was set equal to zero. Later Gakenheimer [6.12] published some important new numerical results for the interior solution. His results are reproduced here in figs. 6.15, 6.16. Respectively they show the response of the dimensionless horizontal and vertical displacements, $v_r(\varphi, \tau)$, $v_z(\varphi, \tau)$, where $\tau = c_d t/\varrho$, $\varrho = (r^2+z^2)^{\frac{1}{2}}$ and $\varphi = \tan^{-1}(r/z)$, $0 \leq \varphi \leq \pi/2$. The displacements v_r, v_z relate to our dimensional displacaments $u_r^H(r, z, t)$, $u_z^H(r, z, t)$ through the expression $u_j^H(\varrho, \varphi, \tau) = (1/\pi^2\mu\varrho)v_j(\varphi, \tau)$, $j = r, z$. The computations are based on $\lambda = \mu$, and a point force of magnitude one. In the figures ϱ is not specified since along any ray (denoted by $\varphi =$ constant) v_j is independent of ϱ. However, through the relation just cited we see the corresponding u_j decays as ϱ^{-1}. *The letters* d, s, sd, *and* R *are used to define the arrival times, along a ray, of the dilatational, equivoluminal, head and Rayleigh waves, respectively.* These times are respectively, $\tau = 1$, 1.73, $\sin \varphi + \sqrt{2} \cos \varphi$ and 1.89.

Interesting in Gakenheimer's results are the jump discontinuities exhibited at the dilatational wavefronts for both displacements along all the interior rays ($0 \leq \varphi < \pi/2$) shown. Similar jumps occur at the regular equivoluminal wavefronts except at $\varphi = 0$. When φ is greater than the critical angle (35.3 degrees), it may be noted that the equivoluminal wavefront becomes sin-

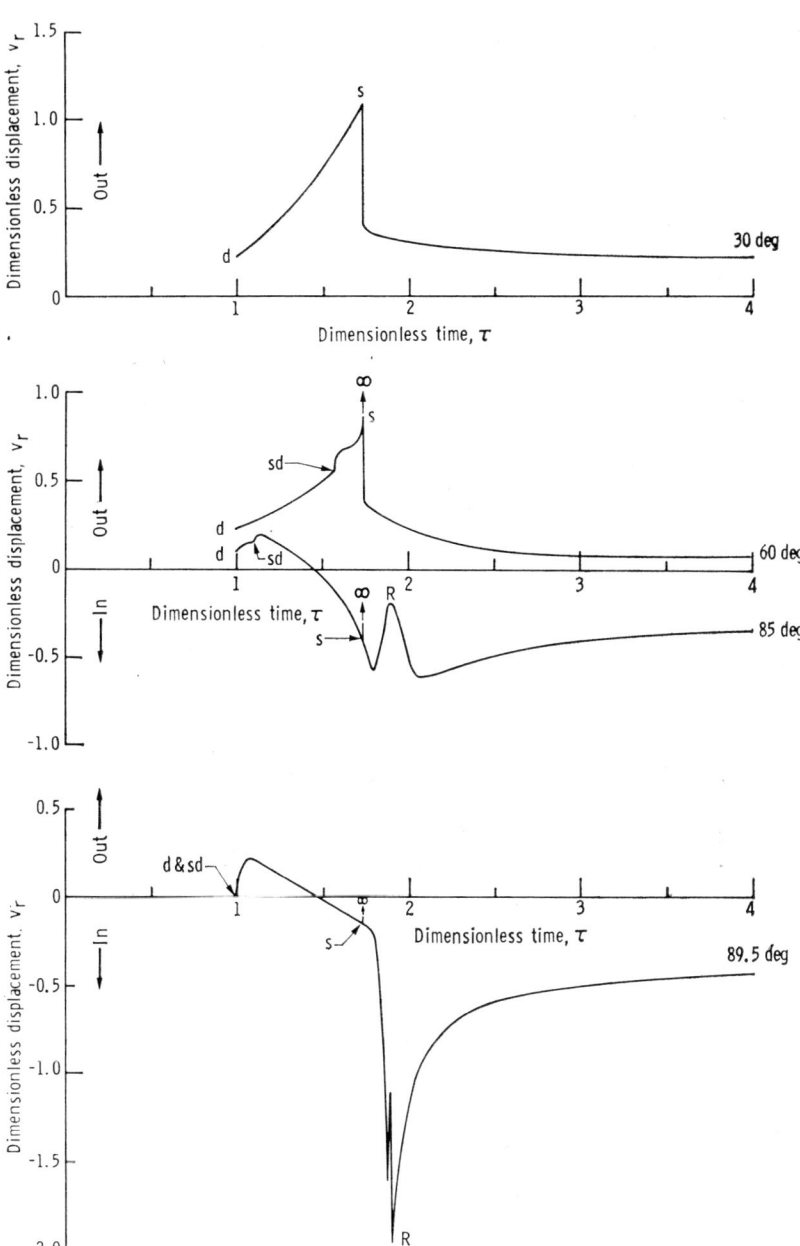

Fig. 6.15. Interior horizontal displacement response in Lamb's problem for surface normal point load source (courtesy Gakenheimer).

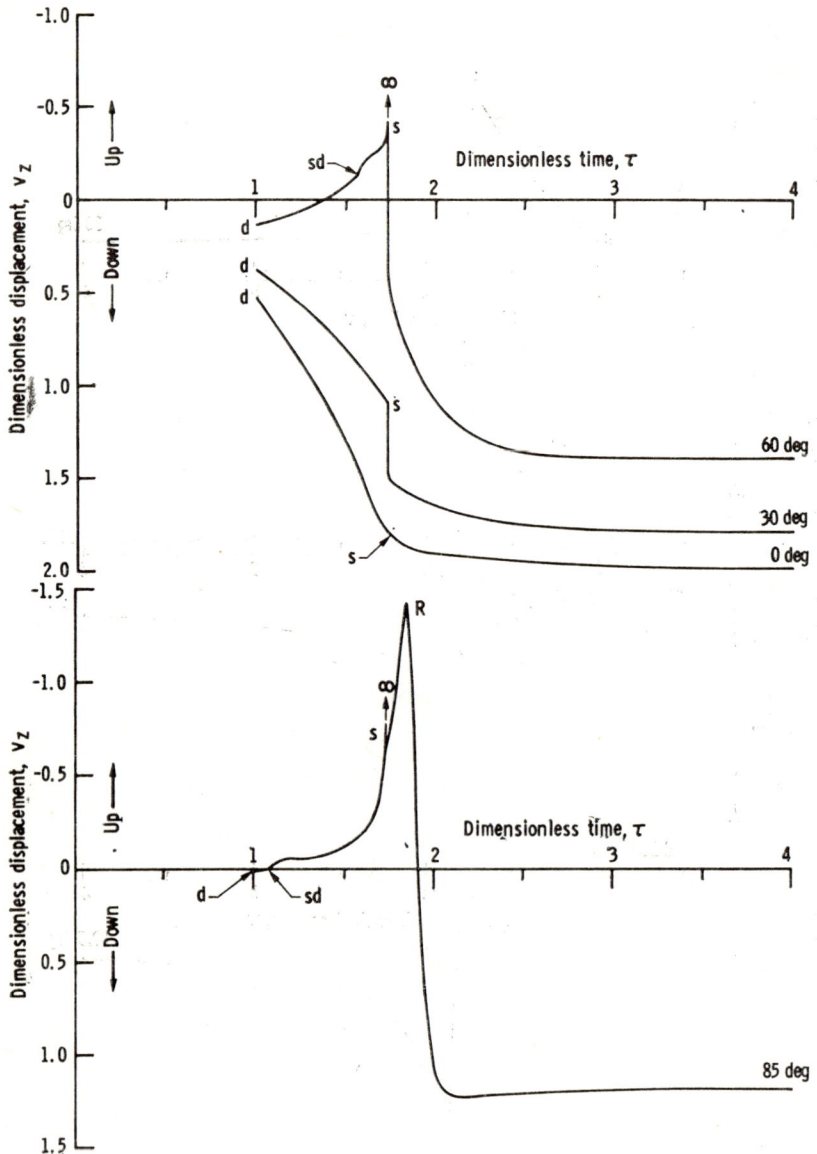

Ch. 6, § 6.3] AXIALLY SYMMETRIC PROBLEMS 343

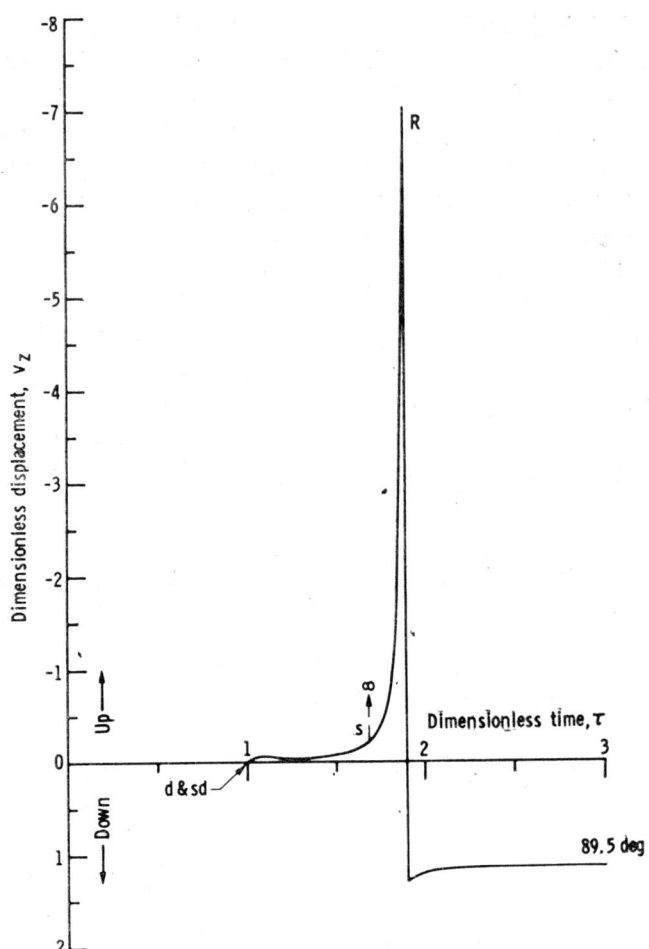

Fig. 6.16. Interior vertical displacement response in Lamb's problem for surface normal point load source (courtesy Gakenheimer).

gular. This corresponds to the logarithmic singularity for this two-sided front discovered recently by Aggarwal and Ablow [6.13], and confirmed in [6.11] by direct wavefront approximations to the exact solution. The jumps at the dilatational and equivoluminal regular wavefronts were noted in Knopoff and Gilbert [6.5] and in [6.11]. The head wave is continuous at its front, having a ramp behavior there as noted in [6.5], [6.11]. Interesting also in figs. 6.15, 6.16, is the growth of the Rayleigh wave exhibited as φ approaches the surface ($\varphi = \pi/2$). It should also be noted, that as the surface is approached the jump discontinuities get smaller in magnitude and finally vanish at the surface, along with the two-sided equivoluminal logarithmic singularity, as Pekeris' results showed (cf. fig. 6.14).

Important is the fact that a step function input in time for the present point load problem has resulted in discontinuities and singularities in the interior displacements, which violates our earlier continuity requirement on these quantities. It is clear, however, that a ramp input instead would not have this discrepancy. Finally, as we remarked on earlier, there is in these wavefront behaviors a more singular nature by a factor of $(t-t_a)^{-\frac{1}{2}}$ (t_a is the arrival time of the front) than the corresponding (with same time input) ones in the line load problem. This holds for the regular dilatational and equivoluminal fronts and the head wavefront. In the case of the two-sided equivoluminal wavefront, it is even more severe, since instead of an $H(t-t_a)$ behavior for the present case we have a $ln|t-t_a|$ behavior compared to a $(t-t_a)^{\frac{1}{2}}$ response for the line load case.

6.3.1.3. Inversion for surface displacements in problem of buried vertical point load source; numerical evaluation and wavefront behaviors. To invert (6.115) for this case Pekeris [6.14] used the Laplace operator integral (5.6b) to write the integral equations for $u_z^H(r, -h, t)$, $u_r^H(r, -h, t)$

$$\int_0^\infty \begin{bmatrix} u_z^H(r, -h, t) \\ u_r^H(r, -h, t) \end{bmatrix} e^{-pt} dt = \begin{bmatrix} \text{right hand sides} \\ \text{of (6.115)} \end{bmatrix}, \qquad (6.124)$$

and then developed a method for solving these equations. Two later works, one by Pekeris [6.15] and the other by Pekeris and Lifson [6.10], were aimed at numerical evaluation of the results obtained in [6.14]. The quoted paper [6.10] is of primary interest here. It contains some important response curves for the surface displacements.

In order to interpret the various events occurring in these response curves, it is an aid to first consider the ray diagrams in fig. 6.17 (fig. 1 of [6.10]) depicting the reflection at the surface of the equivoluminal wave S from the buried source. By considering a vertical diametral plane through the spherical wavefronts emanating from the source here, the action at the free surface can be likened to the plane P and SV waves discussed in Chapter 3. In particular the S wave here would correspond to the SV wave. From § 3.1.4.5, and figs. 3.19, 3.20, on critical angles of incidence for SV waves, and their

Ch. 6, § 6.3] AXIALLY SYMMETRIC PROBLEMS 345

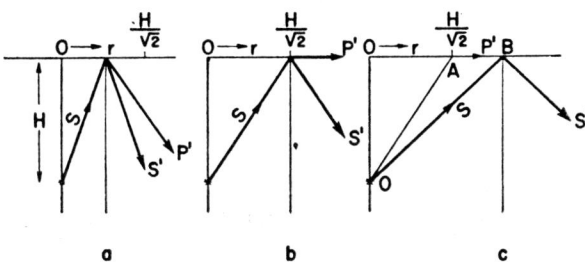

Fig. 6.17. Ray diagram of reflection of an $S(SV)$ wave at a free surface (after Pekeris and Lifson).

corresponding total reflection, we should expect total reflection of the S pulse here for angles of incidence defined by $\beta_{cr} \leq \beta < \pi/2$, where since $\lambda = \mu$, $\beta_{cr} = \cot^{-1}\sqrt{2}$. In fig. 6.17 this range of β corresponds to the reflection points $r \geq H/\sqrt{2}$ (H is our h). Diagram a in fig. 6.17 shows the situation for the ordinary S-wave reflection, generating reflected S' and P' (dilatation) waves. Total reflection starts when the S ray intersects the surface at $r = = H/\sqrt{2}$. Diagram b shows that the first grazing reflected dilatational wave P' is generated. Diagram c depicts the situation of a typical S ray in the region $r > H/\sqrt{2}$. At the observation station B the first signal to arrive is from P'. This event began as an S pulse at the source O, traveled along OA, and then along AB as the P' wave. It is referred to as the SP wave to indicate its P nature was generated by an S wave through reflection. It can be shown that the travel time of the SP wave to B is shorter than that of the direct S wave even though its travel path OAB is longer. There is an infinite number of S rays that intersect AB, hence a continuous SP signal arriving at B between the initial SP signal (emanating from A) and the arrival of the direct S wave from the source.

The response curves for the vertical displacement at various r/H stations on the surface are shown in fig. 6.18 (figs. 2, 3 of [6.10]). The nomenclature corresponding to our $u_z^H(r, -h, t)$ is $w(t, r, H) = (3Z/\pi^2 \mu R)W$, where Z is our P_0, $R = (r^2 + H^2)^{\frac{1}{2}}$, and $W = -W_p + W_s$ is the solution function plotted, W_p giving the dilatational and W_s the equivoluminal contributions to W, respectively. Speed c in the figure equals c_s. The onset of the events P, SP, S and R (for Rayleigh wave) are clearly indicated. The arrival of the SP wave is in accord with our above discussion. Interesting is the fact that as r/H gets larger and larger we progress toward the surface source solution (compare the $H = 0$ case with that in fig. 6.14). It is also interesting to note

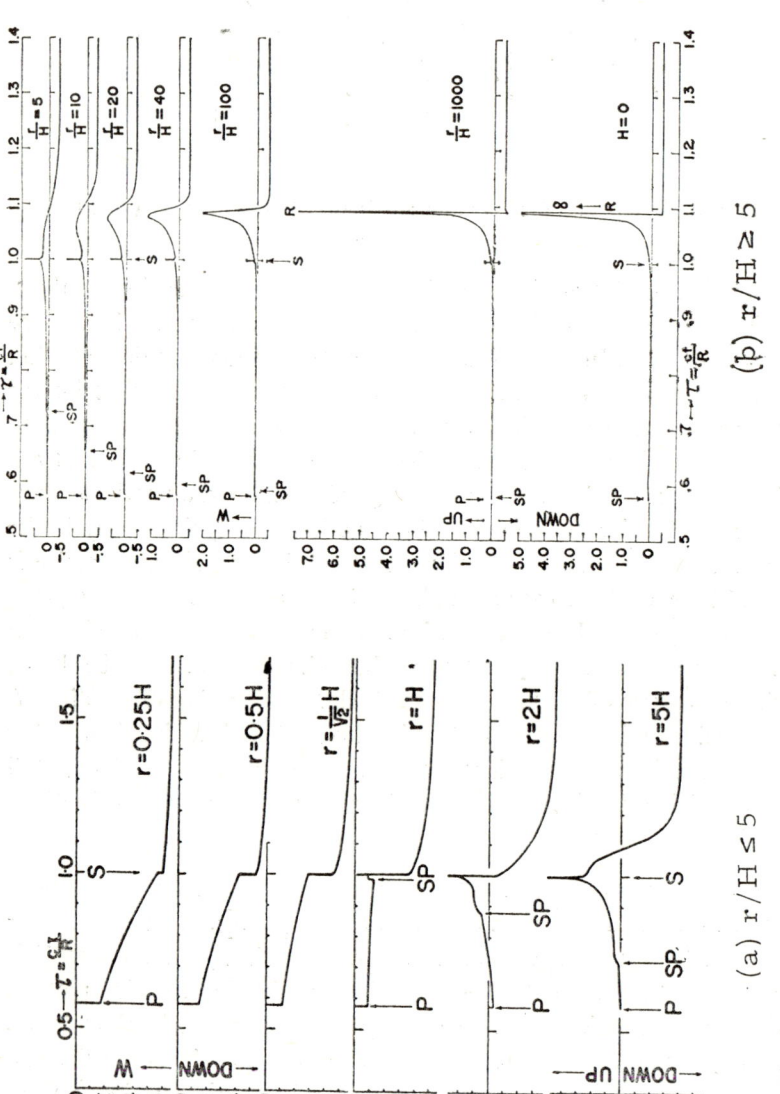

Fig. 6.18. Vertical displacement surface response $w(t, r, H) = u_z^H(r, -h, t)$ in the buried load problem as a function of the epicentral distances r (after Pekeris and Lifson).

Ch. 6, § 6.4]   A NOAXISYMMETRIC PROBLEM   347

the birth and development of the Rayleigh wave which evidently does not exist for $r < H/\sqrt{2}$. It begins to show up at $r = 5H$.

The corresponding response curves for the horizontal displacement are shown in fig. 6.19 (figs. 4, 5 of [6.10]). Here our $u_r^H(r, -h, t)$ is $q(t, r, H)$ $= -(3Z/\pi 2\mu R)Q'$, and $Q' = QR/r$ is the solution function plotted where $Q = -Q_p + Q_s$, Q_p giving the dilatational and Q_s the equivoluminal contributions to Q, respectively. Similar remarks to those discussed for w are obviously applicable here.

Pekeris and Lifson show that (1) the dilatational wavefront behaves as a step discontinuity and (2) for $r < H/\sqrt{2}$, so does the equivoluminal wavefront, but (3) for $r > H/\sqrt{2}$, the equivoluminal wavefront becomes logarithmically singular (two-sided). These might have been expected since one can imagine the P, S spherical waves here, generating reflections at the surface which are analogous to the surface source system of waves shown in fig. 6.7, hence creating the same type of disturbed regions as shown there. It follows then, since $r = H/\sqrt{2}$ is the line separating regular and nonregular wavefronts, that (1) and (2) above correspond to the regular wavefronts, and (3) the analogous two-sided equivoluminal front. Therefore we have *agreement with the wavefront behaviors* of the wave system *for the surface source*. Interesting now though, these discontinuities occur at the surface. Pekeris and Lifson also point out that the SP wave behaves as $(t - t')^{\frac{1}{2}}$ (t' is the arrival time of this event) at its front. It is clear that the Rayleigh wave is singular only when $H = 0$, and hence has the surface response behavior noted earlier for the surface source problem.

6.4. A nonaxisymmetric problem; the suddenly applied normal point load that travels on the surface

6.4.1. Statement of the problem and formal Laplace transformed solution

In [6.11] Gakenheimer and Miklowitz derived the exact solution for the displacements in the subject problem by employing the *Cagniard–de Hoop method*, in effect, extending this technique for the first time to a nonaxisymmetric problem. The problem is very interesting and important, but long and complicated. The plan here is to review the basic method of solution and evaluation of a portion of the results, which will enable the reader to go on further in [6.11] if he wishes. Some of the detail comes from Ga-

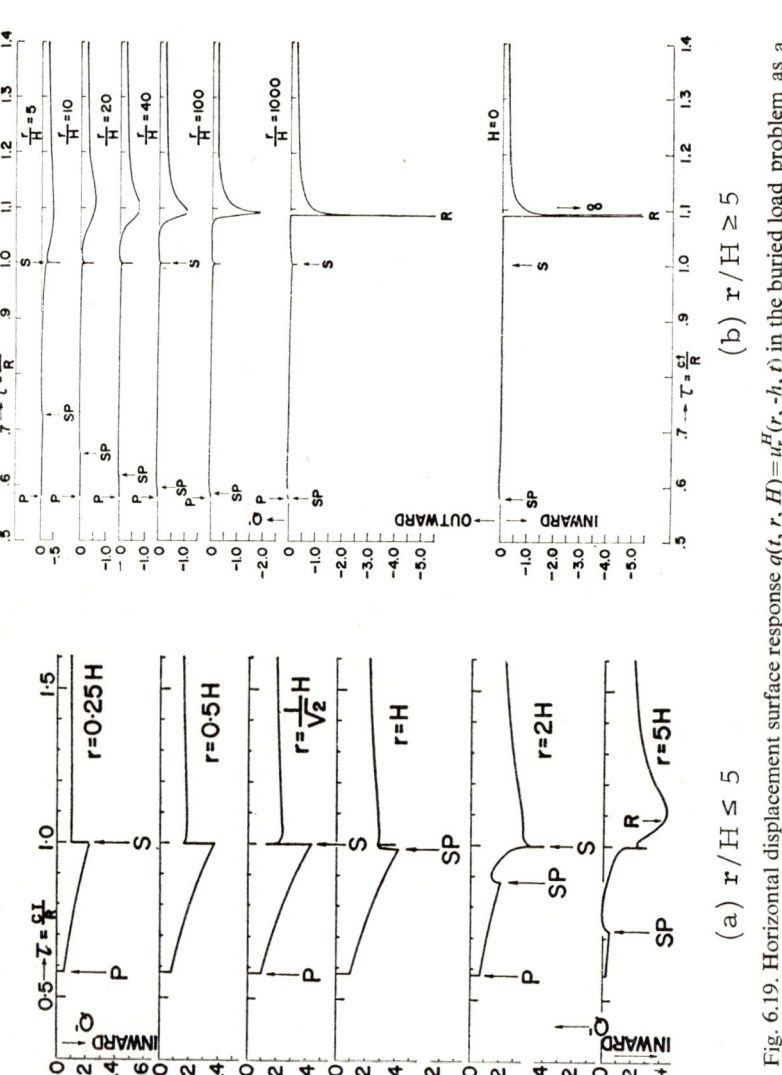

Fig. 6.19. Horizontal displacement surface response $q(t, r, H) = u_r^H(r, -h, t)$ in the buried load problem as a function of epicentral distances r (after Pekeris and Lifson).

Ch. 6, § 6.4]   A NONAXISYMMETRIC PROBLEM   349

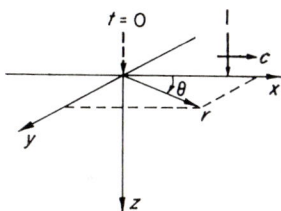

Fig. 6.20. Problem of traveling point load on surface of half space.

k nheimer's thesis [6.16] which underlies [6.11]. Figure 6.20 depicts the
problem. A point normal load of unit magnitude travels along the positive
x-axis with constant speed c, having acquired its velocity instantaneously at
the origin of coordinates (x, y, z) at $t=0$. This boundary-initial value
problem may be stated as follows:

$$\left.\begin{array}{l}\nabla^2\varphi(x,y,z,t)=\ddot{\varphi}/c_d^2 \\ \nabla^2\psi_j(x,y,z,t)=\ddot{\psi}_j/c_s^2\end{array}\right\} \text{in } -\infty<x,y<\infty, z>0, \text{ for } t>0, \text{ with } j=x,y,z, \tag{6.125}$$

$$\varphi(x,y,z,0)=\dot{\varphi}=\psi_j=\dot{\psi}_j=0, \quad \text{in } -\infty<x,y<\infty, z\geqq 0, \tag{6.126}$$

$$\left.\begin{array}{l}\sigma_z(x,y,0,t)=-\delta_s(y)\delta_s(x-ct) \\ \sigma_{zx}(x,y,0,t)=\sigma_{zy}(x,y,0,t)=0\end{array}\right\} \text{for } -\infty<x,y<\infty, t>0, \tag{6.127}$$

$$\lim\; [\varphi(x,y,z,t), \psi_j, \text{ and } \partial\varphi/\partial x, \partial\psi_j/\partial x, \text{ etc.}]=0, \quad \text{for } t>0.$$

$$\begin{array}{lll} x\to\pm\infty, & -\infty<y<\infty, & z\geqq 0 \\ y\to\pm\infty, & -\infty<x<\infty, & z\geqq 0 \\ z\to\infty, & -\infty<x,y<\infty \end{array} \tag{6.128}$$

To derive the solution we first apply the Laplace transform operator integral
to the governing potential equations of motion (6.125), using the initial
conditions (6.126). Next we introduce the *double exponential Fourier transform pairs*, defined through (5.40) by

$$\tilde{\tilde{f}}(\varkappa,\eta)=\int_{-\infty}^{\infty}\int f(x,y)e^{-i(\varkappa x+\eta y)}\mathrm{d}x\mathrm{d}y, \tag{6.129a}$$

$$(x,y)=\frac{1}{4\pi^2}\int_{-\infty}^{\infty}\int \tilde{\tilde{f}}(\varkappa,\eta)e^{i(\varkappa x+\eta y)}\mathrm{d}\varkappa\mathrm{d}\eta, \tag{6.129b}$$

where we have now used ($\tilde{}$) to denote the double exponential Fourier transform. Now, on the basis of the first two radiation conditions in (6.128), we apply (6.129a) to the Laplace transformed equations (6.125) with the results

$$\frac{d^2 \bar{\tilde{\varphi}}(\varkappa, \eta, z, p)}{dz^2} - n_d^2 \bar{\tilde{\varphi}} = 0,$$

$$\frac{d^2 \bar{\tilde{\psi}}_j(\varkappa, \eta, z, p)}{dz^2} - n_s^2 \bar{\tilde{\psi}}_j = 0, \qquad (6.130)$$

where $n_d = (\varkappa^2 + \eta^2 + k_d^2)^{\frac{1}{2}}$ and $n_s = (\varkappa^2 + \eta^2 + k_s^2)^{\frac{1}{2}}$. There are two solutions of each of (6.130) of the form, for example, $\exp(-n_d z)$, $\exp(n_d z)$. But the Laplace, followed by the double exponential Fourier, transform of the last radiation condition in (6.128) shows we must select n_d and n_s with positive real parts. Analogously to our procedure in Lamb's line load problem, with $\varkappa$, η real, we can make p real and positive and select the real positive branches of n_d and n_s to satisfy this condition. It follows the general solutions of the transformed potentials in (6.130) are

$$\bar{\tilde{\varphi}}(\varkappa, \eta, z, p) = A(\varkappa, \eta, p) \exp(-n_d z),$$

$$\bar{\tilde{\psi}}_j(\varkappa, \eta, z, p) = B_j(\varkappa, \eta, p) \exp(-n_s z). \qquad (6.131)$$

Now we expand the boundary conditions (6.127) with the aid of Hooke's law (1.45). After transforming these equations, and the gauge condition $\partial \psi_j / \partial x_j = 0$ (cf. § 2.2.6), we substitute (6.131) in them which leads to four equations for the unknowns A and B_j. The double Fourier inverse transform then yields (6.129b) for the Laplace transformed displacement solution to our problem

$$\bar{u}_j(x, y, z, p) = \bar{u}_{jd} + \bar{u}_{js}, \qquad (6.132)$$

where

$$\bar{u}_{j\alpha}(x, y, z, p) = \frac{1}{4\pi^2 \mu} \int_{-\infty}^{\infty} \int F_{j\alpha}(\varkappa, \eta, p) e^{-n_\alpha z + i(\varkappa x + \eta y)} d\varkappa d\eta, \qquad (6.133)$$

with $\alpha = d, s$ for the dilatational and equivoluminal parts of the displacement respectively, and

$$F_{xd}(\varkappa, \eta, p) = -i\varkappa n_0 G, \qquad F_{xs}(\varkappa, \eta, p) = 2i\varkappa n_d n_s G,$$

$$F_{yd}(\varkappa, \eta, p) = -i\eta n_0 G, \qquad F_{ys}(\varkappa, \eta, p) = 2i\eta n_d n_s G,$$

Ch. 6, § 6.4]   A NONAXISYMMETRIC PROBLEM   351

$$F_{zd}(\varkappa, \eta, p) = n_d n_0 G, \qquad F_{zs}(\varkappa, \eta, p) = -2n_d(\varkappa^2 + \eta^2)G,$$

$$G = 1/(p + ic\varkappa)T, \qquad T = n_0^2 - 4n_s n_d(\varkappa^2 + \eta^2),$$

$$n_0 = [k_s^2 + 2(\varkappa^2 + \eta^2)].$$

6.4.2. Preliminaries in the exact inversion by Cagniard–deHoop method

To invert (6.132) using the Cagniard–deHoop technique we first introduce the *transformation*

$$\varkappa = k_d \beta, \qquad \eta = k_d \sigma, \qquad (6.134)$$

into (6.133), followed by *another important transformation* first used by deHoop [6.17] in a three-dimensional problem in acoustics, namely

$$\beta = q \cos \theta - w \sin \theta, \qquad \sigma = q \sin \theta + w \cos \theta. \qquad (6.135)$$

The resultant new form of (6.133) is

$$\bar{u}_{j\varkappa}(r, \theta, z, p) = \frac{1}{2} \int_0^\infty \int_{-\infty}^\infty K_{jz}(q, w, \theta) e^{-(p/c_d)(m_\alpha z - iqr)} dq dw, \quad (6.136)$$

where

$$K_{xd}(q, w, \theta) = -[iqH \cos \theta + w^2 \sin^2 \theta] m_0 L,$$

$$K_{xs}(q, w, \theta) = 2[iqH \cos \theta + w^2 \sin^2 \theta] m_d m_s L,$$

$$K_{yd}(q, w, \theta) = -\sin \theta [iqH - w^2 \cos^2 \theta] m_0 L,$$

$$K_{ys}(q, w, \theta) = 2 \sin \theta [iqH - w^2 \cos^2 \theta] m_d m_s L,$$

$$K_{zd}(q, w, \theta) = m_0 m_d H L, \qquad K_{zs} = -2H(q^2 + w^2) m_d L,$$

$$L = \{\pi^2 c \mu [H^2 + w^2 \sin^2 \theta] R\}^{-1}, \qquad H = iq \cos \theta + \gamma,$$

$$R = m_0^2 - 4m_d m_s (q^2 + w^2), \qquad m_0 = k^2 + 2(q^2 + w^2),$$

$$m_d = (q^2 + w^2 + 1)^{\frac{1}{2}}, \qquad m_s = (q^2 + w^2 + k^2)^{\frac{1}{2}}, \qquad \gamma = c_d/c,$$

where (r, θ, z) are the cylindrical coordinates shown in fig. 6.20. In getting the form of (6.136) use was made of the evenness of the integrands with respect to w.

In view of the symmetry properties

$$\bar{u}_{x\alpha}(r, \theta, z, p) = \bar{u}_{x\alpha}(r, -\theta, z, p),$$
$$\bar{u}_{y\alpha}(r, \theta, z, p) = -\bar{u}_{y\alpha}(r, -\theta, z, p), \qquad (6.137)$$
$$\bar{u}_{z\alpha}(r, \theta, z, p) = \bar{u}_{z\alpha}(r, -\theta, z, p),$$

$\bar{u}_{j\alpha}$, and hence $\bar{u}_j$, need only be inverted for $0 \leq \theta \leq \pi$. The displacements $u_{j\alpha}$ have different forms depending on the *speed c of the load* relative to the body wave speeds. Therefore the inversion of each $\bar{u}_{j\alpha}$ is separated into three cases. They are the *supersonic case* defined by $c > c_d$, the *transonic case* defined by $c_s < c < c_d$, and the *subsonic case* defined by $c < c_s$. In the next section we carry out the inversion of $\bar{u}_{zd}$, the Laplace transform of the dilatational part of u_z, for the interior of the half space $z > 0$ and for the supersonic case. The related components u_{xd}, u_{yd} can be carried out similarly.

6.4.3. *Exact inversion for the interior and supersonic load motion; dilatational wave*

We consider $\bar{u}_{zd}(r, \theta, z, p)$ in (6.136). The inner integral can be treated on the basis of fixed w in the outer. Note in this that the outer integral can be written as

$$\frac{1}{2} \lim_{M \to \infty} \int_0^M [\ ] \mathrm{d}w,$$

which restricts w to a bounded domain for the q integration. This has certain conveniences in the analysis. The q-integral can be inverted by the Cagniard deHoop transformation

$$t = (1/c_d)(m_d z - iqr), \qquad t \text{ real and positive.} \qquad (6.138)$$

This transformation results in a path over which q is complex, hence the integrand singularities in the q-plane must be studied like in our earlier applications of this method. First there are the branch points of m_d and m_s,

$$Q_d^\pm = \pm i(w^2+1)^{\frac{1}{2}}, \qquad Q_s^\pm = \pm i(w^2+k^2)^{\frac{1}{2}}, \qquad (6.139)$$

respectively. Simple poles arise from the zeros of $H^2 + w^2 \sin^2 \theta$ and R in the denominator of L. They are located at, respectively,

$$Q_c^\pm = (\pm w\sin\theta + i\gamma)/\cos\theta, \qquad Q_R^\pm = \pm i(w^2 + \zeta_R^2)^{\frac{1}{2}}. \qquad (6.140)$$

We note that all these singularities migrate in the q-plane as a function of w. However, since $w \in [0, M)$, they always lie in a domain over which q is finite. For convergence of the inner integral in (6.136) it is clear that the Cagniard path must lie in the upper half of the q-plane. Fig. 6.21

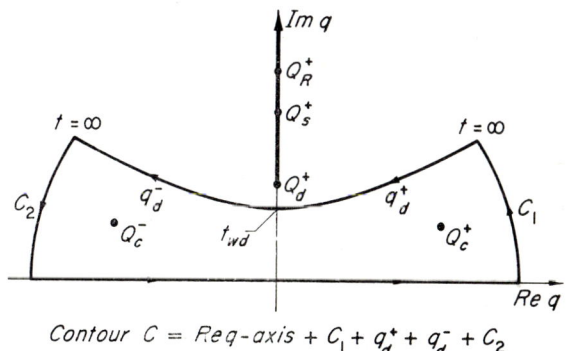

Fig. 6.21. Integration in the q-plane.

depicts this half plane, with its singularities at Q_d^+, Q_s^+, Q_R^+ and $Q_c^\pm$ along with the appropriate cut. Analysis of m_d, m_s is the same as that carried out for $\eta_d'(\zeta)$, $\eta_s'(\zeta)$ (cf. fig. 6.2 and related analysis), since here we lump fixed w^2 with 1 and k^2, respectively. The analysis shows that continuation off the real q-axis results in the required $\operatorname{Re} m_d > 0$, $\operatorname{Re} m_s > 0$ in the cut upper half q-plane. Solving (6.138) for q we find the Cagniard path is given by

$$q_d^\pm = (c_d/\varrho^2)\{\pm z[t^2 - (\varrho, c_d)^2(w^2+1)]^{\frac{1}{2}} + irt\}, \qquad (6.141)$$

where $t \geq (\varrho/c_d)(w^2+1)^{\frac{1}{2}} = t_{wd}$, and $\varrho = (r^2 + z^2)^{\frac{1}{2}}$. This can be shown to be a branch of a hyperbola again, with asymptotes

$$\arg q = \pm \tan^{-1}(r/z) \qquad (6.142a)$$

and vertex, corresponding to $t = t_{wd}$,

$$q_v = ir(w^2+1)^{\frac{1}{2}}/\varrho. \qquad (6.142b)$$

As $t \to \infty$, from this minimum time t_{wd}, we approach the asymptotes on $q_d^\pm$. Since $r/\varrho < 1$, for $z > 0$, from (6.139), (6.142b) it follows that $|q_v| < |Q_i^\pm|$, and therefore the Cagniard path does not intersect the cut. The path is shown in fig. 6.21 based on the assumption that $Q_c^\pm$ lie interior to the closed integration contour C also shown there. Now $Q_c^\pm$ will be interior to this contour if, and only if,

(a) $(-\pi/2) < \theta < \pi/2$ (or $x > 0$),

(b) $c_d tr/\varrho^2 > \gamma \cos\theta$ (or $t > t_L$,   where $t_L = \varrho^2/cx$),   (6.143)

(c) $w \tan\theta > (zc_d/\varrho^2)(t^2 - t_{wd}^2)^{\frac{1}{2}}$.

Condition (a) in (6.143) is necessary for $Q_c^\pm$ to be in the upper half q-plane, and to be finite there for w finite. This condition is related to the fact that the load moves over the $x > 0$ part of the half space. Condition (b) is just a statement that Im $q_d^\pm >$ Im $Q_c^\pm$, so that $Q_c^\pm$ are below $q_d^\pm$, respectively. The alternate statement $t > t_L$, derived from the first, defines t_L. For fixed t and $z > 0$, $t = t_L = \varrho^2/cx$ is a surface of a hemisphere with center at $x = ct/2$, $n = (y^2 + z^2)^{\frac{1}{2}} = 0$. The intersection of this surface with the xz-plane is a circle having its xy-plane intersections at the origin and the point under the traveling load. Surface t_L plays an important role in the form of the solution in this problem. Condition (c) is the statement that $|\text{Re } Q_c^\pm| > |\text{Re } q_d^\pm|$ so that the poles at $Q_c^\pm$ are to the right and left of paths $q_d^\pm$ respectively. The conditions (6.143) therefore insure that $Q_c^\pm$ are interior to C.

Conditions (b) and (c) are equivalent to the condition

$$w^2 > w_{od}^2 = z^2(\gamma^2 \varrho^2 - x^2)/r^2 n^2, \qquad (6.144)$$

which is not difficult to prove. Therefore the *conditions for* $q = Q_c^\pm$ *to lie inside C reduce to satisfying* $x > 0$ *and* $w^2 > w_{od}^2$. From (6.144) we see that for $x > \gamma\varrho = c_d\varrho/c$, $w^2 \geq 0$, since the largest value w_{od}^2 can have is zero. For $x < \gamma\varrho$, we have $w^2 > w_{od}^2$, a positive number. Therefore, the cone $x = \gamma\varrho$, with axis coincident with the positive x-axis, separates the half space into regions where the inversions for the dilatational displacement components will differ, i.e., the contributing dilatational waves will differ. The foregoing conditions on x/ϱ, and condition (a) in (6.143), $x > 0$, define the three regions depicted in fig. 6.22,

Ch. 6, § 6.4]   A NONAXISYMMETRIC PROBLEM   355

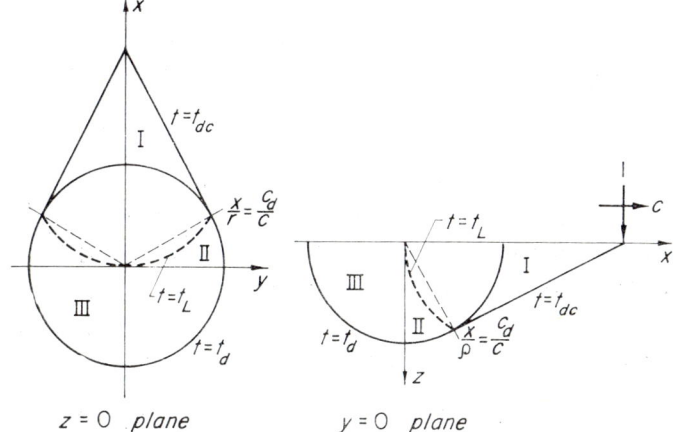

Fig. 6.22. Dilatational waves for supersonic load motion.

I: $x>0$, $\quad x/\varrho > \gamma = c_d/c$.

The poles at $q = Q_c^{\pm}$ lie inside C for $w \in [0, M)$.

II: $x > 0$, $x/\varrho < \gamma$. (6.145)

The poles at $q = Q_c^{\pm}$ lie inside C for $w \in (w_{od}, M)$ and they lie outside C for $w \in [0, w_{od})$.

III: $x < 0$.

The poles at $q = Q_c^{\pm}$ lie outside C for $w \in [0, M)$.

It remains to invert (6.136) for each of these regions. Inversion is carried out much as it was in our treatment of the line load problem. One uses the Cauchy–Goursat theorem and residue theory. Jordan's lemma shows that the contributions from the large circular paths C_1, C_2 are zero. Further, one uses the conjugate nature of the functions over the paths q_d^+, q_d^- to write real integrals. On this basis we have the $\bar{u}_{zd}$ inverses that follow for the regions of (6.145).

Region I

On the basis of the condition on w in (6.145) for this region, (6.136) gives

$$\bar{u}_{zd}(r, \theta, z, p) = \bar{A}_{zd}(r, \theta, z, p) + \bar{B}_{zd}(r, \theta, z, p),\qquad (6.146)$$

where

$$\bar{A}_{zd} = \lim_{M \to \infty} \int_0^M \int_{t_{wd}}^{\infty} \mathrm{Re}\left[K_{zd}(q_d, w, \theta) \frac{\mathrm{d}q_d}{\mathrm{d}t}\right] e^{-pt} \mathrm{d}t \mathrm{d}w,$$

and

$$\bar{B}_{zd} = \lim_{M\to\infty} \int_0^M \mathrm{Re}\{[\hat{K}_{zd}(q,w,\theta)e^{-(p/c_d)(m_d z - iqr)}]_{q=Q_c^+}\}\,dw,$$

where $q_d = q_d^+$ and $\hat{K}_{zd} = m_d m_0 \sec\theta/\pi\mu c R(q,w)$. The integrals in $\bar{A}_{zd}$ converge uniformly for all points (r,θ,z) in the interior $(z>0)$ as $M\to\infty$. This is because t_{wd} [given after (6.141)] also $\to\infty$, hence the exponential $e^{-pt}(p>0)$ in the integrand dominates. Also the integral in $\bar{B}_{zd}$ converges for all points (r,θ,z) in the interior $(z>0)$ because

$$\mathrm{Re}[(m_{dz}-iqr)|_{q=Q_c^\pm}] = \mathrm{Re}[m_d z + (\gamma r/\cos\theta)] > 0,$$

for all w. It follows that we can pass to the limit in $\bar{A}_{zd}$ and $\bar{B}_{zd}$ and they become

$$\bar{A}_{zd} = \int_0^\infty \int_{t_{wd}}^\infty \mathrm{Re}\left[K_{zd}(q_d,w,\theta)\frac{dq_d}{dt}\right] e^{-pt}\,dt\,dw, \qquad (6.147)$$

and

$$\bar{B}_{zd} = \int_0^\infty \mathrm{Re}\{[\hat{K}_{zd}(q,w,\theta)e^{-(p/c_d)(m_d z - iqr)}]_{q=Q_c^+}\}\,dw. \qquad (6.148)$$

The double integral in (6.147) is absolutely convergent since the only singularity in the integrand is the integrable one at $t=t_{wd}$, due to

$$\frac{dq_d}{dt} = \frac{c_d}{\varrho^2}\left[ir + zt(t^2 - t_{wd}^2)^{-\frac{1}{2}}\right],$$

and the exponential in the integrand predominates for large t and/or w (note again that $t_{wd}\to\infty$ as $w\to\infty$). The order of integration in (6.147) can therefore be interchanged (cf. [5.11, § 5.051]). First introducing the step function $H(t-t_{wd})$ in the inner integral and extending its lower limit to zero, (6.147) becomes, after the integration order interchange,

$$\bar{A}_{zd} = \int_0^\infty \int_0^\infty \mathrm{Re}\left[K_{zd}(q_d,w,\theta)\frac{dq_d}{dt}\right] H(t-t_{wd})\,dw\, e^{-pt}\,dt.$$

Now the inner integral must vanish for $t-t_{wd}<0$, because of the H function there. From this inequality we find that $w > [(t/t_d)^2 - 1]^{\frac{1}{2}} = T_d$, where $t_d = \varrho/c_d$. Hence, the upper limit of this integral must be T_d, and with this the H

function can then be set equal to 1. The lower limit $w=0$ gives $t_{wd}=t_d$ which becomes the lower limit of the outer integral. Inversion then gives

$$A_{zd}(r,\theta,z,t) = H(t-t_d) \int_0^{T_d} \mathrm{Re}\left[K_{zd}(q_d, w, \theta) \frac{dq_d}{dt} \right] dw . \quad (6.149)$$

A_{zd} represents a hemispherical dilatational wave for u_{zd}, its front arriving at $t=t_d=\varrho/c_d$. This wave emanates from the initial position of the load. The front is shown in fig. 6.22, marked $t=t_d$ there. Note the wave occurs in region I, which we are dealing with now.

We can also invert $\bar{B}_{zd}$ in (6.148) by the Cagniard-de Hoop method. Branch points of the integrand here are located at the zeros of $m_d(Q_c^+, w)$ and $m_s(Q_c^+, w)$. Poles are located at the zeros of the Rayleigh function $R(Q_c^+, w)$. Hence we have branch points at

$$w = S_d^\pm = i[-\gamma \sin\theta \pm (1-\gamma^2)^{\frac{1}{2}} \cos\theta] ,$$
$$\quad (6.150\mathrm{a})$$
$$w = S_s^\pm = i[-\gamma \sin\theta \pm (k^2-\gamma^2)^{\frac{1}{2}} \cos\theta] ,$$

and poles at

$$w = S_R^\pm = i[-\gamma \sin\theta \pm (\zeta_R^2 - \gamma^2) \cos\theta] , \quad (6.150\mathrm{b})$$

since $\gamma = c_d/c < 1$ for supersonic motion. The Cagniard-de Hoop contour is then obtained by solving

$$t = \frac{1}{c_d}(m_d^z - iqr)|_{q=Q_c^+} ,$$

for w. We find

$$w = w_d^\pm = -i\gamma \sin\theta + (\gamma \cos\theta/n^2)(i\xi y \pm \alpha_d z) , \quad (6.151)$$

for $t \geq t_{dc}$, where

$$t_{dc} = (1/c)\{[(c/c_d)^2 - 1]^{\frac{1}{2}} n + x\} ,$$
$$\alpha_d = \xi^2 - [(c/c_d)^2 - 1]^{\frac{1}{2}} n^2 , \qquad \xi = ct - x .$$

Equation (6.151) defines a branch of a hyperbola with vertex at

$$w_v = i[-\gamma \sin\theta + (1-\gamma^2)^{\frac{1}{2}}(y/n) \cos\theta] , \quad (6.152)$$

and asymptotes $\arg w = \tan^{-1}(y/z)$. Since the integration in (6.148) is defined over just the half range (positive) of w we need only w_d^+ in (6.151) and we drop the super (+) from here on. The singularities, branch cuts and integration contour (marked region I) are shown in fig. 6.23. Again here t varies monotonically from t_{dc} to ∞ which is indicated in the figure.

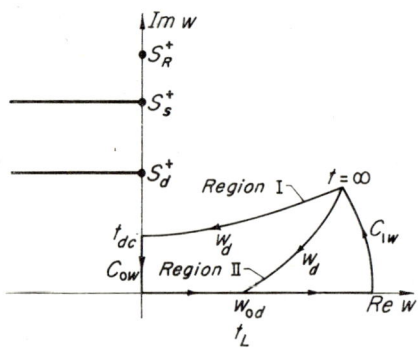

Fig. 6.23. Contour integration in the w-plane.

Since $x/\varrho > c_d/c$ and $y/n < 1$, w_d intersects the imaginary w-axis below S_d^+ and above the real w-axis. The proof is not difficult. Jordan's lemma can be used to show that the contribution of the integral along path C_{1w} is zero. Further, the integral along path C_{0w} vanishes because its real part is zero. Thus the contour integration along the closed path in fig. 6.23 involving the integral in (6.148) reduces the latter to

$$\bar{B}_{zd}(r,\theta,z,p) = \int_{t_{dc}}^{\infty} \mathrm{Re}\left[K_{zd}(w_d,\theta)\frac{\mathrm{d}w_d}{\mathrm{d}t}\right]e^{-pt}\mathrm{d}t,$$

which inverts to

$$B_{zd}(r,\theta,z,t) = H(t-t_{dc})\mathrm{Re}\left[\hat{K}_{zd}(w_d,\theta)\mathrm{d}w_d/\mathrm{d}t\right]. \quad (6.153)$$

This algebraic term represents a *conical wave* generated by and trailing behind the supersonic load. It is shown in fig. 6.22, its front marked by the arrival time $t = t_{dc}$. Note this wave lies in region I. The total contribution to u_{zd} in region I is therefore given by the sum of (6.149) and (6.153).
Region II
  On the basis of the condition on w in (6.145) for this region, (6.136) leads to (6.146), where

Ch. 6, § 6.4] A NONAXISYMMETRIC PROBLEM 359

$$\bar{A}_{zd}(r,\theta,z,p) = \int_0^\infty P.V. \int_{t_{wd}}^\infty \mathrm{Re}\left[K_{zd}(q_d, w, \theta)\frac{\mathrm{d}q_d}{\mathrm{d}t}\right]e^{-pt}\mathrm{d}t\mathrm{d}w ,$$

(6.154)

and

$$\bar{B}_{zd}(r,\theta,z,p) = \int_{w_{od}}^\infty \mathrm{Re}\{[\hat{K}_{zd}(q, w, \theta)e^{-(p/c_d)(m_d z - iqr)}]_{q=Q_c^+}\}\mathrm{d}w .$$

(6.155)

which form the analogous results to those of (6.147), (6.148) for region I. The above are justified since they are based on consideration of letting the upper limit M approach ∞, which turns out to be the same as in the region I case. The *principal value integral* arises in (6.154) because the poles $q = Q_c^\pm$ coalesce on $q_d^\pm$ at $t = t_L$ as $w \to w_{od}$. Gakenheimer expands on this in [6.16, pp. 25–28] by deriving (6.154), (6.155) from (6.136), representing the latter by a sum of four integrals, i.e. those over the integration limits on w of $(0, w_{od} - \varepsilon)$, $(w_{od} - \varepsilon, w_{od})$, $(w_{od}, w_{od} + \varepsilon)$ and $(w_{od} + \varepsilon, M)$. Equation (6.155) differs only from (6.148) for region I by the lower limit here, which is w_{od} instead of zero. This is expected since for the present region the poles $Q_c^\pm$ only lie inside C for $w > w_{od}$.

$\bar{A}_{zd}$ in (6.154) inverts just as it did in region I, so we have

$$A_{zd}(r,\theta,z,t) = H(t - t_d)P.V. \int_0^{T_d} \mathrm{Re}\left[K_{zd}(q_d, w, \theta)\frac{\mathrm{d}q_d}{\mathrm{d}t}\right]\mathrm{d}w . \quad (6.156)$$

A_{zd} here is the continuation of the hemispherical dilatational wave from region I into region II. It is shown in fig. 6.22, its front labeled $t = t_d$.

The contour integration in the w-plane for $\bar{B}_{zd}$ is like that for region I too. Here, however, the contour must start and end with w_{od} [the lower limit of integration in (6.155)] on the positive real w-axis. It is marked by region II in fig. 6.23. Here the singularities S_d^+, S_s^+ and S_R^+ may lie below the real w-axis, but still on the imaginary w-axis, in a way not shown in fig. 6.23. However, since the particular contour here does not intersect the imaginary w-axis, this has no bearing on the contour integration for this region. $\bar{B}_{zd}$ in (6.155) becomes, therefore,

$$\bar{B}_{zd}(r,\theta,z,p) = \int_{t_L}^\infty \mathrm{Re}\left[\hat{K}_{zd}(w_d, \theta)\frac{\mathrm{d}w_d}{\mathrm{d}t}\right]e^{-pt}\mathrm{d}t , \quad (6.157)$$

where t_L corresponds to w_{od} as pointed out earlier. Inversion of (6.157) gives

$$B_{zd}(r, \theta, z, t) = H(t-t_L) \operatorname{Re}\left[\hat{K}_{zd}(w_d, \theta) dw_d/dt\right]. \tag{6.158}$$

B_{zd} has the same algebraic form here as in region I, but the conical front there is replaced with the non-wave hemispherical surface $t=t_L$, shown by the dashed line in fig. 6.22. The dilatational displacement components (u_{zd} here), and their derivatives, are expected to be smooth through $t=t_L$, since this surface is not a characteristic surface of the governing wave equation on φ. Consider fig. 6.22 again. Striking a ray in region II from the origin to the dilatational hemispherical wave front $t=t_d$, we see this ray intersects the leading non-wave surface of B_{zd}. According to (6.158) action goes on in the time interval $t_L < t < \infty$. We note also that $t_L > t_d$. Returning now to $\bar{A}_{zd}$ in (6.154), this same crossing point at $t=t_L$ defines the point in the integration of the inner integral of (6.154), which gives rise to the principal value indicated there, i.e., the integrand behaves as $1/(t-t_L)$ near t_L, where $t_d < t_L < \infty$. The displacement z_{zd} for region II then, is represented by the sum of the right hand sides of (6.156), (6.158). It should be pointed out that B_{zd} in (6.158) has a step discontinuity at $t=t_L$ but a like singularity of opposite sign from the integral of (6.156) renders u_{zd} continuous at this time.

Region III

This region is governed by the third of (6.145), $x<0$. We note there that the poles at $q=Q_c^\pm$ lie outside of the contour C, and therefore $B_{zd} \equiv 0$ here. It follows we are left with only A_{zd} which is the same as it is in region I, i.e., (6.149). This continuation of the hemispherical dilatational wave from region II to III is depicted in fig. 6.22.

A compact expression for u_{zd}, valid for all three regions, is

$$u_{zd}(r, \theta, z, t) = H(t-t_d) P.V. \int_0^{Td} \operatorname{Re}\left[K_{zd}(q_d, w, \theta) \frac{dq_d}{dt}\right] dw$$

$$+ H(t-t_{dc}) H(t-t_L) H(x) \operatorname{Re}[\hat{K}_{zd}(w_d, \theta) dw_d/dt]. \tag{6.159}$$

Equation (6.159) is valid for $0 \leq \theta \leq \pi$, $z > 0$. Striking a few rays in fig. 6.22, from the origin into the particular regions we have dealt with will show (6.159) agrees with our results for the individual regions. Analytical proofs are easily devised too.

6.4.4. Wavefront approximations

In [6.11] wavefront approximations are deduced from the exact solutions. The details will not be given here. However, the results corresponding to (6.159), and others in the problem, are of interest and will be stated. The following remarks hold for $0 \leq r < \infty$, $0 \leq \theta \leq \pi$, $0 < z < \infty$ and all the load speeds, $0 \leq c < \infty$, subject to certain restrictions that will be noted later. At the hemispherical dilatational wavefronts of the displacements (from first term in (6.159) for u_{zd}) there is a step discontinuity, and likewise for the hemispherical equivoluminal wavefronts, except the latter has a two-sided logarithmic singularity when θ is in the domain in which critical reflection occurs. The head wavefront, associated with these hemispherical fronts, behaves as a ramp function. It may therefore be observed that this system of fronts, which stem from the applied load at $t=0$, are in agreement with those of Lamb's problem for the step normal point load. This was anticipated because of the relation of the system to $t=0$. These results do not hold along certain rays defining regions [cf. 6.11].

Interesting are the conical waves for the dilatational displacements, that for u_{zd} being given by the second term in (6.159), which have the half order singularity $(t-t_{dc})^{-\frac{1}{2}}$ at their wavefronts. The conical waves for the equivoluminal displacements behave similarly at $t=t_{sc}$, i.e., like $(t-t_{sc})^{-\frac{1}{2}}$. In addition there is a half order singularity for the equivoluminal displacements in front of $t=t_{sc}$ i.e., like $(t_{sc}-t)^{-\frac{1}{2}}$. There is also a conical head wave, the front of which behaves as $(t-t_{sdc})^{\frac{1}{2}}$. Note this is continuous through its front like the behavior of the ordinary head wave at t_{sd}. These results of course are subject to obvious conditions on the load speed. For example the conical dilatational wave only arises for $c > c_d$ and $x/\varrho \geq c_d/c$. Other cases are discussed in [6.11]. Finally we should note that comparing corresponding waves from the conical system trailing behind the load, and the system emanating from the initial position of the load, the wavefronts of the conical system have a half order stronger singularity, e.g., $u_d \sim O$ $[H(t-t_d)]$, $u_{dc} \sim O[(t-t_{dc})^{-\frac{1}{2}}]$. The two-sided equivoluminal waves are an exception, but the conical wave here does have a stronger singularity.

No numerical evaluations of the solutions for the present traveling point load problem were carried out in [6.11]. However, later related work by Gakenheimer [6.18] on the expanding ring and disk surface loads gives numerical information for several quantities of interest. For example, the dependence of the solution on the load speed relative to the body wave speeds is nicely brought out. Among other things, the reader will also find nteresting plots of particle trajectories for a particle originally just below the half space surface for three different speeds of the expanding disk load, i.e., supersonic, transonic and subsonic.

6.5. Exercises

6.1. For (6.12) of the text, show that since $z \geq 0$,

$$\lim_{x \to \pm\infty} \left\{ \begin{matrix} \tilde{u} \\ \tilde{w} \end{matrix} \right\} = 0,$$

consistent with (5.41). Consider both cases, $z > 0$, and $z = 0$, and note in the latter case, with p fixed, both finite and infinite p must be considered, i.e., p can become infinite on Br too.

6.2 Show, for the branches of $\eta'_d(\zeta)$ and $\eta'_s(\zeta)$ we chose in the derivation of § 6.2.1.1, that

$$-\frac{\pi}{2} < \arg \left\{ \begin{matrix} \eta'_d(\zeta) \\ \eta'_s(\zeta) \end{matrix} \right\} < \frac{\pi}{2} \quad \left(\text{hence } \operatorname{Re} \left\{ \begin{matrix} \eta'_d(\zeta) \\ \eta'_s(\zeta) \end{matrix} \right\} > 0 \right)$$

in the cut lower half plane of fig. 6.2.

6.3. Show, for the branches in exercise 6.2, that

$$R(s) = (k^2 + 2\zeta^2)^2 - 4\zeta^2 \eta'_d(\zeta) \eta'_s(\zeta)$$

has a zero at $\zeta = -i\zeta_R$, identifying it for the case of $\nu = 1/4$ with

$$k_R = c_R/c_s = (2 - 2/\sqrt{3})^{\frac{1}{2}}.$$

Show also in this case, that $R(\zeta)$ does not have for these branches, the other k_R roots, $k_R = 2, (2 + 2/\sqrt{3})^{\frac{1}{2}}$ we noted in our earlier work on Rayleigh waves (cf. § 3.1.4.7). You will find useful (3.65), or (3.66) and (3.68) in your proof.

6.4. Derive (6.41) and (6.42), the contributions to the Laplace transformed surface displacements due to the Rayleigh pole.

6.5. Derive the Rayleigh wave surface displacements corresponding to $F(t) = H(t)$, i.e., $u_{rR}^H(r, \pi/2, t)$ and $u_{\theta R}^H(r, \pi/2, t)$ for $(r/c_R) - \varepsilon < t < (r/c_R) + \varepsilon$, where ε is a small positive number.

6.6. Derive (6.57), the exact solution for the response at the plane of symmetry in Lamb's line load problem for the half space.

6.7. A uniform tangential line load $QF(t)$ (Q is magnitude force per unit length) is suddenly applied to the surface of an elastic half space along the

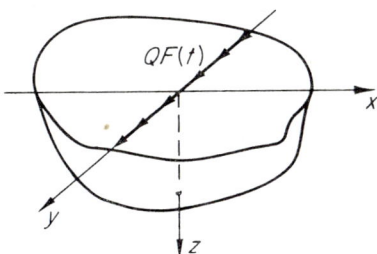

y axis, as depicted in the sketch. Noting that the surface loading here is represented by the stress

$$\sigma_{zy}(x, 0, t) = -Q\delta_s(x)F(t),$$

and that this generates SH waves, formulate the boundary-initial value problem and show with the aid of the Cagniard–de Hoop method it has the solution

$$v^\delta(r, t) = (Q/\pi\mu)[t^2 - (r/c_s)^2]^{-\frac{1}{2}} H(t - r/c_s), \quad 0 < r < \infty, \quad r/c_s < t < \infty,$$

where v is the displacement in the y direction, superscript δ denoting the response to the delta function (the special case $F(t) = \delta(t)$), and $r = (x^2 + z^2)^{\frac{1}{2}}$.

6.8. A strongly related problem to that in exercise 6.7 is that of the uniform v-displacement line source at $r = 0$ in the infinite elastic medium. Noting that the governing equation here is $\nabla^2 v(r, t) = \ddot{v}/c_s^2$, where ∇^2 is the Laplacian for the axially symmetric case, show by a simple Laplace transform inversion that the solution is

$$v^\delta(r, t) = v_0[t^2 - (r/c_s)^2]^{-\frac{1}{2}} H(t - r/c_s), \quad 0 < r < \infty, \quad r/c_s < t < \infty,$$

where v_0 is the magnitude constant of the displacement source. The equivalence of this solution and that of exercise 6.7 stems from the fact that SH waves undergo no mode conversion at a boundary.

6.9. Derive the wavefront approximation for the *regular* equivoluminal waves $u_s^\delta(x, z, \tau)$ and $w_s^\delta(x, z, \tau)$ by (1) the Cagniard-de Hoop method in § 6.2.1.4, and (2) the method of steepest descents in § 6.2.1.5.

6.10. Derive the wavefront approximations for the *head* wave $w_{ds}^\delta(x, z, \tau)$ and the *two-sided* equivoluminal wave $w_s^\delta(x, z, \tau)$ by (1) the Cagniard-de Hoop method in § 6.2.1.4 and (2) the method of steepest descents in § 6.2.1.5.

6.11. Derive the transformed displacement-potential relations (6.112) and transformed stress–potential relations (6.114).

6.12. Referring to Fig. 6.13, verify the values of α and β on the imaginary axis given there.

6.13. Show that the contributions from the paths C_U and C_L are zero, making (6.119) and (6.120) valid.

6.14. In § 6.3.1.3 it was remarked that the solution obtained for the surface displacements, in the problem of the buried vertical point force, exhibited (1) a step discontinuity at the dilatational wavefront and (2) for $r < H/\sqrt{2}$ the same behavior for the equivoluminal wavefront, but (3) for $r > H/\sqrt{2}$ a two-sided logarithmic singularity for the latter. Show with some sketches that the system of reflected P and S spherical waves, generated by the incident P and S waves, is in agreement with the system of waves for the surface normal point load (which is the same as in fig. 6.7.). Noting now in the present problem that the singularities occur on the surface, and for the surface source problem in the interior (cf. figs. 6.15, 6.16), we see our example is a special case of the elastodynamic reciprocity theorem (cf. § 1.13).

6.15. By following the procedure outlined in § 6.4.1 verify that (6.132), (6.133) give the Laplace transformed displacement solution for the traveling point load problem (6.125)–(6.128).

6.16. Show that the transformations (6.134) and (6.135) reduce (6.133) for $\bar{u}_{zd}$ to $\bar{\bar{u}}_{zd}$ in (6.136). Show also that (6.138) has the Cagniard solutions

(6.141) forming a branch of a hyperbola with vertex $q_v = i(w^2+1)^{\frac{1}{2}} r/\varrho$ and asymptotes Im $q = \pm \text{Re } q(r/z)$ [or arg $q = \pm \tan^{-1}(r/z)$].

6.17. Derive the Region I transformed displacement solution $\bar{u}_{zd}$ given by (6.146).

6.18. As shown in fig. 6.23 the Region I path for w_d has its vertex w_v, given in (6.152), on the imaginary axis of w at a point lying between S_d^+ and the real axis. Prove that this is the case.

6.19. The Region II transformed displacement solution $\bar{u}_{zd}$ is given by (6.146) in which $\bar{A}_{zd}$ is given by (6.154). Here a principal value integral arises in (6.154) because the poles $q = Q_c^{\pm}$ coalesce on $q_d^{\pm}$ at $t = t_L$ as $w \to w_{od}$. Prove this. What does it imply about the behavior of the integral of the inner integral in (6.154) near $t = t_L$?

References

[6.1.] H. Lamb, *Philosophical Transactions of the Royal Society (London)* A **203** (1904), 1–42.
[6.2.] A. T. de Hoop, *The Surface Line Source Problem*. Second Annual Report, Seismic Scattering Project; Institute of Geophysics, U. C. L. A., Los Angeles, California (1956–1957).
[6.3.] M. J. Forrestal, L. E. Fugelso, G. L. Neidhardt and R. A. Felder, *Proceedings Engineering Mechanics Division Specialty Conference*, American Society of Civil Engineers (1966), 719–751.
[6.4.] R. L. Rosenfeld and J. Miklowitz, *Proceedings of the Fourth U. S. National Congress of Applied Mechanics*, American Society of Mechanical Engineers, New York, N. Y. (1962), 293–303.
[6.5.] L. Knopoff and F. Gilbert, *Journal of the Acoustical Society of America* **31** (1959), 1161–1168.
[6.6.] R. L. Rosenfeld, *Analysis of Long Compressional Elastic Waves in Rods of Arbitrary Cross Section and Elastic Wave Fronts in Plates and Circular Rods*, Ph. D. Thesis, California Institute of Technology, Pasadena, California (1962).
[6.7.] W. W. Garvin, *Proceedings of the Royal Society (London)* A **234** (1956), 528–541.
[6.8.] E. R. Lapwood, *Philosophical Transactions of the Royal Society (London)* A **242** (1949), 63–100.
[6.9.] C. L. Pekeris, *Proceedings of the National Academy of Sciences* **41** (1955), 469–480.
[6.10.] C. L. Pekeris and H. Lifson, *Journal of the Acoustical Society of America* **29** (1957), 1233–1238.
[6.11.] D. C. Gakenheimer and J. Miklowitz, *Journal of Applied Mechanics* **36** (1969), 505–515.
[6.12.] D. C. Gakenheimer, *Journal of Applied Mechanics* **37** (1970), 522–524.
[6.13.] H. R. Aggarwal and C. M. Ablow, *International Journal of Engineering Science* **5** (1967), 663–679.

[6.14.] C. L. Pekeris, *Proceedings of the National Academy of Sciences* **41** (1955), 629–639.
[6.15.] C. L. Pekeris, *Proceedings of the National Academy of Sciences* **42** (1956), 439.
[6.16.] D. C. Gakenheimer, *Transient Excitation of an Elastic Half Space by a Point Load Traveling on the Surface*, Ph. D. Thesis, California Institute of Technology, Pasadena, California (1969).
[6.17.] A. T. de Hoop, *Applied Scientific Research* B **8** (1959), 349–356.
[6.18.] D. C. Gakenheimer, *Journal of Applied Mechanics*, **38** (1971), 99–110.

CHAPTER 7

TRANSIENT WAVES IN ELASTIC WAVEGUIDES

The present chapter is a natural extension of ch. 4, which through addition of the time variable, will enable us to discuss more physically realistic waveguide problems. In effect here we learn the techniques for integrating over the frequency spectra set down in ch. 4.

7.1. Approximate theories and one dimensional problems

There has been a strong interest in approximate theories in the literature on elastic waveguides, stemming essentially from the complexity of the exact equations of linear elasticity. Attempts are made in these theories to *retain the essential physics of the exact equations along with simplifying features of the elementary theories such as one-dimensionality, and "plane sections remain plane". Comparison is usually made between the exact and approximate theory frequency spectra*, to ascertain the range (frequency and wavelength) of validity of the latter. *The goal is usually a good approximation up to moderately high frequencies and out to moderately short waves.* Our interest here, of course, is the use of these theories in transient wave propagation problems. The literature on approximate theories is vast. We will narrow our interest down to the cylindrical elastic rod on which a good share of the work has been done. In the survey by Miklowitz [4.4] the reader will find sources for further reading in a detailed discussion of the contributions not only for the rod, but for approximate theories for the plate and cylindrical shell.

7.1.1. Approximate theories for axially symmetric compressional waves in a rod

We return to ch. 4 §§ 4.4, 4.4.2 to consider this case. Referring to fig. 4.19, in the well-known *elementary theory* for this case it is assumed (1) that plane sections before deformation remain plane throughout the deformation, and (2) that all stresses vanish except the axial one σ_z which is always uniform across the plane sections. It follows that the stress equations of motion (1.64) for the present case reduce to the one-dimensional equation

$$\frac{\partial \sigma_z(z,t)}{\partial z} = \varrho \ddot{u}_z(z,t) - Z, \qquad (7.1)$$

where Z is the body force term (X_3). Using (4.91), (4.93) and (1.54), where in these last equations x, y are replaced by r, θ, we also find that

$$\varepsilon_z = \partial u_z/\partial z = \sigma_z/E, \quad \varepsilon_r = \partial u_r/\partial r = -v\partial u_z/\partial z, \quad \varepsilon_\theta = u_r/r = -v\partial u_z/\partial z, \qquad (7.2)$$

where the shear displacement gradient $\partial u_r/\partial z$ has been set $=0$ to satisfy $\sigma_{zr}=0$. The last two relations in (7.2) show that we have Poisson's ratio coupling in this theory. Note that with plane sections remaining plane, the existence of u_r means that these sections must extend or contract in their own plane. Substitution of the first of (7.2) in (7.1) gives

$$\frac{\partial^2 u(z,t)}{\partial z^2} = \frac{\ddot{u}(z,t)}{c_b^2} - \frac{Z}{E}, \qquad (7.3)$$

where we have let $u_z = u$, and where $c_b = (E/\varrho)^{\frac{1}{2}}$. Setting $Z=0$, and substituting the infinite wave train

$$u(z,t) = A \exp\left[i\varkappa(z-ct)\right] \qquad (7.4)$$

in (7.3), shows the frequency-wave number relation here is

$$\Omega = (c_b/c_s)\gamma, \qquad (7.5)$$

where Ω and γ were defined in connection with the exact theory rod spectra [cf. discussion following (4.97)]. Noting from the discussion there that (7.5) is the lowest branch (marked 1 in fig. 4.15) asymptote for Ω, $\gamma \to 0$ together, we see (7.5), and hence (7.3), is exact only for $\Omega = \gamma = 0$, i.e., for infinitely long (with respect to a) and vanishing frequency (with respect to ω_s) waves. It can, however, be used as *an approximate theory* for disturbances that are composed of *waves that are very long and of very low frequency*. It should be emphasized, however, that because of (7.5) the theory will not describe dispersion in a rod. It is clear that (7.3) is independent of the rod's cross-sectional shape. Hence, a smooth low ferquency input to the end of a long narrow rod of uniform circular, rectangular, or any other sectional shape, could be treated on the basis of this theory. The reader will find practice in the treatment of these elementary transient wave problems for the rod (and string) in the exercises of Chapter 5.

To improve the elementary theory frequency equation (7.5) Rayleigh [7.1] obtained a correction based on consideration of the *radial inertia* of the rod element. The resultant *second approximation* he found to the lowest branch is given by

$$\Omega = \left(\frac{c_b}{c_s}\right)\gamma\left(1 - \frac{v^2\sigma^2}{4}\gamma^2\right), \qquad (7.6)$$

which is equivalent to the second approximation to the lowest branch of (4.97) *found earlier by Pochhammer* (σ is defined in discussion following (4.96)). A sketch of these approximations is shown in fig. 7.1. It is clear

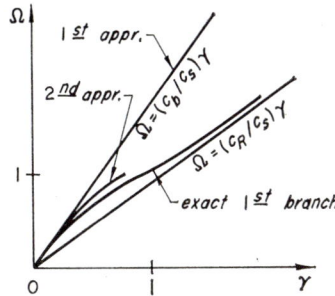

Fig. 7.1. Approximations to lowest branch of Pochhammer frequency equation.

the approximation introduces the needed dispersion in the physics. Later Love [1.2, § 278] derived the equation of motion for the rod which took into account Rayleigh's radial inertia correction. To derive the equation of motion and consistent boundary conditions Love drew on Hamilton's principle. Indeed one of the important uses of the principle is in deriving approximate theories in elasticity. We take out time to derive Hamilton's principle in the next section before putting it to work in deriving our first higher order approximate theory, that for the Love–Rayleigh rod.

7.1.1.1. Hamilton's principle. Consider our elastic body V again. We assume that the body has states which vary continuously between the instants t_0 and t_1. Let the displacements u_i undergo the *variations* δu_i (subject to the rules of the calculus of variations[1]) within the time interval (t_0, t_1) but have prescribed values at the end points t_0 and t_1. The corresponding variation of the potential energy of deformation (strain energy) of the body V, according to (1.77), is

$$\delta U(t) = \int_V \sigma_{ij} \delta \varepsilon_{ij} \mathrm{d}V = \tfrac{1}{2} \int_V \sigma_{ij} (\delta u_{i,j} + \delta u_{j,i}) \mathrm{d}V , \tag{7.7}$$

where it should be noted *in varying this integral only the displacement was subject to variation in agreement with our earlier postulate*. The same will apply to other integrals arising here. Applying (1.57) of the divergence theorem to the first part of the second integral in (7.7) it is not difficult to show that

$$\int_V \sigma_{ij} \delta u_{i,j} \mathrm{d}V = \int_S \sigma_{ij} \delta u_i l_j \mathrm{d}S - \int_V \sigma_{ij,j} \delta u_i \mathrm{d}V , \tag{7.8}$$

where S denotes the surface bounding V. Since $\sigma_{ji} = \sigma_{ij}$, and i, j in (7.7), (7.8) are dummy subscripts, the second term in (7.7) is also represented by the right hand side of (7.8). Using (7.8) in (7.7) then we find that

$$\delta U(t) = \int_S \sigma_{ij} l_j \delta u_i \mathrm{d}S - \int_V \sigma_{ij,j} \delta u_i \mathrm{d}V . \tag{7.9}$$

Now making use of (1.63), (1.64) in (7.9) and integrating it over the time interval (t_0, t_1), we have

$$\int_{t_0}^{t_1} \delta U(t) \mathrm{d}t = \int_{t_0}^{t_1} \delta W(t) \mathrm{d}t - \varrho \int_{t_0}^{t_1} \int_V \ddot{u}_i \delta u_i \mathrm{d}V \mathrm{d}t , \tag{7.10}$$

where $\delta W(t)$ is the variation of the work done on the body V given by

$$\delta W(t) = \int_V X_i \delta u_i \mathrm{d}V + \int_S T_i \delta u_i \mathrm{d}S , \tag{7.11}$$

in agreement with (1.71). The second integral in (7.10) is related to the variation of the kinetic energy of the body K. According to (1.74) this variation is given by

$$\delta K(t) = \varrho \int_V \dot{u}_i \delta \dot{u}_i \mathrm{d}V = \varrho \int_V \frac{\partial}{\partial t} (\dot{u}_i \delta u_i) \mathrm{d}V - \varrho \int_V \ddot{u}_i \delta u_i \mathrm{d}V . \tag{7.12}$$

---

[1] cf. Jeffreys and Jeffreys [5.11, ch. 10].

Ch. 7, § 7.1]   APPROXIMATE THEORIES   371

Integrating (7.12) over (t_0, t_1) we find

$$\int_{t_0}^{t_1} \delta K(t) dt = -\varrho \int_{t_0}^{t_1} \int_V \ddot{u}_i \delta u_i dV dt \tag{7.13}$$

since δu_i vanishes at t_0 and t_1. Hence substituting the left hand side in (7.13) for the right (which appears in (7.10)), we find that (7.10) becomes

$$\delta \int_{t_0}^{t_1} [K(t) - U(t)] dt + \int_{t_0}^{t_1} \delta W(t) dt = 0, \tag{7.14}$$

where K and δW are given by (1.74) and (7.11), respectively, and U by

$$U(t) = \int_V \hat{U} dV = \tfrac{1}{2} \int_V \sigma_{ij} \varepsilon_{ij} dV. \tag{7.15}$$

Equation (7.14) states Hamilton's principle. Our proof is essentially that given by Nowacki [1.3, § 1.10]. *Whenever a strain energy function $\hat{U}$ exists, we can derive the corresponding equations of motion from Hamilton's principle* (7.14). Further, as we shall see the *principle generates the associated boundary conditions*.

7.1.1.2. Love–Rayleigh rod. Except for the addition of radial inertia, the assumptions in this theory are the same as those in the elementary theory discussed in § 7.1.1. Following Love, we consider the *lateral inertia* of an element of a uniform rod of arbitrary-shaped cross section. The lateral displacements of a point in the section, with coordinates (x, y) measured from axes through the section centroid, are, noting the form of u_r in the last of (7.2),

$$u_1 = -\nu x \partial u / \partial z, \qquad u_2 = -\nu y \partial u / \partial z.$$

Note again here since plane sections remain plane, these sections must extend or contract in their own planes. The lateral velocities follow from the displacements through a simple differentiation with respect to t. These, and the axial velocity, define $K(t)$ through (1.74), i.e.,

$$K(t) = \frac{\varrho A}{2} \int_0^l \left[(\dot{u})^2 + \nu^2 r_{gp}^2 \left(\frac{\partial \dot{u}}{\partial z} \right)^2 \right] dz, \tag{7.16}$$

where A, r_{gp}, and l are, respectively, the area, polar radius of gyration, and length of the rod. In getting (7.16) we have made use of the fact that u, and its derivatives, are independent of the cross sectional variables, and hence the integration over the section. From (7.15), U here is simply

$$U(t) = \frac{EA}{2} \int_0^l \left(\frac{\partial u}{\partial z}\right)^2 dz. \tag{7.17}$$

Since the only stress that exists here is σ_z, only those body and surface forces exist which are compatible with σ_z. It follows that only the axial body force Z, and the end applied stresses $\sigma_z(0,t)$ and $\sigma_z(l,t)$ exist. From (7.11) we therefore have

$$\delta W(t) = \int_V Z \delta u \, dV + \int_S \sigma_z \delta u \, dS = A\left[\int_0^l Z \delta u \, dz + [\sigma_z \delta u]_0^l\right]. \tag{7.18}$$

Now (7.16), (7.17) and (7.18) are substituted into (7.14) which gives

$$\delta \int_{t_0}^{t_1} dt \int_0^l \frac{1}{2}\left\{\varrho\left[(\dot{u})^2 + v^2 r_{gp}^2 \left(\frac{\partial \dot{u}}{\partial z}\right)^2\right] - E\left(\frac{\partial u}{\partial z}\right)^2\right\} dz$$
$$+ \int_{t_0}^{t_1} \left\{\int_0^l Z \delta u \, dz + [\sigma_z \delta u]_0^l\right\} dt = 0. \tag{7.19}$$

Consider the first term in the first integral of (7.19). We interchange the integrations on t and z, and then our variation reduces to consideration of

$$\int_{t_0}^{t_1} \delta(\dot{u})^2 dt = 2 \int_{t_0}^{t_1} \dot{u} \frac{\partial(\delta u)}{\partial t} dt.$$

A parts integration on this gives

$$\int_{t_0}^{t_1} \delta(\dot{u})^2 dt = 2\left\{[\dot{u} \delta u]_{t_0}^{t_1} - \int_{t_0}^{t_1} \ddot{u} \delta u \, dt\right\}. \tag{7.20}$$

But $\delta u = 0$ at t_0 and t_1, so our result is the second term in (7.20). The third term in the first integral is carried out in the same manner, except that here we do not interchange integrations on t and z. We have for this term then,

$$\int_0^l \delta\left(\frac{\partial u}{\partial z}\right)^2 dz = 2\int_0^l \frac{\partial u}{\partial z} \frac{\partial(\delta u)}{\partial z} dz = 2\left\{\left[\frac{\partial u}{\partial z} \delta u\right]_0^l - \int_0^l \frac{\partial^2 u}{\partial z^2} \delta u \, dz\right\}. \tag{7.21}$$

Again, the process is similar for the mixed derivative term in (7.19) except that here we carry out the integration on t, followed by one on z. We find

$$\int_0^l dz \int_{t_0}^{t_1} \delta \left(\frac{\partial \dot{u}}{\partial z}\right)^2 dt = -2 \int_{t_0}^{t_1} dt \int_0^l \frac{\partial \ddot{u}}{\partial z} \frac{\partial (\delta u)}{\partial z} dz$$

$$= -2 \int_{t_0}^{t_1} dt \left\{ \left[\frac{\partial \ddot{u}}{\partial z} \delta u\right]_0^l - \int_0^l \frac{\partial^2 \ddot{u}}{\partial z^2} \delta u \, dz \right\}.$$
(7.22)

Substituting the surviving term of (7.20) and the right hand sides of (7.21), (7.22) into (7.19) we find

$$\int_0^l \int_{t_0}^{t_1} \varrho \left[c_b^2 \frac{\partial^2 u}{\partial z^2} - \ddot{u} + v^2 r_{gp}^2 \frac{\partial^2 \ddot{u}}{\partial z^2} + \frac{Z}{\varrho} \right] \delta u \, dt \, dz$$

$$- \int_{t_0}^{t_1} \left[\left(E \frac{\partial u}{\partial z} - \sigma_z + v^2 \varrho r_{gp}^2 \frac{\partial \ddot{u}}{\partial z} \right) \delta u \right]_0^l dt = 0.$$
(7.23)

Since δu is arbitrary over the interval t_0 to t_1, vanishing only at t_0 and t_1, for the first term to vanish in (7.23) the integrand there must vanish. This generates Love's displacement equation of motion for the theory

$$\frac{\partial^2 u(z, t)}{\partial z^2} = \frac{1}{c_b^2}\left(\ddot{u} - v^2 r_{gp}^2 \frac{\partial^2 \ddot{u}}{\partial z^2}\right) - \frac{Z}{E}.$$
(7.24)

The second integral in (7.23) vanishes if the term in parentheses or δu in the integrand vanishes at the ends $z = 0, l$. These are the boundary conditions.[2] The quantity in parentheses is a *stress type condition, a* free end if $\sigma_z(l, t)$ or $\sigma_z(0, t)$ vanish, or forced otherwise. If $\delta u(l, t)$ or $\delta u(0, t)$ vanishes, the end is either fixed (built-in, with $u(l, t)$ or $u(0, t)$ zero), or it has a certain admissible forcing function $u(t)$ there, e.g., $u(0, t) = v_0 t$, v_0 a constant. This is clear since with u specified to be such functions for all t, there is no possible variation of u, i.e., $\delta u = \delta[u(0, t)] = 0$. We note the stress condition is unusual in form. The $\ddot{u}$ term is extraneous, being related to our inertia correction term. It is one failing of this approximate theory making *problems involving stress type end conditions in the theory ill-posed. The longitudinal impact problem* (for the semi-infinite rod), however, *is well posed*, i.e.,

---

[2] When the semi-infinite rod is treated the boundary condition at l is replaced with a radiation condition at $z \to \infty$.

$u(0, t) = v_0 t$, equivalent to $v(0, t) = \partial u/\partial t = v_0 H(t)$. We will treat this problem later and show that *this theory does a remarkably good job of representing the far-field response of the rod*, when compared with the exact theory and experiment.

It should be pointed out that the present theory reduces to the elementary theory, (7.3) and related boundary and initial conditions[3], when the terms corresponding to the radial inertia are deleted. This can be interpreted as the rod radius of gyration r_{gp} becoming smaller for fixed $\varkappa$, hence the wavelengths becoming longer relative to r_{gp}. It is easy to show when $Z=0$, (7.24) has the dispersion relation (7.6).

It is of further interest to point out that a *similar higher order theory for torsional waves and vibrations in a rod exists*. Indeed, as Love shows [1.2, § 279], if one accounts for the *inertia of the motion created by torsional warping of the cross sections into curved surfaces*, one gets an equation of motion of the form (7.24). For the present equation dependent variable, u is replaced by the relative angular displacement of two cross sections. Further, the coefficients of both terms on the right hand side involve the torsional rigidity of the rod, and that of the last term also the integral (over the section) of the square of the torsional function for the section. When the section is circular there is no correction and the equation degenerates into the one-dimensional wave equation in (z, t) for the angular displacement. The solutions then correspond to the nondispersive fundamental axially symmetric torsional mode of the infinite medium [cf. § 4.4.1, (4.87)] having the speed c_s for all wavelengths and frequencies. Indeed it is this simple fundamental mode that has resulted in less need, and hence work, for a higher order rod theory for torsion. This is reflected in the literature (cf. [4.4, pp. 813-814]).

7.1.2. Approximate theories for flexural (nonaxially symmetric) waves in a rod or beam

7.1.2.1. Bernoulli–Euler or elementary bending theory.
To derive the elementary theory consider the bent rod or beam element in fig. 7.2. We assume that the motion takes place in one of principal planes of flexure, taken

---

[3] Initial conditions in both are on u and $\dot{u}$ dictated by the order of the highest derivative in t in the equation of motion, as they usually are.

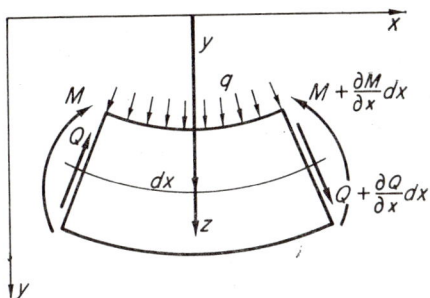

Fig. 7.2. Elementary bending of a beam element.

here as the xy-plane. The beam is assumed to have a uniform cross section of area A, the dimensions of which are small compared to its length. In the *unstrained state* the central axis of the beam (i.e., the line through the centroids of its sections) is the x-axis. When the beam bends it is assumed that the *cross sectional planes remain plane* and normal to the central axis. Thickness normal stresses are assumed zero. As shown in fig. 7.2, *y is a measure of the deflection of the central axis* from its original position. In addition to the *moments M*, and *transverse shear forces Q*, acting on the element, it is acted on by a *distributed loading, say q* (force/unit length), the resultant of possible distributed face loads and/or transverse body forces in the positive y direction. Appealing to *D'Alembert's principle for dynamic equilibrium* we require the sum of the moments, acting on the element about the axis normal to the xy-plane through the element's centroid, to be zero. Neglecting the rotatory inertia of the element we find

$$Q = \partial M/\partial x . \tag{7.25}$$

Likewise requiring dynamic equilibrium of the vertical forces gives

$$q + \partial Q/\partial x = \varrho A \ddot{y} , \tag{7.26}$$

where we have accounted for the vertical translational inertia of the element. Now we introduce (7.25) into (7.26), and in turn the *Bernoulli–Euler curvature theorem*

$$M = -EI \partial^2 y/\partial x^2 , \tag{7.27}$$

where $\partial^2 y/\partial x^2$ *is the curvature* of the central axis of the beam, and *EI the flexural rigidity, I being the moment of inertia* of the beam section about the centroidal axis perpendicular to the *xy*-plane. (The negative sign in (7.27) stems from choosing positive *M* to correspond to negative curvature, à la fig. 7.2.) This produces the governing deflection equation of motion

$$\frac{\partial^4 y(x,t)}{\partial x^4} + \frac{\ddot{y}}{c_b^2 r_g^2} = \frac{q(x,t)}{EI}, \qquad (7.28)$$

where r_g *is the sectional radius of gyration.* This *elementary theory* is commonly referred to as the *Bernoulli–Euler bending theory.* Equation (7.27) is the moment-deflection relation, and substitution of this in (7.25) gives the transverse shear force-deflection relation

$$Q = -EI\,\partial^3 y/\partial x^3. \qquad (7.29)$$

Boundary conditions are the same as in the static problems for this theory. They represent a pinned, built-in, free, or "roller-skate" end (the last having shear force and deflection slope zero), where time dependent end inputs can be treated, e.g., the pin-ended beam would require statements on *M* and *y*. From the moment given by (7.27) one can calculate the *normal stress* on the cross section at *x* according to

$$\sigma_x(x,t) = M(x,t)z/I, \qquad (7.30a)$$

where, as fig. 7.2 shows, *z* measures the distance along the section from the centroidal axis perpendicular to the *xy*-plane. Note positive *z* is downward in the figure giving extensional and compressional stresses, respectively consistent with the longitudinal fiber stretching or contracting indicated there. Note also the maximum stresses occur at the outer fibers. Similarly having *Q* from (7.29), one can calculate, for a particular cross sectional shape, the shear stress along the section at *x* (and its mate along the fibers). For example, for the rectangular section, 2h in depth and b in width, we have

$$\sigma_{xz}(x,t) = \sigma_{zx}(x,t) = \frac{Q(x,t)}{2I}(h^2 - z^2), \qquad I = \frac{2bh^3}{3}. \qquad (7.30b)$$

Most strength of materials books have the derivations of (7.30). For *b* small with respect to h this expression agrees with the corresponding one for the exact theory two-dimensional problem of the end loaded cantilever beam of narrow cross section (cf. Timoshenko and Goodier [7.2]). This parabolic behavior, for the shear stress distribution over the section depth, is fundamental to beam theory, holding for other sectional shapes too.

Setting $q=0$ in (7.28) it is simple to show the dispersion relation is

$$\Omega = \pm c_b \zeta^2 / 2c_s, \qquad (7.31)$$

parabolic in the wave number, where $\Omega = \omega/\omega_s$, $\zeta = 2r_g \varkappa$, and $\omega_s = c_s/2r_g$. Note dimensionless wave number ζ may be real or imaginary for Ω real and positive. This nomenclature is consistent with the Pochhammer frequency equation (4.102), and fig. 4.20 for the circular section beam. In this case (7.31) is also the *first approximation to the lowest mode* in Pochhammer's frequency equation, representing *low frequency-long wave dispersion*. It follows that the *Bernoulli–Euler theory is a good approximation* to the exact theory when used for the analysis of such bending waves, i.e., those that would be generated by low frequency inputs.

7.1.2.2. Timoshenko bending theory. Rayleigh also contributed a lowest branch impoved bending theory (for shorter wavelengths) by taking into account the *rotatory inertia of the element* that was neglected in deriving the Bernoulli–Euler (elementary) theory (cf. [7.1], §§ 161, 162). Subsequently, Timoshenko [7.3] showed that it was equally important to take account of the *shear deformation of the element*. As we shall see, his approximate theory, accounting for both higher order effects, is the best to use if one wants to treat a problem with more accuracy than the elementary theory offers. In our treatment of *Timoshenko's beam theory* we shall see it *also models the exact theory first flexural thickness shear mode for long waves*. Consider the element of the beam shown in fig. 7.3, assuming again the elementary theory assumptions hold here too, except as otherwise stated. As shown, the *beam deflection* y is now assumed to be

$$y = y_b + y_s,$$

the sum of a bending and shear component, where the motion of the element now involves not only a vertical translation, but a rotation and a simple shear. The *rotation* moves a cross-sectional plane through the angle $\partial y_b / \partial x$,

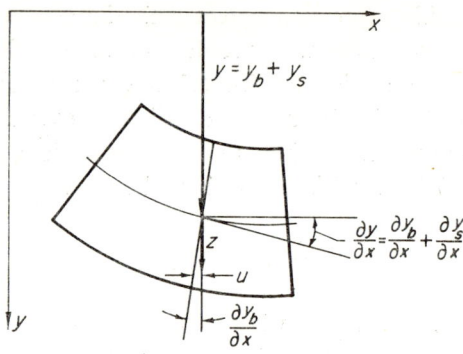

Fig. 7.3. Bending of beam element in Timoshenko's theory.

and the transverse shear force produces the *shear* $\partial y_s/\partial x$ measured at the beam central axis. It follows that we have the slope $\partial y/\partial x = \partial y_b/\partial x + \partial y_s/\partial x$ along this surface. We see therefore, *a cross-sectional plane before deformation remains plane during deformation, but does not remain perpendicular to the central axis.* The displacement in the *x*-direction of points along the cross-sectional plane is given by

$$u = -z\partial y_b/\partial x . \tag{7.32}$$

It follows the corresponding normal strain ε_x is

$$\varepsilon_x = -z\, \partial^2 y_b/\partial x^2 , \tag{7.33}$$

and shear strain ε_{xz} at the central axis is

$$\varepsilon_{xz} = \partial u/\partial z + \partial y/\partial x = \partial y_s/\partial x . \tag{7.34}$$

As in the elementary theory only σ_x and σ_{xz} (and σ_{zx}) are assumed to exist. It follows from this, and (7.33) and (7.34), that the strain energy function is

$$\hat{U} = -\frac{1}{2}\left(\sigma_x z \frac{\partial^2 y_b}{\partial x^2} - \sigma_{xz}\frac{\partial y_s}{\partial x}\right). \tag{7.35}$$

The potential energy of deformation for our uniform beam of length *l* and cross sectional area *A* is given, therefore, by

Ch. 7, § 7.1]   APPROXIMATE THEORIES   379

$$U = -\frac{1}{2}\int_0^l M \frac{\partial^2 y_b}{\partial x^2} dx + \frac{1}{2}\int_0^l Q \frac{\partial y_s}{\partial x} dx, \qquad (7.36)$$

where

$$M = \int_A \sigma_x z \, dA, \qquad Q = \int_A \sigma_{xz} \, dA \qquad (7.37)$$

are the moment and transverse shear force, respectively. *M and Q are the so-called bar (average or weighted) stresses* in this theory by virtue of their integration of stresses across the section. It is through such definitions that we can have a *one-dimensional theory* when there are variations in the stresses across the section. Since the bending here is that of the elementary theory the moment-deflection relation is

$$M = -EI \partial^2 y_b / \partial x^2. \qquad (7.38)$$

We assume a parabolic distribution of the shear stress σ_{xz} across the section, and integrate the second of (7.37) to get the shear force-deflection relation

$$Q = k' A \mu \partial y_s / \partial x, \qquad (7.39)$$

where *k' relates the shear force over the section to the shear strain at the central axis.* k' is easy to get from (7.37) for the common shapes like the circle (3/4), the rectangle (2/3), etc. For unusual cross sections one integrates (7.37) numerically to get $k'A$. Then a calculation for A gives k'. Using (7.38) and (7.39), (7.36) becomes

$$U = \frac{1}{2}\int_0^l \left[EI\left(\frac{\partial^2 y_b}{\partial x^2}\right)^2 + k'A\mu\left(\frac{\partial y_s}{\partial x}\right)^2 \right] dx. \qquad (7.40)$$

By considering the translation of the beam element vertically, and its rotation, we find using (1.74) the kinetic energy of the element to be

$$K = \frac{1}{2}\varrho \int_0^l \left[A(\dot{y}_b + \dot{y}_s)^2 + I\left(\frac{\partial \dot{y}_b}{\partial x}\right)^2 \right] dx. \qquad (7.41)$$

Again $\delta W(t)$ can be written here by noting that the surface and body forces, acting on the beam, must be compatible with the existing stresses

σ_x and σ_{xz}. We note that here $\delta u = -z\delta(\partial y_b/\partial x)$, $\delta w = \delta(y_b + y_s)$, and $\delta v = 0$. It follows from (7.11) that

$$\delta W(t) = \int_0^l q\delta(y_b + y_s)dx - \left[M\delta\left(\frac{\partial y_b}{\partial x}\right) - Q\delta(y_b + y_s) \right]_0^l, \quad (7.42)$$

where $q(x, t)$ is as it was in the elementary theory, the positive sum of transverse distributed loads in the z-direction. As we noted there it can include both face loads and transverse body forces.[3] Upon substituting (7.40), (7.41) and (7.42) into (7.14), and carrying out the indicated variations there, one finds for the equations of motion

$$EI \frac{\partial^3 y_b(x, t)}{\partial x^3} + k'A\mu \frac{\partial y_s}{\partial x} - \varrho I \frac{\partial \ddot{y}_b}{\partial x} = 0,$$

$$k'A\mu \frac{\partial^2 y_s(x, t)}{\partial x^2} - \varrho A(\ddot{y}_b + \ddot{y}_s) + q = 0, \quad (7.43)$$

and the boundary conditions, by making use of the first of (7.43),

$$\left[\left(Q - k'A\mu \frac{\partial y_s}{\partial x}\right) \delta(y_b + y_s) \right]_0^l - \left[\left(M + EI \frac{\partial^2 y_b}{\partial x^2}\right) \delta\left(\frac{\partial y_b}{\partial x}\right) \right]_0^l = 0. \quad (7.44)$$

The first of (7.43) represents rotational, and the second, translational equilibrium. Four possible ends are represented:

i) pin-end: $M=0$, or $M=M_0 f(t)$, and $y=0$, or $y=y_0 g(t)$,

ii) built-in end: $\partial y_b/\partial x = 0$, or $\partial y_b/\partial x = y_0' f(t)$, and $y=0$, or $y=y_0 g(t)$,

iii) free end: $M=0$, or $M=M_0 f(t)$, and $Q=0$, or $Q=Q_0 g(t)$,

iv) "roller skate" end: $\partial y_b/\partial x = 0$, or $\partial y_b/\partial x = y_0' f(t)$, and $Q=0$, or $Q=Q_0 g(t)$.

---

[3] It should be pointed out that a body force moment compatible with σ_x could also be included in this, and the elementary theory, but it is of less interest.

The conditions in (iv) turn out to be the appropriate ones to use for the basic problem of an infinite beam, subjected to a sudden concentrated shear load at the center $x=0$. Note from a solution for y_b and y_s, the moment and shear force are determined by (7.38) and (7.39), respectively. These in turn yield the stresses σ_x and $\sigma_{xz}=\sigma_{zx}$, from (7.30a) and (7.30b), respectively.

The frequency equation is obtained by setting $q=0$ in (7.43), and substituting

$$y_b(x,\ t) = A \exp\left[i\varkappa(x-ct)\right],$$
$$y_s(x,\ t) = B \exp\left[i\varkappa(x-ct)\right], \tag{7.45}$$

in these. The determinant of the two equations for A and B, set equal to zero, gives the Timoshenko beam frequency equation. Figure 7.4 shows how its two branches and the Bernoulli–Euler theory branch, compare for

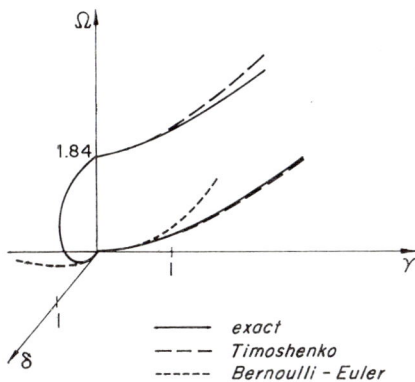

Fig. 7.4. Comparison of approximate and exact theories of bending for relatively low frequency-long waves.

relatively low frequency-long waves, with the corresponding exact theory modes for the circular section rod. *The basic advantages of the Timoshenko theory over the Bernoulli–Euler are (1) it has closer agreement with the exact over a greater domain of the lowest branch, and (2) it has a second branch (flexural thickness-shear) which compares very favorably with the exact theory, at and near cutoff and over the imaginary loop* (shown in fig. 7.4).

We will return to this improved theory later to show how one solves boundary-initial value problems based on it. It is left to the exercises to prove that when the rotatory inertia of the element is neglected, and y_b is taken equal to y, (7.43) and (7.44) reduce to the Bernoulli-Euler theory.

Extensions of the Timoshenko theory to a plate were given by Uflyand [7.4] and Mindlin [7.5]. Only the latter has the correct boundary conditions for the theory.

7.1.3. Boundary-initial value problems based on approximate theories

Since our approximate theories are one dimensional, we can draw on the Laplace transform as we did in ch. 5 § 5.11, based on the material in §§ 5.3, 5.4, 5.10.2.3 and 5.10.2.4. We will find it a powerful method for getting solutions to the boundary-initial value problems based on these higher order partial differential equations.

7.1.3.1. Longitudinal impact problem based on Love-Rayleigh rod theory; exact solution. Consider the problem of a semi-infinite rod of uniform cross section with *polar radius of gyration* r_{gp}, subjected to a step velocity on its end. Using (7.24), with $Z=0$, this boundary-initial value problem may be stated as follows:

$$\frac{\partial^2 u(z,t)}{\partial z^2} = \frac{1}{c_b^2}\left(\ddot{u} - \frac{1}{\alpha^2}\frac{\partial^2 \ddot{u}}{\partial z^2}\right), \qquad \text{for } z>0, \text{ and } t>0, \qquad (7.46)$$

$$u(z, 0) = \dot{u} = 0, \qquad \text{for } z>0, \qquad (7.47)[4]$$

$$\dot{u}(0, t) = \begin{cases} 0, & \text{for } t<0, \\ v_0, & \text{for } t>0, \end{cases} \qquad (7.48)[4]$$

$$\lim_{z \to \infty} u(z, t) = 0, \qquad \text{for } t>0, \qquad (7.49)$$

---

[4] Note, the initial conditions (7.47) are imposed only for $z>0$. However, the boundary condition at $z=0$ (7.48) shows $u=0$ there also, at $t=0$, but the corresponding $\dot{u}$ is $v_0/2$.

where $\alpha^2 = 1/v^2 r_{gp}^2$, and v_0 is a positive magnitude constant. On the basis of (7.47) the Laplace transform of (7.46) is

$$\frac{d^2 \bar{u}(z, p)}{dz^2} - \frac{\alpha^2 p^2}{p^2 + b^2} \bar{u}(z, p) = 0 , \qquad (7.50)$$

where $b^2 = c_b^2 \alpha^2$. It follows that

$$\bar{u}(z, p) = A(p) \exp \left[-n(p) z \right] , \qquad (7.51)$$

where

$$n(p) = \alpha p / (p^2 + b^2)^{\frac{1}{2}} \qquad (7.52)$$

is the one of the two branches of n that will satisfy the transform of (7.49), for $\operatorname{Re} p > 0$. Substituting (7.51) in the transform of (7.48) shows $A(p) = v_0/p^2$, and therefore the transformed solution for the velocity in our problem is

$$\overline{\dot{u}(z, p)} = (v_0/p) \exp \left[-n(p) z \right] . \qquad (7.53)$$

Now this transform has a simple pole at $p = 0$, and branch points at $p = \pm ib$. Further

$$\overline{\dot{u}(z, p)} \text{ is of } O(e^{-\alpha z}/p), \text{ as } |p| \to \infty . \qquad (7.54)$$

It therefore follows that the formal solution for the velocity can be represented by

$$\dot{u}(z, t) = \frac{v_0}{2\pi i} \int_{Br} \frac{e^{pt - n(p)z}}{p} dp , \qquad \text{for } t > 0 , \qquad (7.55)$$

where Br may lie anywhere in $\operatorname{Re} p > 0$, and by

$$\dot{u}(z, t) = 0, \quad \text{for} \quad t < 0 . \qquad (7.56)$$

This solution discloses the *diffusive nature of the Love–Rayleigh theory*, i.e., *the wavefront of the disturbance travels with infinite speed*. To invert the integral in (7.55) an integration is carried out over the contour shown in fig. 7.5. In addition to providing an exact solution for the problem, this contour is a convenient one for obtaining the *long time–far field approxi-*

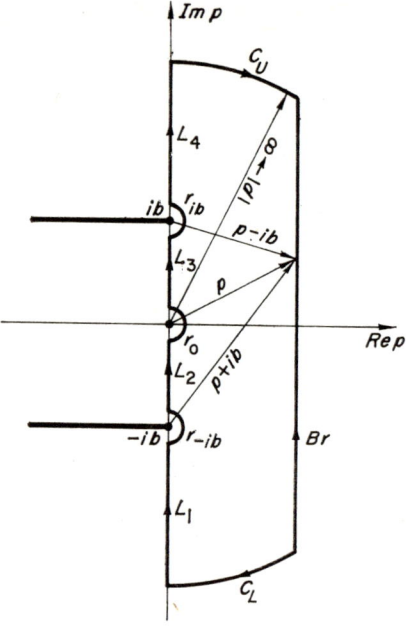

Fig. 7.5. Contour integration for longitudinal impact problem:

mation to this solution. The integration is carried out as it was in ch. 5 § 5.11.2 (cf. also the *s*-plane analysis associated with fig. 6.2 in ch. 6 § 6.2.1.1.). On *Br* we let

$$p = \varrho \exp(i\varphi), \qquad -\pi/2 < \varphi < \pi/2, \qquad (7.57)[5]$$

$$p - ib = \varrho_1 \exp(i\varphi_1), \quad p + ib = \varrho_2 \exp(i\varphi_2), -\pi/2 < \varphi_i < \pi/2,$$

$$i = 1, 2$$

hence

$$n(p) = \alpha \varrho (\varrho_1 \varrho_2)^{-\frac{1}{2}} \exp(i\psi), \qquad \psi = \varphi - (1/2)(\varphi_1 + \varphi_2). \qquad (7.58)$$

---

[5] The range of φ satisfies Re $p > 0$ on *Br*. Since *Br* can be moved arbitrarily as far to the right as we please it follows the ranges of φ_i are the same as φ.

Now having (7.54) we know the integrations over C_L and C_U contribute nothing to the solution. Since $\psi=0$ on the Re p-axis for $p>0$, n is real there. It follows the integrand in (7.55) is real there. Hence for a symmetric region of the p-plane about, and including, a portion of this positive Re p-axis, *the principle of reflection* [cf. Churchill et al, 7.6] applies to the integrand in (7.55), since it is an analytic function in such a region, excluding its singularities. The cross hatched region shown in fig. 7.6 satisfies this requirement,

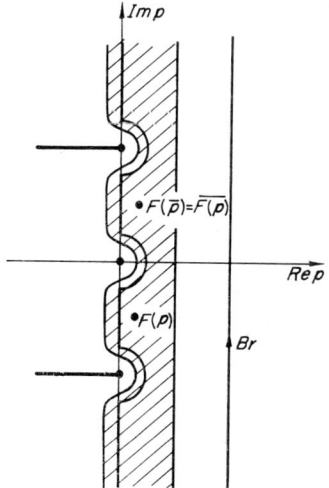

Fig. 7.6. Application of the principle of reflection.

and we note it includes all of the contours along the imaginary axis and indentations. The principle of reflection establishes that the integrand function $F(p)=\exp[pt-n(p)z]/p$ for a point p, say in the lower half of the cross hatched region in fig. 7.6, and $F(\bar{p})$ in the upper half of this region, are complex conjugate functions. It follows therefore that the integration up the imaginary axis in fig. 7.5 gives

$$\frac{v_0}{2\pi i}\left[\int_\Gamma F(p)\mathrm{d}p + \int_{-\Gamma} \overline{F(p)\mathrm{d}p}\right],$$

where $\Gamma = L_1 + r_{-ib} + L_2$. This reduces to

$$\frac{v_0}{\pi}\int_\Gamma \mathrm{Im}\,[F(p)\mathrm{d}p].$$

Hence our solution for $t>0$ is contained in

$$\dot{u}(z,t) = \frac{v_0}{\pi}\left\{\int_\Gamma \text{Im}\left[\frac{\exp[pt-n(p)z]}{p}dp\right] + \frac{1}{2i}\lim_{p\to 0}\int_{r_0}\frac{\exp[pt-n(p)z]}{p}dp\right\}. \tag{7.59}$$

To evaluate the Γ integral we continue the variables (7.57), (7.58) to this path. For L_1 we find $\varphi = \varphi_1 = \varphi_2 = -\pi/2$, $\varrho_1 = \varrho + b$, and $\varrho_2 = \varrho - b$. It follows with $p = -i\varrho$, $n = \alpha\varrho/(\varrho^2-b^2)^{\frac{1}{2}}$ there. By substituting $p = -ib + \varrho_2\exp(i\varphi_2)$ in the integral for the path r_{-ib}, it is easy to show that integral vanishes as $\varrho_2 \to 0$. On path L_2, we find $\varphi = \varphi_1 = -\pi/2$, $\varphi_2 = \pi/2$, $\varrho_1 = b+\varrho$, and $\varrho_2 = b-\varrho$. Since $p = -i\varrho$, and $n = -i\alpha\varrho/(b^2-\varrho^2)^{\frac{1}{2}}$. Approximating (7.52) for small p, it becomes

$$n(p) \simeq (p/c_b)[1-\beta p^2], \tag{7.60}$$

where $\beta = 1/2b^2$. This analysis therefore reduces (7.59) to the solution

$$\dot{u}(z,t) = \frac{v_0}{\pi}\left\{\text{Im}\left[\int_0^b \exp[itf(\varrho)]\frac{d\varrho}{\varrho}\right.\right. \tag{7.61}$$
$$\left.\left. + \int_b^\infty \frac{\exp[-\alpha\varrho z/(\varrho^2-b^2)^{\frac{1}{2}}]}{\varrho}\exp(it\varrho)d\varrho\right] + \frac{\pi}{2}\right\}$$

for $t>0$, where $f(\varrho) = \varrho[1-z\alpha/t(b^2-\varrho^2)^{\frac{1}{2}}]$. The integrals in (7.61) are convergent for all z and $t>0$, and can be evaluated directly by numerical integration.[6]

---

[6] In the present problem there was no real need for the principle of reflection, since L_3, L_4, and r_{ib} lead in a simple way to conjugate integrands of those on L_1, L_2, and r_{-ib}. Hence (7.61) follows easily. On the other hand, the method here, using the principle, requires evaluation of the variables on only one half of the contour, which saves a lot of work in more complicated problems.

One of the reasons for choosing the contour along the imaginary axis (in. fig. 7.5) is that the so-generated integrals, are types that can be readily approximated through certain of the well-known asymptotic expansion techniques we discussed in ch. 5 §§ 5.10.3, 5.10.6. Indeed, because the first integral in (7.61) is of the type that can be approximated by Kelvin's stationary phase method, and the second is a Fourier integral, it is not difficult to find a large time–far field approximation to (7.61). We shall see that the approximation represents an important non-decaying (in z and t) disturbance, which is the same one found from the corresponding approximation in the exact theory.

7.1.3.2. Long time–far field approximation in the longitudinal impact problem. We assume that t is large. The first integral in (7.61) meets all the requirements set down for the stationary phase approximation (5.126), except that $\varsigma(\varrho) = 1/\varrho$ is obviously singular at $\varrho = 0$. Therefore, the first integral is written as

$$\operatorname{Im} \int_0^b \exp\left[itf(\varrho)\right] \frac{d\varrho}{\varrho} = \operatorname{Im}\left[\int_0^\varepsilon \left(\exp\left\{it\varrho\left[1 - \frac{z}{tc_b}(1+\beta\varrho^2)\right]\right\}\right)\frac{d\varrho}{\varrho}\right.$$

$$\left. + \int_\varepsilon^b \exp\left[itf(\varrho)\right] \frac{d\varrho}{\varrho}\right] \tag{7.62}$$

where positive $\varepsilon \ll 1$. Now the second integral on the right-hand side here is one that can be approximated by (5.126). For fixed z/t, and large t, it contributes terms of $O(1/\sqrt{t})$. It can be shown that these groups arrive at station z in the time domain $(z/c_b)(1 + 3\beta\varepsilon^2) \leq t < \infty$, where since z is $\sim c_b t$, it must also be large with t. Now we transform the first integral in (7.62) with

$$\tau = t - z/c_b, \text{ then}$$

$$\frac{\beta z \varrho^3}{c_b} = \frac{r^3}{3}, \quad \text{and } \varrho\tau = r\tau_1, \tag{7.63}$$

so $\tau_1 = \tau/(3\beta z/c_b)^{\frac{1}{3}}$,

hence it becomes,

$$\int_0^\infty - \int_\varepsilon^\infty \frac{\sin(r\tau_1 - r^3/3)}{r} dr. \qquad (7.64)$$

Again the second integral here can be approximated by (5.126), and shown for fixed z/t, and large t, to contribute terms of $O(1/\sqrt{t})$ at a station z, in the same large t- and z-domain as stated for the second integral on the right-hand side of (7.62).

Now the second integral in (7.61), a Fourier integral, is easily seen to have an integrand of bounded variation. It follows by the Riemann–Lebesgue lemma (5.99) this integral contributes a term of $O(1/t)$ for large t and all z. Hence, according to our analysis, (7.61) reduces to

$$\dot{u}(z,t) = \frac{v_0}{\pi} \left[\int_0^\infty \frac{\sin(r\tau_1 - r^3/3)}{r} dr + \frac{\pi}{2} + O\left(\frac{1}{\sqrt{t}}\right) \right] \qquad (7.65)$$

for large t and z. Now

$$\frac{d}{d\tau_1} \left[\frac{v_0}{\pi} \int_0^\infty \frac{\sin(r\tau_1 - r^3/3)}{r} dr \right]$$

$$(7.66)$$

$$= \frac{v_0}{\pi} \int_0^\infty \cos[r(-\tau_1) + r^3/3] dr = v_0 Ai(-\tau_1),$$

where $Ai(-\tau_1)$ is the Airy function. Integrating (7.66) with respect to τ_1 we find

$$\frac{v_0}{\pi} \int_0^\infty \frac{\sin(r\tau_1 - r^3/3)}{r} dr = v_0 \left[\int_0^{\tau_1} Ai(-\tau_1') d\tau_1' + C \right]. \qquad (7.67)$$

Setting $\tau_1 = 0$ in (7.67) it is easily shown that $C = -1/6$. Hence the long time–far field solution in (7.65) reduces to

$$\dot{u}(z,t) = v_0 \left[\int_0^{\tau_1} Ai(-\tau_1') d\tau_1' + \frac{1}{3} \right] + O\left(\frac{1}{\sqrt{t}}\right). \qquad (7.68)$$

The Airy function and its integral in (7.68) are well-known, and tabulated, in the literature, which makes it easy to evaluate this solution (cf. Abramowitz and Stegun [4.10, § 10.4, Tables 10.11, 10.12[7]]. The sketch of (7.68) in fig. 7.7 shows its nature. The solution, sometimes referred to as the "*head*

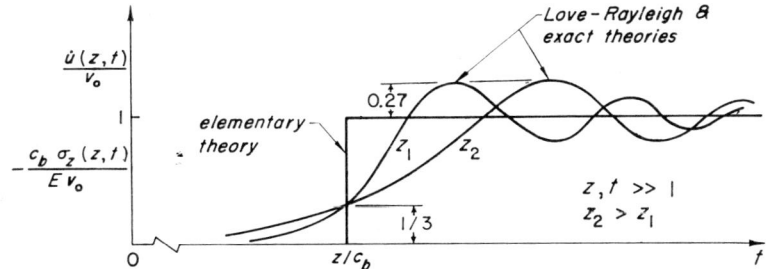

Fig. 7.7. Long time – far field solution of rod longitudinal impact problem ($\dot{u}$ axial velocity, σ_z axial stress).

of the pulse", is valid for small τ_1 (which can be positive or negative, i.e., after or before z/c_b), consistent with the arrival times of lowest branch waves on which the approximation is based (note that the first integral on the right-hand side of (7.62) contains essentially Pochhammer's second approximation (7.6)). Note also, since (7.68) depends only on τ_1, this means that for a larger z, a corresponding increase in τ, to keep τ_1 the same according to the third of (7.63), results in the same disturbance. We see, however, the increase in τ results in a larger time scale or in other words, there is dispersion, where for a certain time interval an increase in station z corresponds to longer period waves in the interval. The response at stations z_1 and z_2 in fig. 7.7 demonstrates this. We note, however, since τ_1 doesn't change, the two response curves exhibit the same amplitudes. *Hence this disturbance is non-decaying in space z and time t.* By approximating the integral in (7.65) for large τ_1, and $(-\tau_1)$, i.e., large time after, and before z/c_b, respectively, one finds agreement of this solution with the elementary theory solution based on (7.3) with $Z=0$.

---

[7] The reader should change the $(-)$ signs in Table 10.12 (for the integral in (7.68)) to $(+)$, i.e., the $(-)$ signs in the second column should all be $(+)$.

By writing $\partial u(z, p)/\partial z$, and comparing it for small $|p|$ with $\overline{\partial u(z, p)/\partial t}$, it is easy to see that (7.65), and therefore (7.68), will result again, except that v_0 is replaced by $(-v_0/c_b)$. It follows that the axial stress is also given by (7.68) as indicated in fig. 7.7, i.e.,

$$\sigma_z(z, t) = E(\partial u/\partial z) = -(E/c_b)\dot{u} = -c_b\varrho\dot{u}, \qquad (7.69)$$

which just expresses the fact that *the one-dimensional momentum equation for the elementary rod theory holds for the present approximation also*.

The foregoing approximate solution was first found by Skalak [7.7] who derived it on the basis of the exact linear theory for the circular sectioned rod (cf. ch. 4 § 4.4.2). As fig. 7.7 shows, *the approximate Love–Rayleigh theory solution* (7.68), found by Miklowitz [7.8], *and Skalak's result are equivalent*. In [7.8] comparison was made of this response on the basis of exact and approximate theories, and experiments by Miklowitz and Nisewanger [7.9]. In the experiments use was made of a shock tube as a device to apply a sudden pressure loading to one "free" end of a long rod of uniform circular section. This end of the rod was inserted through the end plate of the expansion chamber of the tube, and kept flush with its inside face. Hence the rod end was subjected to a plane air shock when the diaphragm separating the compression and expansion chambers was pierced. The input closely approximated a step, since the time across it was of the order 10^{-8} sec., and such a rise time in the 24S–T Al alloy rod that was used corresponds to about 2×10^{-3} in. In this experiment the boundary conditions (corresponding to the exact theory) at the rod end, were essentially a step in σ_z along with $\sigma_{zr} = 0$, representing a pressure shock end loading rather than the longitudinal impact case we have been discussing in which a step in $\dot{u}_z$, along with $\sigma_{zr} = 0$, was applied. However, in [7.8], it is shown the long time–far field response is the same for both problems. Hence, the experimental results of [7.9] can be compared with the theoretical results of [7.7], [7.8]. The experimental results are shown in figs. 7.8, 7.9. The response records shown were obtained electronically with the aid of a thin ring condenser microphone for radial displacement, and very short strain gages for axial strain, both at the lateral surface of the rod. Fig. 7.8 shows the response for several stations, starting with $x = 3/4$ in. (x in these records

Fig. 7.8. Shock tube response records for (*a*) surface radial displacement and (*b*) surface axial strain at various *x*-stations along one inch diameter rod of 24S–T *Al* alloy.

Ch. 7, § 7.1] APPROXIMATE THEORIES 391

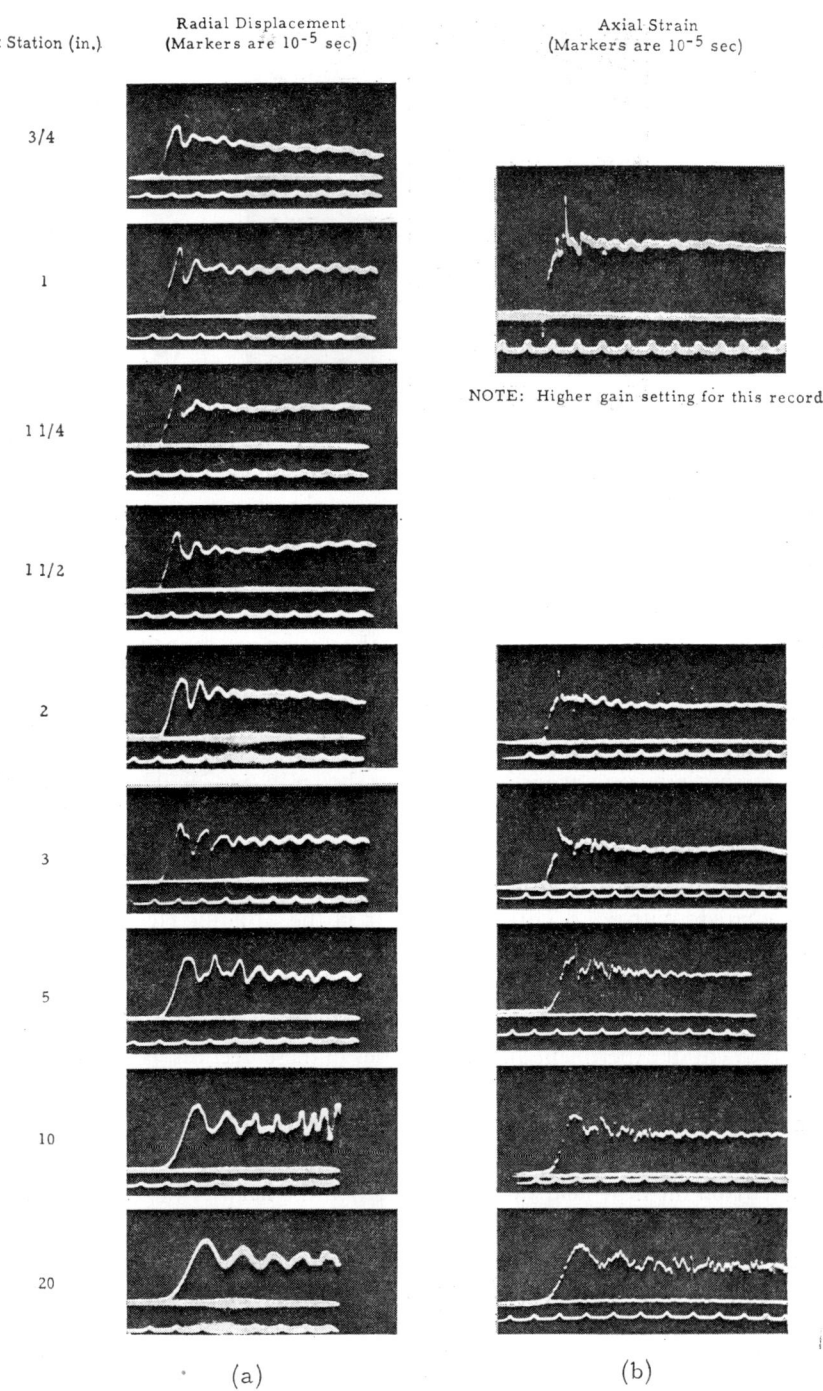

NOTE: Higher gain setting for this record

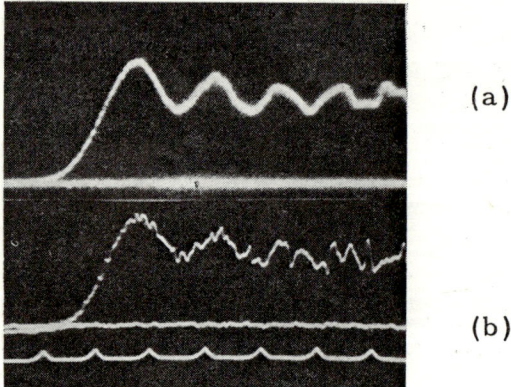

Fig. 7.9. Enlargement of response records in fig. 7.8. at the far field rod station $x = 20$ inches, (a) surface radial displacement and (b) surface axial strain.

corresponds to our z, with the source at $x = 0$; rod diameter was one inch) and out to $x = 20$ inches. The records at the station $x = 20$ inches clearly agree with the form of the head of the pulse solution given by (7.68) and shown in fig. 7.7, where we have noted there is *Poisson's ratio coupling*, given by the second of (7.2), and the momentum equation (7.69) holds. Enlargements of the records at $x = 20$ inches are shown in fig. 7.9. The results of the detailed comparison made in [7.8] for this far field station are shown in fig. 7.10. In the figure the Mindlin–Herrmann higher order, compressional wave rod theory (it considers radial shear as well as radial inertia) is also compared. With $\varkappa_1$, a correction factor in the Mindlin–Herrmann theory, equal to one this theory also exhibits Poisson's ratio coupling. The Love–Rayleigh theory curve is marked Love only. Other nomenclature differences are u'_x for our $\partial u_z/\partial z$, σ for v, w_0 for v_0, and c_0 for c_b. The very close agreement of experiment and the theories in these records is apparent. Essentially it stems from the fact that *all* the theories incorporate Pochhammer's second approximation.

Certain other features in the records of fig. 7.8 are worth mentioning. First note from the near field station records that the head of the pulse begins to develop quite quickly, particularly for the radial displacement. Of further interest is the "ringing" or harmonic oscillations that are apparent for long time in the near field records. This response is that associated with the

Ch. 7, § 7.1] APPROXIMATE THEORIES 393

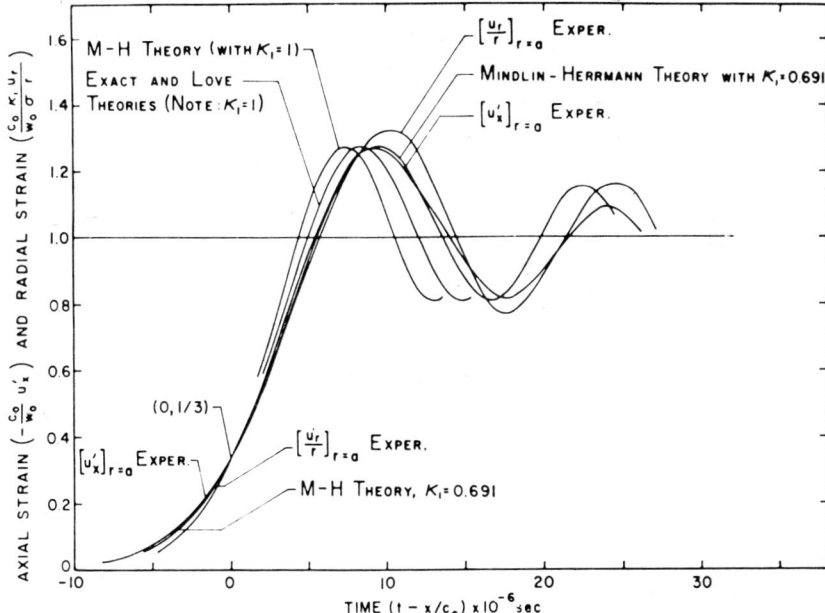

Fig. 7.10. Comparison of theory with experiment for the far field response records of (a) surface radial strain and (b) surface axial strain in fig. 7.9.

rod spectra at and near cutoff. Clearly the axial strain records exhibit much more high frequency (short period) phenomena than the radial displacement records do. This is easy to understand since a strain brings down the wave number as a multiplier. Hence the amplitudes of high frequency waves, associated with short waves (large wave numbers), are relatively larger.

In [7.8] the reader will also be interested in the technique used to obtain solutions for the Mindlin–Herrmann rod theory. This basic technique can be used more generally for impact problems based on higher order approximate theories. It will be discussed in the next section, using the Timoshenko bending theory as an example.

Finally, it should be pointed out that solutions of boundary-initial value problems based on the Love–Rayleigh rod theory must necessarily contain jumps in initial data (at $t=0$) that decay exponentially along the rod. This stems from the characteristic root $n(p)$ given in (7.52). By assuming such data initially, one does arrive at a completely verifiable solution with an unperturbed far field approximation. Since, as has been argued here throughout, the long time–far field solution is the most important use of the theory, the initial jumps would appear unimportant. However, one should be aware that use of the theory in near field responses will certainly be influenced by these jumps. The reader will find an exercise at the end of this chapter that elucidates further on this behavior.

7.1.3.3. Problems based on the Bernoulli–Euler and Timoshenko bending theories; exact solutions.
Boundary-initial value problems for the bending of a beam, governed by the Bernoulli–Euler or Timoshenko theories, can be solved in essentially the same way as discussed for the Love–Rayleigh rod in § 7.1.3.1. In the case of the Timoshenko theory the analysis in the complex p-plane is somewhat more involved, and should be discussed further. Assuming we have problems with lateral loading $q=0$, and initial conditions

$$y_b(x, 0) = y_s(x, 0) = \dot{y}_b(x, 0) = \dot{y}_s(x, 0) = 0, \text{ for } -\infty < x < \infty, \tag{7.70}$$

he Laplace transformed equations of motion are, from (7.43),

$$\left. \begin{aligned} EI \frac{d^3 \bar{y}_b(x, p)}{dx^3} + k'A\mu \frac{d\bar{y}_s}{dx} - \varrho I p^2 \frac{d\bar{y}_b}{dx} &= 0 \\ k'\mu \frac{d^2 \bar{y}_s(x, p)}{dx^2} - \varrho p^2 (\bar{y}_b + \bar{y}_s) &= 0 \end{aligned} \right\}, \quad \text{for } -\infty < x < \infty. \tag{7.71}$$

We let

$$\bar{y}_b(x, p) = \alpha(p) \exp[-n(p)x],$$

$$\bar{y}_s(x, p) = \beta(p) \exp[-n(p)x], \tag{7.72}$$

and substituting these in (7.71), gives the set of equations

$$n\{[\varrho I p^2 - EI n^2]\alpha - k'A\mu\beta\} = 0,$$

$$\varrho p^2 \alpha + (\varrho p^2 - k'\mu n^2)\beta = 0. \tag{7.73}$$

The determinant of the equations (7.73) gives

$$n\{n^4 - [(c_1^2 + c_2^2)/c_1^2 c_2^2] p^2 n^2 + [(p/c_1 c_2)^2 + C] p^2\} = 0, \tag{7.74}$$

APPROXIMATE THEORIES

the characteristic equation of the system of equations (7.71), where

$$c_1^2 = E/\varrho = c_b^2, \qquad c_2^2 = k'\mu/\varrho \qquad \text{with } c_1 > c_2,$$

and

$$C = A\varrho/EI = (c_1 r_g)^{-2}.$$

We discard the root $n=0$, since this corresponds to a rigid body motion. The rest of (7.74) is a bi-quadratic equation in n having the roots

$$n_j(p) = \pm B\sqrt{p}[p \pm D(p^2 - a^2)^{\frac{1}{2}}]^{\frac{1}{2}}, \qquad j = 1, 2, 3, 4, \tag{7.75}$$

where

$$B = (c_1^2 + c_2^2)^{\frac{1}{2}}/\sqrt{2}c_1 c_2, \qquad D = (c_1^2 - c_2^2)/(c_1^2 + c_2^2),$$

$$a = 2\sqrt{C} c_1^2 c_2^2/(c_1^2 - c_2^2).$$

The n_j in (7.75) are four distinct roots, and since (7.73) must hold for all, we find from the two equations there that

$$\frac{\beta_j(p)}{\alpha_j(p)} = \frac{(p/c_1)^2 - n_j^2}{c_2^2 C} = -\frac{p^2}{p^2 - c_2^2 n_j^2} = \varphi_j(p), \tag{7.76}$$

since those equations must be compatible. It follows from (7.72), (7.75) and (7.76) that our transformed displacements are

$$\bar{y}_b(x, p) = \sum_{j=1}^{4} \alpha_j(p) e^{-n_j(p)x},$$

$$\bar{y}_s(x, p) = \sum_{j=1}^{4} \varphi_j(p) \alpha_j(p) e^{-n_j(p)x}. \tag{7.77}$$

Using (7.38), (7.39) the transformed moment $\overline{M}(x, p)$ and shear force $\overline{Q}(x, p)$ can be obtained from (7.77). Note at this stage there are four unknowns $\alpha_j(p)$. For the finite length beam $0 \leq x \leq l$, they are determined by the two boundary conditions at each end à la the cases (i)–(iv) after (7.44). For the infinite or semi-infinite beam the conditions at $x = l$ are replaced by radiation conditions at $|x| \to \infty$, which as we shall see rules out two of the $n_j(p)$ and therefore the corresponding $\alpha_j(p)$. As an example *we treat the case of the infinite beam excited at its center by a suddenly applied shear force*. The problem was first treated by Uflyand [7.4] through essentially the same transform method as that being applied here. Figure 7.11 depicts the problem. As shown in fig. 7.12 the load $Q(0, t)$ can be divided into two equal parts $Q/2$, one acting on the right half of the beam and the other on the

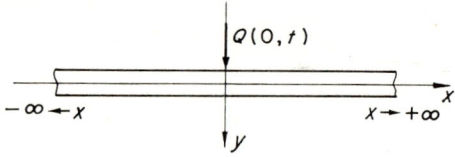

Fig. 7.11. Infinite Timoshenko beam subject to sudden shear load.

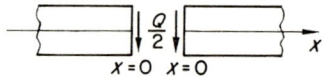

Fig. 7.12. Equivalent problem.

left half. Hence, through this symmetry, we need only treat the right half of the beam $x \geq 0$. The boundary conditions at $x=0$ are

$$Q(0, t) = k'A\mu \partial y_s(0, t)/\partial x = -(Q_0/2)H(t),$$

$$\partial y_b(0, t)/\partial x = 0 \text{ for } t \geq 0, \quad (7.78)$$

where $H(t)$ is the step function and Q_0 a positive magnitude constant. Hence, in writing the first of (7.78), we have used the fact that the applied shear force $Q/2$ is negative, since it is downward on the $x=0$ edge. This is consistent with our definition of a positive shear force in fig. 7.2. The radiation condition is given by

$$\lim_{x \to \infty} [y_b(x, t), y_s, \partial y_b/\partial x, \ldots, Q, M] = 0, \qquad \text{for } t > 0. \quad (7.79)$$

Transforming (7.79) and imposing it on (7.77), it is clear that we must have the two values of $n_j(p)$ that satisfy

$$\text{Re } n_j(p) > 0, \text{ for Re } p > 0 \text{ (on } Br\text{)}. \quad (7.80)$$

Approximating (7.75) for $|p|$ large we find

$$n_j(p) \simeq \pm p/c_j. \quad (7.81)$$

Hence, to satisfy (7.80), we must select from (7.75) the two values of $n(p)$

$$n_j(p) = B\sqrt{p}[p+(-)^j D(p^2-a^2)^{\frac{1}{2}}]^{\frac{1}{2}}, \qquad j=1,2. \tag{7.82}$$

Note from (7.81) that *the Timoshenko theory is hyperbolic, and we will have two wave fronts for every solution based on it*. To determine the remaining two α_j's, use is made of the boundary conditions (7.78). This leads to the formal solutions for the bending rotation and shear force

$$\begin{bmatrix} \dfrac{\partial y_b(x,t)}{\partial x} \\ Q(x,t) \end{bmatrix} = \begin{bmatrix} \dfrac{\partial y_{b1}(x,t)}{\partial x} - \dfrac{\partial y_{b2}(x,t)}{\partial x} \\ Q_1(x,t) - Q_2(x,t) \end{bmatrix}$$

$$= -Y \cdot \frac{1}{2\pi i} \int_{Br} \left\{ \begin{bmatrix} \dfrac{\partial \bar{y}'_{b1}(p)}{\partial x} \\ EI\bar{Q}'_1(p) \end{bmatrix} e^{-n_1(p)x} - \begin{bmatrix} \dfrac{\partial \bar{y}'_{b2}(p)}{\partial x} \\ EI\bar{Q}'_2(p) \end{bmatrix} e^{-n_2(p)x} \right\} e^{pt} dp, \tag{7.83}$$

where

$$\partial \bar{y}'_{bj}(p)/\partial x = p^{-2}(p^2-a^2)^{-\frac{1}{2}}$$

$$\bar{Q}'_j(p) = [Rp - (-)^j B^2 D(p^2-a^2)^{\frac{1}{2}}]p^{-1}(p^2-a^2)^{-\frac{1}{2}}, \qquad j=1,2,$$

with $R = -\sqrt{C}/a$ and $Y = Q_0/4B^2 DEI$. We have chosen to seek the solution for the bending rotation $\partial y_b/\partial x$ rather than the moment, since it (as well as Q) will allow us to verify the boundary conditions (7.78) directly. The moment solution can be obtained later through a simple differentiation of $\partial y_b/\partial x$.

From (7.81) we see for $x/c_1 \leq t < x/c_2$ only the "sub 1" terms in (7.83) survive. For $t \geq x/c_2$ the "sub 2" terms enter. Consider the first case. The function $n_1(p)$ in (7.82) has branch points at $p=0$, $p=\pm a$, and $p=\pm i\gamma$, where $\gamma = aD/(1-D^2)^{\frac{1}{2}}$. These last branch points are at the zeros of $p - D(p^2-a^2)^{\frac{1}{2}}$.[8] Hence the integrands for the first terms in (7.83) have the

---

[8] These zeros are found by the usual squaring process. They can be verified by letting $p = \eta \exp(i\beta)$, $p-a = \eta_1 \exp(i\beta_1)$ and $p+a = \eta_2 \exp(i\beta_2)$, assuming a suitably defined cut for the function $(p^2-a^2)^{\frac{1}{2}}$, and then testing $p - D(p^2-a^2)^{\frac{1}{2}} = \eta \exp(i\beta) - D(\eta_1\eta_2)^{\frac{1}{2}} \exp[i(\beta_1+\beta_2)/2] = 0$ for the zeros $p = \pm i\gamma$.

same singularities. For the second case we note that $n_2(p)$ also has the branch points $p=0$ and $p=\pm a$, but not $p=\pm i\gamma$, since the latter are zeros of $p-D(p^2-a^2)^{\frac{1}{2}}$ and, therefore, not $p+D(p^2-a^2)^{\frac{1}{2}}$. It follows the integrands for the "sub 2" terms in (7.83) have the same singularities. It is not difficult to show that for $|p|\to\infty$, $\partial\bar{y}'_{bj}\partial x = O(p^{-3})$, $\bar{Q}'_1(p)=O(p^{-3})$ and $\bar{Q}_2(p)=O(p^{-1})$. Hence, with path Br anywhere in $\operatorname{Re} p > a$, we can satisfy the conditions (5.19) of the inversion theorem. On this basis it follows that

$$\begin{bmatrix} \dfrac{\partial y_{bj}(x,t)}{\partial x} \\ Q_j(x,t) \end{bmatrix} = 0 \quad \text{for} \quad t < x/c_j, \tag{7.84}$$

which stems from the hyperbolic character of the deflection equations of motion (7.43). For the time domains $t > x/c_j$ we will resort to contour integrations to invert (7.83). *In choosing the contours* inspection of fig. 7.4 gives us some guidance. This figure shows that for physically meaningful results the Timoshenko theory must be restricted to the low frequency- long wave response from its real lowest branch, the long wave response from the region at and near cutoff on its second branch, and edge waves corresponding to the imaginary wave number segments of this second branch. For the first two of these responses fig. 7.4 shows the group velocities are small, approaching zero as the wave number vanishes. *This indicates that a long-time solution would be in order.* The edge waves will be important only very near the origin $x=0$, and even then will involve only the parts of the imaginary segment of the second branch close to the branch points (where the imaginary wave number is smallest). Hence, these waves will be important only in a long-time solution also, i.e., far behind both wave fronts in this theory. To carry out our long-time solution we therefore choose the contours shown in fig. 7.13, which are similar to those that were chosen for the long-time solution of the longitudinal impact problem shown in fig. 7.5. The contour integrations for (7.83) are carried out with the aid of the complex vectors depicted in fig. 7.13, defined by

$$p = \varrho\exp(i\varphi), \quad p-a = \varrho_1\exp(i\varphi_1), \quad p+a = \varrho_2\exp(i\varphi_2), \tag{7.85}$$

where

$$-\pi/2 < \varphi < \pi/2, \text{ and } -\pi/2 < \varphi_i < \pi/2 \quad (i=1,2).$$

These definitions of φ, φ_i are imposed by the Bromwich path condition $\operatorname{Re} p > 0$. From (7.85) we can write

Ch. 7, § 7.1] APPROXIMATE THEORIES 399

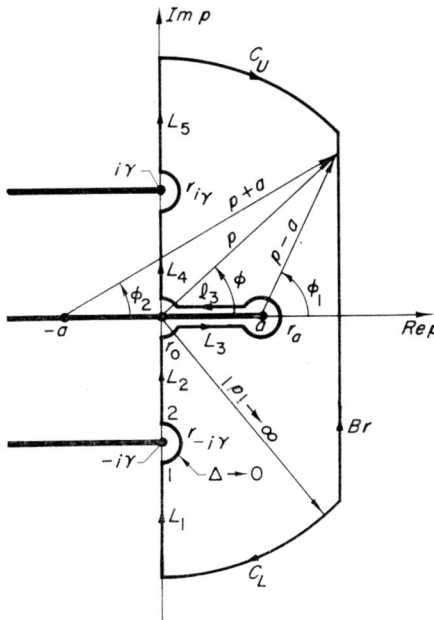

Fig. 7.13. Contour integration for the Timoshenko beam sudden shear force problem.

$$(p^2 - a^2)^{\frac{1}{2}} = (\varrho_1 \varrho_2)^{\frac{1}{2}} \exp(i\psi), \tag{7.86}$$

where $\psi = (\varphi_1 + \varphi_2)/2$ and $\varrho_i > 0$, $-\pi/2 < \psi < \pi/2$. The functions $n_j(p)$ are written as

$$n_j(p) = B(pz_j)^{\frac{1}{2}} = B(\varrho r_j)^{\frac{1}{2}} \exp\left[i(\varphi + \theta_j)/2\right], \tag{7.87}$$

where $z_j = p + (-)^j D(p^2 - a^2)^{\frac{1}{2}} = r_j \exp(i\theta_j)$. We can also let $-\pi/2 < \theta_j < \pi/2$ so that $-\pi/2 < (\varphi + \theta_j)/2 < \pi/2$ [since we have (7.81)], and hence $\operatorname{Re} n_j(p) > 0$ on Br in accord with (7.80).

Since the transforms $\partial \bar{y}_b / \partial x$, $\bar{Q}$ are real on the real p-axis for $p > a$, we can again make use of the principle of reflection and invert (7.83) by evaluating only the lower half plane contours for these integrals, i.e.,

$$\frac{\partial y_b(x,t)}{\partial x} = \frac{\partial y_{b1}(x,t)}{\partial x} - \frac{\partial y_{b2}(x,t)}{\partial x}$$
$$= -\pi^{-1} Y \left\{ \int_\Gamma \operatorname{Im}\left[\frac{\partial \bar{y}'_{b1}(p)}{\partial x} e^{pt - n_1(p)x} dp \right] - \int_\Gamma \operatorname{Im}\left[\frac{\partial \bar{y}'_{b2}(p)}{\partial x} e^{pt - n_2(p)x} dp \right] \right\}, \tag{7.88}$$

$Q(x, t) = Q_1(x, t) - Q_2(x, t)$

$$= -\pi^{-1}EIY \int_\Gamma \text{Im}\{[\bar{Q}_1'(p)e^{-n_1(p)x} - \bar{Q}_2'(p)e^{-n_2(p)x}]e^{pt}dp\}, \quad (7.89)$$

where $\Gamma = L_1 + r_{-iy} + L_2 + r_{0l} + L_3 + r_{al}$, r_{0l} and r_{al} being the lower half plane parts of r_0 and r_a, respectively. Clearly then, since the transforms in (7.88) (7.89) meet the required order conditions, paths C_U, C_L contribute nothing to the integration. Making use then of the vectors in (7.85), and the related functions in (7.86), (7.87) we can continue these functions analytically from their single-valued definitions on Br to the components paths of Γ.

With reference to fig. 7.13, and (7.85)–(7.87), *on the starting path* L_1 we find $\varphi = -\pi/2$, and $\varphi_1 = -\pi + \beta$, $\varphi_2 = -\beta$ so $\psi = -\pi/2$, where β is the acute angle $p-a$, $p+a$ make with the real p-axis. Also we note ϱ varies from ∞ to γ on path L_1, and $\varrho_1 = \varrho_2 = (\varrho^2 + a^2)^{\frac{1}{2}}$ there. It follows that $p = \varrho \exp(-i\pi/2)$, $\sqrt{p} = \sqrt{\varrho} \exp(-i\pi/4)$ and $(p^2 - a^2)^{\frac{1}{2}} = (\varrho^2 + a^2)^{\frac{1}{2}} \exp(-i\pi/2)$, and $z_j = \exp(-i\pi/2)[\varrho + (-)^j D(\varrho^2 + a^2)^{\frac{1}{2}}]$ since $\varrho > \gamma$ on this path. Continuing around the branch point at $p = -i\gamma$ to *path* L_2 we find no change in φ, ψ, ϱ_1, ϱ_2 and therefore p, $\sqrt{p}$ and $(p^2 - a^2)^{\frac{1}{2}}$, except that now ϱ varies from γ to 0. A change does come about in z_1, however, because we have negotiated the branch point of its square root $\sqrt{z_1}$. We can determine this change by first defining the vector $p + i\gamma = \Delta \exp(i\theta)$, which lies along the radius Δ of the semi-circle r_{-iy} (cf. fig. 7.13), where θ varies from $-\pi/2$ at point 1 (the end point of path L_1) to $\pi/2$ at point 2 (the beginning point of path L_2) on this circular path, Δ vanishing as a limit. Substituting p from this vector into z_1, and then using the binomial theorem to expand the $(p^2 - a^2)^{\frac{1}{2}}$ term there, we find $z_1 \simeq [a^2/(a^2 + \gamma^2)]\Delta \exp(i\theta)$ in the neighborhood of $p = -i\gamma$, where use has been made of the fact that $\arg(p^2 - a^2)^{\frac{1}{2}} = -\pi/2$ there. Now at point 1 we know from the path L_1 data that $\theta_1 = -\pi/2$ which is also the value of θ there. It follows that at point 2, where $\theta = \pi/2$, θ_1 is also $\pi/2$. Therefore on L_2, $z_1 = [D(\varrho^2 + a^2)^{\frac{1}{2}} - \varrho] \exp(i\pi/2)$ where the change in form of r_1 is due to the fact that z_1 has gone through its zero (the branch point of $\sqrt{z_1}$ and $\varrho < \gamma$ on this path. Clearly z_2 is as it was on path L_1.

Continuing around the branch point at $p = 0$ to *path* L_3 we find $\varphi = 0$, $\varphi_1 = -\pi$, $\varphi_2 = 0$, $\varrho_1 = a - \varrho$ and $\varrho_2 = a + \varrho$ with ϱ varying from 0 to a. Therefore $p = \varrho$, $\sqrt{p} = \sqrt{\varrho}$ and $(p^2 - \varrho^2)^{\frac{1}{2}} = (a^2 - \varrho^2)^{\frac{1}{2}} \exp(-i\pi/2)$. To determine the forms of z_j on this path we first write z_1 on the basis of the foregoing representations of p and $(p^2 - a^2)^{\frac{1}{2}}$ on this path, i.e.,

$$z_1 = \varrho \exp(i0) - D(a^2 - \varrho^2)^{\frac{1}{2}} \exp(-i\pi/2). \tag{7.90}$$

Since $p=0$ is not a branch point of $\sqrt{z_1}$, we must have continuity in z_1 from the limiting end of L_2 to the limiting beginning of L_3. So at the latter, arg $z_1 = \theta_1 = \pi/2$, i.e., as $\varrho \to 0$ in (7.90), $z_1 = -D(a^2-\varrho^2)^{\frac{1}{2}}\exp(-i\pi/2) \to aD \exp(i\pi/2)$. At the end of L_3, where $\varrho \to a$, (7.90) reduces to $z_1 = \varrho e^{i0}$, or $\theta_1 = 0$. Hence θ_1 varies from $\pi/2$ to 0 on path L_3. A similar procedure for z_2, which is given by

$$z_2 = \varrho \exp(i0) + D(a^2-\varrho^2)^{\frac{1}{2}} \exp(-i\pi/2), \tag{7.91}$$

can be carried out. One finds that θ_2 varies from $-\pi/2$ to 0 on L_3. Since θ_1, θ_2 vary along the path it is important we express this variation in terms of ϱ, the integration variable. Noting from the θ_j variations, that z_1 and z_2 are conjugate functions lying in first and fourth quadrants of the z_1 plane, respectively, we have

$$\begin{aligned} r_1 = r_2 = r = [(1-D^2)(\varrho^2+\gamma^2)]^{\frac{1}{2}}, \\ \sqrt{z_j} = \sqrt{r}[\cos(\theta_j/2) + i\sin(\theta_j/2)] = \lambda(\varrho) \pm i\mu(\varrho), \end{aligned} \tag{7.92}$$

where upper sign is for $\sqrt{z_1}$, and lower $\sqrt{z_2}$, and

$$\lambda(\varrho) = [(r+\varrho)/2]^{\frac{1}{2}}, \qquad \mu(\varrho) = [(r-\varrho)/2]^{\frac{1}{2}}.$$

Shortly we shall use the above determined lineal path variables to write the corresponding line integrals of our inversion for $\partial y_b/\partial x$ and Q. First, however, we assess the contributions to these inversions from the circular paths $r_{-i\gamma}$, r_{0l} and r_{al}, making note that in the first time region $x/c_1 \leq t < x/c_2$, t is bounded, and for all time, x is bounded according to (7.84).

Consider the Γ integral for $\partial y_{b1}/\partial x$ in (7.88) on the first of these paths $r_{-i\gamma}$. We can write $p = \Delta \exp(i\theta) - i\gamma$ on this path. Substituting this into the integrand of this integral and approximating it for small Δ, we find the integral reduces to

$$\frac{\Delta}{\gamma^2(\gamma^2+a^2)^{\frac{1}{2}}} \operatorname{Im}\left[e^{-i\gamma t} \int_{-\pi/2}^{\pi/2} e^{i\theta} d\theta \right] = O(\Delta)$$

which vanishes for $\Delta \to 0$. One can show by a similar procedure that the contribution to $\partial y_{b1}/\partial x$ from the path r_{al} is of $O(\varrho_1^{\frac{1}{2}})$, as $\varrho_1 \to 0$, hence it also vanishes in the limit. It can be similarly shown that the $r_{-i\gamma}$, r_{al} contributions to Q_1 are correspondingly the same as those just found for $\partial y_{b1}/\partial x$. Therefore, since for $\partial y_{b2}/\partial x$, Q_2 there is no path $r_{-i\gamma}$, and the r_{al} contributions of these cancel those of $\partial y_{b1}/\partial x$, Q_1 in $t \geq x/c_2$, we see the circles about the branch points at $p = -i\gamma$ and a contribute nothing to our contour integration, typical of contour integration problems. The same procedure, however, shows that the path r_{0l} does contribute for all four integrals in (7.88), (7.89). This stems from the poles in each at $p=0$, again typical of contour integration problems. We find restricting x, t away from ∞, the r_{0l} integral contributes the terms

$$\left.\frac{\partial y_{b1}}{\partial x}\right|_{r_{0l}} = \pi^{-1}Y\left[\lim_{\varrho \to 0}\left\{\frac{1}{a\varrho} - B\left(\frac{2D}{a\varrho}\right)^{\frac{1}{2}}x\right\}\right],$$

$$\left.\frac{\partial y_{b2}}{\partial x}\right|_{r_{0l}} = \pi^{-1}Y\left[\lim_{\varrho \to 0}\left\{\frac{1}{a\varrho} - (1-\sqrt{2})B\left(\frac{2D}{a\varrho}\right)^{\frac{1}{2}}x\right\}\right], \quad (7.93)$$

to the solution for $\partial y_b/\partial x$, and $B^2 D\pi/2$, $-B^2 D\pi/2$, respectively for Q_1 and Q_2, to the solution for Q. The singular terms in the first of (7.93) are cancelled through contributions of the singular integrals arising on the paths L_2 and L_3. From (7.88), and our path analyses the integrals on these paths lead to the contribution

$$\left.\frac{\partial y_{b1}}{\partial x}\right|_{L_2+L_3} = \pi^{-1}Y\lim_{\delta \to 0}\left\{\int_\delta^\gamma \frac{e^{-B\sqrt{\varrho}\omega_2 x}}{\varrho^2(a^2+\varrho^2)^{\frac{1}{2}}}\sin\varrho t\, d\varrho - \right.$$

$$\left. \int_\delta^a \frac{e^{-B\sqrt{\varrho}\lambda x}\cos B\sqrt{\varrho}\mu x}{\varrho^2(a^2-\varrho^2)^{\frac{1}{2}}}e^{\varrho t}\,d\varrho\right\}, \quad (7.94)$$

where $\omega_2 = [D(a^2+\varrho^2)^{\frac{1}{2}} - \varrho]^{\frac{1}{2}}$. By expanding the integrands here for small ϱ, restricting x, t away from ∞, (7.94) becomes

$$\left.\frac{\partial y_{b1}}{\partial x}\right|_{L_2+L_3} = \pi^{-1}Y\lim_{\delta \to 0}\left\{\frac{t}{a}\int_\delta^\varepsilon\frac{d\varrho}{\varrho} + \int_\varepsilon^\gamma\left[\begin{array}{c}1st\text{ integrand}\\ \text{in (7.94)}\end{array}\right]d\varrho\right.$$

$$\left. -\frac{1}{a}\int_\delta^\varepsilon\frac{1-Bx(aD\varrho/2)^{\frac{1}{2}}+\varrho t}{\varrho^2}\,d\varrho - \int_\varepsilon^a\left[\begin{array}{c}2nd\text{ integrand}\\ \text{in (7.94)}\end{array}\right]d\varrho\right\} \quad (7.95)$$

where ε is a small positive number. The first integral here cancels that for the third term in the third integral, leaving just the first and second terms of the latter. These integrate out to give

$$\pi^{-1}Y\lim_{\delta\to 0}\left[\frac{1}{a}\left(\frac{1}{\varepsilon}-\frac{1}{\delta}\right)-B\left(\frac{2D}{a}\right)^{\frac{1}{2}}x\left(\frac{1}{\sqrt{\varepsilon}}-\frac{1}{\sqrt{\delta}}\right)\right],$$

the singular terms of which cancel those of the first of (7.93). The same type of cancellations occur for the second of (7.93) which enters for the the later time interval $t > x/c_2$. Here, however, the second integral in (7.94) does not occur since in this interval we have $\partial y_{b1}/\partial x - \partial y_{b2}/\partial x$ and it is cancelled by an identical integral contributed by $\partial y_{b2}/\partial x$. However, we have another singular integral, the sum of those over paths L_1 and L_2 (since as pointed out earlier n_2 does not have a branch point at $p = -i\gamma$). Clearly, further cancellations in this later time interval involve the difference of the first and second of (7.93). On the basis of the foregoing analysis the exact solutions for $\partial y_b/\partial x$ and Q, for the time interval $x/c_1 < t < x/c_2$ are

$$\frac{\partial y_b(x,t)}{\partial x} = \pi^{-1}Y\left\{\int_\gamma^\infty \frac{\sin(\varrho t - B\sqrt{\varrho\omega_1}x)}{\varrho^2(a^2+\varrho^2)^{\frac{1}{2}}}d\varrho + \int_\varepsilon^\gamma \frac{e^{-B\sqrt{\varrho\omega_1}x}}{\varrho^2(a^2+\varrho^2)^{\frac{1}{2}}}\sin\varrho t\,d\varrho\right.$$

$$\left.-\int_\varepsilon^a \frac{e^{-B\sqrt{\varrho\lambda}x}\cos B\sqrt{\varrho\mu}x}{\varrho^2(a^2-\varrho^2)^{\frac{1}{2}}}d\varrho + \frac{1}{a\varepsilon} - B\left(\frac{2D}{a\varepsilon}\right)^{\frac{1}{2}}x\right\}, \qquad (7.96)$$

where $\omega_1 = [\varrho - D(a^2+\varrho^2)^{\frac{1}{2}}]^{\frac{1}{2}}$, and

$$Q(x,t) = -\pi^{-1}EIY\left\{\int_\gamma^\infty \frac{R_1}{\varrho(a^2+\varrho^2)^{\frac{1}{2}}}\sin[\varrho t - B\sqrt{\varrho\omega_1}x]d\varrho\right.$$

$$+\int_0^\gamma \frac{R_1}{\varrho(a^2+\varrho^2)^{\frac{1}{2}}}e^{-B\sqrt{\varrho\omega_2}x}\sin\varrho t\,d\varrho$$

$$+\int_0^a \frac{e^{\varrho t - B\sqrt{\varrho\lambda}x}}{\varrho(a^2-\varrho^2)^{\frac{1}{2}}}[R\varrho\cos(B\sqrt{\varrho\mu}x) - B^2D(a^2-\varrho^2)^{\frac{1}{2}}\sin B\sqrt{\varrho\mu}x]d\varrho$$

$$\left. + \frac{B^2D\pi}{2}\right\}, \qquad (7.97)$$

where $R_1 = R\varrho + B^2D(a^2+\varrho^2)^{\frac{1}{2}}$, and for $t > x/c_2$ are

$$\frac{\partial y_b(x,t)}{\partial x} = \pi^{-1}Y\left\{1st\text{ two integrals in (7.96)}\right.$$

$$\left.-\int_\varepsilon^\infty \frac{\sin(\varrho t - B\sqrt{\varrho\omega_3}x)}{\varrho^2(a^2+\varrho^2)^{\frac{1}{2}}}d\varrho - 2B\left(\frac{D}{a\varepsilon}\right)^{\frac{1}{2}}x\right\}, \qquad (7.98)$$

where $\omega_3 = [D(a^2+\varrho^2)^{\frac{1}{2}}+\varrho]^{\frac{1}{2}}$, and

$Q(x, t) = -\pi^{-1}EIY\{1st$ two integrals in (7.97)

$$-\int_0^\infty \frac{R_2}{\varrho(a^2+\varrho^2)^{\frac{1}{2}}} \sin(\varrho t - B\sqrt{\varrho\omega_3}x)d\varrho + B^2D\pi\bigg\}, \qquad (7.99)$$

where $R_2 = R\varrho - B^2D(a^2+\varrho^2)^{\frac{1}{2}}$. The corresponding solution for the moment is easily obtained through differentiation of (7.96), (7.98) with respect to x, according to (7.38). Setting $x=0$ in (7.98), (7.99) shows the solution satisfies the boundary conditions (7.78). The solutions for $t < x/c_1$ in (7.84), and similar solutions which exist for the displacements, velocities and moment, show the radiation condition (7.79) is satisfied. It is easily seen that the solutions (7.84) for $t < x/c_1$ also show that the initial conditions (7.70) are satisfied for $x > 0$. It follows that (7.96)–(7.99) are the exact solution for the posed problem for the Timoshenko theory and can be evaluated through numerical integration of the integrals therein. The admissible parts of such a numerical solution, a la earlier comments on restriction to low frequency-long waves from the lowest branch, and near cutoff frequency-long waves for the second branch, can be determined by comparing predominant period-time of occurence curves (cf. ch. 4 § 4.1.1.3) for the Timoshenko and exact theories. Such a comparison for the related problem of a sudden point normal load on a plate, governed by the Timoshenko theory, was given by Miklowitz [7.10]. This paper will also be of general interest to the reader on this subject. The work of Boley and Chao [7.11] offers further results of interest here. Using a related contour integration technique they derived, and evaluated numerically, a number of solutions for the semi-infinite Timoshenko beam for a variety of end loadings. They compared the results with those obtained from the Bernoulli–Euler theory and found good agreement for long time.

7.1.3.4. *Long-time response for sudden shear load on an infinite Timoshenko beam.* A long-time approximation to the exact solution (7.96)–(7.99) is not difficult to derive. Assuming $t \gg 1$ and $x > 0$, but $x \nrightarrow \infty$, such a solution is obtained from (7.98), (7.99) the exact solution for $t > x/c_2$. Consider first (7.99), the solution for Q. The first integral there is one that can be approximated by the method of stationary phase discussed in § 5.10.6. For t large, and x/t fixed, this integral has first order points of stationary phase at values of ϱ in the range $\gamma \leq \varrho < \infty$, stemming from the real wave number segment of the second branch of the Timoshenko theory (cf. fig. 7.4). These points contribute terms of $O(t^{-\frac{1}{2}})$ to the long-time solution which have the group velocities $0 \leq c_g < c_1$. We can set the second integral in (7.99)

$$\int_0^\gamma = B^2 D \int_0^\varepsilon \frac{e^{-b\sqrt{\varrho}}}{\varrho} \sin \varrho t \, d\varrho + \int_\varepsilon^\gamma \left[\begin{array}{c} 2nd \text{ integrand} \\ \text{in (7.97)} \end{array} \right] d\varrho, \quad (7.100)$$

where ε is a small positive number, and $b = B(Da)^{\frac{1}{2}} x$. The integrals here are Fourier integrals, as we might expect, since they correspond to the imaginary wave number segment of the second branch of the Timoshenko theory. According to the Riemann–Lebesque lemma the second integral on the right of (7.100) contributes a term of $O(t^{-1})$ for large time. The first integral

$$\int_0^\varepsilon \frac{e^{-b\sqrt{\varrho}}}{\varrho} \sin \varrho t \, d\varrho = \int_0^\infty - \int_\varepsilon^\infty \frac{e^{-b\sqrt{\varrho}}}{\varrho} \sin \varrho t \, d\varrho. \quad (7.101)$$

Again here the second integral on the right contributes a term of $O(t^{-1})$. The first integral may be found in the tables (cf. Oberhettinger [5.6] p. 124). Therefore, from (7.100) we have

$$\int_0^\gamma = B^2 D \pi \left\{ \left[\frac{1}{2} - C(\eta) \right]^2 + \left[\frac{1}{2} - S(\eta) \right]^2 \right\} + O(t^{-1}), \quad (7.102)$$

where

$$C(\eta) = \frac{1}{(2\pi)^{\frac{1}{2}}} \int_0^\eta \frac{\cos s}{\sqrt{s}} \, ds, \quad S(\eta) = \frac{1}{(2\pi)^{\frac{1}{2}}} \int_0^\eta \frac{\sin s}{\sqrt{s}} \, ds,$$

are the *Fresnel integrals of the first and second kinds*, respectively, η being $b^2/4t$ here. Similarly we can set the third integral in (7.99)

$$\int_0^\infty = B^2 D \int_0^\varepsilon \frac{\sin(\varrho t - b\sqrt{\varrho})}{\varrho} \, d\varrho + \int_\varepsilon^\infty \left[\begin{array}{c} 3rd \text{ integrand} \\ \text{in (7.99)} \end{array} \right] d\varrho. \quad (7.103)$$

The second integral contributes at most terms of $O(t^{-\frac{1}{2}})$, except for one point which is of $O(t^{-\frac{1}{3}})$, by stationary phase arguments. The term of

$O(t^{-\frac{3}{2}})$ comes from a second order point of stationary phase [cf. (5.121)] due to an inflection point in the lowest branch of the Timoshenko frequency equation at moderately high frequency. The first integral

$$\int_0^\varepsilon \frac{\sin(\varrho t - b\sqrt{\varrho})}{\varrho} \, d\varrho = \int_0^\infty - \int_\varepsilon^\infty \frac{\sin(\varrho t - b\sqrt{\varrho})}{\varrho} \, d\varrho \, . \tag{7.104}$$

The second integral here also behaves as $O(t^{-\frac{1}{2}})$, and again the first may be found in the tables (cf. Oberhettinger [5.6], pp. 26, 138). Therefore from (7.103) we have

$$\int_0^\infty = B^2 D\pi \left\{ \frac{1}{2} - C(\eta) - S(\eta) - [C(\eta)]^2 - [S(\eta)]^2 \right\} + O(t^{-\frac{1}{2}}) \, . \tag{7.105}$$

Substituting the foregoing results into (7.99) we find *the long-time solution for the shear force* to be

$$Q(\xi, \tau)/Q_0 = -(\tfrac{1}{2})\{1 - [C(\xi^2/\tau) + S(\xi^2/\tau)]\} + O(t^{-\frac{1}{2}}) \, , \tag{7.106}$$

where in setting $\eta = \xi^2/\tau$ we have introduced the dimensionless variables $\xi = x/r_g$ and $\tau = 4c_1 t/r_g$. Since $C = S = \frac{1}{2}$ as $x \to \infty$, we see $Q(x, t)$ vanishes there as it should in accord with (7.79). Further, $C = S = 0$ at $x = 0$ so that $Q(x, t)$ satisfies the first of the boundary conditions (7.78). For fixed x, and $t \to \infty$, $C = S \to 0$ and hence $Q(\xi, \tau)/Q_0 \to Q(\xi)/Q_0 = -\frac{1}{2}$, which is the *solution of the static problem*. The numerical evaluation of the response of $Q(\xi, \tau)/Q_0$ is presented in fig. 7.14 for the two stations $\xi = 1, \sqrt{2}$. It is based on the tables for $C(\eta)$, $S(\eta)$ in Abramowitz and Stegun [4.10, table 7.7]. It may be seen that *the shortest period (highest frequency) waves lead in the disturbance, progressively becoming longer and longer as time τ increases at a station*. Comparing the response at $\xi = \sqrt{2}$ with that at $\xi = 1$, shows that *as the disturbance propagates it becomes composed of longer and longer periods over a certain time interval*. These characteristics are in agreement with the predominant period-time of occurrence relation for the low frequency-long wave approximation to the lowest branch of the Timoshenko theory on which our approximation (7.106) is based. But since the approximation is valid only for long time, the early response in these curves are ruled out. We have arbitrarily denoted the interval $\tau_{01} \leq \tau < \infty$

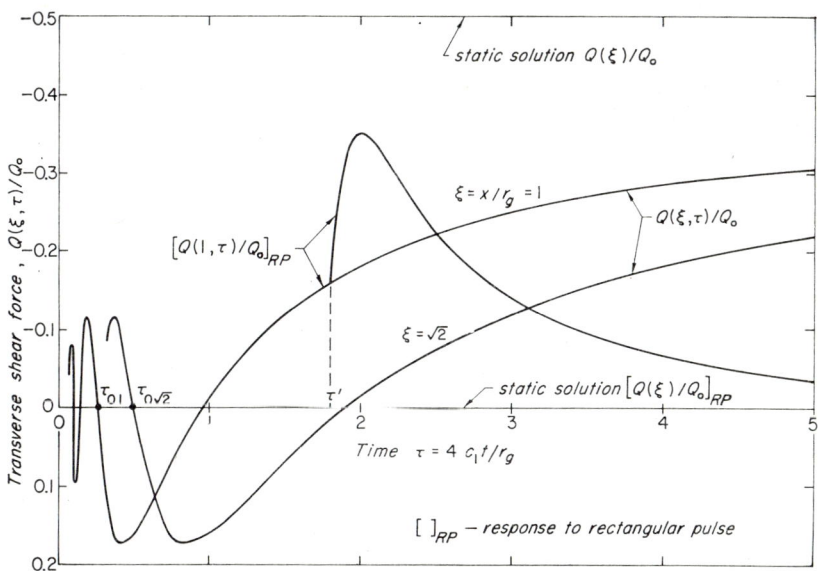

Fig. 7.14. Long-time shear force response of an infinite Timoshenko beam subjected to a (*1*) step and (*2*) rectangular pulse in the shear force.

as admissible for $\xi=1$. The related initial time $\tau_{0\sqrt{2}}$ for $\xi=\sqrt{2}$ is also indicated. It is clear, therefore, that *as ξ grows the approximation becomes better and better.* It may be noted in the figure that $Q(\xi,\tau)/Q_0$ approaches the static solution as time grows, as one would expect. *A further interesting point is that the largest value of the response is that of the static solution.*

To obtain the *long-time solution for* $\partial y_b/\partial x$ we cannot use (7.98) since that was based on restricting t away from ∞. However, we need only return to the original limiting forms for our contour integrals, e.g. (7.93), (7.94), to write the proper solution to be approximated, i.e.,

$$\frac{\partial y_b(x,t)}{\partial x} = \pi^{-1} Y \lim_{\delta \to 0} \left\{ \int_\gamma^\infty \frac{\sin(\varrho t - B\sqrt{\varrho \omega_1} x)}{\varrho^2(a^2+\varrho^2)^{\frac{1}{2}}} \, d\varrho \right.$$

$$+ \int_\delta^\gamma \frac{e^{-B\sqrt{\varrho \omega_2} x}}{\varrho^2(a^2+\varrho^2)^{\frac{1}{2}}} \sin \varrho t \, d\varrho$$

$$\left. - \int_\delta^\infty \frac{\sin(\varrho t - B\sqrt{\varrho \omega_3} x)}{\varrho^2(a^2+\varrho^2)^{\frac{1}{2}}} \, d\varrho - 2B\sqrt{\frac{D}{a\delta}} x \right\} \qquad (7.107)$$

where the last term is the last one in (7.93). Now in approximating the singular integrals in (7.107) by letting $\varrho \to 0$ we must keep the product ϱt intact. The first (nonsingular) integral is of $O(t^{-\frac{1}{2}})$ by stationary phase arguments. The second and third (after expanding their numerators) reduce to integrals of the form

$$\int_0^\infty \varrho^{-\frac{3}{2}} \sin \varrho t \, d\varrho \,, \quad \int_0^\infty \varrho^{-\frac{3}{2}} \cos \varrho t \, d\varrho \,, \tag{7.108}$$

and a term of $O(1)$, again through completing the approximated integrals from ε to ∞ and using the Riemann–Lebesque lemma. The integrals (7.108) can be found in the tables (cf. Oberhettinger [5.6], pp. (110, 2) the first has the value $(2\pi t)^{\frac{1}{2}}$ and the second $\lim_{\delta \to 0}(2/\sqrt{\delta})-(2\pi t)^{\frac{1}{2}}$. Our *long-time solution for* $\partial y_b/\partial x$ is therefore

$$\partial y_b(x,t)/\partial x = -2(2D/\pi a)^{\frac{1}{2}} BYx\sqrt{t} + O(1) \,, \qquad t \gg 1 \,, \qquad x \to \infty \,, \tag{7.109}$$

Making use of the moment-deflection relation (7.38) a differentiation of (7.109) with respect to x gives the long-time moment solution

$$M(x,t) \simeq 2(2D/\pi a)^{\frac{1}{2}} EIBY\sqrt{t} \,. \tag{7.110}$$

So both the *bending rotation and moment become infinite* as $t \to \infty$. For a *rectangular pulse input*, however, they would both approach zero as $t \to \infty$. The responses for this case can be obtained from any of the foregoing solutions using the relation

$$[Z(x,t)]_{RP} = Z(x,t) - Z(x, t-t') \tag{7.111}$$

where the "sub RP" denotes the response to the rectangular pulse input, and Z on the right is the response to the step input (obtained in the foregoing). The time t' is the width of the rectangular pulse. The long-time $[Q(1,\tau)/Q_0]_{RP}$ is shown in fig. 7.14, τ' being the pulse width $4c_1 t'/r_g$ measured from τ_{01}.

It should be pointed out that the long time solutions (7.106), (7.109) and (7.110) in the present case, and related ones according to (7.111), can be derived much more easily by appealing to the asymptotics of the Laplace transform contained in § 5.10.2.4. On the other hand the procedure we have used, fundamental to contour integration methods, is instructional and is indicated in other problems, e.g., cf. § 7.1.3.2. Derivation of (7.106), (7.109) and (7.110) using the asymptotics of § 5.10.2.4 is left to the exercises.

It is also important to point out that had the Bernoulli–Euler theory been used in the present problem, essentially the same long-time results would have been found. This because the two theories agree so closely for the long wave-low frequency part of the lowest branch of the frequency equation (cf. fig. 7.4) on which the long-time approximation is essentially based. Jones pointed this out in his related work [7.12], as did Boley and Chao [7.11], which was pointed out earlier.

7.2. Problems for the infinite plate in plane strain

7.2.1. *Excitation of plate by two symmetric normal line loads; formal solution, numerical evaluation and approximations*

Consider the problem depicted in fig. 7.15. Two symmetrically disposed normal line loads of *magnitude* σ_0, an inherently negative constant of dimensions force/unit length, are suddenly applied to the faces of a plate of $2h$ *thickness*. The mathematical statement of the problem is, using (2.7b),

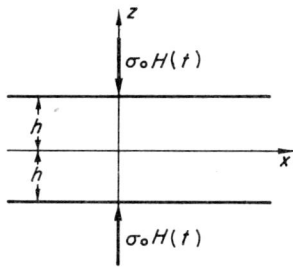

Fig. 7.15. Infinite elastic plate subjected to two symmetric normal line loads.

$$\left.\begin{array}{l}\nabla^2\varphi(x,z,t)=\ddot{\varphi}/c_d^2\\ \nabla^2\psi(x,z,t)=\ddot{\psi}/c_s^2\end{array}\right\}, \quad -\infty<x<\infty, \quad -h<z<h, \, t>0, \quad (7.112)$$

$$\varphi(x, z, 0) = \dot{\varphi} = \psi = \dot{\psi} = 0, \qquad -\infty < x < \infty, \ -h \leq z \leq h, \tag{7.113}$$

$$\left.\begin{array}{l}\sigma_z(x, \pm h, t) = \sigma_0 \delta_s(x) H(t) \\ \sigma_{zx}(x, \pm h, t) = 0\end{array}\right\} \qquad -\infty < x < \infty, \ t > 0, \tag{7.114}$$

$$\lim_{x \to \pm \infty} \left[\varphi(x, z, t), \psi, \frac{\partial \varphi}{\partial x}, \text{etc.} \right] = 0, \qquad -h \leq z \leq h, \ t > 0, \tag{7.115}$$

where $\nabla^2 = \partial^2/\partial x^2 + \partial^2/\partial z^2$, and $\delta_s(x)$ is defined in (6.5). The problem is a direct analog of Lamb's plane problem, i.e., the problem described by (6.1)–(6.5). We derive the formal solution here with essentially the same procedure, using a Laplace transform on t, and an exponential Fourier transform on x. We would therefore have (6.6), (6.7) here again. Now however, we must keep *both branches* of η_d and η_s, since the domain in z is finite. It is more convenient now to rewrite (6.7) as

$$\bar{\varphi}^{\sim}(\varkappa, z, p) = A'(\varkappa, p) \cosh \eta_d z + C'(\varkappa, p) \sinh \eta_d z,$$

$$\bar{\psi}^{\sim}(\varkappa, z, p) = B'(\varkappa, p) \cosh \eta_s z + D'(\varkappa, p) \sinh \eta_s z, \tag{7.116}$$

where A', C', and B', D', are combinations of A, C, and B, D, in (6.7), respectively. Since this problem involves a symmetric loading with respect to the midplane $z = 0$, we drop the C' and B' terms in (7.116). Then a similar procedure to (6.9)–(6.11) is applied to (7.114) for $z = h$ (or $-h$) to determine A' and D'. Applying the double transforms of (3.1) to the now determined (7.116) yields the transformed displacements $\bar{u}^{\sim}$, $\bar{w}^{\sim}$ for the problem.

In the following we will show some interesting techniques that invert directly the underlying double inversion integral for an unknown. We draw on a paper by Lloyd and Miklowitz [7.13], where these techniques are discussed in connection with the derivation and analysis of the solution for the vertical displacement w. The foregoing procedure shows w is given formally by

$$w(x, z, t) = \frac{\sigma_0}{\mu(4\pi^2 i)} \int_{Br} \frac{e^{pt}}{p} \left[\int_{-\infty}^{\infty} \frac{N(\varkappa^2, z, p^2)}{F_s(\varkappa^2, p^2)} e^{i\varkappa x} d\varkappa \right] dp, \tag{7.117}$$

where

$$\frac{N}{F_s} = \frac{\eta_d[(\eta_s^2+\varkappa^2)\sinh\eta_s h\sinh\eta_d z - 2\varkappa^2 \sinh\eta_d h\sinh\eta_s z]}{(\eta_s^2+\varkappa^2)^2 \sinh\eta_s h\cosh\eta_d h - 4\varkappa^2\eta_d\eta_s \sinh\eta_d h\cosh\eta_s h}.$$

We note that $F_s = 0$ is a generalized form of the Rayleigh–Lamb frequency equation (4.48) for symmetric waves, which reduces to the latter if $p = \pm i\omega$.

7.2.1.1. *Inversion of spatial transform first.* Inverting the Fourier transform in (7.117) first is a particularly valuable way to invert (7.117). As we shall see, it leads to *a solution in its most detailed form*, permitting analysis of this transient response in terms of the types of waves in the underlying spectra (4.48). Proceeding with the inversion we *note N/F_s in (7.117) is even in η_d and η_s and, therefore, it is single-valued.* It follows the *inner inversion can be carried out by contour integration and residue theory.* Because of the symmetry of the problem it suffices to consider $x > 0$. Hence, for convergence, we must complete the integration path of the Fourier integral along a large circle in the upper half $\varkappa$-plane. The zeros $\varkappa_j(p)$ of $F_s(\varkappa^2, p^2)$ must therefore satisfy

$$\text{Im } \varkappa_j(p) > 0, \text{ for } \text{Re } p > 0, \qquad (7.118)$$

i.e., on the Br contour. They correspond to poles of N/F_s and are shown in fig. 7.16 along with the integration contours. Such zeros are assumed to exist because their later continuation to the imaginary axis in the p-plane reduces them to the known branches of (4.48) for symmetric waves. With the aid of a Taylor's series expansion it is not difficult to show that the corresponding poles are simple since N and F_s are analytic at $\varkappa = \varkappa_j(p)$, and $N[\varkappa_j^2(p), z, p^2] \neq 0$, for fixed p. The proof is left to the exercises. Approxi-

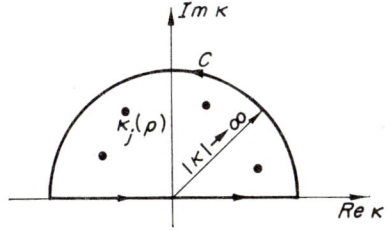

Fig. 7.16. Contour integration in the $\varkappa$-plane.

mating N/F_s for large $|\varkappa|^9$ shows it is of $O(1/\varkappa)$, hence Jordan's lemma applies to the integral along path C in fig. 7.16, and therefore (7.117) reduces to

$$w(x, z, t) = \frac{\sigma_0}{\mu(2\pi)} \int_{Br} \sum_{j=1, 2 \cdots} \frac{N[\varkappa_j^2(p), z, p^2]}{\left.\frac{\partial F_s(\varkappa^2, p^2)}{\partial \varkappa}\right|_{\varkappa^2 = \varkappa_j^2(p)}} \cdot \frac{e^{pt + i\varkappa_j(p)x}}{p} dp,$$

(7.119)

valid for $t > 0$.

Having in mind that inversion of (7.119) will be *restricted to a finite number of branches $\varkappa_j(p)$, since essentially our interest is in the lower modes*, termwise Br integration of the series is valid. This, of course, restricts the problem to inputs ranging from low to only moderately high frequencies. Clearly each of the terms in (7.119) represents the contribution of a branch $\varkappa_j(p)$ of the generalized Rayleigh–Lamb frequency equation $F_s(\varkappa^2, p^2) = 0$. Branch points in these terms are anticipated at places corresponding to the branch points in the spectra given by (4.48) for symmetric waves, i.e., the places where $d\omega/d\varkappa = 0$; at cutoff, and other places on real and imaginary segments of the branches where there is zero slope (cf. figs. 4.12 and 4.15 again). The corresponding condition in (7.119) is where $dp/d\varkappa = 0$, the related branch points in the p-plane lying on the imaginary axis. There are no other branch points of the integrand in (7.119) in the p-plane since by inspection of N in (7.117), and $\partial F_s/\partial(\varkappa^2)$ in (7.126), it can be seen that their quotient is even in η_d and η_s.[10] Since $F_s(\varkappa^2, p^2) = 0$ along the branches $\varkappa_j(p)$, it follows that

$$dF_s = \frac{\partial F_s}{\partial \varkappa} d\varkappa + \frac{\partial F_s}{\partial p} dp = 0,$$

and hence

---

[9] $|\varkappa|$ is always taken larger than $|p|$ in $\varkappa_j(p)$, so that $\varkappa_j(p)$ will always be interior to C.

[10] This is consistent with the fact that a static solution exists for the present problem, which rules out singularities in the right-half of the p-plane (cf. ch. 5, § 5.10.2.4).

$$\frac{dp}{d\varkappa} = -\frac{\dfrac{\partial F_s}{\partial \varkappa}}{\dfrac{\partial F_s}{\partial p}} = -\frac{\varkappa}{p} \frac{\dfrac{\partial F_s(\varkappa^2, p^2)}{\partial(\varkappa^2)}}{\dfrac{\partial F_s(\varkappa^2, p^2)}{\partial(p^2)}} \qquad (7.120)$$

there. By expanding the right side here, about the branch point pairs $(\varkappa_j^*, p^*)$, we find, after a simple integration, that $\varkappa_j(p)$ behaves as

$$\varkappa_j(p) - \varkappa_j^* \sim (p - p^*)^{\frac{1}{2}} \qquad (7.121)^{11}$$

as we stated earlier in (4.52). Note that his expression also shows that $dp/d\varkappa = 0$ at the branch point.

We complete the Br contours for the terms in (7.119) along the imaginary axis, so that $p = \pm i\omega$, and the $\varkappa_j(p)$ become the Rayleigh–Lamb symmetric wave branches of (4.48). As fig. 7.17a shows, the cuts for the branch points of the $\varkappa_j(p)$ are taken to the left. Figure 7.17b shows the analytically conti-

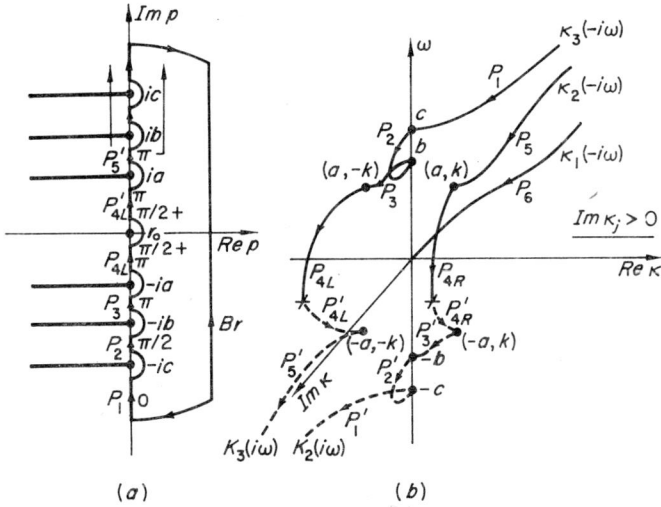

Fig. 7.17. (a) Contour integration in the p-plane and (b) corresponding analytically continued branches $\varkappa_j(p)$ and related paths of integrals I_i.

---

[11] The proof is similar to that carried out in ch. 4, exercise 4.4.

nued spectra for the lowest three branches $\varkappa_1$, $\varkappa_2$ and $\varkappa_3$ for $\omega > 0$ (cf. fig. 4.12), and $\varkappa_2$, $\varkappa_3$ for $\omega < 0$. In the following discussion we verify the character of $\varkappa_2$, $\varkappa_3$ exhibited in fig. 7.17b, showing how the contour integration in fig. 7.17a generates their various segments. The *correspondence between branch points in the $\omega - \varkappa$ space and p-plane is indicated in these two figures*, e.g., $p^* = -ia \rightarrow \omega = a$. For cutoff points, $\varkappa_j^* = 0$, and one of these for $\varkappa_3$ corresponds to $p^* = -ib \rightarrow \omega = b$. The large circles in fig. 7.17a will again give zero contribution here, and *analytical continuation of the integrand in (7.119) for each term there is carried out by observing in fig. 7.17b the nature (real, imaginary, or complex) of the parts of a particular $\varkappa_j$ away from the branch points, and using (7.121) to negotiate the branch points*. For example, suppose we invert the $\varkappa = \varkappa_3(p)$ term in (7.119). As the arrows in fig. 7.17a indicate we progress from the lower end of the imaginary p-axis toward the branch point at $p = -ic$, which corresponds to the P_1 path of the $\varkappa_3$ branch in fig. 7.17b. Figure 7.17b then, gives us the information we need for this portion of $\varkappa_3(-i\omega)$, i.e., it shows it is real and > 0. To establish the change in $\varkappa_3$, as it goes about its cutoff point $p = -ic$, use is made of the vector $p + ic = \varepsilon \exp(i\varphi)$ shown in fig. 7.18. From (7.121) we therefore have

$$\varkappa_3(p) \sim (p+ic)^{\frac{1}{2}} = \sqrt{\varepsilon} \exp(i\varphi/2), \quad (7.122)$$

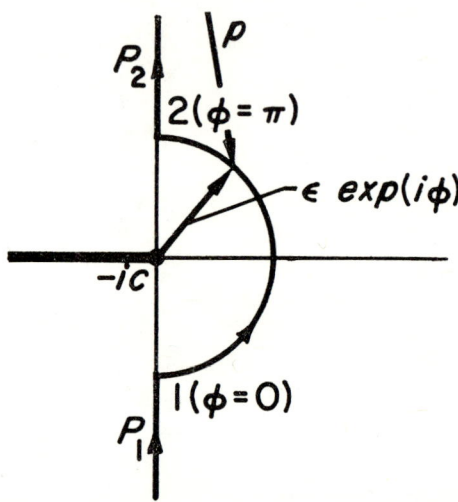

Fig. 7.18. Integration about a branch point at cutoff.

in the neighborhood of $p = -ic$. Since the point 1 in fig. 7.18 corresponds to a point in the approach to c on P_1 in fig. 7.17b, it follows $\varkappa_3(p)$ in (7.122) will have the real nature of P_1 if we set $\varphi = 0$ there. Then letting the range of φ be $0 \leq \varphi \leq \pi$ on the circle in fig. 7.18, at point 2 we have from (7.122), $\varkappa_3(p) \sim i\sqrt{\varepsilon}$. This point corresponds to a point on path P_2 in fig. 7.17b just after leaving c. We note our analysis is in agreement with the positive imaginary nature of P_2. Note also in fig. 7.17a that the corresponding spectral path is indicated to the left of the imaginary axis, along with the corresponding arg $\varkappa_3$ to the right. The next change in $\varkappa_3$ is at the lower cutoff branch point $p = -ib$, corresponding to b in fig. 7.17b. This is handled in the same way as the point $p = -ic$ was. Here the incoming arg $\varkappa_3(p) = \pi/2$ and then there is a rotation to arg $\varkappa_3(p) = \pi$, which is in agreement with fig. 7.17b, i.e. at b, $\varkappa_3(-i\omega)$ changes from positive imaginary (P_2) to negative real (P_3). The change at $p = -ia$ is just slightly different. Here we write

$$\varkappa_3(p) + k \sim (p+ia)^{\frac{1}{2}} = \sqrt{e} \exp(i\varphi/2) \tag{7.123}$$

for the neighborhood of $p = -ia$. In the approach to $p = -ia$ we must have $\varphi = 0$, hence from (7.123), $\varkappa_3(p) \sim -k + \sqrt{\varepsilon}$ in agreement with P_3 in fig. 7.17b. Then with a π change in φ in going around $p = -ia$ we have $\varkappa_3(p) \sim -k + i\sqrt{\varepsilon}$, i.e. the start of P_{4L} (L for left) which is complex with a positive imaginary component. Now $\varkappa_3$ is continuous at the origin $p = 0$ (corresponding to $\omega = 0$ in fig. 7.17b), changing only from $\varkappa_3(-i\omega)$ to $\varkappa_3(i\omega)$. Continuing along the upper half of the imaginary axis of p, $\varkappa_3(p)$ has the branch point $p = ia$. In the neighborhood of this point we have

$$\varkappa_3(p) + k \sim (p-ia)^{\frac{1}{2}} = \sqrt{e} \exp(i\varphi/2) . \tag{7.124}$$

Here the approach is from P'_{4L}, which is the reflection of P_{4L} in the plane $\omega = 0$. So here we have $\varphi = \pi$, i.e. $\varkappa_3(p) \sim -k + i\sqrt{\varepsilon}$ in agreement with the start of P_{4L}. Interesting now, however, is that since φ must again increase by π, after going about the branch point here, $\varkappa_3(p) \sim -k - \sqrt{\varepsilon}$ and we are back to a negative real segment. This is labeled P'_5 in fig. 7.17b, since it appears to be the image of path P_5 in the $\omega - \varkappa$ origin. There are no further branch points for $\varkappa_3(p)$, which is indicated in fig. 7.17a by the upward arrows from P'_5 and arg $\varkappa_3 = \pi$. A similar analysis produces the analytically continued $\varkappa_2(p)$ branch, which is also shown in fig. 7.17b. Proof that it has the features exhibited is left to the exercises.

It is apparent from fig. 7.17 that *complex conjugation here cannot be accomplished for individual branches* $\varkappa_j$. Indeed, as this figure shows, $\varkappa_2$ *and* $\varkappa_3$ *must be taken together for proper conjugation*. Consider first the real segments P_1 and P_1', parts respectively of $\varkappa_3(-i\omega)$ and $\varkappa_2(i\omega)$. On them we have $\varkappa_2(i\omega) = -\varkappa_3(-i\omega) = -\text{Re } \varkappa$, a negative number. Similarly on the real segments P_3 and P_3' we have $\varkappa_2(i\omega) = -\varkappa_3(-i\omega) = \text{Re } \varkappa$, and on the imaginary segments P_2 and P_2', $\varkappa_2(i\omega) = \varkappa_3(-i\omega) = i\text{Im}\varkappa$. All of these are special cases of

$$\varkappa_2(i\omega) = -\bar{\varkappa}_3(-i\omega) . \tag{7.125}$$

It is clear that the complex segments also conjugate in the same way since P_{4L} and P'_{4R} are related through $\varkappa_2(i\omega) = \varkappa_2(-i\omega) = -\bar{\varkappa}_3(-i\omega)$, and P_{4R} and P'_{4L} through the same relations. The relation (7.125) was anticipated since with the kernel $\exp[i\varkappa_j(p)x]$ in (7.119) we must have a relation such as this to obtain a real result. In effect what we are saying then is that *the upper half space* ($\omega > 0$) *spectra is enough information to determine the solution*, i.e., relations like (7.125) must exist, but need not be determined. However, the above analysis shows why.

To write the resultant path integrals for $\varkappa_2$, $\varkappa_3$ in (7.119) we must investigate the integrand there on the basis of our above findings for these branches. We note first that

$$\frac{\partial F_s(\varkappa^2, p^2)}{\partial \varkappa}\bigg|_{\varkappa^2 = \varkappa_j^2(p)} = 2\varkappa_j(p) \frac{\partial F_s(\varkappa^2, p^2)}{\partial(\varkappa^2)}\bigg|_{\varkappa^2 = \varkappa_j^2(p)} \tag{7.126}$$

where

$$\frac{\partial F_s}{\partial(\varkappa^2)} = \frac{(\eta_s^2+\varkappa^2)^2 h}{2} \left(\frac{\sinh \eta_s h \sinh \eta_d h}{\eta_d} + \frac{\cosh \eta_s h \cosh \eta_d h}{\eta_s} \right)$$

$$+ 4(\eta_s^2+\varkappa^2) \sinh \eta_s h \cosh \eta_d h$$

$$- 2\varkappa^2 h(\eta_d \sinh \eta_s h \sinh \eta_d h + \eta_s \cosh \eta_s h \cosh \eta_d h)$$

$$- \left[4\eta_d \eta_s + 2\varkappa^2 \left(\frac{\eta_s}{\eta_d} + \frac{\eta_d}{\eta_s} \right) \right] \sinh \eta_d h \cosh \eta_s h ,$$

and, as previously defined, $\eta_d = (k_d^2+\varkappa^2)^{\frac{1}{2}}$ and $\eta_s = (k_s^2+\varkappa^2)^{\frac{1}{2}}$. Consider then

Ch. 7, § 7.2]   PROBLEMS FOR THE INFINITE PLATE   417

the integral generated by the paths P_1 and P'_1. We have already learned that $\varkappa_2(i\omega) = -\varkappa_3(-i\omega) = -\operatorname{Re}\varkappa$, hence $\varkappa_2^2(i\omega) = \varkappa_3^2(-i\omega) = (\operatorname{Re}\varkappa)^2$. It follows by inspection of (7.119), (7.117) and (7.126) that

(1) $N dp/p [\partial F_s/\partial(\varkappa^2)]$ is the same on P_1 and $P'_1 (p = \pm i\omega)$, taken in opposite directions, and that

(7.127)

(2) when η_d and η_s are real or imaginary (only possibilities here since $\varkappa$ is real and p is imaginary), $N/[\partial F_s/\partial(\varkappa^2)]$ is real since as noted earlier it is even in η_d and η_s.

Hence the integral for these paths is

$$I_1 = -\int_c^\infty \frac{N[\varkappa_3^2(-i\omega), z, -\omega^2]}{\omega\varkappa_3(-i\omega)\left.\dfrac{\partial F_s(\varkappa^2, -\omega^2)}{\partial(\varkappa^2)}\right|_{\varkappa^2 = \varkappa_3^2(-i\omega)}} \cos[\omega t - \varkappa_3(-i\omega)x]d\omega ,$$

$\varkappa_3(-i\omega)$ real and >0 .   (7.128)

The integrals on the real paths P_3 and P'_3 are essentially the same case as those on paths P_1 and P'_1. Here $\varkappa_2(i\omega) = -\varkappa_3(-i\omega) = \operatorname{Re}\varkappa$, and hence $\varkappa_2^2(i\omega) = \varkappa_3^2(-i\omega) = (\operatorname{Re}\varkappa)^2$. It follows that conditions (7.127) apply again, and we are led to the contour integral

$$I_3 = -\int_a^b \frac{N[\varkappa_3^2(-i\omega), z, -\omega^2]}{\omega\varkappa_3(-i\omega)\left.\dfrac{\partial F_s(\varkappa^2, -\omega^2)}{\partial(\varkappa^2)}\right|_{\varkappa^2 = \varkappa_3^2(-i\omega)}} \cos[\omega t - \varkappa_3(-i\omega)x]d\omega ,$$

$\varkappa_3(-i\omega)$ real and <0 .   (7.129)

Now on the positive imaginary paths P_2 and P'_2 we have $\varkappa_2(i\omega) = \varkappa_3(-i\omega) = i\operatorname{Im}\varkappa$, hence $\varkappa_2^2(i\omega) = \varkappa_3^2(-i\omega) = -(\operatorname{Im}\varkappa)^2$. Therefore it follows that

(1) $N\exp(i\varkappa x)dp/p\varkappa[\partial F_s/\partial(\varkappa^2)]$ is the same on P_2 and P'_2 $(p = \pm i\omega)$, taken in opposite directions, and that

(7.130)

(2) since $\varkappa$ and p are both imaginary, $N \exp(i\varkappa x)/[\partial F_s/\partial(\varkappa^2)]$ is real for the reasons given in (2) of (7.127).
These conditions lead to the resultant integral

$$I_2 = \int_b^c \frac{N[\varkappa_3^2(-i\omega), z, -\omega^2] e^{i\varkappa_3(-i\omega)x}}{\omega[-i\varkappa_3(-i\omega)] \dfrac{\partial F_s(\varkappa^2, -\omega^2)}{\partial(\varkappa^2)}\bigg|_{\varkappa^2 = \varkappa_3^2(-i\omega)}} \sin \omega t\, d\omega ,$$

$\varkappa_3(-i\omega)$ imaginary, and Im $\varkappa_3(-i\omega) > 0$. (7.131)

For the complex paths, as we noted earlier, in the case of (1) P_{4L} and P'_{4L} we have $\varkappa_3(i\omega) = \bar{\varkappa}_3(-i\omega)$, and (2) P_{4R} and P'_{4R}, $\varkappa_2(i\omega) = \bar{\varkappa}_2(-i\omega)$. Therefore the first of (7.130) holds for these pairs of paths, leading to

$$\hat{I}_{4L} = \int_0^a \frac{i \sin \omega t}{\omega} \left[\frac{N[\varkappa_3^2(-i\omega), z, -\omega^2] e^{i\varkappa_3(-i\omega)x}}{\varkappa_3(-i\omega) \dfrac{\partial F_s(\varkappa^2, -\omega^2)}{\partial(\varkappa^2)}\bigg|_{\varkappa^2 = \varkappa_3^2(-i\omega)}} \right] d\omega ,$$

$\varkappa_3(-i\omega)$ complex with Im $\varkappa_3(-i\omega) > 0$, (7.132)

for the path $P_{4L} + P'_{4L}$ and

$$\hat{I}_{4R} = \int_0^a \frac{i \sin \omega t}{\omega} \left[\frac{N[\varkappa_2^2(-i\omega), z, -\omega^2] e^{i\varkappa_2(-i\omega)x}}{\varkappa_2(-i\omega) \dfrac{\partial F_s(\varkappa^2, -\omega^2)}{\partial(\varkappa^2)}\bigg|_{\varkappa^2 = \varkappa_3^2(-i\omega)}} \right] d\omega ,$$

$\varkappa_3(-i\omega)$ complex with Im $\varkappa_3(-i\omega) > 0$, (7.133)

for the path $P_{4R} + P'_{4R}$. Now noting that $\varkappa_3(-i\omega) = -\bar{\varkappa}_2(-i\omega)$, the expressions in the brackets of (7.132) and (7.133) are complex conjugate functions which gives for the sum of these integrals

$$I_4 = -2 \operatorname{Im} \int_0^a [\text{in (7.133)}] \frac{\sin \omega t}{\omega} d\omega , \quad \varkappa_2(-i\omega) \text{ as in (7.133)} . \quad (7.134)$$

The paths P_5 and P'_5, as well as P_6 and its mate P'_6 (not shown in fig. 7.17b), give integrals of the same form as (7.128), except that P_6, P'_6 lead to an integrand that is singular at $\omega = 0$. The integral is integrable, however, and for the long time-far field disturbance leads to a result related to that found by Skalak [7.7] for the longitudinal impact problem of the circular cylindrical rod [cf. § 7.1.3.2, (7.68)].

We must also consider the possible contributions from the semi-circular paths about the branch points, and the pole at the origin, in the p-plane (cf. fig. 7.17a). Consider first the path r_0 at $p=0$. From (7.119) we can write for $\varkappa_2$, $\varkappa_3$

$$I_{023} = \sum_{j=2,3} \int_{r_0} \frac{N[\varkappa_j^2(p), z, p^2] e^{pt + i\varkappa_j(p)x}}{2p\varkappa_j(p) \left.\dfrac{\partial F_s(\varkappa^2, p^2)}{\partial(\varkappa^2)}\right|_{\varkappa^2 = \varkappa_j^2(p)}} dp,$$

which by using $p = \varrho \exp(i\varphi)$, $-\pi/2 \leq \varphi \leq \pi/2$, on r_0 becomes

$$I_{023} = \frac{\pi i}{2} \sum_{j=2,3} \lim_{\varrho \to 0} \frac{N[\varkappa_j^2(\varrho), z, \varrho^2] e^{i\varkappa_j(\varrho)x}}{\varkappa_j(\varrho) \left.\dfrac{\partial F_s(\varkappa^2, \varrho^2)}{\partial(\varkappa^2)}\right|_{\varkappa^2 = \varkappa_j^2(\varrho)}}. \tag{7.135}$$

We note that $\varkappa_2$, $\varkappa_3$ in (7.135) must be in the limit, the piercing points (at $\omega = 0$) of the four complex segments P_{4L}, P'_{4L}, P_{4R} and P'_{4R}. Therefore, the conjugation for these segments we carried out to get (7.134) applies to (7.135). Using (7.134) then, we have

$$I_{023} = -\pi \lim_{\varrho \to 0} \left[\operatorname{Im} \left\{ \frac{N[\varkappa_2^2(\varrho), z, \varrho^2] e^{i\varkappa_2(\varrho)x}}{\varkappa_2(\varrho) \left.\dfrac{\partial F_s(\varkappa^2, \varrho^2)}{\partial(\varkappa^2)}\right|_{\varkappa^2 = \varkappa_2^2(\varrho)}} \right\} \right], \tag{7.136}$$

where $\varkappa_2(\varrho) \to \varkappa_2(0)$ is complex with $\operatorname{Im} \varkappa_2(0) > 0$.

Concerning the semi-circular paths about the branch points the situation is much like it was for the Timoshenko beam problem (cf. 7.1.3.3). Consider, for example, the paths about the cutoff branch point $p = -ic$. In the $\varkappa_3(p)$ integral for this path we approximate the integrand with $\varkappa_3(p) = \sqrt{\varepsilon} \exp(i\varphi/2)$, i.e., (7.122), and carry out the resulting simple integration on φ. The result is of $O(\sqrt{\varepsilon})$ for $\varepsilon \to 0$. The same result holds for other paths about cutoff branch points. In the case of the branch points with $\varkappa_j^* \neq 0$ [cf. (7.123)], the same procedure gives a result of $O(\varepsilon)$. Hence these paths contribute nothing to the solution. The *vertical displacement response based on the lowest three Rayleigh-Lamb branches* is therefore given by

$$w_{1+2+3}(x, z, t) = \frac{\sigma_0}{2\pi\mu} \left[I_{01} + I_{023} + \sum_{i=1}^{6} I_i \right], \tag{7.137}$$

where I_{023}, I_1, I_2, I_3 and I_4 are given by (7.136), (7.128), (7.131), (7.129) and (7.134), respectively. The derivations of I_5 (paths P_5+P_5'), I_6 (paths P_6+P_6') and I_{01}, the contribution of the path r_0 for the lowest branch $\varkappa_1(p)$, are left to the exercises.

The *integrals I_1, I_2, and I_4 in (7.137) were evaluated numerically* in [7.13]. They show some *important features that are basic* to the segments of the branches they represent. The time response for I_1, corresponding to the upper real segment, at the station $\xi = x/h = 5$ is shown in fig. 7.19.[12] Note

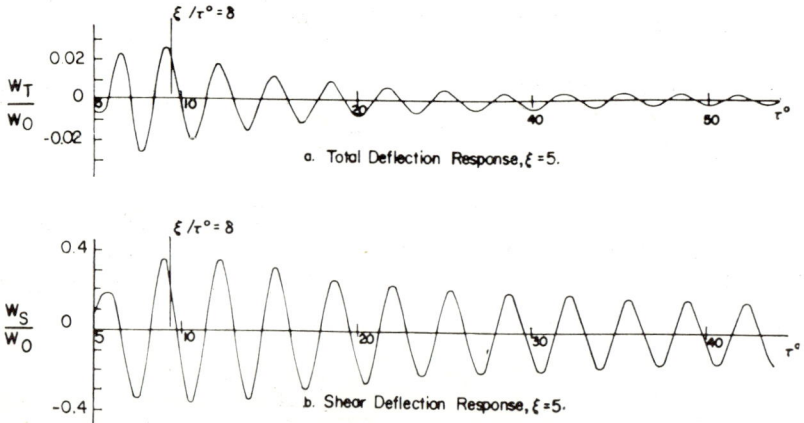

Fig. 7.19. Vertical displacement time response corresponding to the higher real segment P_1 of branch $\varkappa_3$.

that *the long-time cutoff mode has a predominant influence on most of the two responses there.* Figure 7.20 represents I_2 from the imaginary segments of branch $\varkappa_3$. The w_T response shows the *predominance of the cutoff frequency b*. This is *due primarily* to the fact that spatial decay (with x) for this branch almost eliminates entirely everything but the cutoff response itself. Again

---

[12] In the quoted paper these plots were evaluated for the branches and integrals in the corresponding problem of a Timoshenko plate on an elastic foundation, hence the nomenclature for the total deflection $w_T = w_B + w_S$. They, however, are like the symmetric exact theory branches and integrals here, that fig. 2 of the quoted paper shows. δ in the figure corresponds to the arrival of the second wave front in the Timoshenko plate.

for the w_S response, which is much less than w_B (compare magnitudes of w_S and w_T), we see the influence of the other cutoff frequency c (a thickness shear mode) also. The *spatial response for* w_T is also interesting showing the infinitely long waves of this cutoff phenomena. Figures 7.21 and 7.22 are the responses for I_4, representing the complex wave number segments in fig. 7.17b, based respectively, on the Timoshenko beam and present (exact) solutions. Again here the $(a, \pm k)$ *pairs predominate* in the response, because of the more severe spatial decay for other pairs on these segments. Note *one other interesting feature*, namely the nodal points in the spatial response records. They show these *segments govern a non-propagating or standing wave disturbance*. It is also important to note the form of I_3 in (7.129), *representing the real segment* P_3 *of the* $\varkappa_3$ *branch, where* $\text{Re } \varkappa < 0$. This shows the *phase waves* in the integrand *propagate in the negative x-direction*, *whereas the sum of these waves*, i.e., the *integral* I_3, *represents a traveling wave part of the solution that involves propagation in the positive x-direction*. Therefore the integral represents *anomolous dispersion* having the properties $c < 0$, and $c_g > 0$ here, consistent with the nature of the underlying segment in fig. 7.17b.

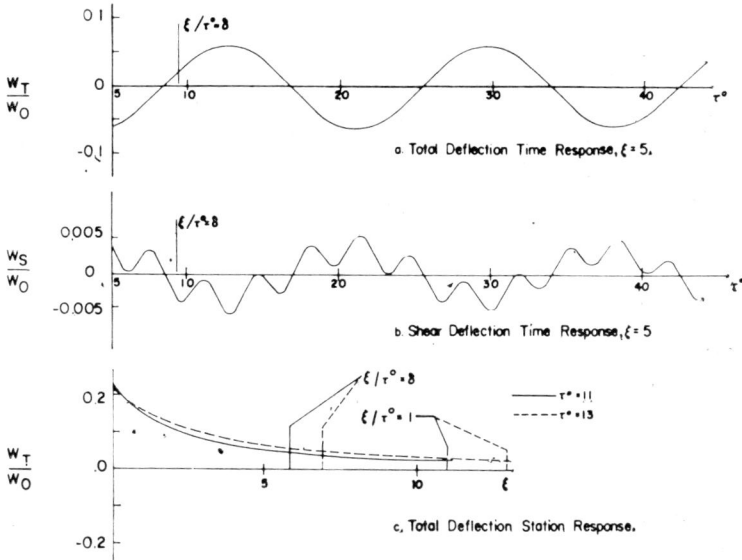

Fig. 7.20. Vertical displacement response corresponding to the imaginary segment P_2 of branch $\varkappa_3$.

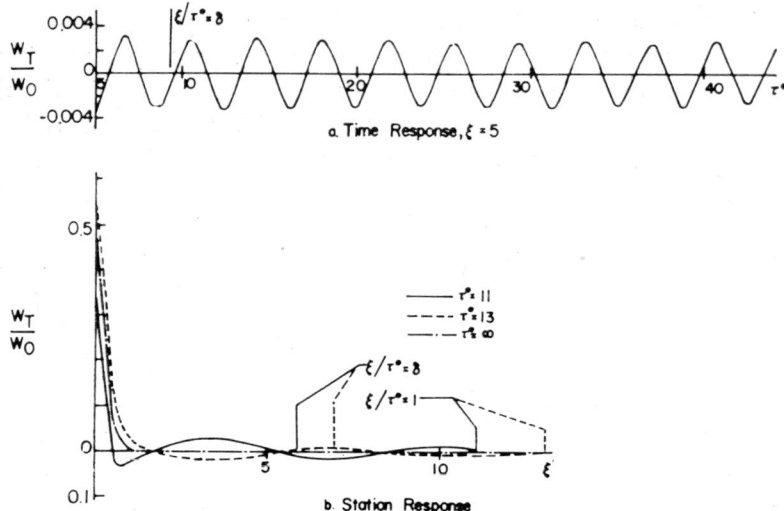

Fig. 7.21. Vertical displacement response corresponding to the complex segments P_{4L}, P_{4R} of branches $\varkappa_2$, $\varkappa_3$ (Timoshenko theory).

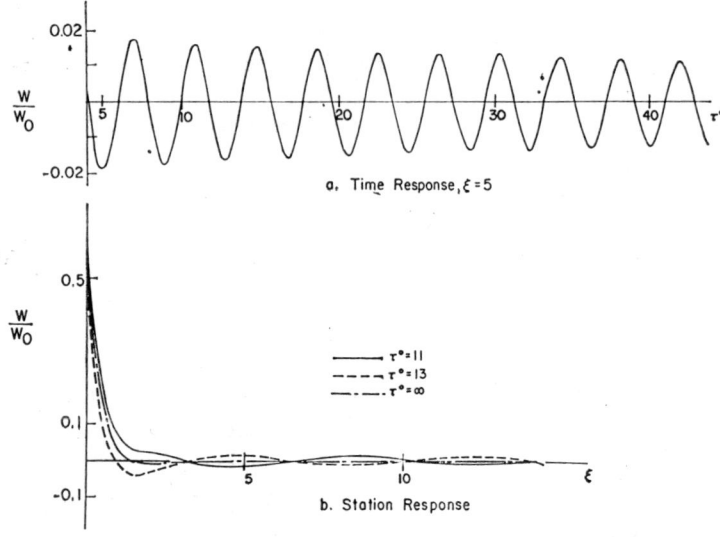

Fig. 7.22. Vertical displacement response coresponding to the complex segments P_{4L} P_{4R} of branches $\varkappa_2$, $\varkappa_3$ (exact theory).

It should be noted again here, as in the preceding approximate theory solutions, *these integrals* (along the imaginary p-axis) *lend themselves nicely to long-time and far-field approximation.* Observation of the integrals in (7.137) shows the following pertinent properties:

I_1, I_3, *and* I_5: These integrals, corresponding to real wave number paths, are improper due to integrand singularities at their finite limits. Such integrals, in our previous approximate theory long-time solutions, were approximated by first piecing off the integration near the singular limit and assessing it separately. The remainder of the integral was approximated by the stationary phase method. In all the present cases, for this part of the integration the method of stationary phase yields at most contributions of $O(t^{-\frac{1}{2}})$ or $O(x^{-\frac{1}{2}})$ for large t or x. Most of the stationary phase contributions are of $O(t^{-\frac{1}{2}})$ but the spectral branches, underlying the present integrals, have inflection points giving rise to the second order points which give the terms of $O(t^{-\frac{1}{3}})$. As the terracing in fig. 4.13 shows an increasing number of such inflection points occur for higher and higher branches. For the integration near the singular limits, for the present integrals which have cutoff type limits stemming from (7.121), consider that for I_1, at and near c. Then (7.128) reduces essentially to

$$I_{1c} \sim \int_c^{c+\varepsilon} \frac{\cos \omega t}{\varkappa_3(-i\omega)} d\omega = \int_c^{\infty} - \int_{c+\varepsilon}^{\infty} \frac{\cos \omega t}{(\omega-c)^{\frac{1}{2}}} d\omega, \qquad (7.138)$$

where the second integral on the right is a Fourier integral and is of $O(t^{-1})$. Through two simple transformations we find the first integral yields

$$I_{1c} \sim (\cos ct - \sin ct)(\pi/2t)^{\frac{1}{2}},$$

which has the same order as most of the contributions from the stationary phase approximation for the rest of I_1. Integrals I_3 and I_5 are similar so that all three contribute at most terms of $O(t^{-\frac{1}{2}})$ for long time, and $O(x^{-\frac{1}{2}})$ for the far field.

I_6: The integrand of this lowest branch integral is continuous everywhere. However, piecing off the integration near $\omega = 0$, the rest of the integral can again be approximated by the stationary phase method. Completion of the pieced off integral from ε to ∞ leads to the following contribution to w

$$w(\xi, \zeta, \tau) \sim \zeta \tau^{-\frac{1}{3}} Ai(-\tau_1'), \qquad (7.139)$$

where $\xi = x/h$, $\zeta = z/h$, $\tau = c_s t/h$, $\tau_1' = (\tau - \xi/b)(3\gamma\tau/b^2)^{-\frac{1}{3}}$ and $b = c_p/c_s$. The approximation technique is like the one that leads to (7.68). It should be pointed out that one second order stationary phase point occurs for this integral at a moderately high frequency. There is therefore another contribution of $O(\tau^{-\frac{2}{3}})$. It should also be pointed out that the stationary phase method breaks down at the ∞ upper limits in I_1, I_5 and I_6. This is because at these limits (where ω, $\varkappa \to \infty$) $c_g^{(n)}(\omega) \to 0$ on the underlying real segments in their monotonic approach to the asymptotes OE and OR in fig. 4.13. The proper way to handle such limiting contributions is through wavefront approximations. We shall discuss this more later.

I_2, I_4: These integrals are Fourier integrals, I_2 being improper at both integration limits. The corresponding pieced off-integrations are the same type as on the L.H.S. of (7.138). Therefore the contributions are of $O(t^{-\frac{1}{2}})$. The Riemann–Lebesque lemma shows the rest of I_2 contributes a term of $O(t^{-1})$ with a small coefficient, since the integrand has spatial decay due to the term $\exp[-\mathrm{Im}\,\varkappa_3(-i\omega)x]$. The integral I_4 has a singular integrand at $\omega = 0$. The pieced-off integral is of the form

$$I_{04} \sim e^{-\mathrm{Im}\,\varkappa_2(0)x} \int_0^\varepsilon \frac{\sin \omega t}{\omega} d\omega = e^{-\mathrm{Im}\,\varkappa_2(0)x} \left[\int_0^\infty \frac{\sin \omega t}{\omega} d\omega + O(t^{-1}) \right]. \quad (7.140)$$

The order term here represents a Fourier integral generated in completing the integral on the left hand side. Since the ∞ integral in (7.140) is equal to $\pi/2$, $I_{04} \sim e^{-\mathrm{Im}\,\varkappa_2(0)x}$. The rest of the I_4 integral yields terms like those of I_2 away from its integration limits.

Finally, we note *solution (7.137) is valid for the H(t) input*, and other moderately sharp inputs, provided we don't ask information at or just behind the initial wavefront. Note the response plots in figs. 7.19–7.22 support this, showing very little high frequency behavior. With all the terms in (7.137) evaluated numerically, like in figs. 7.19–7.22, their sum gives the resultant disturbance. The reader will be interested in the recent work by Viano and Miklowitz [7.14] in which the inversion method of this section was applied to the analogous problem of symmetric normal line load excitation of a symmetrically layered elastic plate. A resultant *near-field response*, based on the integrals from the first few branches evaluated numerically, was compared with a like response involving a few additional higher branches. The *higher branches were found to add very little to the nature of the disturbance*.

7.2.1.2. Inversion of time transform first.

Interchanging the order of integration in (7.117) we have

$$w(x, z, t) = \frac{\sigma_0}{2\pi\mu} \int_{-\infty}^{\infty} e^{i\varkappa x} \left[\frac{1}{2\pi i} \int_{Br} \frac{N(\varkappa^2, z, p^2)}{pF_s(\varkappa^2, p^2)} e^{pt} dp \right] d\varkappa, \quad (7.141)$$

where N/F_s is given in (7.117). Noting again that N/F_s is even in η_s and η_d, there are no branch points of N/F_s in the p plane. The only singularities of the integrand in the Br integral are the simple poles at $p=0$ and $p = \pm i\omega_j(\varkappa)$, $j = 1, 2, 3 \ldots$, where $\omega_j(\varkappa)$ are the real Rayleigh–Lamb branches defined for

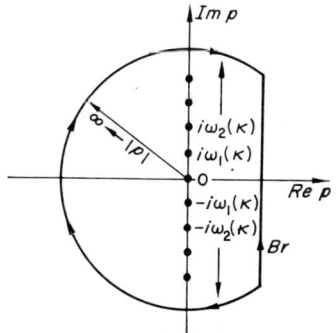

Fig. 7.23. Contour integration in the p-plane.

$\omega \geq 0$. We therefore complete the Br contour to the left as shown in fig. 7.23. It is not difficult to show that for large $|p|$, and all $\varkappa$, the Laplace transform

$$\frac{N}{pF_s} \text{ is of } O\{p^{-2} \exp\left[-k_d(h-z)\right]\}, \quad 0 \leq z \leq h, \quad (7.142)$$

and therefore it decays at least as p^{-2} (the decay at $z = h$). It follows the contour integration over the path indicated in fig. 7.23, along with residue theory, reduces (7.141) to

$$w(x, z, t) = \frac{\sigma_0}{2\pi\mu} \int_{-\infty}^{\infty} \left\{ \lim_{p \to 0} \left[\frac{N'(\varkappa^2, z, p^2)}{F_s'(\varkappa^2, p^2)} \right] \right.$$

$$\left. - \sum_{j=1, 2, \ldots} \left[\frac{N[\varkappa^2, z, -\omega_j^2(\varkappa)]}{\omega_j^2(\varkappa) \frac{\partial F_s(\varkappa^2, p^2)}{\partial(p^2)} \bigg|_{p^2 = -\omega_j^2(\varkappa)}} \right] \cos\left[\omega_j(\varkappa)t\right] \right\} e^{i\varkappa x} d\varkappa, \quad (7.143)$$

where the primes in the first term indicate differentiation with respect to p. This term represents the residue of $N\exp(pt)/pF_s$ at $p=0$. The derivatives are involved here because the limit of $N/F_s = 0/0$ as $p \to 0$. In deriving the series in (7.143), we have made use of the fact that $\partial F_s/\partial(p^2)$ has essentially the same form as $\partial F_s/\partial(\varkappa^2)$ in (7.126), and with $\varkappa$ real (imposed by the real $\varkappa$ path of integration) and p imaginary [i.e., $p = \pm i\omega_j(\varkappa)$] again here, (2) of (7.127) is applicable and therefore the bracketed term in this series is real. Now N'/F_s' in the first term of (7.143) is even in $\varkappa$ and so is the bracketed term in the series there since all terms comprising the latter, including the real branches $\omega_j(\varkappa)$, are even in $\varkappa$ [note $\omega_j(\varkappa)$ for negative $\varkappa$ is just the reflection of $\omega_j(\varkappa)$ for positive $\varkappa$ in the ω-axis ($\varkappa=0$)]. This nature reduces (7.143) to

$$w(x,z,t) = \frac{\sigma_0}{\pi\mu}\left\{\int_0^\infty \lim_{p\to 0}\left[\frac{N'(\varkappa^2, z, p^2)}{F_s'(\varkappa^2, p^2)}\right]\cos\varkappa x\, d\varkappa\right.$$

$$\left. -\frac{1}{2}\int_{-\infty}^\infty \sum_{j=1,2,\ldots} \frac{N[\varkappa^2, z, -\omega_j^2(\varkappa)]}{\omega_j^2(\varkappa)\left.\frac{\partial F_s(\varkappa^2, p^2)}{\partial(p^2)}\right|_{p^2=-\omega_j^2(\varkappa)}} \cos[\omega_j(\varkappa)t - \varkappa x]d\varkappa\right\},$$

(7.144)

valid for $t > 0$. In deriving the series term here we have also used the expansion $\cos[\omega_j(\varkappa)t]\cos\varkappa x = (1/2)\{\cos[\omega_j(\varkappa)t - \varkappa x] + \cos[\omega_j(\varkappa)t + \varkappa x]\}$. The *first term* in (7.144) corresponds to the *static solution* in this problem. The *second term represents the wave propagation*, and as we have already noted is restricted to the real $\omega - \varkappa$ branches. The denominator in this general term is equal to

$$\frac{\omega_j(\varkappa)}{2}\left.\frac{\partial F_s(\varkappa^2, \omega^2)}{\partial \omega}\right|_{\omega^2 = -\omega_j^2(\varkappa)}. \quad (7.145)$$

Hence singularities in the terms of the series (which are the integrands of the associated integrals) would have to be associated with the zeros of the function (7.145). Inspection of the branch functions $\omega_j(\varkappa)$ (cf. the symmetric real segments in fig. 4.13) shows that only $\omega_1(\varkappa)$ has a zero. It occurs at $\varkappa = 0$. Skalak [7.7] has shown that $\partial F_s/\partial \omega$ along the branches $\omega_j(\varkappa)$ cannot vanish. Following Skalak we consider the frequency equation $F(\varkappa^2, \omega^2) = 0$ (the "sub"s is dropped since the proof holds generally for at least Rayleigh–Lamb and Pochhammer frequency spectra). Therefore along any branch $\omega_j(\varkappa)$,

$$\frac{\partial F}{\partial s_t} = \frac{\partial F}{\partial \omega} \frac{\partial \omega}{\partial s_t} + \frac{\partial F}{\partial \varkappa} \frac{\partial \varkappa}{\partial s_t} = 0, \qquad (7.146)$$

where s_t is in the direction tangent to the branch. Let us now assume

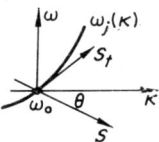

Fig. 7.24. Directions s_t and s at an arbitrary point ω_0 on a branch ω_j.

$\partial F/\partial \omega = 0$ at some arbitrary point ω_0 along a branch ω_j (cf. fig. 7.24). Then from (7.146)

$$\frac{\partial F}{\partial \varkappa} \frac{\partial \varkappa}{\partial s_t} = 0. \qquad (7.147)$$

Now for an arbitrary direction s through ω_0,

$$\frac{\partial F}{\partial s} = \frac{\partial F}{\partial \omega} \frac{\partial \omega}{\partial s} + \frac{\partial F}{\partial \varkappa} \frac{\partial \varkappa}{\partial s} = \frac{\partial F}{\partial \varkappa} \cos \theta, \qquad (7.148)$$

where θ is the angle between the direction s and the $\varkappa$-axis. Therefore, from (7.147), (7.148)

$$\frac{\partial F}{\partial \varkappa} \frac{\partial \varkappa}{\partial s_t} = \frac{1}{\cos \theta} \frac{\partial F}{\partial s} \frac{\partial \varkappa}{\partial s_t} = 0. \qquad (7.149)$$

Now assume $\cos \theta \neq 0$, and observe $\partial \varkappa/\partial s_t$ vanishes only if $\partial \omega_j/\partial \varkappa \to \infty$, i.e., the branches ω_j are vertical. Since there are no such branches, it follows from (7.149) that

$$\partial F(\varkappa^2, \omega^2)/\partial s = 0. \qquad (7.150)$$

From (7.148) we see (7.150) holds for $\cos \theta = 0$ also. Now (7.150) requires that the surface $F(\varkappa^2, \omega^2)$ has a maximum, minimum, or a stationary point (saddle point) at ω_0. However, since $F(\varkappa^2, \omega^2) = 0$ must pass through ω_0 (i.e., ω_0 is on a branch), (7.150) cannot correspond to a maximum or minimum of $F(\varkappa^2, \omega^2)$, since $F(\varkappa^2, \omega^2) = 0$ away from ω_0 too along the branch (cf. fig. 7.25). Now if (7.150) corresponded to a saddle point, there would be two branches ω_j of $F(\varkappa^2, \omega^2) = 0$ passing through ω_0. However, there are

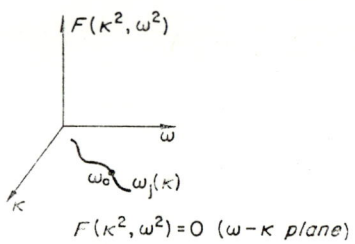

Fig. 7.25. The real branches $\omega_j(\varkappa)$ of $F(\varkappa^2, \omega^2) = 0$.

no such intersecting branches in these real wave number spectra. We conclude that $\partial F/\partial s \neq 0$ and therefore $\partial F/\partial \omega \neq 0$, which was to be proved.

It is also instructive, in the present context, to point out that a similar question arises in the inversion of the spatial transform first, i.e., (7.119). That is, *can $\partial F_s/\partial \varkappa$ there along the branches $\varkappa_j(p)$ vanish?* The answer is that it can, which is exhibited by (7.120). From (7.120) we see near cutoff that $\partial F_s/\partial \varkappa$ and $dp/d\varkappa$ vanish as $\varkappa_j(p) \to 0$, where $\partial F_s/\partial(\varkappa^2)$ and $\partial F_s/\partial p$ do not vanish. The vanishing of $dp/d\varkappa$ means that $\partial F_s/\partial \varkappa$ *vanishes at horizontal slopes on the real and imaginary segments* of the frequency branches. Equation (7.121) shows the related singularities are branch points in the p-plane. For the case of $\varkappa_j^* \neq 0$ one can construct a similar proof. When the segments are real, in the present case, (7.146) applies. Since in this case $\partial \omega/\partial s_l = \partial \omega/\partial \varkappa = 0$, and $\partial F/\partial \omega \neq 0$, (7.146) shows

$$\partial F/\partial s_l = \partial F/\partial \varkappa = 0$$

in agreement with the above discussion.

Having proved that (7.145) is zero only for $\omega_1(\varkappa) = 0$ (at $\varkappa = 0$), all the integrands in (7.144) are continuous over their entire range of integration. This is true also for the $j = 1$ integrand, since it is not singular at $\varkappa = 0$. Hence (7.144) can be integrated numerically or approximated for long time and the far field through the method of stationary phase and the Riemann-Lebesque lemma. We note that *the second integral in (7.144) is comprised of waves that travel in the positive and those that travel in negative x-directions.* For $x > 0$ the latter correspond to the integration interval $-\infty < \varkappa < 0$.

The solutions (7.137), and (7.144) for $j = 1, 2, 3$, are obviously different in form, but the *response they represent must be unique*. In the case where the integrals involve positive group velocities there is a *one to one correspondence* of these integrals in the two solution, i.e., I_1, I_3, I_5 and I_6 in (7.137) from the real segment paths for $\omega > 0$ (cf. fig. 7.17b), and the integrals from the $j = 1, 2, 3$ terms in (7.144) over like paths. Also the *remaining integrals* in (7.137), I_2 and I_4, *from the imaginary and complex wave number segments*, can be identified with those from the $j = 1, 2, 3$ terms in (7.144) corresponding to $\omega_j(\varkappa)$ for $\varkappa < 0$ which *represents waves traveling in the negative x-direction*. For long time the contributions from both of these sets of integrals are of $O(t^{-1})$, according to the Riemann lemma, since they are Fourier integrals. We have shown this earlier in this section for I_2 and I_4 away from their end points of integration. The corresponding integrals in (7.144) also are Fourier integrals and hence are also of $O(t^{-1})$. In this case the integrals are based on the real branches $\omega_j(\varkappa)$ for $\varkappa < 0$ which are the reflections in the ω-axis ($\varkappa = 0$) of the real branches $\omega_j(\varkappa)$ for $\varkappa > 0$. Since the integrals corresponding to the latter branches have points of stationary phase, those of the former cannot, since $\varkappa$ changes its sign, and hence they are *necessarily Fourier integrals* (cf. § 5.10.3). We should also point out that (7.139), the long-time contribution from I_6, has its counterpart that stems from branch $j = 1$ in (7.144). Here, however, a second Airy function is generated with argument showing it corresponds to a wave traveling in the negative x-direction. Finally we should point out that obviously the solution (7.144) is *simpler to obtain*. *On the other hand* it does not bring out as many detailed features as (7.137) does, on the role the frequency spectra play in transient wave solutions.

7.2.2. *Excitation of plate by two antisymmetric normal line loads and by a single normal line load*

By reversing the direction of the input on the face $z = -h$ in fig. 7.15, we have the case of two sudden antisymmetrically disposed normal line loads acting on the infinite elastic plate. This can be solved in the same way as the symmetric problem was in § 7.2.1, by exploiting the antisymmetric spectra of (4.48). Since we have superposition of solutions, because of basic linearity in this subject, summing the two solutions corresponding to one half of the loads gives the solution for the single normal line load applied to one face of the plate. This is demonstrated in fig. 7.26. Direct solution of the single normal line load problem would separate into the two component solutions.

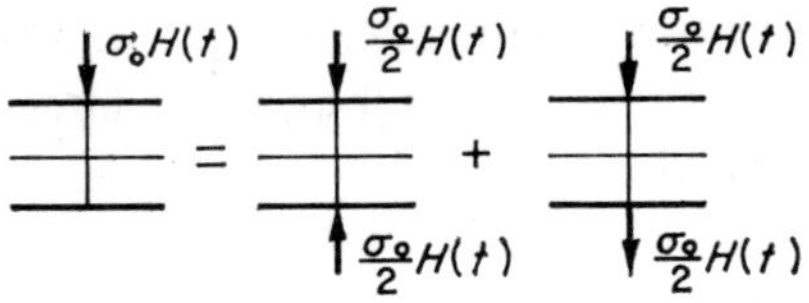

Fig. 7.26. Decomposition of a normal line load on an elastic plate into symmetric and antisymmetric parts.

7.3. Edge load problems for the semi-infinite waveguide

7.3.1. Introduction

The solution of waveguide problems, involving *corners and edges*, and based on the equations of motion from the linear theory for a homogeneous, isotropic, elastic material, is an important topic of long standing interest, difficulty and challenge. The basic difficulty is exhibited in Love's treatment of the free longitudinal vibration of a finite length, circular section, cylindrical rod [1.2, § 201]. The rod here has a stress free lateral surface and ends (or edges), hence stress free corners. Its *edge conditions are of the nonmixed type*, i.e., *stress components* (present case), *or displacement components specified*. Love shows attempts to treat this problem with a classical separation technique, aided by Pochhammer's frequency equation for the infinite rod [4.8] fail, leading to a solution in which the normal stress on the rod ends vanishes, but not the shear stress. This led to the classical approximation that for a long, thin rod (radius ≪ length) this shear stress could be assumed zero because its mate on the neighboring lateral surface is zero. Indeed, as we have discussed earlier, this type of approximation, with the simplifying features brought forth by the low frequency-long waves it represents, has led to a great deal of work on the creation and use of approximate one-dimensional theories for treating wave and vibration problems in the rod, plate, and shell.

The literature shows that contributions to the present subject have been mostly on the problems involving *mixed edge conditions, i.e., a mixture of stress and displacement components*, stemming primarily from their *separable nature*. In the following sections first a review is presented of methods that have been used to study transient wave propagation in waveguide problems of this type and the information they have produced. Then a similar discussion is given of methods and contributions in problems involving *non-*

Ch. 7, § 7.3]   EDGE LOAD PROBLEMS   431

mixed edge conditions. Advances have been made only quite recently in these *nonseparable* problems.

Section 7.2.1 showed that when the formal solution (inversion integrals) for a waveguide problem can be written, inversion procedures based on modal integrals, are available to obtain the solution, which can then be analyzed numerically or through approximations (as outlined there). Obtaining the formal solution in the infinite plate problem of § 7.2.1, for example, was a straight forward application of the indicated transforms. For the semi-infinite waveguide, with its edge or end, the formal solution can be difficult to derive. Some procedures have been established for these cases and a discussion of them, and approximate solutions they have yielded, is worth a bit of our time. We draw on the displacement equations of motion (1.65a), for it will become evident that they offer a certain convenience in treating the edge boundary conditions. We treat the semi-infinite plate in plane strain. Other waveguides, at least for mixed edge conditions can be treated similarly. The displacement equations of motion for the plane strain case are

$$\left. \begin{array}{l} u_{xx}(x, y, t) + k^{-2}(k^2 - 1)v_{xy} + k^{-2}u_{yy} = c_d^{-2}\ddot{u} \\ v_{xx}(x, y, t) + (k^2 - 1)u_{xy} + k^2 v_{yy} = c_s^{-2}\ddot{v} \end{array} \right\}, \quad x > 0, \; -h < y < h, \; t > 0,$$

(7.151)

where we are now using y as the thickness coordinate, and v as the corresponding displacement. The stress-strain relations are

$$\sigma_x(x, y, t)/(\lambda + 2\mu) = u_x + k^{-2}(k^2 - 2)v_y,$$
$$[\sigma_y(x, y, t)/(\lambda + 2\mu) = k^{-2}(k^2 - 2)v_x + u_y, \quad (7.152)$$
$$\sigma_{xy}(x, y, t)/\mu = v_x + u_y, \; \sigma_z = \nu(\sigma_x + \sigma_y) \; .$$

Subscripts attached to displacements indicate differentiation with respect to that coordinate, but when attached to stress indicate the component in the usual way. Initial conditions are taken as

$$u(x, y, 0) = \dot{u}(x, y, 0) = v(x, y, 0) = \dot{v}(x, y, 0) = 0, \; x \geq 0, \; -h \leq y \leq h, \quad (7.153)$$

and conditions at $x \to \infty$ as

$$\lim_{x \to \infty} \begin{bmatrix} u(x, y, t), u_x, \text{etc.} \\ v(x, y, t), v_x, \text{etc.} \end{bmatrix} = 0, \qquad -h \leq y \leq h, \ t > 0. \quad (7.154)$$

7.3.2. Plate in plane strain with mixed edge conditions; formal solutions

The formal solutions for the two problems of this type involving symmetric excitation can be written with the aid of the Fourier sine and cosine transforms (cf. ch. 5 § 5.7). They are discussed in the following sections.

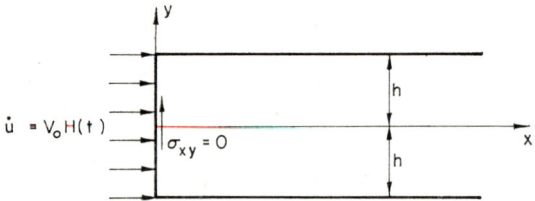

Fig. 7.27. Longitudinal impact problem.

7.3.2.1. Longitudinal impact problem. Figure 7.27 depicts the problem. The plate edge ($x=0$) is subjected to a step uniform normal velocity $\dot{u}$ under zero shear stress σ_{xy}. These edge conditions are given by

$$\left. \begin{aligned} \dot{u}(0, y, t) &= v_0 H(t) \\ \sigma_{xy}(0, y, t) &= \mu[v_x(0, y, t) + u_y(0, y, t)] = 0 \end{aligned} \right\} -h \leq y \leq h, \ t > 0, \quad (7.155)$$

where use has been made of the third of (7.152). First we apply a Laplace transform on t, parameter p, to (7.151), using (7.153). Then since we have a displacement type input for u, and noting the form of (5.55) and (5.56), we further apply a sine transform to the first, and cosine transform to the second, of the Laplace transformed equations (7.151). There results the double transformed equations

$$\tilde{u}^s_{yy}(\varkappa, y, p) - k^2 \eta_d^2 \tilde{u}^s - (k^2 - 1)\varkappa \tilde{v}^c = -k^2 \varkappa \bar{u}(0, y, p),$$
(7.156)
$$\tilde{v}^c_{yy}(\varkappa, y, p) - k^{-2}\eta_s^2 \tilde{v}^c + k^{-2}(k^2-1)\varkappa \tilde{u}^s_y = k^{-2}[\bar{v}_x(0, y, p) + (k^2-1)\bar{u}_y(0, y, p)].$$

Ch. 7, § 7.3] EDGE LOAD PROBLEMS 433

The Laplace transforms of the edge conditions (7.155) are

$$\overline{u}(0, y, p) = p\overline{u}(0, y, p) = v_0/p , \qquad (7.157)$$

since $u(0, y, 0) = 0$ from (7.153), and

$$\overline{\sigma}_{xy}(0, y, p) = \mu[\overline{v}_x(0, y, p) + \overline{u}_y(0, y, p)] = 0 . \qquad (7.158)$$

Now noting from (7.157) that $\overline{u}(0, y, p)$ is independent of y, it follows that $\overline{u}_y(0, y, p)$ in (7.158) vanishes, and hence $\overline{v}_x(0, y, p)$ there also vanishes. Therefore, (7.157), (7.158) reduce to

$$\overline{u}(0, y, p) = v_0/p^2 ,$$
$$\overline{u}_y(0, y, p) = \overline{v}_x(0, y, p) = 0 , \qquad (7.159)$$

and substitution of these in (7.156) determines their right-hand sides. Thus we see the choice of spatial transforms here was made so that the given edge conditions would be *asked for*. The solution of these coupled ordinary differential equations is easily obtained through substitution of the forms

$$\tilde{u}^s(\varkappa, y, p) = A_n(\varkappa, p) \exp\left[-n(\varkappa, p)y\right] + \tilde{u}_p^s(\varkappa, y, p) ,$$
$$\tilde{v}^c(\varkappa, y, p) = B_n(\varkappa, p) \exp\left[-n(\varkappa, p)y\right] + \tilde{v}_p^c(\varkappa, y, p) , \qquad (7.160)$$

where $\tilde{u}_p^s$ and $\tilde{v}_p^c$ are particular integrals needed for the now simple right-hand sides of (7.156). The $n(\varkappa, p)$ are found to be $\pm\eta_d$, $\pm\eta_s$, as might be expected from our earlier work in this chapter. The particular integrals are found to be $\tilde{u}_p^s = v_0\varkappa/p^2\eta_d^2$, and $\tilde{v}_p^c = 0$. It follows that the general solution is

$$\tilde{u}^s(\varkappa, y, p) = A(\varkappa, p) \cosh \eta_d y + B(\varkappa, p) \cosh \eta_s y + v_0\varkappa(p\eta_d)^{-2} , \qquad (7.161)$$

$$\tilde{v}^c(\varkappa, y, p) = -[\varkappa^{-1}\eta_d A(\varkappa, p) \sinh \eta_d y + \varkappa\eta_s^{-1} B(\varkappa, p) \sinh \eta_s y] ,$$

stemming from the symmetry of the loading, and the fact that the two algebraic equations, which yielded $n(\varkappa, p) = \pm\eta_d$, $\pm\eta_s$, must hold for all these values of $n(\varkappa, p)$. The traction free conditions at the plate faces, σ_y

and σ_{xy} in (7.152) vanish at $y = \pm h$, are transformed with an eye on the fact that u involves a sine, and v a cosine transform. Substitution of (7.161) into these transformed conditions, i.e.,

$$\tilde{\sigma}^c_y(\varkappa, h, p)/(\lambda + 2\mu) = k^{-2}(k^2 - 2)[\varkappa \tilde{u}^s(\varkappa, h, p) - v_0 p^{-2}] + \tilde{v}^c_y = 0,$$

$$\tilde{\sigma}^s_{xy}(\varkappa, h, p)/\mu = -\varkappa \tilde{v}^c(\varkappa, h, p) + \tilde{u}^s_y(\varkappa, h, p) = 0,$$

determines A and B, and hence the transformed solution to the problem. The formal solution (inversion integrals) can then be written in the form of (7.117) by making use of the fact that $\tilde{u} = 2i\tilde{u}^s$, and $\tilde{v} = 2\tilde{v}^c$, according to (5.48) and (5.50). The formal solutions for the compressional strains are

$$u_x(x, y, t) = \frac{1}{2\pi i} \int_{Br} e^{pt} dp \left[\frac{1}{\pi} \int_{-\infty}^{\infty} \varkappa^2 \left\{ \frac{N(\varkappa^2, y, p^2)}{F_s(\varkappa^2, p^2)} + \frac{v_0}{p^2 \eta_d^2} \right\} e^{-i\varkappa x} d\varkappa \right], \tag{7.162}$$

$$v_y(x, y, t) = \frac{1}{2\pi i} \int_{Br} e^{pt} dp \left[\frac{1}{\pi} \int_{-\infty}^{\infty} \frac{M(\varkappa^2, y, p^2)}{F_s(\varkappa^2, p^2)} e^{-i\varkappa x} d\varkappa \right],$$

where

$$\left\{ \begin{matrix} N \\ M \end{matrix} \right\} = \frac{(k^2 - 2)v_0}{c_d^2 \eta_d^2} \left\{ \begin{matrix} -[(\varkappa^2 + \eta_s^2) \sinh \eta_s h \cosh \eta_d y - 2\eta_d \eta_s \sinh \eta_d h \cosh \eta_s y] \\ \eta_d[\eta_d(\varkappa^2 + \eta_s^2) \sinh \eta_s h \cosh \eta_d y - 2\varkappa^2 \eta_s \sinh \eta_d h \cosh \eta_s y] \end{matrix} \right\}$$

and F_s is given in (7.117). The technique is basically due to Folk, Curtis et al. [5.16b, 7.15], the first of these references being on the circular cylindrical rod, and second on the plate.

7.3.2.2. *Mixed pressure shock problem.* Consider the problem shown in fig. 7.28. The plate edge is subjected to a step uniform normal stress σ_x, under zero thickness displacement v. These conditions are given by

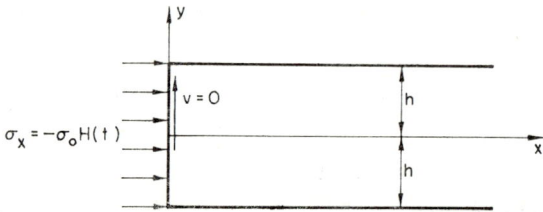

Fig. 7.28. Mixed pressure shock problem.

$$\left.\begin{array}{l}\sigma_x(0, y, t) = -\sigma_0 H(t) \\ v(0, y, t) = 0\end{array}\right\}, \quad -h \leq y \leq h, \, t > 0, \quad (7.163)$$

where σ_0 is the magnitude (a positive constant) of the σ_x input. The method of reducing (7.151) to a set of ordinary differential equations is the same as that leading to (7.156), except that the sine and cosine transform are interchanged so that the given information in (7.163) is asked for. Instead of 7.156), here the analogous transformed equations are

$$\tilde{u}^{\sim c}_{yy}(\varkappa, y, p) - k^2 \eta_d^2 \tilde{u}^{\sim c} + (k^2 - 1)\varkappa \bar{v}^{\sim s}_y = k^2 \bar{u}_x(0, y, p) + (k^2 - 1)\bar{v}_y(0, y, p),$$

(7.164)

$$\bar{v}^{\sim s}_{yy}(\varkappa, y, p) - k^{-2} \eta_s^2 \bar{v}^{\sim s} - k^{-2}(k^2 - 1)\varkappa \tilde{u}^{\sim c}_y = -k^{-2}\varkappa \bar{v}(0, y, p).$$

The Laplace transform conditions at the edge ($x = 0$) are

$$\bar{\sigma}_x(0, y, p)/(\lambda + 2\mu) = \bar{u}_x(0, y, p) + k^{-2}(k^2 - 2)\bar{v}_y(0, y, p) = -\sigma_0/(\lambda + 2\mu)p,$$

$$\bar{v}(0, y, p) = 0, \quad (7.165)$$

where we have used the first of (7.152). From the second of (7.165) it follows that $\bar{v}_y(0, y, p) = 0$, and therefore the first reduces to $\bar{u}_x(0, y, p) = -\sigma_0/(\lambda + 2\mu)p$. Hence the right-hand sides of (7.164) are determined. From here on the procedure is the same as it was in the foregoing case, and a formal solution similar to (7.162) would be obtained.

7.3.2.3. *Mixed edge conditions; problems with antisymmetric excitation.*
Similar procedures exist for plate problems involving mixed edge conditions and antisymmetric excitation. An example is Nigul's treatment [7.16] of the semiinfinite plate subjected to an edge step moment, under zero transverse velocity, i.e., a pinned edge. His formal solution is obtained by essentially the same technique as that devised for the rod in flexure by De Vault and Curtis [7.17], which is a generalization of that in [5.16b].

7.3.2.4. *Formal solutions for other waveguides.*
The reader will be interested in some brief comments on formal solution techniques that have been devised for other waveguide problems involving mixed edge conditions. By developing *exterior domain Hankel transforms* for the infinite plate

with a circular cavity at the plate center, Scott and Miklowitz [7.18] solved the mixed problem of sudden uniform radial displacement of the cavity wall. Jones and Ellis [5.17b], drawing on the Rayleigh–Lamb frequency equation for symmetric waves, and the *plane stress-plane strain analogy*, extended the use of the technique in [5.16b, 7.15] to the mixed pressure shock problem in a semi-infinite, wide, rectangular bar in plane stress, with stress free lateral sides. Rosenfeld and Miklowitz [7.19] also found the technique in [5.16b, 7.15] useful in their work on wave propagation in a rod of arbitrary cross section.

7.3.3. Plate in plane strain with mixed edge conditions; approximate solutions

7.3.3.1. Long-time and/or far-field approximations. It has already been shown in § 7.2.1.1, in the discussion of the lowest branch integral I_6 of solution (7.137), that the pieced off integration of this integral, near $\omega=0$ (and $\varkappa=0$) led to an important contribution to the long-time solution involving an Airy function with wave argument and time decay of $O(t^{-\frac{1}{3}})$. The formal solutions for the present mixed edge condition plate problems, (7.162) for the longitudinal impact problem, and the analogous one for the mixed pressure shock both have in them an I_6 type integral. Here the integrand of I_6 is singular at $\omega=0$, the nature of which is like that in the first integral of (7.61) for the analogous Love–Rayleigh rod problem (cf. § 7.1.3.1). So, the method of stationary phase and the completion of the pieced off part of the I_6 integral here, as discussed in § 7.2.1.1, reduces the formal solution of mixed pressure shock problem to the *long time-far field approximation*

$$u_\xi(\xi, \eta, \tau) \simeq -\hat{\sigma}_0 \left[\int_0^{\alpha'} Ai(-\alpha)d\alpha + \frac{1}{3} \right],$$

$$v_\eta(\xi, \eta, \tau) = -[(k^2-2)/k^2]u_\xi$$

(7.166)

where the dimensionless variables $\xi=x/h$, $\eta=y/h$ and $\tau=c_s t/h$ have been introduced, $\alpha'=(\tau-\xi/b)/(3\delta\xi/b)^{\frac{1}{3}}$, b is the dimensionless plate velocity c_p/c_s, $\delta=(k^2-2)^2/6k^4$, and $\hat{\sigma}_0$ is the constant $k^2\sigma_0/4\mu(k^2-1)$. The solution (7.166) is based on the approximation to the lowest Rayleigh–Lamb frequency branch

$$\varkappa_1(p) \simeq (ip/c_p)[1-\delta(hp/c_p)^2+\ldots] .$$

(7.167)

The second equation in (7.166) expresses the *Poisson's ratio coupling* of the plate. The corresponding approximation for the longitudinal impact problem with formal solution (7.162) differs only trivially from (7.166), i.e., in the constant $\hat{\sigma}_0$. This supports the fact that the long time-far field solution is basically independent of the nature of edge compressional loading. It is not surprising that this type solution is the same for the rod and plate since (7.167) has the same form as (7.60). We should point out that these *exact theory solutions have a first approximation, which corresponds to the elementary theory solution shown in fig. 7.7*. This is *the step function* [in (7.166) it would be $H(\tau-\xi/b)$ instead of the integral of the Airy function appearing there]. *The second approximation (7.166) shows, however, the first should be used with caution*, since having the former we see the *finite jump in the latter cannot be valid*. The first is useful for extremely low frequency-long waves.

Skalak derived his long time-far field solution for the rod [7.7] by inverting the time transform first, and using the stationary phase method (cf. § 7.2.1.2). Folk, Curtis et al. [5.16b, 7.15] derived their solutions for the rod and plate by inverting the spatial transform (Fourier) first, and then approximating the Bromwich integrals by the extended saddle point method (cf. § 5.10.5, refs. 5.15, 5.16a, b, 5.17a, b).

In [5.17b] Jones and Ellis used the extended saddle point technique, as in [5.16b, 7.15], so as to get second-order terms in (7.166) which involve derivatives of the Airy function. These terms account for warping of the cross section of the bar, which is not represented by the plane section nature of (7.166). It was shown by Jones and Ellis [7.20] that the terms accounted for warping of plane sections exhibited in fringe patterns occurring in dynamic photoelasticity shock tube experiments on the rectangular bar. Jones and Norwood [7.21] applied this type of analysis to the semi-infinite, circular section rod, assessing the lateral strains and stresses (the latter corresponding to the rod equivalent of (7.166) are zero). They showed their analysis for the surface radial strain had closer agreement with the corresponding response record in [7.8] than the rod equivalent of the second of (7.166).

In [7.18] the time transform was inverted first, but the approximation (7.166) does not occur, basically because of the spatial decay introduced by the axially symmetric nature of the problem. However, Scott and Miklowitz extended a stationary phase analysis in an earlier related work by Miklowitz [7.22] to the problem in [7.18], and obtained long time-far field, lowest branch approximations for the radial and thickness displacements. The techniques in [7.22] and [7.18] include a time of occurrence-predominant period criterion for restricting the time region of validity for the approximation at a station. This so that higher lowest branch frequencies and the first few thickness branches do not contribute. A comparison of the stationary phase approximation and (7.166) will also be of interest to the reader.

The particular problem solved in [7.17] was that of the semi-infinite, circular section rod, subjected to sudden pressure over half the edge ($-\pi/2 \leq \theta \leq \pi/2$), while this edge begins to move laterally (along $\theta = 0$) with constant velocity. Inversion, evaluation, and experiments in [7.17] followed the theme in [5.16b, 7.15] and the work in Fox and Curtis [7.23]. DeVault and Curtis showed that for long time, or the far field, *the flexural strain disturbance is composed predominantly of lowest branch action, the head stemming from a group velocity maximum at moderately high frequency, and the tail from the low frequency-long wave domain. Amplitudes for the head are proportional to the Airy function, and for the tail, the Fresnel integral* [recall we found the long-time solution for the Timoshenko beam problem treated here was in terms of Fresnel integrals, cf. § 7.1.3.4, eq. (7.106)].

A long time approximation, involving the integrals for some higher antisymmetric real wave number segments, was carried out by Nigul in his work [7.16]. This work is quite detailed, pointing out what segments are needed for various quantities (midplane displacement, moment, etc.) for all long times at a station. Jones contributed similarly [7.24] on a closely related problem to that in [7.16], although not an edge load problem. Important, however, was the fact that he evaluated, for long time, contributions from the first four real segments, for regions on these close to the dilatational wave front [associated with maximum and minimum group velocities $c_g(\varkappa)$].

7.3.3.2. Short-time—near-field, wavefront approximations. Rosenfeld and the author [6.4] have applied the Cagniard-de Hoop technique to the formal solution (7.162), and the analogous solution for the mixed pressure shock problem, establishing the amplitudes and locations of all wavefronts emanating from the edge source in these problems. The technique, and the way to derive wavefront approximations in using it, have already been discussed in ch. 6 §§ 6.2.1.1, 6.2.1.4. We can therefore be brief here. Removing the Bromwich integrals from (7.162) leaves solutions for $\bar{u}_x(x, y, p)$, and $\bar{v}_y(x, y, p)$, i.e., the Laplace transforms of these strains. When

Ch. 7, § 7.3] EDGE LOAD PROBLEMS 439

the transformation $\varkappa = k_d \zeta$, with p real, is introduced into the remaining integrals in (7.162), $\bar{u}_x(x, y, p)$, for example, then takes the form

$$\bar{u}_x(x, y, p) = \frac{V_0}{\pi c_d p} \int_{-\infty}^{\infty} \frac{\zeta^2}{\zeta^2 + 1} e^{-ik_d \zeta x} d\zeta$$

$$-\frac{(k^2 - 2)V_0}{\pi c_d p} \int_{-\infty}^{\infty} \frac{\zeta^2 N(\alpha, \beta, y, \zeta^2)}{R(\alpha, \beta, \zeta^2)} e^{-ik_d \zeta x} d\zeta,$$

(7.168)

where now

$$N = \frac{1}{\zeta^2 + 1} \left[(2\zeta^2 + k^2) \sinh \eta_d h \cosh \eta_d y - 2(\zeta^2 + 1)^{\frac{1}{2}} (\zeta^2 + k^2)^{\frac{1}{2}} \sinh \eta_s h \cosh \eta_s y \right],$$

$$R = (2\zeta^2 + k^2)^2 \cosh \eta_d h \sinh \eta_s h - 4(\zeta^2 + 1)^{\frac{1}{2}} (\zeta^2 + k^2)^{\frac{1}{2}} \cosh \eta_s h \sinh \eta_d h,$$

$$\eta_d = k_d (\zeta^2 + 1)^{\frac{1}{2}}, \quad \text{and} \quad \eta_s = k_d (\zeta^2 + k)^{\frac{1}{2}}.$$

The first integral is easily inverted. It corresponds to a leading plane step wave of u_x, traveling with speed c_d, and is due to the uniform (in y) nature of the edge source. The inversion problem lies in the second integral. The first term in N there corresponds to the dilatational, and the second the equivoluminal waves in u_x (corresponding to the $\bar{\varphi}$ and $\bar{\psi}$ integrals in (14) of [6.4]). Focusing on the integral involving the first term (process is similar for the second), the hyperbolic terms in the integrand are expanded into exponentials. After some algebra, and consideration of the larger exponential terms in R, through use of the binomial theorem the exponential terms of R may be brought into the numerator of the integrands in the form of a series of terms involving exponentials. Further algebra separates this series into one that involves terms with a single exponential. The result is a series of integrals, each of which represents the disturbance following a single wavefront. The integrals are of the form

$$\bar{u}_{xd}(x, y, p) = \int_{-\infty}^{\infty} U_d(\zeta) e^{-p g_d(\zeta)} d\zeta,$$

(7.169)

where sub d denotes dilatational, and

$$g_d(\zeta) \equiv g(\zeta, \gamma, \delta, x) = \frac{1}{c_d} ((ix\zeta + \gamma \eta'_d(\zeta) + \delta \eta_s(\zeta))),$$

$$\gamma = (2n+1)h \pm y, \quad \delta = 2mh,$$

and U_d, an involved function of ζ, also depends on n, and m, which are determined with the aid of geometric ray theory As we have already discussed in ch. 2 § 2.6, *discontinuities in either the geometry or load are sources of wavefronts, and rays, along which wavefronts, normal to the rays, propagate.* The corners $x=0$, $y=\pm h$, here are such points. Figure 7.29 shows the basic

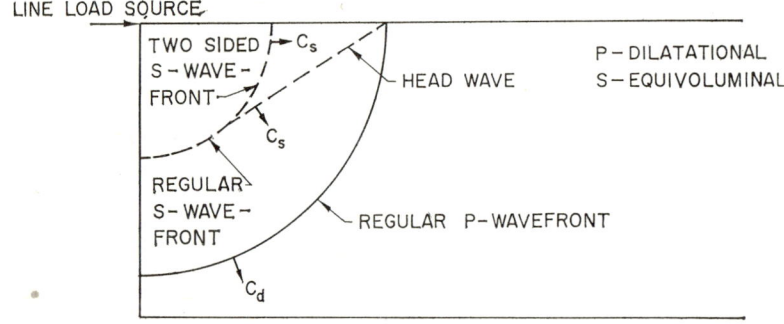

Fig. 7.29. Basic wavefronts in a plate.

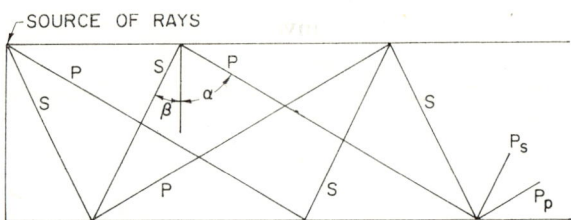

Fig. 7.30. A family of rays.

system of wavefronts that are generated at the upper source $x=0$, $y=h$ (at $x=0$, $y=-h$ there would be another system, symmetric to those in fig. 7.29). Note that the system is that in Lamb's line load problem for the half space. One would expect this. Figure 7.30 shows a *family of multiply reflected rays, these paths being generated by mode conversion at the plate faces* $y=\pm h$, for a pair of P (dilatational) and S (equivoluminal) wavefronts emanating from the source. Letter n measures the number of times a front crosses the plate as a P front, and m the number of times as an S front, ending up (in fig. 7.30) at either P_s or P_p with coordinates (x, y). In fig. 7.30, $n=1$, $m=2$ for any path. Letters γ and δ correspond to the propagation distances along the multiply reflected ray path in the transverse direction y as P and S waves, respectively. The $(-)$ sign is taken in γ in (7.169) for definition at P_p in fig. 7.30, because the last reflection is at a point on the lower plate face ($+$ for upper). With the aid of Snell's law, $\sin \alpha = k \sin \beta$, [6.4] gives expressions for the total distance in the x direction along any path, and the wave front arrival time. These can be used to find the wavefront positions.

Clearly, (7.169) can be inverted by Cagniard's technique. The requirements [which have already been discussed in ch. 6 § 6.2.1.1] establish the analyticity of $g(\zeta)$ in the ζ-plane by suitable cuts for the branch points of $\eta'_d(\zeta)$ and $\eta'_s(\zeta)$. Then the real path integral in (7.169) is traded for one along a path in this region of analyticity, which is determined by $g_d(\zeta) = t$, real and positive. It follows using $\zeta(t)$, defined implicitly by $g_d(\zeta) = t$, (7.169) can be written as

$$\bar{u}_{xd} = \int_{\tau_d}^{\infty} 2\mathrm{Re}\left\{ \frac{U_d(\zeta)}{g'(\zeta)} \bigg|_{\zeta(t)} \right\} e^{-pt} dt, \qquad (7.170)$$

where τ_d is the dilatational wavefront arrival time at a particular station (x, y). Since $2Re\{\}$ in (7.170) is the inverse Laplace transform there, we have

$$\dot{u}_{xd}(x, y, t) = 2\mathrm{Re}\left\{ \frac{U_d(\zeta)}{g'(\zeta)} \right\}_{\zeta(t)} H(t - \tau_d) . \qquad (7.171)$$

A time integration then gives $u_{xd}(x, y, t)$, and this can be approximated near τ_d to give the wavefront time behavior and its amplitude. Results similar to (7.171) exist for the equivoluminal waves, including the head waves. Figure 7.31 shows the wavefronts, that have developed by time $t = 8h/c_d$, and $16h/c_d$, to geometric scale. The wavefronts for the strains, for the edge load problems being discussed here, were all, except for the leading step, found to have the time behavior $(t-\tau)^{\frac{1}{2}}$, where τ is the arrival time of the particular front. The line load problem also treated in [6.4] showed $(t-\tau)^{-\frac{1}{2}}$ for the regular wavefronts. Since the time input for this problem in [6.4] was the step function, this time behavior is in agreement with that for the regular wavefronts in Lamb's line load problem, cf (6.63). This would be expected. In [6.4] numerical data is also given for the longitudinal impact problem for the amplitude coefficients of these wavefronts at $t = 8h/c_d$ and

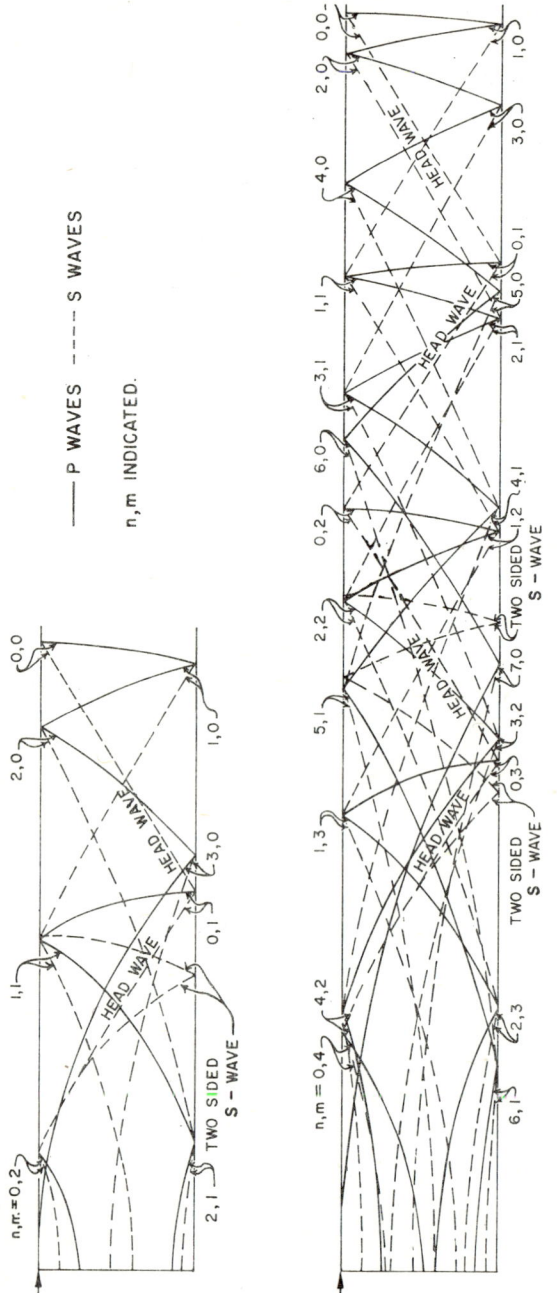

Fig. 7.31. Wavefronts ($t = 8h/c_d$ above, $t = 16h/c_d$ below, $\nu = 0.3$).

$t = 16h/c_d$ (see figs. 6 and 7 there). It was established by these numerical data that *at the surface the strongest wave fronts were created directly by the reflection of the head waves.* Comparison of wavefront arrivals at the surface, with those in the near field records in [7.9], and those in Meitzler's work [7.25], showed definite agreement in the case of the head waves. Other waves are weaker and, since they were poorly defined in the records, were hard to identify positively.

A method for obtaining wavefront approximations, that is applicable to a broader class of waveguide problems than the Cagniard–deHoop method, was recently presented by Randles and Miklowitz [7.26]. The technique is based on high-frequency approximations to the branches of the Rayleigh–Lamb frequency equation, but not in their representation presented in ch. 4. That representation, with its complicating terraces at the dilatational wave speed and superposition of the branches at the equivoluminal wave speed, does not lend itself to high-frequency approximation. In [7.26] the *Rayleigh–Lamb spectra are mapped into a new space*, defined by the change in variables

$$\chi = (\eta_s - \eta_d)/(\eta_s + \eta_d), \quad \eta = h(\eta_s + \eta_d),$$

where η_d, η_s are the thickness wave numbers of ch. 4. The $\chi - \eta$ space offers a more direct approach to the high-frequency response of the plate. In [7.26] the infinite plate, subjected to an impulsive normal line load on one of its faces, is treated (cf. problem in § 7.2.2). The χ, η variables lead to a new form of the formal modal solution of the type given in § 7.2.1.2. The wavefronts are extracted from this solution by exploiting the $\eta - \chi$ branches of the Rayleigh–Lamb equation, and in particular their branch points, about which analytical continuations are made, which uncouple the dilatational and equivoluminal motions. An equivalent modal solution over the dilatational and equivoluminal branches is then derived, and this is approximated using series representations for the branches. Summation over the modes corresponding to these branch approximations then gives the wavefronts. In [7.26] comparison is made of the latter with the known wavefronts from Lamb's line load problem. Randles [7.27] has extended the work to anisotropic plates producing some interesting information on certain *cusp type* wavefronts.

7.3.4. Plate in plane strain with nonmixed edge conditions; formal long-time solutions and their inversions

Because of the *nonseparability in the present class of problems* direct application of integral transforms is not indicated. The right-hand sides of (7.156) and (7.164) point this out clearly, i.e., they do not ask for the given edge conditions which are *both stress or both displacement components*. No analytical procedure has yet been devised that can set down the general formal solutions [like (7.162), or in another form] for both types of problems in this class. Recently, however, Miklowitz [7.28] presented a method that handles such problems for nonmixed edge displacements for long-time. It was extended to the case of nonmixed edge stresses later by Sinclair and Miklowitz [7.29]. The following sections present the technique, application to some of the problems and results found.

7.3.4.1. Inversion integral forms. We return to the displacement equations of motion (7.151) and again apply a Laplace transform on t to them using the initial conditions (7.153). Now, critical in the method, to these a Laplace transform on x, parameter s, is applied. The result is

$$\tilde{\bar{u}}_{yy}(s,y,p) + (k^2-1)s\tilde{\bar{v}}_y(s,y,p) + (k^2s^2 - k_s^2)\tilde{\bar{u}}(s,y,p)$$
$$= k^2[s\bar{u}(0,y,p) + \bar{u}_x(0,y,p)] + (k^2-1)\bar{v}_y(0,y,p) = f(s,y,p),$$
(7.172)
$$\tilde{\bar{v}}_{yy}(s,y,p) + k^{-2}(k^2-1)s\tilde{\bar{u}}_y(s,y,p) + k^{-2}(s^2 - k_s^2)\tilde{\bar{v}}(s,y,p)$$
$$= k^{-2}[s\bar{v}(0,y,p) + \bar{v}_x(0,y,p)] + k^{-2}(k^2-1)\bar{u}_y(0,y,p) = g(s,y,p),$$

where now a tilda over quantities indicates the Laplace transform on x. Note that $f(s,y,p)$ and $g(s,y,p)$ are composed of the *time transformed edge unknowns* $\bar{u}, \bar{u}_x, \bar{v}_y$, and $\bar{v}, \bar{v}_x$, and $\bar{u}_y$, respectively, in a manner similar to the right-hand sides of (7.156) and (7.164). However, (7.172) alone asks for all these edge quantities, a basic factor in why the present technique can handle nonmixed edge problems. Use of terms like the first in (7.160) with $\varkappa$ replaced by s, determine the characteristic roots, and the complimentary functions, $\tilde{\bar{u}}_c$ and $\tilde{\bar{v}}_c$. A further Laplace transform on y, along with a convolution, gives the particular integrals, $\tilde{\bar{u}}_p$ and $\tilde{\bar{v}}_p$. Although certainly not restricted to, particular interest will be symmetric loadings. In addition to edge loadinge-we will also have interest in face loadings in these problems. The face conditions, when the loading there is normal, are

$$\sigma_y(x, \pm h, t) = \sigma_0 F(x) G(t), \qquad \sigma_{xy}(x, \pm h, t) = 0, \qquad (7.173)$$

where $F(x)$, $G(t)$ are Laplace transformable functions, and where σ_0 is an inherently negative magnitude constant of dimensions force/unit length. We of course could have symmetric shear type loadings on the faces as well by interchanging the right-hand sides of (7.173). Now with symmetric loadings,

$$v(x, 0, t) = 0, \qquad \sigma_{xy}(x, 0, t) = 0, \qquad (7.174)$$

for $x > 0$, $t > 0$, and hence from here on we need only consider the upper half of the plate, $x \geq 0$, $0 \leq y \leq h$. Making use of (7.173) then, leads to the inversion integral statements for the displacements

$$\begin{Bmatrix} u(x,y,t) \\ v(x,y,t) \end{Bmatrix} = \frac{1}{2\pi i} \int_{Br_p} e^{pt} \left[\frac{1}{2\pi i} \int_{Br_s} \begin{Bmatrix} \tilde{u}(s,y,p) \\ \tilde{v}(s,y,p) \end{Bmatrix} e^{sx} ds \right] dp , \qquad (7.175)$$

where

$$\tilde{u}(s, y, p) = \tilde{u}_c + \tilde{u}_p ,$$

$$\tilde{v}(s, y, p) = \tilde{v}_c + \tilde{v}_p ,$$

where

$$\tilde{u}_c = 2[A(s,p) \cosh \alpha y + B(s,p) \cosh \beta y],$$

$$\tilde{v}_c = 2[\alpha s^{-1} A(s,p) \sinh \alpha y - s\beta^{-1} B(s,p) \sinh \beta y],$$

$$\tilde{u}_p(s, y, p) = k_s^{-2} \int_0^y \{[s^2\alpha^{-1} \sinh \alpha(y-y') + \beta \sinh \beta(y-y')] f(s, y', p)$$

$$+ k^2 s [\cosh \alpha(y-y') - \cosh \beta(y-y')] g(s, y', p)\} dy' ,$$

$$\tilde{v}_p(s, y, p) = k_d^{-2} \int_0^y \{[\alpha \sinh \alpha(y-y') + s^2\beta^{-1} \sinh \beta(y-y')] g(s, y', p)$$

$$+ k^{-2} s [\cosh \alpha(y-y') - \cosh \beta(y-y')] f(s, y', p)\} dy' ,$$

and

$$A(s,p) = A_N(s,p)/R(s,p), \qquad B(s,p) = B_N(s,p)/R(s,p)$$

$$2A_N(s,p) = -s[k^2(2s^2 - k_s^2)\sinh\beta h \cdot I + 2s\beta\cosh\beta h \cdot J],$$

$$2B_N(s,p) = -\beta[2k^2 s\alpha \sinh\alpha h \cdot I - (2s^2 - k_s^2)\cosh\alpha h \cdot J],$$

$$R(s,p) = (2s^2 - k_s^2)^2 \cosh\alpha h \sinh\beta h + 4s^2\alpha\beta \sinh\alpha h \cosh\beta h,$$

$$I = T(s,p) + k_s^{-2}\int_0^h \{k^{-2}s[\alpha^{-1}(2s^2 - k_s^2)\sinh\alpha(h - y')$$

$$+ 2\beta\sinh\beta(h - y')]f(s, y', p) + [(2s^2 - k_s^2)\cosh\alpha(h - y')$$

$$- 2s^2\cosh\beta(h - y')]g(s, y', p)\}dy' + k^{-2}(k^2 - 2)\bar{u}(0, h, p),$$

$$J = k_s^{-2}\int_0^h \{[2s^2\cosh\alpha(h - y') - (2s^2 - k_s^2)\cosh\beta(h - y')]f(s, y', p)$$

$$+ k^2 s\beta^{-1}[2\alpha\beta\sinh\alpha(h - y') + (2s^2 - k_s^2)\sinh\beta(h - y')]g(s, y', p)\}py' - \bar{v}(0, h, p),$$

$$\alpha = (k_d^2 - s^2)^{\frac{1}{2}}, \qquad \beta = (k_s^2 - s^2)^{\frac{1}{2}}, \qquad T(s,p) = \sigma_0 \bar{F}(s)\bar{G}(p)/(\lambda + 2\mu),$$

and where Br_p and Br_s are, respectively, the Bromwich contours in the p- and s-planes. It should be noted at this point, however, (7.175) is not yet the formal solution to the problem, since it contains the edge unknowns, through $f(s, y, p)$, $g(s, y, p)$, $\bar{u}(0, h, p)$ and $\bar{v}(0, h, p)$. Once these are determined for a specific problem, (7.175) becomes the formal solution. Then it can be inverted by the techniques discussed in § 7.2.1, assuming also the inverses of $\tilde{u}_p$ and $\tilde{v}_p$ can be found.

7.3.4.2. Boundedness condition; integral equations for the edge unknowns. Recall from ch. 4 that the complex segments of branches of the Rayleigh–Lamb frequency equation have images in all the coordinate planes of the $\omega - \varkappa$ space (cf. § 4.1.3.1 and fig. 4.12). This makes a total of four such segments for a branch $\varkappa_j(-i\omega)$ for $\omega \geq 0$, which are also known to pierce the $\omega = 0$ plane, with vertical tangents. There are an infinite set of such segments and their images. Assuming a harmonic wave form of $Ae^{i(\varkappa x - \omega t)}$, those segments satisfying Im $\varkappa_j(-i\omega) > 0$, which are $\varkappa_j^c(-i\omega)$ (super c for complex) and $-\bar{\varkappa}_j^c(-i\omega)$, we have already shown lead to *exponentially decaying (in x) edge waves*. A few of these segments, in the low-frequency domain, are shown as dashed lines in the sketch to the left in fig. 7.32. The two other segments satisfy Im $\varkappa_j(-i\omega)<0$. They are $-\varkappa_j^c(-i\omega)$ and $\bar{\varkappa}_j^c(-i\omega)$, which are the solid lines in the figure. These lead to *exponentially unbounded waves as x grows*. Since $R(s,p)$ in (7.174), set$=0$, is another generalized form of

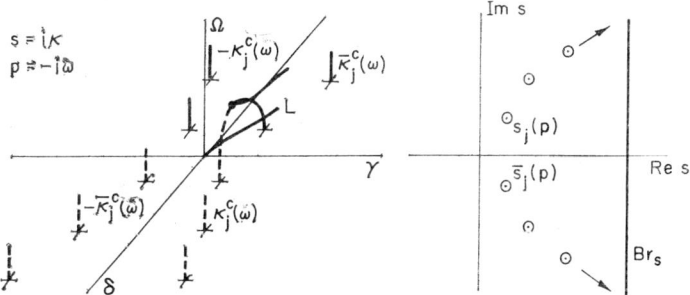

Fig. 7.32. Frequency spectra and related $s_j(p)$ for unbounded waves.

the Rayleigh–Lamb frequency equation, which if $s=i\varkappa$ and $p=-i\omega$, becomes the latter, $R(s,p)=0$ must have the roots $s_j(p)=i\varkappa_j(p)$ on analytical continuation arguments. It follows there are four segments, corresponding to those just discussed, two of which would give, through residue evaluation of the inner integral in (7.175), exponentially unbounded waves at $x\to\infty$, hence violating condition (7.154). These segments satisfy

$$\text{Re } s_j(p)>0, \tag{7.176}$$

i.e., complex $s_j(p)$ and $\bar{s}_j(p)$ (where we have now dropped the super c), which are shown in the sketch to the right in fig. 7.32 (the edge waves $-s_j(p)$, and $-\bar{s}_j(p)$, lying in the left-half plane, are not shown there). It follows that these unbounded waves can be eliminated by requiring the residues in the Br_s integral in (7.175), associated with them, to vanish i.e.,

$$A_N(s,p)|_{s=s_j(p)} = B_N(s,p)|_{s=s_j(p)} = 0, \tag{7.177}$$

where $s_j(p)$ are the two sets of roots of $R(s,p)$ in (7.175) satisfying (7.176). Substitution of A_N and B_N from (7.175) into (7.177) gives two equations for I and J, for each $s_j(p)$. Compatibility of these equations requires that

$$J_j - P_j I_j = 0, \tag{7.178}$$

where J_j and I_j are J and I in (7.175) for $s = s_j(p)$, and

$$P_j = \frac{2k^2 s_j \alpha_j A_j}{2s_j^2 - k_s^2} = -\frac{k^2(2s_j^2 - k_s^2)B_j}{2s_j \beta_j}, \qquad \begin{Bmatrix} A_j \\ B_j \end{Bmatrix} = \frac{\sinh\begin{Bmatrix}\alpha_j \\ \beta_j\end{Bmatrix} h}{\cosh\begin{Bmatrix}\alpha_j \\ \beta_j\end{Bmatrix} h},$$

$$\alpha_j = (k_d^2 - s_j^2)^{\frac{1}{2}}, \qquad \beta_j = (k_s^2 - s_j^2)^{\frac{1}{2}},$$

where we have denoted $s_j(p)$ by s_j. Now arguing on the basis of Lerch's theorem (cf. § 6.2.1), that for the present p can be assumed to be real, and noting the conjugate nature of the roots $s_j(p)$ satisfying (7.176), the boundedness condition (7.178) can be expanded into

$$\begin{Bmatrix}\text{Re} \\ \text{Im}\end{Bmatrix} \Bigg\{ k_s^{-2} \int_0^h \Bigg[\Bigg\{ 2s_j^2 \frac{\cosh \alpha_j y}{\cosh \alpha_j h} - (2s_j^2 - k_s^2) \frac{\cosh \beta_j y}{\cosh \beta_j h} \Bigg\} f(s_j, y, p)$$

$$- \frac{k^2 s_j}{\beta_j} \Bigg\{ 2\alpha_j \beta_j \frac{\sinh \alpha_j y}{\cosh \alpha_j h} + (2s_j^2 - k_s^2) \frac{\sinh \beta_j y}{\cosh \beta_j h} \Bigg\} g(s_j, y, p) \Bigg] dy \quad (7.179)$$

$$- \bar{v}(0, h, p) - P_j \{ T(s_j, p) + k^{-2}(k^2 - 2)\bar{u}(0, h, p) \} \Bigg\} = 0.$$

Equations (7.179) are *two coupled integral equations for the edge unknowns* for each set of parameters p and $s_j(p), j = 2, 4, 6 \ldots$[13] For a particular problem the edge conditions reduce the number of edge unknowns in (7.179). A solution of these equations for the remaining edge unknowns complete the formal solutions given by (7.175). Evidently a boundedness condition like (7.177) was first used, in a simple strip problem by Doetsch [7.30]. (It was learned that M. Picone also used one quite early). More recently, Benthem [7.31] used it effectively in the solution of static problems in elastic strips governed by the biharmonic equation for the Airy stress function for plane stress. The work [7.28, 7.29] extended this thinking to elastodynamics. The applications in the following sections demonstrate that (7.179) can be solved for the edge unknowns in problems involving nonmixed edge conditions.

---

[13] These values of j are consistent with the branch numbering we have used in ch. 4 and this chapter, § 7.2.1.1. Note $j = 3, 5, 7 \ldots$ for the present complex segments, enter only before the conjugation that has now been completed (cf. figs. 4.12, 4.15 and 7.17b).

7.3.4.3. Problem A: Nonmixed pressure shock; formal long-time solution. The problem is illustrated in fig. 7.33. The plate edge is subjected to a step

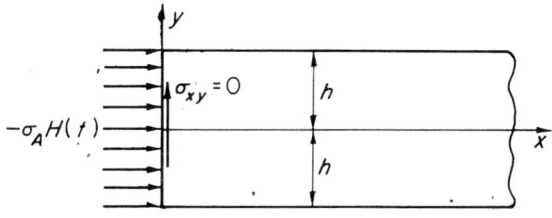

Fig. 7.33. Nonmixed pressure shock problem.

uniform normal stress σ_x, under zero shear stress σ_{xy}. These conditions are given by

$$\left.\begin{array}{l}\sigma_x(0, y, t) = -\sigma_A H(t), \\ \\ \sigma_{xy}(0, y, t) = 0\end{array}\right\}, \quad -h \leq y \leq h, \qquad t > 0, \qquad (7.180)$$

where σ_A is the magnitude of the stress input. From (7.152) the corresponding Laplace transforms on time t for these edge conditions are

$$\begin{array}{l}\bar{\sigma}_x(0, y, p) = \mu[k^2\bar{u}_x(0, y, p) + (k^2-2)\bar{v}_y(0, y, p)] = -\sigma_A/p, \\ \\ \bar{\sigma}_{xy}(0, y, p) = \mu[\bar{u}_y(0, y, p) + \bar{v}_x(0, y, p)] = 0.\end{array} \qquad (7.181)$$

Note that the edge unknowns $\bar{u}_x$, $\bar{v}_y$, $\bar{u}_y$ and $\bar{v}_x$ occur explicitly in (7.181). The long-time formal solution for the present problem can be derived from (7.175) by approximating the double transforms $\tilde{u}(s, y, p)$ and $\tilde{v}(s, y, p)$ appearing there for small p. It follows that a *small p-determination of the edge unknowns* in (7.181), *through the boundedness condition* (7.179), will consistently *define $\tilde{u}$ and $\tilde{v}$ for the analogous approximation of the latter.* To accomplish this we must find *suitable forms for the edge unknowns, for small p,* that we can substitute into (7.179) to "*open up*" this condition, i.e., reduce these integral equations to algebraic ones. As a guide in constructing these forms we *first make use of the elementary theory for compressional waves in a plate,* to enable us to estimate those components of the edge unknowns corresponding to the long-time dependence of the displacements and displacement gradients at the edge. The governing equations for the elementary theory may be written as

$$u_{xx}^e(x,t) = c_p^{-2} \ddot{u}^e(x,t),$$

$$v_y^e(x,t) = -k^{-2}(k^2-2)u_x^e(x,t), \qquad (7.182)$$

$$\sigma_x^e(x,t) = 4\mu k^{-2}(k^2-1)u_x^e(x,t).$$

where superscript e indicates elementary theory, and $c_p = c_s[4(k^2-1)/k^2]^{\frac{1}{2}}$. For the analogous problem to that we are treating here (cf. fig. 7.33), based on the elementary theory (7.182), we take the semi-infinite plate (of unit width) subjected to a step edge force of magnitude $2\sigma_A h$. The Laplace transform easily produces the solution for this problem giving at the edge $x=0$,

$$\bar{u}^e(0,p) = \hat{\sigma}_A/pk_p, \quad \bar{u}_x^e(0,p) = -k^2(k^2-2)^{-1}\bar{v}_y^e(0,p) = -\hat{\sigma}_A p^{-1}, \qquad (7.183)$$

for $-h \leq y \leq h$, $\operatorname{Re} p > 0$, where $k_p = p/c_p$, and $\hat{\sigma}_A$ is the dimensionless stress defined by $\hat{\sigma}_A = \sigma_A k^2/4\mu(k^2-1)$. To the estimates in (7.183) we add a supplementary set of edge unknowns. We represent these *additional contributions by Fourier series in y*, with the p-dependence incorporated in the coefficients of the series. Guided by the symmetric nature of our problem we choose for the displacement gradients

$$\bar{u}_y^a(0,y,p) = \sum_{n=1,3,5,\ldots}^{\infty} A_n(p) \sin(n\pi y/2h),$$

$$\bar{v}_y^a(0,y,\mathrm{p}) = \sum_{n=1,3,5,\ldots}^{\infty} B_n(p) \cos(n\pi y/2h), \qquad (7.184)^{14}$$

for $0 \leq y \leq h$, $\operatorname{Re} p > 0$. These *quarter-range Fourier series*[15] are a good choice

---

[14] Benthem first used series like these in his work [7.31] on related static problems. There the coefficients were not p dependent.

[15] It should be noted that these series representations can be obtained from the half-range series on $[0,2h]$ in the same way the half-range series on $[0, h]$ is obtained from the full Fourier series on $[-h, h]$. Therefore a Fourier theorem holds true for the series of (7.184), and quarter range is an appropriate name for them.

since they preclude the possibility of Gibb's phenomena at the end points of the interval of representation $[0, h]$. This in turn ensures the coefficients A_n, B_n must decay faster than n^{-1} as $n \to \infty$. A proof of this is given by Sinclair [7.32]. Such decay will be of value in any subsequent numerical calculations.

We now integrate (7.184) which, invoking the first of the symmetry relations (7.174), gives

$$\bar{u}^a(0, y, p) = \bar{u}^c(p) + \sum_{n=1,3,5,\ldots}^{\infty} a_n(p) \cos(n\pi y/2h),$$

(7.185)

$$\bar{v}^a(0, y, p) = \sum_{n=1,3,5,\ldots}^{\infty} b_n(p) \sin(n\pi y/2h),$$

for $0 \leq y \leq h$, $\operatorname{Re} p > 0$, where we have introduced $\bar{u}^c(p)$ *for the transformed supplemental corner displacement in the x-direction*, $a_n(p) = -(2h/n\pi)A_n(p)$ and $b_n(p) = (2h/n\pi)B_n(p)$. We note that as a consequence of the large n behavior of $A_n(p)$, $B_n(p)$ noted earlier, i.e. they decay faster than n^{-1}, the coefficients in (7.185) must obey the order conditions

$$a_n(p) = o(n^{-2}), \qquad b_n(p) = o(n^{-2}), \qquad \text{as } n \to \infty.$$

(7.186)

Now these additional terms (7.185) are combined with those from the elementary theory. To do this we (1) integrate the last of (7.183), again invoking the first of (7.174), which yields

$$\bar{v}^e(0, y, p) = \hat{\sigma}_A(k^2 - 2)y/k^2 p,$$

(7.187)

and (2) take u_x^a, v_x^a so that when added to the corresponding elementary theory terms, satisfaction of the edge conditions (7.181) is assured. It follows then, from (7.183), (7.185) and (7.187), the edge unknowns for the present problem can be represented by

$$\bar{u}(0, y, p) = \hat{\sigma}_A/pk_p + \bar{u}^c(p) + \sum_{n=1,3,5,\ldots}^{\infty} a_n(p) \cos(n\pi y/2h),$$

$$\bar{v}(0, y, p) = (k^2 - 2)\hat{\sigma}_A y / k^2 p + \sum_{n=1, 3, 5, \ldots}^{\infty} b_n(p) \sin(n\pi y/2h),$$
(7.188)

$$\bar{u}_x(0, y, p) = -\hat{\sigma}_A/p - k^{-2}(k^2 - 2) \sum_{n=1, 3, 5, \ldots}^{\infty} (n\pi/2h) b_n(p) \cos(n\pi y/2h),$$

$$\bar{v}_x(0, y, p) = \sum_{n=1, 3, 5, \ldots}^{\infty} (n\pi/2h) a_n(p) \sin(n\pi y/2h),$$

for $0 \leq y \leq h$, $\operatorname{Re} p > 0$. We note differentiation of the first two equations here gives $\bar{u}_y(0, y, p)$ and $\bar{v}_y(0, y, p)$.

Substituting the right hand sides of (7.172) into the *boundedness condition* (7.179) (with $T=0$), and in turn (7.188), and carrying out the simple integrations, reduces (7.179) to the *infinite set of linear algebraic equations*

$$\begin{Bmatrix} \operatorname{Re} \\ \operatorname{Im} \end{Bmatrix} \left\{ \sum_{n=1, 3, 5, \ldots}^{\infty} [C_a(n; s_j, p) a_n(p) + C_b(n; s_j, p) b_n(p)] \right.$$
(7.189)
$$\left. + C_c(s_j, p) \bar{u}^c(p) - F_A(s_j, p) \right\} = 0,$$

where

$$C_a(n; s_j, p) = (-)^{(n-1)/2} \left(\frac{k^2 - 1}{k^2} \right) \frac{n\pi s_j}{h} \left[\frac{n^2 \pi^2}{2h^2} + k_s^2 \right] / \alpha_n^2 \beta_n^2,$$

$$C_b(n; s_j, p) = (-)^{(n-1)/2} \left[\left(\frac{k^2 - 1}{k^2} \right) \frac{n^2 \pi^2 s_j^2}{h^2} - \alpha_n^2 k_s^2 \right] / \alpha_n^2 \beta_n^2,$$
(7.190)

$$C_c(s_j, p) = (k^2 - 2) k_d^2 P_j / k^2 \alpha_j^2,$$

$$F_A(s_j, p) = \frac{(k^2 - 2)\hat{\sigma}_A k_d^2}{p \alpha_j^2 \beta_j^2} \left[\left(\frac{\beta_j^2}{k_p} - \frac{s_j k_s^2}{2k_p^2} \right) \frac{P_j}{k^2} + \alpha_j^2 h \right],$$

and $\alpha_n^2 = \alpha_j^2 + (n\pi/2h)^2$, $\beta_n^2 = \beta_j^2 + (n\pi/2h)^2$, for $\bar{u}^c(p)$ and $a_n(p)$, $b_n(p)$ ($n=1, 3, 5, \ldots$). We seek a solution of this set for *small* p, considering first the $s_j(p)$ as $p \to 0$. These limiting roots of $R(s, p)$ in (7.175) are the zeros of

Ch. 7, § 7.3] EDGE LOAD PROBLEMS 453

$$\lim_{p \to 0} \left[\frac{R(s, p)}{k_d^2} \right] = -i(k^2 - 1)s^2 r(s), \qquad (7.191)$$

where $r(s) = \sin 2sh + 2sh$. According to (7.189) we select the one infinite set of these complex zeros of $r(s)$ that correspond to the piercing points of $\bar{\varkappa}_j^c(-i\omega)$ in the plane $\Omega = 0$, and hence the first quadrant $s_j(p)$, shown in fig. 7.32. Robbins and Smith [7.33] list (in order of increasing real part) the first ten of these constant values $s_j(p)$, i.e., $2sh$, satisfying $r(s) = 0$. It can be shown (cf. [7.32], pp. 22–23) that

$$\lim_{p \to 0} \left[\frac{\mathrm{d}s_j(p)}{\mathrm{d}p} \right] = 0, \qquad j = 2, 4, 6, \ldots, \qquad (7.192)$$

corresponding to the fact that the $\varkappa_j^c(\pm i\omega)$ are normal to the $\Omega = 0$ plane in fig. 7.32. It follows that the *zeros of $r(s)$* in (7.191) *are a good approximation to the $s_j(p)$ for a range of p, small, but > 0*. This is obviously *important to the validity of the long-time solution* we are attempting to derive with the present technique.

We now must determine the behavior of the unknowns $\bar{u}^c$, $a_n(p)$ and $b_n(p)$ for $p \to 0$. On the basis of the premise that the elementary theory will describe the dominant time variation for very long time we require $\mathrm{ord}[\bar{u}^c(p)] \leq \mathrm{ord}[p^{-2}]$, $\mathrm{ord}[a_n(p)] \leq [p^{-1}]$ and $\mathrm{ord}[b_n(p)] \leq \mathrm{ord}[p^{-1}]$ for $p \to 0$.[16] Moreover, for these terms to have significant contributions to the long-time solution the order of all three quantities have to be greater than 1. We therefore seek $\bar{u}^c(p)$, $a_n(p)$ and $b_n(p)$ such that

$$\mathrm{ord}[1] < \mathrm{ord}[\bar{u}^c(p)] \leq \mathrm{ord}[p^{-2}], \qquad \text{as } p \to 0^{16},$$

(7.193)

$$\mathrm{ord}[1] < \begin{Bmatrix} \mathrm{ord}[a_n(p)] \\ \mathrm{ord}[b_n(p)] \end{Bmatrix} \leq \mathrm{ord}[p^{-1}], \qquad \text{as } p \to 0.$$

Expanding the terms in (7.190) for $p \to 0$, and substituting the results (given in exercise 7.27d) into the equations (7.189) for $\bar{u}^c(p)$, $a_n(p)$ and $b_n(p)$, yields

---

[16] ord [ ] for $p \to 0$, is being used here in the sense of large order O. For example, the first of (7.193) means $\bar{u}^c(p) = \mathrm{O}(p^{-k})$, $0 < k \leq 2$.

in view of the order requirements (7.193) and the retaining of only the largest compatible terms,

$$\sum_{n=1,3,5,\ldots}^{\infty} (-)^{(n+1)/2} \frac{n^2 s_j}{s_n^4} \left[\frac{n\pi}{2h} a_n(p) + s_j b_n(p) \right] = 0$$

as $p \to 0$, for $j = 1, 2, 3, \ldots$ (7.194)[17]

where $s_n^2 = s_j^2 - (n\pi/2h)^2$. Clearly (7.194) admits the solution $a_n(p) = b_n(p) = 0$ for $p \to 0$. This solution insists that for the present small-p approximation any contributions to the edge unknowns, other than those derived from the elementary theory, must be confined to $\bar{u}^c(p)$. Further since the boundedness condition (7.194) is free of $\bar{u}^c(p)$, this remains an unknown at this point. This indeterminancy can be attributed to the problem in the near field-long time domain asymptotically approaching a second boundary value problem in elastostatic (stresses prescribed), since this type of static problem admits an arbitrary rigid displacement field. Since, however, the elastodynamic problem of the second type has no such arbitrary displacement field we turn to the far-field response and consider the boundedness condition as s and $p \to 0$ together. For this limit, $R(s, p) = 0$ gives

$$s_1(p) = \pm k_p + O(p^3) \text{ as } p \to 0., \qquad (7.195)$$

the lowest branch of R, corresponding to $\varkappa_1(p)$ in (7.167) (note the agreement). The positive sign defines an $s_1(p)$ which for $\operatorname{Re} p > 0$, and a range of p small but not equal to zero, satisfies (7.176). Hence the boundedness condition applies for this $s_1(p) = k_p + O(p^3)$, and we substitute it into (7.189), (7.190) and then let $p \to 0$, and find, in view of the order conditions (7.193) and the solution of (7.194),

---

[17] This equation results by noting that $C_a = C_b = O(1)$, $C_c = O(p^2)$ and $F_A = O(1)$, cf. Exercise 7.27 d. It follows the bracketed expression in (7.189) gives $O(1) \cdot a_n(p) + O(1) \cdot b_n(p) + O(p^2) \cdot \bar{u}^c(p) - O(1) = O$. From the order conditions (7.193) we see that the complete term $O(p^2) \cdot \bar{u}^c(p)$ is at most $O(1)$. Hence, since a_n, b_n are greater than $O(1)$, according to (7.193), it follows only these terms survive.

Ch. 7, § 7.3] EDGE LOAD PROBLEMS 455

$$\bar{u}^c(p) = \sum_{n=1,3,5,\ldots}^{\infty} (-)^{(n-1)/2}(2/n\pi)a_n(p) = 0, \quad \text{as } p \to 0. \quad (7.196)$$

Equation (7.196) completes the determination of the small-p edge unknowns for the present problem. With the unknowns determined it is an easy step to the formal solution. Substituting the reduced version of (7.188) ($\bar{u}^c(p) = 0$, $a_n(p) = b_n(p) = 0$) into the inversion integral statements (7.175) for u and v, through $f(s, y, p)$, $g(s, y, p)$ given in (7.172), and performing the simple integrations indicated there, yields the double transforms

$$\tilde{u}(s, y, p) = \frac{-\hat{\sigma}_A}{p\alpha^2\beta^2}\left[\frac{s}{k_p}(\beta^2 + k_p s) - \frac{k_d^2}{k^2}(3k^2 - 2)\right.$$

$$\left. + \frac{1}{R(s,p)}\{\Theta^u(s, y, p) + \Phi^u(s, y, p)\}\right] \quad (7.197)$$

$$\tilde{v}(s, y, p) = \frac{-\hat{\sigma}_A}{p\alpha^2\beta^2}\left[\left(\frac{k^2-2}{k^2}\right)s\alpha^2 y + \frac{1}{R(s,p)}\left\{\frac{\alpha}{s}\Theta^v(s, y, p) + \frac{s}{\beta}\Phi^v(s, y, p)\right\}\right],$$

with $\Theta^u = (k^2-2)k_d^2 s(\beta^2/k_p)[\gamma \sinh \beta h \cosh \alpha y + 2\alpha\beta \sinh \alpha h \cosh \beta y]$ and similar expressions for Φ^u, Θ^v and Φ^v which are not given here in the interests of brevity.[18] *Equation (7.197) completes the formal long-time solution (7.175) for the present problem. It holds for small p.*

7.3.4.4. Problem A: Nonmixed pressure shock; long-time solution. In tackling the inversion of (7.197) we treat three ranges of s separately: $s/p \to \infty$ as $p \to 0$; $s/p \to c$ as $p \to 0$, c a nonzero constant; and $s/p \to 0$ as $p \to 0$.

For the *first of these ranges*, which corresponds to the *near-field*, (7.197) becomes

$$\tilde{u}(s, y, p) = \frac{\hat{\sigma}_A}{ps}\left[\frac{c_p}{p} - \frac{1}{s}\right] + O(1) \quad \text{as } p \to 0,$$

$$\tilde{v}(s, y, p) = \frac{k^2-2}{k^2}\frac{\hat{\sigma}_A}{ps} + O(1) \quad \text{as } p \to 0, \quad (7.198)$$

---

[18] These quantities are given, respectively, by the last three of equations (1.20) in [7.32], except for multipliers similar to $(k^2-2)k_d^2 s\beta^2/k_p$ in Θ^u here.

for $0 \leq y \leq h$, $\operatorname{Re} p > 0$. Inversion of (7.198) then gives

$$u(x, y, t) \sim \hat{\sigma}_A[c_p t - x] \qquad \text{as } t \to \infty,$$

$$v(s, y, t) \sim k^{-2}(k^2 - 2)\hat{\sigma}_A y \qquad \text{as } t \to \infty, \qquad (7.199)$$

for $0 < x < X$, $0 \leq y \leq h$, where X demarks the extent of the near field. In view of the forms found for the edge unknowns [cf. (7.188) with $\bar{u}^c(p)$, $a_n(p)$, $b_n(p)$ set equal to zero therein] the region of validity for (7.199) may be extended to include $x = 0$. Further, due to the uniformity in s and y of the asymptotics on p for this case, (7.199) may be differentiated with respect to x and y to produce the near-field long-time strains.

Turning to the second case where s and p tend to zero concurrently, which corresponds to being in the vicinity of the *wavefront*, (7.197) subject to this limiting procedure gives

$$\bar{\tilde{u}}(s, y, p) = \frac{\hat{\sigma}_A}{p k_p (s + k_p)} [1 + \mathrm{O}(p^2)] \qquad \text{as } s, p \to 0,$$

$$(7.200)$$

$$\bar{\tilde{v}}(s, y, p) = \left(\frac{k^2 - 2}{k^2}\right) \frac{\hat{\sigma}_A y}{p(s + k_p)} [1 + \mathrm{O}(p^2)] \qquad \text{as } s, p \to 0,$$

for $0 \leq y \leq h$, $\operatorname{Re} p > 0$. Inversion of the terms in (7.200) then produces

$$u(x, y, t) \sim \hat{\sigma}_A [c_p t - x] H(c_p t - x) \qquad \text{as } t \to \infty,$$

$$(7.201)$$

$$v(x, y, t) \sim k^{-2}(k^2 - 2)\hat{\sigma}_A y H(c_p t - x) \qquad \text{as } t \to \infty.$$

Comparison of (7.201) with (7.199) demonstrates that the former applies everywhere in the plate, $x \geq 0$, $-h \leq y \leq h$. The analogous asymptotics on $s\bar{\tilde{u}}$ and $\partial \bar{\tilde{v}}/\partial y$ furnish

$$u_x(x, y, t) \sim -[k^2(k^2 - 2)^{-1}]v_y(x, y, t) \sim -\hat{\sigma}_A H(c_p t - x) \qquad \text{as } t \to \infty,$$

$$(7.202)$$

again applying everywhere in the plate. Equations (7.201), (7.202) constitute our *first far-field approximation* for Problem A and exhibit waves that are non-decaying in space and time with (7.202) attesting to the long-time *Poisson's ratio coupling* of the *longitudinal* and *thickness strains*. Note that although (7.201), (7.202) could have been obtained from the elementary theory directly, this is *not* what we have done in the present treatment. Rather, we have shown that the elementary theory solution is obtained as the *first long-time approximation of the exact theory for the present problem*.

A higher order approximation than (7.202), based on a second approximation to $R(s, p)$, is available for both s and p tending to zero. It can be derived from the lowest Rayleigh–Lamb branch, as discussed in § 7.3.3.1 for I_6, resulting in (7.166) again based on (7.167), where now $\hat{\sigma}_A$ replaces $\hat{\sigma}_0$ there. This result was anticipated earlier but first proved in [7.29] by the present method. Again, the nature of both approximations is exhibited in fig. 7.7. Computation of these approximations for the plate (mixed and nonmixed edges) appears in fig. 4 of [7.28].

To conclude the inversion of the small-p formal solution for Problem A we consider the third s-range, namely p small with $s \to 0$. Under such a limiting procedure, (7.197) establishs that $\bar{u}^\sim$, $\bar{v}^\sim$ are O(1) in s. Hence we have no contributions to the displacements for t large, $x \to \infty$. Similarly it may be shown that the strains are zero as $x \to \infty$ and thus our solution is in accord with the conditions at infinity (7.154).

7.3.4.5. Problem B: Nonmixed line load; formal long-time solution. Following the procedures adopted in § 7.3.4.3 we now seek long-time information for the line-load impact on the edge of the waveguide depicted in fig. 7.34. The edge conditions for this problem are

$$\sigma_x(0, y, t) = -\sigma_B \delta(y) H(t),$$

$$\sigma_{xy}(0, y, t) = 0,$$

(7.203)

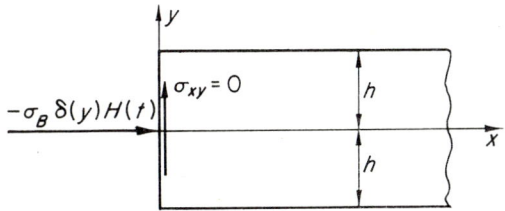

Fig. 7.34. Nonmixed line-load problem.

where σ_B is the magnitude of the stress input (force per unit length). Using (7.152), the Laplace-time transforms of (7.203) give

$$\bar{\sigma}_x(0, y, p) = \mu[k^2 \bar{u}_x(0, y, p) + (k^2-2)\bar{v}_y(0, y, p)] = -\sigma_B p^{-1}\delta(y),$$

$$\bar{\sigma}_{xy}(0, y, p) = \mu[\bar{u}_y(0, y, p) + \bar{v}_x(0, y, p)] = 0,$$

(7.204)

for $0 \leq y \leq h$, Re $p > 0$.

As in § 7.3.4.3 we must now postulate forms for the edge unknowns in order to open up the boundedness condition (7.179). The only differences encountered between the pattern established for Problem A and that required for Problem B will be due to the singular nature of the latter problem. To appraise such singular nature we resolve Problem B, for the long-time, into three problems as depicted in fig. 7.35. The first of these, Problem B1, is the line-load on the elastostatic half-space, sometimes referred to as the *Flamant problem*. Problem B1 will provide the long-time *singular parts of the edge unknowns*. The second problem, Problem B2, is the residual associated elastostatic problem. The stresses applied on the plate faces here (σ'_y and $\pm\sigma'_{xy}$) are of the same magnitude as those acting on the corresponding sections of the half-space in Problem B1 but opposite in sign. Problem B2 is made self-equilibrating by the introduction of a uniform normal stress, $\sigma_B/2h$, acting on the plate edge. The attendant edge values for this problem will contribute to the *regular parts of the edge unknowns*. The third problem, Problem B3, is the uniform normal load applied to our waveguide, recognizable as Problem A with a modified stress input. Problem B3 will furnish the dominant time-dependence of the edge quantities in the long-time, thereby completing the selection of the representations for the regular parts of the edge unknowns.

We now consider each problem in this decomposition individually; first, the singular problem. For the Flamant problem at $x = 0$ we have[19]

$$\bar{u}^s(0, y, p) = -(\hat{\sigma}_B h/p)ln(y/h), \qquad \bar{u}^s_y(0, y, p) = -\bar{v}^s_x(0, y, p) = -\hat{\sigma}_B h/py,$$

(7.205), (cont. next pg.)

---

[19] The appendix at the end of the paper [7.29] gives a sketch proof of (7.205).

Ch. 7, § 7.3] EDGE LOAD PROBLEMS... 459

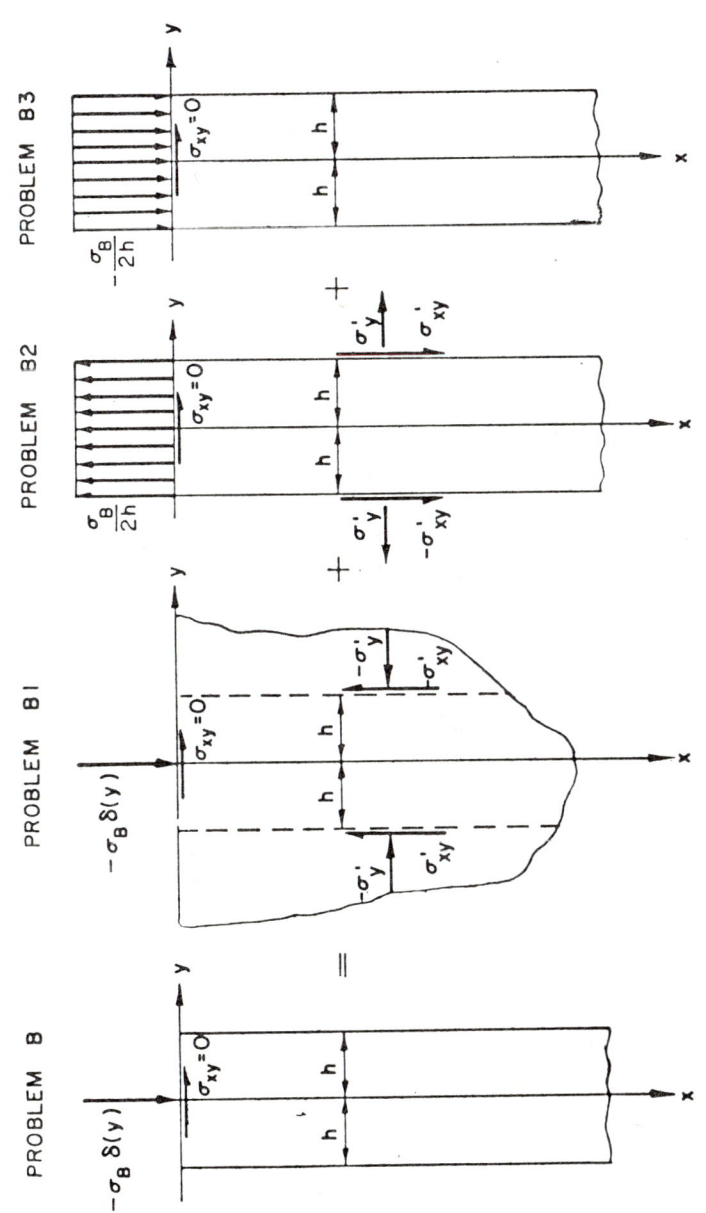

Fig. 7.35. Decomposition of nonmixed line-load problem.

$$\bar{v}^s(0, y, p) = -(\pi\hat{\sigma}_B/2pk^2)\mathrm{sgn}(y),$$

$$\bar{v}^s_y(0, y, p) = \bar{u}^s_x(0, y, p) = -(\pi\hat{\sigma}_B h/pk^2)\delta(y),$$

for $0 < y \leq h$, $\mathrm{Re}\, p > 0$, with extension to include $y = 0$ in an integrable sense whenever this is asked for by the boundedness condition, (7.179). Here $\mathrm{sgn}(y) = 2H(y) = 2H(y) - 1$ is the *signum function* [$\mathrm{sgn}(0) = 0$ by definition] and $\hat{\sigma}_B$ is the dimensionless stress input for Problem B defined by

$$\hat{\sigma}_B = \sigma_B k^2 / 2\pi\mu h(k^2 - 1). \tag{7.206}$$

Note that $\bar{u}^s(0, y, p)$ in (7.205) is determined to within an arbitrary rigid-body displacement; such indefiniteness is consistent with a second boundary-value problem in elastostatics and will be incorporated in a $\bar{u}^c(p)$ term in the same manner as in the previous section.

One should bear in mind that currently we are merely postulating the forms for the singular parts of the edge unknowns. Accordingly (7.205) is only a reasonable guess as to what these terms might be, based on the thesis that, in the near-field, the long-time singular nature of a problem involving, exclusively, outward propagating disturbances is the same as the singular behavior of the corresponding elastostatic problem.

For use in Problem B2 we set $y = h$ in the stress distribution associated with the Flamant problem to obtain

$$\sigma'_{xy} = (x/h)\sigma'_y = \frac{2\sigma_B}{\pi h}\frac{(x/h)^2}{[1+(x/h)^2]^2}, \qquad (x \geq 0). \tag{7.207}$$

In choosing edge forms for Problem B2 we observe the similarity of the prescribed stresses at $x = 0$ for this problem with those in Problem A. This similarity suggests that suitable representations for the edge unknowns in Problem B2 would be the expressions in (7.188) with $\hat{\sigma}_A$ therein replaced by $-\pi\hat{\sigma}_B/4$. However, Problem B2 has in addition to its edge stresses the equilibrating plate-face stresses σ'_y and σ'_{xy}.[20] This fact further suggests that

---

[20] Since the plate-face stresses σ'_y, σ'_{xy} of (7.207) decay as $x \to \infty$, Problem B2 is amenable to a finite-element treatment. Such a treatment is outlined in Appendix 1 of [7.32].

we modify the forms arrived at subsequent to the replacement by dropping the $-\pi\hat{\sigma}_B/4pk_p$ term.

Now turning to Problem $B3$ we note its complete equivalence to Problem A subject to $\sigma_B/2h$ being equal to σ_A. It follows that the appropriate forms for Problem $B3$ are obtained from (7.188) on exchanging $\hat{\sigma}_A$ there for $\pi\hat{\sigma}_B/4$.

Combining the edge unknown representations for these last two problems will cancel the elements in $\bar{v}(0, y, p)$, $\bar{u}_x(0, y, p)$ which give rise to $\pm\pi\sigma_B/2h$ and thus produce the compact forms,

$$\bar{u}^r(0, y, p) = \pi\hat{\sigma}_B/4pk_p + \bar{u}^c(p) + \sum_{n=1,3,5,\ldots}^{\infty} a_n(p) \cos(n\pi y/2h),$$

$$\bar{v}^r(0, y, p) = \sum_{n=1,3,5,\ldots}^{\infty} b_n(p) \sin(n\pi y/2h),$$

(7.208)

$$\bar{u}_x^r(0, y, p) = -k^{-2}(k^2-2) \sum_{n=1,3,5,\ldots}^{\infty} (n\pi/2h)b_n(p) \cos(n\pi y/2h),$$

$$\bar{v}_x^r(0, y, p) = \sum_{n=1,3,5,\ldots}^{\infty} (n\pi/2h)a_n(p) \sin(n\pi y/2h),$$

for $0 \leq y \leq h$, $\text{Re } p > 0$. Here, the superscript r on quantities denotes their *regular nature*; the $\bar{u}^c(p)$ is once again the transformed supplemental corner displacement.

Contingent upon the validity of the thesis that the forms in (7.205) will describe the singular contributions to the edge unknowns in their entirety, the terms in (7.208) will in fact be regular and consequently the $a_n(p)$ and $b_n(p)$ there will comply with the large-n order condition of (7.186). Moreover, such regularity then guarantees the validity of term-by-term differentiation of $\bar{u}^r$, $\bar{v}^r$ in (7.208). Thus, on adjoining (7.205) to (7.208) and the terms $\bar{u}_y^r$, $\bar{v}_y^r$ obtained therefrom, the postulation of the edge unknowns for Problem B is concluded.[21]

---

[21] Details of the forms for the edge unknowns for this problem and of the ensuing opening up of the boundedness condition utilizing the forms are given in [7.32].

We are now in a position to open up the boundedness condition and evaluate the edge unknowns for small p. To achieve this we closely follow the method established for Problem A in § 7.3.4.3. The forms for the edge unknowns for Problem B are substituted into the boundedness condition (7.179) and the expanded (7.179) integrated. This results in an infinite system of linear equations for the unknowns $\bar{u}^c(p)$, $a_n(p)$ and $b_n(p)$ which is similar to (7.189), the only differences encountered being contained in an $F_B(s, p)$ term. The system has one equation for each $s_j(p)$ satisfying (7.176). For p small (as argued in § 7.3.4.3) $s_j(p) \to s_j$ ($j = 1, 2, ...$), the roots of $r(s)$ in (7.191), and $\bar{u}^c(p)$, $a_n(p)$ and $b_n(p)$ obey the order requirements (7.193). In view of these order requirements, the boundedness condition for Problem B reduces to,

$$\sum_{n=1,3,5,...}^{\infty} (-)^{(n+1)/2} \frac{n^2 \hat{s}}{\hat{s}^4} \left[\frac{n\pi}{2} \hat{a}_n + \hat{s} \hat{b}_n \right] = \frac{2}{\pi^2 \cos \hat{s}} [(1 + \hat{s} \tan \hat{s}) si(\hat{s}) + \sin \hat{s}],$$

(7.209)

for $p \to 0$, $\hat{s} = \hat{s}_j$ ($j = 1, 2, ...$); wherein all quantities have been rendered dimensionless by the introduction of $\hat{a}_n = p a_n(p)/\sigma_B h$, $\hat{b}_n = p b_n(p)/\sigma_B h$, $\hat{s} = \hat{s} h$, $\hat{s}_n^2 = \hat{s}_n^2 h^2$, with $\sin 2\hat{s}_j + 2\hat{s}_j = 0$. Here $si(\hat{s})$ is the sine integral (cf. [4.10], p. 232).

As in [7.28], [7.31], we now employ the *method of reduction*[22] to evaluate $\hat{a}_n$, $\hat{b}_n$. Results for the first ten $\hat{a}_n$ and $\hat{b}_n$ found, using twenty-four roots $\hat{s}_j$, are displayed in the following table.

The numerical decay of the $\hat{a}_n$, $\hat{b}_n$ values in the table is faster than $1/n^2$ for $n > 3$, in agreement with our large-n order condition (7.186). Such numerical decay supports our thesis that in the long-time near-field the singular nature of Problem B is the same as for the corresponding static problem.

Using the $\hat{a}_n$, $\hat{b}_n$ values of the table, the edge displacements associated with Problem B2 can be evaluated, enabling comparison with the finite-element treatment given in Appendix 1 of [7.32]. Such a comparison shows agreement to within 1%.

---

[22] That is, solving the finite $2N \times 2N$ system of linear equations associated with the first N roots $\hat{s}_j$, then increasing the size of the finite system solved until stable estimates of the desired number of $\hat{a}_n$ and $\hat{b}_n$ have been found.

Fourier Coefficient Values

n	$\hat{a}_n$	$\hat{b}_n$
1	−0.1129	0.2891
3	−0.0242	0.0036
5	0.0077	−0.0026
7	−0.0030	0.0011
9	0.0014	−0.0005
11	−0.0007	0.0003
13	0.0004	−0.0001
15	−0.0003	0.0001
17	0.0002	−0.0001
19	−0.0001	0.0000

The edge displacements associated with Problem B, however, cannot be evaluated at this juncture since presently $\bar{u}^c(p)$ is an unknown. To ascertain $\bar{u}^c(p)$ we proceed as in Problem A and consider the boundedness condition for the case of s and p tending to zero together. This limiting process defines s_1 as given in (7.195) and the boundedness condition associated with s_1 then furnishes the necessary additional equation for the determination of $\bar{u}^c(p)$, namely

$$\bar{u}^c(p) = \frac{\hat{\sigma}_B h}{p}\left[1 + \sum_{n=1,3,5,\ldots}^{\infty} (-)^{(n-1)/2} \frac{2}{n\pi} \hat{a}_n\right] \quad \text{as } p \to 0. \tag{7.210}$$

Substituting our numerical values for $\hat{a}_n$ into (7.210) then gives $\bar{u}^c(p) = \hat{\sigma}_B \hat{u}^c h/p$ with $\hat{u}^c = 0.935$. We have now completed the determination of the edge unknowns for Problem B.

7.3.4.6. Problem B: Nonmixed line load; long-time solution. For the purposes of exhibiting our results we remove the $\pi\hat{\sigma}_B/4pk_p$ term, carry out the simple inversion of the remaining terms and then define the *static edge displacements* (i.e., time independent displacements) and their *gradients* as follows:

$$\hat{u} = \left[u(0, y, t) - \frac{\pi}{4}\hat{\sigma}_B c_p t\right]/\hat{\sigma}_B h = \hat{u}^c - \ln(y/h) + \sum_{n=1,3,5,\ldots}^{\infty} \hat{a}_n \cos(n\pi y/2h),$$

$$\hat{v}=v(0,y,t)/\hat{\sigma}_B h=-(\pi/2k^2)\,\text{sgn}\,(y)+\sum_{n=1,3,5,\ldots}^{\infty}\hat{b}_n\sin(n\pi y/2h)$$
(7.211)

$$\hat{u}_y=-\left[(h/y)+\sum_{n=1,3,5,\ldots}^{\infty}(n\pi/2)\hat{a}_n\sin(n\pi y/2h)\right],$$

$$\hat{v}_y=\sum_{n=1,3,5,\ldots}^{\infty}(n\pi/2)\cos(n\pi y/2h),$$

for $0<y\leq h$, with $\hat{v}_y$ having a symmetric delta function at $y=0$. Using our numerical values of $\hat{a}_n$, $\hat{b}_n$ and setting $k^2=7/2$ affords a means of calculating $\hat{u}$, $\hat{v}$, $\hat{u}_y$ and $\hat{v}_y$ of (7.211). The results of this calculation are plotted in fig. 7.36.

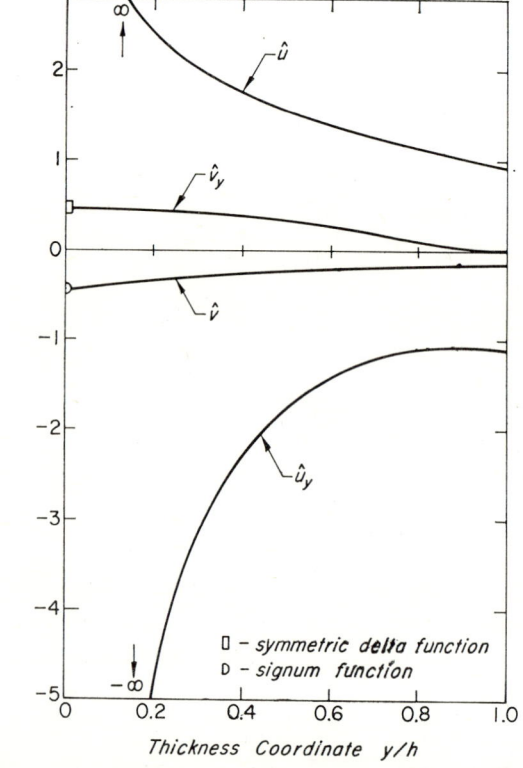

Fig. 7.36. Static edge displacements and their y-derivatives for line load problem.

Ch. 7, § 7.3]   EDGE LOAD PROBLEMS   465

In proceeding to the small-p formal solution for Problem B, and thus to the long-time solution, we substitute the now determined edge unknowns into the formal solution (7.175), with the aid of f and g in (7.172), then the inversion process in undertaken for the pertinent three ranges of s in exactly the same manner as in § 7.3.4.4. For the *near-field* this process produces the edge displacements and their gradients. In the *far-field*, the process gives

$$u(x, y, t) \sim (\pi \hat{\sigma}_B/4)[c_p t - x] H(c_p t - x),$$

$$v(x, y, t) \sim [(k^2 - 2)\pi \hat{\sigma}_B y / 4k^2] H(c_p t - x), \qquad (7.212)$$

$$u_x(x, y, t) = -k^2(k^2 - 2)^{-1} v_y(x, y, t) \sim -\pi (\hat{\sigma}_B/4) H(c_p t - x),$$

for $x, t \to \infty$ as the *first approximation*. The *second approximation* (for $x, t \to \infty$) is again (7.166), with K there now $\pi \hat{\sigma}_B/4$. Comparison of (7.201) (7.202) with (7.212), and the analogous comparison for the second approximations, demonstrates that the long-time far-field approximations for Problems A and B are the same if equal normal forces act on the edge, $x=0$, in both problems, i.e., if $\sigma_A = \sigma_B/2h$ or equivalently $\hat{\sigma}_A = \pi \hat{\sigma}_B/4$.

7.3.4.7. Problems involving nonmixed edge displacements; comments. In [7.28] the problem of the plate with a built-in edge, subjected to two symmetric suddenly applied normal loads on its faces, a short distance from the edge, was treated by the foregoing methods. Interestingly the head of the pulse solution (7.166) was found, but here it represented the long time-far field response of the reflection of the incident pulse from the edge. The magnitude of this response (instead of $\hat{\sigma}_0$ in (7.166)) involved the Fourier coefficients ($\hat{a}_n$ in (7.209)) and another coefficient $\hat{a}_0$ for a term in the edge unknowns representing the singularities at the corners $(0, \pm h)$ of the built-in edge.

Cooper and Craggs [7.34] treated a related problem with a finite difference-numerical method. The edge conditions were

$$\dot{u}(0, y, t) = V_0 H(t), \qquad \dot{v}(0, y, t) = 0,$$

a velocity shock problem. The results for u_x on the plate faces were exhibited in two figures. One shows the time response at a station in the near field

($x \simeq h$), which has some resemblance to the grosser features of a corresponding response record in [7.9] (see axial strain record for $x=1$ in fig. 7.8 here). The second is a spatial response record which exhibits features like those of (7.166).

Mention should also be made of Bertholf's finite difference-numerical integration work [7.35] on the nonmixed pressure shock problem for the rod. He shows some interesting *results for near field*, radial and axial strain, response, which compare favorably with corresponding records from the work [7.9] (again see fig. 7.8 here).

7.4. Axially symmetric problems for the infinite plate

7.4.1. Excitation by two symmetric normal point loads

The problem was treated by Miklowitz [7.22]. It and its equivalent half space problem are illustrated in figs. 7.37 a, b, respectively. Mathematically

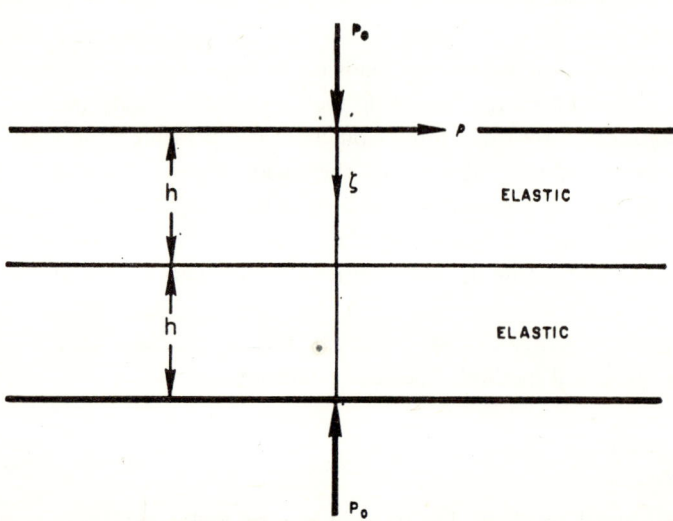

Fig. 7.37a. Infinite plate subjected to sudden symmetric point-loads on faces.

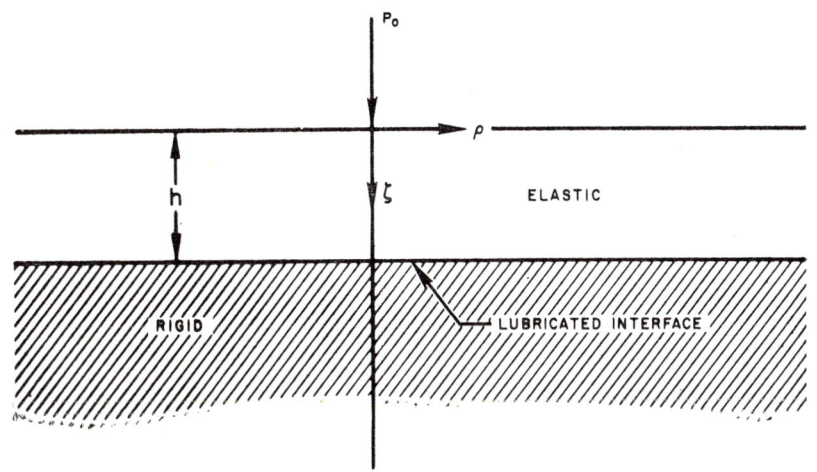

Fig. 7.37b. Equivalent layer-half space problem.

the problem can be stated as follows:

$$\left.\begin{array}{l}\nabla^2\varphi(\varrho,\zeta,\tau)=a^{-2}\ddot{\varphi}\\ \nabla^2\psi(\varrho,\zeta,\tau)=\ddot{\psi}\end{array}\right\} \text{in } \varrho>0, \quad 0<\zeta<1, \quad \text{for } \tau>0,$$

$$\varphi(\varrho,\zeta,0)=\dot{\varphi}=\psi=\dot{\psi}=0, \quad \text{in } \varrho\geqq 0, \quad 0\leqq\zeta\leqq 1,$$

$$\left.\begin{array}{l}\sigma_\zeta(\varrho,0,\tau)=P_0\delta_s(\varrho)H(\tau)/\pi\varrho\\ \sigma_{\zeta\varrho}(\varrho,0,\tau)=0\\ u_\zeta(\varrho,1,\tau)=\sigma_{\zeta\varrho}(\varrho,1,\tau)=0\end{array}\right\}, \quad \text{for } \varrho\geqq 0, \quad \tau>0, \quad (7.213)$$

$$\lim_{\varrho\to\infty}\left[\varphi(\varrho,\zeta,\tau),\psi,u_\varrho, \text{etc.}\right]=0, \quad 0\leqq\zeta\leqq 1, \quad \tau>0,$$

where ϱ, ζ and τ are the dimensionless radial, axial and time coordinates, respectively, defined by $\varrho=r/h$, $\zeta=z/h$ and $\tau=\omega_s t (\omega_s=c_s/h)$, and $a=c_d/c_s$. The displacements-potential relations are

$$u_\varrho=h\,u_r=\frac{\partial\varphi}{\partial\varrho}-\frac{\partial\psi}{\partial\zeta},$$

$$u_\zeta = h\, u_z = \frac{\partial \varphi}{\partial \zeta} + \frac{\partial(\varrho\psi)}{\varrho\partial\varrho}, \tag{7.214}$$

and the stress-potential relations

$$\sigma_\zeta = h^2 \sigma_z = \lambda \nabla^2 \varphi + 2\mu \frac{\partial}{\partial \zeta}\left[\frac{\partial \varphi}{\partial \zeta} + \frac{\partial(\varrho\psi)}{\varrho\partial\varrho}\right], \tag{7.215}$$

$$\sigma_{\zeta\varrho} = h^2 \sigma_{zr} = \mu\left\{\frac{\partial}{\partial\zeta}\left[\frac{\partial\varphi}{\partial\varrho} - \frac{\partial\psi}{\partial\zeta}\right] + \frac{\partial}{\partial\varrho}\left[\frac{\partial\varphi}{\partial\zeta} + \frac{\partial(\varrho\psi)}{\varrho\partial\varrho}\right]\right\},$$

where $\nabla^2 = \dfrac{\partial}{\partial\varrho^2} + \dfrac{\partial}{\varrho\partial\varrho} + \dfrac{\partial}{\partial\zeta^2}$.

The problem is related to Lamb's problem for the surface point load source in § 6.3.1. As in that problem we use a Laplace transform on time and Hankel transforms (cf. § 5.8) on the radial coordinate ϱ. We find the formal solutions for the displacements to be

$$u_\varrho(\varrho, \zeta, \tau) = \int_0^\infty \varkappa J_1(\varkappa\varrho)\left[\frac{1}{2\pi i}\int_{Br} \bar{u}_\varrho^1(\varkappa, \zeta, p)e^{p\tau}dp\right]d\varkappa,$$

$$u_\zeta(\varrho, \zeta, \tau) = \int_0^\infty \varkappa J_0(\varkappa\varrho)\left[\frac{1}{2\pi i}\int_{Br} \bar{u}_\zeta^0(\varkappa, \zeta, p)e^{p\tau}dp\right]d\varkappa, \tag{7.216}$$

where $\varkappa$ is the dimensionless wave number in the ϱ-direction. The double transforms $\bar{u}_\varrho^1$ and $\bar{u}_\zeta^0$ are left to [7.22] to conserve space here. In [7.22] the Laplace transform in (7.216) was inverted first (cf. § 7.2.1.2). Residue theory, exploiting the poles $p_n = \pm \Omega_n(\varkappa)$, where $\Omega_n = \omega_n/\omega_s$, leads to the solutions

$$u_\varrho(\varrho, \zeta, \tau) = \int_0^\infty \varkappa J_1(\varkappa\varrho)\left\{\lim_{|p|\to 0}\left[\frac{N'(\varkappa, \zeta, p^2)}{D'(\varkappa^2, p^2)}\right]\right.$$

$$\left. + \sum_{n=1}^\infty \frac{N(\varkappa, \zeta, -\Omega_n^2)}{\Omega_n^2 D_n(\varkappa^2, -\Omega_n^2)} \cos\Omega_n\tau\right\}d\varkappa, \tag{7.217}$$

Ch. 7, § 7.4]   AXIALLY SYMMETRIC PROBLEMS   469

with a similar expression for u_ζ. The term N in (7.217) is $\tilde{u}_\varrho^{-1}$ except for pD in its denominator, where D is essentially the Rayleigh–Lamb frequency equation F_s in (7.117). The term D_n is the derivative of D with respect to p^2 evaluated for the branches Ω_n. The primes in the first term of (7.217) indicate differentiation with respect to p. Note the similarity in forms of (7.217) and (7.117).

As we have discussed earlier, for the long time-far field solution we need only the limit term, corresponding to the static solution, and the lowest branch $n=1$ term in (7.217), with corresponding terms for u_ζ. The static solution is given by

$$\frac{\pi\mu}{2P_0} u_{\varrho 0}(\varrho, \zeta) = \int_0^\infty \varkappa J_1(\varkappa\varrho)\tilde{u}_{\varrho 0}^1(\varkappa, \zeta)d\varkappa,$$

$$\frac{\pi\mu}{2P_0} u_{\zeta 0}(\varrho, \zeta) = \int_0^\infty \varkappa J_0(\varkappa\varrho)\tilde{u}_{\zeta 0}^0(\varkappa, \zeta)d\varkappa,$$

(7.218)

and the $n=1$ term inverse by

$$\frac{\pi\mu}{2P_0} u_{\varrho 1}(\varrho, \zeta, \tau) = \int_0^\infty \varkappa J_1(\varkappa\varrho)\tilde{u}_{\varrho 1}^1(\varkappa, \zeta) \cos \Omega_1\tau d\varkappa,$$

$$\frac{\pi\mu}{2P_0} u_{\zeta 1}(\varrho, \zeta, \tau) = \int_0^\infty \varkappa J_0(\varkappa\varrho)\tilde{u}_{\zeta 1}^0(\varkappa, \zeta) \cos \Omega_1\tau d\varkappa,$$

(7.219)

where $\tilde{u}_{\varrho 0}^1$, $\tilde{u}_{\zeta 0}^0$, $\tilde{u}_{\varrho 1}^1$ and $\tilde{u}_{\zeta 1}^0$ are given in [7.22]. For large ϱ we can replace the Bessel functions in (7.218), (7.219) by the leading terms of their asymptotic expansions,

$$J_0(\varkappa\varrho) \sim (2/\pi\varkappa\varrho)^{\frac{1}{2}} \cos[\varkappa\varrho - (\pi/4)], \quad J_1(\varkappa\varrho) \sim (2/\pi\varkappa\varrho)^{\frac{1}{2}} \cos[\varkappa\varrho - (3\pi/4)],$$

provided we restrict the integration away from $\varkappa=0$. Equations (7.219) then become

$$-\frac{\pi\mu}{2P_0} u_{\varrho 1}(\varrho, \zeta, \tau) = \frac{1}{2(\pi\varrho)^{\frac{1}{2}}} (\text{Re} - \text{Im}) \int_\varepsilon^\infty \sqrt{\varkappa}\tilde{u}_{\varrho 1}^1(\varkappa, \zeta)e^{i\varrho f(\varkappa)}d\varkappa + O(\varepsilon),$$

(7.220)

$$\frac{\pi\mu}{2P_0} u_{\zeta 1}(\varrho, \zeta, \tau) = \frac{1}{2(\pi\varrho)^{\frac{1}{2}}} (\text{Re} + \text{Im}) \int_{\varepsilon}^{\infty} \sqrt{\varkappa} \tilde{u}_{\zeta 1}^0(\varkappa, \zeta) e^{i\varrho f(\varkappa)} d\varkappa + O(\varepsilon^2) ,$$

where $0 < \varepsilon \ll 1$, and the function

$$f(\varkappa) = \varkappa - \Omega_1(\tau/\varrho) . \tag{7.221}$$

The order terms correspond to the 0 to ε range of integration of (7.219) for large ϱ. The integrals in (7.220) are of the form that can be approximated by the *method of stationary phase* (for large ϱ). In anticipation of this, we have written them for only the outgoing waves in $\varrho > 0$, i.e., their incoming wave-counterparts will have no points of stationary phase, hence are Fourier integrals which are negligible with respect to (7.220), cf. § 5.10.6. The amplitude real functions $\sqrt{\varkappa}\tilde{u}_{\varrho 1}^1$, $\sqrt{\varkappa}\tilde{u}_{\zeta 1}^0$ are continuous functions, and $f(\varkappa)$ is twice continuously differentiable in $0 < \varepsilon \leq \varkappa < \infty$ (Skalak's argument that $D' \to 0$ enters into establishing the continuity of the amplitude functions, cf. § 7.2.1.2). For large ϱ and fixed τ/ϱ the points of stationary phase $\varkappa = s$ for the integrals of (7.220) are those in $\varepsilon \leq \varkappa < \infty$ satisfying

$$[f'(\varkappa)]_{\varkappa = s} = 0 , \tag{7.222}$$

where the prime denotes differentiation with respect to $\varkappa$. From (7.221) and (7.222) we find

$$\Omega_1'(s) = \varrho/\tau , \tag{7.223}$$

the well-known stationary phase condition. Using the stationary phase approximation (5.126) then, we find the far field (and long time) displacements corresponding to the lowest branch $\Omega_1(\varkappa)$ to be

$$-\frac{\pi\mu}{2P_0} u_{\varrho 1}(\varrho, \zeta, \tau) = \frac{1}{\varrho}$$

$$\left[\frac{s\Omega_1'(s)}{|\Omega_1''(s)|}\right]^{\frac{1}{2}} \tilde{u}_{\varrho 1}^1(s, \zeta) \sin\left\{\varrho\Omega_1(s)\left[\frac{1}{\Omega_1'(s)} - \frac{s}{\Omega_1(s)}\right]\right\} + O(\varrho^{-\frac{3}{2}}) ,$$

$$\tag{7.224}$$

$$\frac{\pi\mu}{2P_0} u_{\zeta 1}(\varrho, \zeta, \tau)$$

$$= \frac{1}{\varrho} \left[\frac{s\Omega_1'(s)}{|\Omega_1''(s)|} \right]^{\frac{1}{2}} \tilde{u}_{\zeta 1}^0(s, \zeta) \cos\left\{\varrho\Omega_1(s) \left[\frac{1}{\Omega_1'(s)} - \frac{s}{\Omega_1(s)} \right] \right\} + O(\varrho^{-\frac{3}{2}}),$$

provided that $f''(s) = -(\tau/\varrho)\Omega''(s) \neq 0$. This last condition is satisfied for $0 < \varkappa = s < 1.38$, a domain governing *low frequency-long waves*.

The static solution (7.218) can be similarly approximated. There, however, we are dealing with Fourier integrals in the far field. These lead to terms of $O(\varrho^{-\frac{3}{2}})$, hence, can be lumped in with the order term in (7.224). It follows that (7.224) is our solution. The displacements it represents were evaluated numerically in [7.22]. Fig. 7.38 shows the response of the horizontal displacement $u_{\varrho 1}(\varrho, \zeta, \tau)$ of (7.224) for the three stations $\varrho = 20$, $\zeta = 0$, $1/2$ and 1. The initial time is $\tau = \zeta/b$ where $b = c_p/c_s$, c_p being the "plate velocity". Reference [7.22] shows that the approximation in the region from initial time to A corresponds to the very long waves of very low frequency from the lowest symmetric branch of the Rayleigh–Lamb frequency equation [(4.48) with +sign], i.e., it is free of the high frequencies of this branch, and the complete second and third real branches as shown in a corresponding predominant period-time of occurrence plot. Note that the maximum response occurs in this region. The other regions in fig. 7.38 out to E, C, D and B are discussed further in [7.22]. For further detail on the derivation of (7.224), the reader can consult [7.22].

7.4.2. Mixed edge condition problem; sudden normal displacement on circular cavity wall

Scott and Miklowitz [7.18] extended the work in the previous section to a problem of the infinite plate with a *circular cylindrical hole* or cavity. The plate is excited by a uniform step radial displacement on the cavity wall along with zero shear stress. As such, the problem is one of mixed edge conditions, related to the longitudinal impact problem of § 7.3.2.1. The displacement equations of motion for cylindrical coordinates (r, z) were employed. To write the formal solution *extended Hankel transforms* were developed for the exterior domain $a \leq r < \infty$, a being the radius of the hole.

472 TRANSIENT WAVES IN ELASTIC WAVEGUIDES [Ch. 7, § 7.4

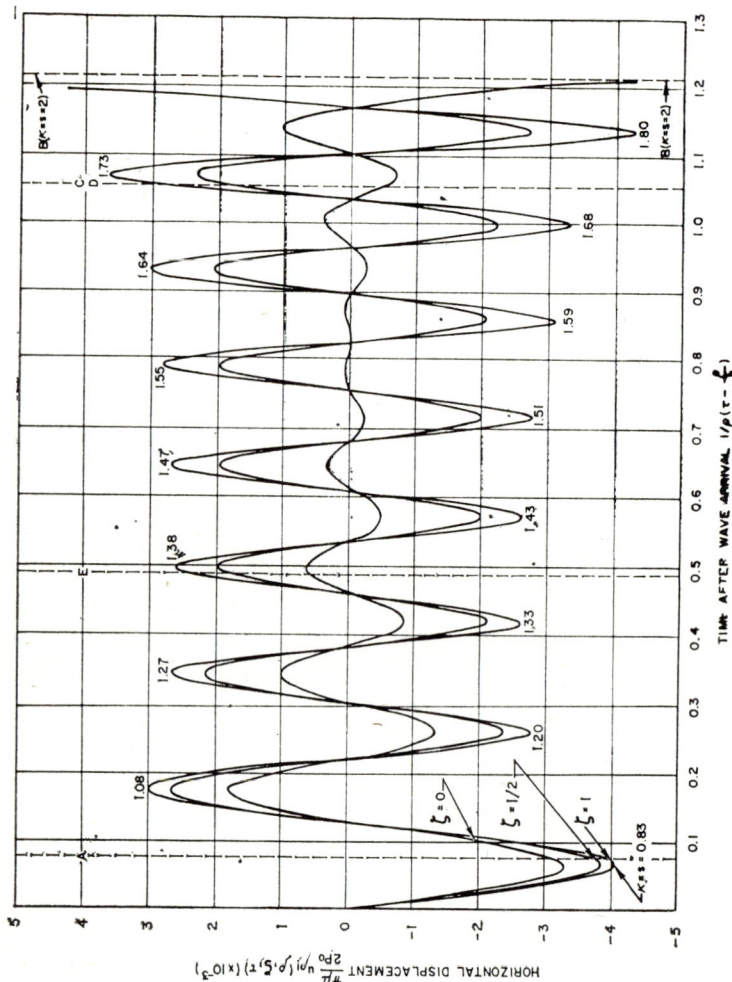

Fig. 7.38. Horizontal displacement $u_{p1}(\rho, \zeta, \tau)$ at a layer station (ρ, ζ) as function of time after wave arrival $(1\varrho)(\tau - \varrho/b)$ (stations $\varrho = 20$, $\zeta = 0, 1/2, 1$).

These transforms can be obtained from the following expansion formulas given in Titchmarsh [7.36, § 4.10]:

$$f(r) = \int_0^\infty \frac{\zeta C_0(\zeta, r, a)}{J_1^2(\zeta a) + Y_1^2(\zeta a)} \, d\zeta \int_a^\infty \xi f(\xi) C_0(\zeta, \xi, a) d\xi ,$$

$$f(r) = \int_0^\infty \frac{\zeta C_1(\zeta, r, a)}{J_1^2(\zeta a) + Y_1^2(\zeta a)} \, d\zeta \int_a^\infty \xi f(\xi) C_1(\zeta, \xi, a) d\xi ,$$

(7.225)

where

$$C_0(x, y, a) = J_0(xy) Y_1(xa) - J_1(xa) Y_0(xy) ,$$

$$C_1(x, y, a) = J_1(xy) Y_1(xa) - J_1(xa) Y_1(xy) ,$$

and $f(r)$ is a suitably restricted arbitrary function. An alternative formal derivation of (7.225) is given by Scott [7.37]. Defining the zero and first-order transforms by

$$\bar{f}^0(k) = \int_a^\infty rf(r) C_0(k, r, a) dr ,$$

$$\bar{f}^1(k) = \int_a^\infty rf(r) C_1(k, r, a) dr ,$$

(7.226)

respectively, the corresponding inverse transforms are, from (7.225),

$$f(r) = \int_0^\infty \frac{k C_0(k, r, a) \bar{f}^0(k)}{J_1^2(ka) + Y_1^2(ka)} dk ,$$

$$f(r) = \int_0^\infty \frac{k C_1(k, r, a) \bar{f}^1(k)}{J_1^2(ka) + Y_1^2(ka)} dk .$$

(7.227)

Again as in our other transforms, integration by parts produced transforms of derivatives and derivative combinations of $f(r)$ (analogous to those of the

Hankel transform, cf. § 5.8), and their relations to the zero and first order transforms $\tilde{f}^0(k)$, $\tilde{f}^1(k)$, and $f(r)$ and df/dr at the cavity wall $r=a$. Two mixed edge problems can be solved, namely the conditions

$$u_r(a, z, t) = u_0 g(t),$$
$$\sigma_{rz}(a, z, t) = 0,$$
(7.228)

or

$$\sigma_{rz}(a, z, t) = \sigma_0 g(t),$$
$$u_r(a, z, t) = 0.$$
(7.229)

The conditions (7.228), governing the problem treated in [7.18], excite compressional waves, and (7.229) flexural waves. It should be pointed out that a mixed pressure shock problem, like that solved in § 7.3.2.2, did not separate in the present coordinates.

The inversion, far field—long time approximation, and numerical evaluation was carried out by essentially the same methods used in the previous section. Further detail is left to [7.18] and [7.36], and [7.22].

7.4.3. *Excitation by a time-dependent thermal field; numerical evaluation of modal solution*

Sve and Miklowitz [7.38] recently solved an axially symmetric infinite plate problem in which the excitation was due to a time-dependent thermal field. The governing equations are the potential wave equations in (7.213) in which the φ equation becomes inhomogeneous due to a term representing the temperature. The governing equations are therefore those of *uncoupled dynamic thermoelasticity*. The temperature distribution was chosen to simulate effects of rapidly absorbed electromagnetic radiation. The radiation was assumed to have a radial Gaussian distribution on one face of the plate and to partially or completely penetrate the plate. Heating time was accounted for, but heat conduction was neglected, the latter being based on the short times of the prime interest (wave propagation effects) in comparison with the time required to reach thermal equilibrium. The temperature distribution was selected to reflect these features. The problem has application in the determination of *laser effects on structures*, and the *design of laser components* such as mirrors and windows subjected to intense radiation.

With the thermal field known, and temperature-independent material properties assumed, the formal solution was derived by the Laplace–Hankel transform method described in § 7.4.1. Assuming complete penetration of the radiation only the symmetric modes of propagation are required. The solution therefore has the form of (7.217). The midplate stresses of the solution were evaluated by numerical integration for near-field stations, summing these modes of propagation up to $n=8$, and comparing the responses based on different sums n, e.g., $n=5$ with $n=8$. The curves, based on zero heating time t_0, are instructive and are shown in fig. 7.39. Part (a) of the figure is for the station $r=0$ (plate center) for five and eight mode sums, and the one-dimensional solution for σ_z, (b) is for $r=4$ and two and eight mode sums, (c) is for $r=4$ and one and five mode sums and (d) is for $r=8$ and two and eight mode sums. In (a) we note only a small difference between the five and eight mode sums. The same is true for the two and eight mode sums of (b). Comparing the results for the five and eight mode sums in (c) and (b) shows the essential features of the solution have been determined. More modes would yield only finer details, even for these close in stations r. This is to be expected for these stations close to the center of the loading, since the plate is responding like a one-dimensional slab (in thickness direction z) for short time with a slowly convergent modal solution. One can see in part (d) for station $r=8$ the response at this more remote (from the source) station becomes more characteristic of a plate. It should be pointed out that the oscillations in these figures can be identified with the cutoff frequencies of the Rayleigh–Lamb spectra. We have noted this type of strong influence on plate response records earlier (cf. fig. 7.19, 7.20 and related discussion in § 7.2.1.1). Further analysis of the results in fig. 7.39, and the response records for a case of non-zero heating time, may be found in [7.38].

7.5. Exercises

7.1. By a derivation similar to that used to obtain the Love–Rayleigh approximate theory for compressional waves in a rod, show that the analogous theory for the plate in plane strain has the governing equation of motion

$$\frac{\partial^2 u(x,t)}{\partial x^2} = \frac{1}{c_p^2}\left[\ddot{u} - \frac{v^2 h^2}{3(1-v)^2}\frac{\partial^2 \ddot{u}}{\partial x^2}\right] - \frac{(1-v^2)X}{E}$$

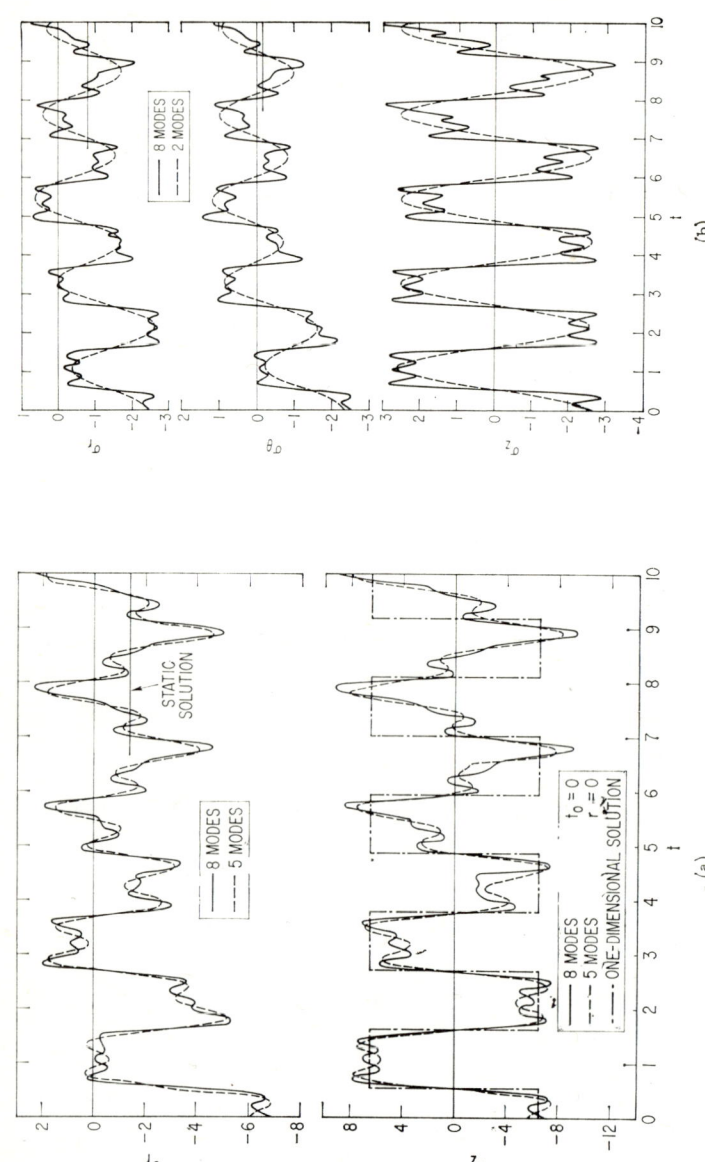

Fig. 7.39. Thermally induced midlayer stress waves in an elastic plate or layer as a function of time. (a) $r=0$, five and eight modes, (b) $r=4$, two and eight modes, (c) $r=4$, one and five modes and (d) $r=8$, two and eight modes.

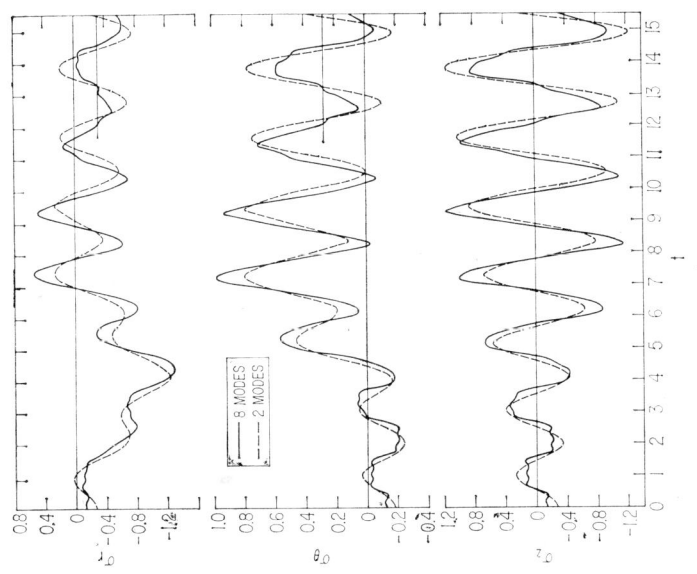
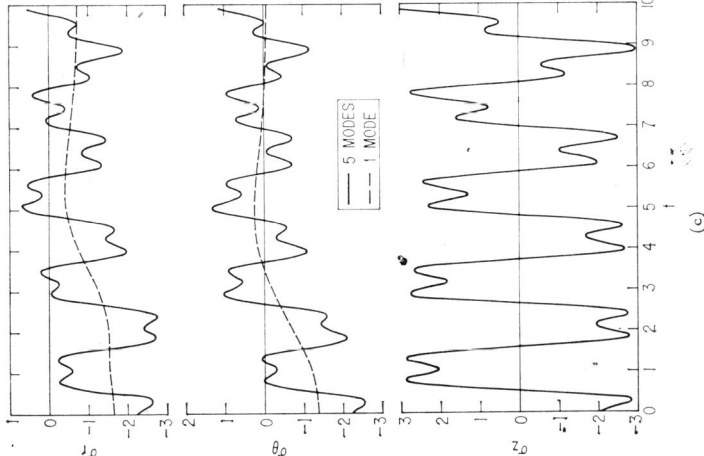

Fig. 7.39. continued

where $c_p = [E/\varrho(1-v^2)]^{\frac{1}{2}}$, X is the body force/unit volume in the x-direction, and h is half the plate thickness. What are the associated initial and boundary conditions for the theory?

7.2. In the Bernoulli–Euler theory for flexural waves in a beam, show that the boundary conditions for the roller skate end are compatible in the case of the time dependent shear force, $Q(0, t) = Q_0 H(t)$ along with $\partial y(0,t)/\partial x = 0$. Give the value of the moment M at $x=0$. Hint: You can write

$$\lim_{x \to 0} \frac{\partial^3 y}{\partial x^3} = -\frac{Q_0}{EI} H(t).$$

7.3. Prove for the Timoshenko beam theory that $\delta W(t)$ is given by (7.42).

7.4. Derive the governing equations (7.43), (7.44) for the Timoshenko beam theory and show they contain the Bernoulli–Euler theory as a special case.

7.5. Derive the frequency equation for the Timoshenko beam theory and approximate each of its two roots (branches) for Ω, $\zeta \to \infty$ and $\zeta \to 0$. Show the latter approximations agree generally with the character depicted in fig. 7.4. What are the phase and group velocities associated with the Ω, $\zeta \to \infty$ approximations?

7.6. In the longitudinal impact problem of § 7.1.3.1 the inversion of the formal solution (7.55) can be carried out alternately by cutting the p-plane along the imaginary axis in the interval $-b \leq \text{Im} p \leq b$ together with an integration over the dumbell contour surrounding this cut. Show that this integration produces another form for the velocity solution (7.61), given by

$$\dot{u}(z, t) = \begin{cases} 0, & t < 0, \\ -(2v_0/\pi) \int_0^b \cos \varrho t \sin [\varrho a z/(b^2 - \varrho^2)^{\frac{1}{2}}] d\varrho/\varrho + v_0, & t > 0. \end{cases}$$

7.7. Prove that the second integral on the R.H.S. of (7.62) contributes terms of $O(1/\sqrt{t})$ for fixed z/t, and large t, in the domain $(z/c_b)(1+3\sigma\varepsilon^2) \leq t < \infty$.

7.8. Making use of the procedure in § 7.1.3.2 obtain the long time-far field approximation from the solution in exercise 7.6 showing it is the same as (7.68). Discuss the differences in the two derivations.

7.9. Making use of the formal solution (7.55) for the longitudinal impact problem, based on the Love–Rayleigh rod theory, show that the velocity $\dot{u}(z, t)$ experiences a jump at $t=0$ that decays exponentially with z. Note therefore it violates the assumed vanishing initial velocity in (7.47). Show with the aid of the governing equation (7.46), however, that such a non-vanishing initial velocity is admissible in this theory, resulting in the same solution as that based on the vanishing initial velocity. Indeed, assuming the nonvanishing initial velocity at the outset, ensures that the solution will completely satisfy the posed boundary value problem.

7.10. The sketch shows a semi-infinite beam suddenly subjected to a step moment on its pin end (at $x=0$), $M(0, t) = M_0 H(t)$. Assume the beam to be

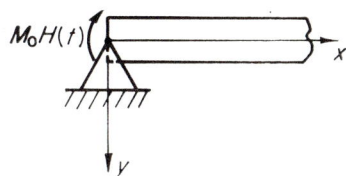

governed by the Euler–Bernoulli displacement equation of motion, (7.28) with $q(x, t) = 0$. Assume further that the beam is at rest initially, and undisturbed at infinity ($x \to \infty$) for all time. Note that the moment-displacement relation in this theory is given by (7.27). Then with the aid of the Laplace transform,

(a) derive the *formal* solutions for the displacement $y(x, t)$ and moment $M(x, t)$, i.e., inversion integral statements for the problem. Show all necessary arguments leading to the definition of this unique solution, including the points necessary to insure that the transform $\bar{y}(x, p)$ or $\bar{M}(x, p)$ can be inverted through this integral statement,

(b) show that the inversion integral solution you obtained in (a) satisfies the boundary, initial, and radiation conditions of the problem.

7.11. (a) Complete the *Br* contour of your $M(x, t)$ integral solution in

Exercise 7.10 (a), down the imaginary p-axis, and carry out the contour integration to derive another more convenient form of solution for evaluating $M(x, t)$. State your arguments in the derivation.

(b) Approximate your solution in (a) of this exercise for long-time. Again state the arguments in your derivations.

7.12. Using the long-time asymptotics of the Laplace transform and its inverse (cf. § 5.10.2.4) and the Laplace transform tables, show that you can derive the long-time solution (7.106), (7.109) and (7.110).

7.13. Prove, for $F_s(\varkappa^2, p^2)$ in (7.117), that

$$\lim_{p \to 0} \frac{F_s(\varkappa^2, p^2)}{k_s^2} = \frac{\varkappa^2}{12(1-v)} (\sinh 2\varkappa h + 2\varkappa h), \qquad p \text{ on } Br,$$

hence establishing that there are corresponding general branches $\varkappa_j(p)$ of $F_s = 0$, $\varkappa$ and p complex, p small, which are the zeros of $\sinh 2\varkappa h + 2\varkappa h$. These correspond to the small ω approximations (4.56) to the complex segments of the Rayleigh–Lamb frequency equation [(4.48) for symmetric waves] that we studied earlier.

7.14. Prove that the branches $\varkappa_j(p)$ of $F_s(\varkappa^2, p^2) = 0$ are simple poles of the integrand function in the inner integral of (7.117), hence residue theory gives (7.119). You may assume

$$N[\varkappa_j^2(p), z, p^2] \qquad \text{and} \qquad \frac{\partial F_s}{\partial \varkappa}[\varkappa^2, p^2]\bigg|_{\varkappa = \varkappa_j(p)}$$

are nonvanishing [at $\varkappa_j(p)$].

7.15. Use the same method that was used to carry out the contour integration for $\varkappa_3(-i\omega)$, along its lower half imaginary p-axis contour in fig. 7.17a, to prove that the branch $\varkappa_2(\mp i\omega)$ is as shown in fig. 7.17b.

7.16. By the methods described in the text derive the integral in (7.119) corresponding to the lowest branch $j = 1$, i.e., $\varkappa_1(-i\omega)$ and $\varkappa_1(i\omega)$.

7.17. Show that the integral I_1 in (7.128) yields (7.138) for the integration near the cutoff point $\omega = c$, and hence the result $I_{1c} \sim (\cos ct - \sin ct)(\pi/2t)^{\frac{1}{2}}$.

7.18. Show that (7.139) results from approximating for long time and the far field the integral derived in Exercise 7.16.

7.19. In (7.117), for the technique of inverting the time transform first, show that N/pF_s behaves at least as $O(p^{-2})$ for large p in the domain $0 \leq z \leq h$.

7.20. Two uniform and symmetric tangential line loads $QF(t)$ (Q is magnitude force per unit length) are suddenly applied to the faces of an infinite plate of $2h$ thickness parallel to the y-axis and through the z-axis as shown in the sketch. Noting that the surface loadings here are represented by the stress $\sigma_{zy} = \pm Q\delta_s(x)F(t)$ at $z = \pm h$, and that this generates SH waves, formulate the boundary-initial value problem and obtain the formal solution

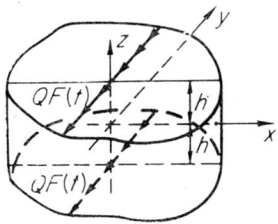

for the displacement v in a form like (7.117). Making use of the latter, obtain a solution through inversion of the spatial transform first and one through inversion of the time transform first.

7.21. Show that for large time and the far field the solution for the displacement w in (7.144) can be reduced to the form

$$w(\xi, \zeta, \tau) \sim K\zeta\tau^{-\frac{1}{3}}[Ai(-\tau_1') + Ai(-\tau_1'')],$$

where Ai is the Airy function, $\xi = x/h, \zeta = z/h, \tau = c_s t/h, \tau_1' = (\tau - \xi/b)(3\gamma\tau/b^2)^{-\frac{1}{3}}$, $\tau_1'' = (\tau + \xi/b)(3\gamma\tau/b^2)^{-\frac{1}{3}}, b = c_p/c_s$ and K is a constant. Note that the lowest symmetric Rayleigh–Lamb branch ω_1 can be approximated by $\omega_1 \simeq (c_p\gamma/h)(1 - \delta\gamma^2)$, where $\gamma = \varkappa h$, $\delta = (k^2 - 2)^2/6k^4$. Note also the discussions in § 7.1.3.2 leading to (7.68), and § 7.2.1.1 on the I_6 integral leading to (7.139). Show that the solution given above reduces to (7.139) for ξ large and positive with τ in the neighborhood of ξ/b.

7.22. Show that the values of $n(\varkappa, p)$ in (7.160) must be $\pm\eta_d$, $\pm\eta_s$. Then show that the general transformed solution of the longitudinal impact problem is (7.161).

7.23. Derive the formal solution for the mixed pressure shock problem in § 7.3.2.2 analogous to (7.162).

7.24. Derive $\tilde{u}_p$ in (7.175). Hint: use a Laplace transform on y in (7.172), and a convolution.

7.25. Starting with the general formal solution (7.175) for the nonmixed edge condition problem, derive in detail the boundedness condition (7.179).

7.26. Consider the semi-infinite plate problem depicted in fig. 7.33. Suppose now the governing equation is that of the plane stress approximation

$$u_{xx}(x, t) = \ddot{u}/c_p^2, \qquad c_p = [E/\varrho(1-v^2)]^{\frac{1}{2}}.$$

Assuming that the initial conditions are $u(x, 0) = \dot{u}(x, 0) = 0$ for $x > 0$, $u(x, t)$ vanishes at $x \to \infty$ for $t > 0$, and $\partial u(0, t)/\partial x = -\hat{\sigma}_A H(t)$, $\hat{\sigma}_A = \sigma_A(1-v^2)/E$, solve this problem by using Laplace transforms on x and t in the theme of § 7.3.4. Here you will find the boundedness condition to be a simple algebraic one rather than the integral equations (7.179). The solution is given by (7.201) as one would expect. Note the technique is an alternate to that usually used in one-dimensional problems (cf. exercises 5.9, 5.10). The finite (in length) waveguide problems can also be solved by employing a finite Laplace transform on x. Then the boundedness condition is replaced with a condition that stems from the required entirety of the finite Laplace transform. This has not been discussed in the text, but the reader can work this out for the present simple problem without great difficulty.

7.27. (a) verify that (7.188) are the edge unknowns for Problem A posed in § 7.3.4.3,

(b) show that the boundedness condition (7.179) reduces to the infinite set of algebraic equations (7.189), (7.190),

(c) show that along the complex arms $s_j(p)$, $ds/dp \to 0$, as $p \to 0$, corresponding to the $\varkappa_j^c(\omega)$ branches in fig. 7.32 being normal to the plane $\omega = 0$, as $\omega \to 0$. Hint: you have (7.191) for the $s_j(p)$.

(d) with the aid of the expansions $\alpha_j(s_j, p)$, $\beta_j(s_j, p)$

$$\begin{Bmatrix} \alpha_j(s_j, p) \\ \beta_j(s_j, p) \end{Bmatrix} = is_j \begin{Bmatrix} 1 - k_d^2/2s_j^2 + O(p^4) \\ 1 - k_s^2/2s_j^2 + O(p^4) \end{Bmatrix}$$

for $p \to 0$, show that

$$C_a(n; s_j, p) = (-)^{(n-1)/2} \left(\frac{k^2 - 1}{2k^2}\right) s_j \left(\frac{n\pi}{h}\right)^3 / s_n^4 + O(p^2),$$

$$C_b(n; s_j, p) = (-)^{(n-1)/2} \left(\frac{k^2 - 1}{k^2}\right) s_j^2 \left(\frac{n\pi}{h}\right)^2 / s_n^4 + O(p^2),$$

$$C_c(s_j, p) = (k^2 - 2) k_d^2 \tan s_j h / s_j^2 + O(p^4),$$

$$F_A(s_j, p) = \frac{\sigma_A c_p (k^2 - 2)}{4\mu(k^2 - 1)c_s^2 s_j^2} \tan s_j h + O(p),$$

where $s_n^2 = s_j^2 - (n\pi/2h)^2$. These lead to the infinite set of algebraic equations for the unknowns $a_n(p)$, $b_n(p)$ given in (7.194).

7.28. Derive the formal solution (7.216) for the axially symmetric infinite plate problem of § 7.4.1., its inversion (7.217) and long time-far field approximation (7.224).

References

[7.1.] Lord Rayleigh, *The Theory of Sound, Vol. 1*, 2nd Edition. Dover Publications, Inc., New York (1945), 251–252.
[7.2.] S. Timoshenko and J. N. Goodier, *Theory of Elasticity*, 3rd Edition (1970), 41–46.
[7.3.] S. Timoshenko, *Philosophical Magazine*, Ser. 6 **41** (1921), 744–746, **43** (1922), 125–131.
[7.4.] Ya. S. Uflyand, *Prikladnaia Matematika I Mekhanika* **12** (1948), 287–300.
[7.5.] R. D. Mindlin, *Journal of Applied Mechanics* **18** (1951), 31–38.
[7.6.] R. V. Churchill, J. W. Brown and R. F. Verhey, *Complex Variables and Applications*, 3rd Edition. McGraw–Hill Book Company, Inc., New York (1974), 290–292.
[7.7.] R. Skalak, *Journal of Applied Mechanics* **24** (1957), 59–64.
[7.8.] J. Miklowitz, *Proceedings of the Third U. S. National Congress of Applied Mechanics*, American Society of Mechanical Engineers, New York, New York (1958), 215–224.
[7.9.] J. Miklowitz and C. R. Nisewanger, *Journal of Applied Mechanics* **24** (1957), 240–244.
[7.10.] J. Miklowitz, *Journal of Applied Mechanics* **27** (1960), 681–689.
[7.11.] B. A. Boley and C. C. Chao, *Journal of Applied Mechanics* **22** (1955), 579–586.
(7.12.] R. P. N. Jones, *Quarterly Journal of Mechanics and Applied Mathematics* **8** (1955), 373–384.

[7.13.] J. R. Lloyd and J. Miklowitz, *Proceedings of the Fourth U. S. National Congress of Applied Mechanics, vol 1*, American Society of Mechanical Engineers, New York, New York (1962), 255–267.
[7.14.] D. C. Viano and J. Miklowitz, *Journal of Applied Mechanics* **41** (1974), 684–690.
[7.15.] C. W. Curtis, *Propagation of Elastic and Plastic Deformation in Solids*, OOR Report, Contr. DA–36–034–ord–1456, Sup. 2, Proj. TB 2–0001 (187), Lehigh University, Sept. 1956.
[7.16.] U. K. Nigul, *Journal of Applied Mathematics and Mechanics (P. M. M.)* **27** (1963–1964), 1602–1620.
[7.17.] G. P. DeVault and C. W. Curtis, *Journal of the Acoustical Society of America* **34** (1962), 421–432.
[7.18.] R. A. Scott and J. Miklowitz, *Journal of Applied Mechanics* **31** (1964), 627–634.
[7.19.] R. L. Rosenfeld and J. Miklowitz, *Journal of Applied Mechanics* **32** (1965), 290–294.
[7.20.] O. E. Jones and A. T. Ellis, *Journal of Applied Mechanics* **30** (1963), 61–69.
[7.21.] O. E. Jones and F. R. Norwood, *Journal of Applied Mechanics* **34** (1967), 718–724.
[7.22.] J. Miklowitz, *Journal of Applied Mechanics* **29** (1962), 53–60.
[7.23.] G. Fox and C. W. Curtis, *Journal of the Acoustical Society of America* **30** (1958), 559–563.
[7.24.] R. P. N. Jones, *Quarterly Journal of Mechanics and Applied Mathematics* **17** (1964), 401–421.
[7.25.] A. H. Meitzler, *Propagation of Elastic Pulses Near the Stressed end of a Cylindrical Bar*, Ph. D. Thesis, Lehigh University, Bethlehem, Pennsylvania (1955).
[7.26.] P. W. Randles and J. Miklowitz, *International Journal of Solids and Structures* **7** (1971), 1031–1055.
[7.27.] P. W. Randles, *International Journal of Solids and Structures* **9** (1973), 31–52.
[7.28.] J. Miklowitz, *Analysis of Elastic Waveguides Involving an Edge*. In: Wave Propagation in Solids, ed. J. Miklowitz. American Society of Mechanical Engineers, New York, New York (1969), 44–70.
[7.29.] G. B. Sinclair and J. Miklowitz, *International Journal of Solids and Structures*, **11** (1975), 275–294.
[7.30.] G. Doetsch, *Theorie und Anwendung der Laplace-Transformation*. Springer-Verlag, Berlin, Germany (1937), 378–383.
[7.31.] J. P. Benthem, *Quarterly Journal of Mechanics and Applied Mathematics* **16**, Part 4 (1963), 413–429.
[7.32.] G. B. Sinclair, *On Nonmixed Symmetric End–Load Problems in Elastic Waveguides*, Ph. D. Thesis, California Institute of Technology, Pasadena, California (1973).
[7.33.] C. I. Robins and R. C. T. Smith, *Philosophical Magazine, Ser 7*, **39** (1948), 1004–1005.
[7.34.] G. J. Cooper and J. W. Craggs, *Journal of the Australian Mathematics Society* **6** (1966), 55–64.
[7.35.] L. D. Bertholf, *Journal of Applied Mechanics* **34** (1967), 725–734.
[7.36.] E. C. Titchmarsh, *Eigenfunction Expansions, Part I*, 2nd Edition. Oxford University Press, London, England (1962).
[7.37.] R. A. Scott, *Transient Wave Propagation in Elastic Plates with Cylindrical Boundaries, Studied with Aid of Multi–Integral Transforms*, Ph. D. Thesis, California Institute of Technology, Pasadena, California (1964).
[7.38.] C. Sve and J. Miklowitz, *Journal of Applied Mechanics* **40** (1973), 161–167.

CHAPTER 8

PULSE SCATTERING BY HALF-PLANE, CYLINDRICAL AND SPHERICAL OBSTACLES

8.1. Introduction; Wave features in a scattering problem

The scattering of a wave by an obstacle in its path is a topic of long standing interest in the study of wave propagation in optics, acoustics and electromagnetics. Attention has been given related problems in elastic wave propagation only in recent years. Sommerfeld's optics problem [8.1] of the diffraction of an incident plane harmonic wave by a wedge, with the half-plane screen as a special case, is a classical example. In order to bring out the basic nature of the waves involved in scattering problems it is instructive to examine the relatively simple wave system in Sommerfeld's *half-plane problem*. Consider fig. 8.1, where we now assume that the incident and other waves have fronts. Shown are two positions of the *incident wave* which propagates along the ray defined by the angle θ_i measured from the positive x-axis (and screen) as is the angular coordinate θ. The angle θ_i is referred to as the *angle of incidence*.[1] In the first position at the lower left (dashed line) in the figure, time is assumed to be $t < -t_1$, the earliest time. In the second position in the upper left and lower right of the figure (solid line) time is $t = t_1$, which holds for the rest of the figure too. A third position of the incident wave (not shown), in which it would intersect the screen edge, would correspond to $t = 0$, when *wave scattering* by the screen begins. Clearly from the figure the *diffracted wave*, a part of the scattered wave system, is generated by the geometrical discontinuity at the screen edge.

---

[1] This is for mathematical convenience. θ_i is really the complement of the angle of incidence which is defined as the angle between the incident ray and the normal to a reflecting surface (cf. ch. 3).

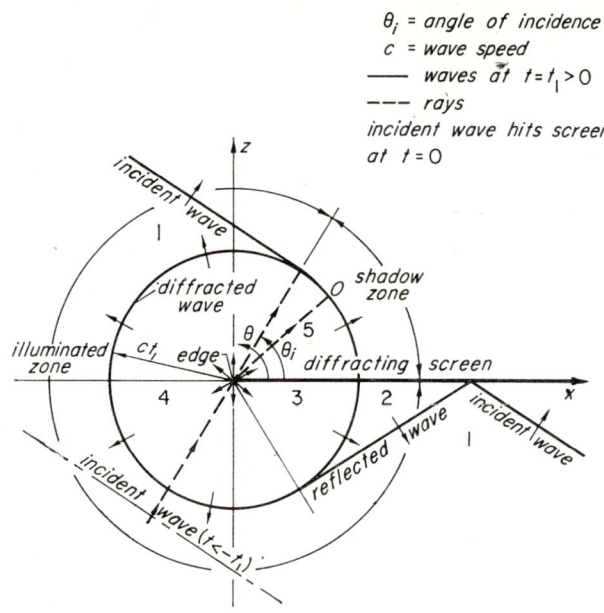

Fig. 8.1. Wave field in Sommerfeld optics problem: diffraction of plane wave from half-plane screen.

The reflected wave completes the scattered wave system.[2] The numbered regions in the figure are distinguished by the wave disturbances present (at $t=t_1$). Region 1 has the incident wave only, 2 the incident and reflected waves, 3 the incident, reflected and diffracted waves, 4 the incident and diffracted waves and 5 the diffracted wave only. As indicated in the figure, 5 lies in the *shadow zone*. This zone is distinguished by the fact that an observer in the zone (at O) receives the signal, not only along a direct ray from the source (normal to the plane wave in this case) but along the direct ray and a deflection of this ray bent by the scatterer (screen here). On the

---

[2] The literature in wave scattering exhibits what appears to be a rather loose use of the word "scattering" or "diffraction" to categorize a problem. Both are scattering problems in which the total wave field, less the incident wave, is the scattered disturbance. When "diffraction" is used in the present sense it usually means the diffracted disturbance is the one of major importance in the problem.

other hand, in the *illuminated zone* shown in the figure (everywhere outside of the shadow zone) the observer receives the signal only by direct rays from the source wave, its reflection and diffracted wave. It should also be noted that the surface of the diffracted wave in the shadow zone is normal to the scatterer. We also note that whereas only the diffracted wave occurs in the shadow zone, all the waves occur in the illuminated zone.

As we shall see later elastic wave diffraction problems are, in general, more complicated primarily because, given an incident wave of one type (say P or SV for example), the scatterer generates, through mode conversion, the same system of waves that occurred in our earlier boundary value problems plus the diffracted waves. Further complexity is introduced by the fact that some classes of the present problems fall into the category of the *mixed boundary value problem*, i.e. the *third of the fundamental elastodynamic problems* [cf. § 1.8, (1.70)]. As we shall see these problems usually involve the *solution of an integral equation*. In spite of these extra complications the reader will note later that the wave features in these problems are basically the same as those just discussed for Sommerfeld's problem.

The plan in this chapter is to first treat the two-dimensional problems of the diffraction of an elastic pulse by a half-plane obstacle, i.e., those of the type in Sommerfeld's problem. The second part of the chapter will be devoted mostly to the two-dimensional problems of the diffraction of the elastic pulse by a circular cylindrical obstacle. A brief treatment of the analogous diffraction by a spherical cavity will conclude the chapter. These *latter diffraction problems fall into the class of either the first or second boundary value problem of elastodynamics*, and as such they do not require the solution of integral equations. As we shall see they are handled in a manner strongly related to some of the techniques in waveguide problems set down in Chapter 7. This is not surprising since they do involve a characteristic length, e.g., the radius of the circular cylindrical obstacle.

8.2. Plane-elastic pulse diffraction by half-plane obstacles

8.2.1. Introduction

As noted earlier the present class of problems are mixed boundary value problems. In solving them one usually is confronted by an *integral equation* or *dual integral equations* that can be reduced to algebraic equations for the unknowns by the *Wiener–Hopf technique*. Noble [8.2] discusses the various methods by which this can be done. In the present work we draw on Clemmow's approach [8.3], involving dual integral equations (in electromagnetic wave problems). Noble [8.2, pp. 152–153] discusses Clemmow's method and Ang [8.4] and Ang and Williams [8.5] have used it successfully in crack propagation studies. The method has certain simplifying features. Alternate treatments in the present problems may be found in (1) deHoop [2.19], who, as we indicated earlier, draws on integral representations for the displacements, and (2) Achenbach [2.14, ch. 9]. The subject is quite important in the theoretical studies of crack propagation in brittle materials.

8.2.2. Diffraction of a plane horizontally polarized shear pulse by a traction free half plane

8.2.2.1. Statement of the problem. This, and the corresponding problem of the rigid (displacement zero) half plane, are the simplest cases of the present class of problems. Being based on a single wave equation they are in effect the Sommerfeld problems of elasticity. The present problem is depicted in fig. 8.1 where use is made of the cartesian coordinates (x, y, z) and the corresponding theory of *SH* waves in ch. 3 § 3.2. The diffracting screen is replaced by the traction free cut or slit which is defined by $z=0$, $0<x<\infty$. The total displacement field v is given by

$$v(x, z, t) = v_i(x, z, t) + v_s(x, z, t), \qquad (8.1\text{a})$$

where v_i is the incident wave and v_s is the scattered wave. *Equation* (8.1a) *is typical of scattering problems.* The incident wave v_i is given by

$$v_i(x, z, t) = f[t - c_s^{-1}(x \cos \gamma + z \sin \gamma)], \qquad (8.1\text{b})$$

where as noted earlier γ (θ_i in fig. 8.1) is the angle of incidence of this pulse and $f(t)=0$ for $t<0$. The angle γ is restricted to the range $0 \leq \gamma \leq \pi/2$. It follows that the *scattered wave v_s must satisfy $v_s=0$ for $t<0$, everywhere in space.* From (8.1a) with v_i known, the boundary-initial value problem reduces to one on the unknown displacement v_s. It is stated as

$$\nabla^2 v_s(x, z, t) = \ddot{v}_s/c_s^2, \qquad -\infty<x<\infty, \ -\infty<z<\infty, \ t>0, \qquad (8.2)$$

$$v_s(x, z, 0) = \dot{v}_s(x, z, 0) = 0, \qquad -\infty<x<\infty, \ -\infty<z<\infty, \qquad (8.3)$$

Ch. 8, § 8.2]   PULSE DIFFRACTION BY HALF-PLANE   489

$$\sigma_{zys}(x, 0, t) = -\sigma_{zyi}(x, 0, t), \qquad 0<x<\infty, \ t>0, \qquad (8.4)$$

$$v_s(x, 0, t) = 0, \qquad -\infty < x \leq 0, \ t>0, \qquad (8.5)$$

$$\lim_{r\to\infty} [v_s(x, z, t), \sigma_{zys}, \sigma_{xys}] = 0, \qquad t>0, \qquad (8.6)$$

where $r = (x^2+z^2)^{\frac{1}{2}}$, *and the condition that the kinetic and strain energy densities for v_s are integrable everywhere in space*. This last condition *limiting the stresses to integrable singularities at $x=0$, is one of four on v_s* that must be met to have a unique solution, as pointed out by deHoop [2.19, § 7]. The three other conditions are that v_s satisfy (8.2), (8.4) and $v_s = 0$ everywhere in space for $t<t_0$ (the initial time). DeHoop's proof is applicable generally to the elastic wave diffraction from half plane scatterers of interest in this chapter. It should be pointed out that for plane elastostatics problems involving cracks, recently Knowles and Pucik [8.6] have shown that uniqueness is obtained by specifying boundedness of the displacement in addition to the hypothesis in the standard uniqueness theorem of Kirchhoff.

The boundary condition (8.4) is a statement that the total stress $\sigma_{zy} = \sigma_{zyi} + \sigma_{zys} = 0$ on both sides of the cut. Equilibrium considerations of elements above and below the cut show that the stresses σ_{zys} are equal in magnitude, but oppositely directed on the two sides of the cut. This leads to *scattered wave motion that is antisymmetric* with respect to the x-axis ($z=0$), and hence the boundary condition (8.5). Equation (8.6) is the radiation condition on v_s and associated quantities.

8.2.2.2. Formal solution of the problem. To solve the boundary value problem we employ the Laplace transform on time t and exponential Fourier transform (with real argument) on x as we did in Chapter 6. The transformed equation (8.2) is

$$\frac{d^2 \tilde{v}_s(\varkappa, z, p)}{dz^2} - \eta_s^2 \tilde{v}_s = 0, \qquad (8.7)$$

which has the solutions

$$\tilde{v}_s(\varkappa, z, p) = A(\varkappa, p) e^{\mp \eta_s z} \qquad (8.8)$$

with $\eta_s = (k_s^2 + \varkappa^2)^{\frac{1}{2}}$. The Fourier transform inverse therefore gives the Laplace transform of v_s

$$\bar{v}_s(x, z, p) = \frac{1}{2\pi} \int_{-\infty}^{\infty} A(\varkappa, p) e^{-(\pm \eta_s z + i\varkappa x)} d\varkappa . \tag{8.9}$$

We first seek the solution to the problem for the domain $z \geq 0$. Then the process for selecting the appropriate branch of η_s is like that in § 6.2. That is, with $\varkappa$ real, and taking p real and positive, we can select the real positive branch of η_s to satisfy the double transform of (8.6) [cf. (6.7) and discussion following there]. Since ultimately use will be made of the Cagniard–deHoop method to invert $\bar{v}_s$, it is convenient to impose the transformation $\varkappa = k_s \zeta$ on (8.9) now. It becomes

$$\bar{v}_s(x, z, p) = \frac{1}{2\pi} \int_{-\infty}^{\infty} A(\zeta) e^{-pg(\zeta)} d\zeta , \tag{8.10}$$

where $g(\zeta) = (1/c_s)[(\zeta^2 + 1)^{\frac{1}{2}} z + i\zeta x]$. The function $(\zeta^2 + 1)^{\frac{1}{2}} = \eta_s/k_s$ can be continued analytically off the real ζ-axis in the manner the like-functions of § 6.2.1.1 were [cf. fig. 6.2, and associated analysis]. We find that $\text{Re}(\zeta^2 + 1)^{\frac{1}{2}} \geq 0$ *for the entire cut ζ-plane shown in fig. 8.2.*[3] This is needed in these diffraction problems as will become apparent later. The unknown function

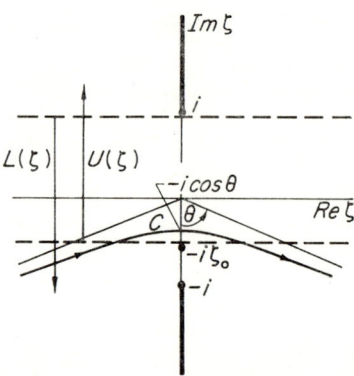

Fig. 8.2. The cut ζ-plane, regions of regularity of $L(\zeta)$ and $U(\zeta)$ functions and Cagniard path for *SH*-pulse diffraction problem.

---

[3] Note the like-functions of § 6.2.1.1 have this property too. However, there the property was required only over the cut lower half ζ-plane.

$A(\zeta)$ is required to behave algebraically at $|\zeta|\to\infty$. From (8.10) note this condition ensures that $\bar{v}_s(x, z, p)$ and its derivatives will be continuous at $x=0$, $z>0$. Therefore the condition gives boundedness to v_s in the neighborhood of $x=0$, relating to uniqueness of the solution as discussed after (8.6).

To determine $A(\zeta)$ we have the mixed boundary conditions (8.4), (8.5). From the first, using (8.10) and $\bar{\sigma}_{zy}=\mu\partial\bar{v}/\partial z$, we find

$$\int_{-\infty}^{\infty} (\zeta^2+1)^{\frac{1}{2}}A(\zeta)e^{-ik_sx\zeta}d\zeta = -2\pi\bar{f}(p)\sin\gamma e^{-k_sx\cos\gamma}, \qquad 0<x<\infty, \tag{8.11}$$

and from the second and (8.10),

$$\int_{-\infty}^{\infty} A(\zeta)e^{-ik_sx\zeta}d\zeta = 0, \qquad -\infty<x\leq 0. \tag{8.12}$$

Equations (8.11), (8.12) are dual integral equations for the unknown $A(\zeta)$.

To solve for $A(\zeta)$ we first note that by simply applying Cauchy's integral formula (cf. [7.6], § 51) to (8.11) it can be reduced to an algebraic equation for $A(\zeta)$. We find

$$(\zeta^2+1)^{\frac{1}{2}}A(\zeta) = -i\frac{\bar{f}(p)\sin\gamma L(\zeta)}{(\zeta+i\zeta_0)L(-i\zeta_0)}, \tag{8.13}$$

where $L(\zeta)$ *is a function having no zeros or singularities in the lower half-plane* $\operatorname{Im}\zeta<1$, *and behaves algebraically at infinity therein*, cf. fig. 8.2. Similarly we can satisfy (8.12) with

$$A(\zeta) = U(\zeta), \tag{8.14}$$

where $U(\zeta)$ *is a function having like properties in the upper half-plane* $\operatorname{Im}\zeta > -\zeta_0 = -\cos\gamma$, also shown in fig. 8.2. Eliminating $A(\zeta)$ in (8.13), (8.14) we find

$$\frac{U(\zeta)}{L(\zeta)} = -i\frac{\bar{f}(p)\sin\gamma}{(\zeta+i\zeta_0)(\zeta^2+1)^{\frac{1}{2}}L(-i\zeta_0)}. \tag{8.15}$$

Now we can *factor* $(\zeta^2+1)^{\frac{1}{2}}$ into $(\zeta-i)^{\frac{1}{2}}(\zeta+i)^{\frac{1}{2}}$, the first term being an $L(\zeta)$ type, and second a $U(\zeta)$ type, function. Hence, in (8.15) if we set

$$L(\zeta) = (\zeta - i)^{\frac{1}{2}},$$

(8.16)

$$U(\zeta) = -i \frac{\bar{f}(p) \sin \gamma}{(\zeta + i\zeta_0)(\zeta + i)^{\frac{1}{2}}[-i(\zeta_0 + 1)]^{\frac{1}{2}}},$$

these satisfy our stated requirements on $L(\zeta)$, $U(\zeta)$. Substituting the second of (8.16) into (8.14) then, gives

$$A(\zeta) = -i \frac{\bar{f}(p) \sin \gamma}{(\zeta + i\zeta_0)(\zeta + i)^{\frac{1}{2}}[-i(\zeta_0 + 1)]^{\frac{1}{2}}},$$

(8.17)

which is the solution of the dual integral equations (8.11), (8.12). Note in getting $A(\zeta)$ that analytical continuation, needed in most applications of the Wiener–Hopf technique, was not involved. This is an appealing feature of Clemmow's method. *The crucial step leading to the solution* (8.17) *was the decomposition of $U(\zeta)/L(\zeta)$ in* (8.15) *which gave* (8.16). *In problems of the present type this decomposition is the core of the Wiener–Hopf procedure, whatever solution technique is used* [8.2, § 1.7]. In the *SH*-wave problem we are dealing with, inspection easily led to the needed factorization. However, *in more complicated problems one has to draw on a more general decomposition technique.* We treat a problem of this type in the § 8.2.4.

8.2.2.3. Inversion of the formal solution and discussion of its nature. With $A(\zeta)$ known we can now proceed with the inversion of $\bar{v}_s$ in (8.10). Having discussed the Cagniard–deHoop method at length in Chapter 6, we can be economical here (essentially following deHoop's treatment [2.19]). Indeed the integral in (8.10) is of the form of the first integral of $\bar{u}$ or $\bar{w}$ in (6.15). Hence, the procedure in § 6.2.1.1 leads here to a change from the path of integration in (8.10) to the hyperbola

$$\zeta = \pm [(c_s t/r)^2 - 1]^{\frac{1}{2}} \sin\theta - i(c_s t/r) \cos\theta, \quad r/c_s \leq t < \infty, \quad 0 \leq \theta \leq \pi,$$

(8.18)

where r, θ are the polar coordinates, $r = (x^2 + z^2)^{\frac{1}{2}}$, $0 \leq r \leq \infty$, $\theta = \tan^{-1}(z/x)$, $0 \leq \theta \leq 2\pi$, with $\theta = 0, 2\pi$ the sides of the cut (cf. fig. 8.1). The Cagniard path (8.18) is shown in fig. 8.2 marked C, with asymptotes defined by θ measured from the lower half of the Im ζ-axis. The minimum time $t = r/c_s$ corresponds to the vertex $\zeta = -i \cos\theta$ which, according to the θ interval specified in

(8.18), shows this intersection of C and the imaginary axis always lies between the branch points $\zeta = i$ and $-i$. That this is admissible stems from $\text{Re}(\zeta^2+1)^{\frac{1}{2}} \geq 0$ everywhere in the cut ζ-plane (stated earlier). *Hence C in the upper or lower half ζ-plane is a valid equivalent path for the original integration path along the real ζ-axis* (C is in the lower half ζ-plane for $0 \leq \theta < \pi/2$ and in the upper half for $\pi/2 < \theta \leq \pi$). Note further that this stated θ interval in (8.18) is consistent with our present analysis which is restricted to $z \geq 0$ (i.e., $0 \leq \theta \leq \pi$). We also note when $0 \leq \theta < \gamma$, the simple pole at $\zeta = -i\zeta_0 = -i\cos\gamma$ lies between C and the real ζ-axis in which case its contribution to the solution must be accounted for. *This contribution gives the scattered wave associated with the geometrical solution of the diffraction problem* (i.e. the solution composed of the incident wave and its reflection). Analysis of (8.10), (8.17) in the complex ζ-plane is like that in § 6.2.1.1 [cf. (6.20)–(6.24)], and the contour integration results in

$$\bar{v}_s(r, \theta, p) = \bar{v}_s^d(r, \theta, p) + \bar{v}_s^g(r, \theta, p), \qquad (8.19)$$

where

$$\bar{v}_s^d(r, \theta, p) = \pi^{-1} \int_{C_4} \text{Re}\left[A(\zeta)e^{-pg(\zeta)}d\zeta\right], \qquad 0 \leq \theta \leq \pi,$$

and

$$\bar{v}_s^g(r, \theta, p) = \begin{cases} -\bar{f}(p)\exp\left[-k_s r \cos(\gamma-\theta)\right], & \text{for } 0 \leq \theta < \gamma, \\ 0, & \text{for } \gamma < \theta \leq \pi, \end{cases}$$

where C_4 is the fourth quadrant path of C in fig. 8.2, and superscript d and g indicate, respectively, the transformed diffracted and geometrical wave components of the scattered disturbance. The result for $\bar{v}_s^g$ in (8.19) stems from the residue evaluation at the simple pole $(\zeta + i\zeta_0)^{-1}$. Introducing $t = g(\zeta)$, and $\omega_s(t) = \zeta(t)$ given by (8.18) (with the $+$ sign there), into $\bar{v}_s^d$ in (8.19) gives

$$\bar{v}_s^d(r, \theta, p) = \pi^{-1} \int_{r/c_e}^{\infty} \text{Re}\left\{A[\omega_s(t)]\frac{d\omega_s(t)}{dt}\right\}e^{-pt}dt. \qquad (8.20)$$

It can be shown that

$$(\omega_s^2+1)^{-\frac{1}{2}}(d\omega_s/dt) = c_s r^{-1}[(c_s t/r)^2 - 1]^{-\frac{1}{2}}, \qquad (8.21)$$

and with this we can write (8.20) as

$$\bar{v}_s^d(r, \theta, p) = \bar{f}(p)\bar{V}(r, \theta, p), \tag{8.22}$$

where

$$\bar{V}(r, \theta, p) = D \int_{r/c_s}^{\infty} [(c_s t/r)^2 - 1]^{-\frac{1}{2}} \operatorname{Re}\left\{\frac{(1-i)(\omega_s - i)^{\frac{1}{2}}}{\omega_s + i \cos \gamma}\right\} e^{-pt} dt,$$

with $D = c_s(\pi r)^{-1} \sin (\gamma/2)$. As we have seen in earlier examples, by inspection the inverse of $\bar{V}(r, \theta, p)$ is given by

$$V(r, \theta, t) = D[(c_s t/r)^2 - 1]^{-\frac{1}{2}} \operatorname{Re}\left\{\frac{(1-i)(\omega_s - i)^{\frac{1}{2}}}{\omega_s + i \cos \gamma}\right\} H(t - r/c_s), \quad 0 \leq \theta \leq \pi. \tag{8.23}$$

Then noting the product form of (8.22), the convolution theorem shows the diffracted wave is given by

$$v_s^d(r, \theta, t) = \left\{\int_{r/c_s}^{t} f(t-\tau) V(r, \theta, \tau) d\tau\right\} H(t - r/c_s), \quad 0 \leq \theta \leq \pi. \tag{8.24}$$

The θ domain here can be extended to $\pi \leq \theta \leq 2\pi$ by recalling that

$$v_s(r, \theta, t) = -v_s(r, 2\pi - \theta, t), \tag{8.25}$$

i.e., the motion is antisymmetric with respect to the x-axis ($z = \theta = 0$) [cf. remarks after (8.6)]. It can be seen that the *diffracted wave is a cylindrical wave generated by the edge of the cut* with wavefront velocity c_s.

By carrying out the algebraic operations indicated in (8.23) deHoop finds the particularly convenient form

$$V(r, \theta, t) = \frac{c_s}{\sqrt{2\pi r}} \left\{\frac{\sin \frac{1}{2}(\gamma - \theta)}{c_s t/r - \cos(\gamma - \theta)} + \frac{\sin \frac{1}{2}(\gamma + \theta)}{c_s t/r - \cos(\gamma + \theta)}\right\} \frac{H(t - r/c_s)}{(c_s t/r - 1)^{\frac{1}{2}}},$$

$$0 \leq \theta \leq 2\pi. \tag{8.26}[4]$$

---

[4] This result can be derived by working with the square of Re{} in (8.23). The numerator of this is a cubic in $c_s t/r$. After expressing the cubic in its factors, and taking the square root of the so-formed [Re{}]^2, a simple application of partial fractions yields (8.26).

Note that (8.26) is valid over the entire θ-domain which stems from (8.25). This result is in agreement with the result found by Sommerfeld [8.7] for this problem.

Noting that

$$v^g(r, \theta, t) = v_s^g(r, \theta, t) + v_i(r, \theta, t),$$

the simple inverse of v_s^g in (8.19), the symmetry relation (8.25) and (8.1b) yield the geometrical solution

$$v^g(r, \theta, t)$$

$$= \begin{cases} 0, & 0 \leq \theta < \gamma, \\ f[t - (r/c_s) \cos(\gamma - \theta)], & \gamma < \theta < 2\pi - \gamma, \\ f[t - (r/c_s) \cos(\gamma - \theta)] + f[t - (r/c_s) \cos(\gamma + \theta)], & 2\pi - \gamma < \theta \leq 2\pi. \end{cases}$$

(8.27)

Noting the critical point $\theta = \gamma$ (and its mate $2\pi - \gamma$) is the place where the pole at $\zeta = -i \cos \gamma$ sets on the Cagniard contour, the geometrical solution for these points is defined through

$$v^g(r, \theta, t) = (\tfrac{1}{2})[v^g(r, \theta-, t) + v^g(r, \theta+, t)], \tag{8.28}$$

where the terms on the right are given by (8.27). Finally then, our solution is given by

$$v(r, \theta, t) = v_s^d(r, \theta, t) + v^g(r, \theta, t), \tag{8.29}$$

valid for all θ, where v_s^d is given by (8.24), and its component parts (8.22), (8.26), and v^g by (8.27), (8.28). The wave system is shown in fig. 8.1.

Of further interest here is the note by Handleman and Rubenfeld [8.8] who investigated the *singularity of the stress* $\sigma_{\theta y}$ *in the neighborhood of the edge of the cut*. They found it *behaved as* $r^{-\frac{1}{2}}$. Such behavior would be admissible for solution uniqueness as discussed after (8.6). The result is of importance in *crack propagation* studies.

8.2.3. Diffraction of a plane horizontally polarized shear pulse by a rigid half plane

The derivation of the solution for this problem closely follows that for the preceding case. We again have (8.1)–(8.3), (8.6), but (8.4) and (8.5) are replaced by

$$v_s(x, 0, t) = -v_i(x, 0, t), \qquad 0 < x < \infty, \qquad t > 0, \qquad (8.30)$$

$$\sigma_{zys}(x, 0, t) = 0, \qquad -\infty < x \leq 0, \qquad t > 0, \qquad (8.31)$$

respectively, and again the remarks on "behavior near $x=0$" following (8.6) apply here. The solution is derived by deHoop [2.19, § 9]. It is given by (8.29), where for v_s^d we again have (8.24) with component parts (8.22) and (8.26), with now a change in sign of the second term in the latter. Similarly for v^g, we have (8.27) and (8.28), where now in the last equation of the former there is a change in sign. The wave system is again that in fig. 8.1. Derivation of the solution for this case by the method of the previous section is left to the exercises.

8.2.4. Diffraction of a plane dilatational pulse by a traction free half plane

8.2.4.1. Statement of the problem.
We now consider the diffraction of a plane dilatational pulse by the semi-infinite cut $0 < x < \infty$, $z = 0$ (coordinates same as in fig. 8.1). The incident displacement for this plane strain problem is represented by its components

$$\begin{Bmatrix} u_i(x, z, t) \\ w_i(x, z, t) \end{Bmatrix} = \begin{Bmatrix} \cos \alpha \\ \sin \alpha \end{Bmatrix} f\left[t - c_d^{-1}(x \cos \alpha + z \sin \alpha)\right], \qquad (8.32)$$

in the x and z directions, respectively, where now α is the angle of incidencn restricted to $0 \leq \alpha \leq \pi/2$, and again $f(t) = 0$ for $t < 0$. Here we can draw oe the theory of P waves in ch. 3 § 3.1, and ch. 6 § 6.2.1. The total displacement field is given by its components

$$u(x, z, t) = u_i(x, z, t) + u_s(x, z, t),$$
$$w(x, z, t) = w_i(x, z, t) + w_s(x, z, t),$$
(8.33)

where $u_s, w_s = 0$ for $t < 0$. With u_i, w_i in (8.32) known, according to (8.33) the boundary initial value problem reduces to one on the unknown displacement components u_s, w_s. Now there are two components of the stress that must vanish on the cut, namely σ_z, σ_{zx}. Again equilibrium considerations of the elements above and below the cut show each of these stresses are equal in magnitude, but oppositely directed, on the two sides of the cut. This leads to a natural decomposition of the problem into one of symmetric motion and one of antisymmetric motion with respect to the x-axis as noted by deHoop [2.19, §§ 12, 14] and Achenbach [2.14, § 9.6]. These motions relate to the vanishing of σ_z and σ_{zx} on both sides of the cut, respectively. The component boundary-initial value problems can therefore be written by making use of the governing equations (6.1), (6.2) and (6.4). For both problems we have

$$\left. \begin{array}{l} \nabla^2 \varphi_s(x, z, t) = \ddot{\varphi}_s/c_d^2 \\ \nabla^2 \psi_s(x, z, t) = \ddot{\psi}_s/c_s^2 \end{array} \right\} \quad -\infty < x < \infty, \ -\infty < z < \infty, \ t > 0, \quad (8.34)$$

$$\varphi_s(x, z, 0) = \dot{\varphi}_s = \psi_s = \dot{\psi}_s = 0, \quad -\infty < x < \infty, \ -\infty < z < \infty, \quad (8.35)$$

$$\lim_{r \to \infty} \left[\varphi_s(x, z, t), \psi_s(x, z, t), \text{etc.} \right] = 0, \quad t > 0, \quad (8.36)$$

along with

$$\left. \begin{array}{ll} \sigma_{zs}(x, 0, t) = -\sigma_{zi}(x, 0, t), & 0 < x < \infty, \\ w_s(x, 0, t) = 0, & -\infty < x \leq 0, \\ \sigma_{zxs}(x, 0, t) = 0, & -\infty < x < \infty, \end{array} \right\} t > 0, \quad (8.37)$$

for the symmetric problem, and

$$\left. \begin{array}{ll} \sigma_{zxs}(x, 0, t) = -\sigma_{zxi}(x, 0, t), & 0 < x < \infty, \\ u_s(x, 0, t) = 0, & -\infty < x \leq 0, \\ \sigma_{zs}(x, 0, t) = 0, & -\infty < x < \infty, \end{array} \right\} t > 0, \quad (8.38)$$

for the antisymmetric problem. Again here we must have the conditions stated after (8.6) to ensure a unique solution for u_s, w_s.

8.2.4.2. Formal solution of the problem. On the basis of (8.34)–(8.36) the double transform technique and the Cagniard–deHoop method leads to the Laplace transformed displacements for the scattered disturbance in $z \geqq 0$

$$\bar{u}_s(x, z, p) = -\frac{k_d}{2\pi} \int_{-\infty}^{\infty} [i\zeta A(\zeta) e^{-pg_d(\zeta)} - \eta'_s(\zeta) B(\zeta) e^{-pg_s(\zeta)}] \, d\zeta ,$$
(8.39a)

$$\bar{w}_s(x, z, p) = -\frac{k_d}{2\pi} \int_{-\infty}^{\infty} [\eta'_d(\zeta) A(\zeta) e^{-pg_d(\zeta)} + i\zeta B(\zeta) e^{-pg_s(\zeta)}] \, d\zeta ,$$

where g_d and g_s are given in (6.15), and $\eta'_d(\zeta) = (\zeta^2 + 1)^{\frac{1}{2}}$, $\eta'_s(\zeta) = (\zeta^2 + k^2)^{\frac{1}{2}}$ from (6.14). Here the functions $\eta'_d(\zeta)$ and $\eta'_s(\zeta)$ can be continued analytically off the real ζ-axis such that

$$\operatorname{Re} \begin{bmatrix} \eta'_d(\zeta) \\ \eta'_s(\zeta) \end{bmatrix} \geqq 0 \qquad (8.39b)$$

for the entire cut ζ-plane shown in fig. 8.3 [cf. fig. 6.2, and associated analysis]. The unknown functions $A(\zeta)$, $B(\zeta)$ are required to behave algebraically at $|\zeta| \to \infty$. As (8.39a) shows this condition ensures that $\bar{u}_s(x, z, p)$, $\bar{w}_s(x, z, p)$ and their derivatives will be continuous at $x = 0$, $z > 0$. This boundedness of $\bar{u}_s$, $\bar{w}_s$ in the neighborhood of $x = 0$ is related to the uniqueness of the solutions u_s, v_s [cf. discussion following (8.6)]. The transformed stresses at $z = 0$ corresponding to (8.39a) are

$$\bar{\sigma}_{zs}(x, 0, p) = \frac{\mu k_d^2}{2\pi} \int_{-\infty}^{\infty} [(k^2 + 2\zeta^2) A(\zeta) + 2i\zeta \eta'_s(\zeta) B(\zeta)] e^{-ik_d x \zeta} \, d\zeta ,$$
(8.40)

$$\bar{\sigma}_{zxs}(x, 0, p) = \frac{\mu k_d^2}{2\pi} \int_{-\infty}^{\infty} [2i\zeta \eta'_d(\zeta) A(\zeta) - (k^2 + 2\zeta^2) B(\zeta)] e^{-ik_d x \zeta} \, d\zeta .$$

Ch. 8, § 8.2]   PULSE DIFFRACTION BY HALF-PLANE   499

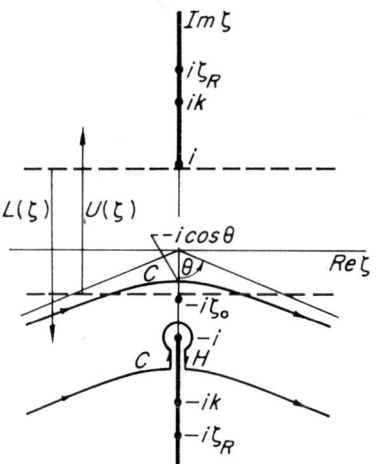

8.3. The cut ζ-plane, regions of regularity of $L(\zeta)$ and $U(\zeta)$ functions and Cagniard paths for P-pulse diffraction problem.

We consider first the *symmetric problem*. Substituting the second of (8.40) into the Laplace transform of the third of (8.37) yields

$$B_1(\zeta) = i2\zeta \eta'_d(\zeta) A_1(\zeta)/(k^2 + 2\zeta^2), \qquad (8.41)$$

where the *subscript 1 denotes the A, B of the symmetric problem*. Now substituting (8.41) into the first of (8.40), and using this in the Laplace transform of the first of (8.37), we have

$$\int_{-\infty}^{\infty} \frac{R(\zeta)}{k^2 + 2\zeta^2} A_1(\zeta) e^{-ikax\zeta} d\zeta = -(2\pi k^2/k_d) \cos 2\beta \bar{f}(p) e^{-kax\cos\alpha}, \quad 0 < x < \infty, \qquad (8.42)$$

where $R(\zeta)$ is the Rayleigh function given in (6.15). The right hand side here represents $-\overline{\sigma_{zi}}(x, 0, p)$ which is calculated from the Laplace transforms of (8.32) where use has been made of (3.17), α and β here (related through $\cos \alpha = k \cos \beta$) being the compliments of α and β there. Finally, substituting the second of (8.39a) into the Laplace transform of the second of (8.37), and using (8.41) again, gives

$$\int_{-\infty}^{\infty} \frac{\eta'_d(\zeta)}{k^2 + 2\zeta^2} A_1(\zeta) e^{-ikax\zeta} d\zeta = 0, \qquad -\infty < x \leq 0. \qquad (8.43)$$

Equations (8.42), (8.43) are the dual integral equations for $A_1(\zeta)$. Applying the Cauchy integral formula to (8.42) (as we did in § 8.2.2) yields

$$\frac{R(\zeta)}{k^2+2\zeta^2} A_1(\zeta) = -\frac{ik^2 \cos 2\beta \bar{f}(p) L(\zeta)}{k_d(\zeta+i\zeta_0)L(-i\zeta_0)}, \qquad (8.44)$$

where $L(\zeta)$ *is a function having no zeros or singularities in the lower half plane* $\operatorname{Im}\zeta < 1$, *behaving algebraically at infinity therein*, cf. fig. 8.3. Similarly we can satisfy (8.43) with

$$\frac{\eta'_d(\zeta)}{k^2+2\zeta^2} A_1(\zeta) = U(\zeta), \qquad (8.45)$$

where $U(\zeta)$ *is a function having like properties in the upper half plane* $\operatorname{Im}\zeta > -\zeta_0 = -\cos\alpha$, also shown in fig. 8.3. Eliminating $A_1(\zeta)$ from (8.44), (8.45) we find

$$\frac{U(\zeta)}{L(\zeta)} = -\frac{ik^2 \cos 2\beta \bar{f}(p)\, \eta'_d(\zeta)}{k_d L(-i\zeta_0)(\zeta+i\zeta_0)(\zeta^2+\zeta_R^2)F(\zeta)}, \qquad (8.46)$$

where $F(\zeta)=R(\zeta)/(\zeta^2+\zeta_R^2)$ with $\zeta_R = c_d/c_R$. Now we factor (8.46) letting

$$L(\zeta) = (\zeta - i\zeta_R)F_L(\zeta)(\zeta-i)^{-\frac{1}{2}},$$

$$U(\zeta) = k_d^{-1} k^2 \cos 2\beta \bar{f}(p) D(\zeta)(\zeta+i)^{\frac{1}{2}},$$

$$\qquad (8.47)$$

$$D(\zeta) = [-i(\zeta_0+1)]^{\frac{1}{2}}/(\zeta_0+\zeta_R)(\zeta+i\zeta_0)(\zeta+i\zeta_R)F_L(-i\zeta_0)F_U(\zeta),$$

$$F(\zeta) = F_U(\zeta)F_L(\zeta),$$

where F_U, F_L must be U- and L-type functions, respectively. The factors F_U, F_L are derived in the next section through a general factorization technique in the Wiener–Hopf method. As will be shown there $F_U(\zeta)$ is a U-type function in the upper half plane $\operatorname{Im}\zeta > -1$ and $F_L(\zeta)$ an L-type function in the lower half plane $\operatorname{Im}\zeta < 1$. It therefore follows that $L(\zeta)$ and $U(\zeta)$ in (8.47) satisfy the requirements stated earlier for these functions in their respective domains shown in fig. 8.3. Now substitution of U in (8.47) into (8.45) gives

$$[k_d/\bar{f}(p)]A_1(\zeta) = A_1'(\zeta) = k^2 \cos 2\beta (k^2 + 2\zeta^2)(\zeta - i)^{-\frac{1}{2}} D(\zeta) \qquad (8.48)$$

which through (8.41) determines $B_1(\zeta)$. These then determine $\bar{u}_s$, $\bar{w}_s$ of the symmetric problem through (8.39a).

Solution of the antisymmetric problem is handled in the same way. Substituting the first of (8.40) into the Laplace transform of the third of (8.38) gives

$$B_2(\zeta) = -(k^2 + 2\zeta^2)A_2(\zeta)/2i\zeta\eta_s'(\zeta), \qquad (8.49)$$

where the *subscript 2 denotes the A, B of the antisymmetric problem*. Now substituting (8.49) into the second of (8.40), and using this in the Laplace transform of the first of (8.38), yields

$$\int_{-\infty}^{\infty} \frac{R(\zeta)}{\zeta\eta_d'(\zeta)} A_2(\zeta) e^{-ik_d x \zeta} d\zeta = (4\pi i/k_d) \sin 2\alpha \bar{f}(p) e^{-k_d x \cos \alpha}, \qquad 0 < x < \infty, \qquad (8.50)$$

where the right hand side here is obtained in a manner similar to that in (8.42). Substituting the first of (8.39a) into the Laplace transform of the second of (8.38), and using (8.49) again, gives

$$\int_{-\infty}^{\infty} \zeta^{-1} A_2(\zeta) e^{-ik_d x \xi} d\zeta = 0, \qquad -\infty < x < 0. \qquad (8.51)$$

Equations (8.50), (8.51) are the dual integral equations for $A_2(\zeta)$. The solution for $A_2(\zeta)$ is obtained as $A_1(\zeta)$ was for (8.42), (8.43). The result is

$$-[k_d/\bar{f}(p)]A_2(\zeta) = A_2'(\zeta) = i2 \sin 2\alpha \zeta(\zeta + ik)^{\frac{1}{2}} D(\zeta), \qquad (8.52)$$

which through (8.49) determines $B_2(\zeta)$. These then determine $\bar{u}_s$, $\bar{w}_s$ of the antisymmetric problem through (8.39a). Therefore the Laplace transformed displacement components for the present P-pulse diffraction problem are, from (8.39a),

$$\bar{u}_s(x, z, p) = -\frac{k_d}{2\pi} \int_{-\infty}^{\infty} \{i\zeta[A_1(\zeta) + A_2(\zeta)]e^{-pg_d(\zeta)}$$

$$- \eta'_s(\zeta)[B_1(\zeta) + B_2(\zeta)]e^{-pg_s(\zeta)}\} d\zeta,$$

(8.53)

$$\bar{w}_s(x, z, p) = -\frac{k_d}{2\pi} \int_{-\infty}^{\infty} \{\eta'_d(\zeta)[A_1(\zeta) + A_2(\zeta)]e^{-pg_d(\zeta)}$$

$$+ i\zeta[B_1(\zeta) + B_2(\zeta)]e^{-pg_s(\zeta)}\} d\zeta,$$

where A_1, A_2 are given by (8.48), (8.52), and B_1, B_2 through (8.41), (8.49), respectively, which when the factorization of $F(\zeta)$ has been carried out completes the Laplaced transformed solution.

8.2.4.3. Factorization of the function $F(\zeta)$. As Noble [8.2] points out, *solving problems with Wiener–Hopf methods ultimately boils down to a successful factorization of a kernel function.* In the SH-wave diffraction problem of § 8.2.2 the kernel function $(\zeta^2 + 1)^{\frac{1}{2}}$ was easily factored (by inspection). In the more general diffraction problems of elastodynamics the kernel function is the present $F(\zeta)$ defined in (8.46). It is essentially the Rayleigh function $R(\zeta)$. To factor this more complicated function we draw on a known general technique based on the Cauchy integral formula. We note first the only singularities of $F(\zeta)$ are its branch points at $\zeta = \pm i$, $\pm ik$. By introducing the branch cuts Re $\zeta = 0$, $1 < |\text{Im}\zeta| < k$, $F(\zeta)$ is made regular everywhere in this cut ζ-plane. There are no zeros of $F(\zeta)$ since the zeros of its numerator $\pm i\zeta_R$ are also those of its denominator. For $|\zeta| \to \infty$, $F(\zeta) = 1 + O(\zeta^{-2})$. These properties of $F(\zeta)$ make the logarithmic function $\ln F(\zeta)$ regular everywhere in the same cut ζ-plane as for $F(\zeta)$. Hence Cauchy's integral formula gives

$$\ln F(\zeta) = \frac{1}{2\pi i} \int_C \frac{\ln F(z)}{z - \zeta} dz,$$

(8.54)

where the contour C is shown in fig. 8.4, and the root functions η_d, η_s in the single-valued integrand function $\ln F(z)$ must be chosen in accord with (8.39b). Assuming first that ζ is not an imaginary number over the intervals $1 < |\zeta| < k$, since $\ln F(z) = O(z^{-2})$ as $|z| \to \infty$, the path C in (8.54) can be deformed into the sum of the paths C_U and C_L surrounding the z-plane cuts (same as those for the ζ-plane) in fig. 8.4. Then from (8.47)

Ch. 8, § 8.2] PULSE DIFFRACTION BY HALF-PLANE 503

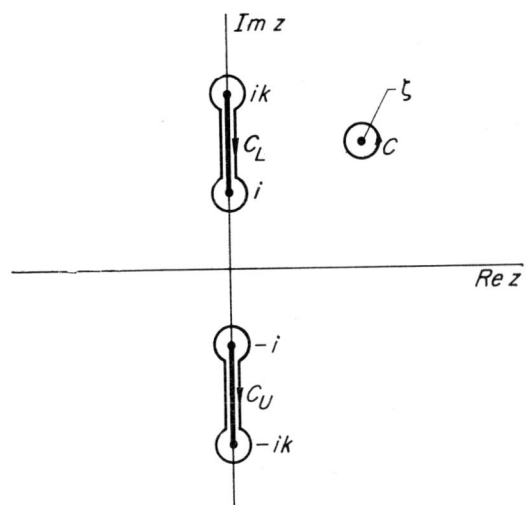

Fig. 8.4. The cut z-plane for $\ln F(z)$ and associated integration contours for the factorization of kernel function $F(\zeta)$.

$\ln F(\zeta) = \ln F_U(\zeta) + \ln F_L(\zeta)$ and we can identify each of these component functions with their respective component paths C_U and C_L. That is,

$$\ln F_U(\zeta) = \frac{1}{2\pi i} \int_{C_U} \frac{\ln F(z)}{z - \zeta} \, dz,$$

$$\ln F_L(\zeta) = \frac{1}{2\pi i} \int_{C_L} \frac{\ln F(z)}{z - \zeta} \, dz,$$

(8.55)

which completes the factorization of $F(\zeta)$. It follows from (8.55) that $F_U(\zeta)$ is a U-type function in the upper half plane $\operatorname{Im} \zeta > -1$, and $F_L(\zeta)$ an L-type function in the lower half plane $\operatorname{Im} \zeta < 1$.

It is easy to show from (8.55) that $F_L(\zeta) = F_U(-\zeta)$ and hence we need only evaluate the first of (8.55). Further, carrying out the indicated contour integration in this integral, employing complex vectors $z = \varrho \exp(i\varphi)$, $z - i = \varrho_1 \exp(i\varphi_1)$, $z + i = \varrho_2 \exp(i\varphi_2)$, etc. (cf. ch. 6 fig. 6.2) subject to (8.39b), it reduces to

$$\ln F_U(\zeta) = -\frac{1}{\pi} \int_1^k \tan^{-1} \left[\frac{4\varrho^2 (\varrho^2 - 1)^{\frac{1}{2}} (k^2 - \varrho^2)^{\frac{1}{2}}}{(k^2 - 2\varrho^2)^2} \right] \frac{d\varrho}{\varrho - i\zeta},$$

(8.56)

which can be handled by numerical integration. For the cases when ζ is just to the right or just to the left of the imaginary axis of ζ where $-k<\mathrm{Im}\ \zeta = -|\zeta|<-1$, and hence is just to the right or left of the cut in the lower half of the z-plane, we can carry out the contour integration of the first of (8.55) assuming ζ approaches the cut as a limit from the right or from the left. The integration is similar to that producing (8.56), except now the simple pole at $z=\zeta$ (on the cut in the z-plane) gives rise to a residue term and a principal value integral for each of the limiting cases. The results are

$$\ln F_U(\zeta) = \pm i \tan^{-1}\left[\frac{4|\zeta|^2(|\zeta|^2-1)^{\frac{1}{2}}(k^2-|\zeta|^2)^{\frac{1}{2}}}{(k^2-2|\zeta|^2)^2}\right]$$

$$-\frac{1}{\pi}\,\mathrm{P.\,V.}\int_1^k \tan^{-1}\left[\frac{4\varrho^2(\varrho^2-1)^{\frac{1}{2}}(k^2-\varrho^2)^{\frac{1}{2}}}{(k^2-2\varrho^2)^2}\right]\frac{d\varrho}{\varrho-i\zeta}, \qquad (8.57)$$

where the upper sign is for the case where ζ has approached the cut from the right and the lower sign for the approach from the left. Comparing (8.57) and (8.56) we see $F_L(\zeta)=F_U(-\zeta)$ holds for (8.57) too.

It is clear from the above fundamental procedure for exact factorization of $F(\zeta)$ that the use of (8.55)–(8.57) and subsequent numerical integrations can become complicated. Approximate factorization is therefore desirable when possible. Koiter's work [8.9] is of note. It's theme is the replacement of $F(\zeta)$ by an approximation, say $F_a(\zeta)$, where F and F_a have similar properties but F_a can be factored algebraically. Noble [8.2, § 4.5] discusses approximate factorization, including further detail on Koiter's work and his examples.

8.2.4.4. Inversion of formal solution and discussion of its nature.
Inversion of $\bar{u}_s$, $\bar{w}_s$ in (8.53) follows that for $\bar{v}_s$ in (8.10). We treat first the scattered dilatational waves, symmetrical and antisymmetrical (with respect to $z=0$) associated with the A_1, A_2 terms respectively, in (8.53). An analogous treatment of the B_1, B_2 terms follows, representing the symmetrical and antisymmetrical scattered equivoluminal waves. Drawing on the procedure in § 6.2.1.1 again and § 8.2.2.3, the path of integration for the A_1, A_2 integrals in (8.53) is transformed into the hyperbola

$$\zeta = \pm[(c_d t/r)^2-1]^{\frac{1}{2}}\sin\theta - i(c_d t/r)\cos\theta,\ r/c_d \le t < \infty,\ 0 \le \theta \le \pi. \qquad (8.58)$$

The path (8.58) is the upper one marked C, shown in fig. 8.3 with asymptotes defined by θ. The minimum time $t = r/c_d$ corresponds to the vertex $\zeta = -i\cos\theta$. Therefore the intersection of C and the imaginary axis always lies between $\zeta = i$ and $-i$, for the same reasons as stated in the analogous case of § 8.2.2 [cf. remarks after (8.18)]. Consider first then, the symmetric diffracted dilatational waves. To determine this contribution to $\bar{w}_s$ in (8.53), for example, $\eta_d(\zeta)A_1(\zeta)$ and $\eta_d(-\bar{\zeta})A_1(-\bar{\zeta})$, the ntegrand functions on the fourth and third quadrant parts of the path C, respectively, must be conjugate functions. Inspection of $A_1'(\zeta)$ in (8.48) shows they are conjugates if $F_L(-i\zeta_0)$ $\times F_U(\zeta)$ and $F_L(-i\bar{\zeta}_0)F_U(-\bar{\zeta})$ are. Making use of the first of (8.55), it is easily shown that $F_U(\zeta)$, $F_U(-\bar{\zeta})$ are conjugate functions. Further, making use of the relations $F_L(\zeta) = F_U(-\zeta)$ and (8.56) it can be shown that $F_L(-i\zeta_0)$ is real. Hence $\eta_d'(\zeta)A_1(\zeta)$ and $\eta_d'(-\bar{\zeta})A_1(-\bar{\zeta})$ are conjugates. Since the asrguments for $\bar{u}_s$ are similar we are led to

$$\left\{ \begin{array}{l} {}_{d1}\bar{u}_s^d(x,z,p) \\ {}_{d1}\bar{w}_s^d(x,z,p) \end{array} \right\} \quad (8.59\text{a})$$

$$= -\frac{c_d \bar{f}(p)}{\pi r} \int_{r/c_d}^{\infty} [(c_d t/r)^2 - 1]^{-\frac{1}{2}} \mathrm{Re} \left\{ \begin{bmatrix} i\omega_d \\ \eta_d'(\omega_d) \end{bmatrix} \eta_d'(\omega_d) A_1'(\omega_d) \right\} e^{-pt} dt,$$

written in terms of the fourth quadrant path so that

$$\zeta = \omega_d(t) = [(c_d t/r^2 - 1]^{\frac{1}{2}} \sin\theta - i(c_d t/r) \cos\theta \quad (8.59\text{b})$$

from (8.58). In deriving (8.59a) we have also made use of a relation like (8.21) involving

$$\eta_d'(\omega_d) = (c_d t/r) \sin\theta - i[(c_d t/r)^2 - 1]^{\frac{1}{2}} \cos\theta, \quad 0 \leq \theta \leq \pi. \quad (8.59\text{c})$$

In ${}_{d1}\bar{u}_s^d$, for example, the leading subscripts d, 1 stand for dilatational and symmetric, respectively, and the following superscript d for diffracted wave. Similar nomenclature will be used in the sequel. The corresponding displacements u_r, u_θ can be written from (8.59) with the aid of

$$u_r = u \cos\theta + w \sin\theta,$$

$$u_\theta = -u \sin\theta + w \cos\theta. \quad (8.60)$$

We find

$$\begin{Bmatrix} {}_{d1}\bar{u}^d_{rs}(r,\theta,p) \\ {}_{d1}\bar{u}^d_{\theta s}(r,\theta,p) \end{Bmatrix}$$

(8.61a)

$$= \frac{c_d \bar{f}(p)}{\pi r} \int_{r/c_d}^{\infty} \begin{Bmatrix} -[1-(r/c_d t)^2]^{-\frac{1}{2}} \\ 1 \end{Bmatrix} \mathrm{Re}\left\{ \begin{bmatrix} 1 \\ i \end{bmatrix} \eta'_d(\omega_d) A'_1(\omega_d) \right\} e^{-pt} dt.$$

Now (8.61a) have the product forms

$${}_{d1}\bar{u}^d_{rs}(r,\theta,p) = \bar{f}(p){}_{d1}\bar{U}_r(r,\theta,p),$$

$${}_{d1}\bar{u}^d_{\theta s}(r,\theta,p) = \bar{f}(p){}_{d1}\bar{U}_\theta(r,\theta,p),$$

(8.61b)

where

$$\begin{bmatrix} {}_{d1}\bar{U}_r(r,\theta,p) \\ {}_{d1}\bar{U}_\theta(r,\theta,p) \end{bmatrix} = -\frac{c_d}{\pi r} \int_{r/c_d}^{\infty} [\text{Integrands in (8.61a)}] dt.$$

(8.61c)

The symmetrical parts of the dilatational diffracted displacement waves are therefore

$$\begin{bmatrix} {}_{d1}u^d_{rs}(r,\theta,t) \\ {}_{d1}u^d_{\theta s}(r,\theta,t) \end{bmatrix} = \left\{ \int_{r/c_d}^{t} f(t-\tau) \begin{bmatrix} {}_{d1}U_r(r,\theta,\tau) \\ {}_{d1}U_\theta(r,\theta,\tau) \end{bmatrix} d\tau \right\} H(t-r/c_d), \quad 0 < \theta \leq \pi.$$

(8.62)

where ${}_{d1}U_r(r,\theta,t)$, ${}_{d1}U_\theta(r,\theta,t)$ are obtained from (8.61c), (8.61a) by inspection. Note that $\theta=0$ *(surface cut)* has not been included in (8.62), (8.61c). *Hence the present solution is the interior solution* [cf. § 6.2.1.1, (6.37), (6.38) and following remarks].

The antisymmetric counterparts of (8.62) are obtained similarly. We find

$$\begin{Bmatrix} {}_{d2}\bar{u}^d_s(x,z,p) \\ {}_{d2}\bar{w}^d_s(x,z,p) \end{Bmatrix} = \frac{c_d \bar{f}(p)}{\pi r} \int_{r/c_d}^{\infty} [(c_d t/r)^2 - 1]^{-\frac{1}{2}} \mathrm{Re}\left\{ \begin{bmatrix} i\omega_d \\ \eta'_d(\omega_d) \end{bmatrix} \eta'_d(\omega_d) A'_2(\omega_d) \right\} e^{-pt} dt.$$

(8.63a)

Ch. 8, § 8.2] PULSE DIFFRACTION BY HALF-PLANE 507

and
$$\left\{\begin{matrix}_{d2}\bar{u}^d_{rs}(r,\theta,p)\\ _{d2}\bar{u}^d_{\theta s}(r,\theta,p)\end{matrix}\right\}$$

(8.63b)

$$=\frac{c_d\bar{f}(p)}{\pi r}\int_{r/c_d}^{\infty}\left\{[1-(r/c_dt)^2]^{-\frac{1}{2}}\right\}\text{Re}\left\{\begin{bmatrix}1\\-i\end{bmatrix}\eta'_d(\omega_d)A'_2(\omega_d)\right\}e^{-pt}\mathrm{d}t.$$

The product forms for (8.63b) are

$$_{d2}\bar{u}^d_{rs}(r,\theta,p)=\bar{f}(p)_{d2}\bar{U}_r(r,\theta,p),$$

(8.64)

$$_{d2}\bar{u}^d_{\theta s}(r,\theta,p)=\bar{f}(p)_{d2}\bar{U}_\theta(r,\theta,p),$$

where

$$\begin{bmatrix}_{d2}\bar{U}_r(r,\theta,p)\\_{d2}\bar{U}_\theta(r,\theta,p)\end{bmatrix}=\frac{c_d}{\pi r}\int_{r/c_d}^{\infty}[\text{Integrands in (8.63b)}]\mathrm{d}t.\quad(8.65)$$

The antisymmetric parts of the dilatational diffracted displacement waves are therefore

$$\left\{\begin{matrix}_{d2}u^d_{rs}(r,\theta,t)\\ _{d2}u^d_{\theta s}(r,\theta,t)\end{matrix}\right\}$$

$$=\left\{\int_{r/c_d}^{t}f(t-\tau)\begin{bmatrix}_{d2}U_r(r,\theta,\tau)\\_{d2}U_\theta(r,\theta,\tau)\end{bmatrix}\mathrm{d}\tau\right\}H(t-r/c_d),\quad 0<\theta\leq\pi,\quad(8.66)$$

where $_{d2}U_r(r,\theta,t)$, $_{d2}U_\theta(r,\theta,t)$ are obtained from (8.65), (8.63b) by inspection.

The path of integration for the integrals containing B_1, B_2 representing the equivoluminal waves is transformed into the hyperbola

$$\zeta=\pm[(c_st/r)^2-1]^{\frac{1}{2}}k\sin\theta-i(c_dt/r)\cos\theta,\qquad r/c_s\leq t<\infty,0\leq\theta\leq\pi.\quad(8.67)$$

The minimum time $t=r/c_s$ corresponds to the vertex $\zeta=-ik\cos\theta$. Since $\cos\theta$ can be positive or negative ($0\leq\theta\leq\pi$), in the case of $0\leq \pm k\cos\theta<1$ the corresponding θ interval is $\cos^{-1}(1/k)<\theta<\cos^{-1}(-1/k)$ and the Cagniard path will again intersect the imaginary axis between $\zeta=i$ and $-i$ free of the cuts. Hence this is again a path like the upper one marked C in fig. 8.3 leading to regular equivoluminal waves. When $-k<k\cos\theta<-1$, which corresponds to $\cos^{-1}(-1/k)<\theta\leq\pi$, it would appear that the cut is intersected between $\zeta=i$ and ik ($\cos\theta$ is negative). However, from (8.41), (8.48) and (8.49), (8.52), we see the integrands in (8.53), associated with $B_1(\zeta)$, $B_2(\zeta)$, are regular in $-\zeta_0=-\cos\alpha<\operatorname{Im}\zeta<k$. Therefore we have the regular Cagniard path C here too. The remaining θ-interval (for $0\leq\theta\leq\pi$) is $0\leq\theta<\cos^{-1}(1/k)$ corresponding to $1<k\cos\theta\leq k$. Hence the path C would intersect the cut between $\zeta=-i$ and $-ik$. Here the integrands in (8.53), associated with $B_1(\zeta)$, $B_2(\zeta)$, have $\zeta=-i$ as a branch point due to $F_U(\zeta)$ appearing in both of these, as well as $(\zeta+i)^{\frac{1}{2}}$ in the case of $B_1(\zeta)$. Thus, as in the analogous case of § 6.2.1.1 [cf. (6.25)–(6.32), fig. 6.5], the Cagniard path has to be supplemented by the path H that approaches, but circumvents the cut as shown in fig. 8.3.

On the basis of this analysis, considerations regarding conjugation are similar to those leading to the transformed diffracted dilatational waves (8.59) and (8.63a). For the symmetric equivoluminal wave case we find for the regular waves

$$\begin{Bmatrix} {}_{s1}\bar{u}_s^d(x,z,p) \\ {}_{s1}\bar{w}_s^d(x,z,p) \end{Bmatrix}$$

$$=\frac{2c_s\bar{f}(p)}{\pi r}\int_{r/c_s}^{\infty}[(c_st/r)^2-1]^{-\frac{1}{2}}\operatorname{Re}\left\{\begin{bmatrix} i\eta_s'(\omega_s) \\ \omega_s \end{bmatrix}\frac{\omega_s\eta_d'(\omega_s)\eta_s'(\omega_s)}{k^2+2\omega_s^2}A_1'(\omega_s)\right\}e^{-pt}dt, \tag{8.68a}$$

in which

$$\omega_s(t)=[(c_st/r)^2-1]^{\frac{1}{2}}k\sin\theta-i(c_dt/r)\cos\theta, \tag{8.68b}$$

and for the head waves

$$\begin{Bmatrix} {}_{ds1}\bar{u}_s^d(x,z,p) \\ {}_{ds1}\bar{w}_s^d(x,z,p) \end{Bmatrix} = \frac{2c_s\bar{f}(p)}{\pi r}$$

$$\int_{t_{ds}}^{r/c_s}[1-(c_st/r)^2]^{-\frac{1}{2}}\operatorname{Re}\left\{\begin{bmatrix} \eta_s'(\omega_{ds}) \\ -i\omega_{ds} \end{bmatrix}\frac{\omega_{ds}\eta_d'(\omega_{ds})\eta_s'(\omega_{ds})}{k^2+2\omega_{ds}^2}A_1'(\omega_{ds})\right\}e^{-pt}dt, \tag{8.69a}$$

in which

$$\omega_{ds}(t) = i\{[1-(c_s t/r)^2]^{\frac{1}{2}} k \sin\theta - (c_d t/r)\cos\theta\}, \tag{8.69b}$$

$$t_{ds} = (r/c_d)[\cos\theta + (k^2-1)^{\frac{1}{2}}\sin\theta], \tag{8.69c}$$

where t_{ds} is the minimum time (for the head waves). In deriving (8.68a), (8.69a) we have made use of

$$\eta'_s(\omega_s) = (c_d t/r)\sin\theta - ik[(c_s t/r)^2 - 1]^{\frac{1}{2}}\cos\theta, \tag{8.70a}$$

and

$$\eta'_s(\omega_{ds}) = (c_d t/r)\sin\theta + k[1-(c_s t/r)^2]^{\frac{1}{2}}\cos\theta, \tag{8.70b}$$

respectively. For the regular equivoluminal waves corresponding to (8.68a), θ is presently restricted to $0 \le \theta \le \pi$. The head waves from (8.69a), however, only occur in the interval $0 \le \theta < \cos^{-1}(1/k)$. Now using (8.60) and (8.68), (8.69) we find for the associated polar components of the displacements

$$\begin{Bmatrix} {}_{s1}\bar{u}^d_{rs}(r,\theta,p) \\ {}_{s1}\bar{u}^d_{\theta s}(r,\theta,p) \end{Bmatrix}$$

$$= \frac{2c_d \bar{f}(p)}{\pi r}\int_{r/c_s}^{\infty}\left\{[1-(r/c_s t)^2]^{-\frac{1}{2}}\right\}\mathrm{Re}\left\{\begin{bmatrix} 1 \\ -i \end{bmatrix}\frac{\omega_s \eta'_d(\omega_s)\eta'_s(\omega_s)}{k^2+2\omega_s^2}A'_1(\omega_s)\right\}e^{-pt}dt, \tag{8.71}$$

and

$$\begin{Bmatrix} {}_{ds1}\bar{u}^d_{rs}(r,\theta,p) \\ {}_{ds1}\bar{u}^d_{\theta s}(r,\theta,p) \end{Bmatrix}$$

$$= \frac{2c_d \bar{f}(p)}{\pi r}\int_{t_{ds}}^{r/c_s}\left\{-[(r/c_s t)^2-1]^{-\frac{1}{2}}\right\}\mathrm{Re}\left\{\frac{\omega_{ds}\eta'_d(\omega_{ds})\eta'_s(\omega_{ds})}{k^2+2\omega_{ds}^2}A'_1(\omega_{ds})\right\}e^{-pt}dt. \tag{8.72}$$

The product forms for (8.71) and (8.72) are

$${}_{s1}\bar{u}^d_{rs}(r,\theta,p) = \bar{f}(p)\,{}_{s1}\bar{U}_r(r,\theta,p),$$

$${}_{s1}\bar{u}^d_{\theta s}(r,\theta,p) = \bar{f}(p)\,{}_{s1}\bar{U}_\theta(r,\theta,p), \tag{8.73}$$

510 PULSE SCATTERING BY OBSTACLES [Ch. 8, § 8.2

and
$$_{ds1}\bar{u}_{rs}^d(r, \theta, p) = \bar{f}(p)_{ds1}\bar{U}_r(r, \theta, p),$$
$$_{ds1}\bar{u}_{\theta s}^d(r, \theta, p) = \bar{f}(p)_{ds1}\bar{U}_\theta(r, \theta, p),$$
(8.74)

respectively, where

$$\begin{bmatrix} _{s1}\bar{U}_r(r, \theta, p) \\ _{s1}\bar{U}_\theta(r, \theta, p) \end{bmatrix} = \frac{2c_s}{\pi r} \int_{r/c_s}^{\infty} [\text{Integrands in (8.71)}] \, dt,$$
(8.75)

$$\begin{bmatrix} _{ds1}\bar{U}_r(r, \theta, p) \\ _{ds1}\bar{U}_\theta(r, \theta, p) \end{bmatrix} = \frac{2c_s}{\pi r} \int_{t_{ds}}^{r/c_s} [\text{Integrands in (8.72)}] \, dt.$$
(8.76)

The symmetrical parts of the equivoluminal diffracted displacement waves are therefore

$$\begin{bmatrix} _{s1}u_{rs}^d(r, \theta, t) \\ _{s1}u_{\theta s}^d(r, \theta, t) \end{bmatrix} = \left\{ \int_{r/c_s}^{t} f(t-\tau) \begin{bmatrix} _{s1}U_r(r, \theta, \tau) \\ _{s1}U_\theta(r, \theta, \tau) \end{bmatrix} d\tau \right\} H(t-r/c_s), \quad 0 < \theta \leq \pi,$$
(8.77)

and

$$\begin{bmatrix} _{ds1}u_{rs}^d(r, \theta, t) \\ _{ds1}u_{\theta s}^d(r, \theta, t) \end{bmatrix} = \left\{ \int_{t_{ds}}^{t'} f(t-\tau) \begin{bmatrix} _{ds1}U_r(r, \theta, \tau) \\ _{ds1}U_\theta(r, \theta, \tau) \end{bmatrix} d\tau \right\} H(t-t_{ds}),$$

$$0 < \theta < \cos^{-1}(1/k), \quad (8.78)$$

where t' is the minimum of t and r/c_s, and $_{s1}U_r(r, \theta, t)$, $_{s1}U_\theta(r, \theta, t)$ and $_{ds1}U_r(r, \theta, t)$, $_{ds1}U_\theta(r, \theta, t)$ are obtained from (8.75), (8.71) and (8.76), (8.72), respectively, by inspection. Note, as indicated in (8.78), that the head waves only occur in $0 \leq \theta < \cos^{-1}(1/k)$ as pointed out earlier.

The antisymmetric counterparts of (8.77), (8.78) are obtained in the same manner. We find

$$\begin{Bmatrix} _{s2}\bar{u}_s^d(x, z, p) \\ _{s2}\bar{w}_s^d(x, z, p) \end{Bmatrix}$$
(8.79)

$$= -\frac{c_s \bar{f}(p)}{2\pi r} \int_{r/c_s}^{\infty} [(c_s t/r)^2 - 1]^{-\frac{1}{2}} \text{Re} \left\{ \begin{bmatrix} i\eta_s'(\omega_s) \\ \omega_s \end{bmatrix} \frac{k^2 + 2\omega_s^2}{\omega_s} A_2'(\omega_s) \right\} e^{-pt} \, dt,$$

$$\left\{ \begin{matrix} {}_{ds2}\bar{u}_s^d(x,z,p) \\ {}_{ds2}\bar{w}_s^d(x,z,p) \end{matrix} \right\}$$

(8.80)

$$= -\frac{c_s \bar{f}(p)}{2\pi r} \int_{t_{ds}}^{r/c_s} [1-(c_s t/r)^2]^{-\frac{1}{2}} \mathrm{Re}\left\{ \begin{bmatrix} \eta_s'(\omega_{ds}) \\ -i\omega_s \end{bmatrix} \frac{k^2 + 2\omega_{ds}^2}{\omega_{ds}} A_2'(\omega_{ds}) \right\} e^{-pt} \mathrm{d}t,$$

where (8.68b) again holds for (8.79), and (8.69b, c) for (8.80), and use has again been made of (8.70). We find further that

$$\left\{ \begin{matrix} {}_{s2}\bar{u}_{rs}^d(r,\theta,p) \\ {}_{s2}\bar{u}_{\theta s}^d(r,\theta,p) \end{matrix} \right\}$$

$$= \frac{c_d \bar{f}(p)}{2\pi r} \int_{r/c_s}^{\infty} \left\{ \begin{matrix} -1 \\ [1-(r/c_s t)^2]^{-\frac{1}{2}} \end{matrix} \right\} \mathrm{Re}\left\{ \begin{bmatrix} 1 \\ i \end{bmatrix} \frac{k^2+2\omega_s^2}{\omega_s} A_2'(\omega_s^2) \right\} e^{-pt} \mathrm{d}t, \quad (8.81)$$

and

$$\left\{ \begin{matrix} {}_{ds2}\bar{u}_{rs}^d(r,\theta,p) \\ {}_{ds2}\bar{u}_{\theta s}^d(r,\theta,p) \end{matrix} \right\}$$

$$= \frac{c_d \bar{f}(p)}{2\pi r} \int_{t_{ds}}^{r/c_s} \left\{ \begin{matrix} -1 \\ [(r/c_s t)^2-1]^{-\frac{1}{2}} \end{matrix} \right\} \mathrm{Re}\left\{ \frac{k^2+2\omega_{ds}^2}{\omega_{ds}} A_2'(\omega_{ds}) \right\} e^{-pt} \mathrm{d}t. \quad (8.82)$$

The product forms for (8.81) and (8.82) are

$${}_{s2}\bar{u}_{rs}^d(r,\theta,p) = \bar{f}(p) {}_{s2}\bar{U}_r(r,\theta,p),$$

(8.83)

$${}_{s2}\bar{u}_{\theta s}^d(r,\theta,p) = \bar{f}(p) {}_{s2}\bar{U}_\theta(r,\theta,p),$$

and

$${}_{ds2}\bar{u}_{rs}^d(r,\theta,p) = \bar{f}(p) {}_{d;2}\bar{U}_r(r,\theta,p),$$

(8.84)

$${}_{ds2}\bar{u}_{\theta s}^d(r,\theta,p) = \bar{f}(p) {}_{d;2}\bar{U}_\theta(r,\theta,p),$$

respectively, where

$$\begin{bmatrix} {}_{s2}\overline{U}_r(r, \theta, p) \\ {}_{s2}\overline{U}_\theta(r, \theta, p) \end{bmatrix} = \frac{c_d}{2\pi r} \int_{r/c_e}^{\infty} [\text{Integrands in (8.81)}] dt, \qquad (8.85)$$

$$\begin{bmatrix} {}_{ds2}\overline{U}_r(r, \theta, p) \\ {}_{ds2}\overline{U}_\theta(r, \theta, p) \end{bmatrix} = \frac{c_d}{2\pi r} \int_{t_{ds}}^{r/c_s} [\text{Integrands in (8.82)}] dt. \qquad (8.86)$$

The antisymmetrical parts of the equivoluminal diffracted displacement waves are therefore

$$\begin{bmatrix} {}_{s2}u_r^d(r, \theta, t) \\ {}_{s2}u_\theta^d(r, \theta, t) \end{bmatrix} = \left\{ \int_{r/c_e}^{t} f(t-\tau) \begin{bmatrix} {}_{s2}U_r(r, \theta, \tau) \\ {}_{s2}U_\theta(r, \theta, \tau) \end{bmatrix} d\tau \right\} H(t-r/c_s), \qquad 0 \leq \theta \leq \pi,$$
(8.87)

and

$$\begin{bmatrix} {}_{ds2}u_r^d(r, \theta, t) \\ {}_{ds2}u_\theta^d(r, \theta, t) \end{bmatrix} = \left\{ \int_{t_{ds}}^{t'} f(t-\tau) \begin{bmatrix} {}_{ds2}U_r(r, \theta, \tau) \\ {}_{ds2}U_\theta(r, \theta, \tau) \end{bmatrix} d\tau \right\} H(t-t_{ds}), 0 \leq \theta < \cos^{-1}(1/k),$$
(8.88)

where again here t' is the minimum of t and r/c_s, and ${}_{s2}U_r(r, \theta, t)$, ${}_{s2}U_\theta(r, \theta, t)$ and ${}_{ds2}U_r(r, \theta, t)$, ${}_{ds2}U_\theta(r, \theta, t)$ are obtained from (8.85), (8.81) and (8.86), (8.82), respectively, by inspection. Here again the head waves occur only in $0 \leq \theta < \cos^{-1}(1/k)$ as indicated in (8.88). This completes the interior solution for the diffracted waves for $0 \leq r < \infty$, $0 < \theta \leq \pi$. To extend it to $\pi \leq \theta < 2\pi$ we draw on the underlying symmetry relations, i.e.,

$$\begin{aligned} {}_1U_r(r, 2\pi-\theta, t) &= {}_1U_r(r, \theta, t), \\ {}_1U_\theta(r, 2\pi-\theta, t) &= -{}_1U_\theta(r, \theta, t), \end{aligned} \qquad (8.89)$$

for symmetric displacements with respect to $z=0$, and

$$\begin{aligned} {}_2U_r(r, 2\pi-\theta, t) &= -{}_2U_r(r, \theta, t), \\ {}_2U_\theta(r, 2\pi-\theta, t) &= {}_2U_\theta(r, \theta, t), \end{aligned} \qquad (8.90)$$

for antisymmetric displacements with respect to $z=0$.

The surface response [for $\theta=0$, with $\theta=2\pi$ by symmetry relations (8.89), (8.90)] is derived like in § 6.2.1.1 [cf. (6.39)–(6.44) and discussion therein]. We note again that with $\theta=0$, (8.59b), (8.68b) and (8.69b) degenerate to $\omega_d=\omega_s=\omega_{ds}=-ic_d t/r$ thereby wrapping the Cagniard contour upper C, and lower C plus H, in fig. 8.3 about the cut along the imaginary axis, $-\infty<\text{Im }\zeta<-1$. Hence again here the Rayleigh pole at $\zeta=-i\zeta_R$ creates the differences between the surface and interior solution, i.e. the surface solution is the interior solution evaluated at $\theta=0$, in which the integrals of (8.61c), (8.65), (8.75) and (8.85) become principal value integrals and where in addition we have residue terms from the Rayleigh pole [cf. (6.47), (6.48)]. One would expect the Rayleigh waves here to be similar to those in Lamb's line load problem [cf. (6.47), (6.54)].

To complete the solution we need the geometrical solution of the diffraction problem which stems from the pole at $\zeta=-i\zeta_0=-i\cos\alpha$. Adding the incident waves to this gives the complete geometrical solution for the problem, e.g., $u^g=u_s^g+u_i$. Again assuming $0\le\theta\le\pi$ at the outset, we treat first the geometrical dilatational waves. These are contained in

$$\begin{Bmatrix} {}_d\bar{u}_s(x,z,p) \\ {}_d\bar{w}_s(x,z,p) \end{Bmatrix} = -\frac{k_d}{2\pi}\int_{-\infty}^{\infty}\begin{Bmatrix} i\zeta \\ \eta'_d(\zeta) \end{Bmatrix}[A_1(\zeta)+A_2(\zeta)]e^{-pg_d(\zeta)}d\zeta, \tag{8.91}$$

from (8.53). Assuming completion along the upper path C in fig. 8.3 (which gave our diffracted dilatational waves) we will have a nonvanishing residue term if $\cos\alpha<\cos\theta$, or $0\le\theta<\alpha$, i.e., the pole lies inside the closed contour. The residue terms yield

$$\begin{Bmatrix} {}_du_s^g(r,\theta,t) \\ {}_dw_s^g(r,\theta,t) \end{Bmatrix} = -\begin{Bmatrix} \cos\alpha \\ \sin\alpha \end{Bmatrix} f[t-(r/c_d)\cos(\theta-\alpha)], \tag{8.92}$$

which cancels the incident waves (8.32), hence the complete geometrical solution in this region is

$$_du^g(r,\theta,t) = {}_dw^g(r,\theta,t) = 0. \tag{8.93}$$

Now for $-1\le\cos\theta<\cos\alpha$ or $\alpha<\theta\le\pi$, the pole lies outside the closed contour. Therefore in this region the complete geometrical solution is given by

$$_du^g(r,\theta,t)=u_i, \qquad _dw^g(r,\theta,t)=w_i. \tag{8.94}$$

Making use of the basic symmetry relations

$$u_s(r, 2\pi-\theta, t) = \pm u_s(r, \theta, t),$$
$$w_s(r, 2\pi-\theta, t) = \mp w_s(r, \theta, t), \tag{8.95}$$

we find for $\pi \leq \theta < 2\pi - \alpha$ that again (8.94) holds. Lastly for $2\pi - \alpha < \theta \leq 2\pi$, from (8.95) we have

$$\begin{Bmatrix} _du_s^g(r, 2\pi-\theta, t) \\ _dw_s^g(r, 2\pi-\theta, t) \end{Bmatrix} = \begin{Bmatrix} _{d1}u_s^g(r, \theta, t) - _{d2}u_s^g(r, 0, t) \\ -_{d1}w_s^g(r, \theta, t) + _{d2}w_s^g(r, \theta, t) \end{Bmatrix}, \tag{8.96}$$

and identifying each of the terms on the right with each of terms in (8.91), and the roles the latter play in the residue calculation leading to (8.92), we have

$$\begin{Bmatrix} _du_s^g(r, 2\pi-\theta, t) \\ _dw_s^g(r, 2\pi-\theta, t) \end{Bmatrix} = \begin{Bmatrix} \cos\alpha \\ \sin\alpha \end{Bmatrix} \{\pm R_{dd}\} f[t - (r/c_d)\cos(\theta+\alpha)], \tag{8.97a}$$

where R_{dd} is the reflection coefficient of the reflected P wave from a P wave incident at a free plane boundary given by

$$R_{dd} = -\frac{k^2\cos^2 2\beta - \sin 2\alpha \sin 2\beta}{k^2\cos^2 2\beta + \sin 2\alpha \sin 2\beta}, \tag{8.97b}$$

in agreement with A_2/A_1 in (3.21a), (3.18). We note also the incident waves u_i, w_i are present in this θ interval.

The geometrical equivoluminal waves are obtained similarly. Here residue evaluation at $\zeta = -i\zeta_0$ is applied to the integrals associated with $B_1(\zeta) + B_2(\zeta)$ in (8.53). Assuming completion along the upper path C in fig. 8.3 vanishing residue terms show $_su_s^g = _sw_s^g = 0$ in $0 \leq \theta < \beta$, i.e., there is no reflected equivoluminal wave in the shadow zone as would be expected. Now for $\beta < \theta \leq \pi$, the pole is outside the closed contour and therefore gives no contributions to the geometrical solution, i.e., just u_i and w_i exist here. The symmetries of the displacements show contributions occur only in $2\pi - \beta < \theta \leq 2\pi$. In this θ-interval, using relations like (8.96), we find

Ch. 8, § 8.2] PULSE DIFFRACTION BY HALF-PLANE 515

$$\begin{Bmatrix} {_s}u_s^g(r, 2\pi-\theta, t) \\ {_s}w_s^g(r, 2\pi-\theta, t) \end{Bmatrix} = \begin{Bmatrix} \sin\beta \\ \cos\beta \end{Bmatrix} R_{ds} f[t-(r/c_s)\cos(\theta+\beta)], \qquad (8.98\text{a})$$

where R_{ds} is the displacement reflection coefficient of the reflected SV wave from a P wave incident at a free plane boundary given by

$$R_{ds} = -\frac{2k\sin 2\alpha \cos 2\beta}{k^2 \cos^2 2\beta + \sin 2\alpha \sin 2\beta}, \qquad (8.98\text{b})$$

in agreement with $-kA_4/A_1$ in (3.21b), (3.19). When the closed contour involves the lower path C in fig. 8.3, then $1 < k\cos\theta \leq k$ which leads to the same results as above for $0 \leq \theta < \beta$ and $2\pi-\beta < \theta \leq 2\pi$, i.e., ${_s}u_s^g = {_s}w_s^g = 0$ and (8.98), respectively. On the basis of the above results the complete geometrical solution $u^g(r, \theta, t)$, $w^g(r, \theta, t)$ is

$$\begin{Bmatrix} u^g(r, \theta, t) \\ w^g(r, \theta, t) \end{Bmatrix} \qquad (8.99)$$

$$= \begin{cases} 0, & 0 \leq \theta < \alpha, \\ \begin{bmatrix} u_i(r, \theta, t) \\ w_i(r, \theta, t) \end{bmatrix} = \begin{bmatrix} \cos\alpha \\ \sin\alpha \end{bmatrix} f[t-(r/c_d)\cos(\theta-\alpha)], & \alpha < \theta < 2\pi-\beta, \\ \begin{bmatrix} u_i(r, \theta, t) \\ w_i(r, \theta, t) \end{bmatrix} + R_{ds} \begin{bmatrix} \sin\beta \\ \cos\beta \end{bmatrix} f[t-(r/c_s)\cos(\theta+\beta)], & 2\pi-\beta < \theta < 2\pi-\alpha, \\ \begin{bmatrix} u_i(r, \theta, t) \\ w_i(r, \theta, t) \end{bmatrix} + R_{ds} \begin{bmatrix} \sin\beta \\ \cos\beta \end{bmatrix} f[t-(r/c_s)\cos(\theta+\beta)] \\ \qquad + R_{dd} \begin{bmatrix} \cos\alpha \\ -\sin\alpha \end{bmatrix} f[t-(r/c_d)\cos(\theta+\alpha)], & 2\pi-\alpha < \theta \leq 2\pi, \end{cases}$$

where R_{dd}, R_{ds} are given in (8.97b), (8.98b), respectively, and $f(\tau) = 0$ for $\tau < 0$. The cases $\theta = \alpha$, $2\pi-\beta$ and $2\pi-\alpha$ are handled by using the arithmetical mean of the values on each side of these special θ values [cf. (8.28)]. The polar displacement components u_r^g, u_θ^g are obtained by substituting (8.99) into (8.60). The complete solution for the problem in terms of the polar displacement components $u_r(r, \theta, t)$, $u_\theta(r, \theta, t)$ is

$$u_r(r, \theta, t) = u_{rs}^d(r, \theta, t) + u_r^g(r, \theta, t),$$
$$\qquad\qquad\qquad\qquad\qquad\qquad\qquad\qquad\qquad (8.100)$$
$$u_\theta(r, \theta, t) = u_{\theta s}^d(r, \theta, t) + u_\theta^g(r, \theta, t),$$

where u_{rs}^d, $u_{\theta s}^d$ are sums obtained from like terms in (8.62), (8.66), (8.77), (8.78), (8.87) and (8.88) for $0<\theta\leq\pi$ (the interior solution). Incorporating the modifications generated by the Rayleigh pole [discussed after (8.90)] extends the interior solution to the surface $\theta=0$. Then extension to the complete θ interval $0\leq\theta\leq2\pi$ is accomplished with the aid of the symmetry relations (8.89), (8.90). Hence the solution (8.100) is valid for $0\leq r<\infty$, $0\leq\theta\leq2\pi$ and $0\leq t<\infty$.

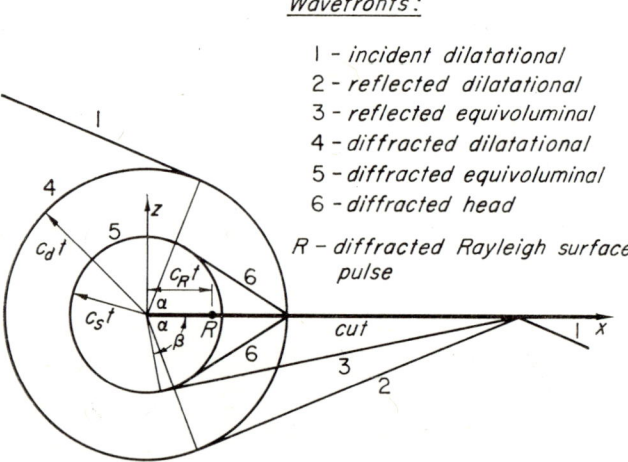

Fig. 8.5. Wavefronts in the diffraction of a plane dilatational pulse by a traction free half plane (α-angle of incidence and reflection of dilatational waves, β-angle of reflection of equivoluminal wave).

Figure 8.5 depicts the wavefronts for the events of the solution. We see the edge of the cut generates a diffracted wave system similar to that in Lamb's line load problem (cf. fig. 6.7), having regular dilatational and equivoluminal waves, a two-sided equivoluminal wave, a head wave and a Rayleigh pulse. Note also that the plane reflected dilatational and equivoluminal waves are supersonic with respect to the diffracted dilatational wave.

Much algebraic and numerical evaluation work remains to be done in the present solution to obtain detailed information on the nature of the

Ch. 8, § 8.3]   SCATTERING BY CYLINDRICAL OBSTACLES   517

waves involved. For the special cases $\theta=0$, $\theta=\pi/2$ and π evaluation is simpler as it was in Lamb's line load problem. For other θ-values one can obtain wavefront expansions without much difficulty using the techniques discussed in §§ 6.2.1.4, 6.2.1.5.

DeHoop [2.19, §§ 12,13] also treats the case of diffraction of a plane dilatational pulse by a rigid half plane. Derivation of the solution to this problem can be carried out by the present technique. The basic nature of this derivation is left to the exercises. Of course the technique is also applicable to the analogous problems involving an incident plane equivoluminal pulse.

An important *alternate method for solving the two-dimensional diffraction problems* of interest here has been given by Miles [8.10]. He has formulated Buseman's *method of conical flows* (more generally referred to as the method of self-similarity) *for two-dimensional elastic wave propagation problems*. The *method employs homogeneous solutions* which are applicable in problems not having a characteristic length, or if one is involved it must be derived from a parameter to which the solution must be proportional. (Note therefore that Lamb's line load problem for the half space can be handled by the method.) In [8.10] Miles treats the present problem of the plane dilatational pulse diffracted from a traction free cut, and the similar problem for the plane equivoluminal pulse. The reader will be interested in his numerical results in both problems for the distribution (with θ) of (1) radial velocity along the diffracted dilatational wavefront, and (2) tangential velocity along the diffracted equivoluminal wavefront, resulting from normal incidence. The reader should also note that the recent book by Pao and Mow [8.11] on diffraction of elastic waves contains a treatment of the present class of problems based on homogeneous solutions. Pao and Mow present a history of elastic wave diffraction which will also be of interest to the reader.

8.3. Elastic pulse scattering by cylindrical and spherical obstacles

8.3.1 Introduction

The present class of problems have their background in *acoustics and seismology*, where in the latter a good share of the work has been on time-harmonic excitation. Gains in the related more difficult problems of transient excitation have been made only in recent years, due primarily to the quest for information in *protective construction* problems (cf. [4.4], pp. 830–32). Pao and Mow [8.11] present a comprehensive survey and treatment of the subject. The present treatment focuses on the transient problems with a detailed look at the solutions and the methods that produce them.

8.3.2 Scattering of an elastic pulse by a circular cylindrical cavity

8.3.2.1. Line load source; general features of the wave system. Consider the infinite elastic solid with a circular section cylindrical cavity that shown in fig. 8.6. The cavity is infinitely long, of radius a and has its axis along $r=0$. We assume a line source S to the right of the cavity at $x=x_0(x_0>a)$. The problem is one of plane strain with coordinates (r, θ, t). In fig. 8.6 it is assumed that one wave system is active with a wavefront velocity c. The wavefronts involved are depicted for two different fixed times t_1, t_2 where $t_1<t_2$. In the figure the numbers 1, 2 and 3 indicate, respectively, the incident, reflected and diffracted wavefronts (solid lines). The associated rays (dashed lines) are indicated by the numbers $\hat{1}$, $\hat{2}$ and $\hat{3}$, respectively. Since there is

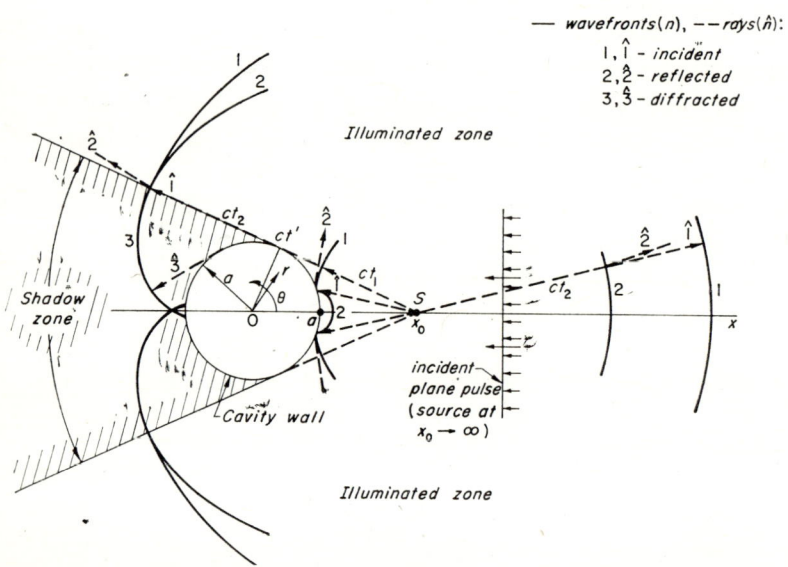

Fig. 8.6. Wavefronts, rays and wave regions in the scattering of an elastic pulse from a circular cylindrical cavity.

symmetry with respect to the x-axis, only the wavefronts and rays in the upper half of the figure have been numbered. The *shadow zone,* as fig. 8.6 shows (cross hatched in figure) is bounded by the rays $\widehat{1}$, outward from their point of tangency with the cavity wall at $ct' = (x_0^2 - a^2)^{\frac{1}{2}}$, and the part of the cavity wall defined by $\cos^{-1}(a/x_0) \leq \theta \leq 2\pi - \cos^{-1}(a/x_0)$. The illuminated zone covers the rest of the plane for $r \geq a$. For the time $t_1 (t_1 < t')$ it may be seen that the wavefronts and rays are 1, $\widehat{1}$ and 2, $\widehat{2}$, the incident and reflected pairs just to the left of S. Note these are confined to the illuminated zone because time t_1 is too short for a disturbance to be created in the shadow zone (note the incident wavefront 1 is a complete circle, but only that part to the left of S has been shown for the sake of simplicity in the figure. For time $t_2(t_2 > t')$, we have the outer system of wavefronts and rays 1, $\widehat{1}$, 2, $\widehat{2}$ and 3, $\widehat{3}$ to the left of S and 1, $\widehat{1}$ and 2, $\widehat{2}$ to the right (the latter are really continuations of those to the left, the portions between again not being shown for simplicity). The new wavefronts and rays 3, $\widehat{3}$ created here are due to diffraction by the cavity. The source of these diffracted wavefronts and rays is the point ct' on the cavity wall, i.e., when 1, 2 touch ct'. The diffracted wavefront 3 is the involute of the disturbed portion of the cavity wall between ct' and the leading edge of this wavefront. The rays $\widehat{3}$ of course must be perpendicular to the wavefront and hence the disturbed portion of the cavity wall is the envelope of these rays. Since the exciting source of the diffracted wavefront is the line source at the point ct', one would expect it to experience two-dimensional spatial decay. It is of interest to point out further that as time grows the edge of 3, and its lower half-plane mate, progress further along the cavity wall, out into the illuminated zone and back into the shadow zone periodically ad infinitum. This creates a *continuous spiraling diffracted wave* with its edge always on the cavity wall and its tail approaching r infinite. These are the fundamentals of wave propagation in the cylindrical cavity scattering problem. In the elastic case two basic wave systems are at play but fundamentally they both behave as in the system we have just discussed.

Two important limiting cases for the position of the line source S are indicated in fig. 8.6, namely when $x_0 \to a$ or $x_0 \to \infty$. The first case is the line source on the cavity wall at $x = a$. This is basically a *Lamb type problem for the exterior space* $r \geq a$. It was treated by Miklowitz [8.12], and will be discussed in detail in the sequel. The case of $x_0 \to \infty$ is equivalent to the plane incident pulse, traveling toward the cavity. This problem was treated by Baron and Matthews [8.13], Baron and Parnes [8.14] and later by

Miklowitz [8.15] and Peck and Miklowitz [8.16]. It will be discussed in detail later, including comparisons of the methods and results used in the quoted works. The general case where S is at x_0, but the obstacle is the rigid cylinder, has been treated by Gilbert and Knopoff [8.17] for wavefront approximations. We will discuss this in a later section on approximations in the incident plane wave case.

8.3.2.2. Friedlander's representation of solution. It is known for harmonic wave diffraction problems, involving a circular cylindrical or spherical scatterer, that in the shadow zone a solution based on *Fourier series converges more and more slowly as the frequency is increased*. The related difficulty in transient excitation occurs in the response at early, and out to moderately long, times which are associated with high frequencies. A method of solution suitable for short-time response was developed by Friedlander [8.18, 2.10 ch. 6]. His representation of the solution can be found through an application of *Poisson's summation formula*. This formula can be written as

$$\sum_{n=-\infty}^{\infty} g(n) = \sum_{m=-\infty}^{\infty} \int_{-\infty}^{\infty} g(\eta) e^{i2m\pi\eta} d\eta . \tag{8.101}$$

As proved in Titchmarsh [5.2 § 2.8], sufficient conditions for the formula are that (1) $g(\eta)$ is of bounded variation in $(-\infty, \infty)$, (2) $g(\eta)$ tends to 0 as $\eta \to \pm\infty$, and (3) the integral

$$\int_{-\infty}^{\infty} g(\eta) d\eta \tag{8.102}$$

exists. The condition of bounded variation on $g(\eta)$ can be relaxed to functions of square integrability, i.e., functions that satisfy

$$\int_{-\infty}^{\infty} |g(\eta)|^2 d\eta < \infty, \tag{8.103}$$

as proved in Morse and Feshbach [2.7, Part I, pp. 466-467].

Applied to the Fourier series representation of the response function $f(r, \theta, t)$, (8.101) gives

$$f(r, \theta, t) = \sum_{n=-\infty}^{\infty} F(r, n, t) e^{in\theta} = \sum_{m=-\infty}^{\infty} \int_{-\infty}^{\infty} F(r, \eta, t) e^{i\eta(\theta + 2m\pi)} d\eta$$

$$= \sum_{m=-\infty}^{\infty} f^*(r, \theta+2m\pi, t), \qquad (8.104)$$

where

$$f^*(r, \theta, t) = \int_{-\infty}^{\infty} F(r, \eta, t)e^{i\eta\theta}d\eta.$$

We refer to f^* as *the wave form of f*, and to the *sum on m in (8.104) as the wave sum*.

The wave form of the response, f^*, has a clear physical interpretation. This response is the disturbance propagating outward in θ with the wavefronts behaving geometrically as discussed earlier, i.e., the diffracted fronts of f^* wind around the cavity. From his wavefront expansions, Friedlander found that f^* is identically zero for θ's beyond the wavefront, therefore, for finite t, the sum on m is finite. Thus, as we noted earlier f^* overlaps itself as it winds around the cavity, and the wave sum on m is simply the sum of the overlapping responses. Both the wave sum f and the wave form f^* are defined on $-\infty < \theta < \infty$, but f^* is not periodic in θ; however, the wave sum on m gives the total solution the 2π periodicity in θ that is phycsially required. This may be seen by asking for the value of f at a θ that is outside the usual physical range, say $\theta = 4\pi$ instead of $\theta = 0$. All this means is that in (8.104) the terms that contribute to the total solution differ (from the solution for $\theta = 0$) by 2 in their value of m, but the number of terms and the values of the individual terms are identical to the $\theta = 0$ case. It is of further interest to note that one can interpret (8.104) as giving definition to solution f on a Riemann surface having the origin as branch point with sheets $(2m-1)\pi < \theta < (2m+1)\pi$, $m = 0, \pm 1, \pm 2,...$ In this the source, say at $(a, 0)$, is represented by an infinity of sources at $(a, 2m\pi)$ which corresponds physically to the already discussed fact that a single source can signal to a receiver not only through a direct ray but by rays corresponding to the diffracted waves that wrap themselves around the cavity. Finally, it is important to note that *to solve for the total response f one needs to solve only for f^* corresponding to the physical plane $m=0$*. Simple substitutions for the higher m terms, together with (8.104), then give f. This will be demonstrated in the sequel.

8.3.2.3. *Normal line load source on cavity wall; formal solution.* Consider in fig. 8.6 the case of a normal line load $PF(t)$ applied suddenly, at time $t=0$, to the cavity wall at $r(=x)=a$, $\theta = 0$; P is a magnitude constant of dimensions force per unit length and $F(t)$ prescribes the time behavior of the input. The problem is one of plane strain ($u_z = \partial/\partial z = 0$, where u_z is the displacement component in the z-direction). The boundary value problem may be stated as

$$\nabla^2 \varphi(r, \theta, t) = \ddot\varphi/c_d^2, \qquad \nabla^2 \psi(r, \theta, t) = \ddot\psi/c_s^2,$$
(8.105)
$$r > a, \qquad -\infty < \theta < \infty, \qquad t > 0,$$

$$\varphi(r, \theta, 0) = \psi(r, \theta, 0) = \dot\varphi(r, \theta, 0) = \dot\psi(r, \theta, 0) = 0, \qquad r \geq a, \ -\infty < \theta < \infty,$$
(8.106)

$$\sigma_r(a, \theta, t) = -PF(t)\delta_s(\theta)/a, \qquad \sigma_{r\theta}(a, \theta, t) = 0, \qquad -\infty < \theta < \infty, \ t > 0,$$
(8.107)

$$\lim_{r \to \infty, \text{ and/or } \theta \to \pm\infty} [\varphi(r, \theta, t), \psi, u_r, u_\theta, \text{etc.}] = 0, \qquad t > 0, \qquad (8.108)$$

where ∇^2 is given by (4.75) without the last term there. The equations (8.104)–(8.108) represent, respectively, the governing wave equations on the potentials (2.7b), initial conditions, boundary conditions for the cavity wall, and the radiation condition. The associated displacement-potential relations are

$$u_r = \partial\varphi/\partial r + \partial\psi/r\partial\theta, \qquad u_\theta = \partial\varphi/r\partial\theta - \partial\psi/\partial r, \qquad (8.109\text{a})$$

and the stress-potential relations

$$\sigma_r = \frac{\lambda}{c_d^2}\ddot\varphi + 2\mu\left[\frac{\partial^2\varphi}{\partial r^2} + \frac{\partial}{\partial r}\left(\frac{\partial\psi}{r\partial\theta}\right)\right],$$

$$\sigma_\theta = \frac{\lambda}{c_d^2}\ddot\varphi + \frac{2\mu}{r}\left[\frac{\partial^2\varphi}{r\partial\theta^2} + \frac{\partial\varphi}{\partial r} - \frac{\partial^2\psi}{\partial r\partial\theta} + \frac{1}{r}\frac{\partial\psi}{\partial\theta}\right],$$
(8.109b)

$$\sigma_{r\theta} = \mu\left[\frac{\ddot\psi}{c_s^2} + 2\left(\frac{\partial^2\varphi}{r\partial r\partial\theta} - \frac{\partial\varphi}{r^2\partial\theta} - \frac{\partial^2\psi}{\partial r^2}\right)\right],$$

$$\sigma_z = \nu(\sigma_r + \sigma_\theta).$$

Noting we have the radius of the cavity a as a characteristic length here, the problem (8.105)–(8.108) can be solved with the double integral transform methods of § 7.2.1 for two-dimensional waveguide problems. In particular,

SCATTERING BY CYLINDRICAL OBSTACLES

we employ the Laplace transform pair (5.6), and the real argument exponential Fourier transform pair (5.39) with $\varkappa$ and x there replaced by v and θ respectively. The procedure in § 7.2.1 then produces the transformed governing equations (8.105) as

$$r^2 \frac{d^2\bar{\varphi}^{\sim *}(r, v, p)}{dr^2} + r \frac{d\bar{\varphi}^{\sim *}}{dr} - (r^2 k_d^2 + v^2)\bar{\varphi}^{\sim *} = 0,$$

$$r^2 \frac{d^2\bar{\psi}^{\sim *}(r, v, p)}{dr^2} + r \frac{d\bar{\psi}^{\sim *}}{dr} - (r^2 k_s^2 + v^2)\bar{\psi}^{\sim *} = 0,$$

(8.110a)

where the super asterisk ( )* implies these quantities are the wave forms of ( ). General solutions of the Bessel equations (8.110a) are $I_v(k_{d,s}r)$, $K_v(k_{d,s}r)$ but the radiation condition (8.108) requires we select the K_v functions. Hence we have the general solutions

$$\bar{\varphi}^{\sim *}(r, v, p) = \alpha(v, p) K_v(k_d r),$$

$$\bar{\psi}^{\sim *}(r, v, p) = \beta(v, p) K_v(k_s r),$$

(8.110b)

where $k_d = p/c_d$, $k_s = p/c_s$. Making use of (8.109) then produces the formal solution for the displacements in the physical plane ($m = 0$)

$$\left.\begin{array}{c} u_r^*(r, \theta, t) \\ u_\theta^*(r, \theta, t) \end{array}\right\} = \frac{1}{4\pi^2 i} \int_{Br} e^{pt} \left[\int_{-\infty}^{\infty} \left\{\begin{array}{c} \bar{u}_r^{\sim *}(r, v, p) \\ \bar{u}_\theta^{\sim *}(r, v, p) \end{array}\right\} e^{-iv\theta} dv \right] dp, \quad (8.111a)$$

where

$$\bar{u}_r^{\sim *}(r, v, p) = \Gamma(v, p)[k_d A K_v'(k_d r) + (v/r) B K_v(k_s r)], \quad (8.111b)$$

$$\bar{u}_\theta^{\sim *}(r, v, p) = -i\Gamma(v, p)[(v/r) A K_v(k_d r) + k_s B K_v'(k_s r)], \quad (8.111c)$$

$$A = c_d^2[K_{v+2}(v) + K_{v-2}(v)], \qquad B = c_s^2[K_{v+2}(u) - K_{v-2}(u)],$$

$$\Gamma(v, p) = \frac{Pf(p)}{\mu a p^2 K_v(u) K_v(v) C(v, p)},$$

$$C(v, p) = \frac{K_{v-2}(u)}{K_v(u)} \cdot \frac{K_{v+2}(v)}{K_v(v)} + \frac{K_{v+2}(u)}{K_v(u)} \cdot \frac{K_{v-2}(v)}{K_v(v)}$$

$$- (1 - k^2) \left[\frac{K_{v+2}(v)}{K_v(v)} + \frac{K_{v-2}(v)}{K_v(v)} \right],$$

where $f(p)$ is the Laplace transform of $F(t)$, $u=k_d a$, $v=k_s a$, $k=c_d/c_s$, and the prime denotes differentiation with respect to the arguments of the K_ν functions. Note that the form of the transformed solutions (8.111b, c) restricts $K_\nu(u)$ and $K_\nu(v)$ to nonvanishing values.

8.3.2.4. Normal line load source on cavity wall; exact inversion. Since the K_ν functions are entire functions of their order ν (cf. Erdelyi *et al.* [2.8] §§ 7.2.1, 7.2.2, 7.2.5), (8.111a) can be inverted exactly through residue theory and contour integration a la the method of § 7.2.1.1 for inverting the spatial transform first. Since the problem is one of symmetry about $\theta=0$ it suffices to consider $\theta>0$. We then complete the integration path of the inner integral in (8.111a) in the lower half ν-plane, hence restricting the roots or branches of $C(\nu,p)$ to Im $\nu_j(p)<0$ for Re $p>0$ to insure convergence. Residue theory then reduces (8.111a) to

$$\begin{Bmatrix} u_r^*(r,\theta,t) \\ u_\theta^*(r,\theta,t) \end{Bmatrix} = -\frac{1}{2\pi i}\int_{Br} e^{pt}\left\{ i\sum_{j=1,2,\ldots}\left\{\begin{bmatrix} N_r(r,\nu,p) \\ N_\theta(r,\nu,p) \end{bmatrix}\frac{e^{-i\nu\theta}}{\frac{\partial C}{\partial \nu}}\right\}_{\nu=\bar{\nu}_j(p)} \right\}dp, \tag{8.112}$$

where $\nu_j(p)=(\nu_r+i\nu_i)_j$ in terms of its real and imaginary parts with $\nu_i>0$, and $N_r=C\bar{u}_r^*$, $N_\theta=C\bar{u}_r^*$. It is known that *residue series*, such as that in (8.112), have *usefulness in the present class of diffraction problems only in the shadow zone*. Friedlander [8.18, pp. 717, 728–731, also 2.10 pp. 147–154] for example, shows in the related general source problem in acoustics, that the terms of an analogous series, when approximated for large p with θ in the *illuminated zone*, do not exist as Laplace transforms. Friedlander treats this case then, not by residues, but by approximating the Laplace transform asymptotically and using the *method of steepest descents*, to get the reflected wavefront. Gilbert and Knopoff [8.17] did similarly for their elastic wave problem.

Restricting (8.112) to shadow zone response we can invert the Br integral there by termwise contour integration, as we did in waveguide problems earlier, by completing the Bromwich contour to the left. Since the K_ν functions and their derivatives in the integrands of (8.112) have a common (logarithmic) branch point at $p=0$, and since the branches $\nu_j(p)$ of the frequency equation $C(\nu,p)=0$ also have a similar branch point there and no other singularities in the p-plane (as will be shown shortly), we can cut this plane along the negative real axis. Then the Br contour can be completed up the imaginary axis ($p = \pm i\omega$) as in § 7.2.1.1. Here, however, with a branch point only at $p=0$, the integration is like that in § 5.11.2 (the pressurized cylindrical cavity problem) and the contour is that of fig. 5.7. Noting that conjugation for the paths L_1, L_2 can involve only the branches $-v_j(-i\omega)$ and $\bar{v}_j(i\omega)$, we find using the basic relations

$$K_{-\nu}(z) = K_\nu(z), \qquad K_{\bar{\nu}}(\bar{z}) = \overline{K_\nu(z)}, \tag{8.113a}$$

which follow from inspection of the integral representation for $K_\nu(z)$ (cf. Copson [8.19]), and

$$K_\nu(-ix) = i(\pi/2) \exp(i\pi\nu/2) H_\nu^{(1)}(x), \tag{8.113b}$$

that (8.112) reduces to

$$\begin{Bmatrix} u_r^*(r, \theta, t) \\ u_\theta^*(r, \theta, t) \end{Bmatrix}$$

(8.114a)

$$= -\frac{1}{\pi} \int_0^\infty \operatorname{Re} \sum_{j=1,2,\ldots} i \begin{bmatrix} N_r[r, -v_j(-i\omega), -i\omega] \\ N_\theta[r, -v_j(-i\omega), -i\omega] \end{bmatrix} \frac{e^{i[v_j(-i\omega)\theta - \omega t]}}{\left.\frac{\partial C}{\partial \nu}\right|_{\substack{\nu = -v_j(-i\omega) \\ p = -i\omega}}} d\omega,$$

(8.114a)

where the roots $v_j(-i\omega)$ are found from

$$C\big|_{p=-i\omega} = \frac{H_{\nu-2}^{(1)}(\eta)}{H_\nu^{(1)}(\eta)} \cdot \frac{H_{\nu+2}^{(1)}(\zeta)}{H_\nu^{(1)}(\zeta)} + \frac{H_{\nu+2}^{(1)}(\eta)}{H_\nu^{(1)}(\eta)} \cdot \frac{H_{\nu-2}^{(1)}(\zeta)}{H_\nu^{(1)}(\zeta)}$$

(8.114b)

$$-(k^2-1) \left[\frac{H_{\nu+2}^{(1)}(\zeta)}{H_\nu^{(1)}(\zeta)} + \frac{H_{\nu-2}^{(1)}(\zeta)}{H_\nu^{(1)}(\zeta)} \right] = 0,$$

where $\eta = \omega a/c_d$ and $\zeta = \omega a/c_s$. The integrands of (8.114a) correspond to component diffracted and radiated harmonic waves, one for each mode of propagation pair $[\omega, v_j(-i\omega)]$ traveling in the positive θ-direction, and outward in r, the latter stemming from the $H_\nu^{(1)}(k_d r)$, $H_\nu^{(1)}(k_s r)$ character of the functions N_r, N_θ. Figure 8.7 shows the position of two such diffracted waves (i.e., the negative θ-traveling wave from the $\theta<0$ solution is also shown) corresponding to a time t when these waves have already begun their second trip around the cavity. It is clear that the Riemann surface sheets $m = \pm 1$ in addition to $m = 0$ are involved here, hence the corresponding terms in (8.104).

As we have seen in the waveguide problem of § 7.2.1, a solution such as (8.114a) can be evaluated by numerical integration once the lower branches of C in (8.114b) have been found numerically. As we have discussed for waveguides this permits a detailed study to be made of the various waves in the problem with an eye on both wave number v and frequency ω. In a later application to the problem of diffraction of a plane dilatational pulse by the circular cylindrical cavity, we shall see these numerically

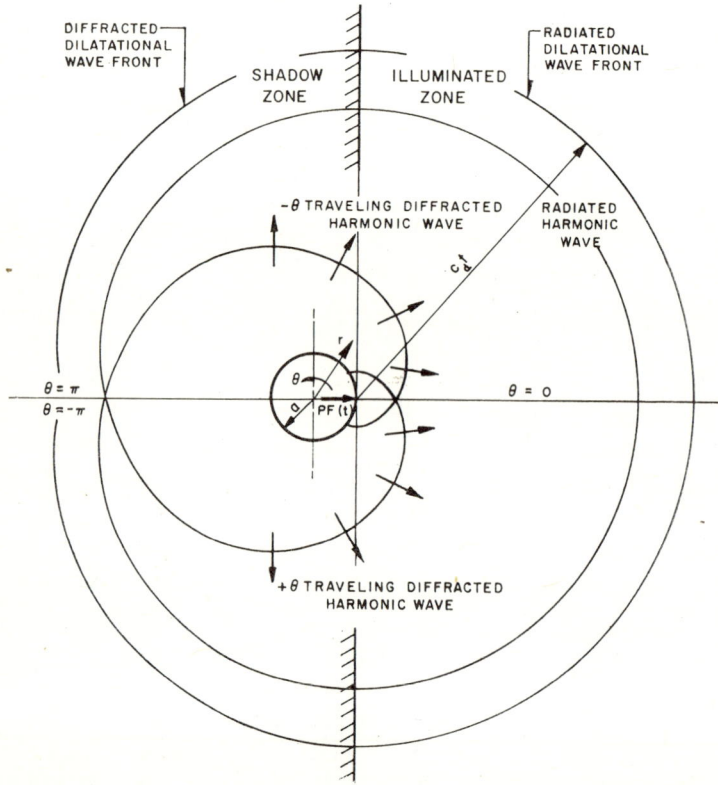

Fig. 8.7. Wave propagation from a surface line load source in a circular cylindrical cavity

evaluated branches are the key to very accurate responses in the shadow zone for the relatively early times. Further they are the key to *long time-far field* (*in θ*) *approximations* which will also be demonstrated shortly.

The frequency equation (8.114b) was first studied by Viktorov [8.20]. He noted that with η and ζ real, C could have only complex roots, but offered no proof of this. He did disclose there was one branch for large $|v|$, large ω and vanishingly small imaginary part v_i, that had the limiting real value $v_R = \omega a/c_R$ corresponding to the Rayleigh surface wave on the free half space. Later, Gilbert [8.21] obtained approximations for an infinity of roots having the limiting velocities c_d as p approaches infinity on the real p-axis.

In his thesis Peck [8.22] made a thorough analytical and numerical study of the branches of C. An abstracted version of this work appears in [8.16]. With the aid of (8.113a) it is easy to prove that if $C(v, p)$ in (8.111) has the root (or branch) $v_j(p)$ then it also has the roots $-v_j(p)$ and $\pm \bar{v}_j(\bar{p})$. It follows therefore that all the branches for real η (imaginary p) can be found from the branches having Im $v > 0$ and $\eta > 0$. Numerical results for seven of the branches, taken from [8.22, 8.16], are shown in fig. 8.8. They were calculated for Poisson's ratio $\frac{1}{4}$. The three roots designated $P1$, $P2$ and $P3$ are the first three of an infinity of roots whose phase velocities approach c_d as $\eta \to \infty$. Their asymptotic approximation, to two terms, is

$$v_j(\eta) \sim \eta + a_j \left(\frac{\eta}{2}\right)^{1/3} e^{-2\pi i/3}, \qquad (8.115a)$$

where the a_j are the roots of the Airy function ($a_1 = -2.338...$, etc.). The three roots designated $S1$, $S2$ and $S3$ are the first three of an infinity of roots whose phase velocities approach c_s, their asymptotic approximation being (8.115a) with η replaced by ζ. The branch marked R is a single branch whose phase velocity approaches the Rayleigh wave velocity c_R. The asymptotic approximation for this branch is

$$v_R(\eta) \sim \frac{c_d}{c_R}\eta + \gamma_1 + i\gamma_2 \eta e^{-\gamma_3 \eta}, \qquad (8.115b)$$

where γ_i are real positive functions of c_d and c_s. The derivations of (8.115), as well as expressions for the γ_i may be found in [8.22], the latter also being given by Viktorov [8.20]. It is important to note that *all the branches*

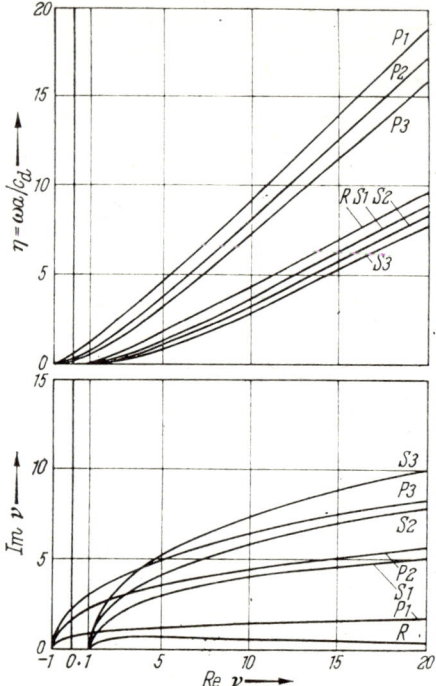

Fig. 8.8. Projections of branches of frequency equation.

$v_j(\eta)$ are complex, as the lower half of fig. 8.8 shows. *As* Re v, hence Ω, grows, Im v also grows for the P and S branches. For the R root, Im v first grows but then decays to zero. It follows, observing (8.114a), that *root R will contribute the predominant disturbance for large θ*, i.e., for large $\theta + 2m\pi$.

It may be observed in fig. 8.8 that as $\eta \to 0$ the P roots approach $v = -1$, and the S roots and R root approach $v = +1$. This behavior was analyzed by Peck [8.22] by first expressing $C(v,p)$ in (8.111) as a power series in p. This series was then used to generate the approximations for the roots found numerically, which show that $v \to \pm 1$ as $p \to 0$ along the imaginary axis of p. Through the symmetry properties of the roots it is only necessary to investigate the region $v \simeq 1$. Setting $v = 1 + \varepsilon(p)$ then, with $\varepsilon \to 0$, it is found that the low frequency approximations for the roots are contained in

$$v_j(p) = 1 - \tfrac{1}{2}\left[\ln\frac{k^2+1}{k^2-1} + (2j+1)\pi i\right]\left(\ln\frac{pa}{c_d}\right)^{-1} + O[(\ln p)^{-2}]. \quad (8.116)$$

Ch. 8, § 8.3]   SCATTERING BY CYLINDRICAL OBSTACLES   529

For small p, $\ln p$ is approximately real and negative, therefore cancelling the negative sign in (8.116). It follows the roots $v_j(p)$ leave $v=1$ with the slope $(2j+1)\pi/\ln[(k^2+1)/(k^2-1)]$ in the v-plane (cf. lower part of fig. 8.8). For $j=0$ (root R) and $j>0$ (S roots), Im $v_j>0$ in agreement with these roots in fig. 8.8. Denote these as $v_j(p)$. The P roots correspond to those in (8.116) having $j<0$ and hence Im $v_j<0$. These then are $\bar{v}_j(p)$, and do not appear in fig. 8.8. Through the symmetry of the roots in $v=0$, however, the P roots appearing in fig. 8.8 must be $-\bar{v}_j(p)$ with $j<0$. These roots leave $v=-1$ with Im $v_j>0$. It may be seen that these results are consistent with the low frequency character of the v_j in fig. 8.8. This information on the branches $v_j(-i\omega)$ of the frequency equation C in (8.114b) enables one to evaluate the solution (8.114a), or like solutions, numerically or through approximations as we shall demonstrate in the following sections.

8.3.2.5. Normal line load source on cavity wall; Rayleigh waves and the long time-far field solution. It has already been pointed out that one would expect the *Rayleigh waves* from the high frequenay-short wave limit of branch R to predominate in a far field solution, since the limiting wave numbers on this branch are the only real wave numbers in the spectra. Such a solution was derived in [8.12]. The derivation of the approximate solution and its evaluation form an instructive example in the present class of problems.

The time dependent Rayleigh waves are obtained by first picking out of (8.114a) the term corresponding to the branch $v_j(-i\omega)$ containing the Rayleigh wave real pair (η, v) as a high frequency-short wave limit $(\eta, v \gg 1)$. The pertinent branch of C in (8.114b) is defined by this limit, i.e., R. The desired approximation of C and other compatible terms in (8.114a) were found by using the appropriate Debye asymptotic expansion for the $H_v^{(1)}$ functions in (8.114b) and (8.114a). The general expansion needed is one in which both order v and argument η(or ζ) are large and positive. The present work further imposes that

$$v/\eta = c_d/c_R > v/\zeta = c_s/c_R > 1 \, . \tag{8.117}$$

The general expansion may be found in Erdelyi *et al.* [2.8, p. 87]. (8.117) means that through the substitution $v/\eta = \cosh \alpha$ this expansion for the first order term can be written as

530   PULSE SCATTERING BY OBSTACLES   [Ch. 8, § 8.3

$$H_\nu^{(1)}(\eta) = -i\left(\frac{2}{\pi\nu \tanh \alpha}\right)^{\frac{1}{2}} e^{-\nu(\tanh\alpha - \alpha)}\{1 + O(1/\nu)\},\qquad (8.118)$$

where $\alpha = \tanh^{-1} q$, $q = (1-b^2\varkappa^2)^{\frac{1}{2}}$, $\varkappa^2 = c_R^2/c_s^2$ and $b^2 = c_s^2/c_d^2$. $H_\nu^{(1)}(\zeta)$ is just like (8.118) except that β is substituted for α, where $\beta = \tanh^{-1} s$, and $s = (1-\varkappa^2)^{\frac{1}{2}}$.

Making this first order approximation to (8.114b) yields

$$(2/b^2\varkappa^4)[(\varkappa^2-2)^2 - 4qs] = 0, \qquad 0 < \varkappa < 1, \qquad (8.119)$$

the well-known equation for the speed of a Rayleigh wave on the free surface of an elastic half space. Taking into account the continuity of (8.118) in ν and ω, the derivation of (8.119) proves that the branch of C containing the Rayleigh wave pair has the real asymptote

$$\nu = c_d \eta / c_R \qquad (8.120)$$

in agreement with (8.115b). The latter equation shows that the branch approaches this asymptote (for $\eta, \nu \gg 1$) through a vanishingly small Im ν.

We have already pointed out that $\theta + 2m\pi$ can ultimately be taken large (the far field nature of the present approximation) by invoking the Friedlander representation of the solution (8.104). Then if we argue that the contributions of the branches of the frequency equation for $\eta \to 0$, $\nu_j \to \pm 1$ (hence real ν_j) will give the static solution (which will be accounted for later), (8.114a) reduces to

$$\begin{Bmatrix} u_{rR}^*(r,\theta,t) \\ u_{\theta R}^*(r,\theta,t) \end{Bmatrix} \qquad (8.121)$$

$$= -\frac{1}{\pi} \int_{\omega_L}^{\infty} \mathrm{Re}\left\{ i \begin{bmatrix} N_r(r,\nu,p)/\frac{\partial C}{\partial \nu} \\ N_\theta(r,\nu,p)/\frac{\partial C}{\partial \nu} \end{bmatrix}_{\substack{\nu = -c_d\eta/c_R \\ p = -i\omega}} e^{i\{[c_d\eta/c_R + i\,\mathrm{Im}\,\nu_R(\mathrm{Re}\,\nu)]\theta - \omega t\}} \right\} d\omega,$$

where subscript R denotes Rayleigh wave response of the displacements, ω_L is an arbitrarily large but finite positive frequency, and Im $\nu_R(\mathrm{Re}\,\nu)$

Ch. 8, § 8.3] SCATTERING BY CYLINDRICAL OBSTACLES 531

is given by the third term in (8.115b). The behavior of Im v_R is shown in fig. 8.8. Note that the first term in the exponential is the first term in (8.115b), the real part of v_R. The bracketed expressions containing N_r, N_θ in (8.121) are approximated by using (8.118) and related expressions. Then setting $F(t) = \delta(t)$, the delta function, (8.121) reduces to

$$\begin{Bmatrix} u_{rR}^{\delta*}(r,\theta,t) \\ u_{\theta R}^{\delta*}(r,\theta,t) \end{Bmatrix} = \int_{v_L}^{\infty} e^{-\operatorname{Im} v_R (\operatorname{Re} v)\theta} \begin{Bmatrix} \tilde{u}_{rR}^{\delta*}(r,v) \sin\left[(D-\theta)v\right] \\ \tilde{u}_{\theta R}^{d*}(r,v) \cos\left[(D-\theta)v\right] \end{Bmatrix} dv , \quad (8.122)$$

where $v_L = c_d \eta_L / c_R = \omega_L a / c_R$, $D = c_R t/a$ and $\tilde{u}_{rR}^{\delta*}$, $\tilde{u}_{\theta R}^{\delta*}$ are given by

$$\tilde{u}_{rR}^{\delta*}(r,v) = M\left[U_A(r) e^{-A(r)v} + U_B(r) e^{-B(r)v}\right], \quad (8.123a)$$

$$\tilde{u}_{\theta R}^{\delta*}(r,v) = -\frac{M}{\varkappa}\left[V_A(r) e^{-A(r)v} + V_B(r) e^{-B(r)v}\right], \quad (8.123b)$$

where

$$M = \frac{Pc_R \varkappa}{\pi \mu a L}, \quad L = 2\left\{\frac{8[2-(1+b^2)\varkappa^2] + 4(\varkappa^2-2)^3 + (\varkappa^2-2)^4}{(\varkappa^2-2)^2}\right\},$$

$$U_A(r) = b(2-\varkappa^2)\left[\frac{a}{b\varkappa r} - \left(\frac{1+Q}{1-Q}\right)^{\frac{1}{2}}\right]\sqrt{\frac{q}{Q}}, \quad U_B(r) = \frac{2aq}{\varkappa r}\sqrt{\frac{s}{S}},$$

$$V_A(r) = (2-\varkappa^2)\frac{a}{r}\sqrt{\frac{q}{Q}}, \quad V_B(r) = 2\varkappa q\left[\frac{a}{\varkappa r} - \left(\frac{1+S}{1-S}\right)^{\frac{1}{2}}\right]\sqrt{\frac{s}{S}},$$

$$A(r) = Q - q - \tanh^{-1} Q + \tanh^{-1} q ,$$

$$B(r) = S - s - \tanh^{-1} S + \tanh^{-1} s ,$$

$$Q(r) = (1 - b^2 \varkappa^2 r^2 / a^2)^{\frac{1}{2}}, \quad S(r) = (1 - \varkappa^2 r^2 / a^2)^{\frac{1}{2}},$$

in which r has been restricted to the *neighborhood of the cavity wall* ($a \leq r < < a/\varkappa$) *where the major effects occur*, rendering $A(r)$ and $B(r)$ both real and ≥ 0. Both approach zero as $r \to a$. Note that $D - \theta = 0$ gives the Rayleigh surface wave arrival time $t = a\theta/c_R$.

From (8.122), taking into account the exponential decay of Im v_R(Re v), we find the bounds on the magnitudes of $u_{rR}^{\delta*}$ and $u_{\theta R}^{\delta*}$ are given by

$$e^{-\text{Im}\nu_{RL}\theta}\left|\int_{\nu_L}^{\infty}\begin{Bmatrix}\tilde{u}_{rR}^{\delta*}(r,\nu)\sin\left[(D-\theta)\nu\right]\\ \tilde{u}_{\theta R}^{\delta*}(r,\nu)\cos\left[(D-\theta)\nu\right]\end{Bmatrix}d\nu\right|<\left|\begin{Bmatrix}u_{rR}^{\delta*}(r,\theta,t)\\ u_{\theta R}^{\delta*}(r,\theta,t)\end{Bmatrix}\right|$$

$$<\left|\int_{\nu_L}^{\infty}\begin{Bmatrix}\tilde{u}_{rR}^{\delta*}(r,\nu)\sin\left[(D-\theta)\nu\right]\\ \tilde{u}_{\theta R}^{\delta*}(r,\nu)\cos\left[(D-\theta)\nu\right]\end{Bmatrix}d\nu\right| \quad (8.124)$$

where $\text{Im}\,\nu_{RL}=\text{Im}\,\nu_R(\text{Re}\,\nu)=\text{Im}\,\nu_R(\nu_L)$. Since ν_L is an arbitrarily large, but finite, number, $\text{Im}\,\nu_{RL}$ can be a corresponding arbitrarily small but nonvanishing number (cf. fig. 8.8). Therefore the upper bounds in (8.124) correspond to spatially (θ) nondecaying waves, and the lower bounds in general spatially (θ) decaying waves. From (8.122), (8.124) we can write (8.122) as the approximation.

$$\begin{Bmatrix}u_{rR}^{\delta*}(r,\theta,t)\\ u_{\theta R}^{\delta*}(r,\theta,t)\end{Bmatrix}\simeq e^{-\text{Im}\nu_{Ra}\theta}\int_{\nu_L}^{\infty}\begin{Bmatrix}\tilde{u}_{rR}^{\delta*}(r,\nu)\sin\left[(D-\theta)\nu\right]\\ \tilde{u}_{\theta R}^{\delta*}(r,\nu)\cos\left[(D-\theta)\nu\right]\end{Bmatrix}d\nu \quad (8.125)$$

where $\text{Im}\,\nu_{Ra}$ is an intermediate value of $\text{Im}\,\nu_R(\text{Re}\,\nu)$ between $\text{Im}\,\nu_{RL}$ and 0. Note that $\text{Im}\,\nu_{Ra}$ can approach but cannot equal zero. Noting (8.123) we see the integrals in (8.125) are simple and can be found in most integral tables. Equations (8.125) therefore reduce to

$$u_{rR}^{\delta*}(r,\theta,t)\simeq Me^{-\text{Im}\nu_{Ra}\theta}$$

$$\times\left\{U_A(r)\frac{e^{-\nu_L A(r)}}{A^2(r)+(D-\theta)^2}\left[A(r)\sin(D-\theta)\nu_L+(D-\theta)\cos(D-\theta)\nu_L\right]\right.$$
$$\left.+U_B(r)\frac{e^{-\nu_L B(r)}}{B^2(r)+(D-\theta)^2}\left[B(r)\sin(D-\theta)\nu_L+(D-\theta)\cos(D-\theta)\nu_L\right]\right\}, \quad (8.126a)$$

$$u_{\theta R}^{\delta*}(r,\theta,t)\simeq -\frac{M}{\varkappa}e^{-\text{Im}\nu_{Ra}\theta}$$

$$\times\left\{V_A(r)\frac{e^{-\nu_L A(r)}}{A^2(r)+(D-\theta)^2}\left[A(r)\cos(D-\theta)\nu_L-(D-\theta)\sin(D-\theta)\nu_L\right]\right.$$
$$\left.+V_B(r)\frac{e^{-\nu_L B(r)}}{B^2(r)+(D-\theta)^2}\left[B(r)\cos(D-\theta)\nu_L-(D-\theta)\sin(D-\theta)\nu_L\right]\right\}, \quad (8.126b)$$

based on $A(r)$ and $B(r) > 0$. Equations (8.126) represent Rayleigh waves in the near surface interior $a < r \leq a/\varkappa$ traveling with the velocity c_R. It is clear they are continuous through the arrival time $D = \theta$. Note they decay exponentially with r and θ. Since v_L is a large real number the r decay is severe for $r > a$. The station $\theta + 2m\pi$ (for fixed r) can of course be large but it must be finite since it must always be behind the dilatational wavefront. It follows, since Im v_{Ra} can approach zero, that the exponential decay with θ is much less severe than with r in the present approximation.

The maximum responses in (8.126) occur at the surface as one would expect from our study of the Rayleigh waves in the strongly related Lamb's line load problem. To find these responses we let $A(r)$ and $B(r) \to 0+$ (as $r \to a+$) in (8.126) which results in

$$u_{rR}^{\delta*}(a+, \theta, t) \simeq M[U_A(a+) + U_B(a+)] e^{-\text{Im } v_{Ra}\theta} \frac{\cos (D-\theta)v_L}{D-\theta}, \quad (8.127\text{a})$$

$$[u_{\theta R}^{\delta*}(a+, \theta, t) \simeq -\frac{M}{X} e^{-\text{Im } v_{Ra}\theta}$$

$$\times \left[V_A(a+) \left\{ \lim_{A(r) \to 0+} \left[\frac{A(r)}{A^2(r) + (D-\theta)^2} \right] \cos (D-\theta)v_L - \frac{\sin (D-\theta)v_L}{D-\theta} \right\} \right.$$
(8.127b)

$$\left. + V_B(a+) \left\{ \lim_{B(r) \to 0+} \left[\frac{B(r)}{B^2(r) + (D-\theta)^2} \right] \cos (D-\theta)v_L - \frac{\sin (D-\theta)v_L}{D-\theta} \right\} \right].$$

Noting that

$$\int_{-\infty}^{\infty} \frac{A(r)}{A(r)^2 + (D-\theta)^2} \, d(D-\theta) = \pi, \quad \text{for } A(r) > 0,$$

and that the integrand function here is zero for $A(r) \to 0+$ and $D \neq \theta$, we conclude the limits in (8.127b) are π times the symmetric delta function $\delta_s(D-\theta)$. Imposing $D-\theta$ small then, (8.127) reduce to the limiting singular surface responses

$$u_{rR}^{\delta*}(a+, \theta, t)/u_0 = \frac{c_R \varkappa^2 q}{\pi a R} e^{-\text{Im} v_{Ra}\theta} \left[\frac{1}{D-\theta} + 0\{v_L^2(D-\theta)\} \right],$$
(8.128a)

$$u_{\theta R}^{\delta*}(a+, \theta, t)/u_0 = -\frac{c_R \varkappa^2 (2-\varkappa^2)}{2aR} e^{-\text{Im} v_{Ra}\theta} [\delta_s(D-\theta) + 0(v_L)],$$
(8.128b)

where $u_0 = P/\mu$. The work [8.12] shows the corresponding stress $\sigma_{\theta R}^{\delta *}(a+, \theta, t)$ behaves as $\delta_s'(D-\theta)$ where the prime indicates differentiation with respect to $(D-\theta)$. It should be pointed out that *these singular displacements are essentially* (as Im $\nu_{Ra} \to 0+$) *those found in Lamb's line load problem* [cf. eqs. (6.52), (6.54) and fig. 6.8], in effect then showing that they represent high frequency-short waves that cannot "see" the curvature of the cavity wall.

Identifying the displacements in (8.128) with $f^*(r, \theta, t)$ in (8.104), and then substituting them into the latter equation, extends the solution to all of finite $\theta + 2m\pi$, giving then the *wave sums* u_{rR}^δ, $u_{\theta R}^\delta$. The observer, standing at a station θ in the physical plane, sees each of the waves in (8.128) as a periodic phenomenon in t (of period $D = c_R t/a = 2\pi$). This agrees with the diffracted wavefronts 3 depicted in fig. 8.6, noting now that (8.128) represent two-sided discontinuities.

The far field-long time solution in the present problem for $\theta + 2m\pi > 1$ is obtained by imposing $\theta + 2m\pi \gg 1$ and $D \gg 1$, along with $|D - (\theta + 2m\pi)| < 1$, on the extension (8.104) of (8.128), and projecting the result into the physical plane. It follows that we have the responses shown in fig. 8.9 for the positive

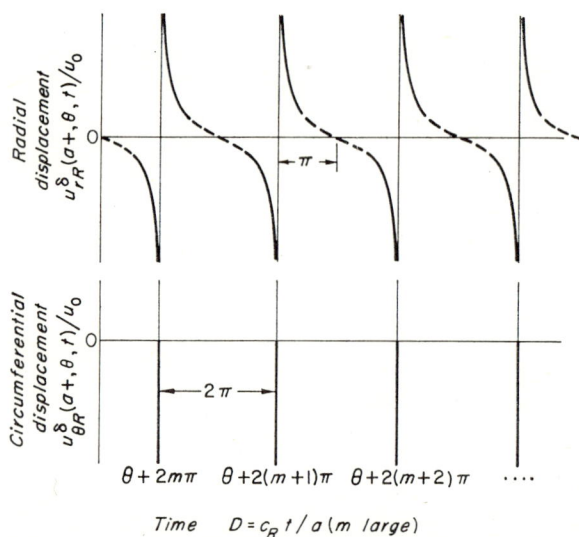

Fig. 8.9. Response (wave sum) for long time at cavity wall due to positive θ-traveling Rayleigh waves (cavity wall line load delta function source).

Ch. 8, § 8.3]   SCATTERING BY CYLINDRICAL OBSTACLES   535

θ-traveling waves. The solid lines in the figure represent the approximations (8.128). The dashed line connection (in the case of the radial displacement) represents an assumption that the heads and tails of these Rayleigh waves cancel away from $|D-(\theta+2m\pi)|$ small. Note that the head associated with a $D-(\theta+2m\pi)=0$ event can extend out to the dilatational wavefront, and similarly the tail to time infinite. Clearly then, near neighboring events are chiefly responsible for the dashed portions of the wave sum u_{rR}^{δ}.

The only further consideration necessary for the long time solution is the static solution. Since the input is the delta function $\delta(t)$, the static solution must be zero. To *complete the solution it is only necessary to have a system like that in fig. 8.9 for the negative θ-traveling waves*, i.e., from the $\theta<0$ solution. The total response at the station $\theta=\pi/2$, for example, would have the same figure for u_{rR}^{δ} as that in fig. 8.9, except the period would be π, and $u_{\theta R}^{\delta}$ would change its sign every π.

The infinite discontinuities in (8.128) and fig. 8.9, of course, are directly dependent on the nature of the input function $F(t)$ [here $\delta(t)$]. It is therefore of interest to get the response to the step $H(t)$, which offers a much less severe high frequency input (i.e., $|\delta(i\omega)|=1, |\tilde{H}(i\omega)| \simeq 1/\omega$). A solution corresponding to (8.128), and its extension by (8.104), can be obtained for the step case throught the use of the Duhamel integral operating on a series of the first terms on the right hand side of (8.128). This yields the Rayleigh positive θ-traveling wave contributions to the long time solution as

$$u_{rR}^{H}(a+,\theta,t)/u_0 = \frac{\varkappa^2 q}{\pi R} \sum_{m=M_0}^{M} e^{-\operatorname{Im} \nu_{Ra}(\theta+2m\pi)} \log |D-(\theta+2m\pi)|,$$
(8.129a)

$$u_{\theta R}^{H}(a+,\theta,t)/u_0 = -\frac{\varkappa^2(2-\varkappa^2)}{2R} \sum_{m=M_0}^{M} e^{-\operatorname{Im} \nu_{Ra}(\theta+2m\pi)} H[D-(\theta+2m\pi)],$$
(8.129b)

where now M and M_0 are large and are determined by the number of periodic waves occuring in a certain domain of large time, and $H[D-(\theta+2m\pi)]$ is the step function

$$H[D-(\theta+2m\pi)] = \begin{cases} 0, & D<\theta+2m\pi, \\ \tfrac{1}{2}, & D=\theta+2m\pi, \\ 1, & D>\theta+2m\pi. \end{cases}$$

The terms in (8.129a) are valid in the vicinity of $D-(\theta+2m\pi)=0$. The full curve can be obtained by numerically integrating (w. r. t. time away from $D-(\theta+2m\pi)=0$) the u_{rR}^δ curve in fig 8.9. Again here the negative θ-traveling waves must be superposed on (8.129) for the full dynamic contribution to the long time solution in the physical plane. At $\theta=\pi/2$ in the present case the oddness of $u_{\theta R}^H/u_0$ (w. r. t. θ) leads to an alternating series of square waves of duration π, decaying (with $\theta+2m\pi$) amplitude, and occurring with periodicity 2π. In the present case the static solution has existing constant (w. r. t. time) terms which can be neglected with respect to the terms in (8.129a), hence the latter are again the predominant disturbances in the long time solution. Note that u_{rR}^H are still represented by infinite discontinuities. In the case of $u_{\theta R}^H$ the static solution would have to be added to the dynamic for the total long time solution.

The reader will be interested in the extension of the present approximation to the *linear viscoelastic case* given in [8.12]. In the work a correspondence principle leads to similar results at the infinite Rayleigh wave discontinuities, but to further spatial decay away from these times.

8.3.2.6. Diffraction of plane compressional pulse by cavity; formal solution.

Figure 8.6 depicts the problem. Recall this is the case which is equivalent to the line source S at $x_0 \to \infty$, resulting in the incident plane pulse shown in the figure. Here the incident plane pulse is an elastic compressional one, which we can write as

$$\varphi_i(r,\theta,t) = \frac{\sigma_0 c_d^2}{2(\lambda+2\mu)}\left(t+\frac{r\cos\theta}{c_d}\right)^2 H\left(t+\frac{r\cos\theta}{c_d}\right). \qquad (8.130)$$

corresponding to a step in stress σ_r of amplitude σ_0 (inherently negative). The governing wave equations (8.105), where $t>0$ is replaced by $t>-a/c_d$ (the time at which the incident wave strikes the cavity), hold here as well as the displacement—and stress-potential relations (8.109). The solution is separated into incident and scattered parts so that the total solution is given by

$$\varphi(r,\theta,t) = \varphi_i(r,\theta,t) + \varphi_s(r,\theta,t),$$
$$\psi(r,\theta,t) = \psi_s(r,\theta,t), \qquad (8.131)$$

in the nomenclature of § 8.2.2.1. It follows, since the incident stresses are known, that the boundary conditions at the cavity wall for the scattered wave solution are

$$\left.\begin{array}{l}\sigma_{rs}(a, \theta, t) = -\sigma_{ri}(a, \theta, t), \\ \sigma_{r\theta s}(a, \theta, t) = -\sigma_{r\theta i}(a, \theta, t),\end{array}\right\} \quad -\infty < \theta < \infty, \qquad t > -a/c_d, \quad (8.132)$$

so that the total stress there is zero. Since the incident wave strikes the cavity at $t = -a/c_d$, we require that

$$\varphi_s(r, \theta, -a/c_d) = \dot{\varphi}_s = \psi_s = \dot{\psi}_s = 0, \qquad r \geq a, \qquad -\infty < \theta < \infty. \quad (8.133)$$

Radiation conditions for the scattered waves can be stated as

$$\lim_{r \to \infty \text{ and/or } \theta \to \pm\infty} [\varphi_s(r, \theta, t), \psi_s, u_{rs}, \text{etc.}] = 0, \qquad t > -a/c_d, \quad (8.134)$$

completing the statement of the problem.

The formal solution of the problem can again be written in the wave sum form (8.104). We argue since the only given function in the problem is the incident potential, once its expression in wave-sum form has been found, we need only require that each term of the wave sum for φ_s and ψ_s satisfy the wave equations (8.105). Correspondingly from (8.132)–(8.134) we have

$$\sigma_{rs}^*(a, \theta, t) = -\sigma_{ri}^*(a, \theta, t), \quad (8.135)$$

$$\sigma_{r\theta s}^*(a, \theta, t) = -\sigma_{r\theta i}^*(a, \theta, t), \quad (8.136)$$

$$\varphi_s^*(r, \theta, -a/c_d) = \dot{\varphi}_s^* = \psi_s^* = \dot{\psi}_s^* = 0, \quad (8.137)$$

$$\lim_{r \to \infty \text{ and/or } \theta + 2m\pi \to \pm\infty} [\varphi_s^*(r, \theta, t), \psi_s^*, u_r^*, \text{etc.}] = 0, \quad (8.138)$$

i.e., the boundary conditions, initial conditions and radiation conditions are also satisfied term by term. Since action in the present problem begins at $t = -a/c_d$ we use the *bilateral Laplace transform on time t* [cf. § 5.5 and eq. (5.38)], and again the exponential Fourier transform (with real argument) on the circumferential coordinate θ [pair (5.39)]. We find again the transformed governing equations (8.10a) and their admissible general solutions (8.110b), but both now for $\bar{\varphi}^{\sim*}(r, \nu, p)$ and $\bar{\psi}^{\sim*}(r, \nu, p)$.

To complete our statement of the general transformed solutions for $\bar{\varphi}^{\sim*}$ and $\bar{\psi}^{\sim*}$ we must derive $\bar{\varphi}_i^{\sim*}(r, \nu, p)$. We derive it in the following by applying the Poisson summation formula to the Fourier series [cf. eq. (8.104)] of the Laplace transform of φ_i. From (8.130), $\bar{\varphi}_i = \bar{\varphi}_0(p) \exp(k_d r \cos\theta)$ $\bar{\varphi}_0(p) = \sigma_0 c_d^2 / 2(\lambda + 2\mu) p^3$. Noting that the exponential function in $\bar{\varphi}_i$ is a *generating function* for a Fourier–Bessel series, we find

$$\bar{\varphi}_i(r,\theta,p) = \sum_{n=-\infty}^{\infty} \bar{\varphi}_0(p) I_{|n|}(k_d r) e^{in\theta}, \qquad (8.139)$$

cf. Erdelyi *et al.* [2.8, p. 7]. The absolute value sign is permissible because $I_{-n}(z) = I_n(z)$ for integral n. We now apply the integral for f^* in (8.104) to $\bar{\varphi}_i$, and take its Fourier exponential transform, with the result

$$\bar{\varphi}_i^{\sim *}(r,v,p) = \bar{\varphi}_0(p) \int_{-\infty}^{\infty}\int_{-\infty}^{\infty} I_{|\eta|}(k_d r) e^{i\theta(\eta+v)} d\eta d\theta . \qquad (8.140)$$

Since I_v approaches zero exponentially as $v \to +\infty$, the inner integral here is uniformly convergent and we may interchange the integrations writing (8.140) as

$$\bar{\varphi}_i^{\sim *}(r,v,p) = \bar{\varphi}_0(p) \int_{-\infty}^{\infty} I_{|\eta|}(k_d r) \left[\int_{-\infty}^{\infty} e^{i\theta(\eta+v)} d\theta\right] d\eta . \qquad (8.141)$$

The inner integral here may be recognized as the inverse of the Fourier exponential transform of the delta function $2\pi\delta(\eta+v)$. Hence

$$\bar{\varphi}_i^{\sim *}(r,v,p) = 2\pi \bar{\varphi}_0(p) I_{|v|}(k_d r) , \qquad (8.142)$$

and from this equation and (8.110b) our present general transformed solution is

$$\bar{\varphi}^{\sim *}(r,v,p) = 2\pi \bar{\varphi}_0(p) I_{|v|}(k_d r) + \alpha(v,p) K_v(k_d r) ,$$
$$\bar{\psi}^{\sim *}(r,v,p) = \beta(v,p) K_v(k_s r) . \qquad (8.143)$$

Substituting (8.143) into the double transforms of the boundary conditions (8.135, 8.136) determines $\alpha(v,p)$, $\beta(v,p)$. They are

$$\alpha(v,p) = 2\pi \bar{\varphi}_0(p) [D(v,p)]^{-1} \{[(2v^2 + k_s^2 a^2) I_{|v|}(k_d a) - 2k_d a I'_{|v|}(k_d a)]$$

$$\cdot [(2v^2 + k_s^2 a^2) K_v(k_s a) - 2k_s a K'_v(k_s a)]$$

$$- 4v^2 [I_{|v|}(k_d a) - k_d a I'_{|v|}(k_d a)][K_v(k_s a) - k_s a K'_v(k_s a)]\} ,$$

$$\beta(v,p) = i 4\pi \bar{\varphi}_0(p) [D(v,p)]^{-1} v [k_s^2 a^2 + 2(v^2 - 1)] , \qquad (8.144)$$

Ch. 8, § 8.3]   SCATTERING BY CYLINDRICAL OBSTACLES   539

where

$$D(v, p) = -\{[(2v^2+k_s^2a^2)K_\nu(k_da) - 2k_daK'_\nu(k_da)][(2v^2+k_s^2a^2)K_\nu(k_sa)$$
$$- 2k_saK'_\nu(k_sa)] - 4v^2[K_\nu(k_da) - k_daK'_\nu(k_da)][K_\nu(k_sa) - k_saK'_\nu(k_sa)]\}.$$

The Wronskian $K_\nu(z)I'_\nu(z) - K'_\nu(z)I_\nu(z) = z^{-1}$ has been used to simplify the expression for $\beta(v, p)$. The function $D(v, p)$, which is used in [8.16, 8.22], is another form for $C(v, p)$ of (8.111). Their zeros are identical. We can now write the formal solution for the displacements u_r^*, u_θ^* which are again given by the double inversion integrals (8.111a), except that now $\tilde{u}_r^{~*}$, $\tilde{u}_\theta^{~*}$ are given by

$$\tilde{u}_r^{~*}(r, v, p) = r^{-1}[k_d r \alpha(v, p) K'_\nu(k_d r) - iv\beta(v, p) K_\nu(k_s r)] + 2\pi k_d \bar{\varphi}_0(p) I'_{|\nu|}(k_d r),$$
(8.145)
$$\tilde{u}_\theta^{~*}(r, v, p) = -r^{-1}\{[iv\alpha(v, p) K_\nu(k_d r) + k_s r \beta(v, p) K'_\nu(k_s r)] - i2\pi \bar{\varphi}_0(p) v I_{|\nu|}(k_d r)\}.$$

The differences in the expressions for $\beta(v, p)$ in (8.144), and the bracketed terms in (8.145), and the corresponding terms in [8.16, eqs. (16), (18a, b)], stem from the use of the Fourier transform pairs (5.40) instead of (5.39) used here; i.e., this introduces a change in sign for the quantities that are odd in v.

8.3.2.7. Diffraction of plane compressional pulse by cavity; exact inversion.
The diffracted displacement waves in the present problem, as in the preceding line load case, can be obtained by inverting the bracketed terms in (8.145) in the manner of § 8.3.2.4. However, there is an important difference in the inversion of the Fourier transform. It stems from $I_{|\nu|}(k_d r)$ in the double transform of the incident potential $\bar{\varphi}_i^{~*}(r, v, p)$ in (8.142). The Bessel function $I_{|\nu|}(z)$ is not an analytic function of complex v by itself, so that continuation off the real axis of v (hence contour integration) for terms involving this function (cf. $\alpha(v, p)$ in (8.144)] would not be possible. As pointed out in [8.22, 8.16], however, *the sum of the incident and scattered dilatational potentials*, the first of (8.143), *is analytic*, since it is even in real v, hence the absolute value signs on order v may be dropped. The proof is left to the exercises. It is clear that this statement also applies to other transformed response functions, e.g., the displacements in (8.145), since they

are associated with $\alpha(v, p)$ and the transforms of the incident pulses there. Then, since I_v, K_v, are *entire functions of their order* v, inversion procedes as in the line load case, cf. § 8.3.2.4. Again considering $\theta > 0$, the analog of (8.112) is

$$\begin{Bmatrix} u_{rs}^*(r, \theta, t) \\ u_{\theta s}^*(r, \theta, t) \end{Bmatrix} = -\frac{1}{2\pi i} \int_{Br} e^{pt} \left[i \sum_{j=1,2,\cdots} \left\{ \begin{bmatrix} N_{rs}(r, v, p) \\ N_{\theta s}(r, v, p) \end{bmatrix} \frac{e^{-iv\theta}}{\frac{\partial D}{\partial v}} \right\}_{v=\bar{v}_j(p)} \right] dp,$$
(8.146)

based on the branches of $D(v, p)$ satisfying $\text{Im } v_j(p) < 0$ for $\text{Re } p > 0$ for convergence, where $v_j(p) = (v_r + iv_i)_j$, $v_i > 0$ [as in (8.112)], and $N_{rs} = D\tilde{u}_{rs}^*$ $N_{\theta s} = D\tilde{u}_{\theta s}^*$, subscript s denoting scattered[4]. Contour integration in the p-plane follows that in § 8.3.2.4, reducing (8.146) to to the analog of (8.114a)

$$\begin{Bmatrix} u_{rs}^*(r, \theta, t) \\ u_{\theta s}^*(r, \theta, t) \end{Bmatrix}$$

$$= -\frac{1}{\pi} \int_0^\infty \text{Re} \sum_{j=1,2,\cdots} i \begin{bmatrix} N_{rs}[r, -v_j(-i\omega), -i\omega] \\ N_{\theta s}[r, -v_j(-i\omega), -i\omega] \end{bmatrix} \frac{e^{i[v_j(-i\omega)\theta - \omega t]}}{\frac{\partial D}{\partial v}} \bigg|_{\substack{v=-v_j(-i\omega) \\ p=-i\omega}} d\omega,$$
(8.147)

where again here the roots $v_j(-i\omega)$ are those depicted in fig. 8.8. Recall they are those of $C(v, -i\omega) = 0$ in (8.114b) [or equivalently of $D(v, -i\omega) = 0$ of (8.144)]. Note that the last terms in (8.145) represent the incident wave and are easily inverted through the known pair (8.130), (8.142).

Again here it may be seen that the integrand of (8.147) corresponds to component diffracted and reflected harmonic waves {one for each mode of propagation pair $[\omega, v_j(-i\omega)]$} traveling in the positive θ-direction, and outward in r, the latter stemming from the $H_v^{(1)}(k_d r)$ and $H_v^{(1)}(k_s r)$ character of the functions $N_{rs}, N_{\theta s}$. These waves are generated by incident wave-

---

[4] Again here we will ultimately be interested only in the diffracted wave parts of these scattered wave representations for the reasons pointed out in the discussion following (8.112).

Ch. 8, § 8.3]   SCATTERING BY CYLINDRICAL OBSTACLES   541

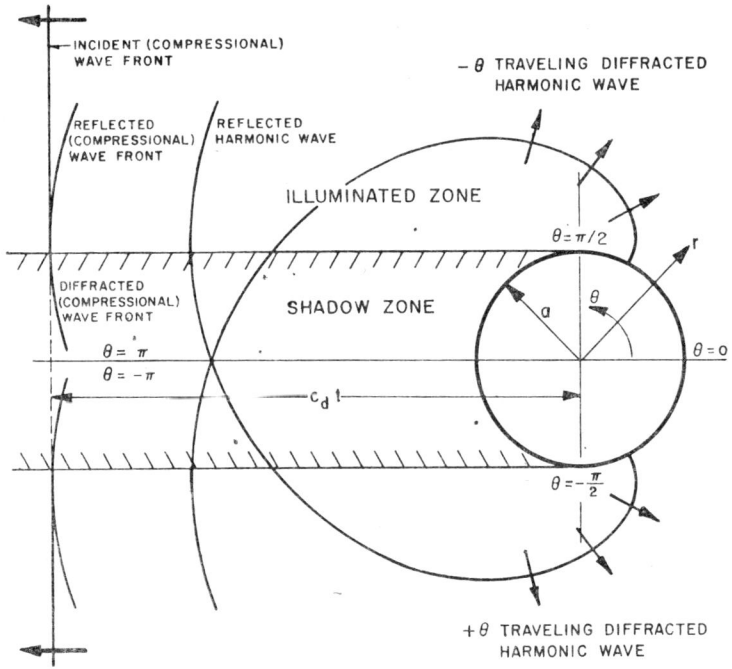

Fig. 8.10. Scattering of a plane compressional pulse by a circular cylindrical cavity.

cavity interaction. Fig. 8.10 shows the position of two such diffracted and reflected waves (i.e., the negative θ-traveling waves from the $\theta<0$ solution are also shown) corresponding to a time t when the incident wavefront has already enveloped the cavity and gone past it. It is clear that the Riemann surface sheets $m=\pm 1$, in addition to $m=0$, are involved here, hence the corresponding terms in (8.104).

Figure 8.10 points out that no diffraction starts until the incident wavefront reaches the vertical line, $\theta=\pm\pi/2$, through the cavity center, i.e., diffracted waves have their origin at $\theta=\pm\pi/2$, $r=a$, propagating into the shadow-zone. This occurs when time $t=0$ (note the incident wavefront reaches the cavity at $\theta=0$ at time $t=-a/c_d$). The reflected harmonic wave, and wavefront, only partially shown in the figure, surround the cavity except in the shadow zone where they do not occur.

It should be pointed out that singularities occur in the integrands of the Bromwich integrals of (8.146) at p=0 causing the small circular paths in the direct contour integration in the p-plane to give infinite contributions. Peck studied these in [8.22]. The results are also given in [8.16]. Noting that our interest is restricted to the shadow zone, where there is quiescence for $t<0$, he made use of the convolution theorem of the one-sided Laplace transform (cf. § 5.4.4) which led to zero contributions from the small circular paths, and associated path integrals that were proper at $p=0$. Details are given in [8.16].

8.3.2.8. Diffraction of plane compressional pulse by cavity; numerical evaluation of solution. In [8.22, 8.16] the solution (8.147) and associated velocities, were evaluated by numerical integration. The results presented here (figs. 8.11–8.14, taken from [8.16]) are for the velocities since they have the most interesting pulse behavior. The velocities were evaluated at the cavity wall $r=a$, at the points $\theta=(3/4)\pi$ and $\theta=(5/4)\pi$ for $0<c_d t/a<10$.[5] The integrals for the modes of propagation (modes for short) associated with the frequency branches P1, P2, P3, R, S1 and S2 (cf. fig. 8.8) were summed. Poisson's ratio was taken to be 1/4. The normalization constant $\dot{u}_0 = \sigma_0 c_d/(\lambda+2\mu)$ used in the figures is the particle velocity behind the incident step-stress dilatational pulse.

Figure 8.11 shows the radial velocity in the wave form of solution, i.e. $\dot{u}^*$ (our $\dot{u}_r^*$), at $\theta=(3/4)\pi$ (both individual mode contributions $\dot{u}_j^*$ and the mode sum $\dot{u}^*$ are shown). The largest contribution comes from the P1 mode, which also exhibits strong impulsive behavior at the arrival time of diffracted P waves. Note that the P1 branch in fig. 8.8 has the smallest Im ν except for R. The second largest contribution comes from the R mode. It is noteworthy that no significant impulsive behavior occurs at the diffracted Rayleigh (R) wave arrival time. As one would expect from their higher Im ν (cf. fig. 8.8), contributions from the higher P modes (P2 and P3) are seen to decrease rapidly as the mode number is increased. The S1 mode has a very small contribution, and the S2 mode response is too small to be plotted. The mode sum is essentially zero ahead of the arrival time of the diffracted P wave. The plot in fig. 8.11 is too crowded to show this, but it can be seen in fig. 8.13. The slight oscillations about zero ahead of the P-wave arrival time (visible in fig. 8.13) are probably caused by truncation

---

[5] Results for velocities at $\theta=\pi$, and displacements at $\theta=(3/4)\pi$ and $\theta=\pi$, are presented in [8.22].

Ch. 8, § 8.3]   SCATTERING BY CYLINDRICAL OBSTACLES   543

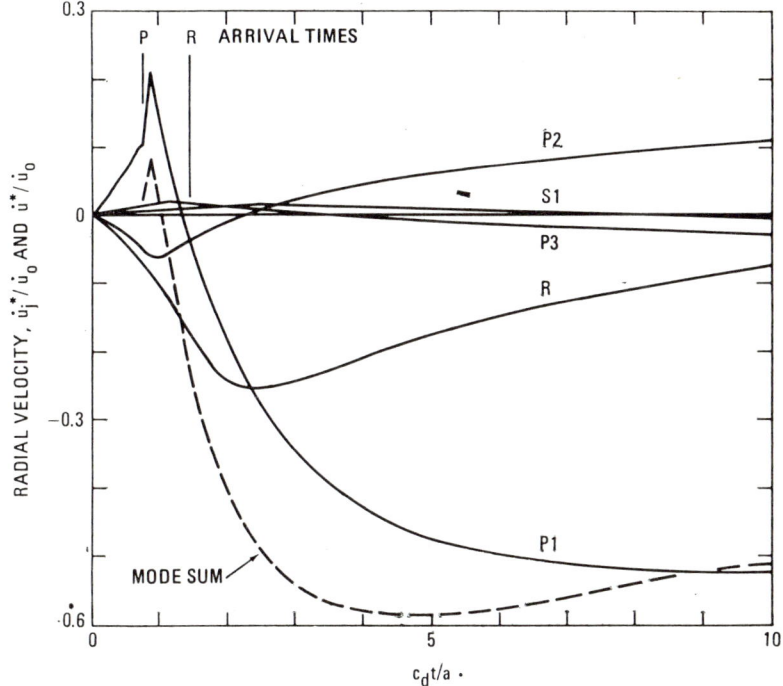

Fig. 8.11. Modal response $\dot{u}_j^*$ and mode sum $\dot{u}^*$ at $r=a$, $\theta=\frac{3}{4}\pi$.

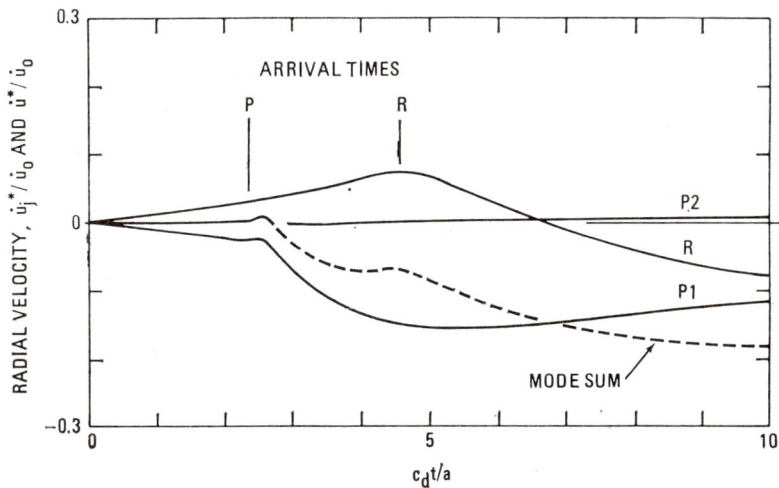

Fig. 8.12. Modal response $\dot{u}_j^*$ and mode sum $\dot{u}^*$ at $r=a$, $\theta=\frac{2}{4}\pi$.

544   PULSE SCATTERING BY OBSTACLES   [Ch. 8, § 8.3

of the infinite integrals at $\eta=40$, since the convergence becomes quite slow as $c_d t/a$ is decreased.

In fig. 8.12, the mode contributions and mode sum are shown for the positive θ propagating $\dot{u}^*$ wave at $\theta=(5/4)\pi$. The additional propagation of $\pi/2$ in θ has effected some striking changes. The most obvious change is the decrease of amplitude of the wave. Second, the mode convergence has become even more rapid than at $\theta=(3/4)\pi$. Third, the mode sum is clearly zero ahead of the P arrival (the infinite integrals converge much more rapidly at higher

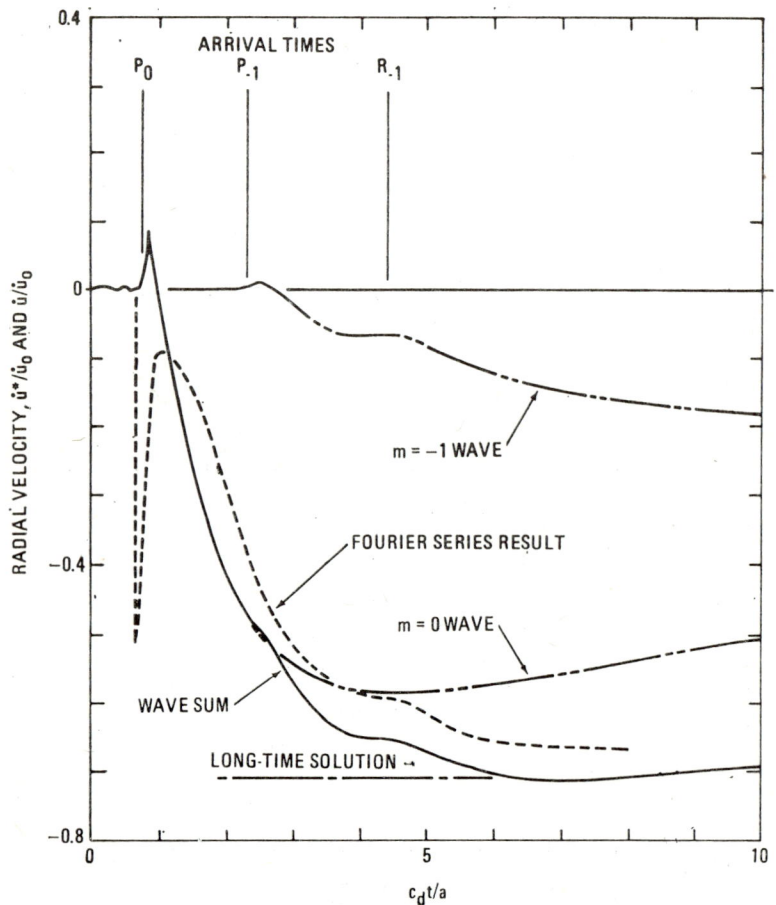

Fig. 8.13. Waves $\dot{u}^*$ and wave sum $\dot{u}$ at $r=a$, $\theta=\frac{3}{4}\pi$
(Fourier series result after Baron and Parnes).

θ's because of the imaginary part of v_j) and the slow, smooth rise of the pulse from zero at the front, which is characteristic of diffracted waves (cf. [8.18] §§ 7.8), has become apparent in the numerical results. Finally, a pulselike behavior has begun to emerge at the R wave arrival time. This delayed emergence of the Rayleigh-type pulse is similar to the behavior in the half-space problem with a buried source disturbance [cf. ch. 6, § 6.2.2, fig. 6.11]. This behavior is also an early indication of the long-time dominance of the Rayleigh pulse shown by Miklowitz [8.15].

To obtain the total response at the $\theta = (3/4)\pi$ point, the wave sum of the $\dot{u}^*$ waves is obtained from (8.104). This is illustrated in fig. 8.13. The first wave to arrive is the $m = 0$ wave, which is just the $\dot{u}^*$ wave at $\theta = (3/4)\pi$ given in fig. 8.11. The second wave to arrive is the $m = -1$ wave, propating in the negative θ direction. By virtue of the symmetry of $\dot{u}^*$ in θ, this second wave is identical to the $m = 0$ wave at $\theta = (5/4)\pi$, given in fig. 8.12. The $m = 1$ and $m = -2$ waves also arrive at $\theta = (3/4)\pi$ for $c_d t/a < 10$, but their contributions are negligible for this time interval. Thus, the wave sum is essentially the sum of the $\dot{u}^*$ waves for $m = 0$ and $m = -1$, as shown in fig. 8.13 by the solid curve. The correspondence with the long-time solution is seen to be good; the slight deviation for $c_d t/a$ approaching 10 is probably due to the neglect of the $P4$ mode in the $m = 0$ wave.

The waves and wave sum for the circumferential velocity $\dot{v}^*$(our $\dot{u}_\theta^*$) at $\theta = (3/4)\pi$ are shown in fig. 8.14. Some comparisons with the radial velocity results in fig. 8.13 worthy of comment are as follows: (1) a much larger disturbance is contributed by the $m = -1$ wave at the P_{-1} arrival time, (2) a barely detectable Rayleigh pulse occurs in the $m = -1$ wave, and (3) the short-time oscillations in the $m = 0$ wave, caused by truncation of the infinite integrals, are slightly larger than they were for the radial velocity.

8.3.2.9. Diffraction of plane compression pulse by cavity; Fourier series solution, comparison of the two methods and results. Let us assume that at time $t = 0$ the incident wave, φ_i in (8.130), strikes the cavity at $r = x = a$. Then on the basis of the initial conditions (8.133), taken now at $t = 0$, we apply the Laplace transform to the wave equations (8.105). These transformed partial differential equations have as solutions the *classical separable forms* for the scattered potentials

$$\bar{\varphi}_{ns}(r, \theta, p) = \bar{\varphi}_n(r, p) \cos n\theta ,$$

$$\bar{\psi}_{ns}(r, \theta, p) = \bar{\psi}_n(r, p) \sin n\theta ,$$

(8.148)

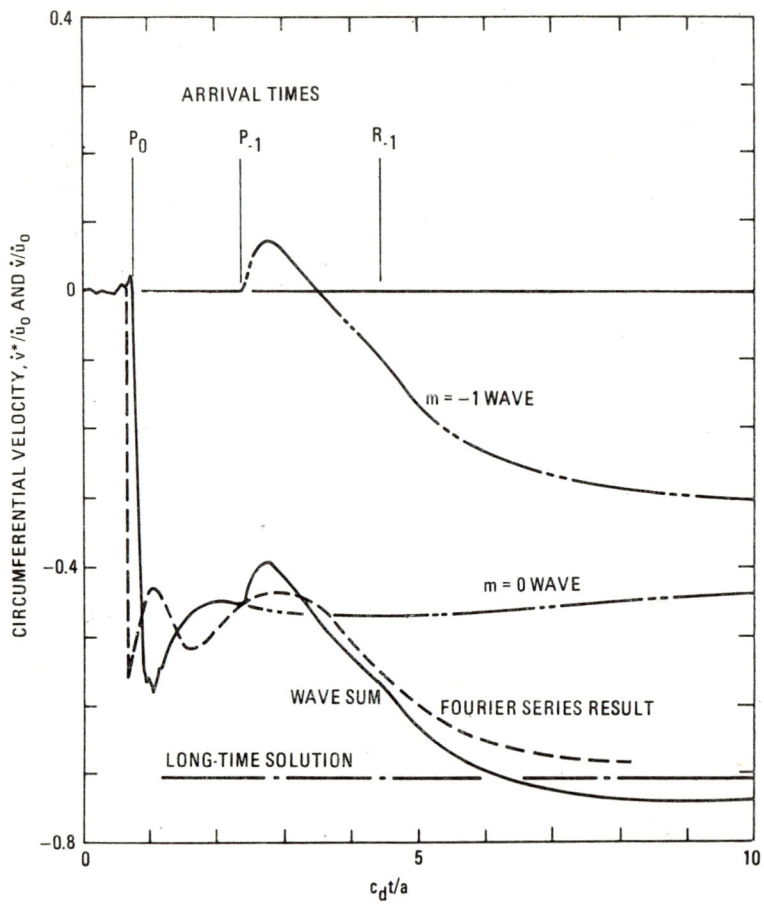

Fig. 8.14. Waves $\dot{v}^*$ and wave sum $\dot{v}$ at $r=a$, $\theta=\tfrac{3}{4}\pi$ (Fourier series result after Baron and Parnes).

where $\bar{\varphi}_n$, $\bar{\psi}_n$ must be solutions of the equations

$$r^2 \frac{d^2 \bar{\varphi}_n}{dr^2} + r \frac{d\bar{\varphi}_n}{dr} - (r^2 k_d^2 + n^2)\bar{\varphi}_n = 0 ,$$

$$r^2 \frac{d^2 \bar{\psi}_n}{dr^2} + r \frac{d\bar{\psi}_n}{dr} - (r^2 k_s^2 + n^2)\bar{\psi}_n = 0 .$$

(8.149)

Noting these equations are of the same form as (8.110a), having solutions (8.110b), it follows that (8.148) become

$$\bar{\varphi}_{ns}(r,\theta,p) = A_n(n,p)K_n(k_d r)\cos n\theta,$$
$$\bar{\psi}_{ns}(r,\theta,p) = B_n(n,p)K_n(k_s r)\sin n\theta, \tag{8.150}$$

making use of the radiation condition (8.134) once again, but just for $r\to\infty$. The general transformed solutions for the scattered potentials $\bar{\varphi}_s$, $\bar{\psi}_s$ are then written as

$$\bar{\varphi}_s(r,\theta,p) = \bar{\varphi}_0(p)\sum_{n=0}^{\infty} e_n A_n(n,p)K_n(k_d r)\cos n\theta,$$

$$\bar{\psi}_s(r,\theta,p) = \bar{\varphi}_0(p)\sum_{n=1}^{\infty} e_n B_n(n,p)K_n(k_s r)\sin n\theta, \tag{8.151}$$

through superposition, where $e_n = 1$ for $n=0$, and 2 for $n \geq 1$. The transformed incident potential can be expressed similarly as

$$\bar{\varphi}_i(r,\theta,p) = \bar{\varphi}_0(p)e^{k_d r\cos\theta} = \bar{\varphi}_0(p)\sum_{n=0}^{\infty} e_n I_n(k_d r)\cos n\theta. \tag{8.152}$$

Making use of the Laplace transforms of (8.131), from (8.151), (8.152) we have the transformed general solutions $\bar{\varphi}(r,\theta,p)$, $\bar{\psi}(r,\theta,p)$. The coefficients $A_n(n,p)$, $B_n(n,p)$ are determined by using the conditions of a stress free boundary (8.132), now with $-\pi < \theta < \pi$ and $t > 0$, leading to the formal solutions for the displacements

$$\begin{Bmatrix} u_r(r,\theta,t) \\ u_\theta(r,\theta,t) \end{Bmatrix} = \frac{1}{2\pi i}\int_{Br} \begin{Bmatrix} \bar{u}_r(r,\theta,p) \\ \bar{u}_\theta(r,\theta,p) \end{Bmatrix} e^{pt}\,dp, \tag{8.153}$$

where

$$\bar{u}_r(r,\theta,p) = \bar{\varphi}_0(p)\sum_{n=0}^{\infty} e_n\{k_d I'_n(k_d r)$$
$$+ r^{-1}[k_d r A_n(n,p)K'_n(k_d r) + nB_n(n,p)K_n(k_s r)]\}\cos n\theta,$$

$$u_\theta(r,\theta,p) = -r^{-1}\bar{\varphi}_0(p)\sum_{n=1}^{\infty} e_n\{nI_n(k_d r)$$
$$+ [nA_n(n,p)K_n(k_d r) + k_s r B_n(n,p)K'_n(k_s r)]\}\sin n\theta, \tag{8.154}$$

where

$$A_n(n, p) = \frac{\alpha(v, p)}{2\pi\bar{\varphi}_0(p)D(v, p)}\bigg|_{|v|=v=n},$$

$$B_n(n, p) = -\frac{n[k_s^2 a^2 + 2(n^2 - 1)]}{\pi\bar{\varphi}_0(p)D(v, p)_{|v=n}},$$

$\alpha(v, p)$, $D(v, p)$ being given in (8.144). This is the Fourier series representation of the solution. Inversion of (8.153) is accomplished by termwise contour integration of the bracketed terms in (8.154), corresponding to the scattered waves. The terms involving the $I_n's$ there correspond to the incident wave and can be inverted by inspection.

We note that when $n=0$ in (8.154), u_r is the only displacement, and this term corresponds to axially symmetric deformation. Substitution of $n=0$ into the A_n term of (8.154) shows, as we might expect, that $\bar{u}_r$ takes the form of $\bar{u}(r, p)$ in (5.161), the transformed solution of the pressurized cylindrical cavity problem of § 5.11.2, in which $D(0, p)$ here $=F(p)$ there. Recall that the possible zeros of $F(p)$ was of concern in our inversion in § 5.11.2, and we proved that $F(p)$ had no zeros for $\operatorname{Re} p \geq 0$, leading to the contour integration over the contour shown in fig. 5.7. It follows the same contour can be employed for the other n values in the present problem based on similar arguments to those in § 5.11.2. That is, (1) since the integer order $I_n(p)$ functions are entire functions of p, we have only the branch point at $p=0$, the one common to all the $K_n(p)$ functions, and (2) there are no zeros of $D(n, p)$ in $\operatorname{Re} p \geq 0$ because the present problem has a static solution. The technique would therefore yield a series of line integrals from the integration up the imaginary axis, plus a series of residue terms from the corresponding paths about $p=0$. This solution can be evaluated by integrating the line integrals numerically.

Baron and Matthews [8.13] used essentially the foregoing technique[6] for solving the present problem. In their work they completed the path of the Fourier transform inversion integral in the upper half plane which necessitated locating the zeros of the denominator function [corresponding to our $D(n, p)$] there for each value of n. Completion of the path along the

---

[6] Instead of the Laplace transform on time t they used a half range exponential Fourier transform with complex argument; cf. § 5.6.2.

real axis in their technique would have eliminated the need for these zeros, as it did in the technique leading to the inversion of (8.153) (with its analogous integration along the imaginary axis in the p-plane). Since the number of these zeros increase with increasing n it is not a trivial numerical problem to locate these zeros, aggravated further by the complicated representations of the Bessel functions making up $D(n, p)$. For the present step incident stress pulse, Baron and Matthews found only $n=0$, 1 and 2 (a three term Fourier series) were needed for the moderately longer time response of the circumferential (hoop) stress σ_θ at a station (a, θ) where the maximum of the stress was found. This is because the contributions from the circumferential modes for $n>2$ are important for early time, but negligible for the later times. In [8.13, figs. 5,6] numerical results are presented for the σ_θ response at stations $(a, 0)$ and $(a, \pi/2)$, restricted away from early time (up to approximately the time it takes the incident pulse to cross the cavity). For Poisson's ratio $= 1/4$ peak stresses were found to be about 10 percent higher than the long time static values.

In a later paper Baron and Parnes [8.14] made a similar analysis for the response of the radial and circumferential velocities $\dot{u}_r$, $\dot{u}_\theta$. Again with $n=0$, 1, 2 they evaluated these velocities for several cavity wall stations (a, θ) selected from the range $0 \leq \theta \leq \pi$. The response curves appear in figs. 7, 8 of [8.14]. Comparison of these results at the shadow zone station $[a, (3/4)\pi]$ with those of the wave sum method is made in figs. 8.13, 8.14 here. Comparing the curves marked "wave sum" and "Fourier series result" one sees the three term Fourier series results obtained by Baron and Parnes are in fairly good correspondence except for the early times, where the three terms are not enough to give even qualitatively correct results. The close correspondence of the wave sum representation with the physics of the problem is brought out by the fact that pulselike disturbances at the P_{-1} and R_{-1} arrival times are a natural part of the wave sum representation. In fig. 8.13 for the radial velocity note also that the results are in better agreement just a bit after the P_0 arrival time. In his thesis [8.22] Peck made further comparisons of results at other cavity wall stations. Also, using a three term Fourier series for the incident wave, instead of the exact one used by Baron and Parnes, Peck found their results were in better agreement with those of the wave sum method. Conceivably this may be due to cancellations of higher order n contributions to the incident and scattered waves. The reader will also be interested in the discussion of the Fourier series technique in the book by Pao and Mow [8.11, ch. III § 5]. They also apply it to the present problem, throwing further light on the influence of the higher order n values, through numerical evaluations of these modes. For sharper inputs than the step, of course, the higher n modes would certainly be needed.

In conclusion *we emphasize the following on the two methods:* the approach based on the wave-sum method provides better physical insight into the early-time, near-field wave motions than does the Fourier series method. The wave-sum form of solution converges rapidly at short times where the Fourier series solutions are ineffective. The convergence at long time is still fast enough so that seven modes of propagation provide an accurate solution. It must be kept in mind that, for general loading functions, the comparative convergence properties of the two methods depend on the time constants of the load. For load histories that are "more impulsive" than the step-function, the wave-sum method would be even more rapidly convergent; however, for more gradually applied loads, the Fourier series method would become more advantageous.

It is also important to recognize the disadvantages of the wave sum method. First, numerical evaluations of the type presented using this method, are restricted to the shadow zone. Second, the relatively difficult mathematics of Bessel functions of complex order come heavily into play. Finally, the roots of a complicated transcendental equation must be evaluated before the evaluation of the inversion integrals themselves, making the overall numerics for the present approach considerably more involved than those for the Fourier series approach.

8.3.2.10. Diffraction of plane compressional pulse by cavity; approximations and comments. It is the purpose here to discuss briefly approximation techniques that have been applied to the present problem. Gilbert [8.23] contributed high frequency information for both normal (present case) and oblique incidence of the plane compressional pulse on the cavity. He considered only the portion of the cavity in the illuminated zone, restricted to $|\theta| < \pi/3$. Gilbert used a *geometric optics method* with further ray theory considerations to determine the reflected wavefronts (cf. § 2.6.4 and ch. 4 § 2 of Friedlander's book [2.10]). He found for the present problem (incident plane step in σ_r) that the reflected stresses also behaved as a step at their wavefronts. He also found that the presence of the cavity produced a maximum amplification of the field (total) stresses of approximately two, and that the maximum stress was the circumferential stress σ_θ. Grimes [8.24] also studied the wavefronts in the illuminated zone using the method of steepest decents applied to the Friedlander representation of the solution ($m=0$). The approach, of course, is of interest here, however, Grimes finds a linear rise in time for the total σ_θ at $r=a$ which does not agree with the findings of Gilbert discussed above. Finally, the work by Gilbert and Knopoff [8.17] on the *rigid cylindrical cavity* will be of interest to the reader in the present context since they derived *wavefront approximations for both the illuminated and shadow zones*. They use Friedlander's representation of the solution. Wavefront approximations for the illuminated zone are then written through the method of steepest descents. For the early motion in the shadow zone, integration was accomplished through a method developed for high-frequency approximations in analogous scattering problems in acoustic and electromagnetic wave theory.

Approximations for long time in the present problem have also been contributed. Miklowitz [8.15] studied the *Rayleigh waves* for long time and the far field a la the method he used on the line load problem (cf. § 8.3.2.5). In the present case of plane wave impingement on the cavity the Rayleigh surface waves are not singular at their arrival time, but they still are dominant in the dynamic long-time solution. The fact that they are nonsingular at their arrival times makes them experience spatial decay with θ at these times too, and further gives them heads and tails about their arrival time which must be summed to get an accurate description of their periodic (in θ) behavior (cf. the radial displacement in fig. 8.9). The figures 8.15–8.18 show the qualitative nature of these waves (n in these figures is the m we have used in the wave sum here; $c_0 = c_R$ here)[7]. The resultant response for the radial velocity $v_{rR}^H(a, \theta, t)/v_i$, where $v_i = c_R \sigma_i / \mu$ and σ_i is minus the σ_0 input in (8.130), is shown by the solid line in fig. 8.15. It represents the sum of the component waves, where because the heads and tails of these waves level off fairly rapidly, and are of opposite sign, only one or two neighboring waves need be accounted for in addition to the main wave at a certain arrival time for a reasonable approximation, e.g., to the right of the k-th wave arrival, in addition to this wave, the $k-1$, $k+1$, and $k+2$ are used in the sum.

A similar procedure yields the circumferential velocity Rayleigh surface wave $v_{\theta R}^H$ in the response to the plane step wave source. Here, however, the

---

[7] The analysis underlying these figures is essentially the same as that described in § 8.3.2.5. It has been left to the exercises.

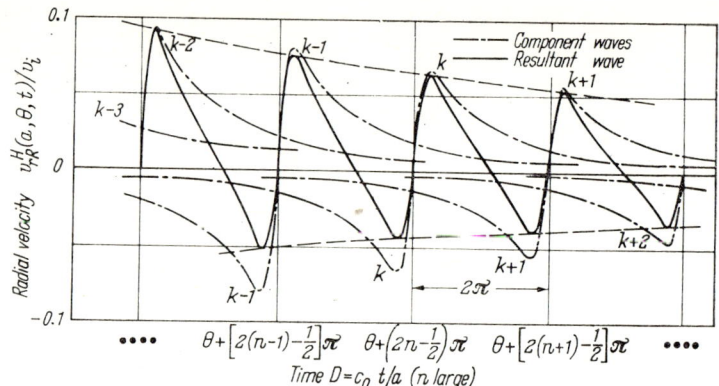

Fig. 8.15. Long time radial velocity response at the cavity wall due to the positive θ-traveling Rayleigh waves (plane step wave source). Note $v_{rR}^H = u_{rR}^{\delta}$.

evenness of $v_{\theta R}^H$ w. r. t. the Rayleigh wave arrival time requires consideration of the positive θ-traveling waves (from the $\theta + 2n\pi > 0$ domain) and negative θ-traveling waves (from the $\theta + 2n\pi < 0$ domain) together. In this manner the oddness of $v_{\theta R}^H$ w. r. t. θ assures convergence. The situation is shown in fig. 8.16a for the station $\theta = \pi/2$, where the component positive θ-traveling waves are shown above and component negative θ-traveling waves below. The resultant $v_{\theta R}^H(a, \theta, t)/v_i$ wave, obtained by summing the component waves in fig. 8.16a, is shown in fig. 8.16b. Here again because the heads and tails of the component waves level off rapidly, and the k and i waves are of opposite sign, only a few neighboring waves in addition to the main wave, are needed in the sum, e.g., in the π domain to the right of the k-th wave arrival, in addition to this wave, the $k-1$, $k+1$, $i+1$, i, and $i-1$ waves are used. In both figs. 8.15 and 8.16a the 2π period of these resultant waves, as well as their spatial attenuation, should be noted. For other θ stations in the physical plane, similar figures can easily be constructed.

The positive θ-traveling component and resultant accelerations $a_{rR}^H(a, \theta, t)/a_i$ and $a_{\theta R}^H(a, \theta, t)/a_i$ are shown in figs. 8.17a, b, respectively, where $a_i = c_R^2 \sigma_i / a\mu$. The figures show that only the neighboring wave to the right or left the main wave, at a particular arrival time, need be considered in addition to the latter in the summation.

Finally in figs. 8.18a, b we show the circumferential stress $\sigma_{\theta R}(a, \theta, t)$ response at the cavity wall due to the component and resultant positive θ-traveling waves for both the delta function (fig. 8.18a), and step (fig. 8.18b) wave sources.

Ch. 8, § 8.3] SCATTERING BY CYLINDRICAL OBSTACLES 553

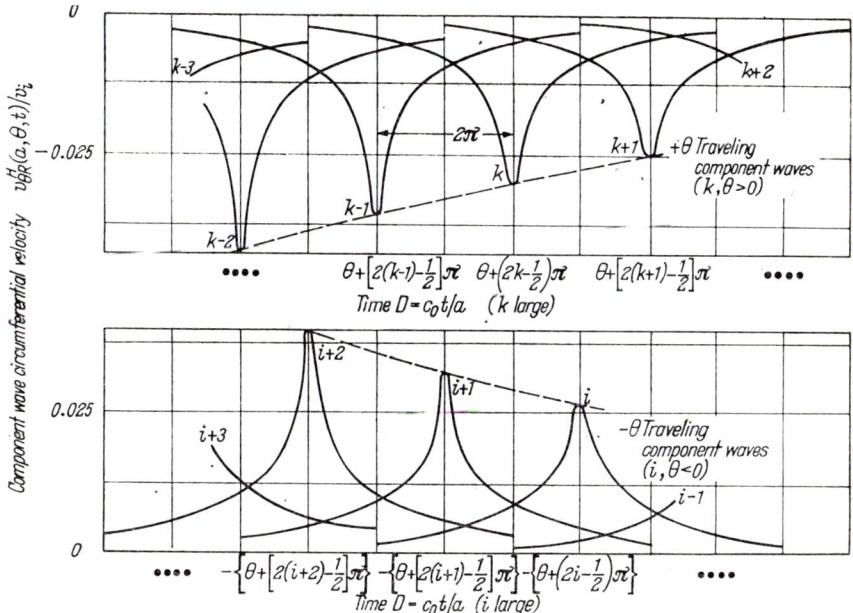

Fig. 8.16a. Component Rayleigh surface waves of circumferential velocity at long time shown in their relative positions of time at $\theta = \pi/2$ (plane step wave source). Note $v_{\theta R}^H = u_{\theta R}^\delta$

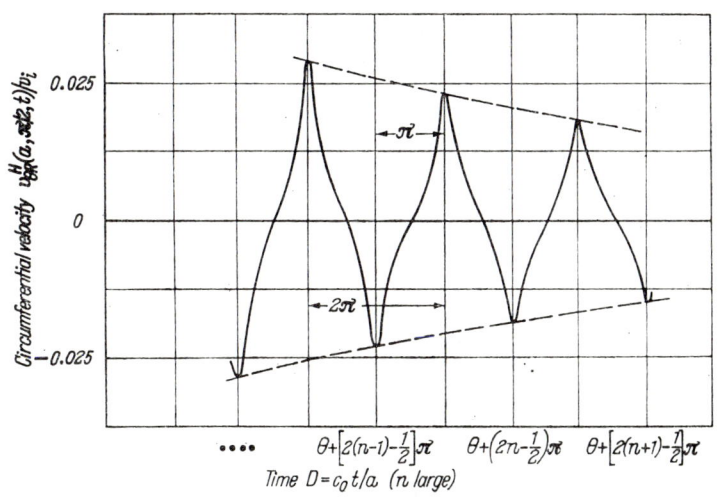

Fig. 8.16b. Rayleigh surface waves of circumferential velocity at $\theta = \pi/2$ at long time (plane step wave source). Note $v_{\theta R}^H = u_{\theta R}^\delta$.

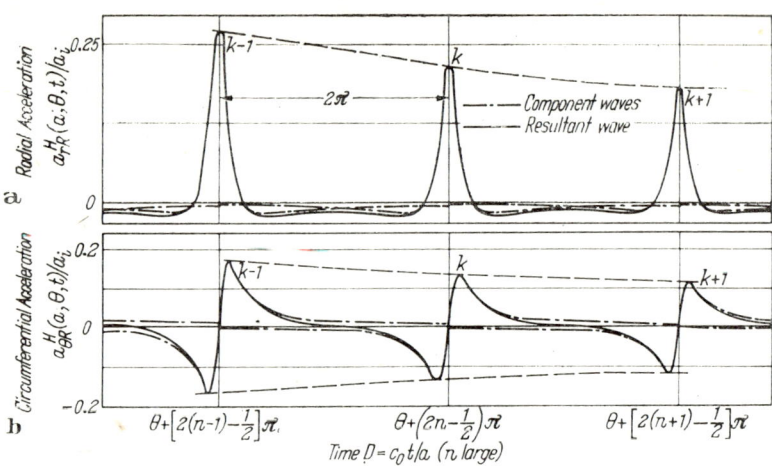

Fig. 8.17. Long time (a) radial and (b) circumferential acceleration response at cavity wall due to positive θ-traveling waves (plane step wave source). Note $a_{rR}^H = v_{rR}^\delta$ and $a_{\theta R}^H = v_{\theta R}^\delta$.

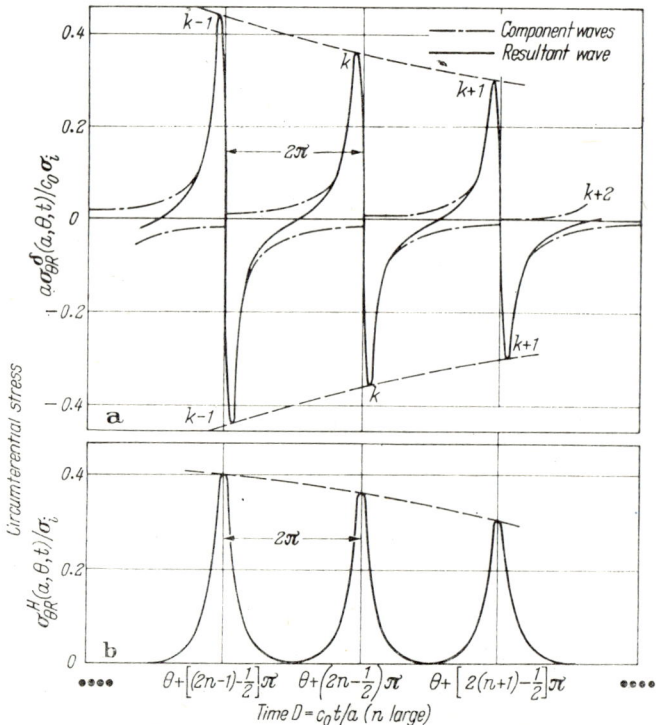

Fig. 8.18. Long time circumferential stress response at cavity wall to plane wave (a) delta function and (b) step function source due to positive θ-traveling Rayleigh waves.

Soldate and Hook [8.25] derived the *long-time response* at the cavity wall ($r=a$) in the present problem. They used the Fourier series form of the solution (8.153), (8.154) applying the asymptotics of the Laplace transform and its inverse (discussed in § 5.10.2.4) to the transformed displacements (8.154) and the corresponding hoop stress $\bar{\sigma}_\theta$ for $r=a$. Specifically these Laplace transforms are expanded in power series in p and inverted term by term. With our step input it is found that only a few terms are needed to give adequate results for the very long time approximation sought. As the input sharpens, more and more terms are needed, and further the accuracy of the results is harder to assess. Soldate and Hook found the leading terms of the long-time velocities to be the rigid body velocities of the cavity wall

$$v_{rB}^H(a,\theta)/v_i = -(c_s^2/c_R c_d)\cos\theta,$$

$$v_{\theta B}^H(a,\theta)/v_i = (c_s^2/c_R c_d)\sin\theta,$$

(8.155)

and the leading term of the hoop stress to be

$$\sigma_{\theta S}^H(a,\theta)/\sigma_i = -(2/k^2)(k^2-1-2\cos 2\theta).$$

(8.156)

Proof of (8.155), (8.156) is left to the exercises. As it should be (8.156) is the same as the solution for the elastostatic hoop stress that can be obtained through superposition of Kirsch's classical solution for the rectangular plate with a circular hole subjected to uniform compressions on one set of the edges (cf. Timoshenko and Goodier [7.2, pp. 90–97]). Of course (8.155), (8.156) are the limiting solutions for time infinite. As such they will be the only responses present, the periodic Rayleigh waves having long since died out through spatial decay. However, since the decay to the static solution is quite slow, one would expect, say for moderately large times, to have the effects of both the static solution and the Rayleigh surface waves acting at cavity wall stations. At the stations $\theta=0$ and π, where $\sigma_{\theta S}^H$ in (8.156) is zero (for $\lambda=\mu$, $k^2=3$), one would still have the Rayleigh waves $\sigma_{\theta R}^H$ (depicted in fig. 8.18b) acting there. This adds an interesting new consideration in the determination of dynamic stress concentrations in the cylindrical cavity problem.

8.3.3. Scattering of a plane compressional pulse by a circular cylindrical elastic inclusion; a brief discussion

8.3.3.1. The problem. Figure 8.6 again depicts the problem with the incident plane pulse shown there, but now the cavity is replaced by the cylindrical elastic inclusion. The *inclusion* which is in the interior region $0 \leq r < a$, is an elastic solid of different properties than that in the exterior region $r > a$ ($r = a$ is the interface). To state the problem mathematically we first define the potentials φ_α, ψ_α, $\alpha = 1, 2$, where 1 corresponds to the inner solid and 2 the outer solid. The outer solid then is governed by the same equations as in the plane pulse cavity problem of § 8.3.2.6, i.e., governing wave equations (8.105), displacement- and stress-potential relations (8.109) and solution forms (8.131), all with subscript 2 on φ and ψ and related quantities. We note φ_i is again given by (8.130). Added here then would be (8.105), (8.109) with subscript 1 on φ and ψ to represent the inner solid. Boundary conditions at $r = 0$ would now be that this interface has continuous displacements and stresses. Quiescent initial conditions of the type (8.133) on the outer solid potentials, and also now on the inner solid potentials, are assumed. In turn the radiation conditions (8.134) apply again to the outer solid potentials, and consistently, we require the inner solid potentials to be bounded. This completes the statement of problem.

This more general case of transient wave scattering from a cylindrical elastic inclusion is *more complicated* than the preceding cases treated for the cavity because of the *refracted waves* that are generated at the interface of the two solids. As fig. 8.19 shows, for example, for an incident *P*-wave ray, the refracted *P* wave is responsible for a multitude of other rays that are due to its *external refractions to the outer solid and reflections within the inclusion*. The problem is of interest, for example, in the dynamic response of *fiber-reinforced composite materials*. In these materials one asks whether stress wave singularities, arising from *wave focusing* can cause separation at the fiber (inclusion)-matrix (outer solid) interface, hence weakening the composite.

8.3.3.2. The literature; methods and results. The problem has been attacked in recent years by Ting and Lee [8.26], Ko [8.27], Achenbach et al. [8.28] and Griffin and Miklowitz [8.29]. The objective in the first two of these works was to ultimately determine the dispersive effect of an array of inclusions on the incident stress pulse. The latter two are concerned mostly with focusing.

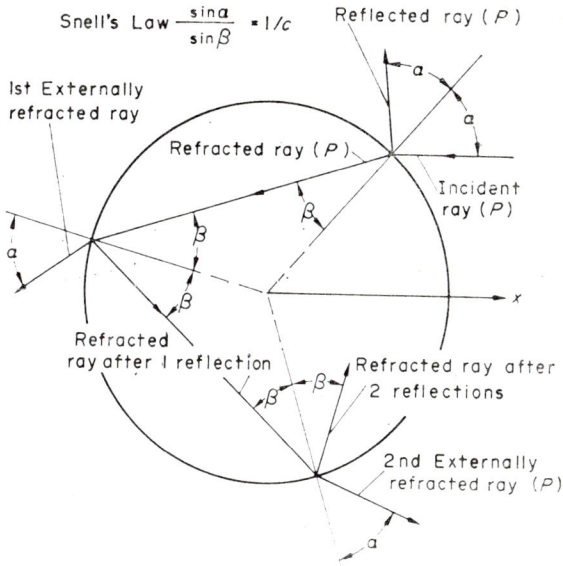

Fig. 8.19. Ray geometry of the refracted dilatational waves.

Focusing occurs when a ray touches a *caustic*, a caustic *being an envelope of converging rays*. Figure 8.20 (taken from [8.29]) shows one example of a caustic that is generated within the inclusion. As one can see, the caustic is formed by the family of refracted *P*-wave rays in the interior, once reflected

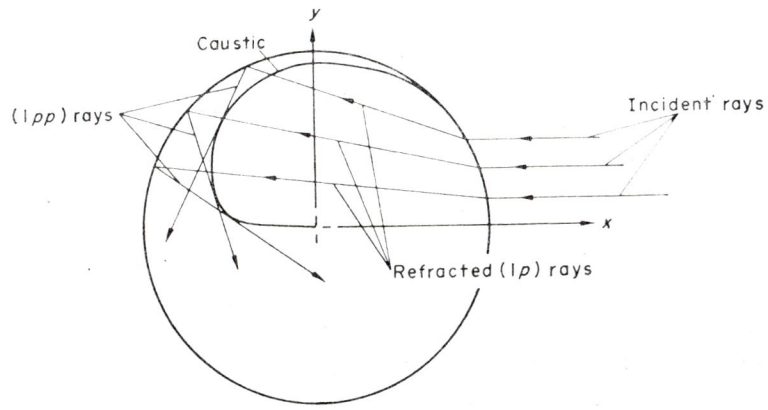

Fig. 8.20. (1 *pp*) rays and caustic for $c=1.5$, $0 \leq \theta < 2\pi$.

from the interface (the 1 pp rays in the figure). This case is the important practical one in which the fiber is a stiffer material than the matrix ($c = c_{d1}/c_{d2} > 1$).

The papers by Ting and Lee and by Ko show that *caustics can occur and have wavefront singularities there*. However, they do not bring out the important wavefront singularities that occur after focusing. Ting and Lee studied the interaction of an incident plane dilatational pulse with a circular cylindrical (or spherical) elastic inclusion using the wavefront analysis of geometrical acoustics, cf. § 2.6.4 and Friedlander [2.10, ch. 3 § 6]. They determine the pressure field for the times that include the incident wavefront's reflection at the interface, its transmission (refraction) into the inclusion and its emergence into the outer medium. Curves are given for the dilatational (and equivoluminal) wavefront positions (with time) and for the corresponding magnitudes at the pressure wavefronts.

Making use of *integral representations* of the Kirchhoff type [cf. (2.117)], Ko determined wavefront stresses and displacements for the interior, exterior and the interface fields. He presents the dilatational wavefront positions for the cases where the inclusion is either stiffer ($c > 1$) or less stiff ($c < 1$) than the matrix material. Further he presents numerical results for the wavefront magnitudes of the stresses and displacements along the interface as a function of circumferential angle.

In their work Achenbach *et al.* were mainly interested in *focusing* effects. They carried out experiments on dynamically edge loaded (explosive charges of lead azide) Plexiglas sheets having a single circular aluminium inclusion and, interestingly, obtained photographic evidence of separation of the inclusion from the Plexiglas. As the magnitude of the charge was increased the amount of separation increased, with total separation of the inclusion occuring for the largest charge used. Analytical work was carried out to try to correlate the regions of separation with the focusing of the first wavefront (dilatation). They used a geometric acoustics approximation for this wavefront up to its arrival at a focus point on a caustic. Then, since this approximation breaks down on the caustic, they followed Friedlander's work in acoustics [cf. 2.10, p. 67] in which he used Poisson's integral formula [cf. (2.146)] to carry the approximation beyond the caustic point. In the elastic case of [8.28], analogously Love's integral representation for the displacement field (cf. [1.2], p. 303 and [2.16] §§ 14–17 for derivation) was the tool. Correlation was found for the experimental results.

The work by Griffin and Miklowitz corroborates, and considerably extends, the findings of Achenbach et al. through a more *general method of analysis for treating singular wavefronts*. In this method a Watson-type lemma is used. The lemma relates the asymptotic behavior of the solution at its wavefronts to the corresponding asymptotic behavior of its Fourier transform on time for large values of the transform parameter. Again, as in the foregoing work on the cylindrical cavity, the Friedlander representation is used to handle the θ variable, here however finding application also to the interior region (inclusion) through the aforementioned Watson-type lemma. The lemma not only handles the first wavefront to arrive but also the later arriving ones. This property is quite important in *focusing problems* since they often have later arriving waves which have focused and are singular.

In [8.29] both of the cases $c>1$ and $c<1$ are analyzed in detail (Fig. 8.21

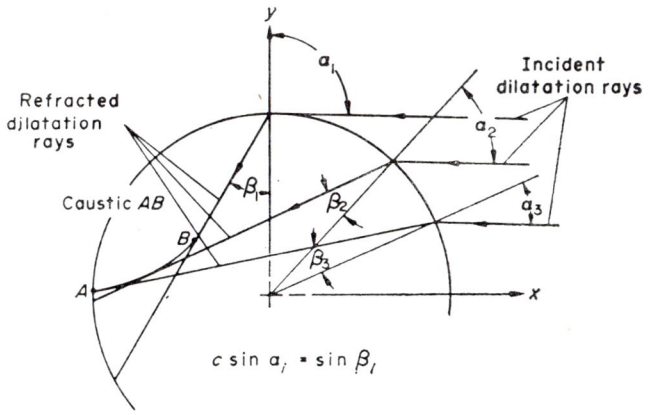

Fig. 8.21. Refracted dilatational rays and caustic for $c=0.5$, $0 \leq \theta < 2\pi$.

shows the rays and caustic for the latter case). Careful studies of the ray geometry involved in focusing, aided by the Watson-type lemma, bring out the nature of this phenomenon. In the *case $c<1$* (fiber softer than matrix) it is shown that along a ray the incident step stress pulse remains a step pulse upon refraction. However, upon reaching the caustic the *wavefront singularity* becomes of the type $(t-t_{d1})^{-1/6}$, where t_{d1} is the dilatational wave arrival time. Further propagation takes the wavefront past the caustic where it becomes logarithmic, i.e., $ln|t-t_{d1}|$.

The *case $c>1$* shows logarithmic singularities develop in the stress wavefronts here too, much as in the manner of the previous case, i.e., after the ray touches and goes beyond the caustic. As fig. 8.20 shows, this happens to the once reflected family of refracted rays ($1pp$). It is clear from this figure therefore, taking into account the fact that there are two systems of wavefronts traveling in this problem (the positive- and negative-θ traveling systems), that *every point on the interface $r=a$ experiences a logarithmic singularity* (from refracted waves that have reflected once).

Other points of interest found in [8.29] are that (1) the interior refracted wavefronts have logarithmic singularities that are refracted into the exterior solid unchanged, (2) the interior wavefronts reflect n number of times, a process which results in a decay in the magnitude of their coefficients as $|d|^n$, $d<1$ and (3) the effects of the diffracted waves in the problem were negligible with respect to those of the focused refracted waves.

8.3.4. Diffraction of an elastic pulse by a spherical cavity; a brief discussion

8.3.4.1. The literature. Important to the literature in the present topic was the relatively early paper by Nagase [8.30] that appeared in 1956, and the paper by Nussenzveig [8.31] in 1965. Nagasse treated the problem of the diffraction of harmonic in time waves from the cavity, generated by an exterior point source, say at $r_1 > r_0$, with r_0 the radius of the cavity. He treated both dilatational and equivoluminal sources and obtained important high-frequency approximations. With a similar interest Nussenzveig treated the related case of the high-frequency scattering of an acoustic harmonic plane wave from an impenetrable sphere. Later work by Norwood and Miklowitz [8.32] extended the results of Nagase and Nussenzveig to transient elastic wave (or pulse) diffraction from the spherical cavity. Figures 8.22, 8.23 depict the two problems treated in [8.32], respectively a *sudden normal point load on the cavity wall, and the impingement of a plane dilatation pulse on the cavity.* Approximations at the cavity wall (in the shadow zone) for the displacements at the dilatational wavefronts, and for the Rayleigh surface waves, were obtained for both problems. The next section reviews briefly the essential features of the method of solution used in [8.32] and the results obtained.

8.3.4.2. Method of solution; results. The problem of the diffraction of a pulse from a spherical cavity is *closely related* to that for the circular cylindrical cavity. Indeed the geometrical optics of the former case is similar to that

Ch. 8, § 8.3] SCATTERING BY CYLINDRICAL OBSTACLES 561

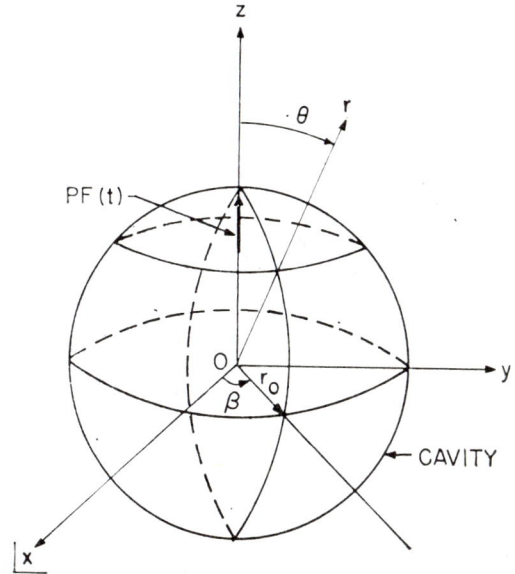

Fig. 8.22. Problem of cavity surface normal point load source.

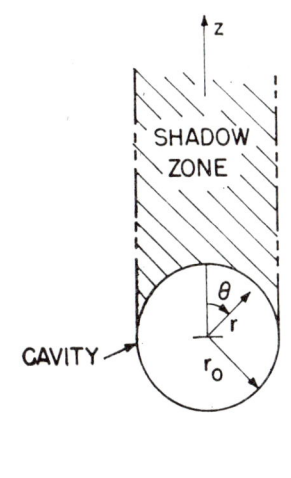

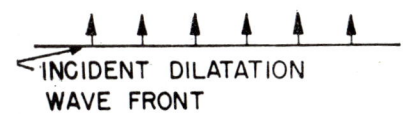

Fig. 8.23. Problem of incident plane dilatational pulse.

of the latter. In fact the wavefronts for the circular cylindrical cavity case are the meridional sections of those in the spherical cavity case, the full fronts in the latter being obtained by rotation about their axes of symmetry. One would expect then, the present two problems to be closely analogous to the problems of §§ 8.3.2.3, 8.3.2.6 with methods of analysis being analogous to those used on the latter. We bring out here therefore only the *essential differences in the methods and features* of the two classes of problems, further detail being left to [8.32].

The Laplace transform is again used on time t. This reduces the governing potential equations of motion to

$$\nabla^2 \bar{\varphi}(r, \theta, p) = k_d^2 \bar{\varphi}, \qquad \nabla^2 \bar{\chi} = k_s^2 \bar{\chi}, \qquad \bar{\psi} = \partial \bar{\chi}/\partial \theta \qquad (8.157)$$

where $\bar{\varphi}$ is the transformed dilatational, and $\bar{\chi}$, $\bar{\psi}$ equivoluminal scalar potentials, r, θ are the radial and latitudinal angle coordinates (cf. fig. 8.22) and $k_d = p/c_d$, $k_s = p/c_s$, p being the Laplace transform parameter. Further spherical harmonics separation of (8.157), and invoking the radiation condition, produces the general solutions

$$\bar{\varphi} = \sum_{n=0}^{\infty} A_n(k_d r)^{-\frac{1}{2}} K_{n+\frac{1}{2}}(k_d r) P_n(\cos \theta),$$

$$\bar{\psi} = \sum_{n=0}^{\infty} B_n(k_s r)^{-\frac{1}{2}} K_{n+\frac{1}{2}}(k_s r) \mathrm{d}P_n(\cos \theta)/\mathrm{d}\theta, \qquad (8.158)$$

where $K_{n+\frac{1}{2}}(z)$ is the modified Bessel function of the second kind of order $n+\frac{1}{2}$, and $P_n(\cos \theta)$ is the *Legendre polynomial of order n*.

Point load problem

This problem is analogous to the line load problem of the circular cylindrical cavity. Consideration of the boundary conditions at the cavity wall produces the Laplace transformed solution $\bar{\varphi}$, $\bar{\psi}$ for the problem. The Bromwich integral then gives the formal solutions for $\varphi(r, \theta, t)$, $\psi(r, \theta, t)$. Singularities of $\bar{\varphi}$, $\bar{\psi}$ in the p-plane are simple poles, steming from the frequency equation $\triangle (r_0, n, p) = 0$ and the transform of the time input function of the point load. Such poles cannot be in the right half p-plane, $\operatorname{Re} p > 0$, since the problem has a static solution. It follows the *Br* contour can be

Ch. 8, § 8.3]   SCATTERING BY CYLINDRICAL OBSTACLES   563

traded for a path up the imaginary axis. The dynamic solution for φ then becomes

$$\varphi(r, \theta, t) = \frac{1}{2\pi i} \int_0^\infty \bar{\varphi}(r, \theta, i\omega) e^{i\omega t} d\omega, \quad (8.159)$$

where $\bar{\varphi}$ has the form

$$\bar{\varphi}(r, \theta, i\omega) = \sum_{n=0}^\infty (2n+1) \frac{f(r_0, n, i\omega)}{\Delta(r_0, n, i\omega)} \frac{K_{n+\frac{1}{2}}(i\omega r/c_d)}{(i\omega r/c_d)^{\frac{1}{2}}} P_n(\cos\theta)$$

with a similar expression for $\psi(r, \theta, t)$.

As we have noted earlier, a series such as that in (8.159) would have slow convergence properties at the high frequencies. Our remedy for this in the cylindrical cavity and inclusion problems was the use of Poisson's summation formula (8.101) and related theorem given in § 8.3.2.2. The analogous treatment here exploits *Watson's transformation*, a technique introduced initially and used widely in the study of the diffraction of electric waves by the earth [8.33]. It was also used in [8.30, 8.31].

Watson's transformation is based on the formula

$$\sum_{n=0}^\infty g(n+1/2) = \frac{1}{2} \int_{C_w} g(v) \frac{\exp(-i\pi v)}{\cos v\pi} dv, \quad (8.160)$$

where C_w is the contour shown in fig. 8.24. The formula is easily checked

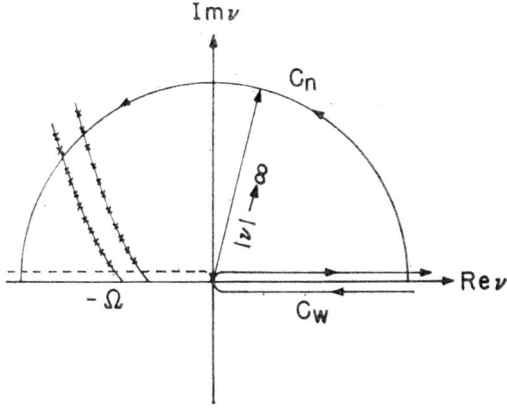

Fig. 8.24. Integration in the v-plane.

by evaluating the residues at $v = n + \frac{1}{2}$. On applying (8.160) to $\bar{\varphi}$ in (8.159) (i.e., identifying the series in (8.159) with that in (8.160)) and using the expression $P_n(\cos\theta) = \exp(i\pi n) P_n[\cos(\pi - \theta)]$, one finds

$$\bar{\varphi}(r, \theta, i\omega) = -i \int_{C_w} \frac{\varphi_{v-\frac{1}{2}}(r_0, r, v-\frac{1}{2})}{\Delta(r_0, i\omega, v-\frac{1}{2})} P_{v-\frac{1}{2}}[\cos(\pi - \theta)] \frac{v \, dv}{\cos v\pi}, \qquad (8.161)$$

where

$$\varphi_{v-\frac{1}{2}}(r_0, r, v-\frac{1}{2}) = f(r_0, i\omega, v-\frac{1}{2}) \frac{K_v(ibv)}{(ibv)^{\frac{1}{2}}}, \quad v = \frac{\omega r}{c_s}, \quad b = \frac{c_s}{c_d}.$$

By substituting $-v$ for v in the integrand of (8.161), using the identities $P_{-\lambda-\frac{1}{2}}(\cos\theta) = P_{\lambda-\frac{1}{2}}(\cos\theta)$, $K_{-v}(z) = K_v(z)$ it can be shown that this integrand is an odd function of v. It follows that the lower half of C_w in fig. 8.24 may be replaced by its reflection in the origin, the dashed line in the figure. This contour and the upper half of C_w are equivalent to a straight line located just above the real axis on which the expansion

$$\frac{1}{\cos v\pi} = 2 \sum_{s=0}^{\infty} (-)^s \exp[i\pi v(2s+1)] \qquad (8.162)$$

is valid. Substitution of this result into (8.161) yields

$$\bar{\varphi}(r, \theta, i\omega) = -2i \sum_{s=0}^{\infty} (-)^s \int_{-\infty}^{\infty} v \frac{\varphi_{v-\frac{1}{2}}}{\Delta} P_{v-\frac{1}{2}}[\cos(\pi-\theta)] \exp[i\pi v(2s+1)] dv$$
$$(8.163)$$

with a similar result for $\bar{\psi}$.

Next in the method one exploits the zeros of Δ (and corresponding simple poles) in the integrand of (8.163). They are the branches of the frequency equation $\Delta(r_0, i\omega, v-\frac{1}{2}) = 0$, which give the modes of propagation through residue theory. This is done through the sequence of paths C_n passing between the zeros of Δ, as shown in fig. 8.24. The paths C_n can be shown to give a zero contribution to the contour integrations (cf. Norwood's thesis [8.34] for the proof). It follows that (8.163) reduces to

$$\bar{\varphi}(r, \theta, i\omega)$$
$$= 4\pi \sum_{s=0}^{\infty} (-)^s \sum_{j=1, 2\ldots} \left\{ v \varphi_{v-\frac{1}{2}} P_{v-\frac{1}{2}}[\cos(\pi-\theta)] \exp[i\pi v(2s+1)] \left(\frac{\partial \Delta}{\partial v}\right)^{-1} \right\}_{v=v_j}$$
$$(8.164)$$

for $0 < \theta \leq \pi$, v_j being the zeros of $\Delta(r_0, i\omega, v-\tfrac{1}{2})$ in the second quadrant. Similar expressions for $\bar{\psi}$ and displacements $\bar{u}_r$, $\bar{u}_\theta$ are given in [8.32]. At this point it should be emphasized that Watson's transformation was the important tool that enabled us to get a solution in terms of the frequency branches for the stress free spherical cavity. Simply put, it has allowed us to *trade one set of poles for a much more important set*, just as Poisson's summation formula did for us in the cylindrical cavity problem. In [8.34, pp. 17,52] it is proved that the two techniques are equivalent for the present problems. The usefulness of (8.164) in obtaining the high frequency-wavefront and Rayleigh surface wave approximations will become apparent in the sequel.

The dilatational wavefronts for the displacements u_r, u_θ are derived in [8.32, 8.34] for the step in time load. They are obtained from expressions like (8.164) by utilizing high frequency-large wave number approximations to the P branches of $\Delta(r_0, i\omega, v-\tfrac{1}{2})=0$, say v_{2j} (nomenclature of [8.32]), which are given in Nagase's work [8.30]. They are of the same form as (8.115a), which means the higher the branch the larger its imaginary part. Note this is indicated by one of the series of poles in the second quadrant of v in fig. 8.24. Seeking the response on the surface of the cavity we first substitute the v_{2j} into $\bar{u}_r(r_0, \theta, i\omega)$ which gives

$$\bar{u}_r(r_0, \theta, i\omega)$$
$$= M\left(\frac{i\omega a}{c_d \sin\theta}\right)^{-\tfrac{1}{2}} \sum_{s=0}^{\infty} (-)^s \sum_{j=1,2,\ldots} \left\{e^{iv[2\pi(s+1)-\theta]} + e^{iv[2\pi s+\theta]}\right\}_{v=v_{2j}}, \quad (8.165)$$

valid for $0 < \theta < \pi$, where M is a constant. $\bar{u}_\theta$ is similar. Letting $s=0$ we see the terms in (8.165) correspond to two diffracted waves, one which reaches the station (r_0, θ) directly from the point source at the north pole $(r_0, 0)$, and the other through a reflection from the south pole (r_0, π). This *reflection process goes on ad infinitum at the north and south poles*, times permitting, the reflecting waves being represented by $s \geq 1$. They are waves that encircle the cavity $2s$ times, so that the corresponding angular paths are increased by $2\pi s$. Note here our wave sum is the sum on s in (8.165).

Recall the inversion path in the p-plane that led to (8.165) was the imaginary axis, and this path is equivalent to the Br path. This permits setting $i\omega = p$ in (8.165), and inverting it through the asymptotics of this inversion integral for large p and short time. Leaving the details to [8.32, 8.34], it was found to determine the wavefronts for u_r, u_θ only the terms $s=0, j=1$ needed to be taken into account for $\theta > 1$. These *wavefronts* have the forms

$$\begin{Bmatrix} u_r(r_0, \theta, t)/u_0 \\ u_\theta(r_0, \theta, t)/u_0 \end{Bmatrix} = (\sin \theta)^{-\frac{1}{2}} \begin{Bmatrix} M_r T^{-\frac{1}{2}} \\ M_\theta T^{\frac{5}{2}}/\theta^3 \end{Bmatrix} \exp(-b\theta^{\frac{3}{2}}T) H(T), \quad (8.166)$$

where $u_0 = P/r_0(\lambda + 2\mu)$ (P is the magnitude constant of the point load), $T = tc_d/r_0 - \theta$, $H(t)$ is the Heaveside step function and the inequalities

$$1 < \theta < \pi, \qquad (tc_d/r_0) < \min(k\theta, 2\pi - \theta)$$

must hold and M_r, M_θ and b are constants. These wavefront reponses for Poisson's ratio 1/3 and the two stations $(r_0, \pi/3)$, $(r_0, 3\pi/4)$ are plotted in

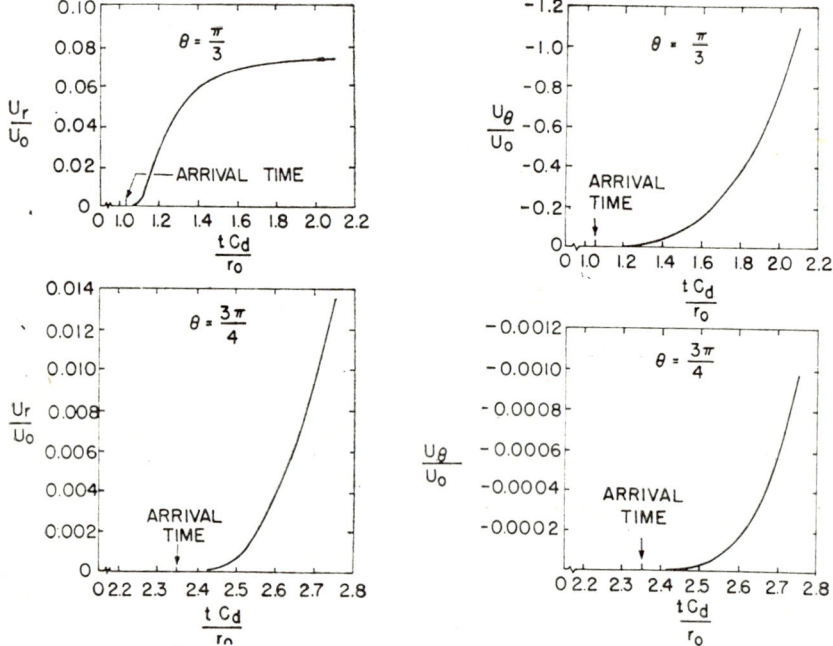

Fig. 8.25. Radial displacement response to point load at cavity wall $r = r_0$, for $\theta = \pi/3$ and $3\pi/4$.

Fig. 8.26. Tangential displacement response to point load at cavity wall $r = r_0$, for $\theta = \pi/3$ and $3\pi/4$.

figs. 8.25, 8.26, respectively for u_r and u_θ. Their behaviors are typical of the diffraction of scalar pulses by curved boundaries (cf. [2.10], ch. 6, § 7).

The Rayleigh surface waves were also calculated using (8.115b) by essentially the method given in [8.12]. The results are similar to those in fig. 8.9.

Ch. 8, § 8.3]   SCATTERING BY CYLINDRICAL OBSTACLES   567

Incident plane dilatational pulse
This boundary value problem is analogous to the corresponding one of the cylindrical cavity [cf. § 8.3.2.6]. The technique for solving it is essentially the same as that just used to solve the point load problem (cf. [8.32, 8.34] for further detail). The resulting dilatational wavefronts for u_r, u_θ at the station $\pi/6$ for Poisson's ratio again $1/3$ are similar to (8.166). They are shown in fig. 8.27 where $U_1 = \tau_0 r_0/(\lambda+2\mu)$, τ_0 being the magnitude constant of the axial stress associated with the incident potential pulse. The Rayleigh

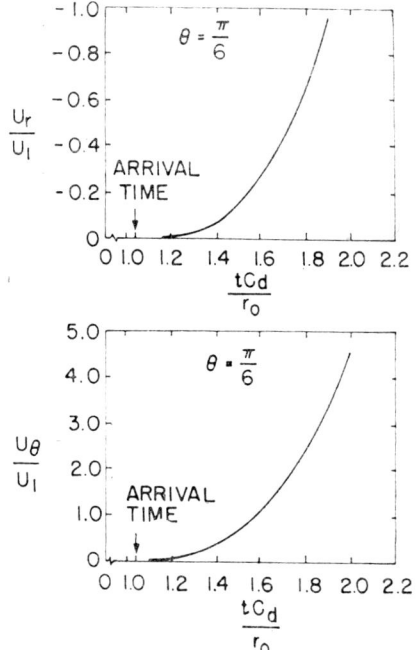

Fig. 8.27: Response of radial and tangential displacement at cavity wall $r=r_0$, and $\theta = \pi/6$, plane pulse diffraction problem.

surface waves here are similar to those found in [8.15]. The latter are discussed in § 8.3.2.10 and exhibited in figs. 8.15–8.18.

8.4. Exercises

8.1. In the SH-pulse diffraction problem of § 8.2.2. derive (a) the dual integral equations (8.11), (8.12) for the unknown $A(\zeta)$, and (b) their solution given by (8.17).

8.2. Making use of $A(\zeta)$ in (8.17) derive the solution (8.29) for the displacement $v(r, \theta, t)$ in the SH-pulse diffraction problem of § 8.2.2.

8.3. Derive the solution for the diffraction of a plane horizontally polarized shear pulse by a rigid half plane (§ 8.2.3) by the method of § 8.2.2 used on the analogous problem for the traction free half plane.

8.4. In the dilatational-pulse diffraction problem of § 8.2.4 derive (a) the dual integral equations (8.42), (8.43) for the unknown $A_1(\zeta)$, and (b) their solution given by (8.48). Through (8.41) $B_1(\zeta)$ is then also determined, and so the transformed displacements $\bar{u}_s$, $\bar{w}_s$ of the symmetric problem through (8.39a). Similarly, derive (c) the dual integral equations (8.50), (8.51) for the unknown $A_2(\zeta)$, and (d) their solution given by (8.52) with $B_2(\zeta)$ then being given by (8.49). These then determine $\bar{u}_s$, $\bar{w}_s$ of the antisymmetric problem also through (8.39a).

8.5. Derive the factorization of $F(\zeta)$ given by (8.55). Then carrying out the indicated contour integrations, derive (8.56) and (8.57).

8.6. For the symmetric part of the dilatational pulse diffraction problem the Laplace transforms of the dilatational diffracted wave contributions, to the scattered displacements $\bar{u}_s$, $\bar{w}_s$ in (8.39a), are given by (8.59). Verify the latter relations by first showing that $\eta'_d(\zeta)A_1(\zeta)$ and $\eta'_d(-\bar{\zeta})A_1(-\bar{\zeta})$ are conjugate functions on the fourth and third quadrant parts of the upper path C in fig. 8.3, respectively, path C being represented by (8.58).

8.7. Making use of the discussion following (8.66), verify that (8.68) give the Laplace transformed diffracted regular equivoluminal waves for the symmetric part of the dilatational pulse diffraction problem, and (8.69) the Laplace transformed head waves. Note the derivation of (8.68a), (8.69a) makes use of (8.70a), (8.70b), respectively.

8.8. Derive in detail the complete geometrical displacement solution $u^g(r, \theta, t)$, $w^g(r, \theta, t)$, given by (8.99), for the dilatational pulse diffraction problem.

8.9. For the diffraction of a dilatational pulse from a rigid half plane set down the analogous symmetric and antisymmetric problems corresponding to (8.34)–(8.38) for the traction free half plane case. Then deduce the analogous dual integral equations for these problems corresponding to (8.42), (8.43) and (8.50), (8.51). Discuss what further differences in the solution procedure, and results, you might expect to find in the present problem when compared to those for the traction free half plane case treated in the text.

8.10. For the problem of the normal line load source on the circular cylindrical cavity wall, derive the formal solution for the wave forms u_r^*, u_θ^* given in (8.111). Then verify the exact inversion for these forms (8.114).

8.11. Approximate the frequency equation (8.114b), for large and positive frequency η, ζ and wave number v, to show it degenerates to (8.119), the equation for the speed of the Rayleigh wave on the free surface of an elastic half space. Note then that this proves (8.114b) contains a branch which has the asymptote $v = c_d \eta / c_R$ (8.120) for the Rayleigh wave pair (η, v).

8.12. Starting with (8.121) derive (8.122)–(8.128) for the Rayleigh wave forms $u_{rR}^{\delta*}(r, \theta, t)$, $u_{\theta R}^{\delta*}(r, \theta, t)$, and then representations on the cavity wall $(r \to a+)$, in the normal line load source problem. Then derive (8.129), the Rayleigh positive θ-traveling surface wave contributions to the long time-far field response in the step normal line load problem.

8.13. For the problem of the diffraction of a plane compressional pulse by a circular cylindrical cavity, derive the formal solution (8.111a) for the wave forms u_r^*, u_θ^* where now $\tilde{u}_r^*$, $\tilde{u}_\theta^*$ are given by (8.145). Then verify the exact inversion for these wave forms (8.147), making sure you prove the sum of the incident and scattered dilatational potentials in the first of (8.143) is analytic, since it is even in order v which permits dropping the absolute value signs on v wherever it occurs.

8.14. Derive the Fourier series formal solution (8.153), (8.154) for the problem of the diffraction of a plane compressional pulse by a circular cylindrical cavity.

8.15. As pointed out in § 8.3.2.10, Rayleigh waves in the problem of the diffraction of a plane compressional pulse by a circular cylindrical cavity were analyzed in [8.15] through essentially the same method described in § 8.3.2.5 for the case of the normal line load source on the cavity wall. Making use of that method, show that the wave forms $u_{rR}^{\delta*}$, $u_{\theta R}^{\delta*}$ are given by

$$\begin{bmatrix} u_{rR}^{\delta*}(r,\theta,t) \\ u_{\theta R}^{\delta*}(r,\theta,t) \end{bmatrix} = \int_{v_L}^{\infty} e^{-\mathrm{Im}\, v_R(\mathrm{Re}\,v)\theta} \begin{bmatrix} \tilde{u}_{rR}^{\delta*}(r,v) \sin\left[(E-\theta)v\right] \\ \tilde{u}_{\theta R}^{\delta*}(r,v) \cos\left[(E-\theta)v\right] \end{bmatrix} dv, \qquad v_L \gg 1,$$

where

$$\tilde{u}_{rR}^{\delta*}(r,v) = (2/\pi)^{\frac{1}{2}}\, M\, \left[Q_F(r) e^{-F(r)v} + Q_G(r) e^{-G(r)v}\right] v^{-\frac{1}{2}},$$

$$\tilde{u}_{\theta R}^{\delta*}(r,v) = -(2/\pi)^{\frac{1}{2}}\, M\left[R_F(r) e^{-F(r)v} + R_G(r) e^{-G(r)v}\right] v^{-\frac{1}{2}},$$

$$M = \frac{2\sigma_i(2-\varkappa^2)}{\rho c_R \sqrt{qL}}, \qquad E(t) = D(t) + \pi/2, \qquad D(t) = c_R t/a,$$

$$Q_F(r) = (2-\varkappa^2)\left[\frac{a}{r} - b\varkappa\left(\frac{1+Q}{1-Q}\right)^{\frac{1}{2}}\right]\sqrt{\frac{q}{Q}}, \qquad Q_G(r) = \frac{2aq}{r}\sqrt{\frac{s}{S}},$$

$$R_F(r) = (2-\varkappa^2)\frac{a}{r}\sqrt{\frac{q}{Q}}, \qquad R_G(r) = 2q\left[\frac{a}{r} - \varkappa\left(\frac{1+S}{1-S}\right)^{\frac{1}{2}}\right]\sqrt{\frac{s}{S}},$$

$$F(r) = A(r) + A_0, \qquad G(r) = B(r) + A_0, \qquad A_0 = \tanh^{-1} q - q,$$

with q, s, $\varkappa$ and b, defined in (8.118) and $Q(r)$, $S(r)$, $A(r)$, $B(r)$ and L defined in (8.123). Constants σ_i, ρ and a are respectively the magnitude of the input pulse, the material density and cavity radius. Based on the reasoning that $u_{rR}^{\delta*}$, $u_{\theta R}^{\delta*}$ will have their maxima on the cavity wall and at or near the arrival time $E-\theta=0$, show that the integrals for these quantities given here reduce to

$$\begin{bmatrix} u_{rR}^{\delta*}(a,\theta,t) \\ u_{\theta R}^{\delta*}(a,\theta,t) \end{bmatrix} \simeq e^{-\operatorname{Im} v_R a \theta} \begin{bmatrix} A_r K_r \\ -A_\theta K_\theta \end{bmatrix} [\{A_0^2+(E-\theta)^2\}^{\frac{1}{2}} \mp A_0]^{\frac{1}{2}}/\{A_0^2+(E-\theta)^2\}^{\frac{1}{2}}$$

representing the maxima. Here $A_r = 1 - \operatorname{erf}(A_0 v_L)^{\frac{1}{2}} + 2(A_0 v_L)^{\frac{1}{2}} e^{-A_0 v_L}/\sqrt{\pi}$
$A_\theta = 1 - \operatorname{erf}(A_0 v_L)^{\frac{1}{2}}$, $K_r = M[Q_F(a) + Q_G(a)]$ and $K_\theta = M[R_F(a) + R_G(a)]$.

In like theme, but for $|E-\theta|$ large (before and after the arrival time $E - \theta = 0$), one can approximate the given Fourier integrals (cf. § 5.10.3) to assess the heads and tails of the Rayleigh surface waves. Show by this means that these approximations give

$$\begin{pmatrix} u_{rR}^{\delta*}(a,\theta,t) \\ u_{\theta R}^{\delta*}(a,\theta,t) \end{pmatrix} = \begin{bmatrix} K_r \\ -K_\theta \end{bmatrix} e^{-\operatorname{Im} v_R a \theta} \left[-(2/\pi v_L)^{\frac{1}{2}} e^{-A_0 v_L} \begin{pmatrix} \sin[v_L(E-\theta)-\pi/2] \\ \cos[v_L(E-\theta)-\pi/2] \end{pmatrix} \right]$$

$$\times \frac{1}{E-\theta} + O[(E-\theta)^{-\frac{3}{2}}],$$

which shows the heads and tails of the Rayleigh waves decay rapidly. The figures 8.15–8.18 of the text are based on the results in this exercise.

8.16. Using the Fourier series form of the solution (8.153), (8.154) for the problem of the diffraction of a plane compressional pulse by a circular cylindrical cavity, derive the long-time solution (8.155), (8.156) with the aid of the asymptotics of the Laplace transform and its inverse (cf. 5.10.2.4).

References

[8.1.] A. Sommerfeld, *Optics–Lectures on Theoretical Physics, Vol. IV* Academic Press, New York (1964), 249.
[8.2.] B. Noble, *Methods Based on the Wiener–Hopf Technique for the Solution of Partial Differential Equations*, Pergamon Press, New York (1958).
[8.3.] P. C. Clemmow, *Proceedings of the Royal Society of London* **205** (1951), 286–308.
[8.4.] D. D. Ang, *Journal of Mathematics and Physics* **38** (1960), 246–256.
[8.5.] D. D. Ang and M. L. Williams, *Proceedings of the Fourth Annual Conference on Solid Mechanics*, University of Texas, Austin, Texas (1959), 36–52.
[8.6.] J. K. Knowles and T. A. Pucik, *Journal of Elasticity* **3** (1973), 155–160.
[8.7.] A. Sommerfeld, *Zeitschrift für Mathematik und Physik* **46** (1901), 11–97.

[8.8.] G. H. Handelman and L. A. Rubenfeld, *Journal of Applied Mechanics* **36** (1969), 873–874.
[8.9.] W. T. Koiter, *Proceedings Koninklijke Nederlandse Academie van Wetenschappen, Series B57* (1954), 558–579.
[8.10.] J. W. Miles, *Quarterly of Applied Mathematics* **18** (1960), 37–59.
[8.11.] Y. H. Pao and C. C. Mow, *Diffraction of Elastic Waves and Dynamic Stress Concentrations*, Crane, Russik and Company, Inc., New York (1973), 572–595.
[8.12.] J. Miklowitz, *Pulse Propagation in a Viscoelastic Solid with Geometric Dispersion*. In: Stress Waves in Anelastic Solids, IUTAM Symposium, eds. H. Kolsky and W. Prager. Springer–Verlag, Berlin (1963), 255–276.
[8.13.] M. L. Baron and A. T. Matthews, *Journal of Applied Mechanics* **28** (1961), 347–354.
[8.14.] M. L. Baron and R. Parnes, *Journal of Applied Mechanics* **29** (1962), 385–395.
[8.15.] J. Miklowitz, Proceedings of the Eleventh International Congress of Applied Mechanics, Springer–Verlag, Berlin (1966), 469–483.
[8.16.] J. C. Peck and J. Miklowitz, *International Journal of Solids and Structures* **5** (1969), 437–454.
[8.17.] F. Gilbert and L. Knopoff, *Journal of the Acoustical Society of America* **31** (1954), 1169–1175.
[8.18.] F. G. Friedlander, *Communications of Pure and Applied Mathematics* **7** (1954), 705–732.
[8.19.] E. T. Copson, *An Introduction to the Theory of Functions of a Complex Variable*, Oxford University Press, London (1935), 344.
[8.20.] I. A. Viktorov, *Soviet Physics–Acoustics* **4** (1958), 131–136.
[8.21.] F. Gilbert, *Journal of the Acoustical Society of America* **32** (1960), 841–857.
[8.22.] J. C. Peck, *Plane Strain Diffraction of Transient Elastic Waves by a Circular Cavity*, Ph. D. Thesis, California Institute of Technology, Pasadena, California (1965).
[8.23.] J. F. Gilbert, *Elastic Wave Interaction with a Cylindrical Cavity*, Report to WDIE, AFBMD, Contract AF 04(647)–342, December 1959, E. H. Plesset Associates Inc., Los Angeles, California.
[8.24.] C. K. Grimes, *Studies on the Propagation of Elastic Waves in Solid Media*, Ph. D. Thesis, California Institute of Technology, Pasadena, California (1964).
[8.25.] A. M. Soldate and J. F. Hook, *A Theoretical Study of Structure–Medium Interaction*, Report to AFSWC, Kirtland Air Force Base, Contract AF29(601)–2838, National Engineering Science Company Report AFSWC–TN–61–6, November 1960.
[8.26.] T. C. T. Ting and E. H. Lee, *Journal of Applied Mechanics* **36** (1969), 497–504.
[8.27.] W. L. Ko, *Journal of Applied Mechanics* **37** (1970), 345–355.
[8.28.] J. D. Achenbach, J. H. Hermann and F. Ziegler, *Journal of Applied Mechanics* **37** (1970), 299–304.
[8.29.] J. H. Griffin and J. Miklowitz, *International Journal of Solids and Structures* **10** (1974), 1333–1356.
[8.30.] M. Nagase, *Journal of the Physical Society of Japan* **11** (1956), 279–301.
[8.31.] H. M. Nussenzveig, *Annals of Physics* **34** (1965), 23–95.
[8.32.] F. R. Norwood and J. Miklowitz, *Journal of Applied Mechanics* **34**, (1967),735–744.
[8.33.] G. N. Watson, *Proceedings of the Royal Society of London, Series A* **95**, (1918), 83–99.
[8.34.] F. R. Norwood, *Diffraction of Transient Elastic Waves by a Spherical Cavity*, Ph. D. Thesis, California Institute of Technology, Pasadena, California (1967).

SUPPLEMENTARY READING

The purpose of this addendum on further reading is two-fold. First it is appropriate to guide the reader to other important works that are *natural extensions of the text material*. Secondly, attempts have not been made here to develop topics on elastic waves *dealing with additional effects in the linear elastic medium*, because of limitations on time and space. Hence, it is also appropriate to guide the reader to suitable introductory material on waves in anisotropic and inhomogeneous media, as well as in other complicated media.

On text material

Reflection and refraction of waves at a plane interface

Here an alternate method to that given in Chapter 3 is contained in the papers by Crandall [S.1] and McNiven and Mengi [S.2] on the use of *slowness (reciprocal velocity) diagrams* to study wave reflection and refraction.

Transient waves in the half space

We first note that Chao [S.3] with the aid of Cagniard's method derived the solution for the suddenly applied *surface tangential point load source*. He evaluated the surface displacements and the horizontal displacement along the normal axis below the load. Afandi and Scott [S.4] solved the case of half space excitation by a *time-dependent dipole*, a *nonaxisymmetric problem*. They evaluated numerically the surface displacements due to a surface dipole.

Of note also are works by Mitra [S.5, S.6] for *finite distributed surface sources*. The first was a uniform pressure over an infinite strip. The second was a *uniform pressure over a circular portion* of the surface. Mitra gives numerical evaluations of the displacements in this work. Mitra [S.7] also derived the displacement solution for the case of *a finite distributed impulsive twisting moment* in the surface plane of the half space. Eason [S.8, S.9] also contributed here. The first of his works was on *torsional impulsive loading over a finite circular surface area.* In the second he treated the case of an *impulsive body force.* Evaluations are given for the displacements. Norwood [S.10] treated the *nonaxisymmetric* transient response due to a distributed load over a rectangular area of the surface of the half space.

Concerning traveling loads on the surface of the half space note should be made of the recent work of Freund [S.11] on *nonuniformly moving loads* under the conditions of plane strain. Other works of interest in the present topic are discussed in [4.4].

Transient waves in waveguides

We mention first the work of Kaul and McCoy [S.12]. They used the higher order approximate circular rod theory of Mindlin-McNiven to study the *head of the pulse response,* taking into account higher order terms in this event. Working with the radial strain they show the magnitude of the first oscillation of the head of the pulse becomes larger as the x-station becomes smaller. Further works of interest are those of exact plate theory involving the *surface normal point load source* by Knopoff, Knopoff and Gilbert, Broberg, Davids and Pytel. They are aimed at the early time response. These works and others, are discussed further in [4.4].

Elastic wave scattering

Datta has recently presented a survey [S.13] that is of interest here. He discusses four topics; the scattering of (1) dilatational waves by a liquid sphere or cylinder, (2) waves by *rigid spheroids* and (3) waves by a *rigid circular disk.* The fourth topic is wave propagation in a *half space containing a cylindrical cavity.* The *nonseparable nature* of this last problem has led to contributions only in recent years with pioneering work by Ben-Menahem and Cisternas (on the spherical cavity) and later important contributions by Thiruvenkatachar and Viswanathan and Gregory. Datta points out that work by him and Sangster, based on matched asymptotic expansions, has applicability to the problem and he demonstrates this.

Of note also is work on wave *diffraction by a finite rigid strip and crack.* The early work by Ang and Knopoff [S.14, S.15] on the diffraction of a *time harmonic* dilatational wave for both cases should be pointed out. More recently Thau and Lu [S.16] have treated the *transient* incident dilatational wave case for the finite crack to establish corresponding stress intensity factors at the crack edges.

Two-dimensional wedge; quarter plane

Progress has been made in the study of wave propagation in a *two-dimensional elastic wedge*. Most of the work has been on the special case of the *quarter plane*. As in the analogous waveguide case the *separable problem* with *mixed boundary conditions* (tractions specified on one edge and a mixture of stress and displacement on the other) can be handled with a double integral transform (cf. ch. 7, § 7.32). Wright [S.17] treated *four such problems*, involving uniform inputs for each of the two stresses and two displacements to one edge (cf. ch. 7, § 7.32). Brock and Achenbach [S.18] treated similarly the *longitudinal impact problem* for *two welded quarter spaces*. Both works present numerical evaluations of the solutions.

Concerning the *quarter plane* with *nonmixed boundary conditions* on its edges (both having either the two stresses or the two displacements specified, or on one edge the two stresses and the other the two displacements) we first note the early work of Lapwood [S.19]. He treated the problem of the *sudden line load on one edge* and studied, through a multi-integral transform and successive approximations, the behavior of the *Rayleigh surface wave* transmitted and reflected by the corner. Later Viswanathan [S.20] treated the analogous but more complicated problem of *two welded quarter spaces*. His analysis modeled Lapwood's which he used to study the various first order events in the problem. These included the body waves radiated from the interface corner, transmitted Rayleigh pulse and the Stoneley pulse traveling along the interface. The work of Alterman and Rotenberg [S.21] on the quarter plane with stress free edges and an internal line compressional source, using a *finite difference method* will also be of interest.

Of note also is Kraut's treatment [S.22, S.23] of the *rigid quarter plane scatterer in the infinite elastic solid, a three-dimensional problem*. He formulated it as a Wiener-Hopf problem in two complex variables and studied the scattered waves generated by an incident plane compressional pulse.

On wave propagation in the *two-dimensional elastic wedge* of *arbitrary angle* we note first the important survey in 1969 by Knopoff [S.24], who discusses a variety of *analytical methods* that have been tried for nonmixed edge conditions but have failed. More recently Achenbach [S.25] has reported success in applying the *method of self-similar solutions* to such a problem, i.e. a line load suddenly applied to one of the edges which then moves with a constant velocity. His method generates a *system of coupled integral equations* which he solves by series expansions and numerics. He shows numerical results for the radial particle velocity on one edge of the wedge.

More recently Wojcik [S.26], also using the *method of self-similar solutions*, has reduced such problems to solving (numerically) a *single Fredholm integral equation* of the second kind in one dimension for one unknown.

Elastodynamic modeling of electrical signal processing devices

Progress has been made recently in elastodynamic wave problem modeling for certain of the subject devices. It is well known, for example, that *thin strips* of a material on a *substrate of another material* can guide surface waves. The great advantage in using such strips and the surface waves they guide, instead of electromagnetic waves, is the extremely large reduction in size of surface-wave devices compared to their electromagnetic counterparts. Tiersten [S.27] discusses elastic *surface waves guided by thin films*. Freund [S.28] treated *guided surface waves on an elastic half space* as a *Wiener-Hopf problem*. Auld has a good discussion of strips and other types of waveguides on a substrate (cf. I. 22, vol. 2, pp. 118–128). Simons studied the *scattering of a Rayleigh wave* [S.29] *and Love wave* [S.30] by the *edge of a thin surface layer* on a half space. Both are Wiener-Hopf problems in which use is made of Tiersten's boundary conditions [S.27] to approximate the effect of the surface layer on the half space.

On additional effects

Anisotropic media; composite materials

Instructive in the subject of wave propagation in homogeneous, anisotropic, linear, elastic media are the surveys of Helbig in 1958 [S.31] and Musgrave in 1959, 1961 [S.32, S.33]. Equally important are the papers by Synge in 1956, 1957 [S.34, S.35] on the more *mathematical features* of the topic. Kraut's 1963 survey [S.36] focuses on advances in *transient wave propagation*. In it he discusses in detail the initial value problem treated first by Duff [S.37] in 1960, as well as certain generalizations that can be introduced into that work. Further, Kraut discusses his thesis work [S.38] on transient waves in *Lamb's problem for a transversely isotropic, elastic half space*. Other notable works on this problem were presented by Abubakar [S.39, S.40] and Buchwald [S.41]. The survey by Scott [S.42] in 1969 is exclusively on advances made in transient wave propagation in the anisotropic solid. It discusses, among other things, progress made in *waveguide problems*.

The monographs [S.43, S.44] on the *dynamics of composite materials* (laminated media) will be of interest to the reader. The survey by Bedford et al [S.45] should also be pointed out. The topic has exhibited strong activity in wave propagation studies recently.

Inhomogeneous media

An introduction to the theory of wave propagation in an inhomogeneous, linear, elastic medium and a review of developments in the subject to 1957 is given in Ewing et al [3.11 ch. 7]. Hook's work [S.46] in 1960 was an important advance. It showed that *separability* of the vector displacement equation of motion could be extended to several types of inhomogeneous media. Karal and Keller's work [2.13] was also of note. They extended the theory of geometrical optics to the inhomogeneous elastic medium.

Recent activity and gains in this subject have been primarily in *seismology*. Two papers by Chapman [S.47, S.48] on the application of *the exact and approximate generalized ray method* (Cagnard-de Hoop technique, cf. ch. 6) to waves in a vertically inhomogenous medium are good sources for this work. He points out that works of Spencer, Helmberger, Müller and others showed the Cagnard-de Hoop technique could be applied to inhomogeneous media provided the model is approximated by plane homogeneous layers. Chapman extended the method to inhomogeneous models where the density and wave speeds are functions of the vertical coordinate.

Of note too is Zemell's work [S.49] on *high frequency cylindrical and spherical waves* for the vertically inhomogeneous half space. His paper also has a good critical account of the subject involving important earlier works.

Thermal effects

An introduction to elastic wave propagation under the influence of thermal effects is provided by Nowacki's book [S.50]. He first discusses the *basic equations of motion for a linear thermoelastic medium*. Chapter I is then a treatment on *harmonic waves,* and II the propagation of *transient waves* in such a medium. The early survey by Chadwick [S.51] and the recent ones by Francis [S.52] and Boley [S.53] can then school the reader on the important papers in this topic.

Anelastic media

For waves in *linear viscoelastic media* the book by Bland [S.54] is a good introduction. That by Christensen [S.55] is also recommended. The early surveys by Kolsky [S.56] and Hunter [S.57] should also be noted.

For waves in *nonlinear elastic media* the monograph by Bland [S.58] can be used as an introduction. It discusses the propagation of discontinuity surfaces and shocks. Also of interest here will be the article by Seymour and Mortell [S.59] on nonlinear geometric acoustics.

For waves in *plastic media* we first note the book by Cristescu [S.60] on the general subject of dynamic plasticity. Surveys of note on *plastic waves* are the early ones by Craggs [S.61] and Hopkins [S.62], and a recent one by Clifton [S.63].

References

[S.1.] S. H. Crandall, *Journal of the Acoustical Society of America* **47** (1970), 1338–1342.
[S.2.] H. D. McNiven and Y. Mengi, *Journal of the Acoustical Society of America* **44** (1968), 1658–1663.
[S.3.] C. C. Chao, *Journal of Applied Mechanics* **27** (1960), 559–567.
[S.4.] O. F. Afandi and R. A. Scott, *International Journal of Solids and Structures* **8** (1972), 1145–1161.
[S.5.] M. Mitra, *Proceedings of the National Institute of Science, India, Part A* **28** (1962), 199–205.
[S.6.] M. Mitra, *Proceedings of the Cambridge Philosophical Society* **60** (1964), 683–696.
[S.7.] M. Mitra, *Zeitschrift für Angewandte Mathematik und Mechanik* **38** (1958), 40–43.
[S.8.] G. Eason, *Quarterly Journal of Mechanics and Applied Mathematics* **17** (1964), 279–292.
[S.9.] G. Eason, *Mathematika* **2** (1964), 75–82.
[S.10.] F. R. Norwood, *Journal of Applied Mechanics* **36** (1969), 516–522.
[S.11.] L. B. Freund, *Journal of Applied Mechanics* **40** (1973), 699–704.
[S.12.] R. K. Kaul and J. J. McCoy, *Journal of the Acoustical Society of America* **36** (1964), 653–660.
[S.13.] S. K. Datta, Scattering of Elastic Waves. In: *Mechanics Today* **4**, ed. S. Nemat-Nasser, Pergamon Press, New York.
[S.14.] D. D. Ang and L. Knopoff, *Proceedings of the National Academy of Sciences* **52** (1964), 201–207.
[S.15.] D. D. Ang and L. Knopoff, *Proceedings of the National Academy of Sciences* **52** (1964), 1075–1081.
[S.16.] S. A. Thau and T-H. Lu, *International Journal of Solids and Structures* **7** (1971), 731–750.
[S.17.] T. W. Wright, *Journal of the Acoustical Society of America* **45** (1969), 935–943.
[S.18.] L. M. Brock and J. D. Achenbach, Wave Propagation in Two Joined Quarter-

REFERENCES

Spaces. In: *Developments in Theoretical and Applied Mechanics* **5**, eds. G. L. Rogers, S. C. Kranc and E. G. Heneke. Pergamon Press, New York (1970), 449–476.

[S.19.] E. R. Lapwood, *The Earth Today*, Oliver and Boyd, Edinburgh (1961), 174–196.

[S.20.] K. Viswanathan, *Geophysical Journal* **11** (1966), 293–322.

[S.21.] Z. S. Alterman and A. Rotenberg, *Bulletin of the Seismological Society of America* **59** (1969), 347–368.

[S.22.] E. A. Kraut, *Bulletin of the Seismological Society of America* **58** (1968), 1083–1096.

[S.23.] E. A. Kraut, *Bulletin of the Seismological Society of America* **58** (1968), 1097–1115.

[S.24.] L. Knopoff, *Elastic Wave Propagation in a Wedge*. In: *Wave Propagation in Solids*, ed. J. Miklowitz, American Society of Mechanical Engineers, New York, New York (1969), 3–43.

[S.25.] J. D. Achenbach, Wave Propagation, Elastodynamic Stress Singularities and Fracture. In: *Theoretical and Applied Mechanics*, ed. W. T. Koiter, North Holland Publishing Company, Amsterdam (1976), 71–88.

[S.26.] G. L. Wojcik, *Self-Similar Elastodynamic Solutions for the Plane Wedge*, Ph. D. Thesis, California Institute of Technology, Pasadena, California (1977).

[S.27.] H. F. Tiersten, *Journal of Applied Physics* **40** (1969), 770–789.

[S.28.] L. B. Freund, *Journal of Applied Mechanics* **38** (1971), 899–905.

[S.29.] D. A. Simons, *Journal of the Acoustical Society of America* **59** (1976), 12–18.

[S.30.] D. A. Simons, *Journal of Applied Mechanics* **42** (1975), 842–846.

[S.31.] K. Helbig, *Gerlands Beitr. Geophysics* **67** (1958), 177–211.

[S.32.] M. J. P. Musgrave, *Rept. Progr. Physics* **22** (1959), 74–96.

[S.33.] M. J. P. Musgrave, Elastic Waves in Anisotropic Media. In: *Progress in Solid Mechanics* **II**, I. N. Sneddon and R. Hill, eds., North Holland Publishing Company, Amsterdam (1961), 63–85.

[S.34.] J. L. Synge, *Proceedings of the Royal Irish Academy A* **58** (1956), 13–21.

[S.35.] J. L. Synge, *Journal of Mathematics and Physics* **35** (1957), 323–335.

[S.36.] E. A. Kraut, *Reviews of Geophysics* **1** (1963), 401–448.

[S.37.] G. F. D. Duff, *Philosophical Transactions of the Royal Society London Series A*, **252** (1960), 249–273.

[S.38.] E. A. Kraut, *Propagation of a Pulse from a Surface Line Source on a Transversely Isotropic Elastic Half-Space*, Ph. D. Thesis. University of California at Los Angeles (1962).

[S.39.] I. Abubakar, *Geophysical Journal* **6** (1961–1962), 337–359.

[S.40.] I. Abubakar, *Geophysical Journal* **7** (1962–1963), 87–101.

[S.41.] V. T. Buchwald, *Quarterly Journal of Mechanics and Applied Mathematics* **14** (1961), 293–317.

[S.42.] R. A. Scott, Transient Anisotropic Waves in Bounded Elastic Media. In: *Wave Propagation in Solids*, ed. J. Miklowitz. American Society of Mechanical Engineers, New York, New York (1969), 71–91.

[S.43.] *Dynamics of Structured Solids*, ed. G. Herrmann, American Society of Mechanical Engineers, New York, New York (1968).

[S.44.] *Dynamics of Composite Materials*, ed. E. H. Lee, American Society of Mechanical Engineers, New York, New York, (1972).

[S.45.] A. Bedford, D. S. Drumheller and H. J. Sutherland, On Modeling the Dynamics of Composite Materials. In: *Mechanics Today* **3**, ed. S-Nemat-Nasser. Pergamon Press, New York (1976), 1–54.

[S.46.] J. F. Hook, *Journal of the Acoustical Society of America* **33** (1961), 302–313.

[S.47.] C. H. Chapman, *Geophysical Journal* **36** (1974), 673–704.

[S.48.] C. H. Chapman, *Geophysical Journal* **46** (1976), 201–233.

[S.49.] S. H. Zemell, *SIAM Journal of Applied Mathematics* **31** (1976), 1-15.
[S.50.] W. Nowacki, *Dynamic Problems of Thermoeleasticity*, Translated by H. Zorski, Noordhoff International Publishing, Leyden, The Netherlands (1975).
[S.51.] P. Chadwick, Thermoelasticity, The Dynamical Theory. In: *Progress in Solid Mechanics* **I**, eds. I. N. Sneddon and R. Hill. North Holland Publishing Company, Amsterdam (1960), 265-328.
[S.52.] P. H. Francis, *Journal of Sound and Vibration* **21** (1972), 181-192.
[S.53.] B. A. Boley *Thermal Stresses Today*, Proceedings of the Seventh U. S. National Congress of Applied Mechanics, American Society of Mechanical Engineers Publications, New York, N. Y. (1974), 99-107.
[S.54.] D. R. Bland, *Theory of Linear Viscoelasticity*, Pergamon Press, London (1960).
[S.55.] R. M. Christensen, *Theory of Viscoelasticity, An Introduction*. Academic Press, New York (1971), 110-145.
[S.56.] H. Kolsky, The Propagation of Stress Waves in Viscoelastic Solids. In: *Applied Mechanics Surveys*, eds. H. N. Abramson, H. Liebowitz, J. N. Crowley and S. Juhasz. Spartan Books, Washington (1966), 841-846.
[S.57.] S. C. Hunter, Viscoelastic Waves. In: *Progress in Solid Mechanics* **I**, eds. I. N. Sneddon and R. Hill. North Holland Publishing Company, Amsterdam (1960), 3-57.
[S.58.] D. R. Bland, *Nonlinear Dynamic Elasticity*. Blaisdell Publishing Company, Waltham, Massachusetts (1969).
[S.59.] B. R. Seymour and M. P. Mortell, Nonlinear Geometrical Acoustics. In: *Mechanics Today* **2**, ed. S. Nemat-Nasser, Pergamon Press, New York (1975), 251-312.
[S.60.] N. Cristescu, *Dynamic Plasticity*. North Holland Publishing Company, Amsterdam (1968).
[S.61.] J. W. Craggs, Plastic Waves. In: *Progress in Solid Mechanics* **II**, I. N. Sneddon and R. Hill, eds., North Holland Publishing Company, Amsterdam (1961), 143-197.
[S.62.] H. G. Hopkins, Dynamic Nonelastic Deformations of Metals. In: *Applied Mechanics Surveys*, eds. H. N. Abramson, H. Liebowitz, J. N. Crowely and S. Juhasz. Spartan Books, Washington (1966), 847-867.
[S.63.] R. J. Clifton, Plastic Waves: Theory and Experiment. In: *Mechanics Today* **1**, ed. S. Nemat-Nasser, Pergamon Press, New York (1972), 102-167.

AUTHOR INDEX

Ablow, C. M. 343, 365
Abramowitz, M. ... 219, 230, 389, 406
Abubakar, I. 576, 579
Achenbach, J. D. ... 17, 78, 79, 82, 83, 85, 94, 96, 118, 148, 172, 192, 488, 497, 557–559, 572, 575, 578–579
Afandi, O. F. 573, 578
Aggarwal, H. R. 343, 365
Alterman, Z. S. 575, 579
Ang, D. D. 488, 571, 574, 578
Arenberg, D. L. 128, 129, 132, 133, 139, 140, 142, 177
Auld, B. A. 17, 18, 576

Baron, M. L. ... 519, 544, 546, 548–549, 572
Bedford, A. 577, 579
Ben-Menahem, A. 574
Benthem, J. P. 448, 450, 462, 484
Bertholf, L. D. 466, 484
Bishop, R. E. D. 136–138, 144, 177
Bland, D. R. 578, 580
Bleich, H. H. 54, 56
Boley, B. A. ... 404, 409, 483, 577, 580
Brekhovskik, L. M. 17, 18, 226
Bremmer, H. 242, 296
Brillouin, L. 9, 17, 187, 229, 271
Broberg, K. B. 574
Brock, L. M. 575, 578
Brockway, G. S. 52, 56
Brown, J. W. 385, 483, 491
Buchwald, V. T. 576, 579
Budiansky, B. 115, 118
Bullen, K. E. 16, 18

Cagniard, L. 5, 6, 11, 12, 17, 167, 168, 169, 177, 302–318
Cerrillo, M. V. 271, 297, 437
Chadwick, P. 577, 580
Chao, C. C. ... 404, 409, 483, 573, 578
Chapman, C. H. 577, 579
Chou, P. C. 115, 118, 290
Christensen, R. M. 578
Churchill, R. V. ... 232, 234, 241, 248, 278, 281, 293, 296, 385, 483, 491
Cisternas, A. 574
Clebsch, A. 3, 17, 61, 167
Clemmow, P. C. 488, 490–492, 499–502, 571
Clifton, R. J. 578, 580
Cooper, G. J. 465–466, 484
Copson, E. T. 525, 572
Courant, R. 17, 96, 105, 107, 112, 113, 118
Craggs, J. W. ... 465–466, 484, 578, 580
Crandall, S. H. 573, 578
Cristescu, N. 578, 580
Curtis, C. W. 271, 297, 434–438, 484

Das Gupta, S. C. 228, 230
Datta, S. K. 574, 578
Davids, N. 574
Davies, R. M. 190–191, 229
de Hoop, A. T. ... 11, 12, 54, 55, 56, 95, 96, 118, 302–318, 351–352, 365–366, 488–489, 492–497, 502–517
DeVault, G. P. 435, 438, 484

DiMaggio, F. L. 54, 56
Dix, C. H. 17, 177
Dobrin, M. B. 150, 151, 177
Doetsch, G. 448, 484
Donnell, L. H. 10, 18
Drumheller, D. S. 577, 579
Duff, G. F. D. 576, 579
Duhem, P. 3, 17, 61

Eason, G. 79, 118, 573–574, 578
Ellis, A. T. ... 271, 297, 436–437, 484
Erdelyi, A. .. 70, 118, 255–257, 261–264, 267, 269, 283, 297, 338, 524, 529, 538
Ergin, K. 169, 170, 177
Eringen, A. C. 17, 18
Ewing, W. M. ... 5, 8, 17, 148–150, 159–161, 162, 165, 168, 177, 211, 214, 226, 298, 577

Felder, R. A. 316–317, 319, 365
Feshbach, H. 63, 118, 215, 520
Flinn, E. A. 17, 177
Folk, R. 271, 297, 434–438
Forrestal, M. J. 316–317, 319, 365
Fox, G. 271, 297, 434–438, 484
Francis, P. H. 577, 580
Freund, L. B. 574, 576, 578, 579
Friedlander, F. G. ... 17, 71, 75, 77, 78, 83, 118, 520–521, 524, 545, 550, 558, 566, 572
Fugelso, L. E. 316–317, 319, 365
Fulton, J. 79, 118

Gakenheimer, D. C. ... 340–344, 347, 349–362, 365–366
Gangi, A. F. 54, 56
Garabedian, P. R. 107, 112, 118
Garvin, W. W. 329–332, 365
Gilbert, F. .. 323, 326–328, 343, 365, 524, 527, 550–551, 572, 574
Goldsmith, W. 10, 18
Goodier, J. N. ... 5, 17, 136–138, 144, 177, 290, 297, 377, 483, 555
Graff, K. F. 17, 18
Graffi, D. 52–55, 56, 96

Gregory, R. D. 574
Griffin, J. H. 556–560, 572
Grimes, C. K. 550–551
Gurtin, M. E. 54, 56
Gutenberg, B. ... 6, 131, 133, 134, 162, 165, 177

Hadamard, J. 96, 118
Handelman, G. H. 495, 572
Havelock, T. H. ... 8, 9, 17, 186–187, 192, 229, 277
Hayes, M. 147, 177
Helbig, K. 576, 579
Helmberger, D. V. 577
Hermann, J. H. 557–559, 572
Hilbert, D. 17, 96, 105, 107, 112, 113, 118
Hook, J. F. 555, 572, 577, 579
Hopkins, H. G. 578, 580
Hopkinson, B. 9
Hunter, S. C. 578, 580

Ingard, K. U. 17, 18

Jahsman, W. E. 290, 297
Jardetsky, W. S. 5, 8, 17, 148–150, 159–161, 162, 165, 168, 177, 211, 214, 226, 298, 577
Jeffreys, B. S. ... 252–255, 261, 297, 370
Jeffreys, H. 6, 18, 147, 252–255, 261, 269, 297, 370
Jones, D. S. 4, 17, 79, 82, 95, 118
Jones, O. E. 271, 297, 436–437, 484
Jones, R. P. N. 409, 438, 484

Karal, F. C. Jr. 78, 118, 577
Kaul, R. K. 574, 578
Keller, H. 71, 76, 118
Keller, J. B. 71, 78, 118, 577
Kellogg, O. D. 43, 56
Knopoff, L. ... 54, 56, 147, 148, 177, 323, 326–328, 343, 365, 524, 551, 572, 574, 575, 578, 579

Knott, C. G. 6, 17
Knowles, J. K. 151, 152, 177, 489, 571
Ko, W. L. 556, 558, 572
Koenig, H. A. 115, 118, 290
Koiter, W. T. 504, 572
Kolsky, H. 16, 127, 128, 177, 578, 580
Kraut, E. A. 575, 576, 579
Lamb, H. ... 4, 5, 7–9, 11, 17, 18, 196–197, 298, 318, 365
Lamé, G. 3, 59, 61, 62, 118
Lapwood, E. R. ... 5, 329–330, 365, 575, 579
Lawrence, P. L. 150, 151, 177
Lee, E. H. 556, 558
Leonard, R. W. 115, 118
Lifson, H. 55, 332, 344–348, 365
Lloyd, J. R. 229, 230, 410–422, 425–428, 484
Love, A. E. H. ... 2–4, 6–10, 19, 56, 71, 79, 86–93, 95, 103, 118, 120, 122, 165, 177, 211–212, 215, 225, 369, 371, 374, 430, 558
Lu, T-H. 574, 578

Magnus, W. ... 70, 118, 283, 338, 524, 529, 536
Matthews, A. T. 519, 548–549, 572
McCoy, J. J. 574, 578
McLachlan, N. W. 251–252, 297
McNiven, H. D. 207–209, 222, 230, 573, 578
Medick, M. A. 198–201
Meitzler, A. H. 443, 484
Mengi, Y. 573, 578
Meres, M. W. 6, 161, 177
Miklowitz, J. ..... VIII, 11, 197, 215, 226, 229, 230, 277, 285, 287, 290, 297, 319–323, 340, 343, 347, 349–361, 365, 374, 390–393, 404, 410–422, 424–428, 436–477, 483–484, 519–546, 549–567, 569–572
Miles, J. W. 517, 572

Mindlin, R. D. ... 133–135, 138, 145, 146, 177, 180, 182, 185, 188–189, 191, 194–209, 222, 224, 230, 382, 483
Mitra, M. 573, 578
Morse, P. M. 17, 18, 63, 118, 215, 520
Mortell, M. P. 578, 580
Mow, C. C. 10, 17, 517–518, 549, 572
Musgrave, M. J. P. 576, 579
Muskat, M. 6, 161, 177

Nagase, M. 560, 563, 565, 572
Nakano, H. 5, 17
Neidhardt, G. L. 316–317, 319, 365
Neumann, F. 3, 51, 56
Nigul, U. K. 435, 438, 484
Nisewanger, C. R. ... 390–392, 443, 466, 483
Noble, B. 488, 492, 502, 504, 571
Norwood, F. R. ... 437, 484, 560–567, 572, 574, 578
Nowacki, W. 19, 20, 56, 371, 577, 580
Nussenzveig, H. M. 560, 563, 572

Oberhettinger, F. ... 70, 118, 248, 283, 296, 338, 405, 406, 408, 524, 529, 536
Onoe, M. ... 198–201, 207–209, 221–222, 230

Pao, Y. H. ... 10, 17, 224–225, 230, 517–518, 549, 572
Parnes, R. 519, 544, 546, 549, 572
Payton, R. G. 54–56
Pearson, K. 59, 118
Peck, J. C. 520, 527–529, 536–546, 549–550, 572
Pekeris, C. L. 55, 332–334, 336–340, 344–348, 365–366
Phillips, H. B. 46, 56, 60, 61
Pochhammer, L. 7, 214–215, 219, 221, 224–225, 230, 369, 430

Poisson, S. D. ... 2, 7, 9, 58, 59, 105, 118
Press, F. 5, 8, 17, 148–150, 159–161, 162, 165, 168, 177, 211, 214, 226, 298, 577
Pucik, T. A. 489, 571
Pytel, A. 574

Randles, P. W. 443, 484
Rayleigh, Lord (J. W. Strutt) .. 4–8, 10, 18, 146, 147, 177, 196–197, 369, 377, 483
Redheffer, R. M. .. 99, 102, 112, 118
Redwood, M. 17, 131, 177
Rivlin, R. S. 147, 177
Robins, C. I. 453, 484
Rosenfeld, R. L. ... 319–323, 325–328, 365, 436, 438–443, 484
Rotenberg, A. 575, 579
Rubenfeld, L. A. 495, 572

Sangster, J. D. 574
Scott, R. A. 215, 230, 437–438, 471, 473–474, 484, 573, 576, 578, 579
Sears, J. E. 9
Selberg, H. L. 282, 285, 287, 290, 296
Seymour, B. R. 578, 580
Shook, C. A. ... 271, 297, 434–438
Simon, R. F. 150, 151, 177
Simons, D. A. 576, 579
Sinclair, G. B. 444, 448–465, 484
Skalak, R. 390, 418, 426–428, 437, 483
Smith, M. G. 260, 297
Smith, R. C. T. 453, 484
Smith, T. L. 43, 56, 88
Sneddon, I. N. 79, 118, 248, 250, 296
Sokolnikoff, I. S. 19, 48, 52, 56, 99, 102, 112, 118
Soldate, A. M. 555, 572
Somigliana, C. ... 3, 17, 61, 62, 118, 147

Sommerfeld, A. 9, 17, 143, 177, 485, 495, 571
Spencer, T. W. 577
Stegun, I. A. 219, 230, 389, 406
Sternberg, E. 51, 52, 54, 56, 59, 61, 95, 96, 118
Sternberg, W. J. 43, 56, 88
Stoker, J. J. 11, 18
Stokes, G. G. 4, 10, 79, 118
Stoneley, R. ... 6, 13, 166, 167, 177, 214, 230
Stratton, J. A. 51, 56, 95
Suhubi, E. S. 17, 18
Sutherland, H. J. 577, 579
Sve, C. 474–477, 484
Synge, J. L. 576, 579

Tedone, O. 3, 17
Thau, S. A. 574, 578
Thiruvenkatachar, V. R. 574
Tiersten, H. F. 576, 579
Timoshenko, S. 8, 377, 483, 555
Ting, T. C. T. 556, 558, 572
Titchmarsh, E. C. .. 232, 296, 473–474, 484, 520
Todhunter, I. 59, 118
Tricomi, F. G. ... 70, 118, 283, 338, 524, 529, 536
Tricomi, F. G. ... 70, 118, 283, 338, 524, 529, 536

Uflyand, Ya. S. 382, 395, 483

van der Pol, B. 242, 296
Van der Waerden, B. L. 270, 297
Verhey, R. F. 385, 483, 491
Viano, D. C. 424, 484
Viktorov, I. A. 17, 149, 150, 152, 177, 527, 572
Viswanathan, K. 574, 575, 579

Walker, G. W. 6, 18
Watson, G. N. ... 248–249, 297, 563, 572

Weinberger, H. F. ... 103, 118, 243–245, 248
Wheeler, L. T. .. 51, 52, 54, 56, 95, 96
Whitham, G. B. 17, 18
Widder, D. V. 237, 296, 300, 328
Williams, M. L. 488, 571
Wilson, E. B. 75, 118
Wojcik, G. L. 576, 579
Wong, P. K. 215, 230
Wright, T. W. 575, 578
Wu, T. Y. T. 251–252, 255–257, 261–269, 271–275, 297

Zemel, S. H. 577, 580
Ziegler, F. 557–559, 572
Ziv, M. 115, 118

SUBJECT INDEX

Acceleration 43
Acoustics 17
Additional effects 576–578
Airy function 388, 423, 429, 438
  derivatives of 437
Amplitude
  of P wave 121
  ratios (see Reflection coefficients; Transmission coefficients)
  of SH wave 154
  of SV wave 121
Analytical continuation of functions, use of 286
Angle
  incident 120–121, 155–156
  critical 132, 139–144, 163–164
  reflection 120–131, 155–156
  refraction 157, 161, 165
Anharmonic overtones, infinite plate (See Modes, thickness)
Anisotropic media, waves in, early work, surveys on 576
  for Lamb's problem of a transversely isotropic half space 576
  recent work, surveys on 576
  surfaces of 3
  for waveguides 576
Anomalous dispersion 187
  in transient response of a plate .. 421
Antiplane
  displacement source problem, infinite solid 363–364
  shear deformation 152
  (see also SH wave, Wave)

displacement-potential relations of 153, 154
  gauge condition in 153
  SH waves of 152–156, 170–174
  strain-potential relations of 154
  stress-potential relations of 154–155
  tangential surface load problem, half space 363
Antisymmetric
  edge load problems,
    plate, semi-infinite 435
    rod, semi-infinite 435
  line loads, problem of,
    plate, infinite 429–430
Approximate theories, waveguide (See Waveguides, approximate theories for; Waveguide problems)
Asymptotic expansions 250–277
  nature of 251–252
  Poincaré's definition of ... 252–253
  properties of 253–255
  differentiation 255
  integration 254–255
  product, quotient, sum 254
  uniqueness 253–254
Asymptotic expansions of integrals 255–277
  integrals in 256
  critical points of 256
  Fourier 256
  integrand conditions of 256
  Laplace 256
  methods for,

Cagnard–deHoop technique, method in, applications of .. 319–323, 438–443
Laplace's method 262–264
  applications of .. 268, 325, 327–328
  parts integration ... 257, 261–262
  application of 251
  stationary phase, method of 271–275
    applications of .. 275–277, 387–388, 404–408, 423–424, 470–474
  steepest descents, method of 264–270
    applications of .. 323–328, 550–551
  Van der Waerden's method . 270
    application of 270
  Watson's lemma 257
    applications of 260, 326, 327–328
Axially symmetric
  compressional waves in circular cylindrical rod 220–223
  diffraction problems for spherical cavity 565–567
  plate problems 470–476
  torsional waves in
    circular cylindrical rod .. 217–220
    thin plate 290
  vertical point load problem, half space, ... 333–348
  waves, ... 68, 338–348, 470–476

Bar velocity 222
Bernoulli-Euler beam (bending theory) 6, 8
  (See also Waveguides, approximate theories for; Waveguide problems)
Bessel functions
  modified ...283, 523–525, 562–565
    asymptotic expansions of 283
    Hankel function relation 287
  ordinary

asymptotic expansion of ... 469
Bilateral Laplace transform . 241–242
  in cylindrical cavity diffraction problem 537–542
  properties of 242
Body force problems 78–94
  center of compression in .. 91–92
  center of rotation in 92–93
  distributed source of potentials in 83–84
    retarded potential solutions for 84, 88, 94
  double forces in 91–93
  early work on 4
  general solution for displacement in 84–85
  governing inhomogeneous wave equations for 80
  line source of potentials in 85
    solution for 86
  point load in 86–91
    dilation solution for 90
    displacement solution for 89
    rotation solution for 90
    stress solution for 90
    wavefronts, ramp input, for 90–91
    waves generated by 90
  point source of potentials, in 83–84
    solutions for 82–83
  time harmonic 93
Body waves
  in bounded media 68–69
    cylindrical wave surfaces of 69
    generation of interface waves by 68–69
    interference of 69
    reflection and refraction of .. 69
  fundamental 57–63
    dilatation 58
    dilatational displacement 58
    equivolumnal displacement 57–58
    governing equations for . 57–63
    propagation speeds for .. 57–58
    rotation 58
    time harmonic 69–70

588 SUBJECT INDEX

axially symmetric 70
frequency of 69
Helmholtz equations for 69
nodal points of 70
standing 70
trains of 70
wavelength of 69
wave number of 69
types of 63–68
axially symmetric 68
nonaxially symmetric 68
plane 63–67
spherically symmetric 67
Boundary-initial value problems fundamental 46–47
uniqueness of solutions of .. 3, 47 50–51
discontinuous loadings 52
finite body 47, 50–51
infinite anisotropic body ... 51
infinite body 51
integral representations in . 94–96
Kirchhoff's formula 95
Love's formula 95–96
integral transform methods in (see Integral transform methods)
method of characteristics in 112–115
Boundedness condition, solution, for semiinfinite plate with nonmixed edge conditions, .. 446–448, 461–462
algebraic 452, 461
approximation corresponding to long time 454–455, 462
integral equations representing . 448
Branches of frequency equations for, plate, approximations, .. 369, 381, 475
plate, infinite .. 7 (See also Rayleigh-Lamb frequency equation)
rod, approximations . 7–8, 219, 368–369, 381
rod, infinite circular cylindrical .. 7 (See also Pochhammer frequency equations)

Bromwich contour 236
Bulk modulus (Modulus of compression) 41
Buried line dilatational source in half space 329–332

Cagniard-deHoop method
in half plane diffraction problem 490, 492–496, 498–499, 504–516
for nonaxymmetric, surface normal point traveling load problem 351–360
displacements in 357–360
integral equations for 356, 358, 359
Laplace transformed 355–356, 359
integration paths in, deformed 353, 357–358
principal value integrals in 359
singularities and controur integrations in 352–355, 357–359
transformations in, basic ... 351, 352, 357, 358–359
wavefront approximations in . 361
for plane-strain problems .. 302–319, 329–332
displacements in 312–314
integral equations for . 310, 312
Laplace transformed . 307, 309, 310, 311–312
integration paths in, deformed 303–306, 307–308, 311
principal value integrals in .. 311
singularities, contour integration in 303–306
transformations in, basic 302, 303
wavefront approximations in 319–323
Cagniard's method, early use of ... 5
Calculus of variations, use of .. 369–373
Cauchy–Goursat theorem, use of 259, 286
Cauchy problems (See Initial value problem)
Carrier wave 187

modulation of 187
  frequency of 187
  of simple wave group 187
Caustic 557, 559
  singular stress at .. 558 (See also Focusing, wave)
Cavity source problems 277–290
  circular cylindrical cavity, decaying uniform pressure on 290
  circular cylindrical cavity in plate, Kromm's solution method .. 290
  uniform pressure on ... 282–290
  uniform shear stress, rotary velocity on 290
  circular cylindrical cavity, uniform pressure in 282–290
    boundary-initial value problem 283
    displacement-potential relation 284
    Laplace transform solution method 283–287
      singularities, contour integration in 284–287
    solution 286–287
    static 289
    stress responses in 287
    verification of 288
    wavefronts in 289–290
  solution by method of characteristics 290
  stress-potential relations 284
  spherical cavity, nonsymmetric sources 282
  spherical cavity, uniform pressure on 277–282
    boundary-initial value problem 278
    displacement-potential relation 279
    Laplace transform solution method 278–281
    solution 281
      response of stresses in 282
      static 281
      verification of 181
      wavefronts of 281
    stress-potential relation 279

Center of
  compression 91
  rotation 92
Characteristic length 1
Characteristics 75
  theory of 17
  (See also, Wavefronts; Surface of discontinuity, propagating)
Characteristics, method of .. 107–115
  for one-dimensional problems 107–115
    boundary-initial value problems, numerical method 112–115
    characteristic condition in ... 109
    characteristic directions in .. 109
    first hyperbolic theorem, initial values, numerical method 110–112
    second hyperbolic theorem, initial values, numerical method . 112
    uniqueness of solution 112
    use of 290
  for two-dimensional problems . 115
Coefficients
  of reflection 123
  of transmission 160
Completeness of Lame solution ... 3, 61–62, 116
Composite materials, waves in, monographs on 577
Compressional waves in
  an infinite plate with circular cylindrical cavity,
    cavity wall normal displacement, generated by 474
  an infinite semi-infinite circular cylindrical rod
    end load, generated by 466
    frequency equations for 221
    frequency spectra for 208–209, 222
    motion generated by 222–223
    potential modes for 221
    shock tube excitation of . 390–393, 437
  an infinite, semi-infinite plate

cylindrical dilatational, equivolumnal wave surfaces in .. 179–180, 210
frequency equations for . 183, 197
    branches of 197–201
frequency spectra for 185, 198, 202
longitudinal impact, generated by 436–437
mixed edge pressure, generated by 436–437
near base normal line loads, cantilever, generated by 465
nonmixed edge line load, generated by 463–465
nonmixed edge pressure, generated by 455–457
nonmixed edge velocities, generated by 465–466
normal line loads, generated by 420–422
normal point loads, generated by 470–472
P, SV waves in 178–180
phase velocity spectra for 185
thermal field, generated by . 475–477
thickness modes in 185
an infinite solid generated by cavity pressure sources,
    circular cylindrical 287
    spherical 282
a thin circular cylindrical rod, approximate theories for, ... 367–369 371–374, 392–393
    longitudinal impact, generated by 388–389, 393
a thin plate with circular cylindrical cavity,
    cavity wall pressure, generated by 287
Conical flows, Buseman's method of 517 (See also Self-similar solutions)
Conical, waves 358, 360–361
Convolution theorem
    of exponential Fourier tranform 245

of Laplace transform 239
use of 313, 541–542
Coordinates
    cylindrical 68
    rectangular Cartesian 19, 42
    spherical 67
Cusp-type wavefronts in anisotropic plates 443
Cutoff frequencies for plate
    with mixed face conditions ... 188
        group velocities at 189
        phase velocities at 189
        P, SV waves at 189
    with traction free faces   203–204
Cutoff waves, influence of in transient response of a plate ... 420, 475–477
Cylindrical shell, circular, time harmonic waves in 226
Cylindrical waves,
    circular .. 68, 70, 77–78, 286–289
    in diffraction ... 485–487, 494–495, 515–517
    in half space ... 314, 316, 319, 331–332
    in plates 69, 180–186, 188–214, 420–422, 440–443, 456–457, 464–465

D'Alembert's
    principle 43, 375
    solution
        for one-dimensional initial value problem 63–64
Deformation 30–38
    finite 31
    infinitesimal 32
    pure 34, 35–36
Deformation and motion, description of Eulerian (spatial) 19
    Lagrangian (material) 19–20
Delta functions, Dirac .. 80–82, 240
    for cartesian coordinates 81
    derivatives of 81
    Laplace transform of 241

for spherical coordinates ... 81–82
symmetrical 300
Diffraction problems
  elastic pulse by circular cylindrical
    obstacle 517–560
    Friedlander's representation of solution 520–521
    Poisson's summation formula 520–521
    Riemann surface in θ, interpreted as 521
    waveform of function in ... 521
    wave sum in 521
  line source by circular cylindrical
    cavity 518–536
    diffracted wavefront, ray . 518–519
    illuminated zone 518–519
    incident wavefront, ray ... 518
    limiting cases of 519–520
    reflected wavefront, ray ... 518
    shadow zone 518
    source of diffracted wavefront 519
    spiraling diffracted wavefront 519
  line source by rigid circular cylinder 521–551
    wavefront approximations in 551
  normal line load on cavity wall 521–536
    Bessel functions, modified, in 523–525
    boundary-initial value problem 522
    displacement-potential relations 522
    frequency equation analysis of wave number branches 525, 527–529 (See also 411–416)
    high, low frequency analysis of branches 527–529
    Laplace-exponential Fourier transform method. ... 523–536 (See also 411–420)
    solution, displacements ... 525
    diffracted, radiated harmonic waves in 525–526
    long time – far field solution; Rayleigh waves, static solution 529–536
    wave systems of 526
    stress-potential relations 522
  plane-compressional pulse by circular cylindrical cavity . 536–555
    bilateral Laplace-exponential Fourier transform method, 537–542 (See also 523–536, 411–420)
    convolution theorem, use of 541–542
    boundary-initial value problem 536–537
    displacement-potential relations 522
    frequency equation, analysis of wave number branches 525, 527–529, 539 (See also 411–416)
    Laplace transform-Fourier series method 545–550
    circumferential stress response 549
    velocity responses 549
    solution, displacements ... 540
    diffracted, reflected harmonic waves in 540–541
    long time – far field solution: Rayleigh waves, static circumferential stress 551–555, 569–571
    rigid body velocities of cavity wall 555
    velocities, numerical mode, mode sum, wave and wave sum, responses of 542–545
    wave system of 541
    stress-potential relations .. 522

two methods of solution, comparison of 549–550, 544, 546
  advantages and disadvantages of each 550
  velocity responses 549
  wavefront approximations for 550–551
  reflected stresses by geometric optics method 550
  reflected stresses by method of steepest descents 550
plane-compressional pulse by circular cylindrical inclusion 556–560
  boundary-initial value problem 556
  decay of refracted wavefronts, reflection 560
  fiber-reinforced composite materials, application to ... 556
  focusing, wave 557–559
    caustics in 557, 559
    rays of 557, 559
  interface singularity in 560
  refracted dilatational waves, rays of 556–557
    inclusion, inside 556–557
    inclusion, outside .. 556–557
    for reflection inside inclusion 556–557
  separation at interface, wavefront analysis, experiments on 558
  transmission of singular refracted wavefronts to outer solid 560
  wavefront approximations,
    by geometrical acoustics . 558
    by integral representations 558
    singular for stress at caustic 558–559
    by Watson-type lemma .. 559
  wavefront changes along refracted ray in vicinity of caustic 559–560

elastic pulse by spherical obstacle 558, 560–567
  normal point load on cavity wall 560–568
  Bessel functions in, modified 562–565
  boundary-initial value problem 562
  frequency equation, analysis of wave number branches 562–565
  high frequency approximations for P branches 565
  Laplace transform-spherical harmonics method 562–566
  Legendre polynomials in 562–565
  relation to circular cylindrical cavity case 562
  solution cavity wall displacement 565–566
  diffracted P, Rayleigh wavefront, by Laplace transform asymptotics 565–566
  reflecting diffracted P, Rayleigh wavefronts in ...565
  wavefront responses, P, Rayleigh, numerical 566
  Watson's transformation in 563–564
plane-compressional pulse by spherical cavity 567
  solution, cavity wall displacements 567
  wavefront responses, P, Rayleigh, numerical 567
point source, exterior time harmonic, spherical cavity 560
  high frequency approximations 560
finite crack, rigid strip,
  incident dilatational wave, harmonic 574
  incident dilatational wave, transient (crack case) 574

SUBJECT INDEX 593

plane-elastic pulse by half-plane 487–517
  acoustic, light or *SH* wave field 485–487
    diffracted wave 485–486
    illuminated zone 487
    incident wave 485–486
    rays 485–486
    reflected wave 486
    regions for 485
    shadow zone 486
    wavefronts 485–487
  dilatational pulse by rigid half plane 517
  dilatational pulse by traction free half plane 496–517
    boundary-initial value problems 497–498
    Cagniard-deHoop method in 498–499, 504–516
    dual integral equations and solution of 499–501
    factorization, Wiener–Hopf 500–504
    Laplace-exponential Fourier transform method 498–516
    solution, displacements, complete 515–517
    solutions, scattered displacement wave, diffracted .. 506, 507, 510, 512, 513
    geometrical 515
    symmetric, antisymmetric components of problem 497
    total displacement field ... 497
    uniqueness of solution, conditions for 498
    velocity responses 517
  equivoluminal pulse by traction free half plane 517
    velocity responses 517
  SH pulse for rigid half plane . 496
    boundary-initial value problem 496

  solution 486, 496
  SH pulse, traction free half plane 488–495
    boundary-initial value problem 488–489
    Cagniard–deHoop method in 490, 492–494
    dual integral equations and solution of 490–492
    Laplace-exponential Fourier transform method .. 489–494
    solution, scattered displacement wave 486, 494–495
    stress singularity of 495
    total displacement field ... 488
    uniqueness of solution, conditions for 489, 491
Diffraction, pulse
  by circular cylindrical obstacle 517–560
  by half-plane 485–517
  by spherical obstacle .. 558, 560–567
  (See also Diffraction problems)
Dilatational wave of
  cylindrical cavity diffraction problem 542–545
  cylindrical obstacle diffraction problem 556–560
  half plane diffraction problem 506, 507
  half space buried line dilatation source problem 331–332
    surface 331–332
  half space surface normal line load problem ... 312–316, 318–320
    grazing incidence 314
    interior 312–314
    plane of symmetry 318–319
    surface 313–316
    wavefronts of 318–320
  half space surface normal point traveling load,
    interior, supersonic load ... 355, 357–360
    conical 358, 360

hemispherical ... 357, 359–360
  wavefronts of 361
half space vertical point load problem 339–348
  buried load case 344–348
    in surface response ... 344–348
    grazing reflection (*SP* event) of 344–348
  surface load case 339–344
    in interior response ... 340–344
    in surface response 339
  wavefronts of . 340–344, 345–348
spherical cavity diffraction problem 565–567
spherical obstacle diffraction problem 558
Dilatational waves in half space .. 119–152, 312–321, 326, 331–332, 339–348, 357–361
  infinite medium 66–67, 78–94, 96–107, 281–282, 286–290, 515–517, 528–529, 534–536, 542–546, 550–560, 566–567
  two welded half spaces 156–170, 175–177
  waveguides 178–186, 188–209, 220–225, 388–394, 403–404, 408, 419–424, 436–438, 440–443, 456–457, 464–465, 470–472, 475–477
Dipole source problem of half space 573
Dispersion relations 183–184
Dispersion, wave
  anomalous 187
  data for 188
  harmonic 178
  normal 187
Dispersive waves 17, 186–188
Displacement
  components of 32
  continuity condition on 32
  dilatational waves 2
  equations of motion 2, 45–46
    integration of 57–117
    Lamé solution of 59
  equivoluminal waves 2

field 3
  gradients 32
    linearization condition on 32
  integral representation for 4
  in *P* wave 120
  relative, near a point 33
  rigid body 32
    motion 34
    rotation 32
      infinitesimal, components of . 34
      nature of 34–35
    translation 32, 35
  in *SH* wave 152
  in *SV* wave 120
  vector 45, 59
Divergence theorem 43
Double forces 91–92
  generalized 93

Eikonal equation 96
Elastic string, transverse displacements of 97
Elastodynamics, linear theory of 19–56
Electric signal processing devices, elastodynamic modeling of .. 576
  Rayleigh, Love wave scattering by edge of a thin surface layer on half space 576
  surface waves guided by thin films for 576
  guided surface waves on half space 576
Electromagnetic waves 17
Elementary bending theory (See Bernoulli—Euler bending theory)
Energy
  density
    kinetic 48–49
    strain 49
    total 128
    velocity of 192
      equality with group velocity 192–193
    waves of 129–130
  flux 130, 142–144

identity of 49
  generalized 51
  kinetic 48
partition under
  P-wave
    reflection 128–131
    refraction ... 161–162, 165, 169–170
  SH-wave
    refraction 172
  SV-wave
    reflection 133–134, 142–144
    refraction ... 161–162, 165, 169
principle of conservation of . 48–50
reflection coefficients 131, 133–134
strain 49
transmission coefficients ... 161–162, 165, 169–170
waves of, time averaged 130
Equations of motion
  displacement 45–46
  stress 44
Equivoluminal wave of
  cylindrical cavity diffraction problem 542–545
  cylindrical obstacle diffraction problem 558
  half plane diffraction problem,
    head 510, 512
    regular 510, 512
    two-sided 510, 512
  half space buried line dilatational
    source 331–332
    surface 331–332
  half space surface normal line load
    problem 312–316, 318–319, 321–323, 326–328
    head 312–316
    interior 312–316
    plane of symmetry 318–319
    surface 313–316
    two-sided 313–316
    wavefronts of ... 321–323, 326–328

half space surface normal point traveling load,
  wavefronts of 361
  (See also Head wave in)
half space vertical point load problem 339–348
  for buried load case 344–348
    surface 344–348
  for surface load case ... 339–344
    interior 340–344
    surface 339
    wavefronts of .. 340–344, 345–348
spherical obstacle diffraction problem 558
Equivoluminal waves in half space
  119–156, 312–319, 321–323, 326–328, 331–332, 339–348, 361
  infinite medium 66–67, 78–94, 96–107, 290, 494–495, 515–517, 528–529, 534–536, 542–546, 551–555, 558–559
  two-welded half spaces ... 156–177
  waveguides 179–186, 188–214, 217–225, 404, 406–407, 419–424, 440–443, 464, 470–472, 475–477
Expanding ring, disc surface load problems of half space 361–362
Exponential Fourier transform,
  complex argument 245
    integrability requirements on
      inverse of 245
    inversion integral of 245
    operator integral of 245
    properties of 245
  real argument 242–245
    applications of 300–302, 329–331, 349–359, 410–420, 489–516, 523–542
    conditions on 243
    conditions on inverse ... 231–233, 244
    of derivatives 244
    inversion integral of 242, 243
    operator integral of 242, 243

pairs 242, 243
product and convolution theorem 245
relation with sine and cosine transforms 246
shift rule 244–245
tables of 248
Extended saddle point technique, use of 437
Extensions of text material, reading on 573–576

Fiber-reinforced solid, waves in .. 556–560
Finite distributed surface source problems of half space 573–574
impulsive twisting moment 573
load over rectangular are 574
torsional loading over circular area 573–574
uniform pressure over circular area 573
uniform pressure over infinite strip 573
Flammant problem 458–459
stress distribution in 460
Flexural waves in,
an infinite plate on elastic foundation,
frequency equation for .. 227–228
frequency spectra for 229
normal line load, generated by 420–422
an infinite, semi-infinite circular, cylindrical rod,
displacement modes for 224
end load, generated by 435
frequency equation for 224
frequency spectra for ... 224–225
motion generated by 224
potential modes for 223
an infinite, semi-infinite plate, cylindrical dilatational, equivoluminal wave surfaces in ... 179–180, 210

edge moment, generated by .. 435
frequency equation for .. 184, 197
branches of 197–201
frequency spectra for ... 185, 198, 202
P, SV waves in 179–180
phase velocity spectra for 185
thickness modes in 185
a thin plate
approximate theory for 404
point shear force, generated by 404
a thin plate on elastic foundation,
approximate theory for 404
line shear force, generated by 420–421
a thin rod (beam)
approximate theories for 374–377
shear force, generated by 404–409
Focusing, wave 557–559
caustic in 557–559
converging rays in 557, 559
Forces
body 21, 45
inertia 43
surface 21–22
Fourier
Bessel series
for incident pulse transform .. 538
generating function for ... 538
Mellin inversion theorem 236
series, quarter-range 450
transform, exponential (see Exponential Fourier transform)
Fourier integral 256
asymptotic expansion of 261
by parts integration 261–262
by Riemann–Lebesgue lemma 261
in Lamb's problem, use of 5
theorem of 231
conditions in, relaxation of .. 232, 236
Fourier sine and cosine transforms 246–248
appropriate selection of 248
conditions on 243, 247

conditions on inverse ... 231–233, 244, 247
of derivatives 247–248
inversion integrals of 246
operator integrals of 246
pairs 246
properties of, other 248
relation to exponential transform 246, 248
tables of 248
use of 432–435
Fracture, second   VIII
Frequency 121
cutoff, for waveguides (see Cutoff frequencies) ....
circular 69
dimensionless, for infinite plate . 183
equations, for waveguides 6–7, 185
(See also Rayleigh-Lamb, Pochhammer frequency equations)
spectra, for waveguides ... 7–8, 11, 185
Fresnel integrals 405, 438
Friedlander's representation of solution
in exterior cylinder diffraction problems 520–521
in interior cylinder refraction-reflection problems 558–559
wave form of function in 521
wave sum in 521

Gauge
conditions 62–63
use of 153, 350
invariance 63
Geometrical acoustics, optics . 71, 78
Geometrical wave solution in halfplane diffraction 515
Governing equations in
cylindrical coordinates ... 215–217
and axially symmetric compression 220–221
plane strain 284

torsion 217–218
and plane strain 522
rectangular Cartesian coordinates 32, 40, 42, 44–46, 59, 66
and antiplane strain 152–155
and plane strain 121–123
spherical coordinates 45–46, 59-60
and axial symmetry 561
and spherical symmetry . 278–279
Grazing
incidence
of P wave 4, 136–137, 146, 163, 166, 314
of SH wave 173
of SV wave 4, 137–138, 146, 163, 166
reflection
of P wave 141, 164
of SV wave 138
refraction
of P wave 164, 165–166
of SV wave 164, 165–166
Green's function
for three dimensions 83
for two dimensions 86
Group, wave
simple 186
carrier wave in 187
modulation of 187
in spectral analysis of wave train 189–192
stationary phase condition for . 187–188
velocity of 186
relation with velocity of energy transmission 192–193
spectra for infinite plate 191
Guided surface waves on a half space 576
Guided waves 178
(See also Waveguide problems; Waveguides, approximate theories for)

Half space problems
  antiplane surface tangential line load 363
    displacement solution 363
  axially symmetric vertical point load 5, 333–348
    boundary-initial value problem 333–334
    displacement-potential relations 335
    displacement response, interior 340–344
      surface load 340–344
      wavefronts in 340–344
    displacement response, surface 338–340, 345–348
      buried load 345–348
      surface load 338–340
      wavefronts in 339–340, 347
    Laplace–Hankel transform method 334–339, 344
      inversion in 336–339
    reflected wave system, buried load analog of surface source system 347
    stress-potential relations 335
    symmetric delta function for 333–334
    wavefronts compared with line load case 339–340, 344
  buried line dilatational source .. 5, 329–332
    boundary-initial value problem 329
    Lapwood's representation for transformed source 330
      inverse of 331
    Laplace-exponential Fourier transform method 329–331
      Cagniard-deHoop method in 330–331
    surface response 331–332
      Rayleigh waves in, development of 331–332
  dipole source 573

  surface displacement response to surface dipole 573
  expanding ring, disk surface loads
    particle trajectory dependence on load speeds 361–362
    solution dependence on load speeds 361–362
  finite distributed surface sources, displacement responses 573–574
    impulsive twisting moment .. 573
    load over rectangular area ... 574
    torsional loading over circular area 573–574
    uniform pressure over circular area 573
    uniform pressure over infinite strip 573
  surface nonuniformly traveling loads 574
  surface normal line load . 4, 299–328
    boundary-initial value problem 299–300
    displacement-potential relations 301
    Laplace-exponential Fourier transform method 300–313
      Cagniard-deHoop method in 302–319
      solution 312–314
      plane of symmetry response in 318–319
      surface displacement response in ....:.... 314–318
      wavefront approximations 319–328
    stress-potential relations 301
  surface normal point traveling load 347, 349–362
    boundary-initial value problem 349
    displacement component, symmetries of 352
    gauge condition in 350
    Laplace-double exponential Fourier transform method .. 349–359

SUBJECT INDEX

Cagniard-deHoop method in, 351–359
  load speed ranges 352
  solution, displacement
    interior, supersonic load, dilatation waves 355, 357–360
    conical 358, 360
    hemispherical ... 357, 359–360
    regions and conditions for 354–355
    principal value integrals in .. 359, 360
  wavefronts
    conical 361
    hemispherical 361
  surface tangential point load ... 573
    horizontal displacement response along axis below load 573
    surface displacement response 573
Half space, superficial layer on, Love waves in 211–214
Hamilton's principle 371
  approximate theories, application in 380, 371–379
  condition of existence of strain energy function 371
  derivation of 369–371
Hankel
  functions 70
    asymptotic expansions for 70, 529–530
  integral theorem 248–249
  transform 248–250
    applications of 250, 334–339, 468–471
    conditions on 249
    conditions on inverse ... 248–249
    of derivatives 249–250
    inversion integral of 248
    operator integral of 249
      convergence of 249
    pairs 248–249
    properties of, other 250
    tables of 250
  transform, extended (exterior domain) 471, 473–474
    application of 471, 474
Head of the pulse approximation 389, 436–437
  effect of station on magnitude of 574
  elementary theory relation to .. 389
  higher order terms in 574
  momentum equation for 390
  nondecaying nature of 389
Head wave in
  diffraction of dilatational pulse by traction free half plane 510, 512, 516
  half space surface normal line load problem .. 312–316, 321–322, 326–327
    interior 312–314
    interior 312–314
    wavefronts of 322, 327
  half space surface normal point load problem 340–344
  half space surface normal point traveling load problem 361
    wavefronts of 361
  semi-infinite plate, mixed edge conditions 440–443
    wavefronts of 440–443
Helmholtz resolution of vector 60–61
Hemispherical waves 357, 359–360, 361
History, early 2–10
Homogeneous solutions 517
  (See also Self-similar solutions)
Hooke's law, generalized 38–39
  homogeneous, isotropic case . 39–40
Huyghen's
  construction of wavefronts 77
  principle 77
Hydrostatic pressure, uniform 41
Hyperbolic equations 17
Illuminated zone 486–487, 518–519, 526, 541
Impact 9–10
Incident
  P wave 120, 121, 157

Rayleigh surface waves ... 146–151
SH wave 154, 170–171
Stoneley interface waves .. 165–168
SV wave 120, 121, 157
wave in diffraction problems .. 485–486, 496, 518, 541
Infinite medium problems, antiplane displacement source, displacement solution of 363–364 (See also Body force problems; Cavity source problems; Initial value problems; Diffraction problems)
Inhomogeneous media, waves in .. 577
early work on 577
recent work on 577
by Cagniard–deHoop method 577
by geometrical optics 577
for high frequency cylindrical and spherical waves 577
Inhomogeneous wave equation, solution of 116
Initial value problems 3–4, 96–112
displacement solutions for two- and three-dimensional 106
Huyghen's principle in ... 106–107
one-dimensional 96–103
characteristics in 98
discontinuities in initial values in 102–103
initial conditions in 96
method of characteristics in . 107–112
point domain of dependence in 98
point domain of influence in .. 98
solution for initial displacement of 98–99
solution for initial velocity of . 98–102
three-dimensional 103–105
Kirchhoff's formula for 103
point domain of dependence in 106
Poisson's solution for potentials in 105
two-dimensional 105–106

method of descent for ... 105–106
point domain of dependence in 106
solution for potentials in 106
stretched infinite membrane in 295
Integral equations
dual, ... 491, 499–500, 501
Clemmow's method for solving 490–492, 499–501
Integral representations
in diffraction problems .. 12, 96, 488
Kirchhoff's formula for potential . 4, 95
Love's formula for displacement .. 4, 95–96
Integral transforms 232–250
Integral transform methods (see Bilateral Laplace, Laplace, Exponential Fourier, Fourier sine and cosine, Hankel, extended Hankel)
Interface, reflection and refraction from (see Reflection, refraction from interface; P- and SV-wave, SH-wave reflection and refraction)
Interface waves 6, 57, 165–168
Inversion integral,
bilateral Laplace transform 242
exponential Fourier transform 242, 243, 245
Fourier cosine transform 246
Fourier sine transform 246
Hankel transform 248
Laplace transform 232
Isotropy 39–42

Jordan's
inequality 259
lemma, application of 259, 305

Kelvin's method of stationary phase 271–277
$K_0(z)$ function,
integral representation for 329
Lapwood's integral representation for 330
Kinetic energy 48

density 48–49
Kirchhoff's integral representation . 95
Kronecker delta 20

Lagrange multipliers, method of ... 29
Lamb's problem
   for buried line dilatational source, 329–332
   early
      techniques for 11
      work on 11
   for surface normal line load 299–328
   for surface normal point traveling load 347, 349–362
   for transversely isotropic half space 576
   for vertical point load 333–348
      buried 333–336, 334–348
      surface 336–344
Lamé solution of displacement equations of motion 58–62
   for body force problems .. 79–80
   completeness of 61–62, 116
   dilatation in 59
   displacement vector field of 59
   potentials in 59
   rotation in 59
   wave equations in 59
Laplace integral 256
   asymptotic expansion of .. 256–261
      by parts integration 257
      by Watson's lemma 257
      for Laplace transform, large parameter–short time (wavefront) approximation 257–258
      for Laplace transform, small parameter–long time approximation and static solution 258–261
Laplace's method
   for asymptotic expansion of integrals 262–264
      critical points in; corresponding approximations 264
Laplace transform 232–241

applications of ... 278–360, 383–420, 432–474, 489–550, 562–566
asymptotic expansions of . 257–261
   for long time (static solution) 258–261
   for short time (wavefront) 257–258
conditions on 236
conditions on inverse .. 231–233, 236
   relaxation of 232, 236
of derivatives 238
division by parameter 239
Fourier–Mellon inversion theorem 236
inversion integral of 232
negative parameter, inversion for 328
operator integral of 232
   convergence of 233, 236–237
   linearity of 237
   use of 233–234
pairs 232
parameter, transformation 233
product and convolution theorem 238–239
   application of 52–54
shift rule 239–240
tables 241
transforms of singular inverses 240–241
   delta function, Dirac 241
uniqueness of 237
Laplacian
   in cylindrical coordinates 217
   in rectangular Cartesian coordinates 45
   of vector 45–46
Laws of
   P-wave
      reflection 124
      refraction 158–159, 164
   SH-wave
      refraction 171
   SV-wave
      reflection 124

refraction 158–159, 164
Layer (see Plates, infinite, semi-infinite; Waveguide problems)
Layered media 16
   time harmonic waves in 226
Legendre polynomials 562–565
Lerch's theorem 300
   use of 300, 448
Longitudinal impact problem
   of semi-infinite plate with mixed edge conditions 431–434, 436–443
   of semi-infinite rod based on Love-Rayleigh theory ... 382–394, 478
Long-time approximation 260
   Watson's lemma–Laplace transform derivation of 258–261
   solution of static problem ... 261
   for cavity source problems 289, 295
   for circular cylindrical cavity diffraction problems 535, 555
Long time–far field approximations in circular cylindrical cavity diffraction problems .. 529–536, 551–555, 569–571
   cavity wall rigid body velocities in 555
   Rayleigh waves in 529–536, 551–555, 569–571
   static solutions in 535, 555
   waveguides (see Waveguide problems)
Long time–near field approximations in a semi-infinite plate 455–456, 463–465
Love-Rayleigh rod theory ... 8, 369, 371–374
   diffusive nature of 383
   initial data jumps governed by 393–394
   (See also Waveguides, approximate theories for; Waveguide problems)
Love's integral representation for displacement 95–96
Love waves 8, 211–214
   phase velocity equation for 213
      branches of 213
      asymptotics of 213
   phase velocity, group velocity, spectra, lowest two branches of 214
   scattering of, by edge of half space surface layer 576

Mass density 43
Membrane, stretched infinite,
   initial value problem of 295
   solution of 295
Mindlin–Herrmann rod theory 392–393
Mindlin–McNiven rod theory 574
   head of the pulse solution for ... 574
   effect of station on magnitude of 574
   higher order terms in 574
Mixed boundary value problems in wave diffraction 487–517
   dual integral equations in .. 491, 499–500, 501
Mixed edge conditions problem, uniqueness of solution of 55
Mode integral sums, waveguide .. 419–422, 424, 426, 428–429, 474–477, 540, 542–544
   solutions based on comparisons of 424, 428–429, 474–477, 542–544
Modern work and reading 10–12
Modes
   conversion of 119
   no 124–125, 155–156
   partial 1, 132, 160–161
   total 127, 132, 149, 140
   coupling of .... 132 (See also "conversion of" here)
   of propagation in infinite plate,
      dilatational 182–183
      equivoluminal 182–183
   thickness- 185
      dilatational 185
      equivoluminal 185

shear 185
stretch 185
Modulus of
compression (bulk modulus) 41
elasticity (Young's modulus) 38
restriction on 42
rigidity (shear modulus) 41
Moment 21

Nodal points
in time harmonic axially symmetric waves 70
in transient excitation of plates 421–422
Nonaxially symmetric
diffraction problems of circular cylindrical obstacles
waves in 518–519, 532–536, 541–546, 550–560
flexural waves in circular cylindrical rod 222–225
traveling point load problem in half space 347, 348–362
waves in 355–361
Nonlinear elastic media 578
discontinuity surfaces and shock in 578
geometric acoustics for waves in 578
Nonlinear waves, theory of 17
Nonmixed edge displacements, problems of, for semi-infinite plate 465–466
cantilevered plate, symmetric face normal line loads near base ... 465
velocity shock 465–466
Nonseparable problems
semi-infinite plate with nonmixed edge conditions 430–431, 444–466
wedge, with nonmixed edge conditions (see Wedge, two-dimensional)
Nonuniformly moving surface load problems of half space 574

Normal line load problem of infinite plate 429–430

Operator integral for
bilateral Laplace transform 232
exponential Fourier transform 242, 243, 245
Fourier cosine transform 246
Fourier sine transform 246
Hankel transform 249
Laplace transform 232
Operators, linear 47
Optics 17, 485–487
Order symbols, definition of 33

Partial differential equation, linear 47
Particle velocity 43
Parts integration in asymptotic expansions of integrals 257, 261–262
application of 251
Period, predominant 190
–time of occurrence relations .. 191
application of 471
dimensionless variables in 190–192
Period, wave 121
Permutation symbols 20
Phase velocity,
infinite plate 179
dimensionless 183
equation of, for mixed face conditions 183–184
spectra for
for mixed face conditions .. 185
for elastically restrained faces 195
–wave number relations,
for mixed face conditions . 183–184
Plane
strain 120
conditions for 120
displacement-potential relations of 121, 123, 158
P and SV waves of 119–152, 156–170

strain-potential relations of . 122 123
stress-potential relations of .. 122, 123, 158
stress 120
(See also Plane strain)
waves,
    constant phase of 65–67
    definition of 64
    displacement 66–67
    harmonic P-, SV-, SH- 67
    one-dimensional 63–64
    phase of 65
    potential 65–66
    three-dimensional 65–67
Plastic media, waves in, book, surveys on 578
Plate,
    vibrations of,
        flexural, thin plate 7
        nodal figures in 7
    velocity 222, 471
Plate, infinite, time harmonic waves in, cylindrical wave surfaces, dilatational, equivoluminal in .. 179–180, 210
    displacement waves in 180, 210
    with elastically restrained faces,
        coupling between P, SV waves in 194–196
        frequency equation for 195
        roots common with mixed face conditions case 194–196
        frequency spectra for 195
        phase velocity spectra for 195
    on an elastic foundation .. 227–229
        frequency equation, DasGupta, for 227–228
        cutoff frequencies from 228–229
        frequency spectra from 229
        sandwich plate equivalent problem 228
    with mixed face conditions ... 181–186, 188–198

anharmonic overtones in 185
branches of frequency equation for, imaginary wave number segments of 185
real wave number segments of 185
cutoff frequencies in 188–189
    P, SV waves at 189
edge waves in 184
frequency equations for . 183–184
frequency spectra for 185
fundamental mode of,
    dilatational 185
    equivoluminal 185
group velocity spectra for 191
high frequency-short waves in 189–190
phase, group velocities for 189–190
modes of propagation in 182–183
phase velocity equations for . 183–184
phase velocity spectra for 185
phase velocity-wave number relations for 183–184
predominant period-time of occurrence relations for 190
standing waves in 184
thickness modes in,
    dilatational 185
    equivoluminal 185
    shear 185
    stretch 185
wavelength-frequency relation for 190
phase velocity of 179
P waves, symmetric, antisymmetric, in 178–180
SH waves, symmetric, antisymmetric in 209–210
stress waves in 181, 210
SV waves, symmetric, antisymmetric in 179–180
with traction free faces 197–209

frequency equation, Rayleigh–Lamb, for 197
frequency spectra, Rayleigh–Lamb for 197
  complex wave number segments in 197–201, 207–209
  general character of .. 197–201
  real, imaginary wave number segments in 197–207
  generalized frequency equation, Rayleigh–Lamb, for 200
    branches of, near branch point behavior of 200
  (See also Rayleigh–Lamb frequency equation)
  wave number
    in propagation direction of .. 179
    in thickness direction of 179
  wave pairs in 180
Plates,
  infinite
    axially symmetric loadings in 466–477
    plane strain loadings in . 409–430
  semi-infinite
    mixed edge condition loadings in 430–443
    nonmixed edge condition loadings in 444–466
  (See also Waveguide problems)
  thin, infinite
    bending theory of Timoshenko–Uflyand–Mindlin 404
    point shear force problem based on 404
Pochhammer frequency equations .. 7, 219, 221, 224
  general branches of .. 219, 221–222, 224–225
  complex wave number segments of 208–209, 222
  further detail of (see Rayleigh–Lamb frequency equations, general branches of)
  imaginary wave number segments of 220, 208–209, 222, 224–225
  real wave number segments of 220, 208–209, 222, 224–225
  (See also Rod, infinite, circular cylindrical; Rayleigh–Lamb frequency equation)
Point
  load problem in infinite medium .. 4, 86–91, 93–94
    time harmonic 93–94
      steady state solution for 94
  shear load problem
    in infinite plate based on Timoshenko–Uflyand–Mindlin theory 404
  source problem
    in two welded half spaces 6
Poisson's
  integral formula 4
  ratio 41
    coupling 393, 436–437
    restriction on 42
  summation formula 520–521
    Fourier series, applied to 520–521
    square integrability, for functions of 520
    sufficient conditions for 520
Potential,
  displacement 59
  energy of deformation 49
  integral representation for 4
  retarded 4, 84
  scalar 3–4, 59
  vector 3, 59
Power input 48
Pressure shock problem of semi-infinite plate,
  with mixed edge conditions .. 431–432, 434–435, 436–443
  with nonmixed edge conditions 431–432, 444–448, 449–457
Principal value integrals .. 311, 359, 360

Principle of reflection in complex variables, use of 286
Propagating surface of discontinuity 71–78
(See also Wavefronts; Characteristics)
Pulse
  diffraction
    Cagniard–deHoop method for 12, 490–516
    from half plane rigid barrier .. 11–12, 496, 517
    from half plane slit . 11–12, 485–517
    from plane finite obstacles 12, 574
    self-similar solutions for .. 12, 517, 575–576
    Wiener–Hopf method for .. 12, 488–504
  propagation 8
  (See also Diffraction problems)
P wave 67
  reflection from
    boundary with mixed conditions 123–125
    fluid-solid interface .. 168–170, 176–177
    free boundary 125–131, 135–137, 139–140, 144–146
      at grazing incidence .. 136–137, 144, 145
      at normal incidence ... 135, 145–146
      total mode conversion in .. 139, 140
    rigid boundary 152, 175
    solid-solid interface ... 156–161, 175–176
  refraction from
    fluid-solid interface ... 168–170, 176–177
      at critical angles .. 169, 176–177
      at grazing incidence 169
      at normal incidence 169

solid-solid interface ... 156–161, 175–176
  at critical angles 163–165, 175–176
  at grazing incidence 163
  at normal incidence .. 162–163
  for wave pairs 166–168
of surface type 3
(See also Dilatational waves in; Dilatational waves of)

Quarter
  plane
    problems (see Wedge, two-dimensional)
    rigid scatterer in infinite solid, incident compressional pulse 575
  planes 575
  two welded, mixed, nonmixed edge condition problems .. 575–576

Rayleigh function 301
Rayleigh–Lamb waves, unbounded 446–447
Rayleigh surface waves
  in diffraction problems ... 513, 529–536, 551–555, 565–571
  in half space buried line dilatational source 331–332
    far-field domination of 332
    nondecaying nature of .. 331–332
    nonsingular nature of .. 331–332
    two-sided nature of 331–332
  in half space surface normal line load problem 313–318
    far-field domination of .. 317–318
    nondecaying nature of 316–317
    response to delta function of 315–317
    response to step function of .. 318
    singular, nonsingular nature of 313, 315–317
    two-sided nature of 313–314, 317

in half space vertical point load problem 339–343, 345–348
  for buried load case 345–348
    development of 345–348
    nonsingular nature of . 346, 348
    two-sided nature of ... 346, 348
  for surface load case 339–343
    interior response of .. 340–343
    one-sided nature of . 339–340
    singular nature of ... 339–340
    spatial decay of 340
  time harmonic 146–151
    displacement amplitudes of . 149–150
    early work on 4–5, 8
    equation for velocity of 147
    experimental studies of . 150–151
    generalized 151–152
    particle motion of 148–149
    scattering of, by edge of half space surface layer 576
    stress amplitudes of 149–150
    ultrasonic, use of 152
    velocity of 147–148
Rayleigh–Lamb frequency equation . 7, 197
  general branches of 197–201
    analytical continuation in 197–200
    branch points in 197–200
    complex wave number segments of 197–201
    complex wave number segments of, further on 207–209
      analytic continuation scheme 207–209
      low frequency behavior 209
    cutoff, behavior near, of .. 197–200, 227   2
    cutoff in 197–200
    edge waves governed by 200
    imaginary wave number segments of 197–201
    modes associated with .. 200–201
    real-, imaginary-wave number segments of, further on .. 201–207

    bounds 195, 196, 201
    critical Poisson's ratio and cutoff 204–205
    cutoff frequencies 203
    frequency spectra 198, 202
    grazing incidence waves 206, 207
    high frequency-short wave behavior 207
    modes, thickness-strech, -shear, simple 204
    phase, group velocities and curvature at cutoff 205
    Rayleigh waves 207
    spectra, related modes and waves 205–207
    terracing 196
    real wave number segments of 197–201
    wave pairs associated with ... 201
  generalized ... 200, 411, 434, 447
    branches of 411–416, 447
    branches of, near branch point behavior of 200
    for complex frequency-complex wave number segments ... 480
    governing waves corresponding to low frequency-complex wave number segments 480
  (See also Plate, infinite, time harmonic waves in)
Rays,
  in diffraction problems ... 485–486, 518–519, 556–567, 559
  equation for 76
  for homogeneous medium 77
  theory of 17
  use in wavefront analysis of 76
Reciprocal theorem, Graffi 52–55
  applications of 54–55
  dynamic equilibrium states in 52
  extensions of 54
Reflection coefficients 123
  in half space for
    displacements 126–127, 132
    energy 131, 133–134, 143–144

P waves 120, 123–131
  potentials 123–125, 131, 133, 135, 141, 155–156
  SH waves 152–156
  stress 127, 132
  SV waves 120, 131–134
Reflection from half space boundary, coefficients in
  complex 132, 141
  for displacements .. 126–127, 132
  for energy 131, 133–134, 143–144
  for potentials ... 123–125, 131, 133, 135, 141, 155–156
  for stress 127, 132
  of P waves 120, 123–131
  of SH waves 120, 131–134
  of SV waves 152–156
Reflection principle in functions of complex variable,
  application of ... 385, 399–400
Reflection, refraction from interface, coefficients in
  complex 164–165, 169, 175–176, 177
  for displacements ... 161, 162, 163, 169, 172
  for energy ... 161–162, 165, 169–170
  for potentials ... 160–161, 164–165, 168, 172, 175–176
  for stress 161, 169
  of P wave ... 156–161, 168–170, 175
  of SH wave 170–174
  of SV wave 156–160, 161–162, 169
Refraction (see Reflection, refraction from interface)
Resolution of vector, Helmholtz ... 60
Retarded potentials 84
  in body force problems .. 84, 88, 94
Riemann–Lebesgue lemma 243
  applications of ... 388, 405, 408, 424, 428–429
Rod, infinite, circular cylindrical,

displacement-potential relations for,
  axially symmetric compression 220
  axially symmetric torsion 217
  general (nonaxially symmetric) 216
  Lamés solution for 215–216
  vector potentials in 215
  gauge conditions for 215
strain-displacement relations for 217
stress-potential relations for,
  axially symmetric compression 221
  axially symmetric torsion 218
  general (nonaxially symmetric) 217
stress-strain relations for 217
time harmonic waves in ... 214–225
  axially symmetric compressional 220–223
    analog with infinite plate .. 221–222
    frequency spectra for . 208–209, 222
    nature of motion generated by 222–223
    potential modes for 221
  axially symmetric torsional . 217–220
    analog with SH waves in plate 220
    displacement modes for ... 220
    frequency spectra for 219–220
    phase velocity spectra for 220
  nonaxially symmetric (flexural) 223–225
    analog with infinite plate .. 224
    displacement modes for ... 224
    frequency spectra for . 224–225
    natire of motion generated by 224
    potential modes for 223
  Pochhammer frequency equation for,
    axially symmetric compression 221
    axially symmetric torsion 219–220

nonaxially symmetric (flexure) 224
(See also Waveguide problems)
Rods,
  approximate theories, one-dimensional for 367–409
    compressional wave,
      elementary 367–369
      end stress problem 292
      Love—Rayleigh .. 369, 371–374
      longitudinal impact problem based on 382–394
      Mindlin–Herrmann .. 392–393
    flexural wave,
      Bernoulli–Euler (elementary) 374–377
      Timoshenko 377–382
      step shear force problem based on 394–409
    torsional wave,
      fundamental nondispersive mode 219
      initial angular twist problem 293
    section warping 374
  arbitrary section 436
  elliptic section 217
  rectangular section,
    photoelasticity-shock tube experiments with 437
    warping of plane sections in .. 437
  semi-infinite, circular cylindrical,
    mixed edge load problems for,
      axially symmetric 437
      nonaxially symmetric 435
    nonmixed pressure shock problem for,
      finite-difference numerical method in 466
      radial, axial strain responses in 466
      comparison with response records 466
    shock tube excitation of circular cylindrical 390–393
    axial strain, radial displacement response in 391–393
    head of pulse in 391–393
    high frequency waves in 391–393
    near field cutoff waves in 391, 393
  vibrations of, early work on,
    approximate theory 6
    exact theory 7
  (See also Waveguides, approximate theories for; Waveguide problems)

Saddle
  point 265–266
  higher order 266
  method, extended 271
Scattering of
  dilatational waves by liquid sphere or cylinder 574
  pulse by
    circular cylindrical obstacle . 517–560
    half plane 485–517
    spherical obstacle ... 558, 560–567
  waves by cylindrical cavity in a half space 574
    asymptotic expansions, matched, in 574
  waves by rigid spheroids, circular disk 574
Scholte wave 170
(See also Stoneley interface waves)
Seismology 16
Self-similar solutions 5, 517
  for wedge problems 575–576
Semi-infinite elastic plate, problems of,
  with mixed edge conditions .. 430, 432–443
  with nonmixed edge conditions 430–431, 444–466
Shadow zone 486, 518, 526, 541, 561
Shear
  force problem of infinite
    rod (beam) based on Timoshenko theory 394–409

thin plate, based on Timoshenko–Uflyand–Mindlin theory .. 404
modulus (Modulus of rigidity) ... 41
Shells, circular cylindrical, waves in 223
Shift rule,
   exponential Fourier transform 244–245
   Laplace transform 239–240
SH wave 67
   diffracted 485–486
   in half space antiplane tangential surface load problem 363
   in infinite medium antiplane displacement source problem . 363–364
   in infinite plate 209–211
   (See also Plate, infinite, time harmonic waves in)
   reflection from
      free boundary 155–156
      rigid boundary 156
   reflection, refraction from solid-solid interface
      at critical angle 173
      at grazing incidence 173
      at normal incidence 173
      for total transmission 174
   in superficial layer of half space ... 8
   (See also Love waves)
   surface 3
Sine integral 462
Simple shear 41
Slowness 167
   diagrams, use of in study of wave reflection and refraction 573
Sommerfeld optics diffraction problem 485–487
Spherically symmetric
   cavity source problem 277–282
   waves 67–68, 281–282
Standing waves 189
   in transient response of plate . 421–422
Static problem, solution of .. 258–261
Stationary phase 186–188
   condition of 187–188

method of 9, 271–277
   applications of 275–277, 291, 387–388, 404–408, 423–424, 428–429, 436–437, 469–472
   asymptotic expansion of integral in 273
   for second order stationary phase point 274, 277
   for continuous distribution of wave groups ... 274, 275–276
   as special case of steepest descents method 274–275
Steady-state solution for time harmonic body force problem 94
Steepest descents, method of ... 264–271
   applications of 269
   for wavefronts 323–328, 550–551
   in waveguides 271
   in asymptotic expansion of integral 269
   extended saddle point method 271
   when singularities occur . 269–271
   deformation of integration path to steepest paths 267, 269
   traverse paths 267
   saddle surfaces in, geometric features of 265–267
      example of 266–267
      hills 265, 266
      level curves 265
      relief surface 265
      saddle point, conditions at 265–266
      saddle points, higher order ... 266
      steepest paths 265
      valleys 265, 266
Step function, Heaviside 86
Stoneley
   interface waves, time harmonic 165–168
   early work on 9
   equation for velocity of 167
   existence of 168
   velocity of 168

SUBJECT INDEX

Strain,
  analysis of 30–38
  dilatation 38
  energy 3, 49
  energy density 39, 49, 55
    Clapyron's form for 49
  finite 31–32
    components of 32
    linearization of 32
  infinitesimal 20, 32
    components of 32
    nature of 35–36
      extension 35–36
      shear 36
  invariants 38
  principal 37–38
  principal directions of 37
  principal planes of 37
  quadric of Cauchy 37–38
  -stress relations, Cartesian 42
  tensor 37
    components of 36–37
    law of transformation of 37
Strained elastic body 30–31
  continuity conditions on 31
Stress,
  analysis of 21–30
  components of 22–25
    normal 22, 24, 26–28
    shear 22, 28
    transformation of 24–25
  compression 22
  equations of motion 44
    early work on 2
  invariants 28
  maximum shear 28–30
  principal 27–28
  principal directions of 27
  principal planes of 27
  quadric of Cauchy 26–28
  tension 22
  tensor 24
    components of 25
    law of transformation of 25
    symmetry of 24

  vector 22–26
  plane of action of 22
Stress-strain relations 38–42
  Hooke's law, generalized ... 38–39
    elastic coefficients for 39
    homogeneous, anisotropic body 39
    symmetry relation for 39
  Hooke's law, homogeneous, isotropic body 39
    Lamé constants for 40–41
    restrictions on 42
    in terms of Young's modulus, Poisson's ratio 41
Stretched elastic string,
  initial value problem of 96–103
Superposition principle 47
Supplementary reading on 573–580
  additional effects 576–578
  text material 573–576
Surface
  load problems of half space,
    antiplane tangential line 363
    normal line 299–328
    normal point 333–344
    normal point traveling .. 347, 349–362
    tangential point 573
  waves 57
    guided by thin film 576
    P- 141
    Rayleigh 146–151
    (See also Rayleigh surface waves)
    SV- 138
Surface of discontinuity, propagating 4 71–78
  dynamical conditions at 74
  for inhomogeneous medium .. 71, 78
  kinematical conditions at 73
  velocities of 74–75
SV-wave 67
  reflection from
    boundary with mixed conditions 123–125

elastically restrained boundary 133–135
fluid-solid interface 169
free boundary .. 131–134, 137–146
  at critical angle 139–144
  at grazing incidence .. 137–138
  at normal incidence .. 145–146
  at $\pi/4$ angle 138
  total mode conversion in 139, 140
rigid boundary 152
solid-solid interface .. 156–160, 161–162
refraction from
  fluid-solid interface 169
    at critical angles 169
    at grazing incidence 169
    at normal incidence 169
  solid-solid interface 156–162
    at critical angles 163–165
    at grazing incidence 163
    at normal incidence .. 162–163
  for wave pairs 166–168
of surface type 3
(See also Equivoluminal waves in; Equivoluminal wave of)
Tables
  exponential Fourier transforms and inverses 248
  sine and cosine transforms and inverses, relation with 246
  Hankel transforms and inverses 250
  Laplace transforms and inverses 241
  sine and cosine transforms and inverses 248
Tauberian theorems 328
  for negative Laplace transform parameter 328
Tensor, Cartesian 20
  first order 25
  law of transformation
    first order 25
    second order 25
  notation of 20
  second order 24

subscripts of 20
summation convention of 20
Thermal effects 577
  equations of motion for linear thermoelastic medium 577
  harmonic waves based on ... 577
  transient waves based on 577
  surveys on 577
Thermoelasticity, uncoupled dynamic, governing equations of . 474, 577
  infinite plate problem, application to 474–477
Time average of harmonic waves . 130–131
Time harmonic body waves ... 69–70
Time harmonic waves in
  half space 119–156, 174–175
  infinite medium 60–70, 93–94
  two welded half spaces 156–177
  waveguides 178–230
Timoshenko
  bending theory ... 8, 377–382, 394–409
    hyperbolicity of 397
    wave speeds in 395
  –Uflyand–Mindlin plate theory 404
  (See also Waveguides, approximate theories for; Waveguide problems)
Timoshenko beam theory (see Timoshenko bending theory)
Torsional waves
  in an infinite, semi-infinite circular cylindrical rod,
    displacement modes for 220
    frequency equation for 219–220
    frequency spectra for ... 219–220
    phase velocity spectra for 220
  in a thin plate with circular cylindrical cavity,
    cavity wall rotary velocity, generated by 290
    cavity wall shear stress, generated by 290
  in a thin rod,

initial angular twist, generated by 293
nondispersive fundamental mode in 219
section warping theory of 374
Total reflection ... 139–144, 164–165, 169–170, 173–174, 177
Transient waves (See Infinite medium problems; Half space problems; Waveguide problems)
Transmission coefficients 160
at interface for
displacements .. 161–162, 163, 169, 172
energy ... 161–162, 165, 169–170
P waves ... 156–161, 168–170, 175
potentials ... 160–161, 164–165, 168, 172, 175–176
SH waves 170–174
stress 161, 169
SV waves ... 156–160, 161–162, 169
Traveling sources .. 5, 347, 349–362, 574–576

Uniqueness of solution, conditions for in half-plane diffraction problems 489, 491, 498
Uniqueness of solution for .. 47, 50–52
discontinuous loadings 52
finite body 47, 50–51
infinite anisotropic body 51
infinite body 51
Unloading waves, VI, VIII

Van der Waerden's method
for asymptotic expansion of integrals 270
Vector
decomposition of 60–61
Laplacian of 45
operators 45–46
potential 59
in tensor notation 20

components of 20
law of transformation of 25
scalar product for 20
vector product for 20
Viscoelastic media, linear,
books, surveys on waves in 578
Maxwell material rod in,
boundary-initial value problem for, 295
solution of 296
Rayleigh waves diffracted from circular cylindrical cavity in .. 536
by correspondence principle 636
von Schmidt wave (See Head wave)

Water waves...................17
Watson's lemma 257
application to asymptotics of Laplace transform and inverse . 257–261
examples of
long time approximations, static solution of 289, 295, 535, 555
(See also Long time -far and -near field approximations)
wavefront approximations . 289, 323–328, 550–551, 565–567
Watson's transformation
in spherical obstacle diffraction problems 563–564
Wave
diffraction 10, 485–567
dispersion 1, 8, 9, 178–230
group
analysis 8, 186–188
composition of 188
simple 186–187
velocity 8, 186–187
guide 1, 6, 16–17, 178–230
(See also Waveguides, approximate theories for; Waveguide problems)
length of
P wave 121
SH wave 154
SV wave 121

number
  along boundary 124
  in infinite plate rod. . 11, 178–225
    in propagation direction. . . 179, 183
    in thickness direction . . 179, 197
    of P wave 121
    of SH wave 154
    of SV wave 121
  pairs,
    reflection of 145–146
    refraction of 166–168
  phase velocity 179
  in piezoelectric-elastic solid 17
  propagation studies,
    early history of 2–10
    diffraction in 10
    fundamental representations in 2–4
    half space in 4–6
    impact in 9
    two welded half spaces in ... 6
    waveguides in 6–9
    modern work in 10–12
  reflection 120, 123–134, 152–162, 168–175
  refraction 156–162, 168–175
  trains, harmonic,
    finite number of 187
    stationary phase condition for 187–188
    infinite 7, 186
    spectral analysis of ... 188–192
  unloading .... VI–VIII
  velocity along boundary 124
Wavefront approximations
  in Cagniard-deHoop method ..319–323
    for response of half space 319–323
    for response of plate 438–443
    steepest descents, by method of 264–271
    for response of half space ..323–328
  by Randles-Miklowitz method .. 443

for response of plate 443
by Watson's lemma—Laplace transform expansions 257–258
for response to cavity diffraction.. 565–567
for response to cavity sources 289
Wavefronts
  approximations 71, 78
  in cavity source problems . 281, 289
  cusp type, in anisotropic plates . 443
  in diffraction 485–487, 516, 550–551, 558–560, 565–567
  (See also "in a half space" here)
  dynamical conditions at 74
  in a half space
    conical 361
    head 322, 327, 341–344, 361
    hemispherical 361
    regular, equivoluminal, dilatational 318, 320–321, 326, 340–348, 361
    SP wave (P wave grazing reflection) 346–348
    two-sided equivolumnal 313–314, 322–323, 327–328, 341–344, 346–348, 361
  in homogeneous medium 76–77
  Huyghen's construction of 77
  kinematical conditions at 73
  magnitude variations of 77–78
  velocities 74–75
  in waveguides 438–443
  (See also "in a half space" here)
Waveguide problems
  approximate theories for rod .. 382–409
    longitudinal impact, Love–Rayleigh theory 382–394
    boundary-initial value problem 382
    Laplace transform method 383–387
    solution 383, 386
    comparison with experiment, exact and approximate theories 393

initial data jumps in 393-394
long time-far field head of
  pulse response . 388-389
effect of station on magnitude of 574
higher order terms in, based on Mindlin-McNiven rod 574
shear force, Timoshenko theory 394-409
boundary-initial value problem 394, 396
Laplace transform method 394-409
solution 398, 403-404
  admissible parts 404
  comparison with Bernoulli-Euler result 404, 409
  long time approximation of 404-409
  verification 404
approximate theory for plate Timoshenko-Uflyand-Mindlin 404
on elastic foundation, line shear load response 420-421
point shear load 404
for infinite plate and axial symmetry 466-477
normal displacement on circular cavity wall with mixed edge conditions 471, 473-474
Laplace-extended Hankel transform method .. 471, 473-474
solution, long-time approximation 474
normal point load 574
  early time response to 574
symmetric normal point loads; equivalent layer—half space problem 466-467
boundary-initial value problem 466-467
displacement-potential relations 467-471

Laplace-Hankel transform method 468-471
solution, displacements .. 468-469
long time-far field 470-472
long time-far field, derivation of 469-471
maximum response 471
predominant period—time of occurrence criterion in long-time solution 471
stress-potential relations... 468
time-dependent thermal field, excitation by 474-477
Gaussian temperature distribution on plate face 474
Laplace-Hankel transform method 475
stresses, near-field response of midplane 475-477
comparisons based on mode integral sums 475-477
cutoff frequencies, influence of 475-477
mode integrals, numerical evaluation of 475-477
non-zero heating time... 475
for infinite plate in plane strain 409-430
antisymmetric normal line loads 429
solution of 429
normal line load
  decomposition of 429-430
  direct solution of 429
solution by addition of symmetric and antisymmetric parts 429-430
symmetric normal line loads . 409-429
boundary-initial value problem 409-410
inversion of spatial transform first 411-420

generalized Rayleigh–Lamb frequency equation, wave number branches and their properties 411–416
path integrals in ... 417–418
inversion of time transform first 425–429
frequency branches, real Rayleigh–Lamb in ... 425–428
path integrals in.... 425–426
Laplace-exponential Fourier transform method .. 410–420
solution by inverting spatial transform first ... 419–420
anomalous dispersion in 421
cutoff frequencies, influence of 420
long time–far field approximations to integrals of .. 423–424
responses from wave number branch segment (real, imaginary, complex) integrals in 420–422
standing waves in .. 421–422
validity of, for near field, restriction to moderately sharp inputs 424
solution by inverting time transform, first 426
branch integrands in, proof of nonsingular nature of 426–428
long time–far field approximations 428–429, 481
negative x-traveling waves in 428–429
static 426
uniqueness of the solution 428
comparison of two forms of solution and inversion techniques 428–429
for semi-infinite plate in plane strain with mixed edge conditions 430–443

anti-symmetric excitation ... 435
longitudinal impact 431–434, 436–443
boundary-initial value problem 431–434
Laplace-sine, cosine transform method 432–434
inversion techniques 411–420 425–429 (detail under "infinite plate" here)
long time–far field approximations 436–437
solution, formal axial, thickness, strains 434
stress-strain relations 431
wavefront approximations 438–443
amplitude coefficients .. 441, 443
by Cagniard-deHoop method 438–443
comparison with experiment 443
by Randles–Miklowitz method 443
rays, family of S, P 440
sources, rays and wavefront 440
for strain............. 441
wavefront positions ... 441–442
mixed pressure shock ... 431–432, 434–435, 436–443
boundary-initial value problem 431–435
Laplace-sine, cosine transform method 435
inversion techniques .. 411–420, 425–429 (detail under "infinite plate" here)
long time–far field approximations 436–437
solution, formal axial thickness strains 435
wavefront approximations 438–

443 (detail under "longitudinal impact" here)
for semi-infinite plate in plane strain with nonmixed edge conditions 444–466
  long-time solution, restriction to 444
  method preliminaries ... 444–448
    boundedness condition on solution and corresponding integral equations 446–448
    double Laplace transform, use of 444
    double transformed displacements 445–446
    edge unknowns, time transformed 444
    quasi-formal solution . 445–446
    Rayleigh–Lamb exponentially unbounded waves and corresponding generalized complex wave numbers in 446–447
  nonseparability of . 430–431, 444
  problem A: nonmixed pressure shock ... 431–432, 444–448, 449–457
    boundary-initial value problem 431–432, 445, 449
    edge unknowns, determination of 449–452
      boundedness condition for coefficients of, algebraic 452
      boundedness condition for small p 454–455
      time transformed .. 449–452
    formal long-time solution . 445–446, 455
    long-time solution, displacement strains 455–457
  problem B: nonmixed line load 431–432, 444–448, 457–466
    boundary-initial value problem 431–432, 445, 457
    edge unknowns, determination of time transformed 458–461

    decomposition of problem into Flammant singular, self-equilibrated residual and nonmixed pressure shock problems 458–461
    edge unknowns, Flammant problem singular . 458–460
    edge unknowns, regular . 461
    edge unknowns, time transformed 461
    boundedness condition for coefficients of, algebraic 461
    boundedness condition for small p 462
    formal long-time solution 445–446
    long-time solution, displacements, strains 463–465
    method of reduction for coefficients of edge unknowns 462–463
    problems of nonmixed displacements 465–466
    cantilevered plate, symmetric face normal line loads near base 465
    head of the pulse long time— far field solution for reflected disturbance 465
    velocity shock problem ... 465
    by finite-difference numerical method 465
    longitudinal strain responses 465–466
(See also Waveguides, approximate theories for)
Waveguides, approximate theories for, one-dimensional, ... 367–382
  compressional wave, rod 367–374
  elementary theory 367–369
    displacement equation of motion 368
    frequency spectrum . 368–369
    use of 368–369
  Love–Rayleigh theory 369, 371–374

boundary conditions 373–374
displacement equation of motion 373
frequency spectrum 369
use of 369, 373–374
flexural wave, rod 374–382
Bernoulli–Euler (elementary) theory 374–377
boundary conditions ... 376
curvature theorem .. 375–376
deflection equation of motion 376
element motion 375
frequency spectrum 377
moment-deflection relation 375
section normal stress-moment relation 376
section shear stress-shear force relation 376
transverse shear force-deflection relation 376
use of................. 377
Timoshenko bending theory 377–382
advantages of 381–382
boundary conditions ... 380
deflection equations of motion 380
deflection slope 378
element motion 377–378
extensions to thin plate .. 382
frequency spectrum, comparison with exact, Bernoulli–Euler theory 381
moment-deflection relation 379
moment, transverse shear force 379
potential energy 379
section normal stress-moment relation 376
section shear stress-shear force relation 376
shear coefficient k' 379
shear force-deflection relation 379
strain energy function ... 378
strains, sectional 378
total deflection, bending and shear components 377
general nature of 367
torsional wave, rod,
fundamental nondispersive mode theory 219
section warping theory 374
thin plate and circular cylindrical shell 367, 404
(See also Waveguide problems)
Wedge, two-dimensional
of arbitrary angle, nonmixed edge conditions 575–576
analytical methods tried, Knopoff's survey of 575
self-similar solutions for, recent work with 575–576
quarter plane,
with mixed edge conditions, all loadings 575
with nonmixed edge conditions, internal source, stress free edges 575
normal line load on one edge 575
two welded quarter planes,
longitudinal impact (mixed edge conditions) 575
normal line load on edge of one quarter plane (nonmixed edge conditions) 575
Wiener–Hopf method 488, 490–492, 499–504
Clemmow's approach 490–492, 499–501
factorization in ... 491, 502–504
approximate 504
regular, nonvanishing functions L, U in ...490–492, 498, 500–501
Young's modulus (Modulus of elasticity) 38
restriction on 42